MODERN MICROBIOLOGY

PRINCIPLES & APPLICATIONS

MODERN MICROBIOLOGY

PRINCIPLES & APPLICATIONS

Edward A. Birge
Arizona State University

Wm. C. Brown Publishers

Book Team

Editor *Kevin Kane*
Developmental Editor *Carol Mills*
Production Editor *Renee A. Menne*
Designer *Eric Engelby*
Art Editor *Margaret Rose Buhr*
Photo Editor *Robin Storm*
Permissions Editor *Gail Wheatley*
Visuals Processor *Jodi Wagner*

Wm. C. Brown Publishers

President *G. Franklin Lewis*
Vice President, Publisher *George Wm. Bergquist*
Vice President, Operations and Production *Beverly Kolz*
National Sales Manager *Virginia S. Moffat*
Group Sales Manager *Vincent R. Di Blasi*
Vice President, Editor in Chief *Edward G. Jaffe*
Executive Editor *Earl McPeek*
Marketing Manager *Paul Ducham*
Advertising Manager *Amy Schmitz*
Managing Editor, Production *Colleen A. Yonda*
Manager of Visuals and Design *Faye M. Schilling*
Production Editorial Manager *Julie A. Kennedy*
Production Editorial Manager *Ann Fuerste*
Publishing Services Manager *Karen J. Slaght*

WCB Group

President and Chief Executive Officer *Mark C. Falb*
Chairman of the Board *Wm. C. Brown*

Copyedited by Kathy Massimini

All line art prepared by Hans and Cassady, Inc.

Library of Congress Catalog Card Number: 91–71937

ISBN 0–697–07628–8

Printed in the United States of America by Wm. C. Brown Publishers, 2460 Kerper Boulevard, Dubuque, IA 52001

10 9 8 7 6 5 4 3 2 1

For Lori, Anna, and Mara

brief contents

Part 3

THE INTERACTIONS OF MICROORGANISMS AND THE WORLD AROUND THEM

Part 4

MICROBIAL TAXONOMY

table of contents

Part 1

THE WORLD OF MICROORGANISMS

Part 2

THE INNER WORLD OF MICROORGANISMS: PHYSIOLOGICAL PROCESSES

Part 3

THE INTERACTIONS OF MICROORGANISMS AND THE WORLD AROUND THEM

Part 4

MICROBIAL TAXONOMY

preface

P

This moderately-sized book introduces bacteria and other selected microorganisms to students who are already familiar with the basic principles of biology and chemistry. In breadth and scope, the material discussed falls in a middle ground between the very basic and the highly detailed encyclopedic. Both microbiology and nonmicrobiology majors can enroll in a course for which this book is intended, provided that they have the necessary chemistry and biology prerequisites.

The undergraduate curriculum proposed by the American Society for Microbiology (ASM) requires a student to have these prerequisites and to take additional, advanced courses in all microbiology subject areas. In accord with that philosophy, I have organized this introductory text to aid students in becoming literate in microbiological history and terminology. Keep in mind that the material herein is not the final word on any particular topic. Rather the goal is to make it possible for a student to continue his or her study of microbiology using contemporary technical review literature. I have assumed that students who plan to continue in the formal study of microbiology will follow the suggested ASM curriculum and take the recommended additional courses in physiology, biochemistry, genetics, and systematics.

Given the audience for which this book is intended, I have tried to provide even-handed coverage of all major subjects traditionally discussed in an introductory microbiology class. This coverage includes health-related issues, food and industrial microbiology, genetic engineering, ecology, immunology, and taxonomy. Because this is not a text for clinicians, I have placed greater emphasis on the epidemiology of diseases and the methods of prevention rather than on the methods of treatment. Moreover, I detail the biology of microorganisms, as opposed to the biology of humans, and stress the causative agents of diseases, not the organ systems that these agents affect.

Part I provides a brief review of the development of microbiology as a science and outlines the general categories and descriptive traits of microorganisms. The physiological processes comprising the inner world of microorganisms constitute part II (chapters 3–11). Interactions of microorganisms with the world around them, covered in part III (chapters 12–19), include society's use of microorganisms in waste management, food preservation and processing, and genetic engineering. The impact of disease-causing agents on humans also is included in part III.Topics covered relate to immunology, epidemiology, and disease etiology. Part IV (chapter 20 and appendix A) provides a detailed treatment of microbial taxonomy.

It seems to me that a good reference book is not necessarily a good text; therefore, I have chosen to emphasize those features that make a sound text. Various topics are developed sequentially, and some are rediscussed in more than one chapter to expand old concepts and highlight new ones. I deliberately have used the repetition of ideas to emphasize their importance to the student.

Terms printed in **boldface** type can be found in the glossary. Specific items of bacteriologic technique are not covered in this text but can be supplied by any standard laboratory manual such as *Microbiological Applications,* (5th edition) by Harold J. Benson. However, "applications" boxes within each chapter relate the material covered to practical uses of microbiology in order to demonstrate the relevance of microorganisms in our lives today.

Individual instructors may find that some material needs to be omitted in order to fit the length of their particular term. Not all the material found in part I, for example, may be essential for students with excellent backgrounds in biology. Other instructors may find that the sequence of topics does not exactly match their own preferences. For instance, the material in chapter 8 probably could be covered any time after chapter 3 with only minimal difficulties. The major problem would be the implicit dependence on chapters 5 through 7.

More significantly, the detailed treatment of microbial taxonomy is positioned at the end of the book, even though many instructors may want to cover that material earlier. This position reflects the lack of general agreement among textbook authors and other instructors as to when during the semester taxonomy is best discussed. I have endeavored to present the taxonomic information in such a way that the reader can comprehend most of the material any time after completing chapters 3–7; however, the molecular biological ideas used in taxonomy are not covered until chapter 11.

I hope that this book will encourage more students of biology to discover and explore the fascinating and enthralling world of microorganisms. If it does so, it will have achieved one of its primary goals.

Edward A. Birge
Tempe, Arizona

acknowledgments

A

A book such as this one is not possible without the cooperation of many people. I am very grateful to all my colleagues who graciously provided illustrations and information to make this a better book. Special thanks to Jean M. Schmidt who so often dropped what she was doing in order to help resolve specific microbiological questions. It has also been a pleasure to work with the fine people at Wm. C. Brown who have been extremely helpful in all phases of production.

Reviewers

L. Rao Ayyagari
Lindenwood College

Susan T. Bagley
Michigan Technological University

Glendon R. Miller
Wichita State University

Charles W. Pratt
University of Illinois

Ralph Rascati
Kennesaw State College

James E. Struble
North Dakota State University

OTHER TITLES OF RELATED INTEREST FROM WM. C. BROWN PUBLISHERS

Aids/Sexually Transmitted Diseases

Viruses, The Immune System and AIDS by Edward Alcamo

The Biology of AIDS and Other Sexually Transmitted Diseases (1992) by Gerald Stine

Biotechnology

Biotechnology: A Guide to Genetic Engineering by Pamela Peters

Cell and Molecular Biology

Introduction to Experimental Cell Biology (1991) by Holly Ahern

Introduction to Experimental Molecular Biology (1992) by Holly Ahern

Laboratory Manual of Cell and Molecular Biology by John Choinski

A Manual of Laboratory Exercises in Cell Biology (1989) by C. Edward Gasque

Genetics

Genetics: A Human Perspective, 3rd Edition (1992) by Linda R. Maxson and Charles H. Daugherty

Genetics, 2nd Edition (1992) by Robert F. Weaver and Philip W. Hedrick

Basic Genetics (1991) by Robert F. Weaver and Philip W. Hedrick

Principles of Genetics, 3rd Edition (1991) by Robert Tamarin

A Guide to Problem Solving in Genetics (1991) by William Wellnitz

Immunology

Fundamental Immunology, 2nd Edition (1992) by Robert M. Coleman, Mary F. Lombard, Raymond E. Sicard, and Nicholas J. Rencricca

Immunology: A Laboratory Manual, 2nd Edition (1992) by Richard L. Myers

Laboratory Manual and Workbook in Microbiology: Applications to Patient Care (1991) by Josephine Morello, Helen Mizer, and Marion Wilson

Microbiology

Microbiology Laboratory Exercises, 2nd Edition (1992) by Margaret Barnett

Microbiology Laboratory Exercises, Short Version (1992) by Margaret Barnett

Microbiological Applications, Complete, 5th Edition (1990) by Harold J. Benson

Microbiological Applications, Short, 5th Edition (1990) by Harold J. Benson

Foundations of Microbiology (1992) by Kathleen Talaro

Virology

Introductory Experiments in Virology Laboratory Manual (1992) by Gerald Goldstein

part

1

THE WORLD OF MICROORGANISMS

Welcome to the study of microbiology, one of the youngest of the biological sciences! Its beginnings in the middle to late nineteenth century were contemporaneous with other great discoveries in biology, such as those of natural selection and Mendelian genetics. Because tremendous advances have been made in this field in but a little over 100 years, microbiology now stands in partnership with the older biological sciences of botany and zoology.

As their name implies, most microorganisms are not readily visible to the unaided eye; yet each of us can see many of their effects no matter where we live. Every living thing on this planet depends on microorganisms, either directly or indirectly, for its nutrition and well-being. The very diversity of these organisms, however, presents a problem in defining exactly what constitutes a microorganism. Although there is probably no one absolutely inclusive definition, microorganisms are single- and multicellular organisms that do not exhibit tissue specialization. Chapters 1 and 2 outline and develop this definition.

The medical, industrial, and commercial impact of microorganisms reflects dynamic processes on which all living things depend. Microorganisms are active at many points in the food chain and used specifically in the foundations of diverse commercial enterprises. They provide a form of nitrogen to fertilize a wide variety of plants and aid in the production of foods such as yogurt, cheese, pickles, wine, and beer. In addition, they are an integral part of both natural and artificially constructed waste disposal systems that recycle nutrients required by all living organisms. Microorganisms produce antibiotics, amino acids for food supplements, and enzymes for a variety of household and scientific uses. Many magazine and newspaper articles describe the exciting developments in genetic engineering, an industry in which microorganisms are used to provide more exotic products such as human growth hormone, tissue plasminogen activator (used for treating heart attack victims), and human insulin.

In this introductory part, chapter 1 provides a brief summary of the historical development of microbiology and a review of the types of microscopes used to visualize microorganisms. A general classification of microorganisms and how bacteria differ from eukaryotic organisms in general and eukaryotic microorganisms in particular are discussed in chapter 2.

chapter

1

Beginnings of Microbiology

chapter outline

GENERAL GOALS

Be able to discuss the lines of research that directly led to the establishment of microbiology as a scientific discipline.

Be able to discuss the theory of operation of microscopes used commonly to study microorganisms.

■ This chapter draws together the various historical trends in microbiology and defines the status of our knowledge in this field approximately to the year 1900. Further historical reviews are highlighted within individual chapters as topics are developed more extensively. Chapter 1 concludes with a brief overview of microscopes so that you will derive an appreciation of the many micrographs that illustrate this book.

Early Observations of Microorganisms

As Western civilization moved out of the Middle Ages, the general perception among biologists was of two great kingdoms of living organisms, plants and animals. The former derived their energy for growth and metabolism from sunlight, and the latter obtained their energy from consumption of other organisms and were capable of independent movement. All life was assumed to be readily visible to the interested observer. Therefore, it was with considerable astonishment that scientists first read the news of unsuspected tiny organisms being found everywhere.

This discovery came about as a natural consequence of the development of the science of optics. During the seventeenth century, revolutionary changes in biology occurred because scientists were able to examine the fine structure of tissues with their magnifying lenses. For example, in 1665 Robert Hooke developed his concept of cells by observing magnified sections of cork. Numerous other microscopists were active at that same time, but, from the viewpoint of a microbiologist, one stands out.

Antonie van Leeuwenhoek, a merchant living in the city of Delft in the Netherlands, was fascinated by the practical study of optics and became an expert at lens preparation (figure 1.1A). The lenses that he crafted were much finer and possessed greater resolution than those of his contemporaries (see section on microscopes at end of this chapter). In a series of reports filed with the Royal Society of London from about 1670–1690, Leeuwenhoek described his "animalcules," little single-celled organisms that he could see in a drop of water from a pond or from an **infusion,** a water extract prepared by boiling a nutritive substance like hay, meat, or a vegetable.

Leeuwenhoek could tell that his animalcules were alive because he observed cell division of the larger forms. Moreover, his infusions had few animalcules present initially, but with the passage of time, he saw as many as 10,000 organisms in a single drop of water.

A

FIGURE 1.1

(**A**) Antonie van Leeuwenhoek, 1632–1723. (**B**) Some of van Leeuwenhoek's animalcules as observed in aqueous solutions of tooth scrapings.

(A) The Bettmann Archive. (B) Redrawn from the *Philosophical Transactions of the Royal Society of London* 14:568–574, 1684.

His drawings illustrate some exceptionally small forms obtained from scrapings of his teeth (figure 1.1B). Leeuwenhoek's observations are now thought to be the very first made of bacteria. The term *bacterium* (plural bacteria) originates from a latinized form of the Greek word meaning little rod and came into general scientific use about the middle of the nineteenth century.

Further evidence indicated that these animalcules were as alive as other living things in that they could be killed by simple boiling of the infusions. Lazzaro Spallanzani, in 1799, was able to show quite clearly that there were at least two types of animalcules, those of the "higher class" that could not survive exposure to temperatures greater than 92° C and those of the "lower class" that could survive temperatures as high as 106 to 108° C. Animalcules of the higher class were consistently larger than those of the lower class, suggesting that some bacteria (as we now know them to be) have a greater heat resistance than others. Animalcules of both classes are considered *micro*organ-

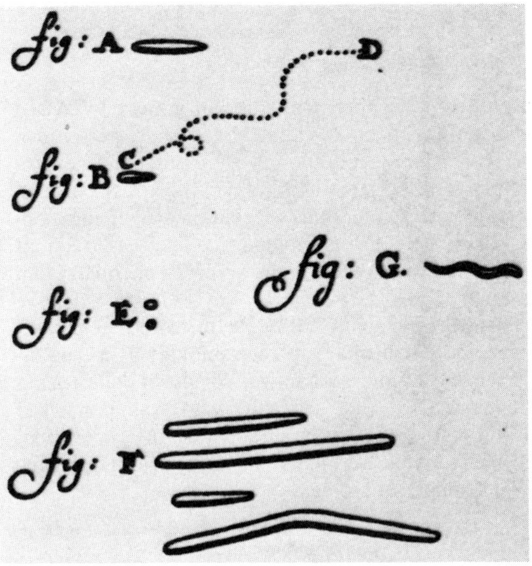

B

FIGURE 1.1 CONTINUED

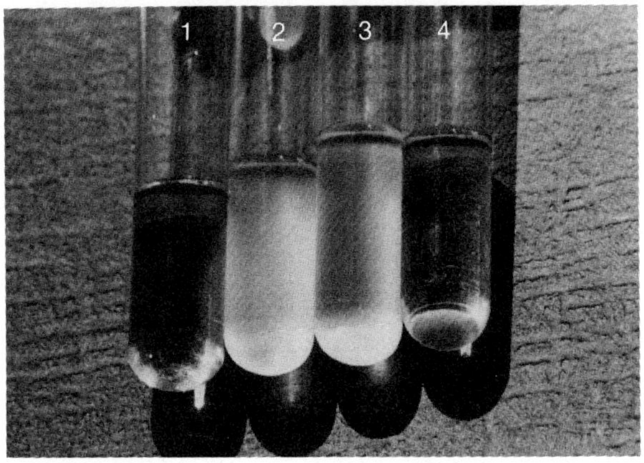

FIGURE 1.2

Growth of bacteria. Identical nutrient solutions were prepared and bacteria added to tubes 2 through 4 at various times. Shortly following the addition of bacteria, the tubes turned cloudy, as illustrated by tube 2. After a week or more, the tops of the tubes began clearing, as depicted in tube 3. Upon standing for a month, the bacteria settled to the bottom of the tubes, leaving a clear solution once again (tube 4).

isms in the sense that a microscope is necessary for the human eye to see them. If the definition of a microorganism is based solely on size, however, what maximum size is appropriate? A more practical definition involving cell structure is considered in chapter 2.

Controversy about Spontaneous Generation

Following the Renaissance in Europe, a great scientific and religious controversy erupted over the subject of spontaneous generation. Advocates of spontaneous generation proposed that living "simpler organisms" could arise from nonliving matter spontaneously, given the availability of appropriate conditions. In 1668 Francesco Redi disproved one specific aspect of the theory of spontaneous generation by covering samples of meat with a cloth that allowed free exchange of air but prevented flies from landing. Because the flies could not land to lay their eggs, no maggots (larval flies) developed even though uncovered meat quickly became riddled with maggots. This demonstration effectively excluded spontaneous generation in the case of macroscopic organisms but left open the possibility of spontaneous generation of still smaller organisms, such as Leeuwenhoek's microorganisms. Proponents of spontaneous generation thought that apparently simple organisms like bacteria could form spontaneously, and they used the existence of microorganisms to bolster their position.

The basic observations of Leeuwenhoek and others certainly seemed at first glance to support a belief in spontaneous generation. They were that an infusion of a substance such as meat went through various stages (figure 1.2). Initially, the nutrient solution was clear, but later it turned cloudy (turbid) and developed various unpleasant odors. A microscopic analysis showed the presence of numerous organisms of various sizes. The larger ones were fungal (filamentous) or protozoan in nature, while the smaller were bacteria. If left to itself, the solution eventually cleared again, but a sediment (the bacteria and other microorganisms) accumulated at the bottom of the container. Because this process occurred even when the container was covered with a cloth, some scientists argued that this sort of putrefaction must exemplify spontaneous generation. In fact one cleric, Turbevill Needham, boiled his infusion briefly and still observed that putrefaction occurred. It was in response to such observations that Spallanzani conducted his careful studies on the temperatures necessary to kill the various types of microorganisms.

The alternative idea of bacteria being all around us and, therefore, present during the preparation of the infusion was an unpopular notion because it seemed to fly in the face of common sense. Nevertheless, Spallanzani as well as the British physicist John Tyndall demonstrated that the entire cycle of putrefaction could be prevented if the infusion was boiled shortly after its

APPLICATIONS BOX

Putrefaction

Ever wonder why our ancestors were so interested in the putrefaction or rotting of various types of infusions? After all, who really cares about what happens to water extracts of hay? Why did many chemists become intrigued with these developments?

One difficulty most of us have in understanding the answers to such questions is that we tend to view historical events from our own perspective in time, with our knowledge of subsequent events influencing our view. When the major work on infusions was undertaken, there was no general agreement that biology was involved at all. What was known was that industrially useful substances such as lactic acid, butyric acid, or ethanol (as in wine or beer) could be obtained from infusions that had putrefied. Within these infusions, insoluble substances called "ferments" could be detected readily. These ferments did not appear to consist of plants or animals, and so classical chemists, personified by Justus von Liebig, maintained that putrefactive processes simply resulted from heating, from contact between different substances, or from a spontaneous change observed before putrefaction had completed. They asserted that ferments had nothing to do with the observed chemistry and were not alive.

The controversy about ferments mirrored a general concern about the legitimacy of microbiology as a true biological science (did microorganisms consist of cells arising from other cells?). When Pasteur, who was himself a chemist, laid to rest the notion of spontaneous generation, he moved the study of "ferments" from the realm of chemistry into the sphere of biology.

preparation and kept in a closed container thereafter. From this observation, they concluded that bacteria were indeed cellular and had to have arisen from preexisting cells that were both invisible to the unaided eye and killed by the boiling process.

Tyndall reinforced his conclusion about the ubiquitous nature of bacteria by demonstrating that they existed among the particles present in air. He showed that a beam of light shining at right angles to the observer's line of sight reflected off small, normally invisible particles in the air and made them readily visible as tiny bright points of light, a variation of the principle now used in dark-field microscopy discussed in the last section of this chapter. Tyndall noticed that if he boiled an infusion inside a closed container with windows in it, waited until all particulate matter had settled out of the air (no scattering of a beam of light), and then uncovered the infusion, no growth occurred.

Skeptics argued, however, that the boiling had inactivated some vital principle in the air and that Tyndall was preventing access of this principle to the infusion by keeping it in a closed container. Such an idea was not deemed particularly strange because scientists of this age still accepted the concept of phlogiston, believing it to be a gas released by burning objects. Louis Pasteur, a French chemist turned biologist, performed the definitive experiment in this controversy (figure 1.3A). He had observed previously that when a so-called souring disease of wine occurred, small, budding microorganisms larger than bacteria appeared in the wine. Moreover, he could prevent souring by heating the wine gently to kill these organisms. Pasteur's successful treatment of the "ailing" wine led him to wonder whether microorganisms also might cause human diseases. If the theory of spontaneous generation were correct, however, there would be no way to prevent such diseases by eliminating the microorganisms. Finally, in 1861 Pasteur thought it essential to disprove spontaneous generation once and for all.

Pasteur used a special container called a swan-necked flask in which he placed his infusion (figure 1.3B). After the infusion was boiled, the open neck would allow air (and any vital principle) to return to the infusion, but the curves in the neck of the flask would cause any particulate matter in the air to settle out on the moist walls of the tubing, thereby preventing it from reaching the infusion. To Pasteur's delight, his boiled infusions remained perfectly sound until the necks of the flasks were broken off, at which time the infusions rapidly putrefied. This result led Pasteur to conclude that microorganisms were always present as particulate matter in normal air. The experiment further reinforced his conviction that they could be the causes of human diseases.

Despite Pasteur's results, there was significant confusion among scientists about the reproducibility of certain experiments with infusions. Pasteur, for instance, never seemed to have any trouble preventing growth of microorganisms in his infusions by boiling them for a few minutes. Other scientists, such as John Tyndall and Turbevill Needham, still found putrefaction occurring even after they boiled their infusions for longer periods of time than Pasteur did.

In 1877 Tyndall finally resolved the difficulty with a series of careful experiments that demonstrated the existence of extremely heat-stable entities in hay in-

A

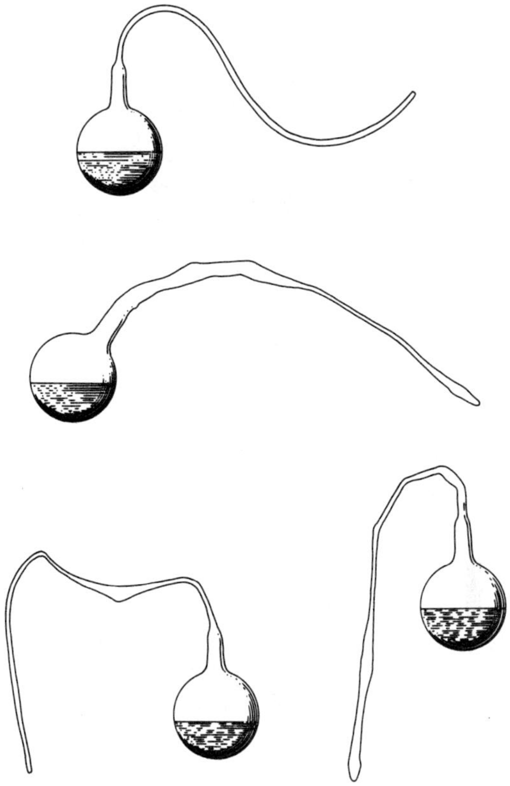

B

FIGURE 1.3

(**A**) Louis Pasteur, 1822–1895. (**B**) The swan-necked flasks used by Pasteur in his experiments disproving spontaneous generation of microorganisms.

(A) Historical Pictures Service. (B) Source: Louis Pasteur, "Mémoire sur les Corpuscles Organisés Qui Existent Dans L'Atmosphère: Examen de la Doctrine des Générations Spontanées; *Annales Sciences Naturelle,* 4th Series, Vol. 16, pp. 5–98, 1861.

fusions. He postulated that there was a "bacterial germ" resistant to boiling temperatures that could develop later into bacteria of the normal sort. Normal bacteria were killed easily by boiling for just a few minutes, but "germs" were not. To demonstrate the validity of his proposition, Tyndall prepared two batches of infusions. He boiled the first batch for 10 minutes and allowed it to stand. The second batch was boiled for one minute, but then reboiled for another minute after 12 hours and for 30 seconds at intervals over the course of the ensuing night and day. Within two days, the first batch of infusions had putrefied, while the second batch remained pure and clear for at least two months. The success of the second batch was traceable to the intervals between boiling episodes. Tyndall's "bacterial germs" are known now as bacterial endospores, and during the intervals between boilings, microscopic observations showed that these endospores germinated to yield vegetative cells easily killed by subsequent heating. The Tyndallization or fractional heating process is still used occasionally today. Figure 1.4 depicts an apparatus from the early 1900s made for this purpose.

Controversy about Microorganisms as Disease-Causing Agents

Our society has come to take for granted the amazing successes of medical science over the past 60 years; therefore, it is difficult to conceive of the state of human knowledge 150 years ago. It is important to realize that even at the beginning of the twentieth century, there was no such thing as a cure for a disease. Physicians had no antibiotics and no antimicrobial drugs with which to work. They primarily treated the superficial symptoms of a patient. No effective and safe treatment for any disease was available until the 1920s when patients with pneumonia were treated successfully.

FIGURE 1.4

Apparatus for killing microorganisms by Tyndallization. The material to be treated is placed inside the cabinet, whose door has been opened to show the internal structure. The lower pan is heated to generate steam that circulates through a space between the walls of the cabinet, thereby heating the contents before escaping through the chimney on top.

Indeed, as the nineteenth century began, there was not even agreement that such a thing as an infectious disease existed.

People often had noticed that some diseases seemed contagious, that is, passed from one person who had the disease to another who did not. The Chinese knew of the infectious nature of smallpox by the tenth century if not before, but the first post-Renaissance suggestion in Western civilization is generally credited to Girolamo Fracastoro in 1546. The idea took hold very slowly, and those who espoused it were shunned frequently.

For example, in 1850 Ignaz Semmelweis, a Hungarian physician, reported to the Vienna medical society his study of two obstetrical clinics that had drastically different rates of childbed (puerperal) fever. In a teaching clinic that trained obstetricians, the death rate was 8.2%, whereas in the other clinic staffed by midwives it was only 2.2%. Semmelweis theorized that the physicians themselves were transmitting the disease by going directly from infected patients to newly admitted patients in the teaching clinic. Infections were rarely seen in the other clinic because the midwives were not exposed to cases of the disease; therefore, they could not transmit it from one patient to another. When Semmelweis instituted a regimen that required physicians and students to wash their hands with lime chloride ($CaOCl_2$) after they saw each patient, the death rate at the teaching clinic dropped essentially to the same level as at the other clinic. Because physicians in the community were so completely offended by the suggestion that they themselves were the source of infection, they ostracized Semmelweis. In the United States Oliver Wendell Holmes, a professor of anatomy and physiology at Harvard Medical School, advanced similar unpopular ideas.

Pasteur was convinced that microorganisms were the source of most if not all diseases that afflicted humans and other members of the animal kingdom. Many other scientists shared his passionate desire to prove this hypothesis, predominant among them was the German physician Robert Koch (figure 1.5). Koch published a nearly definitive proof in 1876 claiming that anthrax, a disease afflicting sheep and other animals as well as humans, was caused by one particular bacterium, *Bacillus anthracis.* Koch and Pasteur soon became bitter rivals, but the work both produced in the course of their feuding was monumental. For instance, in 1884 Koch culminated his laboratory's work with a published paper that demonstrated the causative organisms of anthrax and tuberculosis. In this paper, he also formulated his famous four postulates that, if carried out, would demonstrate whether a particular microorganism was responsible for causing a particular disease. These postulates are outlined as follows.

1. The suspect organism must be found associated with all cases of the disease.
2. The suspect organism must be obtained as a pure culture (no other organism present).
3. The suspect organism must cause the disease if it is then injected into a susceptible animal.
4. The suspect organism must be reisolated from the test animal and shown to be the same as the one used originally to eliminate the possibility of coincidence.

In order to fulfill these postulates, it was necessary for Koch to develop a methodology to isolate individual organisms, somthing that is very difficult to accomplish using liquid medium. He stipulated use of a solid surface as part of a growth medium so that cells descended from one original cell would not have a chance

FIGURE 1.5
Robert Koch, 1843–1910.
Historical Pictures Service

to mingle with other cells but would have to remain near their point of origin.

With the knowledge that specific microorganisms caused specific diseases came the possibility of preventing and/or treating a disease. The preferred technique was to prevent the disease by administering what Pasteur called a **vaccine,** a substance that, when given to an animal, causes the animal to become resistant to the disease. Pasteur and his co-workers developed vaccines consisting of weakened causative agents for anthrax, fowl cholera, and rabies, although they had no real understanding of why the vaccines were effective. Other laboratories became active in this field as well, and in 1890 Emil von Behring and Shibasaburo Kitasato published a paper describing the mechanisms of immunity to diphtheria and tetanus. Their work led directly to the development of the science of immunology.

Meanwhile, in Koch's laboratory at the turn of the century, Paul Ehrlich began his search for chemicals that would specifically kill or retard the growth of disease-causing microorganisms, thereby establishing the science of **chemotherapy.** His first success was with an arsenic-containing compound called Salvarsan, which was reasonably effective against syphilis. A German chemical company later patented this drug. Ehrlich's work stimulated the search for other antimicrobial chemicals and led indirectly to the discovery of **antibiotics,** compounds produced by organisms that kill or inhibit the growth of nonproducers.

At a somewhat earlier time, Joseph Lister, an English physician, was convinced sufficiently by Pasteur's arguments that microorganisms caused disease; therefore, he proposed the concept of **antiseptic surgery** in 1867. He observed that not all surgical intervention failed because of errors in actual surgical technique, but many postoperative patients died from **sepsis,** the subsequent growth of microorganisms in a wound. If microbial growth could be prevented, patient survival would increase. After Lister observed that carbolic acid (phenol) was sufficient to prevent growth of microorganisms in nutrient solutions, he reasoned that if the chemical were used to clean a wound, the complications of infection would be prevented because phenol would kill the microorganisms. Although Lister had some problems with phenol being toxic to normal skin, his antiseptic method worked, especially when large areas were involved. Eventually, he used a paste or putty prepared from calcium carbonate, phenol, and linseed oil for this purpose. Subsequent to the introduction of this methodology, two wards of the Glasgow Royal Infirmary that had had the worst record with respect to infections reported no such cases after Lister's antiseptic had been used for a nine-month period. Modern and less dangerous variations of Lister's work now exist.

As more and more diseases were investigated, it soon became apparent that not all could be prevented or treated with the methods that had been developed. Although a disease clearly was infectious in many instances, no associated microorganism could be seen or grown on artificial medium. In addition, special filters, whose pores were small enough to remove bacteria from a solution, did not reduce the infectivity of the solution. These "poisons" were given the Latin name of viruses (meaning poison) and sometimes referred to as "filterable viruses." In 1891 Dmitri Ivanowski demonstrated that a virus caused a mosaic disease in tobacco plants (figure 1.6). Friedrich Loeffler showed that a virus was responsible for foot and mouth disease in domestic livestock in 1898. However, a real understanding of virus-caused diseases was not possible until the molecular biology revolution in the last half in this century.

APPLICATIONS BOX 1

Disease-Causing Agents

The general public has remarkable faith that medical scientists and microbiologists can quickly pinpoint the cause of any disease and immediately find a cure and/or a preventive treatment. Indeed an application of Koch's postulates would seem to be the solution for any problem. The difficulty with Koch's postulates is their underlying assumption that it is possible to grow the organism. If the organism cannot be grown, the investigator is reduced to experiments such as those used to demonstrate the existence of viruses. Moreover, the possibility of chemical and not biological poisons must be considered.

One example of these difficulties is Legionnaires disease (as it came to be known). In 1976 numerous individuals became sick and several died from a respiratory disease they had contracted at an American Legion convention in Philadelphia. An immediate search was undertaken to determine the cause of the disease, but nearly six months of intensive work were required before a bacterium was grown in the laboratory. Once the organism was isolated, it was comparatively straightforward to demonstrate, using animal hosts, that it was indeed the causative agent of the disease.

An example of a continuing problem with disease treatment is acquired immune deficiency syndrome (AIDS). The virus that causes the disease is now well known. A great deal of effort has been devoted to finding an animal species, other than humans, that is susceptible to the virus and can be used to study its effects. Only within the last few years have inexpensive animal models involving mice been available for AIDS research. These inexpensive models can be used to replace such expensive animal models as chimpanzees and allow more extensive experimentation. At the time of this writing, no reliable treatments have been developed.

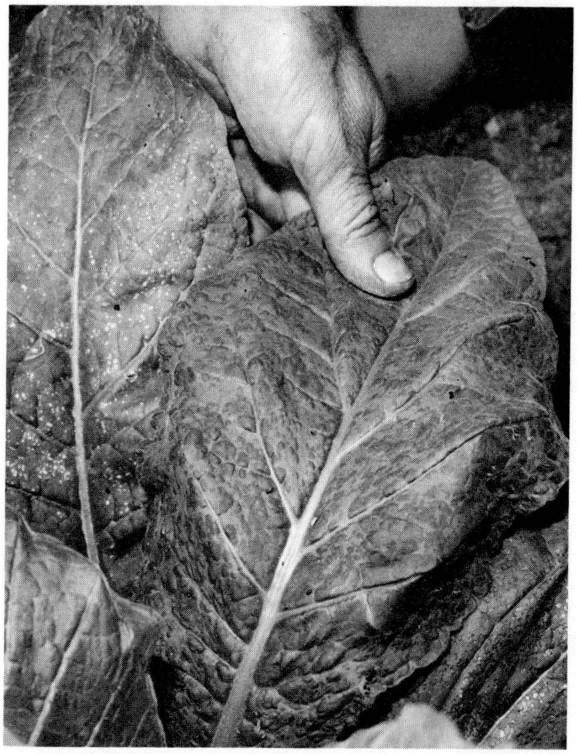

FIGURE 1.6
Tobacco plant infected with tobacco mosaic virus.
© Grant Heilman

Beginnings of Microbial Physiology and Ecology

Perhaps the first clue that there was something really special about microorganisms came from Pasteur's 1861 observations that "infusoria," the organisms he found in infusions, were responsible for the production of butyric acid from a mixture of sugar, ammonium ion, and phosphates. In addition, this production could occur in the absence of air or free oxygen (anaerobic growth), which was a revolutionary idea. It was only 100 years prior to this date that Antoine Lavoisier had shown that animals require oxygen in order to derive the necessary energy for life. If these microorganisms do not require oxygen, then how do they survive? Many biochemists turned their attention to microbial metabolism during the twentieth century.

Microorganisms also were responsible for important portions of the nutrient cycles that occur in the environment, another surprising discovery. Two major, independent investigators working in this area were Martinus Beijerinck in the Netherlands and Sergei Winogradsky in Paris. Together, they were able to demonstrate the role bacteria play in the sulfur and nitrogen cycles, especially in the conversion of atmospheric nitrogen to forms usable by other organisms. For this reason, they are considered to be among the founders of the science of microbial ecology.

The science of food microbiology was built on our increasing knowledge of microorganisms and their activities. Many processes, such as the preparation of alcoholic beverages, olives, bread, cheese, and yogurt, depend on the metabolic activities of microorganisms. Until the existence of microorganisms was proved, food preparation could not be considered on a scientific basis. Similarly, food preservation depends on inhibiting the growth of microorganisms. With modern technology, we have developed many different methods of food processing and preservation.

Microscopes: Basic Tools of the Microbiologist

By the beginning of this century, microbiologists knew much about the general behavior of bacteria and their physiology but somewhat less about their structure and how their structure related to their functions. The discovery of improved microscopic techniques that allowed better resolution of such tiny objects was driven by this lack of knowledge about structure. In order for you to understand how microbiologists obtain the information presented in the following chapters, you need to know how the major types of microscopes function. You then can begin your study of the correlation between the structure and function of microorganisms outlined in chapter 2.

Light Microscopes

The earliest microscope used by Leeuwenhoek was a far cry from the common microscopes of today, being basically a very tiny, perfect lens that was used essentially like a magnifying glass (see the example in figure 1.7). The difference between it and a conventional magnifying glass lies in the amount of useful magnification. A high-quality magnifying glass of today magnifies 10- to 20-fold, whereas Leeuwenhoek's best microscope magnified objects 200-fold. Leeuwenhoek mounted a specimen on the tip of an adjustable screw that he could position to focus the specimen. This is an example of a simple microscope.

The microscope familiar to all biology students is the compound microscope composed of a series of lenses rather than a single lens. The general construction of the compound microscope is shown in figure 1.8. By constructing the individual lenses of this microscope from multiple pieces of glass, the manufacturer has eliminated chromatic aberration, the appearance of colored fringes around the object being viewed, as well as provided a flat field. In a flat-field microscope, all the viewing area is in focus at the same time. In a simple microscope such as Leeuwenhoek's, either the center

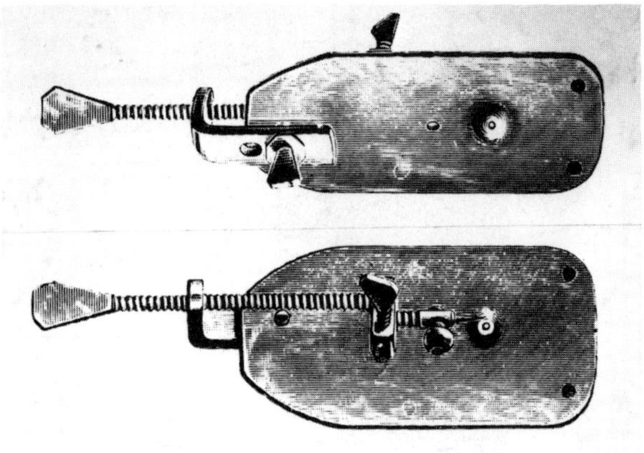

FIGURE 1.7

Model of Leeuwenhoek's original microscope. The observer attached the specimen to the threaded screw that held it at the appropriate distance from the lens. The entire microscope was then held close to the eye like a conventional magnifying glass.

The Bettmann Archive

or the edge of the field can be in focus, not both simultaneously. The Abbé condenser of a compound microscope also greatly contributes to one's ability to visualize microorganisms because it provides a beam of illumination that can be focused precisely on the specimen. Although this type of microscope has been the mainstay for microbiologists over the years, it has its limitations, especially because most cells are transparent and, therefore, difficult to discern against their background.

One solution to the problem of transparency has been to use various dyes to stain the cells to make them visible, a technology that is discussed further in chapter 2. Staining is satisfactory for many purposes, but it does require preparation of the material and usually results in cell death. Microbiologists and other biologists as well have always had an interest in techniques that would allow them to study cells under conditions resembling a normal environment.

The dark-field microscope was an early development permitting the visualization of transparent cells. The principle behind this microscope is based on a common observation that Tyndall used in his experiments: Particles in a beam of light scatter the light at sharp angles to the original beam. For example, a beam of sunlight will highlight dust particles in the air that are not visible normally in a room. Similarly, if a microscope is arranged with a special condenser to block rays of light that normally would hit the objective lens,

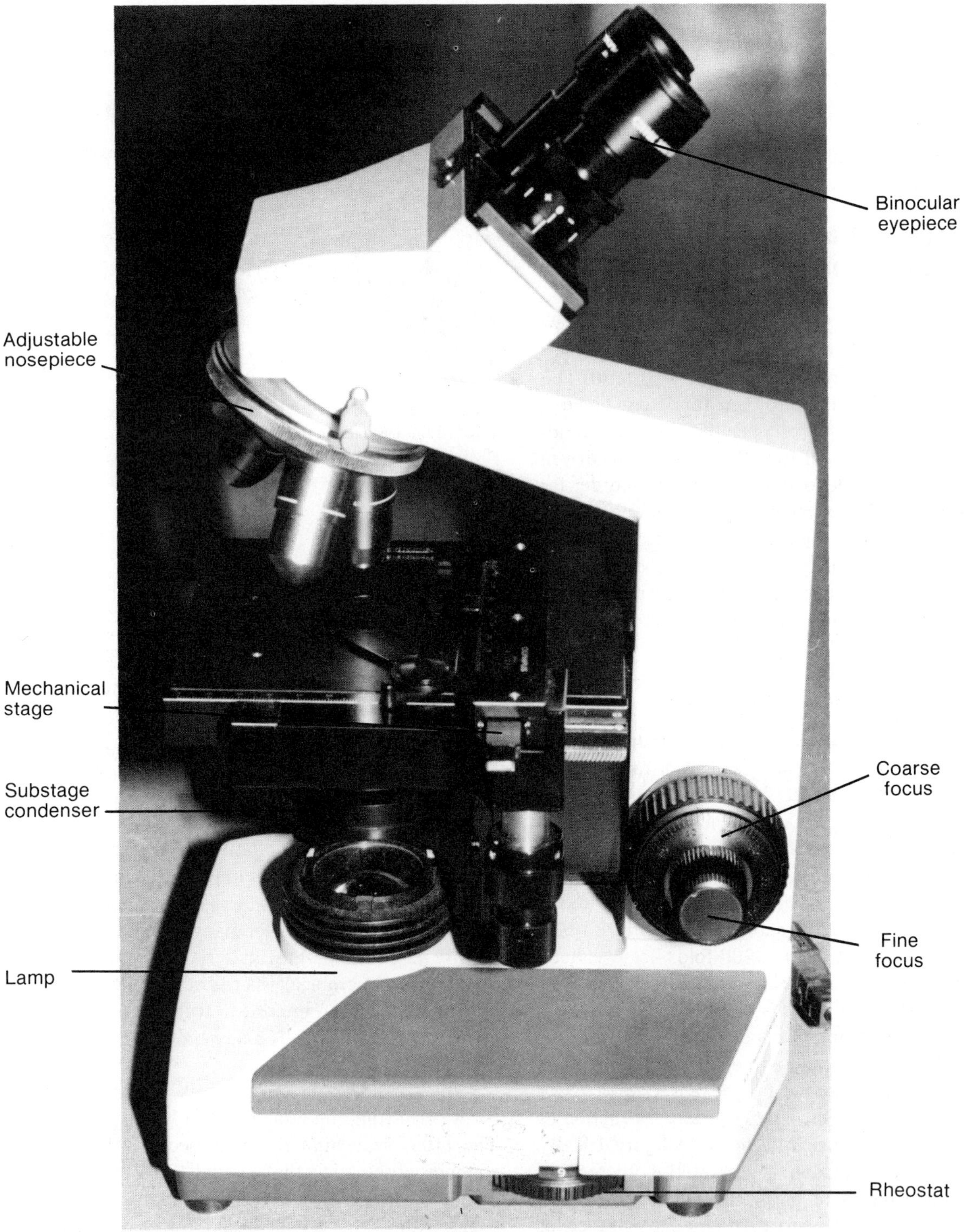

FIGURE 1.8

Photograph of a modern compound microscope. The built-in light source contains a high intensity quartz-halogen bulb. The rheostat regulates the intensity of the light. The substage (Abbé) condenser adjusts vertically to focus the image of the light bulb on the plane of the specimen. With the mechanical stage, the investigator can make very small, reproducible movements of the specimen. An adjustable nosepiece with several different power lenses permits a range of magnification of the specimen. Binocular eyepieces allow specimen examination with both eyes.

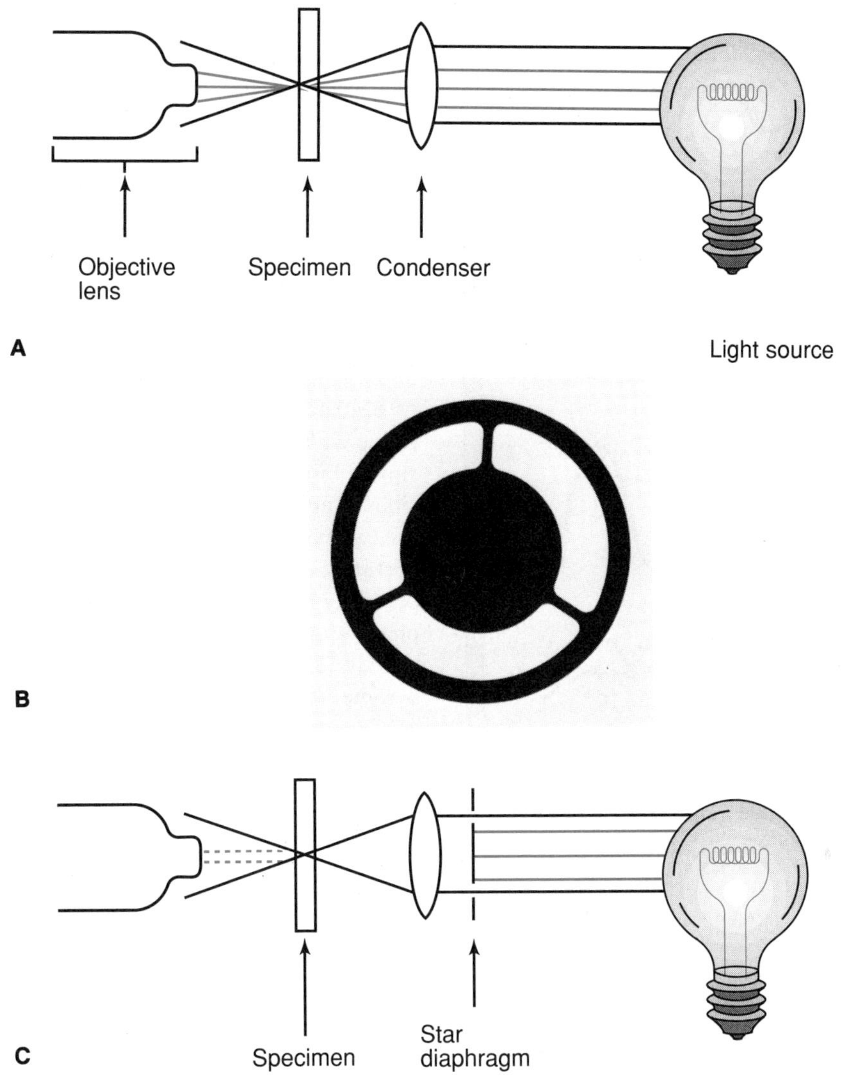

FIGURE 1.9
Dark-field microscopy. (**A**) The normal light path for bright-field microscopy. Note that some of the light rays passing through the specimen do not strike the objective lens. (**B**) A star diaphragm can be inserted between the light source and the condensing lens. (**C**) Dark-field microscope with a cross section of the star diaphragm shown in its proper position. Note that the direct rays of light (those that would enter the objective lens) are completely blocked. Only those indirect light rays that reflect off an object in the specimen plane will be able to enter the objective lens.

the observer sees only a dark field (figure 1.9A–C). Any cells present, however, scatter the light, and the scattered rays of light produce an approximate image of the cell (figure 1.12B).

Other major developments in light microscopy of transparent objects are all based on the phase-contrast principles formulated by the physicist Frits Zernike. He knew that light rays have maximum speed in a vacuum but travel slower when passing through any transparent substance. Furthermore, whenever a beam of light changes speed, it is refracted (bent) at an angle that depends on the magnitude of the velocity change. Therefore, when a specimen on a microscope stage has a different refractive index from the surrounding medium, the light rays are refracted and altered in their phase but unaltered in their wavelength and intensity as they enter and leave the specimen (figure 1.10A–E). Because the human eye is most sensitive to changes in

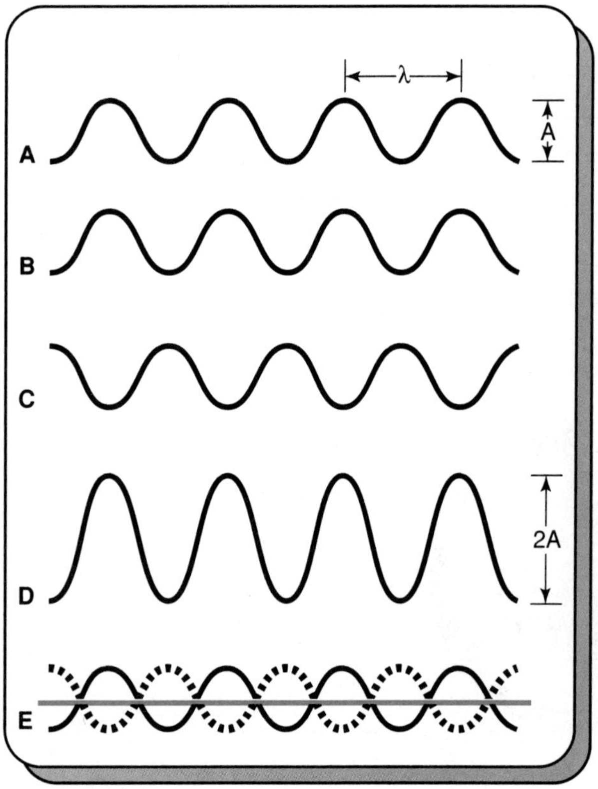

FIGURE 1.10
The wave nature of light. A shift in phase of 1/2 wavelength is often confused with a change in color. (**A**) A ray of light can be represented as a simple sine wave having a definite wavelength (λ) and amplitude (A). The eye interprets variations in λ as changes in color and variations in A as changes in brightness. (**B**) A ray of light identical to that depicted in the top diagram. Note that both light rays are in phase because they go up at the same time and down at the same time. (**C**) A ray of light identical to the one depicted in the top diagram except that it is 180° (1/2 wavelength) out of phase; that is, one wave is going up while the other is going down. By themselves, waves A, B, or C would appear to have the same color and brightness to the eye. (**D**) Constructive interference. If waves A and B arrive at the eye simultaneously, the color is unchanged, but the eye perceives the light to be twice as bright as either individual wave. (**E**) Destructive interference. The two waves exactly cancel (colored line), and the eye of an observer perceives nothing.

wavelength and intensity (color and brightness), Zernike devised an optical system (the phase-contrast microscope) to convert the refraction and phase shift normally provided by the transparent specimen into a change in intensity that the eye readily can detect.

The phase-contrast optical system depends for its function on a special condenser and modified objective lenses. The condenser is designed with a special annular stop (annulus or annular diaphragm) located below the substage condenser so that the specimen can be illuminated with a relatively narrow, hollow cone of light. As the cone of light spreads again after passing through the plane of the stage, it will, in the absence of a specimen, pass through an appropriately sized ring of glass (the phase ring) in the objective lens (figure 1.11A–B). By trial and error, Zernike observed that light passing through a specimen is retarded in phase by 1/4 wavelength. These light rays are also refracted away from the path of the cylinder of light and will miss the phase ring because of this refraction. At the ocular lens, both direct and indirect refracted light rays will converge on the eye of the observer and possibly undergo interference. Interference may be constructive or destructive and occurs when light rays are out of phase with one another (figure 1.10C–E). If the phase ring is built properly into the objective lens, it is possible to ensure that, for light of a specific wavelength, the two types of waves will arrive at the eyepiece either perfectly in phase again (**constructive interference**) or perfectly out of phase (**destructive interference**). The phase shift necessary to accomplish this feat is obviously 1/4 wavelength; for this reason, the phase ring is sometimes referred to as the quarter-wave plate.

The phase-contrast microscope was of such fundamental importance that in 1953 Zernike won the Nobel prize for its development. Since Zernike's time, several modified phase-contrast systems have become available (see micrographs in figure 1.12C–D). The Nomarski differential interference phase-contrast microscope produces an image (figure 1.12C) that appears to have depth, but the theory behind this type of optics is beyond the scope of this book. A recent development is the Hoffman modulated-contrast system (figure 1.12D) that mimics the type of image produced by the Nomarski system but is less complex optically.

Electron Microscopes

In order to learn more about microorganisms, microbiologists need to see their structures defined as clearly as possible. However, the problem is not so much one of magnification as it is of resolution. The term **resolution** refers to an observer's ability to distinguish between two closely spaced objects. The degree of resolution or resolving power is expressed as the minimum separation (distance apart) necessary for two objects to be seen discretely (figure 1.13); therefore, the smaller the number, the better the resolution of the microscope being tested. A light microscope can magnify over 1,500 diameters, but above about 1,000 diameters the resulting image is usually a larger blur.

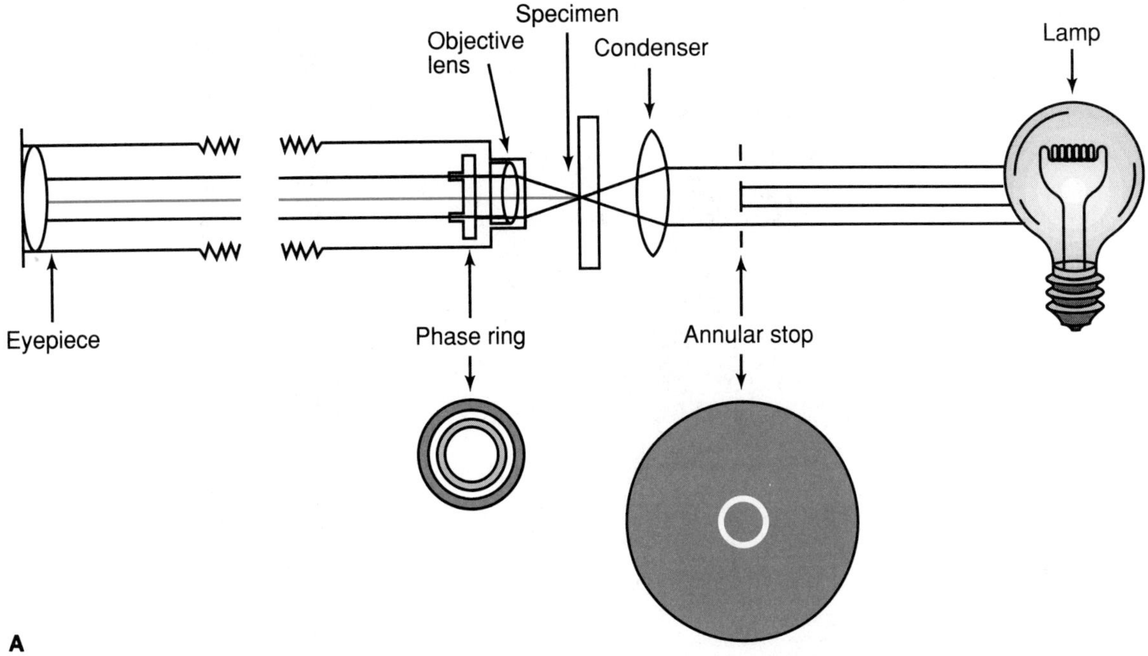

A

FIGURE 1.11

Phase-contrast microscopy. (**A**) The heart of the microscope is the phase ring and the corresponding annular stop. Although the annular stop appears similar in form to a star diaphragm (compare with figure 1.9), its function is quite different. This particular model fits on top of the regular condenser like a cap. Light rays passing through the annular stop are guided to a specific area of the objective lens in which the phase ring is located. Rays passing through the specimen are refracted so that they miss the phase ring (colored lines). In this example, the phase ring retards the direct rays of light so that their phase will match that of the indirect light passing through the specimen. The result is bright phase contrast (a bright image on a darker field) due to interference at the eye of the observer. (**B**) The alga *Bangia fuscopurpuria* photographed with phase contrast microscopy. The smaller entities are bacteria.

Micrograph courtesy of Dr. Pearl H. Lin, Department of Microbiology, Arizona State University.

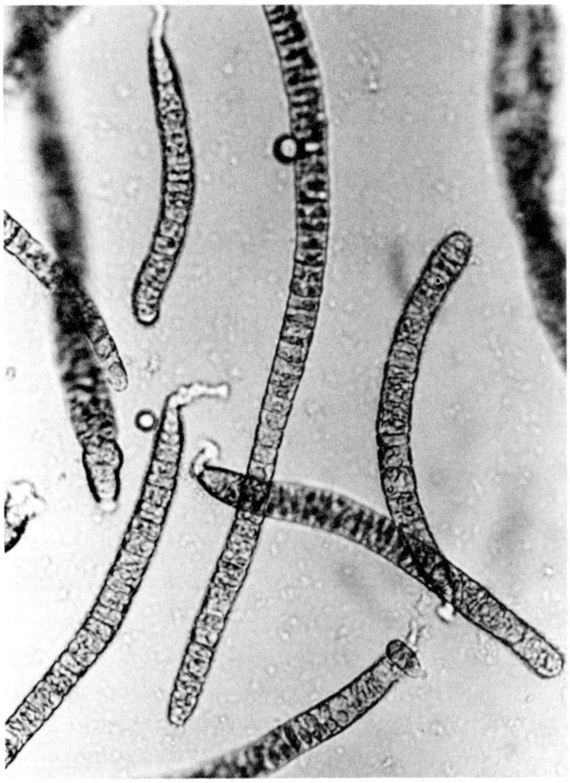

B

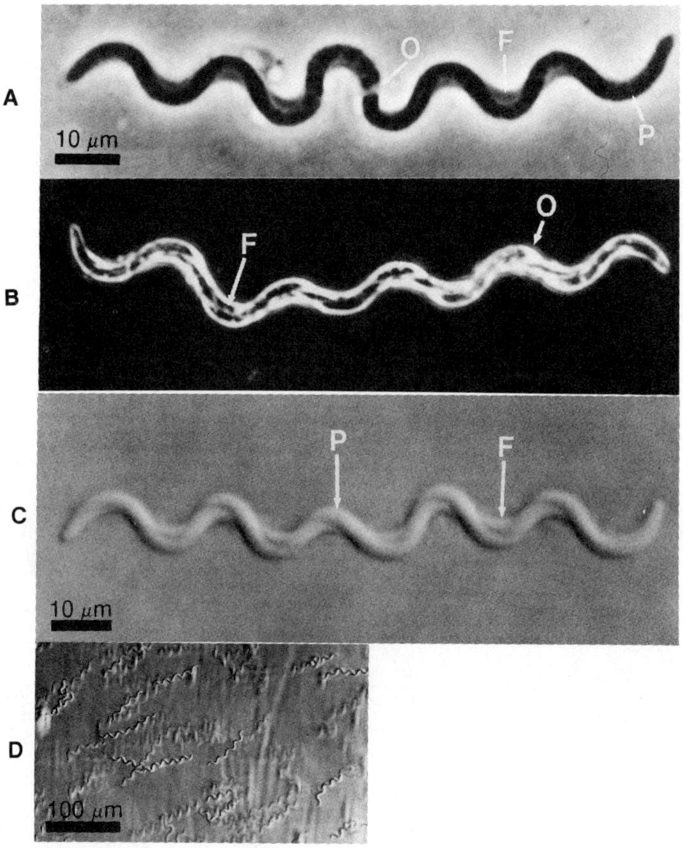

FIGURE 1.12

Images of *Cristispira,* a spirochete isolated from a mollusk, viewed with various types of optical microscopes. (**A**) Phase-contrast microscopy. (**B**) Dark-field microscopy. (**C**) Nomarski optics. (**D**) Hoffman modulated contrast. Labeled structures: protoplasmic cylinder (P); axial filament or crista (F); outer sheath (O)

Reprinted with permission from E. V. Lawry, H. M. Howard, J. A. Baross, R. Y. Morita. 1981. The fine structure of *Cristispira* from the lamellibranch *Cryptomya californica* Conrad. *Current Microbiology* 6:355–360.

The general formula expressing resolution is

Equation 1.1
$r = k \cdot \text{wavelength/NA}$

where r is the resolution (expressed in the same units of distance as the wavelength), k is a proportionality constant that has a value ranging from 0.5 to 0.6, and NA is the numerical aperture of the lens, a measure of its light-gathering capabilities. Numerical aperture values generally fall between 1.4 to 1.5 for very good lenses and below 1.25 for less expensive ones; therefore, a high-quality microscope with a short wavelength of light (blue) should have a resolution of 0.1 to 0.3 μm.

Because the numerical aperture is only slightly variable, the major improvements in resolution have come from changes in the wavelength of radiant energy used for illumination. This is why most light microscopes come equipped with blue filters (wavelength of approximately 480 nm) rather than green (520 nm) or red (640 nm). From the standpoint of physics, still shorter wavelengths are possible, but they are not visible to the human eye. Nevertheless, ultraviolet microscopes have been produced with cameras that record images for later examination.

The most dramatic improvement in resolution has come with the development of the electron microscope, a device that employs beams of energetic electrons inside an evacuated cylinder instead of electromagnetic radiation at longer wavelengths (light) to illuminate the specimen. Electron microscopes use two

APPLICATIONS BOX

Energy-Dispersive Analysis

In later chapters, we consider the surface of a cell and its chemical composition. Obviously, it is not a trivial problem to determine what chemical elements are present within something as small as a bacterial cell, but recent advances in computer and semiconductor technology have made such analyses routine.

When a beam of electrons hits an object being scanned under a microscope, X-ray fluorescence occurs. The energy of the electrons is absorbed by individual atoms and then re-emitted as X rays with an energy spectrum characteristic of the atom's atomic number. If an SEM is fitted with a sufficiently small X-ray detector that can provide information about the energy emitted from a particular location, the chemical composition of the cell and subcellular surface structures can be determined. Because electrons have a limited ability to penetrate any object, only surface layers can be examined in this way. Nevertheless, the technique is a powerful one and is becoming increasingly popular.

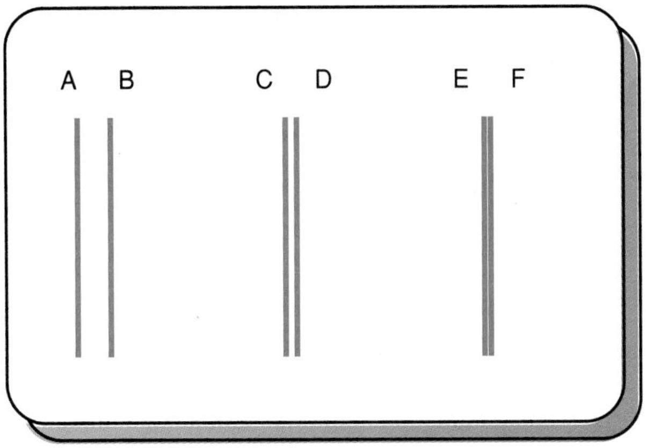

FIGURE 1.13
Test your resolution. At normal reading distance, you should be able to distinguish lines A and B as separate structures. Unless your eyesight is particular acute, however, you are unlikely to be able to separate lines C and D visually without bringing the page closer to your eye. Lines E and F are so close together that even increased magnification is unlikely to permit your eye to resolve them.

basic modes to develop an image (figure 1.14A–D). In the case of the transmission electron microscope (TEM), electrons pass through a specimen, just as photons of light do in a compound light microscope, and produce an image when they strike a fluorescent screen. **Fluorescence** refers to a phenomenon in which a material absorbs energy striking it at one wavelength and re-emits that energy at a different, longer wavelength. Generally, the fluors (chemicals that fluoresce) are chosen so that the emitted radiation is at a yellow-green wavelength to which the human eye is most sensitive. Because electrons have very poor penetrating power, TEM specimens are usually sections rather than whole cells, even for something as small as a bacterium. Electron-dense (absorbing) stains such as osmium tetroxide or uranyl acetate can be used to make particular cellular structures stand out in much the same way that stains for the light microscope absorb specific wavelengths of light to make cells and subcellular structures visible.

In some cases, cells are not sectioned but cracked open. In the **freeze-fracture** process, bacterial cells are suspended as a paste and then deeply frozen to temperatures at which they become brittle. When the frozen cells are struck with a "rammer," they crack along the lines of least resistance, which usually means along a lipid bilayer. While still frozen, the cracked surface is overlaid with a carbon film to create a carbon replica of the cell. Carbon replica preparations allow microbiologists to examine surface and immediate subsurface structures. Examples of freeze-fracture electron micrographs can be seen in the applications box on archaebacteria in chapter 3 and in figure 4.5.

The second mode by which an electron microscope operates is one in which a beam of electrons is scanned across a specimen in much the same way that an electron beam is scanned across the surface of a television tube to produce a picture. In a scanning electron microscope (secondary emission microscopy; SEM), the image is developed from the number and angles of scattered electrons emitted or reflected from the surface of the specimen as the electron beam scans it. For best effect, the specimen must be coated with a thin layer of conducting metal, usually gold. Information derived from the angles of scatter is then converted into a television picture that depicts the three-dimensional relationships of the various parts of the specimen. A monitor displays the picture. Scanning transmission electron microscopes (STEM) also are available and offer some advantages over TEM in that their scanning beam of

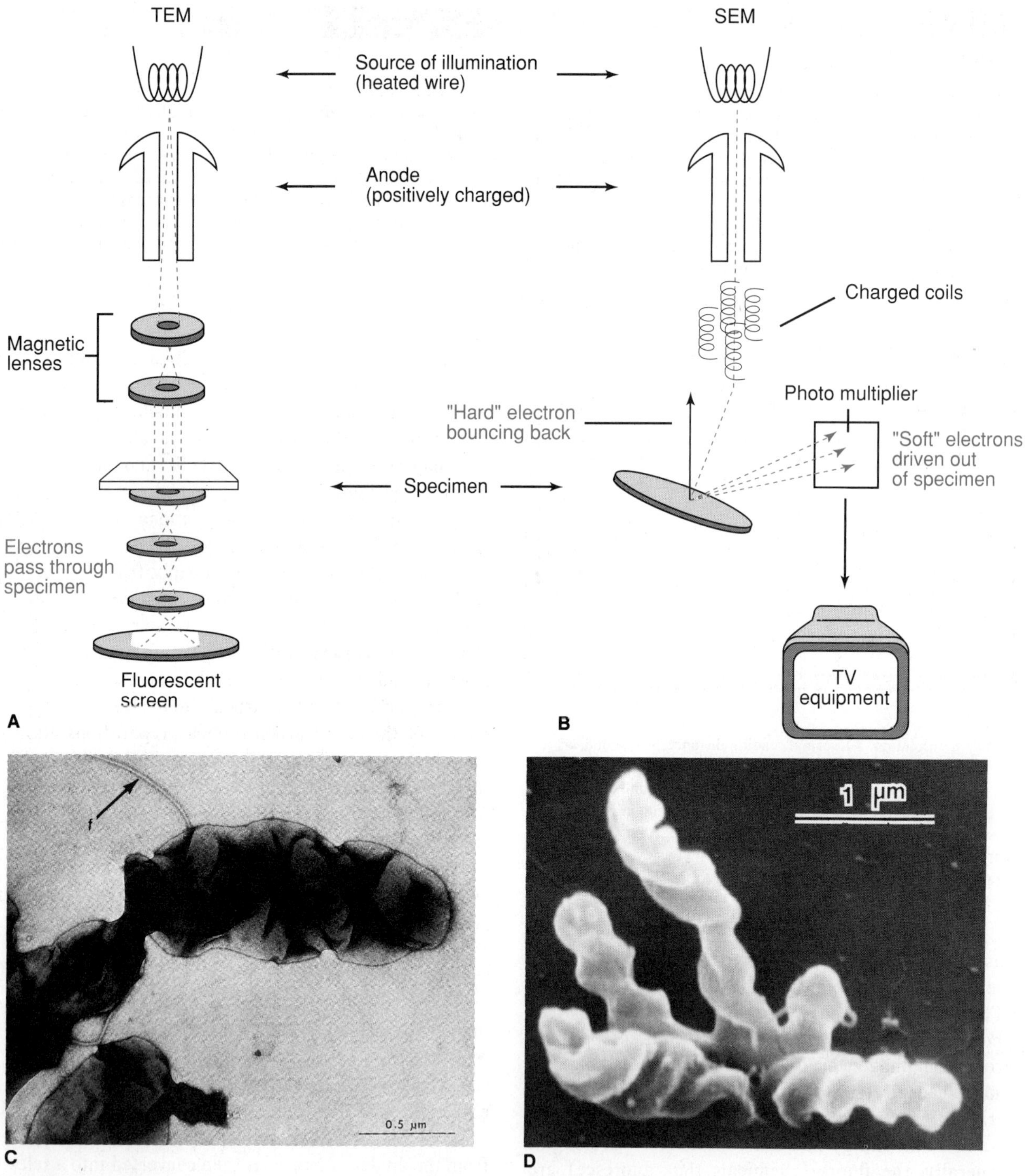

FIGURE 1.14

Types of electron microscopes. (**A**) With the transmission electron microscope (TEM), electrons are accelerated away from the filament by a hollow cathode. They then pass through the specimen to strike a fluorescent screen or piece of photographic film. Although the diameter of the electron beam is variable, it is always present in the center of the field. (**B**) With the scanning electron microscope (SEM), the accelerated electrons are formed into a narrow beam that is scanned across the specimen in a series of parallel lines. The electrons that reflect off the surface of the specimen are used to form an image on a television screen. (**C**) Electron micrograph of the bacterium *Seliberia stellata* stained with phosphotungstic acid and photographed with a TEM. (**D**) Photograph of the same organism taken with an SEM.

Electron micrographs courtesy of James R. Swafford, Brandeis University.

electrons does not stay in one place long enough to degrade the specimen significantly, whereas specimen absorption of energetic electrons causing specimen deterioration can be a serious problem in TEM.

Because electron microscopes have electromagnets instead of glass for their condensing and objective lenses, their optical systems are less precise and do not approach the theoretical maximum resolving power for electrons. Nevertheless, a typical resolution for a high-quality SEM is about 10 nm and that for a good TEM is less than 0.3 nm compared to the resolution of an excellent light microscope that might be around 100 nm. Figure 1.14C–D presents examples of specimens observed with both types of electron microscope.

Summary

The origins of microbiology are bound intimately to the origins of microscopy. The earliest workers in this field had to make do with simple light microscopes. Development of the compound light microscope made better magnification and resolution available to average scientists. Various physical techniques are used to adapt the light microscope for specific functions, such as one to circumvent the lack of contrast between living cells and their surrounding medium. These techniques include dark-field microscopy, phase-contrast microscopy, Nomarski differential interference phase-contrast microscopy, and Hoffman modulated-contrast microscopy.

Nineteenth-century microbiology was fraught with controversy. First, scientists questioned whether microorganisms behaved in the same way as other organisms. Then there was the question of whether they in fact were the causative agents of various diseases. In the course of settling these controversies, not only bacteriology but also immunology were established. Louis Pasteur and Robert Koch are the two scientists most often credited with establishing microbiology as a science.

Questions for Review and Discussion

1. In which situations would it be most advantageous to use each of the following kinds of microscopes: dark field, bright field, phase contrast, TEM, SEM?
2. How could you adapt Tyndall's experiments in order to prepare a solution containing only bacterial spores?
3. What proofs could you offer to a skeptical person about microorganisms being disease-causing agents?

Suggestions for Further Reading

Brock, T. D. (ed. and trans.). 1961. *Milestones in microbiology.* Prentice-Hall, Inc., Englewood Cliffs, N.J. A selection of classic papers and excerpts of papers in microbiology beginning at the time of Leeuwenhoek and continuing to the time of antibiotic discovery. A good source book that includes the works of many scientists discussed in this chapter.

Brock, T. D. 1988. *Robert Koch: A life in medicine and bacteriology.* Science Tech Publishers, Madison, Wis.

Dobell, C. 1958. *Antony van Leeuwenhoek and his "little animals": Being some account of the father of protozoology and bacteriology and his multifarious discoveries in these disciplines.* Russell and Russell, New York.

Postgate, J. 1986. *Microbes and man.* Penguin Books, Ltd., Harmondsworth, England. A short, general introduction to the subject on how microorganisms affect humans.

Shih, G., and R. Kessel. 1982. *Living images: Biological microstructures revealed by scanning electron microscopy.* Jones and Bartlett, Boston. Covers structures of organisms ranging from bacteria to mushrooms, plants, and animals.

chapter

2

Descriptive Traits of Microorganisms

chapter outline

GENERAL GOALS

Be able to describe accurately in standard terms the appearance of any particular microorganism.

Be able to explain how certain laboratory tests can be used to provide the information needed for the description of a microorganism.

At the beginning of this century, scientists had the important problem of identifying what differences, if any, there were between microorganisms and conventional plants and animals. They obtained information both through microscopic examination and through biochemical analyses. This chapter presents selected similarities and differences between eukaryotic and prokaryotic microorganisms in their cell structure, function, and behavior.

Cytology of Microbial Cells

Bacteria were classified first as animals because they could move; later, however, they were classified as plants (specifically fission fungi or *Schizomycetes*) on the grounds that they have a cell wall and do not ingest food like animals. Nevertheless, it seemed to many observers that there were fundamental differences among bacteria, plants, and other microorganisms. One advantage of having access to the modern microscopes described in chapter 1 is that scientists now can view differences that their predecessors only could guess at even as recently as 40 years ago. Figure 2.1A–C displays three transmission electron micrographs, one of a thin section through an algal cell (a plantlike microorganism) and the other two of thin sections through bacterial cells. Notice the striking similarities and differences in your comparison of the three.

Cells in the first two examples (figure 2.1A–B) are filled with cytoplasm containing fibrous DNA and surrounded by a typical unit membrane (the cell or plasma membrane) composed of a lipid bilayer. Notice the additional cell wall that acts to protect these cells against environmental stresses (figure 2.1A–C). When the interiors of the three cells are compared, however, they appear to be organized differently. The algal cell (figure 2.1A) contains numerous cytoplasmic organelles bounded by membranes composed of lipid bilayers having the same appearance as the cell membrane. These include the nucleus, mitochondria, and chloroplasts. By contrast, the bacterial cells (figure 2.1B–C) have no clearly differentiated subcellular organelles and specifically lack a defined, membrane-bound nucleus. This deficit might seem like an obvious point but, in fact, remained an object of controversy as recently as the early 1950s.

Cytologists recognize two general types of cells: (1) plant (and animal) cells as well as all nonbacterial microorganisms called **eukaryotes,** a word taken from the Greek meaning possessed of a true nucleus (karyon) and (2) bacterial cells called **prokaryotes,** meaning lack of a true nucleus. Alternative spellings of both terms are eucaryotes and procaryotes. The names were chosen to suggest that the earliest cells were prokaryotic and eukaryotic cells evolved from prokaryotic ones.

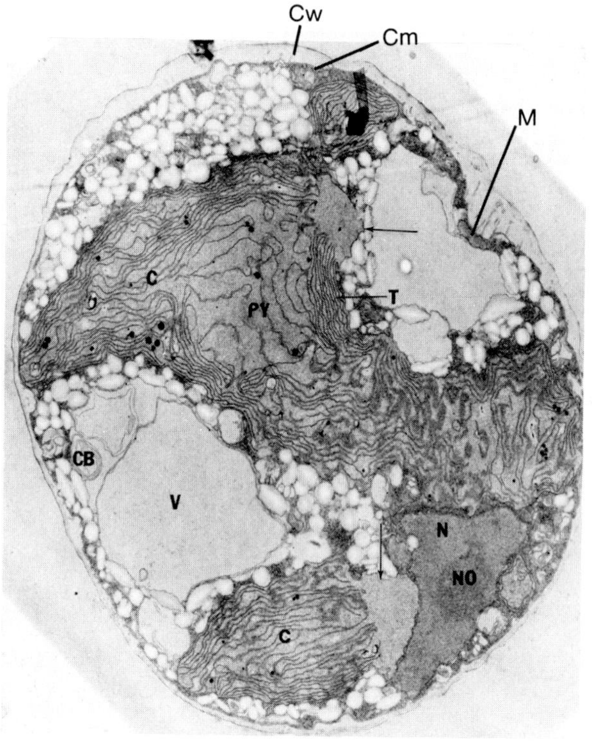

A

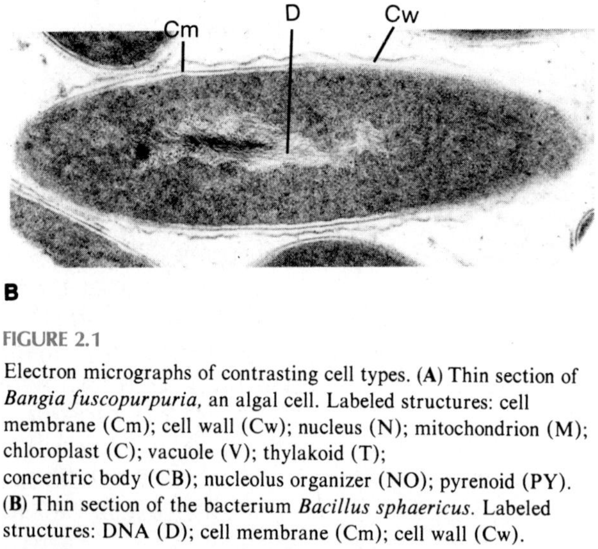

B

FIGURE 2.1

Electron micrographs of contrasting cell types. (**A**) Thin section of *Bangia fuscopurpuria,* an algal cell. Labeled structures: cell membrane (Cm); cell wall (Cw); nucleus (N); mitochondrion (M); chloroplast (C); vacuole (V); thylakoid (T); concentric body (CB); nucleolus organizer (NO); pyrenoid (PY). (**B**) Thin section of the bacterium *Bacillus sphaericus.* Labeled structures: DNA (D); cell membrane (Cm); cell wall (Cw).

(A) Electron micrograph courtesy of Dr. Pearl H. Lin, Department of Microbiology, Arizona State University. (B) Electron micrograph courtesy of Dr. Elizabeth W. Davidson, Department of Zoology, Arizona State University.

APPLICATIONS BOX

The Search for the Bacterial "Nucleus"

Most early bacteriologists assumed that bacteria were similar to eukaryotic cells, only smaller. They applied staining techniques to bacterial cells that demonstrated nuclei in eukaryotic cells, and they reported seeing "nuclei," mitotic figures, etc. Skeptics failed to find convincing evidence of the existence of any nuclei or chromosomes in bacterial cells, however. The controversy raged for decades until advances in electron microscopy and genetics in the 1950s demonstrated that bacteria had no truly comparable structures to nuclei or chromosomes. As T. D. Brock pointed out (see his review article referenced in the suggestions for further reading at the end of this chapter), the conclusions reached by early investigators tended to support the preconceived theory of each individual experimenter. Problems with the bacterial nucleus question serve to emphasize the difficulties inherent in any observational experiment that requires interpretation by an experimenter.

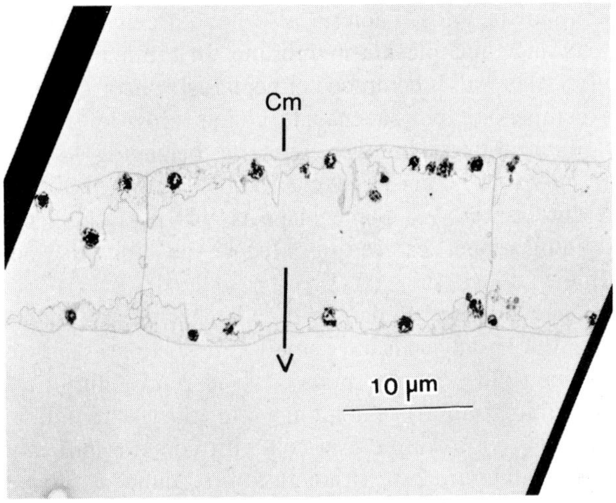

FIGURE 2.1 CONTINUED

(C) Thin section of the bacterium *Beggiatoa*. Note the large central vacuole bounded by a unit membrane (Cm).

(C) Micrograph courtesy of Dr. John Larkin and Margaret C. Henk, Department of Microbiology, Louisiana State University.

The dramatic difference between the two bacterial cells shown in figure 2.1B–C provides an important clue to the nature of cellular organization in prokaryotes. The two cells differ greatly in width, but much of the larger cell (figure 2.1C) is actually vacuole, not cytoplasm. Larkin and Henk have suggested that this organization indicates a fundamentally different approach to cell structure in prokaryotes than in eukaryotes.

Eukaryotic cells can be quite large owing to their internal structures and processes. They have specialized organelles to provide needed functions at appropriate locations within the cell. Mitochondria can cluster at any point at which many ATP molecules are needed. Proteins synthesized deep within the cytoplasm can be exported to the surface through the Golgi apparatus. Cytoplasmic extensions can engulf material outside the cell surface (endocytosis or pinocytosis), and vacuoles can bring this material into the cytoplasm. Cytoplasmic streaming, present in all eukaryotic cells and easily seen in an amoeba, can be used later to move engulfed material to an appropriate region of the cell. The net result is a cell whose surface-area-to-volume ratio can be quite small because diffusion is not the limiting factor in movement of molecules through the cytoplasm.

Contrast the state of affairs in a eukaryotic cell with that of a prokaryotic cell. There are no internal, membrane-bound organelles and no cytoplasmic streaming. Consequently, prokaryotic cells are limited to simple diffusion processes to distribute material within their cytoplasm. The result of this limitation is that cells exceeding a particular surface-area-to-volume ratio appear to develop vacuoles that limit the cytoplasm to an outer cylinder in which the surface area to volume of the cytoplasmic cylinder is such that diffusion is sufficient for purposes of nutrient supply and waste removal. In the *Beggiatoa* cell section shown in figure 2.1C, the thickness of the cylinder wall is about 2 to 3 μm and presumably represents the upper limit for efficient diffusion. In cylindrical bacterial cells, most growth occurs by elongation of the longitudinal axis rather than by elongation of the transverse axis, which correlates to a correct surface-area-to-volume ratio. Cell volume varies linearly with changes in length, but it varies as the square with a change in diameter.

Regardless of the strategy employed to move molecules through their cytoplasm, prokaryotic and eukaryotic cells are composed essentially of the same materials. Consequently, it is reasonable to assume that

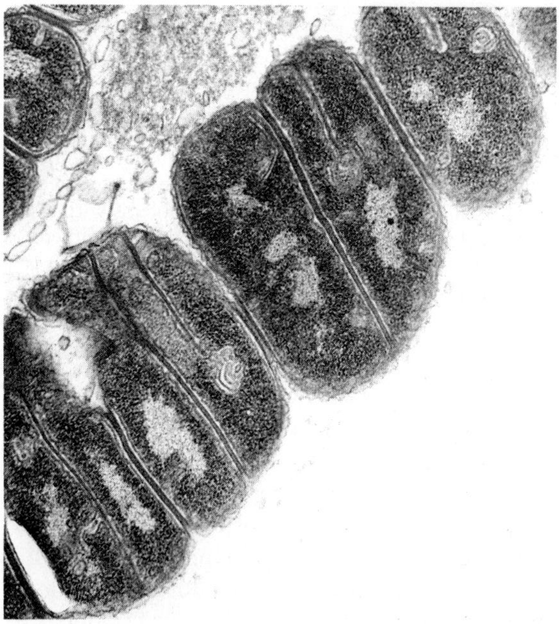

FIGURE 2.2

Mesosomes as seen in a thin section of a fixed bacterium, *Simonsiella muelleri.*

Electron micrograph courtesy of Dr. Jean M. Schmidt, Department of Microbiology, Arizona State University.

all cells, prokaryotic and eukaryotic, must carry out the same core metabolic functions, such as DNA and protein synthesis. Prokaryotes must carry out the same cellular functions as eukaryotic cells, therefore, but they do so without any organelles and possibly with different biochemistry as well.

Cell Surface and Shape

The bacterial cell membrane or plasma membrane, composed of a lipid bilayer, is fundamentally the same as that of a eukaryotic cell. There have been many reports of invaginations of the cell membrane to form narrow channels within the cytoplasm called **mesosomes** (figure 2.2). A variety of functions have been ascribed to them, primarily transport of materials to the center of the cell. However, careful observation with the aid of electron microscopy indicates that mesosomes are probably artifacts of the fixation process used to prepare these cells for electron microscopy because when cells are frozen instead of chemically fixed before they are sectioned, no mesosomes are seen. The possibility that certain regions of the cell membrane have specialized functions that make them susceptible to invagination during fixation has not been ruled out.

The general shape of a eukaryotic cell is maintained by an internal cytoskeleton composed of interwoven networks of actin and protein filaments. Actin fibers may function in the movement of some subcellular components; however, only a few bacteria, such as *Coxiella burnetti,* can be shown to have any sort of intracellular filament that might serve as a cytoskeleton. When the cell wall is removed from most bacteria, they round up into a sphere of minimum volume.

Plants as well as fungi and algae characteristically have cell walls surrounding their cell membrane. Composition of cell walls is extremely variable. It may be chitinous for some fungi but cellulosic for plants and some algae. Most bacteria also have a cell wall surrounding their plasma membrane. In the majority of cases, this wall is composed of **peptidoglycan** or murein, a complex of polysaccharide chains cross-linked by short peptides. However, bacteria belonging to the archaebacterial group have cell walls composed of different materials. Some aspects of peptidoglycan staining properties are presented in the subsection on simple stains.

The surfaces of bacterial cells show considerable variation. Many cells have small, thin structures called common **pili** (singular pilus), type I pili, or fimbriae (singular fimbria) extending into the surrounding medium. An example of a cell with common pili can be seen in figure 2.3. Although at first glance common pili may resemble the cilia seen in protozoa, there is really no comparison. They are much shorter and thinner than cilia. Unlike cilia, they are thought to be used only for adhesion; there is no correlation between cell motility and the presence of common pili. Certain longer, thin, flexible structures that also can be present on the surface of bacterial cells are sex pili associated with special, self-replicating DNA elements called plasmids. Sex pili are involved in some types of DNA transfer. Note the **rugose** or rough surface of the *Escherichia coli* in figure 2.3.

A layer of polysaccharide called a **glycocalyx** or slime layer can surround bacterial cells (figure 2.4A–B). If the glycocalyx is relatively rigid and attached to the cell, it is referred to as a **capsule.** The glycocalyx varies in thickness, and its production can be so great that cells growing on solid medium literally will drip polysaccharide when held vertically. There are two general types of capsules: type I is relatively thick and not heavily charged, whereas type II is thinner, carries a strongly negative charge, and is more likely to be found on disease-causing bacteria. Capsules can be visualized readily under a light microscope if the cell is

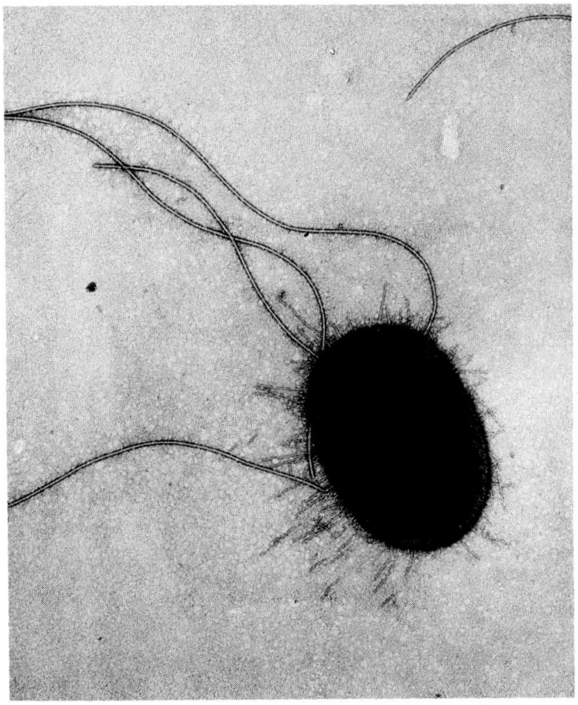

FIGURE 2.3

Bacterial surface structures. Electron micrograph of negatively stained *Escherichia coli* exhibiting short, stubby projections called common pili. The very long, thick, helical filaments are flagella. Note the rugose or rough surface characteristic of a gram-negative cell.

Electron micrograph courtesy of Dr. Margit Olson.

prepared with a negative stain, that is, one in which the background is stained while the object of interest is not (figure 2.4A–B). The function of capsules is apparently to defend the producing cell by making it difficult for other types of cells or proteins to approach the producer.

DNA Organization

Despite differences in form and life-style, cells of eukaryotic microorganisms share a number of common features, and central among these is, of course, the presence of a nucleus. The nucleus is bounded by a unit membrane that has small pores allowing molecular exchange between the nucleoplasm and the cytoplasm. Within the nucleus are the nucleolus, in which ribosomal RNA synthesis occurs, and the individual **chromosomes.** Although a chromosome can be defined in several ways, it is basically a linear molecule of double-strand DNA wrapped around individual cylinders that are composed of two copies each of four different histone proteins. Each individual cylinder plus its DNA corresponds to one **nucleosome.** DNA strands between the nucleosomes can be coated with a fifth histone protein. The structure of a chromosome allows for DNA to be packed into a smaller volume than would otherwise be possible. During cell division, the nuclear membrane breaks down, and the chromosomes condense into characteristic assemblages visible under a light microscope. Each chromosome possesses a centromere, a structure used specifically as an attachment point for spindle fibers (microtubules) during cell division.

One of the few readily observed examples of some localization of function within the prokaryotic cell comes from the bacterial chromosome. In figure 2.1B the DNA appears to be located within a central region of the cell. This DNA aggregation can be isolated as a discrete entity after gentle cell lysis but cannot be considered as a true nucleus because it is not a membrane-bound structure; therefore, it is called the **nucleoid.** DNA within the nucleoid of a prokaryotic cell is not arranged into a true chromosome with nucleosomes as is DNA in a eukaryotic cell, but it does have specific histonelike proteins associated with it and is held in its three-dimensional configuration in part by RNA molecules.

Another way in which bacterial DNA molecules differ from eukaryotic ones is in the absence of a true centromere, which implies that prokaryotic cells do not move their DNA molecules within the cell by means of microtubules (that is, there is no division by mitosis or meiosis). Indeed, examination of thin sections of bacteria with an electron microscope indicates that only bacteria known as spirochetes possess any microtubules. On the other hand, the bacterial DNA still must be partitioned after replication so that each daughter cell has an equal complement of DNA. Hence, there must be a region of bacterial DNA that has the function, but not the structure, of a centromere.

Cytoplasmic Organization

The nucleus is not the only membrane-bound structure in a eukaryotic cell. Mitochondria serve as a source of energy for cells, carrying out reactions of the tricarboxylic acid or Krebs cycle and the electron transport system. Chloroplasts found in photosynthetic cells produce ATP and carbohydrates by using the energy they obtain from light. The endoplasmic reticulum is involved in the production of proteins, and the Golgi apparatus in their transport either to other parts of the cytoplasm or to the exterior of the cell.

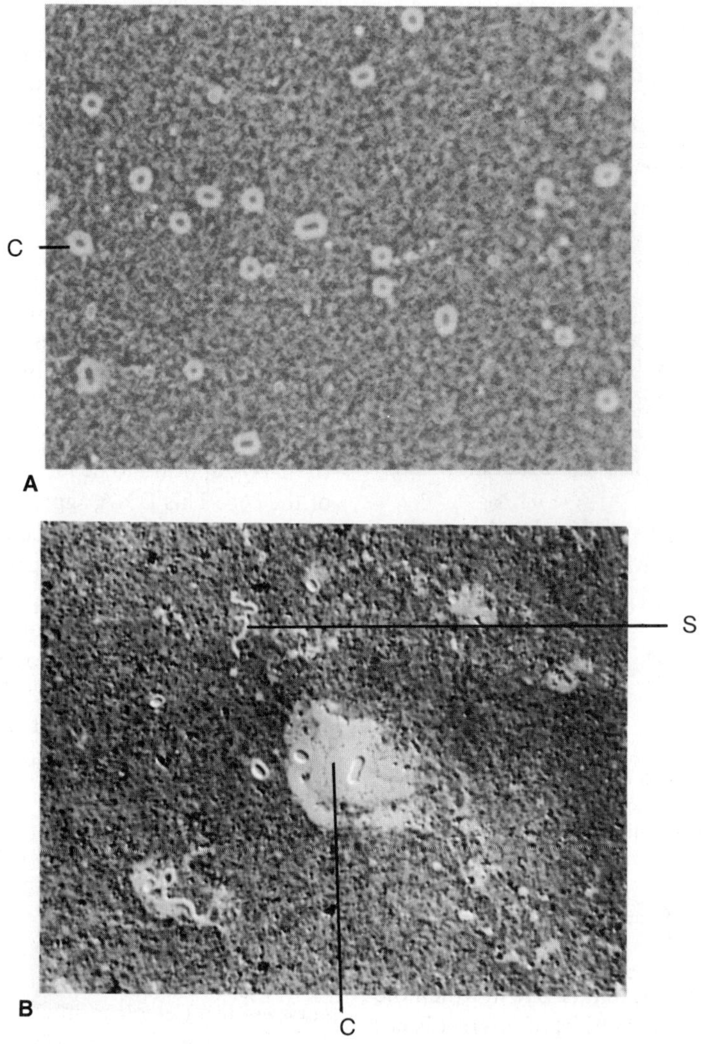

FIGURE 2.4
The bacterial glycocalyx. Two photomicrographs of the anaerobic organism *Bacteroides thetaiotaomicron* demonstrate differences in the appearance of the glycocalyx. The bacteria have been negatively stained with India ink particles. (**A**) Photomicrograph depicts a conventional view of the capsule (C).
(**B**) Photomicrograph was taken with a differential interference phase-contrast microscope. Note how the capsule (C) appears to be much larger and more diffuse. Some detached strands (S) of glycocalyx are also visible.

Reprinted with permission from Lambe, Jr., D. W., K. P. Ferguson, and D. A. Ferguson, Jr., 1988. The *Bacteroides* glycocalyx as visualized by differential interference contrast microscopy. *Canadian Journal of Microbiology* 34:1189–1195.

Protein synthesis within a eukaryotic cell occurs on ribosomes that are either free in the cytoplasm or attached to the walls of portions of the endoplasmic reticulum. Eukaryotic ribosomes are composed of two subunits, designated as 40S and 60S (**sedimentation coefficients**) to reflect their relative sizes and shapes; a fully assembled ribosome (one copy of each subunit) is designated as 80S[1]. Multiple ribosomes engaged in translating a single mRNA molecule into protein can be observed with the electron microscope. This complex is called a polysome.

Prokaryotic ribosomes are similar but not identical in structure to their eukaryotic counterparts. The small subunit is comprised of a 16S RNA molecule and 20 protein molecules to give a 30S ribosomal subunit. The larger 50S ribosomal subunit consists of a 5S RNA molecule, a 23S RNA molecule, and 29 protein molecules. A complete bacterial ribosome has a sedimentation value of 70S. There is no prokaryotic structure

1. A particle that has a sedimentation coefficient of 1S (Svedberg unit) will move 10^{-13} cm/s in the direction of a unit gravitational field.

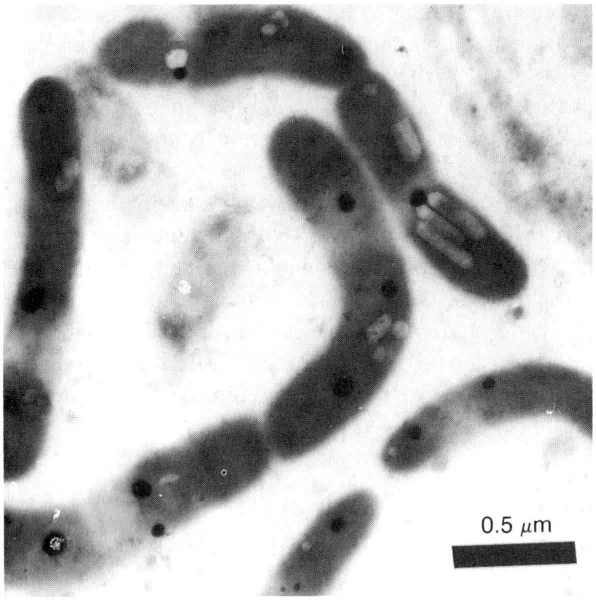

FIGURE 2.5

Gas vacuoles in bacteria isolated from antarctic sea ice. The transparent, rectangular areas in these intact cells are gas vacuoles bounded by a protein membrane. Storage products, probably polyphosphate granules, appear as dark dots.

Photomicrograph courtesy of Dr. J. T. Staley, Department of Microbiology and Immunology, University of Washington.

analagous to a nucleolus, and ribosome assembly occurs wherever the necessary proteins are being transcribed. Because there is no endoplasmic reticulum, ribosomes are free in the cytoplasm or bound as polysomes. The lack of a nuclear membrane in prokaryotes means that polysomes can be assembled before mRNA is synthesized fully.

Sometimes the cytoplasm contains **gas vacuoles,** cylindrical structures observed in many photosynthetic and nonphotosynthetic bacteria (figure 2.5). Gas vacuoles are relatively strong structures that apparently allow an organism to locate itself at a particular level in a body of water. Although they are not surrounded by a true unit membrane, gas vacuoles are encompassed by a layer of proteins oriented so that the surface of the protein layer (or membrane) facing the cytoplasm binds water molecules while the interior surface of the membrane does not.

Cell Division

Cell division among eukaryotic microorganisms occurs primarily by mitosis and meiosis. After replication, condensed chromosomes pair along the midline of a cell. These chromosomes then move to the poles of the cell along a spindle composed of microtubules that attach to the centromere. A furrow develops down the midline of the cell, and two daughter cells are produced when the opposite sides of the furrow touch. If the chromosomes align so that they are identical on opposite sides of the cleavage plane, each daughter cell receives an identical complement of chromosomes in the mitotic process of cell division. If, on the other hand, pairing occurs so that each replicated pair stays together and homologs (chromosomes from different parents) align on opposite sides of the cleavage plane, each replicated homolog migrates to a different pole, resulting in two new cells with only half the normal number of different chromosomes. This process of reduction division is called meiosis I. A second division (meiosis II), following closely on the first, separates each replicated pair of chromosomes without additional DNA replication. Thus, the products of mitosis are two identical cells having a normal chromosomal complement, while the products of meiosis are four not necessarily identical cells each having half the normal number of chromosomes.

Most bacteria reproduce by binary fission, a process in which large cells form a furrow and divide equally in two despite their lack of chromosomes and spindle apparatus. Consequently, there must be some sort of mechanism for partitioning DNA molecules into the two daughter cells. Current theory suggests that the cell accomplishes this task by attaching the DNA molecules to its membrane and then separating the points of attachment by inserting new material into the membrane.

A somewhat less common process of cell division observed in both prokaryotes and eukaryotes is budding. During normal cell division, (mitosis or binary fission), once invagination begins to form the cell septum, the size of the daughter cells does not change. In budding, the daughter cell initially appears as a small protrusion of cytoplasm that gradually enlarges with time (figure 2.6A–B). While the bud is forming, an appropriate DNA complement is transferred into the bud, but how this transfer occurs is not known.

Prokaryotic Cells Seen with Aid of Light Microscopy

A large number of staining procedures have been developed to give biologists additional information about organisms without necessitating their study under an electron microscope. Dyes used to stain cells and make

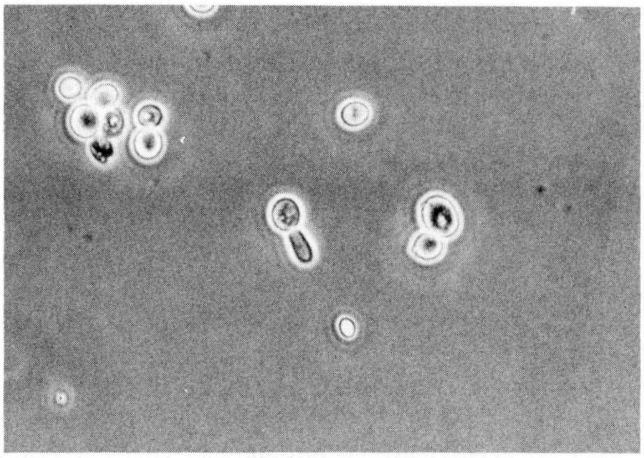

A

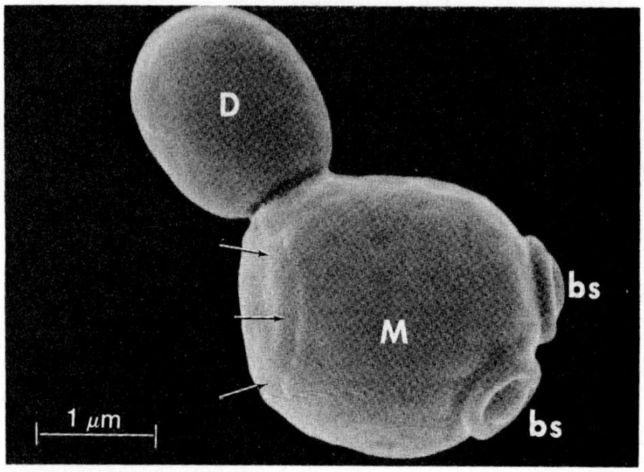

B

FIGURE 2.6
Cell division by budding. (**A**) These *Saccharomyces cerevisiae* cells reproduce by extrusion of the cytoplasm into a bud rather than by fission of the cell. (**B**) Scanning electron micrograph of *Saccharomyces rouxii* showing a mother cell (M), daughter cell (D), and bud scars (bs) from previous divisions. The arrows indicate the rim of a birth scar from where a previous daughter cell has detached itself.

(B) Reprinted from Arnold, W. N., A. T. Pringle, 1980. Scanning electron microscopy of cells and protoplasts of *Saccharomyces rouxii*. *Current Microbiology* 3: 283–285.

them visible in many microbiological procedures provide contrast to the background. Certain staining techniques are useful only with specific organisms, while others have general applicability. Those stains most useful with bacteria are the subject of this section.

Simple Stains

Many dyes, such as methylene blue, result in a general, nonspecific coloration of bacteria and are referred to as simple stains. Some dyes, such as nigrosin (India ink particles), produce a negative stain either because they are too large to enter the cell or they carry a negative charge that is repelled by a negatively charged cell. Consequently, they stain the background but not the cell itself. An example of a simple stain can be seen in color plate 1 and a negative stain is shown in figure 2.4A.

Sometimes a dye reacts with a specific cell structure in an unexpected way. Toluidine blue, for example, imparts a general blue color to the cell, but this dye turns red in the presence of chains of phosphates (polyphosphate granules). An abrupt color change of this type is said to be **metachromatic,** and polyphosphate granules are sometimes referred to as metachromatic granules.

Differential Stains

A **differential stain** is one that reacts only with certain structures or cells. The most fundamental staining procedure bacteriologists use also is a good example of a differential stain. Around 1900 the Danish microbiologist Hans Christian Gram developed this procedure to assist in identifying disease-causing bacteria. He found that bacterial cells stained with crystal violet (the primary stain), treated with an iodide solution, and then destained with an organic solvent responded in one of two ways (figure 2.7). Cells that retained the primary stain and were colored purple were designated as gram positive. Cells that lost the primary stain and could be seen well only if they were counterstained with a secondary stain of contrasting color (pink), such as safranin, were designated as gram negative. Occasionally, an organism will stain positive at certain times and negative at others, depending on its growth conditions. Such an organism is described as gram variable. The Gram stain is still of primary importance in bacteriology because this easy procedure can serve to divide nearly all bacteria into two groups (color plate 1B).

The Gram staining reaction of a bacterial cell depends on the nature of its cell wall and membrane. Electron microscopic studies show that gram-positive cells have a thick cell wall composed primarily of peptidoglycan and the usual cell membrane (figure 2.8A). Gram-negative cells fail to retain the crystal violet dye complex for any of several reasons. A few lack a protective cell wall. Most commonly encountered gram-negative bacteria, however, possess two lipid bilayers surrounding the cell (figure 2.8B). The inner plasma membrane is a conventional cell unit membrane. The outer membrane is a nonconventional membrane whose chemical nature is discussed in chapter 3. Between these two membranes is a thin layer of peptidoglycan.

APPLICATIONS BOX

Importance of the Gram Stain

It is difficult to overemphasize how fundamental Gram stain classification has become in bacteriology. The positive or negative staining reaction of a bacterium is still one of the first items mentioned in its description. Gram stains are also of considerable practical importance in medicine. As a general rule, drugs work optimally either on gram-positive bacteria or on gram-negative ones; therefore, treatment of any infection is often based in part on a rapid determination of the Gram reaction of the disease-causing organism. Once that is known, then an appropriate drug can be administered even if complete identification of the organism has not been made.

Serious medical consequences can result when a laboratory error occurs in the Gram staining procedure. In one such instance, a patient suffering from an unknown disease had a high fever and swollen lymph nodes. After the patient was admitted to the hospital, the infecting organism was identified as gram positive and round in shape. A drug particularly effective against gram-positive organisms was administered, but no special precautions to prevent spread of the infection were taken. Within six hours, the patient had died. Later investigation revealed that the bacteria in question were gram-negative, short rods. In fact, the disease in question was identified as plague. More than 30 people were exposed needlessly to possible infection as a result of this laboratory error.

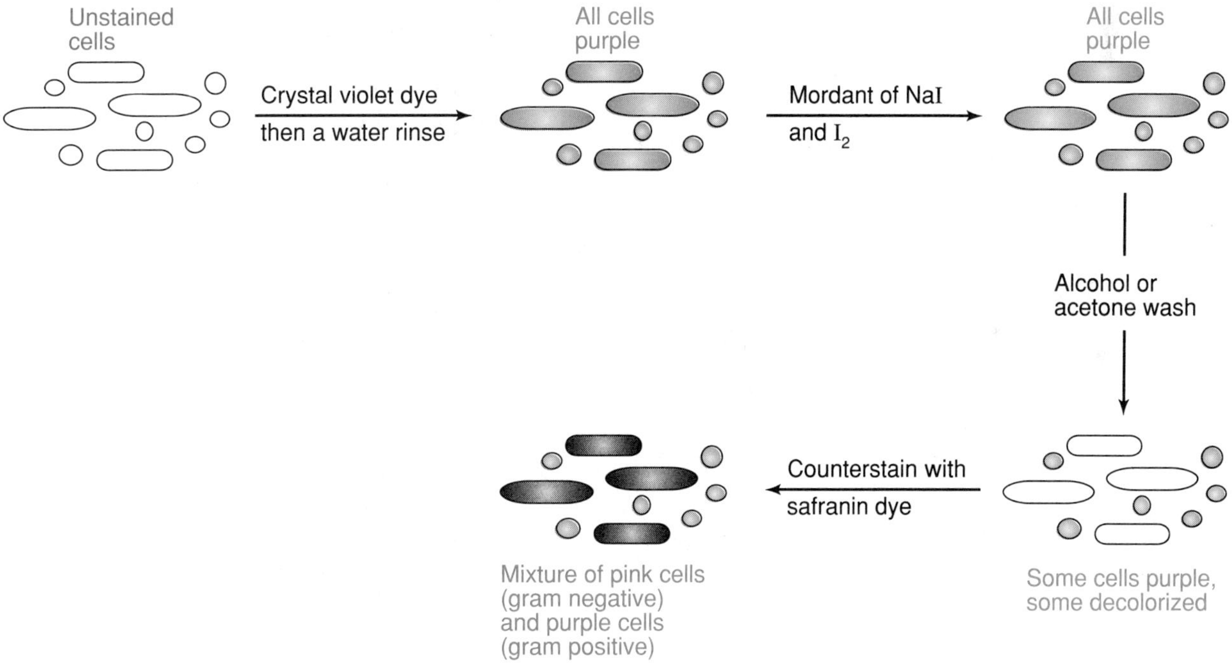

FIGURE 2.7
Differential staining procedure devised by Hans Christian Gram. It is the initial step in nearly all classification procedures (see chapter 20).

The crystal violet used in the Gram stain procedure reacts with the iodide to form an insoluble precipitate. The iodide thus behaves as a mordant by assisting in dye retention. When an organic solvent such as ethanol or acetone is added, the precipitate dissolves from the cell surface and also within the peptidoglycan. In gram-positive cells, the peptidoglycan layer consists of interwoven strands with a pore size so small that even the solubilized crystal-violet-iodide complex permeates with difficulty. In gram-negative cells, however, the solvent attacks the peptidoglycan monolayer beneath the outer membrane. The peptidoglycan becomes porous and can develop actual holes. As a result, the solubilized dye complex can be extracted readily from the cells. Cells that totally lack a cell wall, of course, pose no barrier to dye extraction.

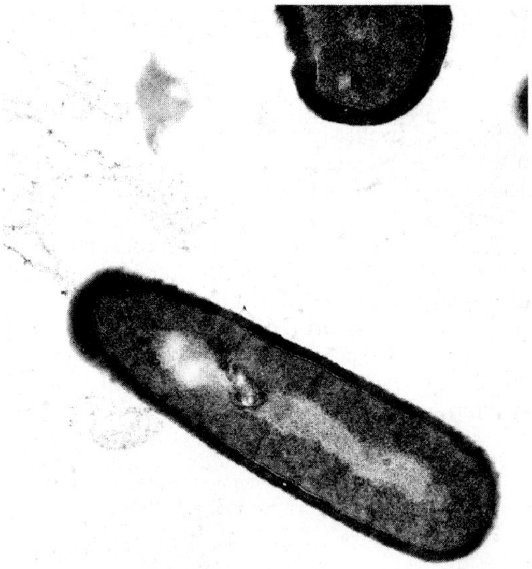

A

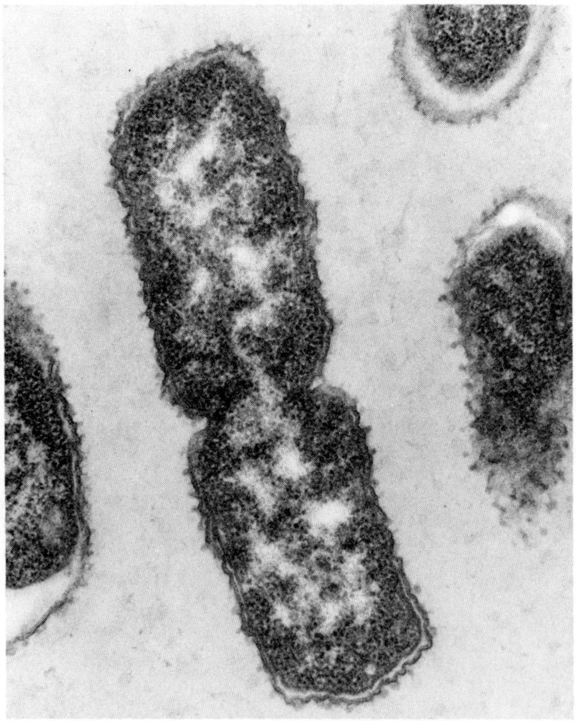

B

FIGURE 2.8

(**A**) Electron micrograph of a thin section of the gram-positive bacterium *Bacillus subtilis.* The light-colored DNA molecule is clearly visible. (**B**) Electron micrograph of a thin section of the gram-negative bacterium *Enterobacter* sp. caught in the act of cell division.

(A) Electron micrograph courtesy of Dr. Jean M. Schmidt, Department of Microbiology, Arizona State University. (B) Electron micrograph courtesy of James R. Swafford, Brandeis University.

Another differential stain whose outcome depends on the nature of the cell surface is the acid-alcohol-fast stain in which carbol-fuchsin is used as the primary stain. This dye is extracted easily from most cells by a solvent mixture of ethanol and hydrochloric acid (acid alcohol). However, certain cells contain in their cell wall a waxy substance called mycolic acid that, under appropriate conditions, will take up the carbol fuchsin dye and protect it from the acid alcohol. This prevents decolorization and makes the dye "fast" so that the cells remain a bright purple (acid-alcohol-fast cells). The other cells then can be counterstained with a dye like methylene blue (nonacid-fast) to give a preparation such as that seen in color plate 1C.

Bacterial Responses to Environmental Changes

Sporulation in Bacteria

Some gram-positive bacteria, primarily members of the genera *Bacillus* and *Clostridium,* are capable of producing the true dormant stages or endospores that Tyndall first described. Unlike certain fungi, bacterial endosporulation is not a true reproductive process but rather one of differentiation because each bacterial cell produces only one spore. The producing cell (sporangium) does not survive the process. The bacterial endospore is the most chemically and thermally resistant form of life that has been discovered. Other resting-stage structures similar to endospores have been observed in the gram-negative genus *Desulfotomaculum.*

Endospores clearly are different in several other ways from the bacterial cells that produce them. They are present only at certain times within the life cycle of the cell, generally during times of nutrient shortage. As figure 2.9A–B shows, they are round or oblong in shape and highly refractile (focusing light so as to appear bright). Some spores are the same width as their producing cell, but others are actually wider so that the sporangium appears distinctly swollen. The sporulation process is highly regulated and considered a classic example of bacterial differentiation.

Bacterial Motility

The unicellular eukaryotic microorganisms can have different mechanical means of locomotion. These include swimming, powered by **flagella** or cilia, and crawling via the extrusion of pseudopods. Two fundamentally similar types of motility are observed among bacteria, although mechanically they are different. Some bacteria possess flagella and move with a rapid

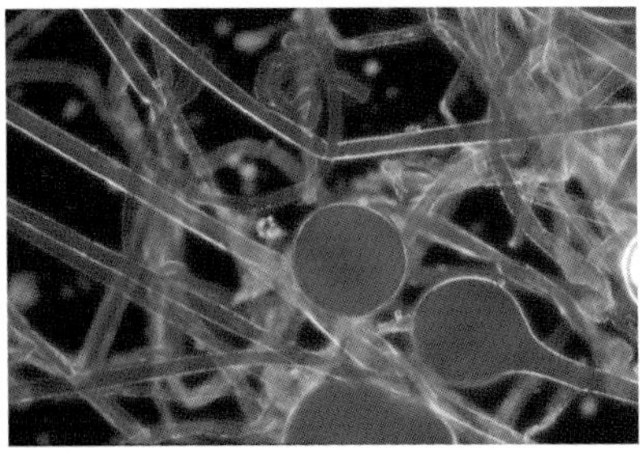

(A)

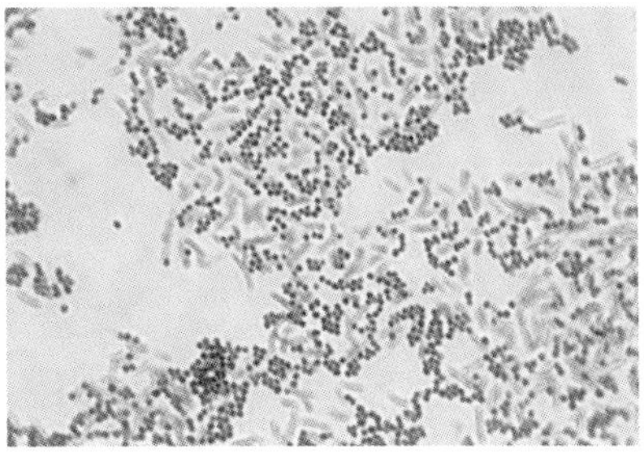

(B)

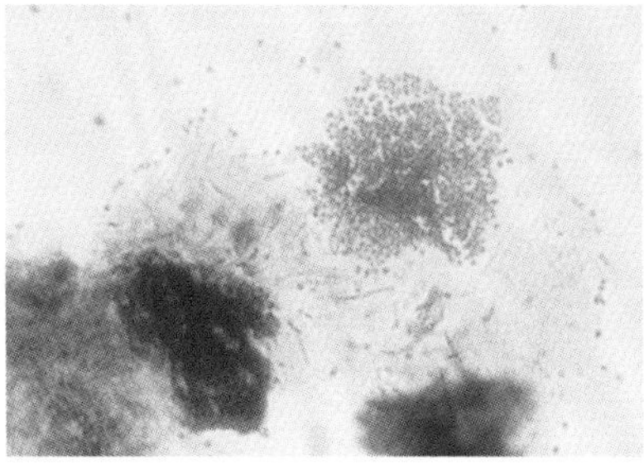

(C)

Color plate 1

Examples of basic staining reactions. (**A**) Simple stain. The fungus *Rhizopus* spp. was stained with cotton blue and photographed under dark-field illumination. (**B**) Gram stain. This differential stain has been applied to a mixture of *Staphylococcus carnosus* (pink indicates gram positive) and *Escherichia coli* (violet-purple indicates gram negative). (**C**) Acid-alcohol-fast stain. This differential stain has been applied to a mixture of *Mycobacterium fortuitum* (pink indicates acid fast) and *S. carnosus* (blue indicates nonacid fast).

(A) Micrograph courtesy of Terry Thomas, Mesa Community College, Mesa, Arizona

(A)

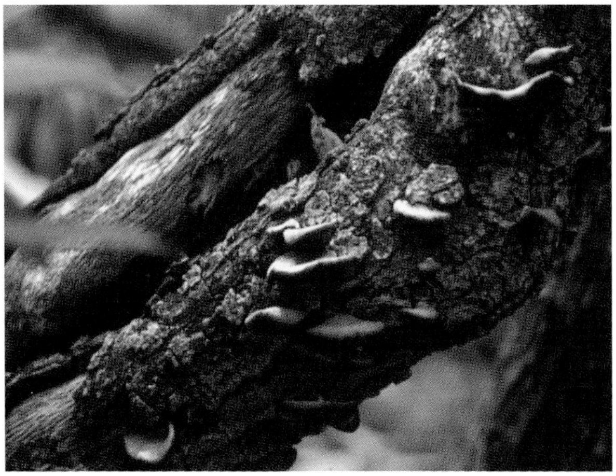

(B)

Color plate 2

Examples of fungi. (**A**) This particular specimen of mushroom is *Amanita* spp. and is poisonous. (**B**) These bracket fungi grow on the sides of tree trunks.

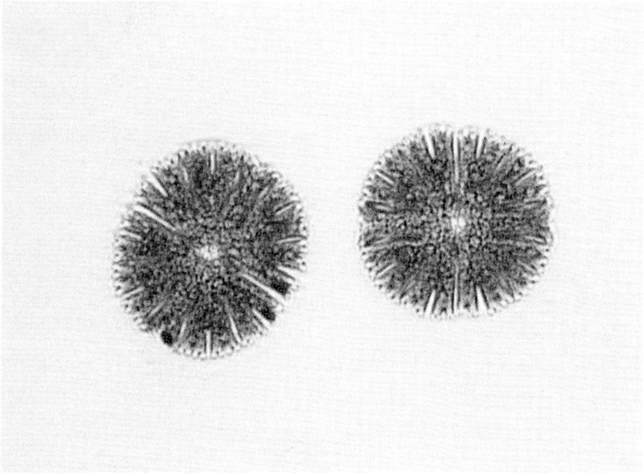

(A)

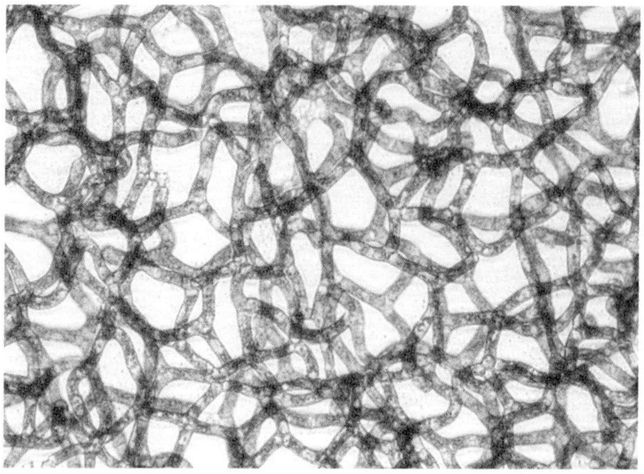

(B)

Color plate 3

Examples of green algae. (**A**) *Micrasterias* spp. (**B**) *Hydrodictium reticulatum.* (**C**) *Zygnema* spp.

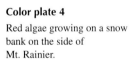

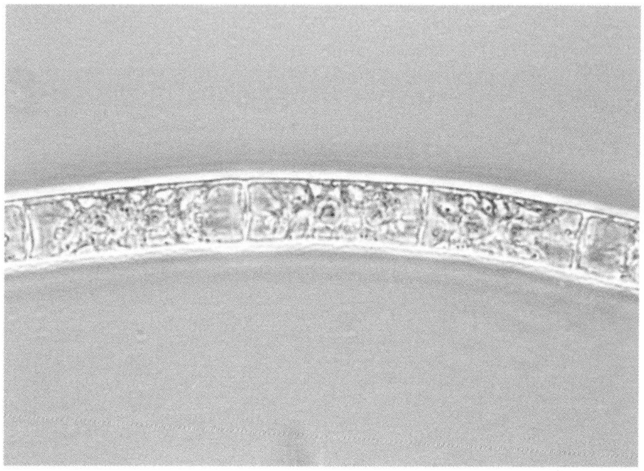

(C)

Color plate 4

Red algae growing on a snow bank on the side of Mt. Rainier.

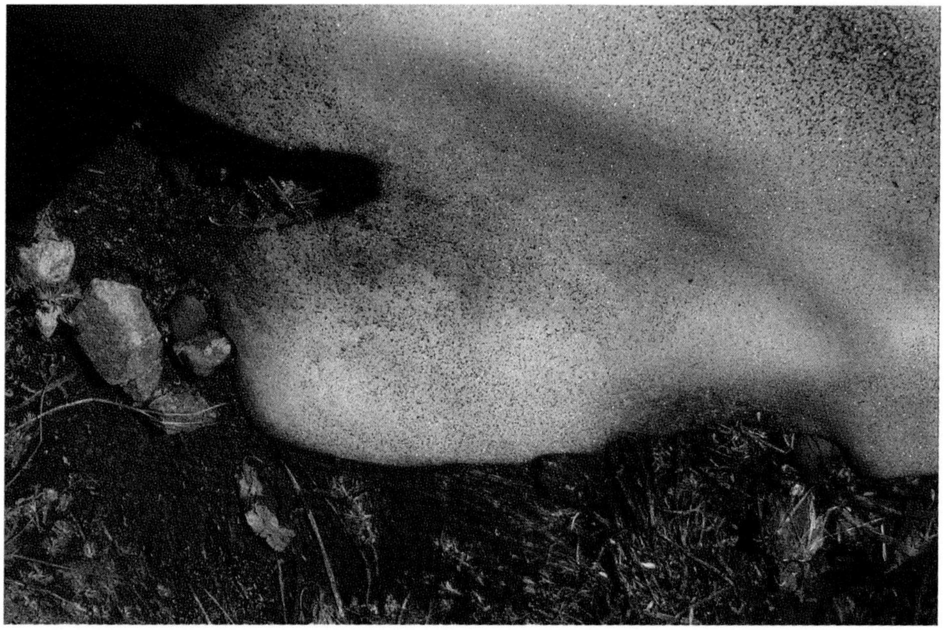

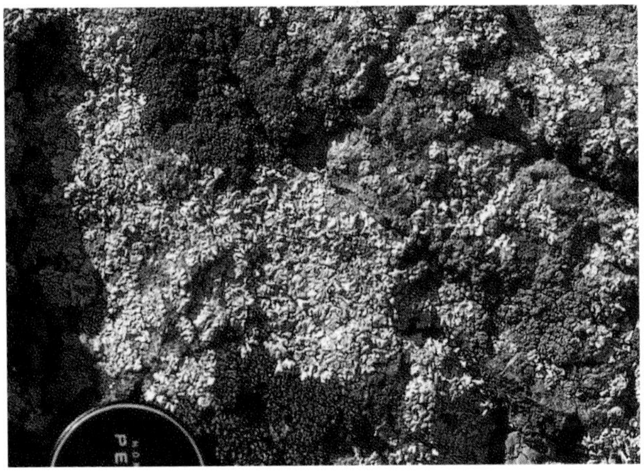

(A)

(B)

Color plate 5

Examples of lichens. (**A**) Crustose lichens growing on a rock in the upper Sonoran desert. (**B**) The foliose lichen *Cladonia alpestris* growing in Anaktuvak Pass, Alaska.
Micrographs courtesy of Thomas H. Nash III, Department of Botany, Arizona State University

Color plate 6

Effects of water pollution. An excess of nutrients in this freshwater lake has allowed extensive algal growth on the water surface. The prevailing winds have concentrated the algae along the beach.

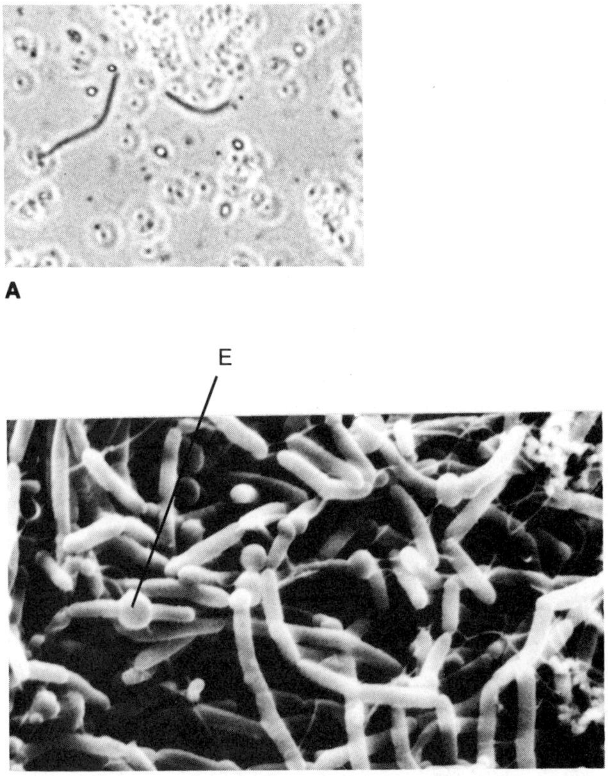

FIGURE 2.9
(**A**) Bacterial endospores from *Bacillus megaterium* as viewed under phase-contrast microscopy. Note their highly refractile (light concentrative) nature when compared to the few remaining cells. (**B**) An endospore that has swollen the producing cell (E). Scanning electron micrograph courtesy of Dr. Joseph M. Madden.

swimlike motion either in liquids or on moist solid surfaces. Bacterial flagella are extremely long (several body lengths), relatively thick filaments built entirely of a protein helix made of flagellin subunits. These flagella are, at first glance, seemingly similar to those found on eukaryotic cells (figure 2.10A–D). However, ultrastructural examination shows that they are similar only in function and not in structure or mechanism. A cross section through a bacterial flagellum reveals that it is composed entirely of protein subunits. No microtubules are present. Moreover, instead of the typical eukaryotic basal body, a prokaryotic flagellum terminates at the cell membrane in a structure called a **flagellar hook.** The filament of the prokaryotic flagellum is not flexible like its eukaryotic counterpart, but it has a permanent helical configuration.

The hook region of the flagellum (figure 2.10D) behaves similarly to a small rotary motor, causing the flagellum to spin like an airplane propeller rather than to wave like a whip as do eukaryotic flagella. The structure of the hook is somewhat variable, depending on the cell type from which it is derived. Opposite rotations of the S and M rings located within the cell membrane are thought to impart the driving force to the hook. A P ring located within the peptidoglycan layer stabilizes the base of the flagellum. If an outer membrane is present, the thinner peptidoglycan layer apparently necessitates the presence of an extra L ring located within the outer membrane itself.

Other motile bacteria that lack flagella use **gliding motility,** a form of motion seen only on solid surfaces. From electron microscopic observations, structures that resemble flagellar hooks without filaments or short wavy filaments themselves on cells exhibiting gliding motility have been reported. There is no consensus on the actual mechanism by which these structures move a cell, however. It may be that bacteria can achieve gliding motion in more than one way.

In the remainder of this section, our focus turns to flagellar motility. Because of the inherent helicity of bacterial flagella (figure 2.3), their effect on the cell depends on the direction of rotation. In a cell that has only a single flagellum at one or both ends (polar flagellation), the direction of movement clearly depends on the direction of rotation. However, in a rod-shaped cell that has a tuft of flagella at one or both ends (lophotrichous flagellation), the situation is more complex. Rotation in a counterclockwise direction will cause the flagella to wind around one another to produce a bundle of flagella all acting in a concerted manner (figure 2.11). The result is a coordinated push that moves a cell rapidly in one direction. On the other hand, if the flagella all rotate in the other direction (clockwise), they tend to act independently, and the cell is less likely to have directed movement. This latter effect is even more pronounced if the cell has flagella located all over its surface (peritrichous flagellation).

Flagellar behavior and cell movement can be visualized under a light microscope. If the cell has flagella acting in concert, it moves across the microscopic field in a more or less straight line rather than in a random walk typical of Brownian motion. This concerted movement is designated as a run. When flagella are not synchronized, the cell seems to spin while remaining in the same place, an action technically called a tumble.

In the past few years, an extensive body of literature has developed describing the ways in which motile bacteria can respond to conditions in their immediate

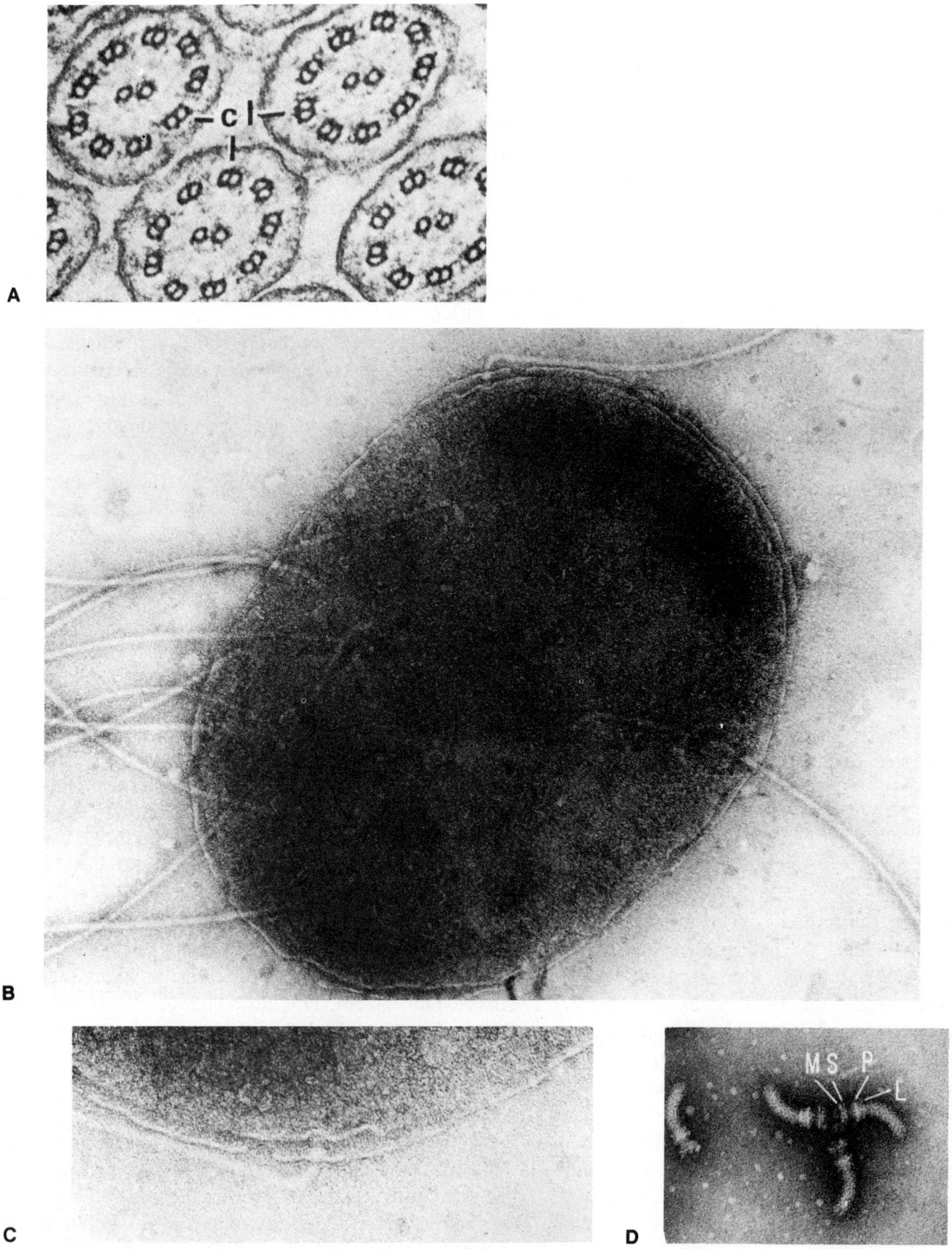

FIGURE 2.10

Flagellar ultrastructure. (**A**) Cross section through a eukaryotic flagellum. Note the 9 + 2 structure. (**B**) *Escherichia coli* cell that has been plasmolyzed (treated with a 25% sucrose solution) to illustrate more clearly the flagellar insertions in the plasma membrane and outer membrane. (**C**) Closer view of the flagellar insertion in the *E. coli* cell. (**D**) Isolated flagellar hooks with an apparent diameter of 17 nm. The M ring inserts in the cell membrane, and the S ring is located just exterior to the membrane. The P and L rings are unique to gram-negative cells. The P ring inserts in the peptidoglycan layer, and the L ring in the outer membrane.

(A) © J. André and E. Favret-Fremiet. (B–D) Reprinted from Hilmen, M., and M. Simon, 1976. Motility and the structure of bacterial flagella, pp. 35–45. *In* Goldman, R., T. Pollard, and J. Rosenbaum, (eds.) *Cell Motility.* Cold Spring Harbor Laboratory: Cold Spring Harbor, NY.

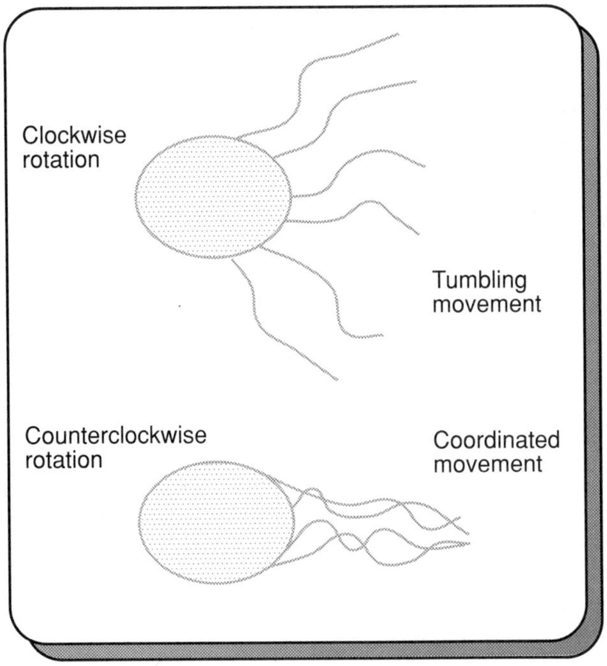

FIGURE 2.11

Effect of flagellar rotation. Diagram illustrates the effect of flagellar aggregation and disaggregation on cellular movement.

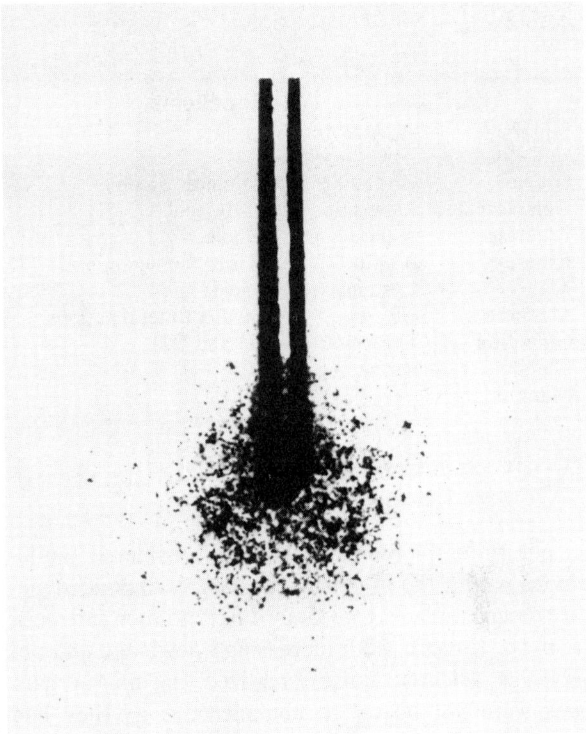

FIGURE 2.12

Demonstration of chemotaxis. The capillary tube contains an attractant chemical. After some period of time has elapsed, bacteria are observed to have concentrated in the vicinity of the tube.

Photograph courtesy of Dr. Jerry Hazelbauer, Washington State University.

vicinity. Much of this behavior seems to be **chemotactic** in nature (movement in response to concentration gradients). Julius Adler and his co-workers demonstrated chemotaxis with an easy technique that they developed. A glass capillary tube containing a solution of a chemical to be tested is dipped into a liquid culture of bacteria (figure 2.12). After some period of time, the number of bacteria per milliliter in the liquid outside the tube is compared to the number of bacteria per milliliter inside the tube. If no taxis has occurred, the numbers are roughly comparable. If positive chemotaxis has occurred, there are significantly more bacteria per milliliter inside the tube than outside. The reverse is true for negative chemotaxis.

It is remarkable that a cell only a few micrometers in length is capable of detecting minute changes in a concentration gradient, but this is obviously the case. Experiments by Howard Berg and his co-workers have shown that the binding of one molecule of an attractant chemical to a cell receptor can give a significant response in terms of flagellar movement.

Microscopic observations have revealed that the mechanism of the response a cell employs is to change the relative amount of time it spends running or tumbling. If by chance a bacterium is moving toward a higher concentration of an attractive chemical, the probability of its shifting to the tumble state is greatly reduced. Conversely, if the bacterium is moving away from the attractive chemical, the frequency of tumbles increases. Bacteria apparently can "remember" the previous concentrations of a chemical for about 4 seconds. Because tumbles result in a random reorientation of the cell, sooner or later the cell will begin, by chance, to move in the correct direction. At that time, the frequency at which the cell tumbles will decrease, and it will continue to move in more or less the correct direction. The inverse of this scheme occurs from the action of repellent chemicals. Further information about chemotaxis can be found in the discussion of membrane biology in chapter 3. Table 2.1 lists some examples of chemotactically active chemicals and the responses they elicit.

TABLE 2.1	Known chemotactic substances for some bacteria	
Attractants		**Repellents**
Sugars	Amino Acids	
Glucose	Glutamate	Aliphatic alcohols
Galactose	Aspartate	Fatty acids
Fructose	Asparagine	Indole
Ribose	Alanine	Hydrophobic amino acids
Maltose	Cysteine	Divalent metal cations
Mannose	Glycine	H^+ and OH^-
Galactitol	Threonine	
Glucitol	Serine	
Mannitol		

Types of Microorganisms

At one time, microorganisms were considered thallophytes within the kingdom *Plantae.* A **thalloid** structure is one that lacks true tissue differentiation (no roots, stems, or leaves). With increasing knowledge and observation, scientists quickly realized that microorganisms were not nearly so homogeneous as they had thought; therefore, additional taxonomic subdivisions were introduced. Eventually, the number of differences between eukaryotes and prokaryotes, some of which have been discussed already, became so great that a third kingdom, *Monera,* was added to *Plantae* and *Animalia.* As more and more information was gathered, additional kingdoms for microorganisms were proposed. Each new taxonomic arrangement was intended to better present our understanding of the world around us within the rigid framework characteristic of any classification system. R. H. Whittaker developed a popular classification concept by subdividing all organisms into five kingdoms (table 2.2). Margulis and Schwartz have prepared a survey of the five kingdoms, which is referenced in the additional reading suggestions at the end of this chapter.

Note that bacteria are assigned their own kingdom, *Monera.* The remainder of the microorganisms, which include fungi, algae, and protozoa, have a cell structure typical of that found in plants and animals but they do not develop from blastulas (like animals) or from preformed embryos (like plants). They are distinguished from plants and animals as members of the kingdoms *Fungi* and *Protista.* The kingdom *Protista* also has been referred to as *Protoctista,* because historically protists are considered to be single-celled organisms and many algae are clearly multicellular. In general, every broad type of eukaryotic microorganism has an approximately similar bacterium, presumably as a result of convergent evolution to use available habitats rather than direct descent. A brief survey of the organismal types that fall into the various kingdoms of microorganisms follows.

TABLE 2.2	Whittaker's five-kingdom taxonomic classification of microorganisms
Kingdom	**Characteristics**
Animalia	Nonphotosynthetic eukaryotic organisms characterized by movement of the entire organism during some stage of the life cycle
Plantae	Photosynthetic eukaryotic organisms producing gametes in multicellular structures with surrounding nonfertile tissue
Fungi	Eukaryotic organisms reproducing by means of spores and having a tubular construction with comparatively rigid cell walls
Protista	Eukaryotic microorganisms not fitting into any of the other kingdoms (i.e., algae and protozoa)
Monera	Prokaryotic organisms of all types

Fungi: Ubiquitous and Versatile Organisms

The fungi (singular fungus) are some of the easiest microorganisms to observe with the unaided eye (see color plate 2A–B). They include the mushrooms seen in the supermarket, the yeast used in baking, and the fuzzy growths found on old food. When these organisms are examined under a microscope, it becomes apparent that the structure of most fungi is based on a cytoplasmic tube known as a **hypha** (plural hyphae). A hypha can be linear or branched and divided into individual cells (**septate**) or have a continuous cytoplasm with multiple nuclei (**coenocytic**), as figure 2.13 illustrates. This difference is not an absolute one because the cross membranes have a pore that permits exchange of cytoplasm between cells instead of complex transport systems used by plants and animals (figure 2.13A–C).

This type of organization exemplifies a thalloid system. Note that the branch points represent an actual splitting of the hypha (true branching). This is a significant point because there are some bacteria that resemble fungi in terms of hyphal organization but do not exhibit true branching. Large masses of hyphae are called **mycelia** (singular mycelium) and form the macroscopically visible structures commonly associated with fungi. Many fungi reproduce by means of spores,

A

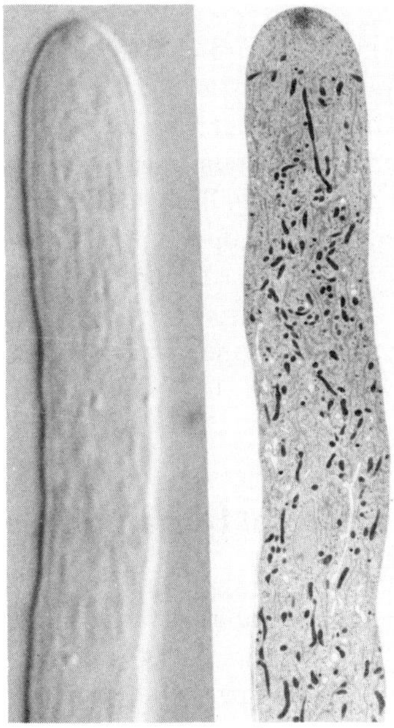
B C

FIGURE 2.13

Fungal hyphae. (**A**) Multiple hyphae of *Sclerotium rolfsii,* a soil-borne plant pathogen. Occasional branches are visible. (**B**) Light micrograph of the tip of a single hypha. (**C**) Transmission electron micrograph of cryofixed hypha. Note the lack of cross walls.

(A–B) Reprinted from Roberson, R. W. 1991. The hyphal tip cell of *Sclerotium rolfsii*: Cytological observations (in press). *In* J. P. Latge and D. Boucias (eds.), *Fungal cell wall and immune response* (NATO Scientific Series). Springer-Verlag, KG, Berlin. (C) Reprinted from Roberson, R. W., and M. S. Fuller. 1988. Ultrastructural aspects of the hyphal tip of *Sclerotium rolfsii* preserved by freeze substitution. *Protoplasma* 146: 143–149.

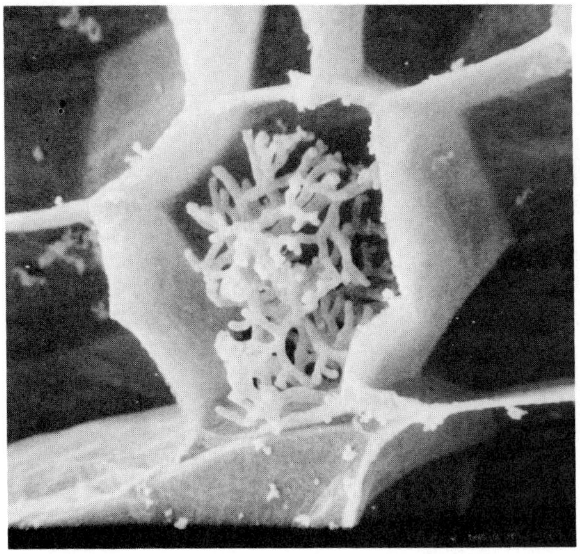

FIGURE 2.14

Haustorium. This hyphal structure permits the fungus to derive nutrients from inside the cell of an adjacent organism.

© R. F. Brown/Visuals Unlimited

resting-stage cells that arise from various types of specialized structures. Asexual spores are produced as a result of mitosis, and sexual spores are the result of meiosis. Yeasts are fungi that have a radically different cell shape. They are composed of ovoid or round cells, some of which reproduce by budding rather than by normal fission.

Fungi that contribute to the decay process are saprophytic. Those that live in association with other organisms have a symbiotic life-style. Mycelia serve to anchor an individual fungus in position and possibly may transport nutrients from one portion of the fungus to another. Frequently, specialized extensions of individual hyphae protrude from the fungal mycelium into the membrane of adjacent host cells and extend into the center of these cells (figure 2.14). In parasitic relationships in which the fungus harms its adjacent cells, these projections called **haustoria** (singular haustorium) drain nutrients from the host cells; however, in other types of fungus-host associations, the haustoria facilitate an exchange of nutrients between the fungus and the other organism for their mutual benefit, as in the case of fungi associated with plant roots.

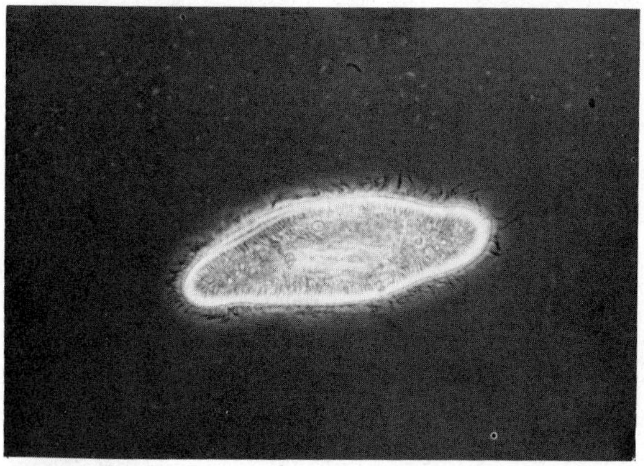

A

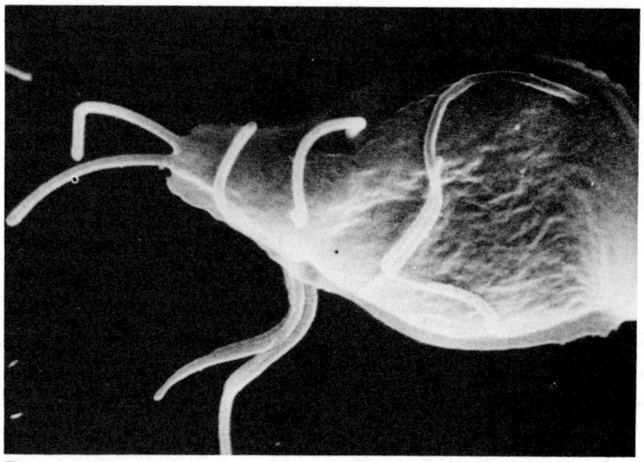

B

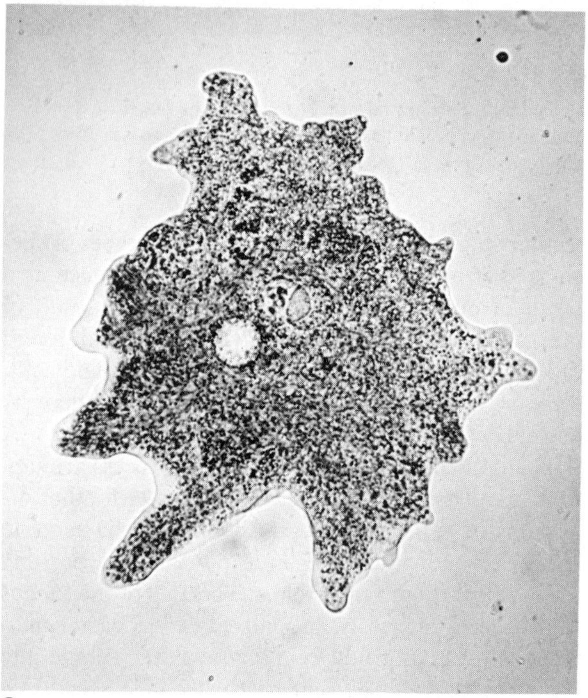

C

FIGURE 2.15

Members of the phylum Protozoa. Phase contrast micrograph of (**A**) *Paramecium,* (**B**) *Giardia.* (**C**) *Amoeba.*

(A) © Paul W. Johnson/BPS. (B) Jerome Paulin/Visuals Unlimited. (C) Runk/Schoenberger/Grant Heilman.

Protozoa: Microscopic Animals

The protozoa, like the fungi, are nonphotosynthetic, but all are unicellular with flexible cell membranes that can be strengthened by an outer covering called a pellicle. They exhibit all the different types of locomotion seen in eukaryotic microorganisms (figure 2.15A–C). Protozoa are members of the kingdom *Protista* primarily because they do not fit into any other kingdom.

As a group, protozoa can be found nearly everywhere. Most are free living and significant to the bacteriologist because bacteria are an important protozoan food source. Parasitic protozoa are frequent contaminants of drinking water in many parts of the world. In addition, they can cause diseases that are spread in other ways, such as by mosquitoes (for example, malaria).

Slime Molds: Combination of Fungal and Protozoan Characteristics

The slime molds are peculiar organisms whose properties seem to fit them best for the role of decay organisms in forests, although some parasitic forms are known. There are a number of varieties, all having similar life cycles. They begin as single-celled organisms (**myxamoebae**) produced by germinating spores. During this stage of their life cycle, slime mold cells are amoeboid like a protozoan and move about feeding on bacteria and yeasts in their vicinity. In some cases, they produce enzymes that break down the cellulose in decaying plant material. After a sufficient period of feeding, some slime molds (*Myxomycetes*) produce a plasmodium, a sluglike structure of varying size that is assembled from individual myxamoebae. The individual cells lose their cell membranes and identity so that the plasmodium is in essence a gigantic bag of nuclei. For this reason, the *Myxomycetes* are designated as acellular slime molds.

Other types of slime molds produce a pseudoplasmodium in which myxamoebae aggregate, but cell identity is not lost and neither are individual mem-

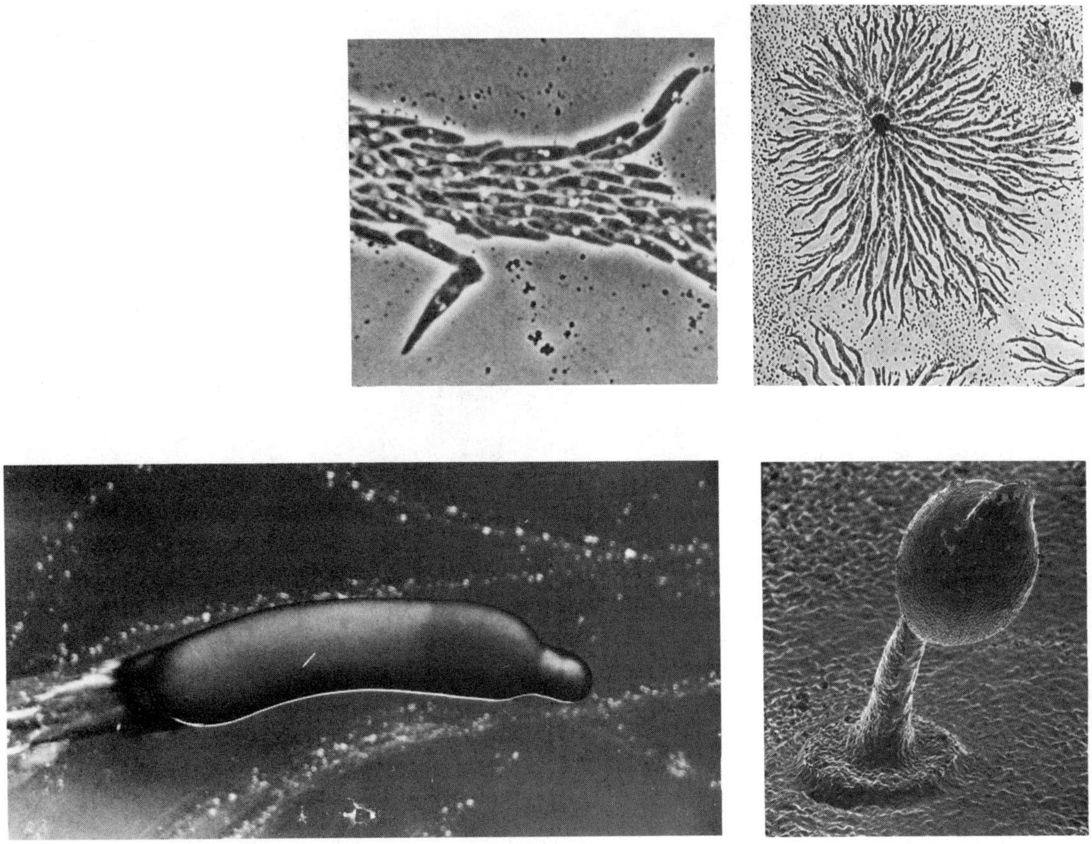

FIGURE 2.16

The cellular slime mold *Dictyostelium discoideum.*

(Top left and right) © Dr. Kenneth B. Roper. (Bottom left) © Dr. David Francis. (Bottom right) © David Scharf/Peter Arnold, Inc.

branes. Organisms belonging to this group are designated as cellular slime molds.

Regardless of whether they are cellular or acellular, all slime molds eventually produce a sporophore, a stalked structure resembling a fungus, at the top of which asexual (mitotic) spores form (figure 2.16). The life cycle then repeats. The slime molds thus form a boundary between the protozoa and the fungi, although they usually are considered as members of the kingdom *Fungi* whose organisms lack cell walls.

Algae: Photosynthetic Microorganisms

The algae (singular alga) are most familiar as the green organisms growing in bodies of water in the summer (see color plate 3A–C). This view, however, is a narrow one that does not consider the diversity in algal size (ranging from single-celled organisms to multicellular kelp many meters in length) and habitat (ranging from the polar areas to the equator). In fact, because of this diversity, it is difficult to develop a comprehensive definition for algae.

For the purposes of this discussion, algae can be defined as unicellular or multicellular photosynthetic protists whose sexual reproduction is based on a unicellular mode. Thus, unicellular algae can function as their own gametes. In the multicellular forms, gametes are produced either by unicellular structures or by multicellular structures (gametangia) in which each cell produces a gamete.

Algae form a conceptual bridge between "higher" plants and some photosynthetic prokaryotes. All algae contain chlorophyll molecules, of course, but they frequently have **accessory pigments** as well that play important roles in photosynthesis. Accessory pigments impart a broad range of colors to algae from which their common names often are derived (for example, red algae, green algae, golden-brown algae).

APPLICATIONS BOX 1

Algal Economics

Algae are of considerable economic importance even to industry. The silicaceous shells of diatoms (see micrograph) can be ground into a powder that is sold under the name of diatomaceous earth, suitable for various types of filters, or used as an absorbent of nitroglycerine in the production of dynamite. Deposits of these tiny shells have formed over millenia as the organisms have lived and died. The layers of shells formed can be of substantial thickness. Commercial deposits are located in the California coastal mountain range.

Some algae are edible. For years, the Japanese have used various forms as wrappers for a variety of foodstuffs. Attempts are presently being made to raise unicellular algae such as *Chlorella* for sale as an inexpensive source of protein for humans and animals. Commercial kelp farms have been established off the coast of California.

Only a few algae affect the health of humans and animals. *Gonyaulax* is a widely distributed organism of this type that produces a toxic substance normally of no consequence because the organism is present in such small quantities. Occasionally, however, *Gonyaulax* grows in such profusion (an **algal bloom**) that the toxin accumulates to levels resulting in the death of fish living in the area affected. Proliferation of this organism produces the notorious red tide of warm coastal waters. At these times, shellfish, which are filter feeders, are most likely to concentrate potentially hazardous levels of the *Gonyaulax* toxin, although the toxin does not affect them adversely as it does humans who consume them. This problem is a recurrent one; therefore, the taking of shellfish is forbidden in coastal areas affected by the red tide during certain months of the year.

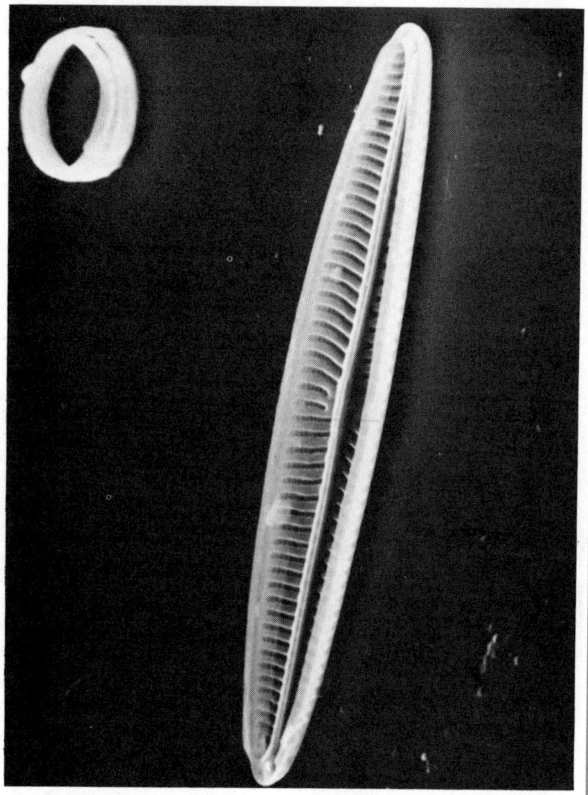

Diatom skeleton.

Electron micrograph courtesy of Diane Apostolakos.

Like other microorganisms, algae are ubiquitous. The greenish subsurface layer found in some porous rocks of the high Arizona deserts is formed by endolithic algae. The red color of a sunlit snow bank in early spring results from the growth of an organism such as *Chlamydomonas rubra* (see color plate 4). The Sargasso Sea just north of the equator contains large mats of algae, and even the more saline waters of the world, such as the Great Salt Lake in Utah, have their algal inhabitants. In fact, so numerous are the algae in the oceans that the majority of photosynthetic products in the world are estimated to be generated in the sea rather than on the land, making algae the major group of primary producers in an ecological sense.

Lichens: Symbiotic Organisms

The lichens are unusual entities in that a single individual is composed of a fungus and an alga or a cyanobacterium (type of photosynthetic bacterium) living as a symbiotic unit. In this relationship, the alga provides nourishment for the fungus from the products of photosynthesis (some cyanobacteria also provide reduced nitrogen), while the fungus anchors the composite organism to the substrate and serves to protect the photosynthesizer and trap moisture. Collectively, the pair comprises a symbiotic organism.

Morphologically, lichens do not resemble their separately grown algal (phycobiont) or fungal (mycobiont) partner. Lichens can be foliose (leaflike), fruticose (shrublike), or crustose (crustlike) and come

FIGURE 2.17
Lichens seen on the rocks in the lower Sonoran desert in southwestern United States.

in a wide variety of sizes and colors (see color plate 5A–B). The frequency of lichen symbiosis can be estimated from the fact that at least 20,000 species are recognized and members of this family occur in all parts of the globe, predominantly in terrestrial habitats. In the arctic, they are large enough to serve as one of the primary food sources for caribou, and in the antarctic, they are the dominant terrestrial organism. On the other hand, lichens can be very small and inconspicuous as in desert areas where they grow in relative profusion (figure 2.17). Genus and species names of the mycobiont are also applied to the symbiotic association.

Reproduction in the lichen is obviously tricky. One simple form of reproduction occurs when a piece of a lichen is broken off and carried to a new location. Special asexual structures (isidia and soredia) also serve as vegetative propagules, allowing the fungus-alga association to be preserved. In other cases, however, the fungus sporulates and disperses its spores in the usual manner. When these germinate, they must find new algal partners; therefore, the form of a lichen may not necessarily be maintained. Similarly, resting-stage algae can be dispersed by the wind over considerable distances.

Monerans: Prokaryotic Organisms

One of the first impressions to strike any person who looks at bacteria through either a light or transmission electron microscope is their very small size. They are, as a general rule, substantially smaller than eukaryotic cells, with a width ranging from 0.5 to 1.0 μm and a length ranging from 1 μm to as much as 100 μm, although a more usual length falls between 1 to 4 μm (table 2.3). The smallest bacterial cells are, therefore, just barely within the limits of resolution of a good light

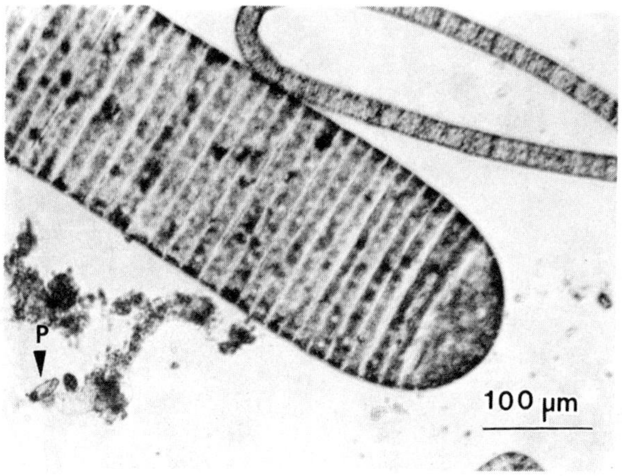

FIGURE 2.18

Largest known bacterium. This *Beggiatoa* was isolated from a hot sulfur spring. The small object labeled (P) is a typical ciliated protozoan.

Micrograph courtesy of Dr. John Larkin, Department of Microbiology, Louisiana State University.

TABLE 2.3 Comparative sizes of some microorganisms and their structures

Organism and Structure	Size
Bacterial cells	0.5 to 1 μm (rarely to 200 μm) wide; 1 to 4 μm (rarely to 100 μm or more) long
Bacterial endospore	0.5 to 1 μm, round or oval
Tetrahymena (a ciliated protozoan)	25 μm wide; 50 μm long
Fungal hypha	2 to 30 μm wide; variable length
Fungal sexual spore	14 to 16 μm wide; 27 to 30 μm long
Chlorella (a green alga)	5 to 6 μm, round or oval

microscope. However, size is not an absolute criterion for differentiation between eukaryotic cells and prokaryotic cells (figure 2.18).

Examination of assorted cultures demonstrates that bacterial cells can have a wide variety of shapes. Generally recognized forms are illustrated in figure 2.19A–D. The most common types of cells are the cylindrical **rod** (figure 2.19A) and the spherical **coccus** (figure 2.19B). In the early days of bacteriology, all rod-shaped cells were classified in the genus *Bacillus,* and the word *bacillus* is still used as a generic term for a rod-shaped cell. In theory, the distinction between rods and cocci should be quite clear-cut because the difference between a long cylinder and a sphere should be easy to detect. However, many bacteria are found to be short, fat rods that are difficult to discern from cocci under a light microscope, although the difference can be visualized more clearly with an electron microscope. Because of this difficulty, the term **coccobacillus** (plural coccobacilli) has been coined to describe those cells that cannot be assigned with certainty to one or the other group on the basis of light microscopy.

The other two common forms of bacterial cells are readily distinguishable. They are the spiral cell or a cell shaped like a portion of a spiral (figure 2.19C) and the stalked cell (figure 2.19D). A stalked cell can have a cellular stalk in which cytoplasm is present or an acellular stalk in which cytoplasm is absent. There have been reports of square bacterial cells, but that shape is rarely encountered.

Microbial Associations

Although bacteria are primarily unicellular, they do associate with other prokaryotes and with eukaryotes as well. Prokaryote-prokaryote associations tend to form for mutual biochemical benefit. They are seen usually between two (or more) organisms with complementary nutritional needs. Some examples of prokaryote-eukaryote associations are seen in certain types of lichen and in nitrogen-fixing root nodules found on legumes. Prokaryote-eukaryote associations also are found in algal communities in areas like the Sargasso Sea as well as in intestinal tracts of animals. In all these cases, the prokaryotic cells are thought to make important contributions to the nutritional state of the associated eukaryotic cells. Of course, not all prokaryote-eukaryote relationships are benign. Some involve real parasitisms that can cause severe illness or death.

Evolutionary Relationships

Although the discussion of bacterial taxonomy is deferred until chapter 20, there is one general subdivision important to discuss in this section. The accumulating body of molecular biological information about various types of bacteria indicates that there is a group of organisms that can be thought of as "living fossils" in the same sense that certain plants and animals are. In all cases, these are bacterial organisms whose evolutionary line of descent diverged from the rest of the kingdom at an early time.

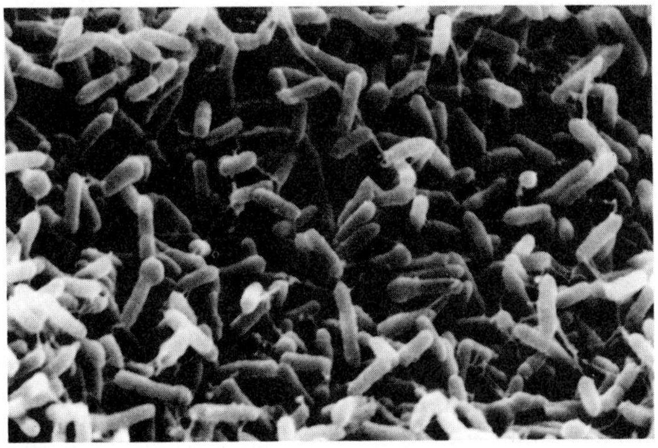

A

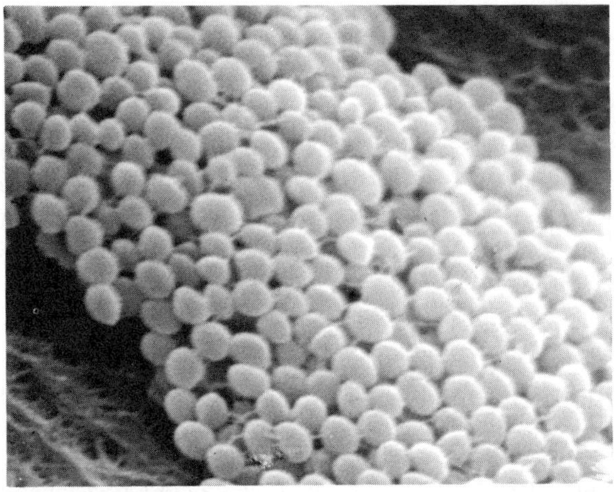

B

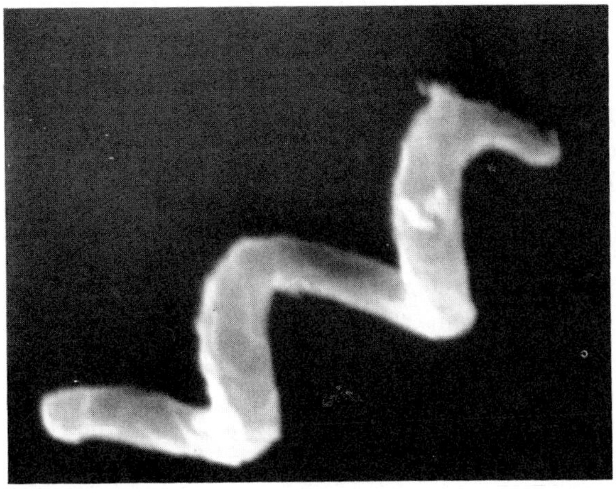

C

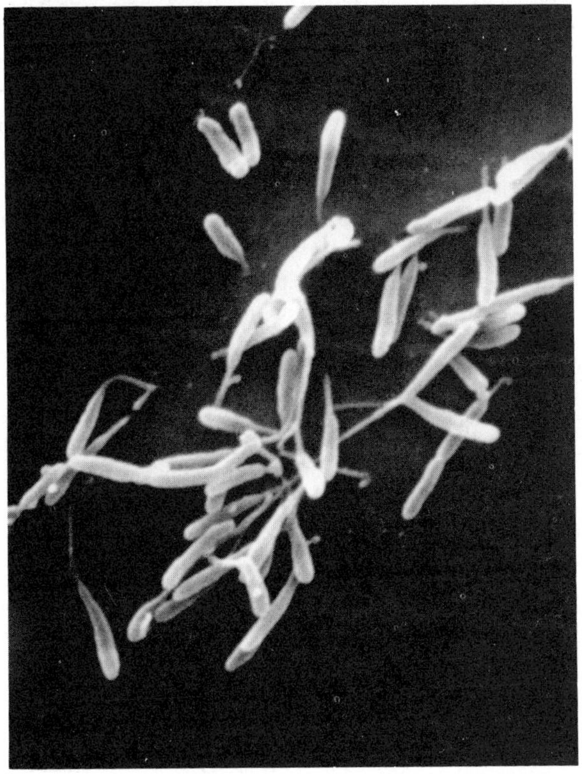

D

FIGURE 2.19

Common shapes of bacterial cells as seen through a scanning electron microscope. (**A**) Rod or bacillus shape exemplified by *Bacillus* sp. (**B**) Round or coccus shape of *Staphylococcus epidermidis*. (**C**) Spiral shape of *Rhodospirillum rubrum*. (**D**) A stalked *Caulobacter crescentus*.

(A) Micrograph courtesy of Dr. Joseph Madden. (B–D) Micrographs courtesy of William P. Sharp, Department of Botany, Arizona State University.

Bacteria that fall into this group are those whose life-styles correspond to environments rarely found on the planet at the present time, although such environments probably were more common in earlier times. These bacteria include the halophiles that tolerate high-salt concentrations, the methanogens that produce methane, the bacteria that grow by oxidizing elemental sulfur, and a number of other, metabolically interesting organisms. The general term assigned to this group of organisms is the *Archaebacteria* (also spelled *Archaeobacteria*).

Carl Woese and his collaborators, who were studying the base sequence of ribosomal RNA, first identified the archaebacteria. Using the argument that similarity of base sequence implies a close relationship in a taxonomic sense, the Woese laboratory was able to show that the monerans fell into two broad groups. The **eubacteria,** whose members possess the properties outlined in the earlier portions of this chapter, is the better-studied group. The **archaebacteria,** whose members possess unusual properties, is the lesser-studied group.

Archaebacteria are no less heterogeneous than eubacteria, but certain generalizations about archaebacteria are possible. Their cell walls, if present, are not composed of peptidoglycan. They can consist of protein, complex polysaccharides, or even what has been called pseudomurein, a cross-linked polysaccharide that has a strong resemblance to peptidoglycan but lacks the specific amino sugar muramic acid normally present in peptidoglycan. Further differences from eubacteria have been observed in the structure of archaebacterial unit membranes, transfer RNA molecules, enzyme cofactors, and such specific enzymes as RNA polymerase.

Detailed comparisons between eubacteria and archaebacteria are made throughout this book as specific topics are considered. A summary of the relationships between the two groups of bacteria can be found in chapter 20.

Noncellular Self-Reproducing Structures

All organisms discussed previously in this chapter have certain features in common, such as cell membranes, DNA and RNA molecules, etc., even if they are not necessarily free-living organisms. Within the biological world, however, there are certain entities that clearly are different in their organization and mode of reproduction. These are not true organisms in the sense that they are not cellular and are only arguably "alive," yet when they are introduced into an appropriate host cell, they are capable of self-reproduction. These are the viruses, the viroids, and the prions or unconventional viruses.

A **virus** is an obligately parasitic entity composed of a single type of nucleic acid, either DNA or RNA, surrounded by a protein coat of greater or lesser complexity. Upon infection of a suitable host, the virus uses the host cell's metabolism to produce more nucleic acid and protein coat. Progeny viral particles are assembled that later are released from the cell as infectious entities. When a virus encounters a suitable host, the cycle begins again. Meanwhile, however, a free virus has no metabolic activity and no ability to reproduce itself.

A **viroid** is a small, uncoated RNA molecule found in the cytoplasm of plant cells, such as those of the potato and tomato. It can reproduce itself and is associated with certain plant diseases. At least 15 different viroids are known. A **prion,** on the other hand, seems to be a protein molecule capable of stimulating its own production when it is introduced into a suitable host cell. Prions are associated with certain "slow" diseases of humans and other animals, such as scrapie (a disease of sheep) and kuru (a disease that formerly occurred in natives of certain Pacific islands). The mode of transmission of viroids and prions is generally unclear and presumably not very efficient.

Summary

Microorganisms are not necessarily so small as to require the use of a microscope in order to see them. Instead, their classification as microorganisms depends on their unicellular organization or, in the case of multicellular microorganisms, on their lack of true organ systems. In the common five-kingdom approach to biological taxonomy, microorganisms are divided among three kingdoms: *Fungi, Protista,* and *Monera.* All eukaryotic microorganisms are classified in the kingdoms *Fungi* or *Protista,* depending on the ways in which they move, reproduce, and derive energy for growth. In addition to the algae, fungi, and protozoa that are commonly recognized microorganisms, there are some groups of microorganisms whose members seem to have properties of an intermediate type. These groups include the slime molds and the lichens.

The kingdom *Monera* includes the organisms called bacteria and encompasses all prokaryotic cells. Bacteria are much smaller than eukaryotic cells, although they carry out the same kinds of functions and seem to have many similar properties. In terms of structure, bacteria have smaller ribosomes and, generally lack internal unit membranes and microtubules. Their motility mechanisms are different structurally and/or functionally from eukaryotic cells. Recent observations have shown that a subgroup, called *Archaebacteria,* are not closely related to the rest of the bacteria, now called *Eubacteria.* Archaebacteria differ from eubacteria as dramatically as bacteria differ from eukaryotes. They have different membranes, cell walls, enzymes, and biochemistry.

Bacterial cells can be seen with the aid of any microscopic process discussed in chapter 1, but additional structural information can be obtained by the use of differential stains. Hans Christian Gram developed the most commonly used differential stain. It colors purple those cells with walls consisting of a thick peptidoglycan layer, and pink those cells that have an outer membrane as a part of the cell wall or those that lack a conventional cell wall. The purple-staining cells are designated as gram positive, and the pink-staining cells as gram negative. Another commonly encountered differential stain is the acid-alcohol-fast stain that imparts a reddish-purple color to cells having mycolic acids in their cell wall (acid-alcohol fast) and a blue color to all other cells.

Some bacteria produce endospores that are highly refractile structures serving as a dormant stage for the cell. They are not truly reproductive because no more than one spore can be produced per cell. Bacterial endospores can be highly resistant to the effects of environmental changes.

Viruses, viroids, and prions are entities that seem to border on life. Viruses are acellular and have as genetic material either DNA or RNA, but they never contain both types of nucleic acid at once. Reproduction is possible only when a virus infects a living prokaryotic or eukaryotic cell and uses that cell's metabolic functions to assemble progeny viral particles. Viroids are naked, self-reproducing RNA molecules found in plant cells. Both viruses and viroids can cause diseases in eukaryotic cells, but only viruses have been observed in prokaryotic cells. Prions are protein molecules associated with certain poorly understood diseases of humans and domestic animals.

Questions for Review and Discussion

1. It has been suggested that prokaryotic cells were the progenitors of eukaryotic cells or that at least they had a common ancestor. Based on your present knowledge, do you see any evidence in favor of the suggestion?
2. In some early taxonomic classifications, the protozoa were grouped with the animals, and the algae, bacteria, fungi, and plants were grouped together. What advantages, if any, can you see to the present scheme? Why?
3. If you were presented with an unknown organism, how would you decide into which taxonomic kingdom it should be placed? If your decision would be to place it in the kingdom *Protista,* how would you decide if the organism were an alga or a protozoan?
4. Microorganisms can have both a positive and a negative impact on our society. Give some examples of microorganisms that are beneficial to our society. Present some examples of microorganisms that can damage our society or its members.
5. What is your definition of a microorganism? How do viruses, prions, and viroids fit into your definition? Why did you include or fail to include them in your definition?
6. What interpretation(s) can be placed on the results of the Gram stain? How might you check whether the interpretation is correct?

Suggestions for Further Reading

Adler, J. 1987. How motile bacteria are attracted and repelled by chemicals: An approach to neurobiology. *Biological Chemistry Hoppe-Seyler* 368:163–173. A general review article by one of the pioneers in the field.

Beveridge, T. J., and J. A. Davies. 1983. Cellular responses of *Bacillus subtilis* and *Escherichia coli* to the Gram stain. *Journal of Bacteriology* 156:846–858. A paper on the mechanism of the Gram stain.

Bold, H. C., and M. J. Wynne. 1985. *Introduction to the Algae,* 2d ed. Prentice-Hall, Inc., Englewood Cliffs, N.J. This is one of the standard texts in the field.

Brock, T. D. 1988. The bacterial nucleus: A history. *Microbiological Reviews* 52:397–411.

Hale, M. E., Jr. 1983. *The biology of lichens.* Arnold, London.

Ingold, C. T. 1984. *The biology of fungi,* 5th ed. Hutchinson, London.

Margulis, L., J. O. Corliss, M. Melkonian, and D. I. Chapman. 1990. *Handbook of Protoctista: The structure, cultivation, habitats, and life histories of the eukaryotic microorganisms and their descendants exclusive of animals, plants, and fungi.* Jones and Bartlett, Boston. An extensive reference book of more than 900 pages.

Margulis, L., and K. V. Schwartz. 1982. *Five kingdoms. An illustrated guide to the phyla of life on earth.* Freeman, San Francisco. This is an excellent, compact overview of contemporary taxonomy written for the nonspecialist.

Rogers, H. J. 1983. *Bacterial cell structure.* American Society for Microbiology, Washington, D.C. Also pertinent to chapter 1.

Segall, J. E., S. M. Block, and H. C. Berg. 1986. Temporal comparisons in bacterial chemotaxis. *Proceedings of the National Academy of Sciences of the United States of America* 83:8987–8991. A somewhat technical article on the extent of bacterial "memory."

Sleigh, M. 1989. *Protozoa and other protists.* Arnold, London.

Whittaker, R. H. 1969. New concepts of kingdoms of organisms. *Science* 163:50–60. The original proposal for five taxonomic kingdoms.

part

2

THE INNER WORLD OF MICROORGANISMS: PHYSIOLOGICAL PROCESSES

Chapters 1 and 2 have provided basic facts about microorganisms and how to study them, so now it is time to consider the metabolic and genetic processes that underlie the enormous versatility of prokaryotes. One important strength of the members of the kingdom *Monera* is their remarkable ability to take advantage of their environment. For example, a bacterium may be able to grow when provided with a medium containing only glucose and inorganic salts, meaning that the bacterial cell can synthesize the molecules necessary to construct its cellular components from this medium. However, the same bacterium, when provided with a medium containing amino acids in addition to the other ingredients, also can shut down its own biosynthetic pathways and use the available molecules instead.

The first section of part 2 (chapters 3–8) introduces basic nutrition, biochemistry, and cellular function. Particular emphasis is placed on the boundary between the prokaryotic cell and the world around it, namely, the plasma membrane and cell wall. Chapter 3 presents current scientific understanding of what makes a membrane semipermeable and of how a cell manages to transport materials into and out of its cytoplasm. The connections between membrane transport and energy metabolism are also introduced in this chapter as well. In chapters 4 through 6, various aspects of prokaryotic biochemistry are highlighted. The conversion of energy to a state usable by cells and the fixation of carbon by cells are considered first, followed by an analysis of the energy-requiring processes of biosynthesis. Chapters 3 through 6 outline all the biochemistry necessary for you to understand the topics covered later in the book. Chapter 7 provides information about how cells can regulate their metabolic processes at various levels. Chapter 8 presents a detailed discussion of bacterial growth and cell division.

The last section of part 2 (chapters 9–11) is devoted to the genetic processes of the bacterial cell and to the viruses that can parasitize it. Animal and plant viruses are discussed briefly as well. Special genetic elements, the plasmids and transposons, that seem to exert a major influence on the rapid evolution of bacteria are presented in chapter 10. Included in chapters 7 through 10 is material essential for the discussion of genetic engineering covered in chapter 11.

chapter

3

Bacterial Nutrition, Membranes, and Transport Processes

chapter outline

■

GENERAL GOALS

Be able to describe models for fluid mosaic membranes and simple chemiosmotic processes.
Be able to give examples of some transport phenomena in bacteria and the kinds of molecules that must be transported.
Be able to describe the minimum nutritional requirements of a bacterium.

This chapter illustrates the point that prokaryotic cells function similarly to eukaryotic cells, but their structures are not necessarily the same. General nutrition, cell surface, and nutrient transport primarily as they occur in bacteria are considered. The chemiosmotic theory that is an important component of chapter 4 is introduced as well.

Feeding a Microorganism

Early Culture Methods

Because all microorganisms, including bacteria, were once considered as variant forms of plants, some terminology applied to them is derived from that used for plants. Specifically, the process of growing microorganisms is referred to as culturing, the organisms grown within a confined area being designated as a **culture.** Thus, when Pasteur allowed his infusions to putrefy, he cultured microorganisms. Such a culture, however, was not terribly useful because it consisted of an unknown mixture of organisms. The development of methods for producing pure cultures, cultures that contain only one type of organism, was a major advance in microbiology.

Credit for developing the methodology to grow pure cultures generally is given to Koch and his co-workers. Koch, like Pasteur, intensely believed in the possibility of microorganisms being the causative agents of many diseases that afflict humans. According to his postulates, listed in chapter 1, it is essential to study organisms in pure culture. In casting about for ways in which to accomplish this task, Koch soon determined that the most practical method was to place microorganisms onto a solid surface.

Microorganisms grown on a solid medium tend to stay more or less in one place, depending on the extent of their motility. Each cell will eventually give rise to a large cluster of cells that form a mound on the surface of the medium. Each cluster of cells is called a **colony,** and unless the colonies overlap as they enlarge, the cells in a colony are all descendents of one original cell. Examples of bacterial colonies can be seen in figure 3.1A–B. If a new culture is **inoculated** (started) from a single truly isolated colony, it is of necessity a pure culture.

Koch almost immediately had problems with his pure culture techniques. He tried growing bacteria on solid substrates, either on the cut surface of a potato or on gelatin (a protein extracted from bone or skin and used to solidify certain desserts) supplemented with various nutrients, but both gave less than ideal results. There were two main difficulties. Many bacteria are capable of degrading starch in the potato or protein in the gelatin and therefore tend to destroy the solid surface necessary for colony formation. Moreover, Koch also found it difficult to maintain a pure, uncontaminated culture when working around the various unprotected cultures that soon accumulated in his laboratory.

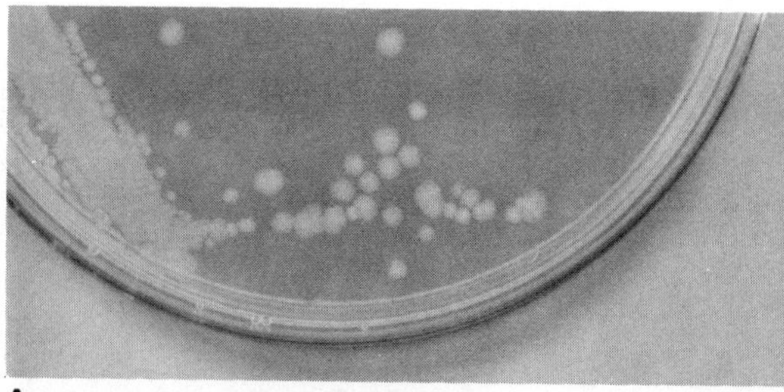

A

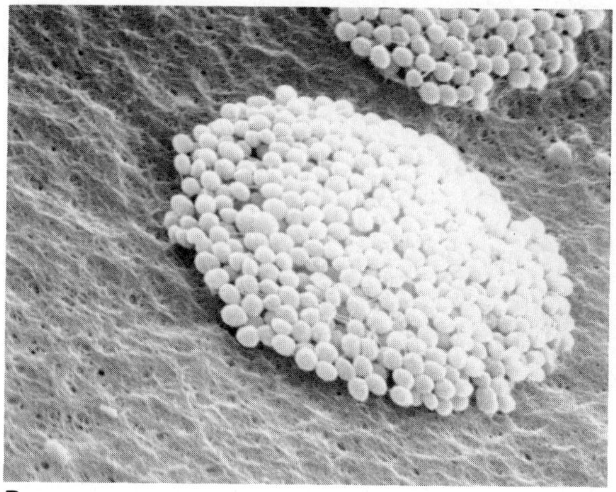

B

FIGURE 3.1

(**A**) Colonies of bacteria growing on the surface of an agar medium. This method of obtaining separated colonies is called an isolation streak. (**B**) SEM photograph of bacterial colonies.

Micrograph courtesy of Dr. Joseph M. Madden.

From Frau Hesse's kitchen came the suggestion to use a substance called **agar** as a solidifying agent. Agar is a mixture of complex polysaccharides (chains of sugars) produced by several types of algae. Originally, it was extracted primarily from the genus *Gelidium,* but more recently the genus *Gracilaria* has predominated. There are two major families of polysaccharides found in agar, agarose (the fraction that will solidify by itself) and agaropectin (the fraction that will not solidify by itself). The basic repeating unit of agarose is outlined in figure 3.2A.

Agar has the property of being solid until it is heated to a temperature greater than 90° C, at which time it liquifies and forms a suspension with water (figure

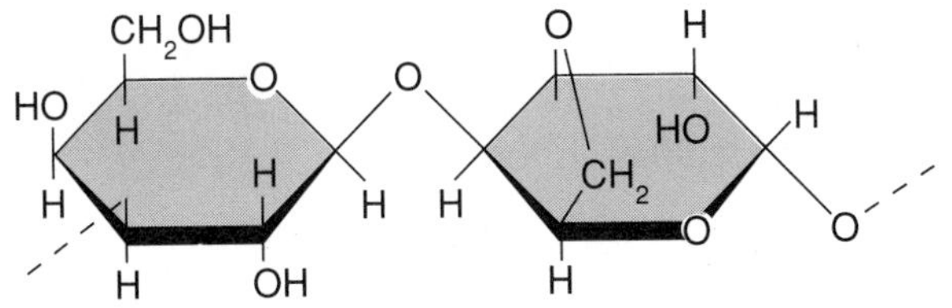

A

B

FIGURE 3.2

(**A**) Agarobiose, the basic repeating unit found in the polysaccharide agarose. Natural preparations contain extra chemical groups (not shown) on the sugar rings. (**B**) Agar solutions. The bottle on the right contains melted agar, whereas the bottle on the left contains a solidified form of the same solution. Note the difference in transparency.

3.2B). It will stay liquid down to about 40° C, at which time it again solidifies. Once an agar medium is prepared, it is generally not necessary to worry about high temperatures melting the medium as occurs, for example, when a dish of gelatin is left in the sun for awhile. Agar has yet another advantage over other solidifying agents. Despite its polysaccharide nature, relatively few bacteria are able to break it down into its component sugars; therefore, it serves as an essentially inert substrate for bacterial growth. Agar is still the

FIGURE 3.3

Petri dishes. Glass and polystyrene dishes are manufactured in several different sizes to accommodate needs for different surface areas. These two Petri dishes are glass. In normal use, the dishes are kept completely closed so that contaminating organisms cannot enter the culture.

primary solidifying agent used today, although various other chemicals, primarily silica derivatives, are used for special purposes.

The other major development in Koch's laboratory was Julius Petri's invention of a special glass dish to hold bacteriologic media. It consisted of nested glass plates with raised edges (figure 3.3). The Petri dish was invaluable because it permitted cultures to be stacked in a small space and transported from one place to another without contamination.

The contributions from Koch's laboratory were critically important in establishing firm scientific footing for bacteriology. On the other hand, they have led to an implicit bias in most studies because it is possible for some nonparasitic bacteria to grow only in the presence of other organisms (symbiotically). Such associations are probably more common than recognized presently, but the possibility of their existence is frequently ignored.

Culture Media and Their Ingredients

A culture **medium** (plural media) is any substance or mixture of substances used to grow one or more kinds of microorganism in the laboratory. Bacteriologic media frequently are classified as complex or defined according to their chemical composition. These terms refer not so much to the number of ingredients present but to our knowledge of their exact chemistry. A complex medium is one that contains such chemically ill-defined substances as yeast or beef extract, while a defined medium can have 10 to 20 different chemical substances present, the identity of each being known

TABLE 3.1 Bacteriologic media

Luria Broth (complex medium)		Pseudomonas Medium (defined medium)	
NaCl	8 g	K_2HPO_4	2.65 g
Yeast extract (water-soluble portion of yeast cell lysate)	5 g	KH_2PO_4	2.08 g
Tryptone (from casein digested with the enzyme trypsin)	10 g	NH_4Cl	1.0 g
		$MgSO_4 \cdot 7H_2O$	0.5 g
		$CaCl_2$	0.5 mg
		Ferric ammonium citrate	5.0 mg
		Succinic acid	8.0 g

precisely. An example of the ingredients present in a typical medium of each type can be seen in table 3.1. The term *synthetic* is sometimes used instead of the word *defined* in order to imply that the substances present have been chemically purified rather than extracted from natural products.

Materials used in complex media are generally water-soluble portions of cells or tissues from various types of organisms. Meat extracts from different animals and/or organs and water-soluble portions of disrupted yeast cells commonly are used. Regardless of the ingredients, a culture medium must fulfill all the nutritional needs of the organisms to be grown on it. These needs are the same as the general nutritional requirements for plants or animals. An organism must have a source of energy to grow as well as various chemical elements to synthesize the molecules of which it is composed. In order of decreasing quantity, these elements generally include carbon, hydrogen, nitrogen, oxygen, phosphorus, potassium, sodium, sulfur, magnesium, calcium, and iron. **Trace metals** such as molybdenum, zinc, and manganese also are required but in such small quantity that the normal contaminants in standard chemicals will often suffice to supply them.

The nutritional requirements of bacteria are diverse. Some need essentially no organic material for growth, whereas others require many amino acids, nucleic acid bases, and even vitamins to grow; therefore, the media necessary to grow microorganisms range from the simple to the very complex. Table 3.1 provides an example of a simple defined medium (*Pseudomonas*) and of a very complicated one (*Methanobacterium*). Both, however, have certain features in common. Phosphorus is supplied in the form of phosphate, sulfur in the form of sulfate, and nitrogen as ammonium ion. Calcium and magnesium are also supplied as salts, although care must be taken not to form insoluble phosphate or sulfate compounds during preparation of the medium. Sometimes a **chelator,** a compound that binds strongly to certain cations, is added to keep the calcium and magnesium tied up as part of a soluble complex until they are needed by the organisms. Commonly used chelating agents include citric acid and ethylenediaminetetraacetic acid (EDTA).

Carbon can be supplied in a wide variety of forms depending on the organism to be grown. **Autotrophs** can take carbon dioxide from the air and fix (reduce) it into carbohydrate compounds that can be used to provide the majority of compounds they need for metabolism. The process of carbon fixation is discussed further in chapter 5. Other organisms called **heterotrophs** use small quantities of carbon dioxide, but for the most part they can use carbon only after it has been reduced at least to the valence level of carbohydrates. Heterotrophs can, however, use many different forms of organic molecules.

A few bacteria have the capability of reducing gaseous nitrogen (N_2) into ammonium ion and, therefore, do not require the addition of any nitrogenous compounds to the medium. This process, called **nitrogen fixation,** is analogous to carbon dioxide fixation and is extremely important environmentally (see detailed discussion in chapter 12). Certain other bacteria are capable of reducing nitrate or nitrite ions to ammonium ions in order to supply their metabolic needs.

The sources of energy used by bacteria are more diverse than those used by eukaryotes, including protozoa and fungi. There are, of course, the familiar processes of photosynthesis and oxidation of simple sugars

***Methanobacterium* Medium (defined medium)**

K_2HPO_4	15 mg	H_3PO_4	0.1 mg
KH_2PO_4	300 mg	Na_2MoO_4	0.1 mg
$(NH_4)_2SO_4$	300 mg	Cysteine HCl	60 mg
NaCl	610 mg	Resazurin	1 mg
$MgSO_4 \cdot 7H_2O$	1,200 mg	Nicotinic acid	0.05 mg
$CaCl_2 \cdot 2H_2O$	800 mg	Folic acid	0.02 mg
Na_2CO_3	400 mg	Pyridoxine HCl	0.1 mg
$MnSO_4$	5 mg	Biotin	0.02 mg
$FeSO_4$	1 mg	Thioctic acid	0.05 mg
$CoCl_2$	1 mg	Thiamine HCl	0.05 mg
$ZnSO_4$	0.1 mg	Riboflavin	0.05 mg
$CuSO_4$	0.1 mg	Vitamin B_{12}	0.001 mg
$AlK(SO_4)_2$	0.1 mg	Calcium pantothenate	0.5 mg

Note: All ingredients are listed per liter.

APPLICATIONS BOX

Petroleum and Bacteria

When people think about the problems facing oil companies, they usually focus on the obvious tasks of locating oil deposits and drilling wells. Generally, pumping is considered a straightforward process, although such is not always the case. Often it is necessary to pump water into the well in order to force the oil (floating on the water) to the surface. This process brings with it the risk of bacterial contamination of the oil deposit.

Bacteria, especially in the genus *Pseudomonas*, pose the potential problem because they have the ability to use short-chain aliphatic or simple aromatic compounds as energy sources. These bacteria are valuable in the environment because they tend to break down spilled oil and grease. However, if they are introduced into a previously uncontaminated oil deposit, they tend to remove the lighter and economically more valuable compounds, leaving behind the so-called heavy crude oil that is much more viscous.

Oil companies working in the Arab states often use very large quantities of seawater as part of their pumping process. This water obviously contains numerous bacteria that must be removed in the most inexpensive way possible. Toward that end, extremely toxic chemicals, such as nitrogen mustards, sometimes are added to kill any bacteria present before the water is pumped underground.

In the aftermath of the massive oil spill in Alaska in 1989, a variety of methods to clean the environment were tried. One microbiological method attempt consisted of providing the environment with additional nitrogen and phosphorus sources (as fertilizer) to encourage growth of existing bacteria so that they would digest the remaining oil.

to provide energy. Organisms that get the energy needed for growth from radiant energy are called **phototrophs,** and those that derive their energy from the oxidation of preexisting chemicals are called **chemotrophs.** However, chemotrophic organisms are more versatile than might be suspected. Although certain bacteria oxidize organic molecules in more or less the same manner as animals do, many bacteria can oxidize inorganic molecules such as sulfur or nitrogen, a phenomenon unknown among eukaryotic organisms. Because many bacteria possess this latter property, chemotrophs have been subdivided into **chemoorganotrophs,** organisms oxidizing organic molecules, and **chemolithotrophs,** organisms oxidizing inorganic molecules such as those found in rocks (lithos is the Greek word for stone). Generally speaking, a given chemotroph has a preferred mode of energy production and does not readily mix organotrophy and lithotrophy. The various possible types of metabolism that organisms can possess are summarized in table 3.2.

TABLE 3.2 Types of bacterial metabolism

Organism	Carbon Source	Energy Source
Chemoautotroph	CO_2	Chemicals
Chemoheterotroph	Organic chemicals	Chemicals
Photoautotroph	CO_2	Radiant energy
Photoheterotroph	Organic chemicals	Radiant energy

Cell Boundaries

Although it is natural to think of the boundary between the bacterial cell and its surrounding medium as the cell membrane, in fact that membrane may not be the sole component. A cell may have additional exterior layers, such as a cell wall or capsule, to insulate it further from the environment. Each extra layer contributes to the overall behavior of the bacterial cell.

The Cell Membrane

Plasma membranes in bacteria, like those in their eukaryotic counterparts, serve to contain the cytoplasm and regulate the entry and exit of chemicals. Because they can be highly variable, this discussion focuses primarily on cell membranes of the most common bacteria, excluding those that resemble eukaryotic organisms (cyanobacteria, funguslike bacteria, protozoanlike bacteria). The cell membrane of bacteria that are clearly alternative forms (anaerobes and archaebacteria) is discussed briefly.

Eubacterial membranes are composed chiefly of phospholipids, such as those shown in figure 3.4A–D, that are arranged in a standard bilayer the same as that found in eukaryotic cells. The typical phospholipid is a glycerol derivative with fatty acid(s) esterified on one side of the molecule and some sort of charged moiety linked via a phosphate on the other side. Common gram-negative bacteria tend to have phosphatidylethanolamine as their primary phospholipid, while common gram-positive organisms usually have phosphatidyl glycerol or cardiolipin (figure 3.4B). Despite differences in structure, all bacterial plasma membranes are semipermeable just like eukaryotic cell membranes.

Lipids from archaebacteria and some anaerobic eubacteria are based on ether linkages rather than on ester bonds (figure 3.4D). Glycerol is still the fundamental molecule, but the side chains are different. Anaerobic eubacteria have plasmalogens (1-*O*-alkenyl-2-acyl glycerophosphatides), while archaebacteria have alkyl ethers. The latter are phytanyl derivatives 20 carbons in length. Diether lipids can be used to form a standard unit membrane that is a lipid bilayer. Members of the genus *Halobacterium* have membranes of this type. Other lipids are tetraethers in which double-length phytanyl chains are linked to glycerol at both ends. Tetraethers form a lipid monolayer with glycerols still on the exterior and phytanyl chains in the middle. Thickness of both mono- and bilayers are 70 nm. A lipid monolayer cannot be split in half like a conventional plasma membrane, however. Members of the genera *Thermoplasma* and *Sulfolobus* have monolayer membranes.

Just as fatty acids can exist in either a solid or liquid state, so too phospholipids and ether lipids can exist in either a fluid or a gel state. That temperature at which lipids change from gel to fluid or vice versa is known as the **transition temperature.** Normal functioning of the cell membrane requires phospholipids in a fluid state because the various components of the membrane must be free to move about. The **fluid mosaic** model of membrane structure, originally proposed by Singer and Nicholson, best represents our understanding to date. In this model, the membrane is envisioned as a dynamic structure in which embedded proteins or other macromolecules are free to move about slowly or to change conformation so as to present different surface views at different times. A diagram of a eubacterial membrane is shown in figure 3.5. The small bulbous structures represent the glycerol "head" of the phospholipid, and the long fibers represent the fatty acids.

The fatty acids define a **hydrophobic** region within the membrane, a region from which water and other charged atoms and molecules are excluded. The hydrophobic layer also serves as an electric insulator because charged compounds cannot pass through it. The outer surfaces of a membrane are **hydrophilic,** however, because they carry charged groups like phosphatidyl derivatives that can form ionic bonds with charged molecules and ions.

The fatty acids used to form bacterial phospholipids are usually saturated or monounsaturated (that is, they have one carbon-carbon double bond), making the phospholipids appropriately flexible. At low temperatures, saturated long-chain fatty acids behave almost like waxes, whereas unsaturated or branched fatty acids tend to remain in a liquid state. Generally speaking, the colder the temperature at which an organism customarily grows, the more unsaturated or branched fatty acids are found in its cell membrane. Conversely, organisms growing at high temperatures preferentially incorporate straight-chain, saturated fatty acids into their membranes.

Both plant and animal cells employ large planar molecules to stabilize their membranes and eliminate certain phase transitions of fatty acids. Animal cells use cholesterol or related molecules, while plants use lanosterol for the same function. In most bacteria, a structurally similar family of compounds, the hopanes, commonly bacteriohopanetetrol (figure 3.4C), seem to serve this stabilizing function. Certain structural similarities between cholesterol and the hopanes are apparent and may provide clues to evolutionary relationships between prokaryotes and eukaryotes. In some archaebacteria, the bacteriorubrins, noncyclic C_{50} compounds related to carotenoid pigments, substitute for the sterols.

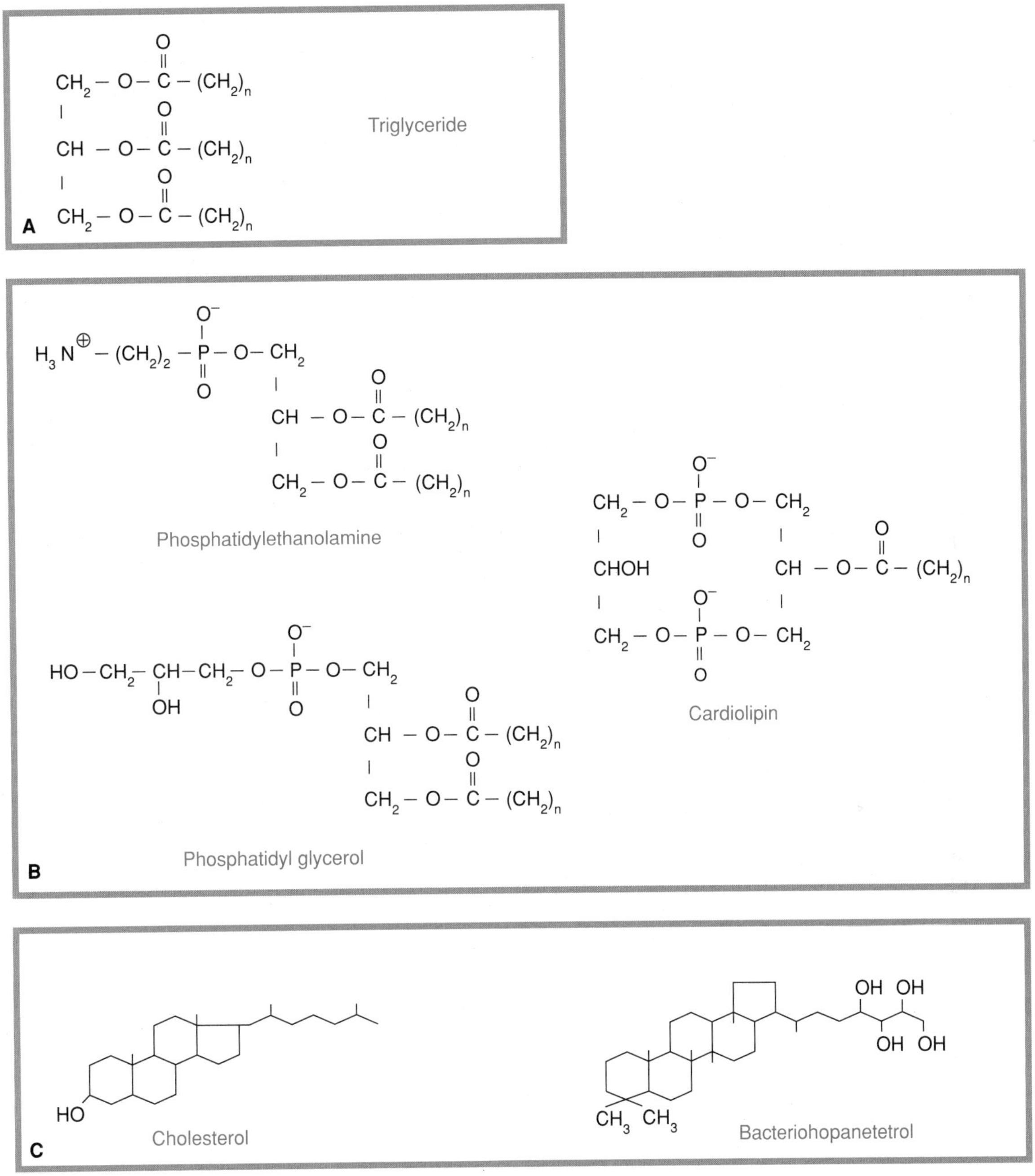

FIGURE 3.4

Continued on next page

Some components of bacterial membranes. (**A**) Eubacterial membranes are composed of lipids consisting of fatty acids esterified to a polyalcohol-like glycerol. Triglycerides have three fatty acids, diglycerides have two, etc. The fatty acids in this example are represented as saturated aliphatic chains of uncertain length. A typical bacterial fatty acid is 14 to 18 carbons in length, and one carbon-carbon double bond (unsaturation) also may be present. (**B**) Some lipids commonly found in eubacterial cell membranes. Phosphatidylethanolamine is the major component of gram-negative cell membranes. Phosphatidyl glycerol and cardiolipin both are found commonly in gram-positive bacterial membranes. (**C**) Most membranes require the presence of large planar molecules to act as stabilizers. Two such molecules are cholesterol, a sterol usually found in eukaryotic cell membranes, and bacteriohopanetetrol, a sterol analog found in many bacterial membranes. (**D**) Some anaerobic eubacteria have ether plasmalogens in their membranes, while archaebacteria have lipids with glycerol linked to phytanyl side chains by an ether bond. The lipids may be diethers that form a typical lipid bilayer or tetraethers that form a monolayer membrane. In either case, the thickness of the membrane is approximately the same.

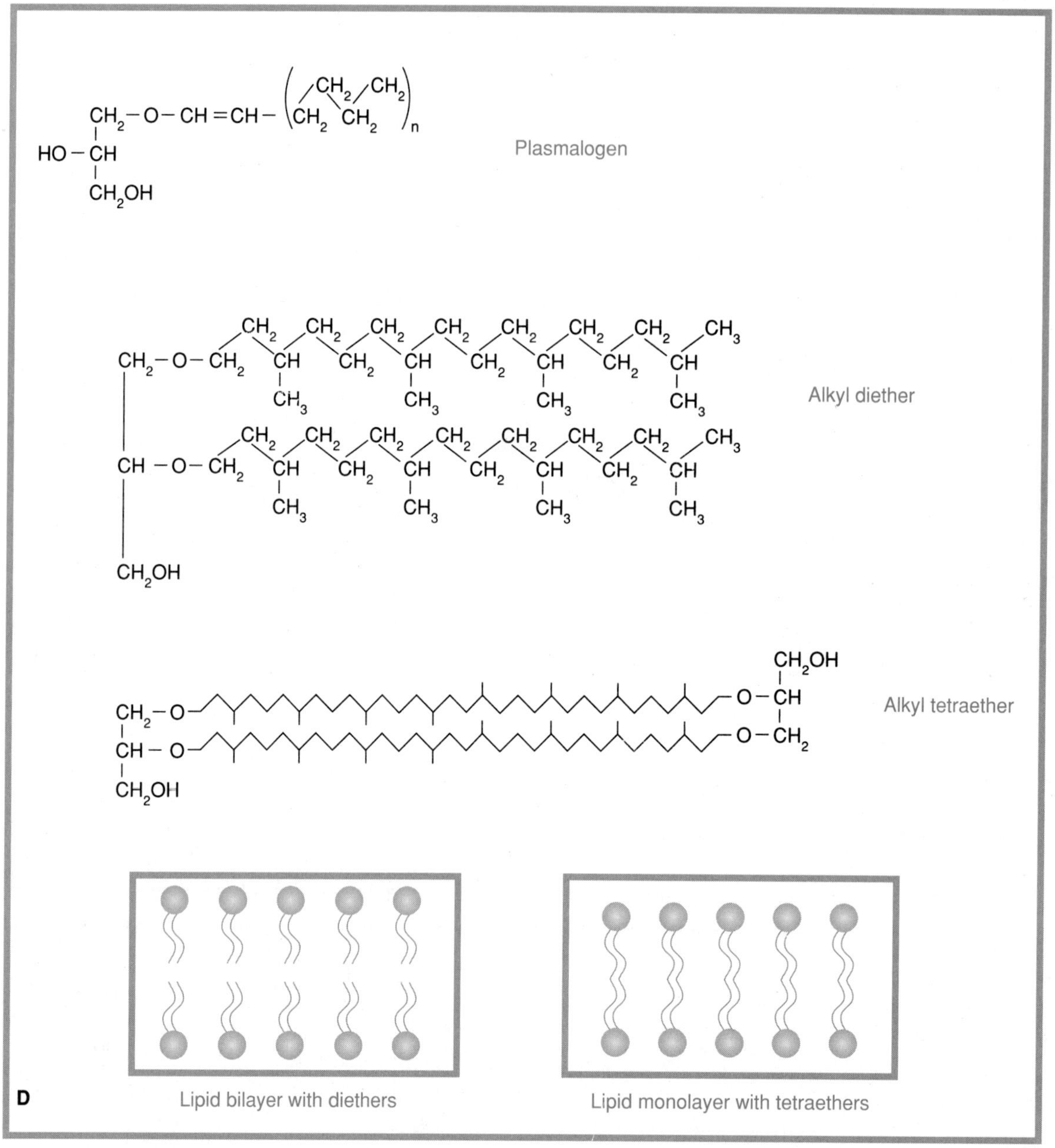

FIGURE 3.4 CONTINUED

The membrane structure shown in figure 3.5 tends to leave the impression that there is a very solid surface separating the cytoplasm from its surroundings, yet this cannot be entirely true because proteins, peptidoglycan, and various complex polysaccharides can be observed exterior to the cell membrane of nearly all bacteria. Each of these macromolecules is characteristic of a particular cell type and, therefore, must be either produced in the cytoplasm and exported to the surface or assembled on the cell surface from specific components exported from the cytoplasm along with enzymes necessary for its assembly. Examples of such activity can be seen in the formation of the cell walls of bacteria.

In addition to the proteins that are exported to the cell surface, there are others that are actually a part of

APPLICATIONS BOX

Crude Oil: From Dinosaurs or Bacteria?

Some time ago, a television commercial for an oil company depicted dinosaurs dissolving into puddles of oil that accumulated underground. Later the oil was pumped to the surface. Just how accurate is such a conception of the origin of crude oil?

Several groups of scientists have been examining the chemical composition of petroleum precursors found in shale rock estimated to be 50 million years old (see the review article by Prince in the suggestions for further reading). Among the major components is bacteriohopanetetrol, accounting for some 5% of soluble organic matter in these sediments. Some estimates place the global amount of bacterial hopanoids at 10^{11} or more tons. This mass is greater than the total mass of organic carbon in all living organisms. Such a large number of these molecules suggests that petroleum is not a direct derivative of dinosaurs and plants but rather is produced from phospholipids and other organic molecules of bacteria that degraded dinosaur and plant tissues once they were covered by layers of soil or mud.

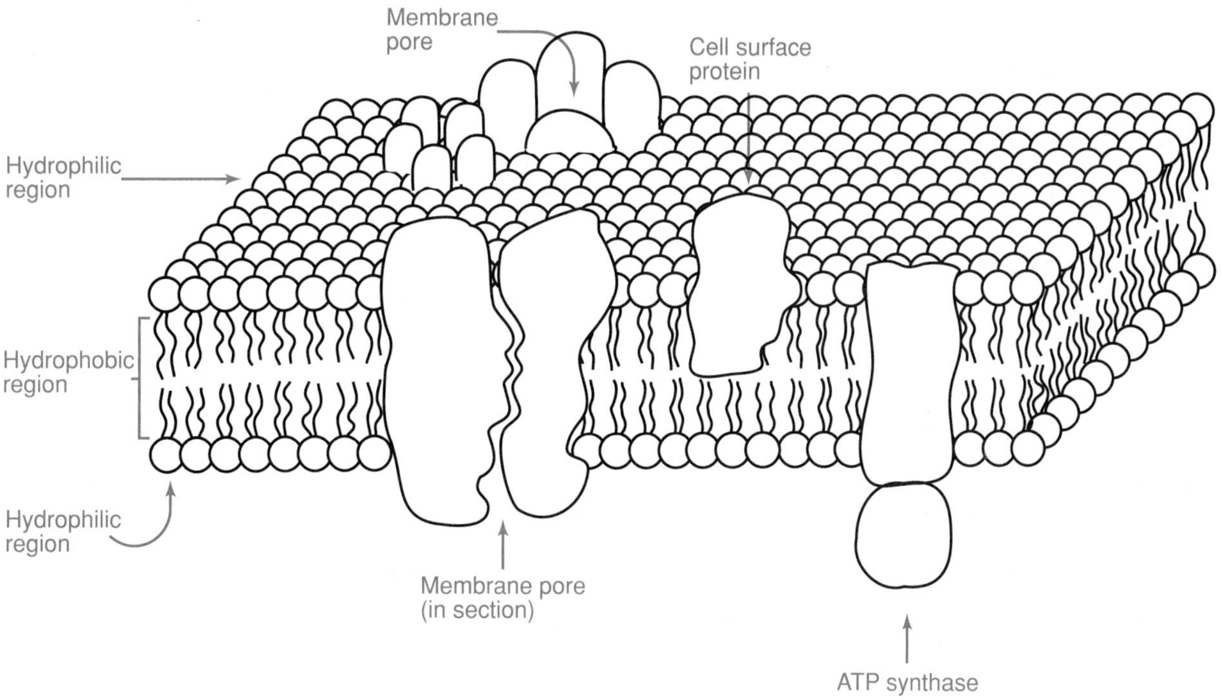

FIGURE 3.5
Singer and Nicholson fluid mosaic model applied to the structure of a bacterial cell membrane. The large irregular objects represent proteins inserted into the membrane. Note that proteins need not protrude all the way through the membrane. Some proteins that do traverse the membrane may define physical pores within the membrane structure. This membrane is fundamentally symmetrical, whereas that shown in figure 3.10 is asymmetrical.

the membrane itself. Proteins embedded in the membrane can be of two types, those that protrude only from one side of the membrane and those that traverse the membrane completely. This means that the inner and outer surfaces of a cell membrane can display different enzymatic functions and can be distinguished on the basis of the exposed proteins.

Because of the chemical nature of the lipid bilayer or monolayer comprising a bacterial cell membrane, membrane proteins generally are thought to have a minimum of two domains, one hydrophobic and one hydrophilic. A hydrophobic region of a protein is not composed of highly charged amino acids; therefore, the protein is soluble in nonpolar solvents, but not in water. A hydrophilic region, on the other hand, is composed of highly charged amino acids and behaves exactly the opposite. In practice, then, the hydrophobic portion of a protein embeds itself in the fatty acid or phytanyl portion of the lipid in the plasma membrane that excludes the aqueous cytoplasm, and the hydrophilic region protrudes from a membrane surface. In many cases, individual molecules of membrane proteins traverse the membrane numerous times, with each hydrophobic region coiled to form a β-pleated sheet (figure 3.6).

The membrane proteins can define actual small pores or holes in the surface of the cell or exhibit other functions relating to enzymatic activity and/or transport of materials into and out of the cell. These types of functions are discussed in more detail later. Presently, it is sufficient to note that one function of a cell membrane is to act as a semipermeable regulator of materials flowing into and out of the cell. Water molecules are free to move in and out of the cell at will, but charged molecules, large molecules, and most other small molecules are not.

One final point of interest is that the impermeability of the cell membrane allows the cell to maintain a cytoplasmic ionic environment different from the ionic composition of its surrounding medium. A bacterial cell maintains a constant internal pH of 7.4 to 7.8 by pumping protons into or out of itself. This is true even for acidophilic cells like *Bacillus acidocaldarius* that grow best in a medium whose pH ranges from 2 to 4. In most situations, therefore, the cell interior is alkaline (low proton concentration) relative to its exterior environment. Because protons are charged (H^+), a charge difference also exists between the cell and its environment (cytoplasm negative with respect to the surrounding medium). The cell can augment the charge difference by pumping sodium ions out of its cytoplasm. The sum of the charge differences from ions being pumped across the cell membrane constitutes an electric potential ($\Delta\Psi$) across the membrane. This electric potential is maintained by insulating properties of the membrane. The difference in proton concentration creates a chemical potential (ΔpH). If there is a sodium ion gradient, it can be expressed as ΔNa^+. These gradients are important for the transport of materials across the membrane, which is discussed in the following section. Mechanisms for generating these gradients are presented in chapter 4.

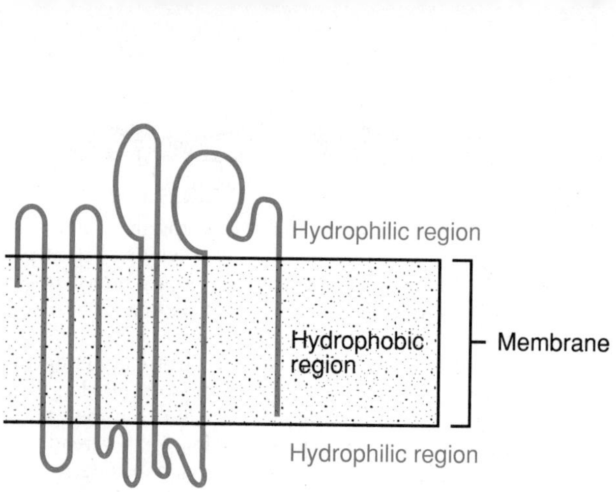

FIGURE 3.6

Example of the folding of a membrane protein. The protein may pass through the membrane more than once, indicating that the protein consists of alternating hydrophobic and hydrophilic regions.

The Cell Wall

Most bacterial habitats are hypotonic, having fewer ions and molecules per unit volume than the cytoplasm of the cells they surround. Therefore, water molecules have a tendency to move through the semipermeable plasma membrane into the cytoplasm. This tendency increases with the difference in solute concentration. The net influx of water results in cell distension and, if isotonicity is not rapidly achieved, the membrane can rupture because of the osmotic pressure produced by the additional water molecules. Most bacteria counteract osmotic pressure with a rigid cell wall that prevents excessive swelling of the cell.

Eubacterial cell walls are either gram positive or gram negative. The gram-positive cell wall is a relatively homogeneous structure composed almost entirely of peptidoglycan with certain teichoic acids woven into a latticework (figure 3.7A–D). The structure produced is a three-dimensional network or **sacculus** whose backbone is filaments composed of the **hexoses** (6–carbon sugars) *N*-acetylglucosamine (NAG) and *N*-acetylmuramic acid (NAM) in a strictly alternating arrangement. Sometimes their chemical derivatives also are used. The length of the polysaccharide chains is variable, averaging 35 to 45 disaccharide units per chain, and each chain is oriented perpendicular to the long axis of a rod-shaped cell (like an incomplete hoop). Each NAM molecule in a disaccharide pair has a lactyl side chain to which a pentapeptide, a chain of five amino acids, is attached initially. In type A peptidoglycan, the sequence of amino acids (beginning at the lactyl group) is L-alanine, D-glutamic acid, a species-specific diamino acid, D-alanine, and finally another D-alanine. In type B peptidoglycan, the first amino acid is replaced by glycine or L-serine.

Note that the pentapeptide consists of alternating stereoisomeric forms of the various amino acids, first an L form, then a D form, etc. This is unlike normal cellular proteins in which all the constituent amino acids are in the L form; therefore, the shape of the pentapeptide chain is unlike that of a normal protein chain (figure 3.8A–B). The pentapeptide is relatively resistant to the action of enzymes called **proteases** that

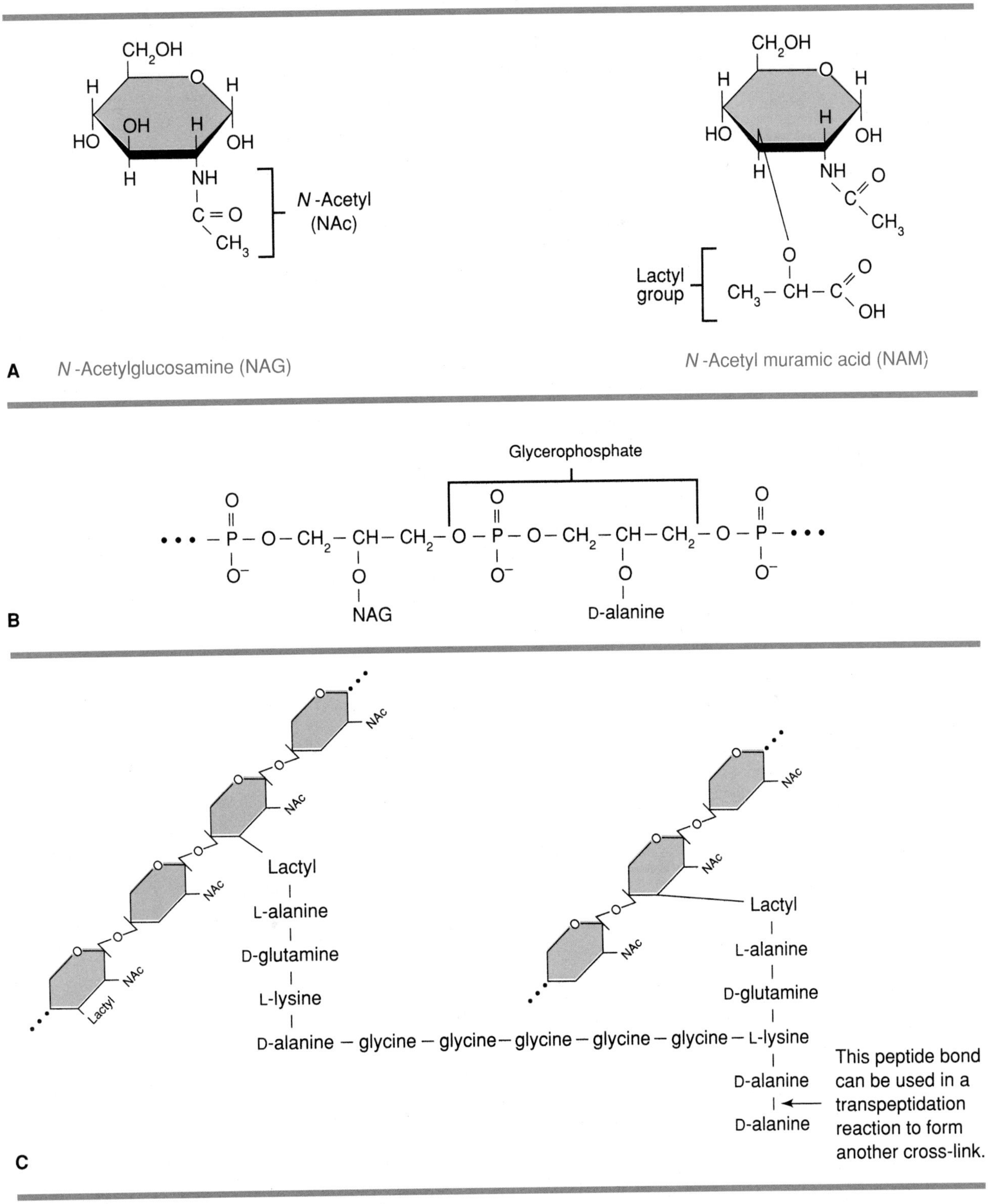

FIGURE 3.7

Chemical composition of a bacterial cell wall. (**A**) Sugar components of the peptidoglycan itself. (**B**) Teichoic acid found in gram-positive cell walls. (**C**) Cross-link between filaments showing a peptide bridge.

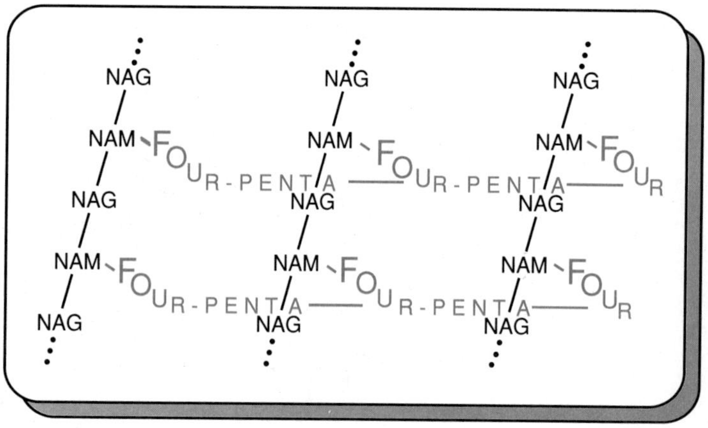

FIGURE 3.7 CONTINUED

(**D**) One layer of peptidoglycan. The peptide cross bridges are shown in color. Each colored letter represents one amino acid in the peptide cross bridge.

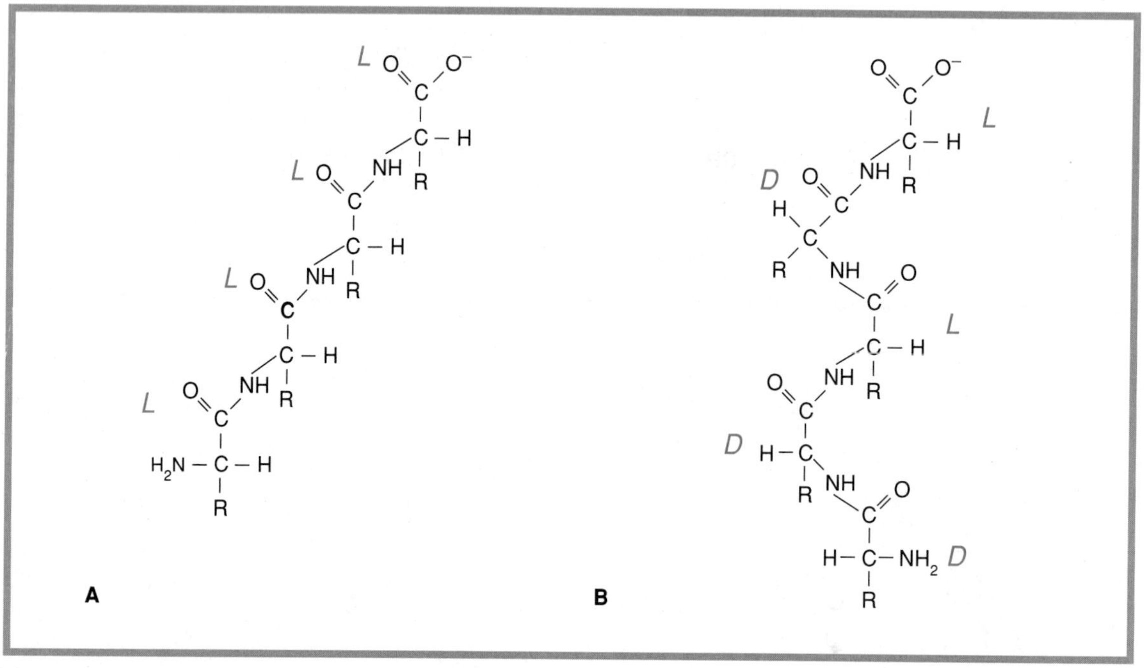

FIGURE 3.8

Stereoisomerism in proteins. (**A**) Protein composed of L-amino acids. (**B**) Protein composed of a mixture of L- and D-amino acids.

cleave peptide bonds because these enzymes generally are specific for the peptide bonds between L-amino acids. The pentapeptides of the peptidoglycan layer are thus more stable than might be expected, particularly in the case of organisms like *Bacillus subtilis* that produce large quantities of extracellular proteases.

Individual polysaccharide polymers are joined to each other by bonds connecting the peptide chain of one filament to the peptide chain of an NAM located on an adjacent filament. The connection in type A peptidoglycan is via peptide bonds produced when a **transpeptidase** enzyme takes the peptide bond joining amino acids 4 and 5 of one pentapeptide and transfers it so as to join amino acid 4 of the first chain to the diamino group of an adjacent chain, releasing an alanine molecule (amino acid 5) in the process (figure 3.9). The connection in type B peptidoglycan is from the extra carboxyl group of amino acid 2 to the diamino acid of

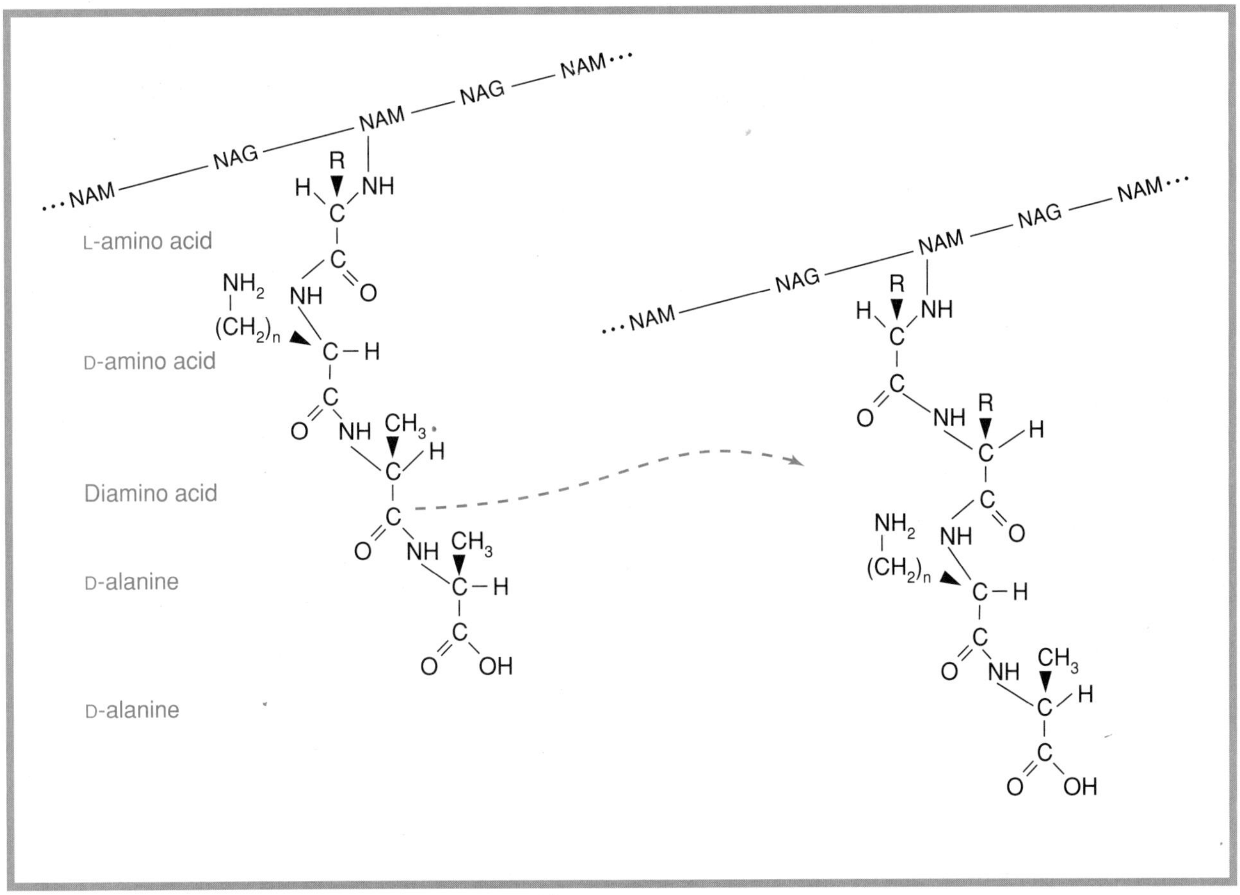

FIGURE 3.9

Transpeptidation reaction for type A peptidoglycan. A preexisting peptide bond is transferred from the nitrogen atom of one amino acid to the nitrogen atom of another.

the adjacent chain. Even if a cross bridge is not formed with a particular peptide chain, amino acid 5 is removed so that the length of an individual peptide chain in mature peptidoglycan is only 4 amino acids.

The direct connection of tetrapeptides is observed in *Bacillus subtilis* and nearly all common gram-negative bacteria. If the distance between the peptidoglycan filaments is too great for the pentapeptide to span, a peptide bridge of 1 to 5 amino acids is added. This bridge serves only to extend the diamino group and does not affect any other aspects of peptidoglycan structure. In *Micrococcus,* the bridge seems to consist of an entire tetrapeptide chain that has been moved as a unit in a transpeptidation reaction. Because the gram-positive cell wall is composed of multiple layers of peptidoglycan, it frequently contains peptide bridges, especially in the outer layers. As a result, it is reasonably porous and presents a significant barrier to molecules of small and medium size as they approach the cell.

Comparatively little is known about the function of the teichoic acids found in the gram-positive cell wall. Members of this family of compounds incorporate some of the same components found in peptidoglycan, but the backbone of teichoic acid is polyglycerol phosphate or ribose phosphate instead of hexose.

A gram-negative staining reaction merely indicates that the crystal-violet-iodine complex is not retained by the cell. There are a variety of cell boundary structures that fail to keep the dye complex, but only one structure is considered a true gram-negative cell wall. It is much more complex than the gram-positive wall and presents a definite barrier to large molecules moving in the vicinity of the bacterial cell. The peptidoglycan layer of the true gram-negative wall does not have teichoic acids present and is much thinner than the gram-positive wall, being only one layer thick (figure 3.10A–B). The lipoprotein-lipopolysaccharide layer surrounding the peptidoglycan is the so-called

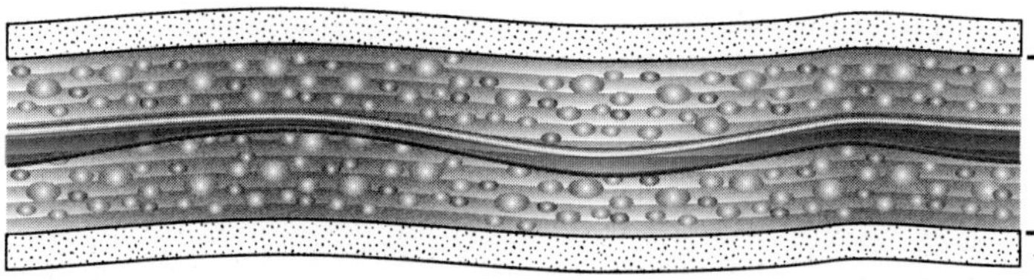

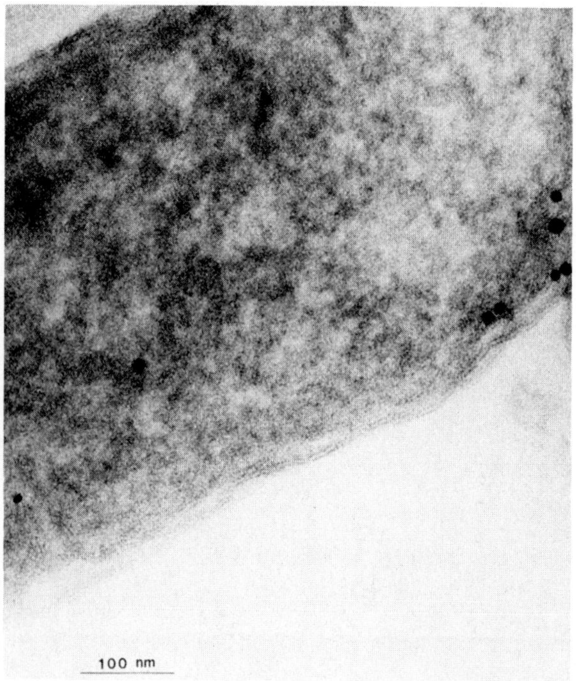

FIGURE 3.10

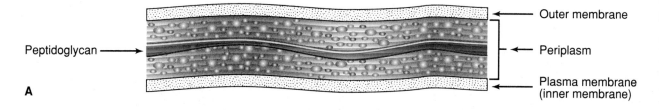

Gram-negative cell wall. (**A**) Diagram of the general composition of the wall. (**B**) Gram-negative cell wall viewed in cross section under an electron microscope. This section was prepared by a cryopreservation technique that shows the peptidoglycan layers more realistically.

Micrograph courtesy of James R. Swafford, Brandeis University.

outer membrane, another lipid bilayer about 75 nm thick and very rich in protein molecules. Roughly 50% of the outer leaflet and 60% of the inner leaflet are composed of protein.

The outer membrane, unlike the cell membrane, is highly asymmetrical in its lipid composition because its inner surface is composed primarily (90%) of phosphatidylethanolamine and lipoprotein, while its outer surface is composed of a lipopolysaccharide based on **lipid A.** Lipid A consists of phosphorylated glucosamines with fatty acids covalently bonded to the hydroxyl groups and amino group (figure 3.11A–B). The fundamental pattern consists of four myristoyl moieties, two on each glucoseamine. Myristic acid is a β-hydroxy fatty acid (hydroxyl group on the second carbon from the carboxyl group). One to three lauroyl moieties are added to complete the lipid A, each one attached to a β-hydroxy group. A principal characteristic of lipid A is, therefore, fatty acids joined to fatty acids.

The outer membrane also interacts with other macromolecular structures in its immediate vicinity. Attached to one lipid A sugar located on the outer surface of the finished membrane is a polysaccharide chain that is characteristic of the species being studied. The polysaccharide chain terminates in a series of sugars called the core that can form a bridge between lipid A and an extended polysaccharide chain or O antigen. The core contains both positive and negative charges, and adjacent chains can bind to one another providing added stability to the cell surface, especially in the presence of a detergent. O antigens, when present, are important in the recognition of invading bacteria by the immune system of a host and in the effect that the bacterium has on the host.

Some scientists have suggested that the outer membrane also may be linked to the cell membrane through peptidoglycan by zones of adhesion that represent pathways for the movement of some materials between the inner and outer membranes. Others have suggested that the zones of adhesion are artifacts caused by the simultaneous budding of material from the inner and outer membranes.

An abundant type of protein found in the outer membrane is a lipoprotein, which serves as a link between the inner leaflet of the outer membrane and the underlying peptidoglycan. At a lipoprotein's amino terminus (cysteine) are attached a fatty acid in an amide linkage and a diglyceride (two fatty acids) via the sulfhydryl (SH) group. These fatty acids have hydrophobic interactions with the fatty acids of lipid A (figure 3.11). The carboxyl terminus of a lipoprotein is lysine, a diamino acid that can be joined via its carboxyl group to the diamino acid found in the peptidoglycan pentapeptide chain. Normally, about 30% of the lipoprotein molecules are so linked. Another abundant *o*uter *m*embrane *p*rotein is OmpA, and it may serve,

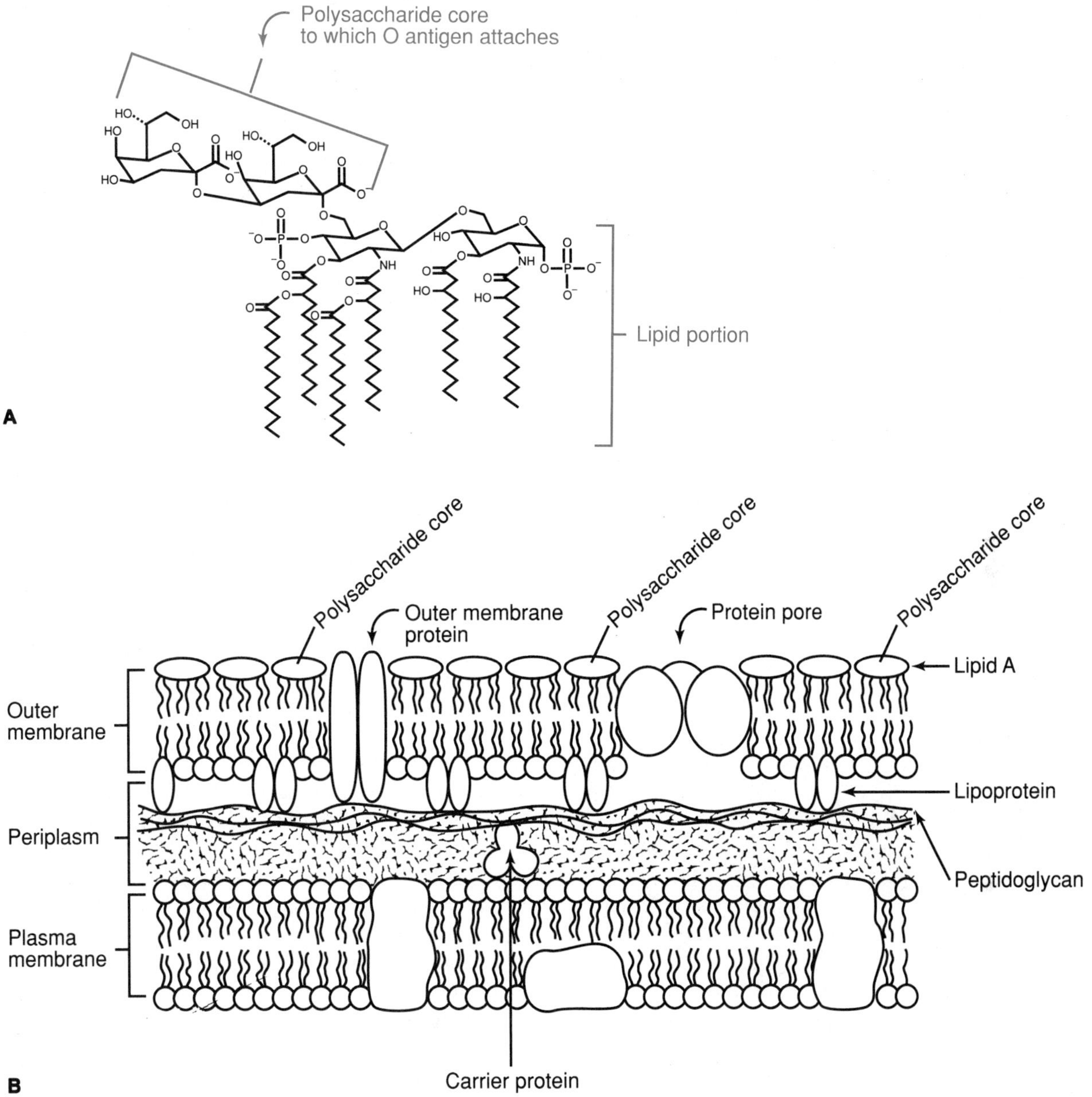

FIGURE 3.11

Model of the structure of the outer membrane. (**A**) The minimal structure of lipid A from *Escherichia coli* K-12, the foundation of the outer surface of the outer membrane. (**B**) One small portion of the outer membrane. Note the periplasmic gel that exists between the outer and plasma membranes. The stippled area represents peptidoglycan, the density of the stippling indicating the degree of cross-linkage. Lipoprotein molecules join the outer membrane to the peptidoglycan layer.

Reproduced, with permission, from the *Annual Review of Biochemistry,* "Biochemistry of Endotoxins," by C. R. H. Raetz, Volume 59, pp. 129–170. Copyright © 1990 Annual Reviews Inc., Palo Alto, CA.

among other things, to make the relatively rigid outer membrane more fluid.

In thin sections of gram-negative cells, a **periplasmic space** has been observed between the plasma membrane and the outer membrane. Recent electron microscopic studies of thin sections prepared from frozen bacteria have indicated that, like the mesosomes, the large periplasmic space often observed in thin sections of conventionally fixed bacteria is an artifact produced by the fixation process. Such an observation does not indicate that the space never occurs but

APPLICATIONS BOX

Archaebacteria as a Third Form of Life

Archaebacteria stain gram negative and were for a long time considered to be somewhat unusual members of that group. In the late 1970s however, Carl Woese and his collaborators suggested that archaebacteria have their evolutionary origins in a primitive form of life that diverged from the rest of the prokaryotes at a very early date.

One piece of evidence that favors their hypothesis is that the cell surface of archaebacteria is unlike the conventional one of gram-negative bacteria. The archaebacterial cell wall lacks both standard peptidoglycan and outer membrane. Instead, there seem to be three variant wall structures. The most similar to the structure seen in gram-negative bacteria is found in *Methanobacterium* and consists of a pseudomurein that has glycan chains linked by peptides, but the glycans do not include muramic acid. In some cases, such as in *Methanosarcina,* the cell wall consists of polysaccharides that have no cross-links at all. In *Halobacterium* the wall appears to be composed entirely of glycoproteins, proteins linked to an oligosaccharide as well as to di- or trisaccharides. These glycoproteins are in addition to the extensive S layer of protein found on the outer surface of all archaebacteria (see micrographs).

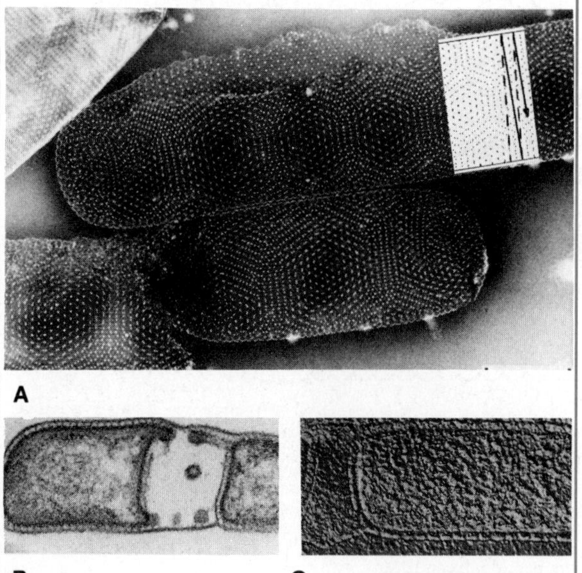

Bacterial S layers. (**A**) Negatively stained isolated S layer from *Thermoproteus tenax.* Prominent white dots represent needlelike spikes protruding from the inner surface of the layer and anchoring it via their distal ends to the cell membrane. Thin sections (**B**) and freeze-fractures (**C**) show the S layer at a distance of approximately 25 nm above the cell membrane, thus a distinct interspace is created.

Reprinted with permission from Baumeister, W., I. Wildhaber, and H. Engelhardt. 1988. Bacterial surface proteins. Some structural, functional, and evolutionary aspects. *Biophysical Chemistry* 29:39–49.

rather suggests that it is filled normally with some material that shrinks during fixation. It is perhaps best to envision the periplasmic space as a hydrated gel filled with membrane-derived oligosaccharide (MDO) located between the inner and outer membranes (figure 3.10B). The outer membrane has firm bonds via lipoproteins to the underlying peptidoglycan layer, whereas the plasma membrane has only tenuous links. Therefore, the periplasmic space is more likely to appear on the inner surface of the peptidoglycan after fixation has taken place. In practice, the distinction between the inner and outer surface of the peptidoglycan is not significant because the peptidoglycan itself has large pores that make it freely permeable to proteins.

The periplasmic space has been estimated to account for as much as 30% of the total cell volume. Located within this region are degradative enzymes, enzymes that shape the peptidoglycan structure, and transport proteins; therefore, it has been suggested that preliminary degradation of materials sometimes occurs in the periplasmic space before they are transported into the cytoplasm of the cell.

Various experiments have shown that the periplasm is iso-osmotic (has the same number of molecules and ions per unit volume) with respect to the cytoplasm. The osmotic determinant appears to be the MDO whose synthesis varies with the osmolality of the medium. The MDO molecules are too large to pass through the outer membrane, but they form ionic bonds with numerous water molecules and act to retain water that would otherwise be drawn into the medium as a result of high ionic strength.

Lying on the very surface of all archaebacteria and many eubacteria are regular arrays of surface proteins called S layers or RS layers that can be seen clearly with the electron microscope (see the micrographs in the applications box on archaebacteria). In gram-negative cells, these proteins are not considered to be outer membrane proteins because they are not inserted substantially into the lipid bilayer. Archaebacterial S layers are more highly organized and located slightly above the cell membrane. Nevertheless, they are attached covalently to the cell membrane (white dots in the micrographs) by specific protrusions from the S layer. Proteins of the S layer can function as receptor sites for viral attachment; possibly they also may act as a type of molecular sieve. In archaebacteria, they may serve to stabilize the cell shape.

Chemiosmotic Theory of Membrane Function

Standard descriptions of membranes often leave the impression that the cell membrane is a static structure full of medium-sized holes. Small molecules fit through the holes whereas large molecules do not, thereby making the membrane semipermeable. Such a mental picture does not do justice to the breadth of membrane function.

Peter Mitchell's **chemiosmotic theory** best describes the function of a semipermeable membrane. It is based on the observations that a membrane is an insulator and that ions may proceed across it regardless of whether the cell is eukaryotic or prokaryotic. Furthermore, many cell functions require the presence of a semipermeable membrane for routine metabolic activity in order to maintain separate cellular and subcellular compartments having different ionic concentrations from normal cytoplasm. Ionic concentration gradients are maintained in a steady state by specific pumps that act continuously. In typical prokaryotes, there is a general excess of Na^+ and H^+ exterior to the cell and a surplus of K^+ in the cytoplasm. The gradient of charge across a membrane is designated as $\Delta\Psi$. If the ion involved is a proton (H^+), there is a corresponding pH gradient (ΔpH) whose directionality is the same as $\Delta\Psi$.

For every ion on the surface of a normal membrane, there is a corresponding counterion on the other side. In the case of H^+, for instance, the counterion is often OH^- or Cl^-. The electric attraction between the ion and counterion is sufficient to prevent the ion located outside the cell from moving any significant distance away from the membrane. The semipermeable membrane prevents the ions from coming together and neutralizing one another. The ionic attraction thus represents potential energy of position. If ions are allowed to come together, energy is released; if ions are moved further apart, energy is required. Conceptually, the movement of ions across a membrane can be imagined to occur through a special reversible pump. If the pump operates in the forward direction, input of energy is required to translocate an ion across the membrane. If the pump operates in the reverse direction, the ion returns to its original side of the membrane and energy is released.

Because chemiosmotic processes are intrinsically reversible, a cell can either use the energy derived from a biochemical reaction to produce an ion gradient or allow the ions to travel down the gradient in order to drive other metabolic reactions such as transport or flagellar rotation (discussed next), or to synthesize energy-containing molecules such as ATP (see chapter 4).

Transport and Membrane Function

Transport refers to the movement of material in *either* direction across a cell membrane. This is a significant point because a cell would commit metabolic suicide if it allowed its intracellular levels of various compounds to rise too high. Although the following discussion focuses on the movement of materials into the cell, remember that for each mechanism bringing material into the cell, there is a counter mechanism for moving the same or similar material back out should the concentration reach dangerous levels. In transport phenomena, ion pumps driven by chemical or photochemical reactions are designated as primary transport systems. All transport systems that use either ion gradients formed during primary transport or energetic compounds, such as ATP, synthesized by the cell as a source of energy are then considered as secondary transport.

For gram-negative bacteria, the problem of transport begins with the outer membrane. Hydrophilic molecules cannot penetrate the phospholipid bilayer directly. Several types of well-defined channels are present within the outer membrane (see figure 3.11B, for an example) that allow small molecules to diffuse toward the peptidoglycan layer but exclude larger ones. The channels are maintained in the outer membrane by special proteins called **porins.** In *Escherichia coli,* the chief porin proteins are OmpC, OmpF, PhoE, maltoporin (LamB), and Tsx. The OmpC and OmpF pores are nonspecific in nature, accepting hydrophilic solutes with a preferred molecular weight of about 250 and excluding molecules with molecular weights larger than 650. Maltoporin seems to be used primarily to transport unbranched polysaccharides derived from starch.

In addition to porins, there are some outer membrane proteins that function to transport certain large molecules. Examples are BtuB, which handles uptake of vitamin B_{12} and the Fec family of proteins, which transport chelated iron complexes.

Regardless of the presence or absence of an outer membrane, all cells must move materials across a plasma membrane. When the cell membrane is carrying out transport functions, its semipermeable nature becomes very important. As noted, charged molecules cannot pass through the membrane unaided, and the passage of uncharged molecules that are much larger than water is also inhibited unless they are lipid soluble. Thus, for all practical purposes, movement of material across a cell membrane is mediated by some sort of carrier.

Carrier proteins span the membrane and can bind to a substance on one side of the cell membrane and transfer it to the opposite membrane surface. For metabolically important compounds, there are frequently at least two transport systems. One involves a carrier that has a relatively low affinity for the substance (works well when there is a high concentration of the substance outside the cell). The other involves a carrier with a high affinity for the same substance, but it functions only when low exterior concentrations are present. The former type of carrier protein generally is present in the cell membrane at all times, while the latter is made only when needed.

At the molecular level, any movement of a compound across a cell membrane requires the input of energy. In a simple solution, molecules move from an area of high concentration to an area of lower concentration. This process of **diffusion** results from the presence of the concentration gradient itself. In chemical terms, the energy of concentration, expressed as the chemical activity *a,* represents the driving force for the movement of molecules from an area of higher concentration to an area of lower concentration. This energy also can be used by cells to drive a process called **facilitated diffusion** (carrier-mediated diffusion) in which molecules at a higher concentration bind to a protein exposed on the membrane surface and then move to the other side of the membrane where they are released (figure 3.12). Facilitated diffusion is advantageous to the cell because this process does not require any metabolic energy. However, it is not as useful as some other mechanisms because it can operate in reverse to remove material from the cell as the ionic concentration outside the cell drops. Facilitated diffusion is common among multicellular organisms, such as mammals, but rare among prokaryotes.

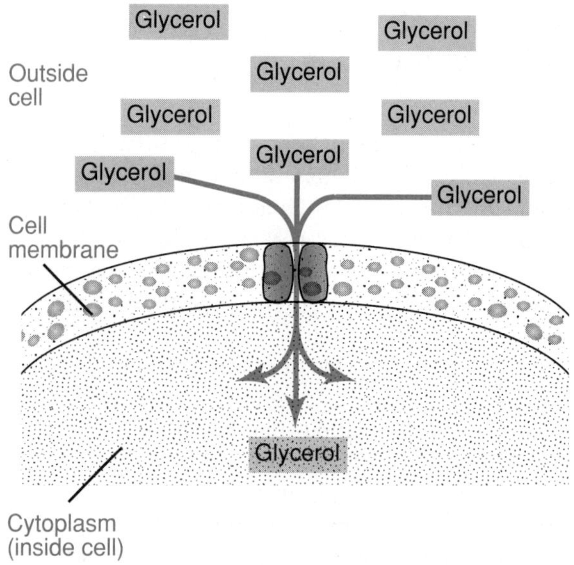

FIGURE 3.12

Facilitated diffusion. Molecules move into the cytoplasm in response to their concentration gradient. Specific carrier proteins within the cell membrane facilitate their entry.

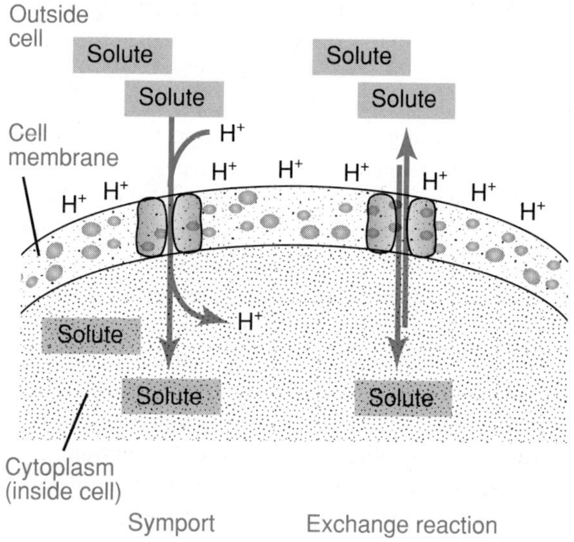

FIGURE 3.13

Symport. Net concentration of the substance requires movement of protons into the cell, while simple exchange does not.

Many bacterial cells grow in areas of relatively low nutrient concentration (10^{-6} M or less); therefore, they need some sort of concentrative transport mechanism in order to grow. This activity is designated as **active transport** and requires an external energy input because the desired molecule actually is moving against its concentration gradient. One source of energy is the **proton motive force** ($\Delta\tilde{\mu}_{H^+}$), equivalent to the sum of the electric ($\Delta\Psi$) and chemical (ΔpH) gradients across the cell membrane. In systems that use this source of energy, the desired substance is cotransported along with a proton. As a proton moves down its electric and chemical potential gradients (from positive to negative and from higher to lower concentration, respectively), it provides the energy necessary to move the cotransported substance against its concentration gradient (see figure 3.13). The transport of a desired substance against its concentration gradient in conjunction with

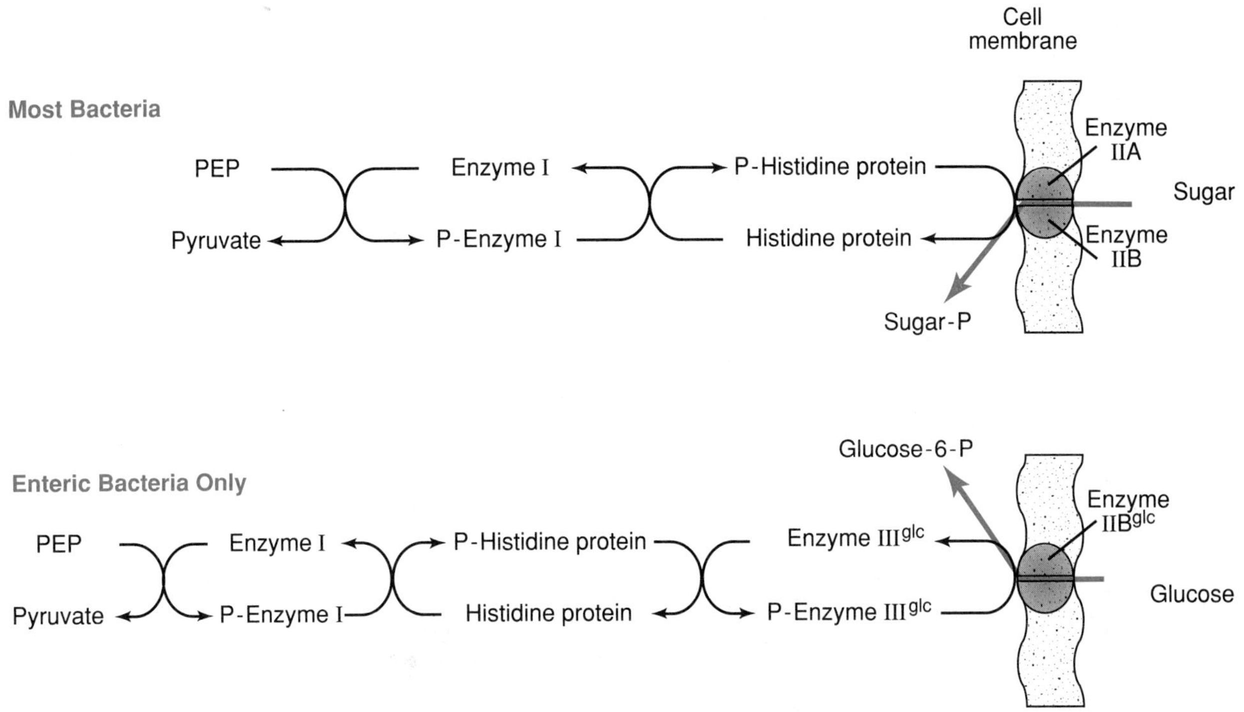

FIGURE 3.14
Glucose phosphotransferase system (PTS). The source of the phosphate group is phosphoenolpyruvate (PEP), and the actual pathway may take either of the two routes shown. In enteric bacteria (*Escherichia coli* and its relatives), glucose uptake proceeds via enzyme III.

the simultaneous transport of an ion down its concentration gradient is called **symport.** The ion is often H^+ but some other positively charged ion such as Na^+ or K^+ can be used instead. Transport of H^+, Na^+, and K^+ ions back to the surface of the cell then occurs by some form of energy-requiring antiporter pump (another example of active transport).

Another type of active transport is binding protein-dependent transport. The molecule to be moved binds to a specific carrier, which can be part of a chain of carriers, to be transported across the membrane. After transport has occurred, the carrier (or some element in the chain) must be reset by an input of energy from ATP or its equivalent.

In bacteria, active transport is often associated with **group translocation,** a process in which a molecule is actively transported into the cell and chemically modified at the same time. A good example of this process is glucose transport that occurs via the carbohydrate phosphotransferase system (PTS) found uniquely in prokaryotes (figure 3.14). Glucose in the cell is found in the form of glucose-6-phosphate, while glucose outside the cell occurs as free glucose. The phosphate group is supplied by phosphoenolpyruvate (PEP), a compound with intrinsic energy equivalent to an ATP molecule. The energetic phosphate is transferred along a series of protein molecules until it is attached to the glucose.

The phosphorylated glucose carries a net negative charge and, consequently, is less likely to be able to cross the cell membrane and escape from the cytoplasm than is an uncharged free glucose molecule. However, the gram-positive bacterium *Streptococcus faecialis* has an anion-anion **antiporter system** that can catalyze the exchange of a cytoplasmic phosphorylated glucose for an external inorganic phosphate molecule. Unlike symport, in which the two molecules being transported are moving in the same direction, the solute molecules are moving in opposite directions during an antiport reaction. This exchange reaction is an example of a transport system that can prevent excessive accumulation of a substance in the cytoplasm.

In the basic PTS pathway, there are four proteins involved. PEP reacts with enzyme I in the cytoplasm to form a phosphorylated enzyme I molecule. This phosphorylated protein then reacts with Hpr (heat-stable protein), another cytoplasmic protein, and transfers the phosphate group to it. Enzyme II, located in the cell membrane, carries out the actual transport of the sugar. It receives a phosphate from either Hpr or from an intermediate protein designated as enzyme III and uses the energy to transport the sugar molecule and phosphorylate it as well. Whether enzyme III is used depends on the sugar to be transported. Enzyme III is involved in certain regulatory pathways.

Specific Examples of Membrane Functions

Transport of the Sugar Lactose by *Escherichia coli*

Lactose is a disaccharide composed of the monosaccharides glucose and galactose (figure 3.15). The enzyme that breaks down the disaccharide to its component monosaccharides is β-galactosidase, which is found only in the cytoplasm. Therefore, use of lactose as an energy source requires that the sugar cross the outer membrane and be transported across the inner membrane. The carrier protein is a galactoside permease found in the inner membrane. It is folded so that it traverses the membrane several times (figure 3.6).

The transport reaction is one of symport in which a proton and lactose bind to the carrier on one side of the cell membrane to be transported to the other side. Thus, the same carrier protein can handle both entry and exit of lactose. The transport mechanism is thought to be one in which the carrier proteins are normally in a "V" configuration, with the opening of the "V" directed toward the outside of the cell. The binding of lactose and a proton cause a momentary change in the shapes of the carrier protein molecules so that the normal opening is closed and the cytoplasmic ends of the proteins separate to release the lactose and proton into the cytoplasm (figure 3.16).

For a cell in which the energy-generating pathways used to maintain the ionic gradients are blocked, symport of lactose actually will deplete the proton gradient, causing the interior of the cell to become more acid than normal. As the ΔpH drops, transport of lactose slows and then stops. However, a physical exchange of lactose molecules between the inside and outside of the cell does not require any energy and can continue regardless of the state of the proton gradient. No energy is required because the same number of molecules are present on each side of the membrane at the end of the process as were present at the beginning, and equilibrium has been maintained. Therefore, a simple exchange of molecules across a membrane can occur spontaneously in the absence of a proton gradient.

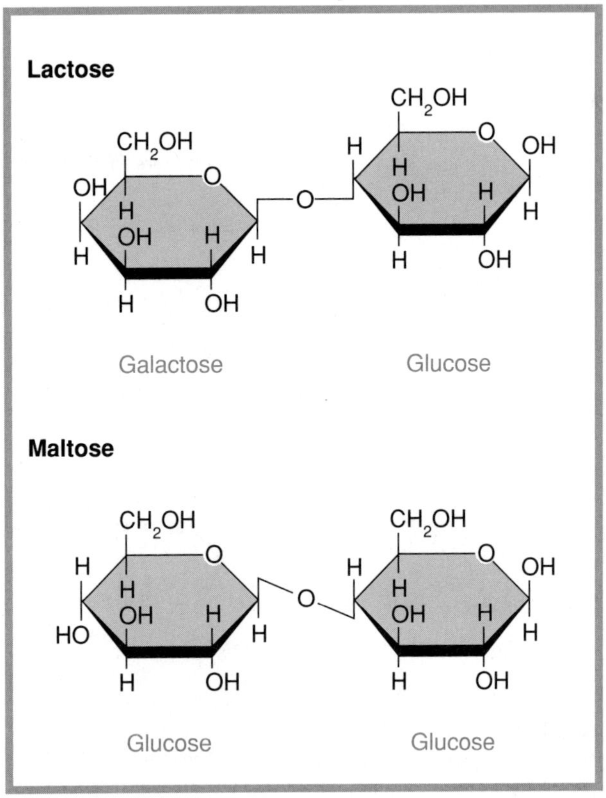

FIGURE 3.15
Structural formulas for the disaccharides lactose and maltose.

Transport of the Sugar Maltose by *Escherichia coli*

Transport of maltose (figure 3.15) is more complicated than that of lactose, but it is also extremely efficient. Maltose transport has been observed only in gram-negative bacteria and exemplifies binding protein-dependent transport. In just a few minutes, *Escherichia coli* can raise its internal maltose concentration 10,000–fold compared to the maltose concentration outside the cell. The general pattern of the transport is shown in figure 3.17.

In a rate-limiting step, maltose passes through a pore in the outer membrane formed by a trimer of maltoporin proteins and enters the periplasmic space. There it adsorbs to the maltose-binding protein (MalE) that will pass through the peptidoglycan layer and deliver it to the cell membrane. In the cell membrane is a potential pore formed by two proteins (MalF and MalG). When a loaded maltose-binding protein arrives at the pore, the pore opens, possibly in a manner similar to the galactoside permease V configuration, allowing the

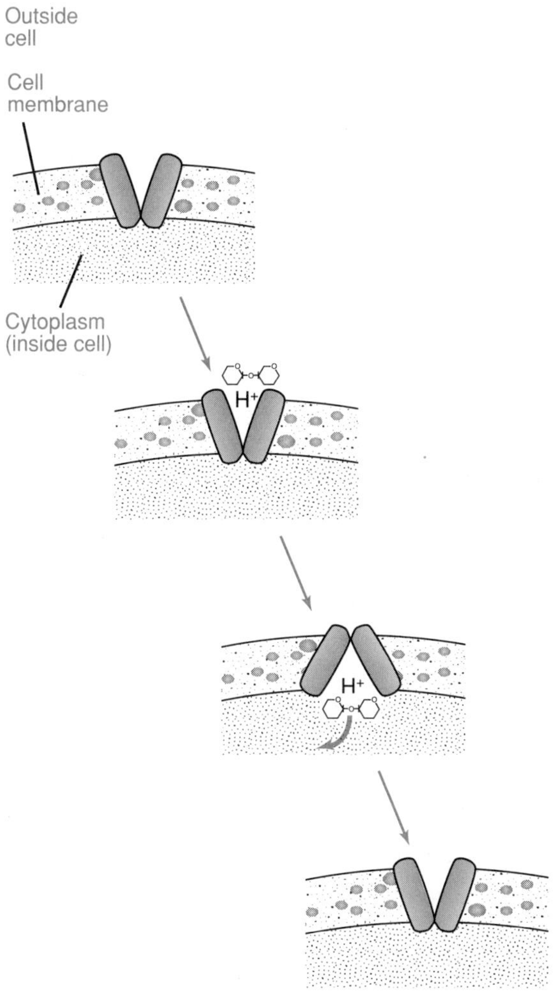

FIGURE 3.16

Model for the function of permease proteins in lactose symport. Permease proteins are oriented normally with a cleft open on the surface of the cell. When lactose and a proton are bound to the permease molecules, the proteins change configuration so as to close the cleft on the outer surface of the membrane and open a new one on the inner surface. The lactose molecule and the proton then are released into the cytoplasm, and the permease proteins return to their original configuration.

From G. A. Scarborough, "Binding Energy, Conformational Change, and the Mechanism of Transmembrane Solute Movements," in *Microbiological Reviews* 49:214–231. Copyright © 1986 by the American Society for Microbiology, Washington, DC. Reprinted by permission.

maltose to enter the cytoplasm and releasing the binding protein. The pore then must be regenerated to its original configuration by another protein (MalK) that uses the energy in an ATP molecule to accomplish its task.

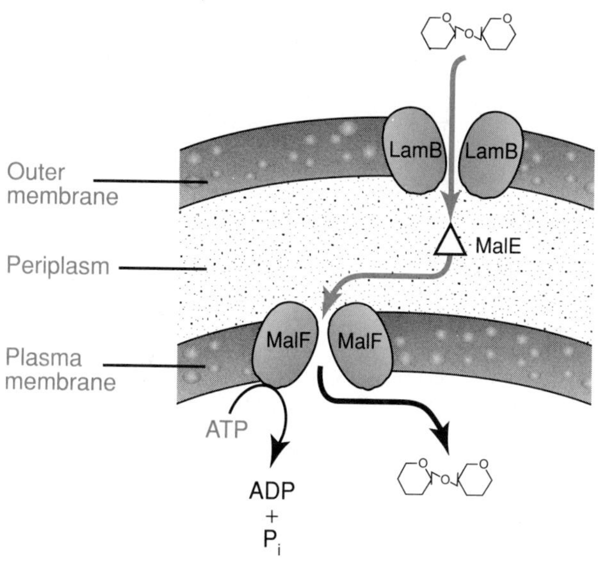

FIGURE 3.17

Model for the transport of maltose in *E. coli*. The MalF proteins form a pore within the cell membrane. Their V-shaped structure can be thought of as opening in or out. When the pore opens outward, a maltose molecule may bind to it and be transported into the cytoplasm. When it opens inward, ATP hydrolysis is necessary to return it to the original outward orientation. P_i designates inorganic phosphate.

Note that this process is not the same as group translocation of glucose because maltose is not modified chemically as it passes through the cell membrane.

One difference between the lactose and maltose systems is clear. The maltose transport system has extra carriers to move the sugar through the outer membrane, periplasmic space, and inner membrane so that it does not depend on diffusion of the free maltose molecule at any point in the process.

Chemotactic Signals

Rotation of bacterial flagella is driven by the proton motive force rather than by ATP, although ATP is necessary for normal chemotactic function. In current models, this force is assumed to be converted somehow into a rotary motion in which the M ring moves with respect to the P ring anchored in the peptidoglycan layer (figure 2.11). Although the direction of the proton motive force is fixed, it is possible for bacteria to reverse the direction of their flagellar rotation; therefore, something equivalent to a gear box must exist. In *Streptococcus*, the speed of the motor has been shown to be proportional to the proton flux across the cell membrane. At low speeds, about 5 Hz, the motor is

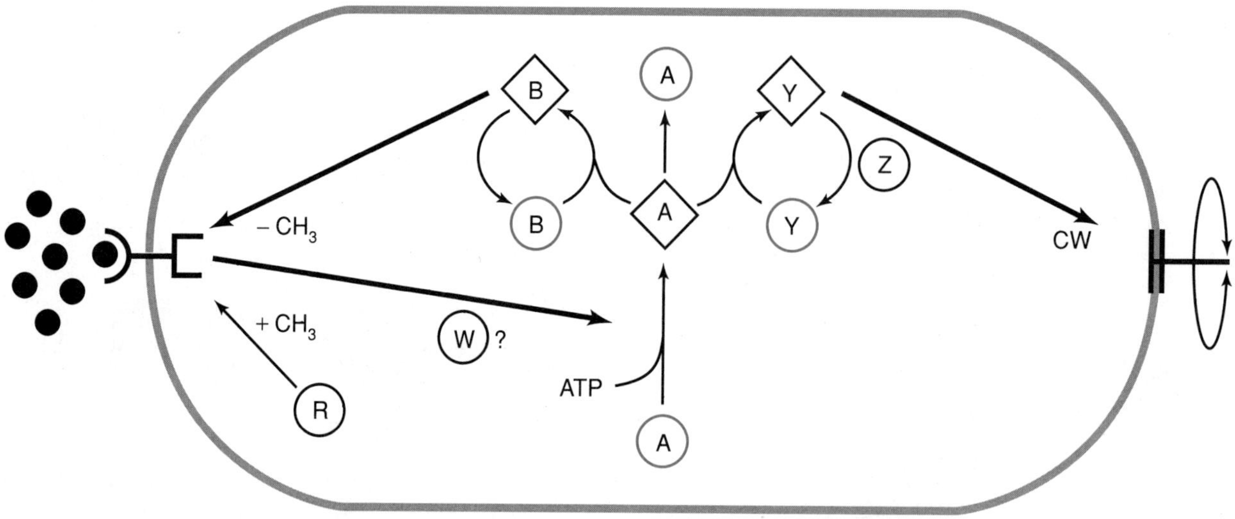

FIGURE 3.18

Signal transduction in bacteria. Lettered geometric figures refer to specific chemotaxis (Che) proteins. Proteins rendered as diamonds are phosphorylated, whereas those in colored circles are not. The receptor molecule is located at the left end of the cell and the flagellar motor at the right.

nearly 100% efficient and can generate a torque of 2.2 $\times 10^{-11}$ dyne-cm.

The chemotactic response of bacteria requires that a cell be able to detect differences in the concentration of a particular chemical with respect to time. The cell measures the concentration of a chemical over a period of about 1 second and compares the value obtained to that obtained some 3 seconds earlier. If the two concentration values differ, the cell alters its behavior according to whether the chemical is an attractant or a repellent (chapter 2). If the direction in which the bacterium is swimming is favorable, counterclockwise rotation of the flagella is maintained. If the direction is unfavorable, clockwise rotation with its concomitant tumbling of the cell is initiated, followed by another period of counterclockwise rotation.

Genetic and biochemical studies have shown that there is a class of **methyl-accepting chemotaxis proteins** embedded in the cell membrane that can affect the frequency with which flagellar rotation is reversed. One possible mechanism by which these proteins act is diagrammed in figure 3.18. Each protein is oriented so that its carboxyl terminus is in the cytoplasm and its amino terminus is outside the cell membrane. The proteins are not associated specifically with the ends of flagella but rather seem to be distributed randomly throughout the plasma membrane. As shown in the diagram, an attractant molecule bound to a receptor stimulates a chemotaxis (Che) protein, perhaps CheW, to cause phosphorylation of CheA proteins by ATP. The phosphate groups are then rapidly transferred to the CheB and CheY proteins. When the CheY protein is phosphorylated, it stimulates the flagella to rotate in a clockwise direction (a tumble).

Any sensory transducing apparatus must have some mechanism for adapting to the presence of a continuing stimulus, and bacteria are no exception. Like many other organisms, they use a system in which adaptation is mediated by a covalent modification of the receptor proteins. The bacteria have a methyl transferase enzyme (CheR) that attaches a methyl group to the carboxyl group of glutamic acids in the sensory receptor protein (figure 3.18). The presence of the methyl group effectively nullifies the signal sent by the receptor protein. Because methyl transferase does not act infinitely fast, there is a time lag before all receptor proteins with ligands are methylated. The greater the amount of chemotactic material present, the more receptors will have to be methylated and, therefore, the longer the cell will continue to exhibit chemotactic behavior. When the concentration of the chemotactic material drops, some methylated receptors will lose their ligand. Subsequently, a demethylase enzyme, phosphorylated CheB, removes the methyl group from the receptor, returning the cell to its original state.

Summary

Bacterial cells can be separated from their environment by several layers: capsules, S layers, cell walls, and cell membranes. Although all cells have semipermeable plasma membranes, these other structures are not necessarily present in every case. Capsules usually are layers of slimy polysaccharide and, when present, represent the most exterior layer of the cell. S layers are a loosely bound protein meshwork located on the surface of the cell wall. Cell walls vary in structure, with the gram-positive cell wall being less complex than the true gram-negative one. Most cell walls contain peptidoglycan, a layer composed of filaments of sugar molecules cross-linked by short peptides; however, some cell walls contain pseudomurein, which has similar structure to peptidoglycan but different components. Less common are cell walls composed of protein. The gram-positive peptidoglycan layer is substantially thicker than the gram-negative one. On the other hand, gram-negative cells have their peptidoglycan layer surrounded by an outer membrane, an asymmetrical lipoprotein-lipopolysaccharide layer that acts as a semipermeable barrier to penetration by large molecules. The cell membrane is a fluid structure composed of lipids and proteins that serves as the osmotic barrier to the cell. Bacterial membranes are essentially similar in structure to those of eukaryotic cells, but they contain different strengthening molecules.

Transport of materials into bacterial cells can occur by a number of different routes. Simple diffusion through protein pores is known to occur in a few cases, but most transport requires the input of energy (active transport). In the case of primary transport, energy input is direct, and ATP-yielding molecules react with some portion of the transport system. Sometimes the compound transported is also modified chemically by the addition of a phosphoryl group (group translocation). A good example of this type of transport is the PEP-dependent phosphotransferase system. In the case of symport, energy is derived from one of the ion gradients that are maintained by the cell across its membrane. When a molecule is transported into the cell, an ion is also transported, with the corresponding reduction in the ionic gradient providing the energy for the process. This exemplifies a secondary transport process.

Sensory transduction is also a property of the cell membrane. In bacteria, it is accomplished by specific membrane proteins that bind to chemotactically active chemicals. When one of these receptor proteins is bound, it sends a signal via one or more phosphorylated proteins to the flagella and, thereby, affects the frequency with which the flagellar motors reverse direction. Methylation of receptor proteins to neutralize their signal is a way in which bacteria adapt to the presence of chemotaxic chemicals. When the receptor is no longer bound to a chemical, its methyl group is removed.

Questions for Review and Discussion

1. What are the essential ingredients in a bacteriologic medium?
2. Why is it so important to work with pure cultures in studying bacteria? What sort of limitation does this concept put on bacteriology?
3. How do cell wall structures of gram-positive and gram-negative cells differ? Why do you think that they give their different staining reactions? What other types of cell walls are possible?
4. What effects might the presence of an outer membrane have on the response of a cell to its environment? What effects might a capsule have? In what ways are these effects similar and in what ways are they different?
5. How would you briefly summarize the chemotactic behavior of bacteria as observed with the aid of a light microscope? How would you summarize the same behavior on the biochemical level? How do the two behaviors interrelate?
6. What are the transport processes used by bacteria? From where is the energy to move the molecules derived in each process? Once a substance has been transported, is it sequestered in the cell or is it free to exchange itself for another similar molecule on the other side of the cell membrane?
7. Would you expect an archaebacterium to exhibit the same kinds of chemiosmotic phenomena as other bacteria? Why? How would you test your predictions?

Suggestions for Further Reading

Baumeister, W., I. Wildhaber, and B. M. Phipps. 1989. Principles of organization in eubacterial and archaebacterial surface proteins. *Canadian Journal of Microbiology* 35:215–227. This review focuses on all types of S layers but primary emphasis is given to archaebacteria.

Brass, J. M. 1986. The cell envelope of gram-negative bacteria: New aspects of its function in transport and chemotaxis. *Current Topics in Microbiology and Immunology* 129:1–92. A very extensive review of a wide variety of transport processes.

Goldfine, H., and T. A. Langworthy. 1988. A growing interest in bacterial ether lipids. *Trends in Biochemical Sciences* 13:217–221. A survery of the structure, function, and biosynthesis of ether lipids in archaebacteria and anaerobic eubacteria.

Hancock, I., and I. Poston. 1988. *Bacterial cell surface techniques.* John Wiley & Sons, New York. Chapter 1 provides a good review of the bacterial cell surface.

de Mendoza, D., and J. E. Cronan, Jr. 1983. Thermal regulation of membrane lipid fluidity in bacteria. *Trends in Biochemical Sciences* 8:49–52.

Nicholls, D. G. 1982. *Bioenergetics: An introduction to the chemiosmotic theory.* Academic Press, Inc., New York. An excellent introduction to some topics covered in this and the next chapter.

Parkinson, J. S. 1988. Protein phosphorylation in bacterial chemotaxis. *Cell* 53:1–2. A short review of signal transduction in bacteria.

Postma, P. W., and J. W. Lengeler. 1985. Phosphoenolpyruvate: carbohydrate phosphotransferase system of bacteria. *Microbiological Reviews* 49:232–269.

Prince, R. C. 1987. Hopanoids: The world's most abundant biomolecules? *Trends in Biochemical Sciences* 12:455–456. A short review.

Randall, L. L., and S. J. S. Hardy. 1989. Unity in function in the absence of consensus in sequence: Role of leader peptides in export. *Science* 243: 1156–1159. A general review of the subject that requires some knowledge of protein synthetic mechanisms.

Russell, N. J. 1984. Mechanisms of thermal adaptation in bacteria: Blueprints for survival. *Trends in Biochemical Sciences* 9:108–112. An article intended for the general scientific reader.

Saier, M. H., Jr., L-F. Wu, and J. Reizer. 1990. Regulation of bacterial physiological processes by three types of protein phosphorylating systems. *Trends in Biochemical Sciences* 15:391–395. A review that includes discussions of chemotaxis sensing and PTS function.

Unwin, N. and R. Henderson. 1985. The structure of proteins in biological membranes. *Scientific American* 252 (2):78–94. A discussion of both pro- and eukaryotic membrane proteins.

chapter

4

Energy Conversion

chapter outline

■

GENERAL GOALS

Be able to describe the operation of chemotrophic and phototrophic energy production systems and their similarities and their differences.
Be able to describe how chemical potential or pH gradients are generated and how their intrinsic energy can be tapped.

The preceding chapter focuses on the nature of the prokaryotic cell envelope and on other structures surrounding it as well as how solutes move into and out of the cell. The chemiosmotic model of membrane function explains many of these secondary transport phenomena by postulating that the energy to carry them out is derived from gradients of electric potential and/or ionic concentration across a semipermeable membrane. The ways in which chemical or light energy can be used in primary transport to form such gradients are one subject of this chapter. The principal focus, however, is on the manner in which cells produce ATP, their chemical storage form of energy, as well as the NADPH reductant necessary for a number of important cellular reactions discussed in chapter 5.

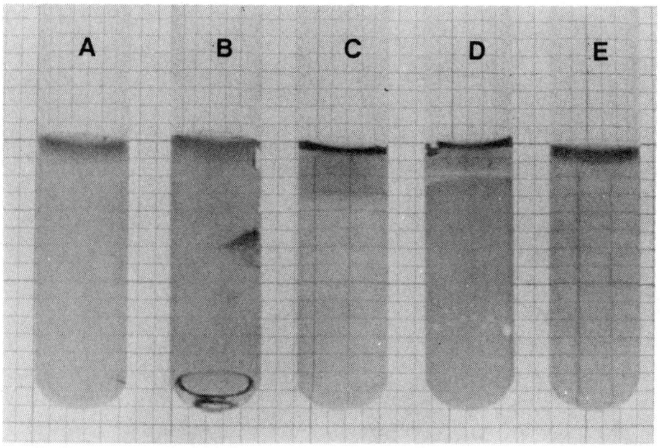

FIGURE 4.1

Oxygen relations. Culture tubes containing thioglycolate agar were inoculated with various bacteria and incubated until growth occurred. Disappearance of background lines indicates areas of growth. The agar contains resazurin, which turns reddish when exposed to oxygen; therefore the dark-colored band at the top of the agar indicates where free oxygen is present. The bacteria and their oxygen relations are: (**A**) *Streptococcus* sp., a facultative anaerobe that is obligately anaerobic. (**B**) *Escherichia coli*, a facultative anaerobe that is oxybiontic. (**C**) An uninoculated control tube with a thick band of oxygenated medium. (**D**) *Clostridium sporogenes*, a strict anaerobe. (**E**) *Bacillus fermis*, a strict aerobe.

Bacterial Oxygen Relations

Just as plants and animals show great diversity in the ways in which they grow and behave so do microorganisms. They are found in all parts of our planet in nearly every conceivable environment, underscoring the fact that conditions suitable for one microorganism may be entirely unsuitable for another. Eukaryotic microorganisms grow with a metabolism that is fundamentally similar to that of plants and animals. Bacteria, however, display a number of unusual metabolic features and are the focus of this chapter.

One important difference between prokaryotes and eukaryotes concerns oxygen relations. An environment that contains gaseous oxygen is said to be **aerobic,** while an environment that lacks oxygen is considered to be **anaerobic.** From our own parochial perspective, it is difficult to imagine living organisms that do not require the presence of gaseous oxygen in order to survive, but that, as Pasteur first observed, is the case for a large number of bacteria. It is possible to perform a very simple test to determine the oxygen preference of any microbial species. The organism to be tested is put into a tube containing a soft agar medium (an agar "deep"). A stabbing motion is used in this process so that the entire length of the agar is inoculated with cells. The medium contains an easily oxidized chemical, such as sodium thioglycolate, that retards the diffusion of oxygen into the depths of the medium by combining with it. After the organism has been allowed to grow, the tube is examined to see where growth has occurred within the agar. Typical growth patterns of specific organisms are shown in figure 4.1A–E.

An organism that is an aerobe will grow in the presence of oxygen and, therefore, will grow on the surface of the agar. On the other hand, an organism that is an anaerobe will not be able to grow on the surface of the agar and will be found only at the very bottom of the agar where all oxygen has been excluded. Some aerobic organisms also are able to grow under anaerobic conditions and are considered to be **facultative anaerobes.** A few organisms are neither aerobic nor anaerobic. These are termed **microaerophilic** because they will grow only in the region of reduced oxygen tension located a few millimeters below the surface of the agar. Occasionally, an organism does not need reduced oxygen tension but rather increased carbon dioxide tension. In this case, the organism is said to be **capneic.** Additional terms sometimes used to describe more precisely the relationship of bacterial metabolism to air are summarized in table 4.1, along with those discussed in this paragraph. The important point to remember is that every bacterial culture has a characteristic response to oxygen and, therefore, must be incubated (allowed to grow) under appropriate conditions.

General Chemical Considerations

In most of the discussion that follows, it is assumed that cells accumulate metabolically useful energy in the form of adenosine triphosphate (ATP). It is important

TABLE 4.1 Terms used to describe the oxygen relations of bacteria

Type of Organism	Observed Growth Properties
General Terms	
Aerobe	Grows on the surface of a solid medium freely exposed to air
Anaerobe	Does not grow on a solid surface freely exposed to air
Metabolic Types	
Microaerophile	Grows only under conditions of reduced oxygen tension; incapable of growth in a normal atmosphere or in an atmosphere lacking oxygen
Facultative anaerobe	Can grow also in the absence of air; can be oxybiontic or obligately anaerobic
Oxybiont	Uses an oxygen-linked respiratory chain (Note that an aerobic organism is not necessarily oxybiontic.)
Obligate anaerobe	Does not use an oxygen-linked respiratory chain; sometimes called an aerotolerant anaerobe
Strict anaerobe	Cannot tolerate the presence of oxygen (inhibited or killed by oxygen or oxidized components of media)

Source: Data from a proposal by R. S. Fulghum in "Definition of Anaerobic Bacteria," *ASM News* 49:432–434. Copyright © American Society for Microbiology, Washington, DC.

to remember, however, that some reactions produce alternative nucleoside triphosphates such as guanosine triphosphate (GTP) or other products, for example, phosphoenolpyruvate (PEP) or acetyl phosphate, that require large amounts of energy for their synthesis. The alternative molecules can undergo reactions yielding ATP in one step (figure 4.2). In practice, once the cell has synthesized some nucleoside triphosphate molecules, it can transfer energy from their chemical bonds to make any other nucleoside triphosphates. For example:

Equation 4.1
$$GTP + ADP \rightleftharpoons GDP + ATP$$

Oxidation-Reduction Reactions

Energy-yielding reactions of all cells, prokaryotic as well as eukaryotic, are **oxidoreduction reactions** that involve the removal of electrons from one compound (the electron donor or substrate that is oxidized) and the transfer of these electrons to another compound (the electron acceptor or substrate that is reduced). The tendency of a substance to lose electrons is expressed by the midpoint electrochemical potential (E_m) for a particular half reaction. A complete reaction is the sum of two half reactions, and the amount of energy involved in the reaction is a function of the difference in E_m for the two half reactions (ΔE_m). If the value for ΔE_m is negative, the reaction releases energy. If it is positive, energy is required in order for the reaction to occur.

Consider the specific case of bacteriochlorophyll (Bchl), a molecule used as a light receptor by photosynthetic bacteria:

Equation 4.2
$$Bchl + light \rightarrow Bchl^+ + e^-$$
$$E_m \cong -1{,}100 \text{ mV}$$

The electron shown in this reaction does not actually exist in the free state but is transferred to an electron acceptor such as a quinone molecule (Q). The half reaction for a quinone is:

Equation 4.3
$$Q + e^- \rightarrow Q^- \qquad E_m = -20 \text{ mV}$$

The overall reaction is then the sum of the half reactions

Equation 4.4
$$Bchl + light + Q \rightarrow Bchl^+ + Q^-$$

and has a total electrochemical change of $\Delta E_m \cong -1{,}120$ mV.

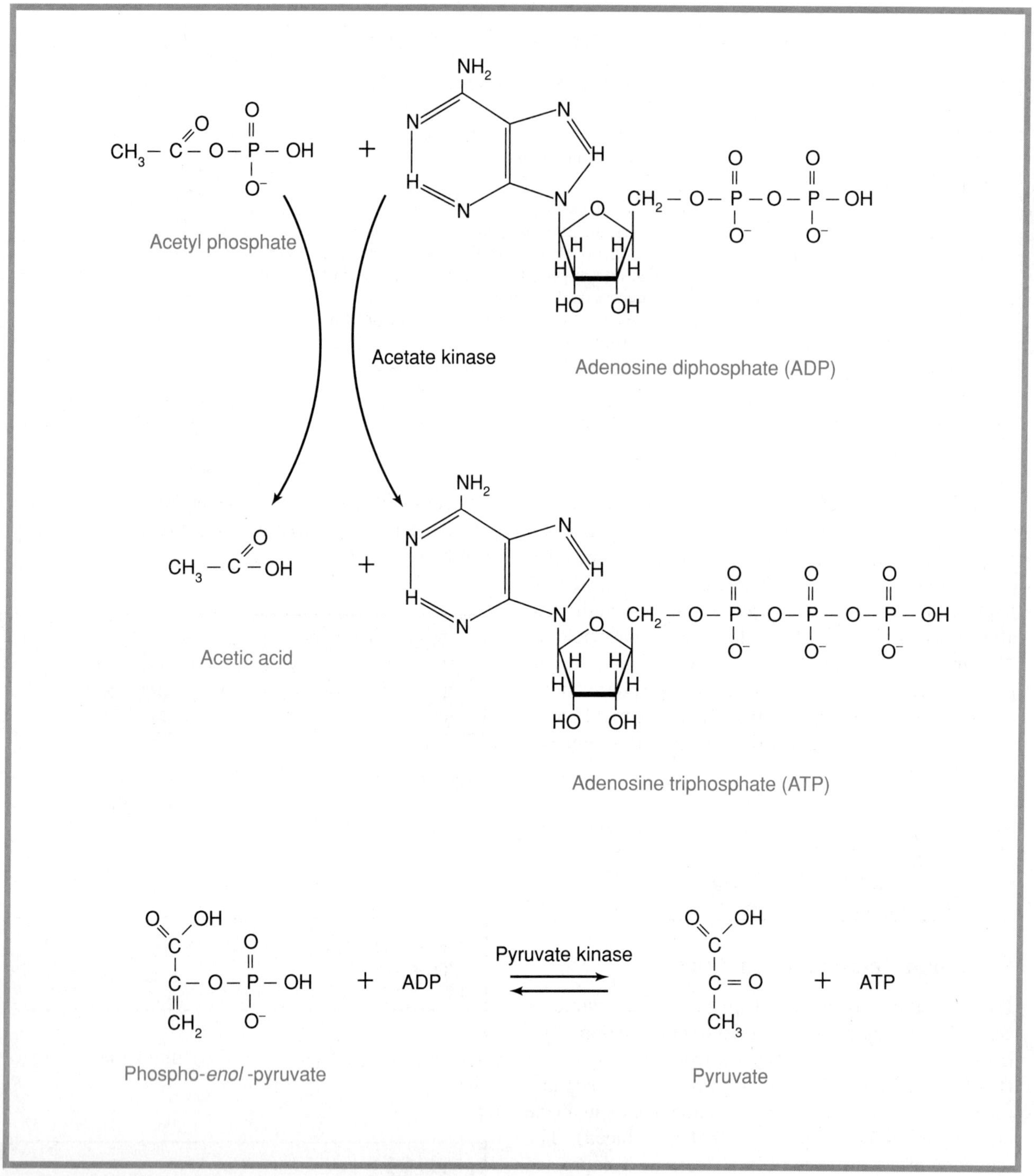

FIGURE 4.2

Certain energetic metabolites can yield ATP in one step (substrate level phosphorylation). Among these are acetyl phosphate and phospho-*enol*-pyruvate.

In oxidoreduction reactions, the greater the change in electrochemical potential of the complete reaction, the more useful energy is available from the reaction. Specifically, a ΔE_m value of 1,000 mV is equivalent to about 23,000 cal, which is substantially more than the 7,300 cal/mol (calories per mole) required for the formation of the high-energy phosphate bond in the conversion of ADP to ATP.

With respect to living organisms, the chemical energy inherent in a molecule to be oxidized can be thought of as being localized in what Stokes has called reduced bonds, such as C — C, C — H, N — H, and N — C (see reference to Stokes's article in suggestions for further reading at the end of this chapter). The amount of energy recovered from oxidizing a molecule depends on the nature of the molecule that is correspondingly reduced. Most often this molecule is oxygen. The intrinsic energy can be tapped by the reactions of metabolism or by the simple burning of the molecule. In either case, the existing reduced bonds are oxidized to such products as CO_2 and NO_2. When coupled to the oxygen half reaction, the energy intrinsic to C — C and C — H bonds is about 220 kJ/mol (kilojoule per mole), a value approximately twice that for C — N and N — H bonds. C — O and O — H bonds contribute nothing to the oxidoreduction reaction; therefore, a fatty acid (lacking in hydroxyl groups) has approximately three times the intrinsic energy value of a carbohydrate with its many hydroxyl moieties.

As electrons are removed from a molecule during oxidation, the energy associated with the chemical bond(s) that they mediated also is removed. During combustion, the energy is released as heat, but in metabolism much of the energy is trapped in one of several ways. During **fermentation,** electrons are removed from one organic molecule and used to reduce another organic molecule that is then discarded. During **respiration,** on the other hand, electrons are removed from an organic or inorganic molecule and passed to an electron transport chain for further processing.

Storing the Energy from Oxidoreduction Reactions

Although ATP production is often thought of as the only energy-storage mechanism in a cell, both prokaryotic and eukaryotic cells can store energy in alternative forms. For certain oxidoreduction reactions, the electrons from the oxidized substrate can be localized in a phosphate bond and later used directly to synthesize ATP as an integral part of a reaction in which ADP reacts with a phosphorylated molecule to yield ATP and a dephosphorylated substrate. This type of reaction is an example of **substrate level phosphorylation;** such reactions are shown in figure 4.2.

An alternative method of ATP production involves the use of chemiosmotic mechanisms to store energy and later release it for ATP synthesis. In the case of chemiosmotic storage, a proton gradient is established with energy from oxidoreduction reactions or photosynthesis. The most common method used by bacteria to establish their proton gradient is electron transport. This is the same electron transport process that occurs in photosynthesis and in chemotrophic energy production in eukaryotic cells. Molecules positioned within the prokaryotic cell membrane are reduced and oxidized sequentially. At various points during the process, protons are pumped from the cytoplasmic side of the membrane to the exterior surface of the membrane. Eventually, electrons are passed to a **terminal electron acceptor** that is reduced and released into the medium as a waste product.

Typical sequences of reactions involved in respiration, including the points at which protons are pumped to the membrane exterior, are shown in figure 4.3. According to chemiosmotic theory, each proton translocation depicted in the figure provides the potential necessary for synthesis of one ATP molecule. Note, however, that organisms may have several possible series of electron transport reactions, depending on the growth conditions.

APPLICATIONS BOX

Oxidase Test

Organisms like the strict aerobe *Pseudomonas* that include cytochrome aa_3 in their chain of electron transport reactions also have the enzyme cytochrome aa_3 oxidase. In an easily performed assay referred to as the **oxidase test,** this enzyme forms a colored product after being incubated with N,N-dimethyl-*p*-phenylenediamine. Many commonly encountered facultatively anaerobic organisms lack the enzyme and hence give a negative oxidase test; therefore, the test often is used as a quick tool for identification of bacteria.

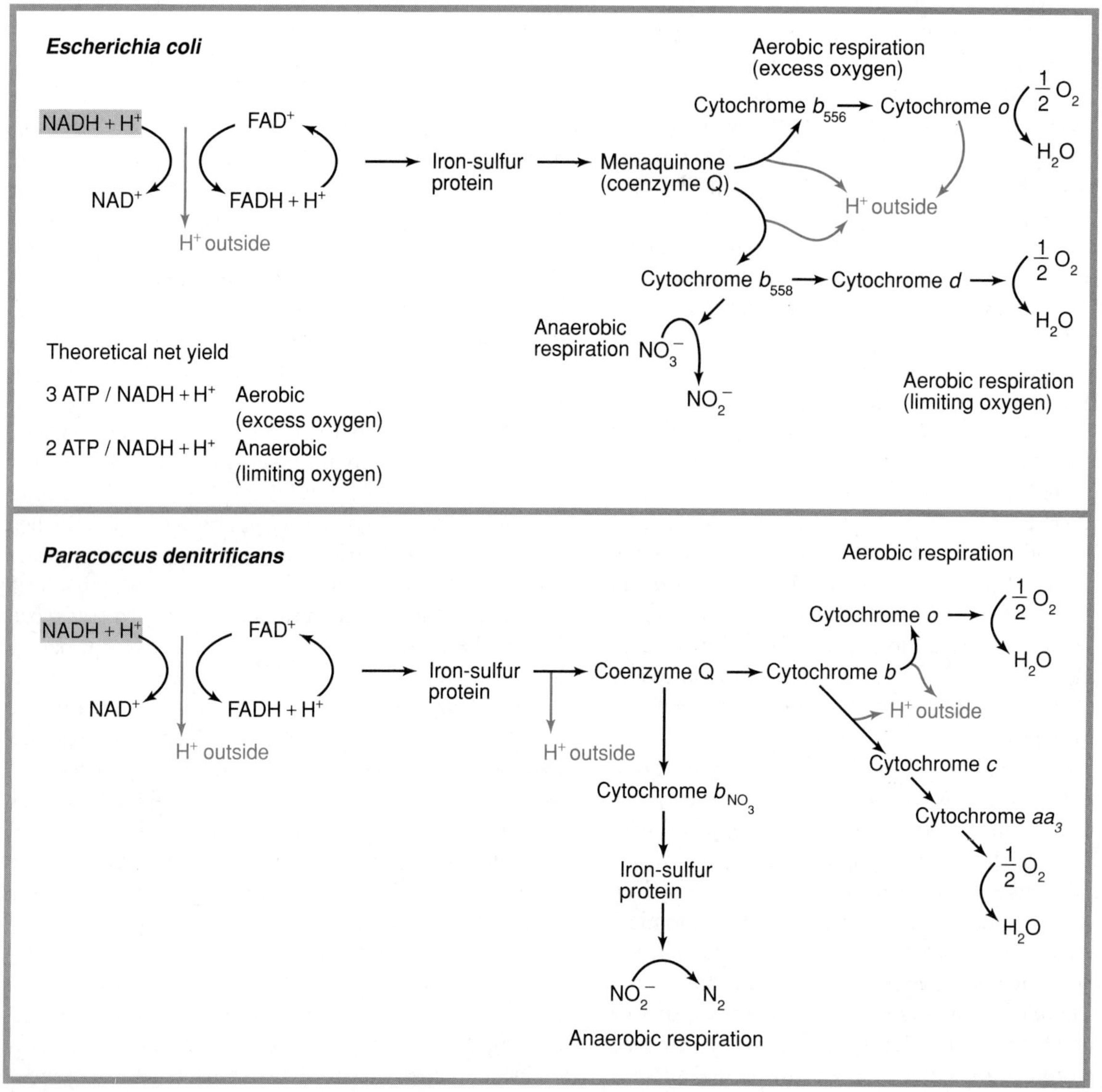

FIGURE 4.3

Possible pathways for electron transport in prokaryotes beginning with reduced NADH + H^+ from an oxidation-reduction reaction. Pathways shown are not necessarily present in all respiratory organisms, but they are characteristic for the organisms listed. Note that the number of protons translocated during the complete oxidation of NADH + H^+ is variable, depending on the organism and terminal electron acceptor used.

Electron transport in chemotrophic eukaryotic cells is usually an aerobic process; that is, oxygen is the terminal electron acceptor in the reaction. From a chemical standpoint, however, there are other potential terminal electron acceptors, although their electrochemical potentials are not as large as oxygen. While not commonly employed by eukaryotes, the alternative acceptors are frequently used by prokaryotes. Biochemists use the term *respiration* in a special sense that has nothing to do with inhalation or exhalation. Whenever the electron transport system is functioning in a chemotroph, the organism is said to be carrying out *respiration,* which is classified as aerobic if the terminal electron acceptor is oxygen and anaerobic if it is

TABLE 4.2 Electrochemical potentials for reactions with alternative terminal electron acceptors

Half Reaction	E_m (V)
$O_2 + 4H^+ + 4e^- \leftrightharpoons 2H_2O$	.820
$NO_3^- + 2H^+ + 2e^- \leftrightharpoons NO_2^- + H_2O$	.430
Adenosine phosphosulfate $+ 2e^- \leftrightharpoons$ AMP $+ SO_3^{2-}$	−.060
$NAD^+ + 2H^+ + 2e^- \leftrightharpoons NADH + H^+$	−.320
$SO_4^{2-} + 3H^+ + 2e^- \leftrightharpoons (6{:}4\ HSO_3^-{:}SO_3^{2-}) + H_2O$	−.510

Note: Any one of the first three half reactions can be added to the fourth one. The more positive the ΔE_m, the more energy is available from the reaction pair.

Source: Data from P. M. Wood, "Chemolithotrophy," pp. 183–230, in *Bacterial Energy Transduction*. Copyright © 1988 Academic Press, Ltd., London.

APPLICATIONS BOX

Origins of Mitochondria and Chloroplasts

The observed similarities between the molecular organization of prokaryotic cells and mitochondria or chloroplasts are truly striking. Both organelles have their own DNA molecules, and prokaryotic-type ribosomes; both carry out essentially similar biochemical reactions to free-living prokaryotes. Although it is true that nuclear DNA in eukaryotes does code for some proteins that chloroplasts require for activity, nuclear DNA certainly does not code for the majority of chloroplast proteins. These observations have led evolutionary biologists to suggest that chloroplasts may have arisen from an event in which a nonphotosynthetic cell, such as a protozoan, engulfed a prokaryotic photosynthesizer but failed to digest it. Such events have been observed. It is presumed that the photosynthetic cell then colonized the host cell over the course of millenia and gradually lost its ability to live as a separate organism.

Although the origin of mitochondria is assumed to be similar but of greater antiquity, there is evidence against this hypothesis in that the majority of mitochondrial proteins are encoded within the nucleus and then exported to the mitochondria. However, striking similarities between the ribosomal RNA sequence of mitochondria and the eubacteria provide favorable evidence. If the endosymbiont hypothesis is assumed to be correct, the numerous chemical differences in the components of the electron transport systems of contemporary prokaryotes and mitochondria may reflect the greater variety of metabolic reactions among chemotrophic prokaryotes.

not. An interesting quirk of terminology is that during aerobic respiration the final enzyme is called an "oxidase," as in the case of cytochrome aa_3 oxidase. When respiration is anaerobic, however, the final enzyme is a "reductase," as in the case of nitrate reductase.

The most commonly used compounds for terminal electron acceptors in anaerobic respiration are nitrate, sulfate, and formate. The electrochemical potential for the nitrate half reaction is less than that for oxygen (less total energy is available), and the number of protons pumped (the ultimate energy yield) is lower in anaerobic respiration than in aerobic respiration (table 4.2). Although the sulfate-sulfite reaction appears more favorable, the sulfate must be be activated first by reacting with ATP to yield adenosine phosphosulfate, which is a good electron acceptor. The net yield of energy is, therefore, lower than might be expected. Cells capable of anaerobic respiration normally regulate their metabolism carefully so that only the terminal electron acceptor yielding the greatest energy is actually used. In *Escherichia coli,* for example, when nitrate is present, the fumarate reductase enzyme is not synthesized.

The similarity in aerobic electron transport observed in prokaryotes and in mitochondria or chloroplasts of eukaryotes suggests that there may be an evolutionary relationship between prokaryotic cells and present-day mitochondria and chloroplasts. It has been proposed that prokaryotic **endosymbionts** (cells living within other cells) had colonized primordial eukaryotic cells at some time in the distant past and have evolved now to the point at which they are no longer capable of independent existence. In other words, eukaryotic cells are really chimeras of a nucleated cell and one or more prokaryotic cells.

After electron transport and proton pumping reactions have been completed, ATP synthesis can occur when protons are allowed to flow back into the cytoplasm via pores maintained by an ATP synthase (ATP phosphohydrolyase) complex. The potential energy inherent in the original position of the proton is trapped within the ATP synthase complex. This enzyme complex can then use that energy to synthesize new ATP molecules from ADP at the cytoplasmic surface of the membrane (figure 4.4).

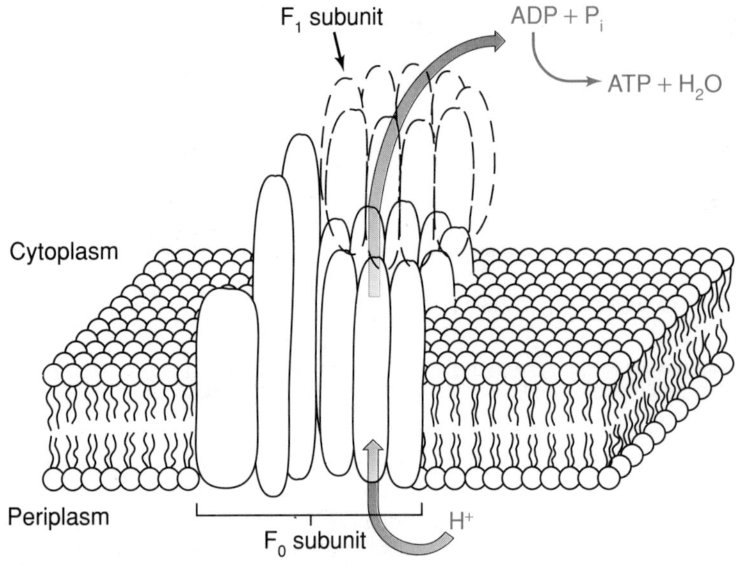

FIGURE 4.4

Chemiosmotic production of ATP. Bacterial and mitochondrial enzymes called ATP synthases are composed of two multiprotein components. The F_1 subunit, located in the cytoplasm, is responsible for synthesis/hydrolysis of ATP. The F_0 subunit, located in the membrane, is a channel for conduction of protons from the surface of the cell to the cytoplasm and vice versa. The reaction can proceed in either direction. Therefore, energy for the synthesis of ATP comes from the return of a proton across the plasma membrane into the cytoplasm. Conversely, if ATP is broken down, protons can be pumped back to the cell surface.

Photosynthesis, the Prime Energy Source

Living organisms derive most of the energy they use from the sun, either directly via photosynthesis or indirectly via ingestion of phototrophs or phototroph consumers. The photosynthetic processes in the macroscopic world are oxygenic, meaning that they produce molecular oxygen. However, geologic theory suggests that the early atmosphere of our planet was anaerobic and strongly reducing (that is, it contained the reduced forms of the major chemical elements for life: H_2, NH_3, and CH_4). Many photosynthetic prokaryotes are described as anoxygenic because they do not produce oxygen as a result of photosynthesis. They have properties that seem to be appropriate to such a predicted environment and may represent the metabolic type of cells that were alive when our atmosphere was anaerobic. These present-day cells have a simple photosynthetic system that offers a good introduction to the basic process of photosynthesis.

Anoxygenic Photosynthesis

The anoxygenic, phototrophic prokaryotes are divided into four broad groups based on the color of their photosynthetic pigments, their Gram stain reaction, and their oxygen relations. There are the gram-negative purple bacteria represented by members of the genera *Rhodospirillum* and *Chromatium;* the gram-negative green bacteria represented by members of the genera *Chlorobium* and *Chloroflexus;* the gram-positive helical bacteria represented by *Heliobacterium chlorum;* and the aerobic, anoxygenic bacteria represented by the genus *Erythrobacter.* In all groups, the photosynthetic apparatus is membrane associated, and therefore the quantity of membrane available per cell limits the amount of photosynthesis that is possible.

Purple and green bacteria as well as *Erythrobacter* create extra membranes to increase their photosynthetic capacity (figure 4.5A–B). By simple invagination of the cell membrane into the cytoplasm, purple bacteria and *Erythrobacter* obtain additional membrane area. Green bacteria have ovoid membranous structures called **chlorosomes** or chlorobium vesicles underlying the cell membrane and contiguous with its inner surface. This is not an exception to the statement made previously that prokaryotic cells do not have membrane-bound organelles because chlorosomes are, in fact, invaginations of the cell membrane. Extra membranes used for photosynthesis should not be confused with structures like the gas vacuoles shown in figure 2.5. The aerobic, anoxygenic photosynthesizers can be orange or pink and seem to be a heterogeneous

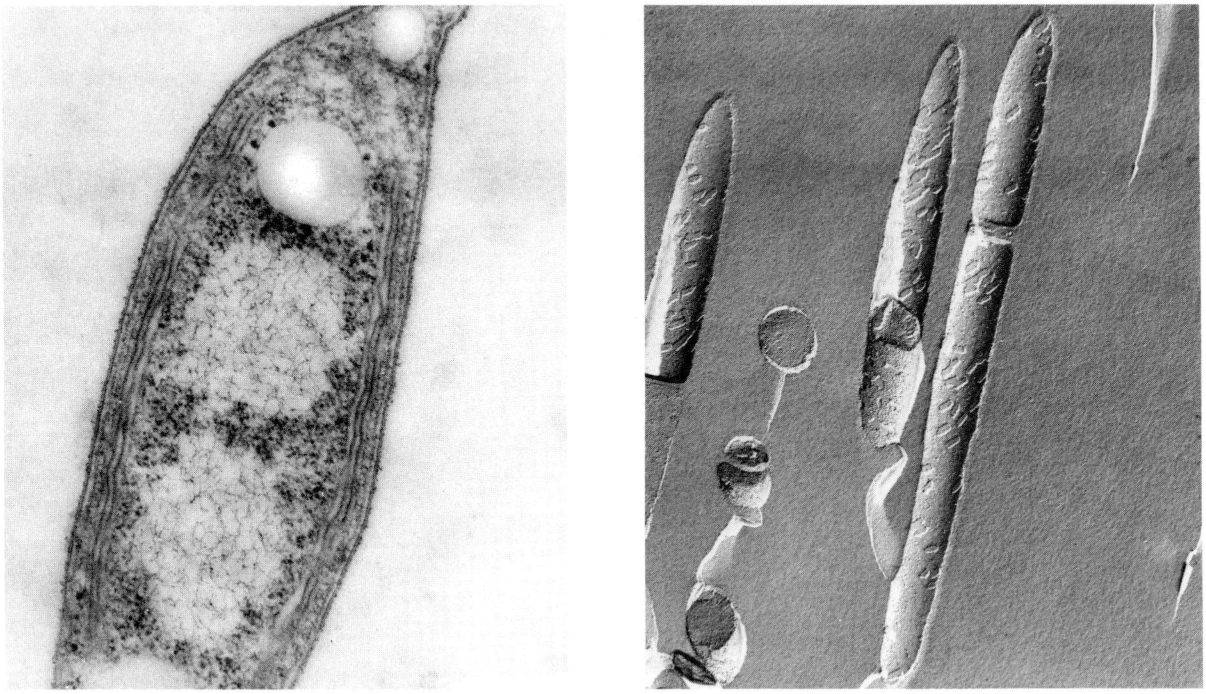

A B

FIGURE 4.5
Photosynthetic membranes observed in anaerobic bacteria.
(**A**) Thin section of *Rhodomicrobium vannielii.* The extra layers of membranes can be seen running parallel to the plasma membrane.
(**B**) Freeze-fracture specimen of *Chloroflexus auranticus.* The dividing cell was frozen and then cleaved along the center of the plasma membrane, making the hydrophobic surface visible. The chlorosome vesicles were budded out from the plasma membrane and appear as depressions. The rough surface of the depressions is the photosynthetic antenna system.

(A) Reprinted from Trentini, W. C., and M. P. Starr. 1967. Growth and ultrastructure of *Rhodomicrobium vannielii* as a function of light intensity. *Journal of Bacteriology* 93:1699–1704. (B) Electron micrograph courtesy of James R. Swafford, Brandeis University.

group that probably will be subdivided into several genera. Presently, they are set apart by their aerobic synthesis of bacteriochlorophyll *a*.

Photosynthetic cells of all types use a variety of pigments to capture photons (quanta of light energy) from sunlight, with each pigment absorbing photons of a particular wavelength. These pigments are referred to as a **photosynthetic antenna** system or light-harvesting complex, and when they absorb a photon of light, the energy from that photon (now called an exciton) can be transferred to other members of the antenna system in a series of equilibrium reactions. The energy is passed eventually to a photochemical reaction center in which the actual energy-yielding reactions begin. These reactions trap energy of the exciton and prevent its return to the antenna system.

The reaction centers contain chlorophyll or bacteriochlorophyll molecules, substances that are identical in function and quite similar in structure (figure 4.6). Minor variations in structure are indicated by letters following each name, so that bacteriochlorophyll *a* has a slightly different structure from bacteriochlorophyll *c*. Such minor structural differences correlate with changes in the absorption maxima of the molecules (the wavelengths of light that they absorb most strongly). The molecules are sometimes designated by an abbreviation including the absorption maximum (for example, $Bchl_{850}$ or P850). Table 4.3 lists some commonly encountered chlorophylls and their absorption maxima.

Purple bacteria use both bacteriochlorophylls *a* and *b* in their reaction centers, while green bacteria use bacteriochlorophylls *a, c, d,* and *e*. If examined carefully, the data in table 4.3 show that the absorption maxima displayed by the various chlorophyll molecules correlate with the expected physical positions of the cells containing them in their natural habitats. For example, aerobic organisms growing directly at an air-water interface tend to absorb all the shorter wavelengths of light, leaving only the longer wavelengths to

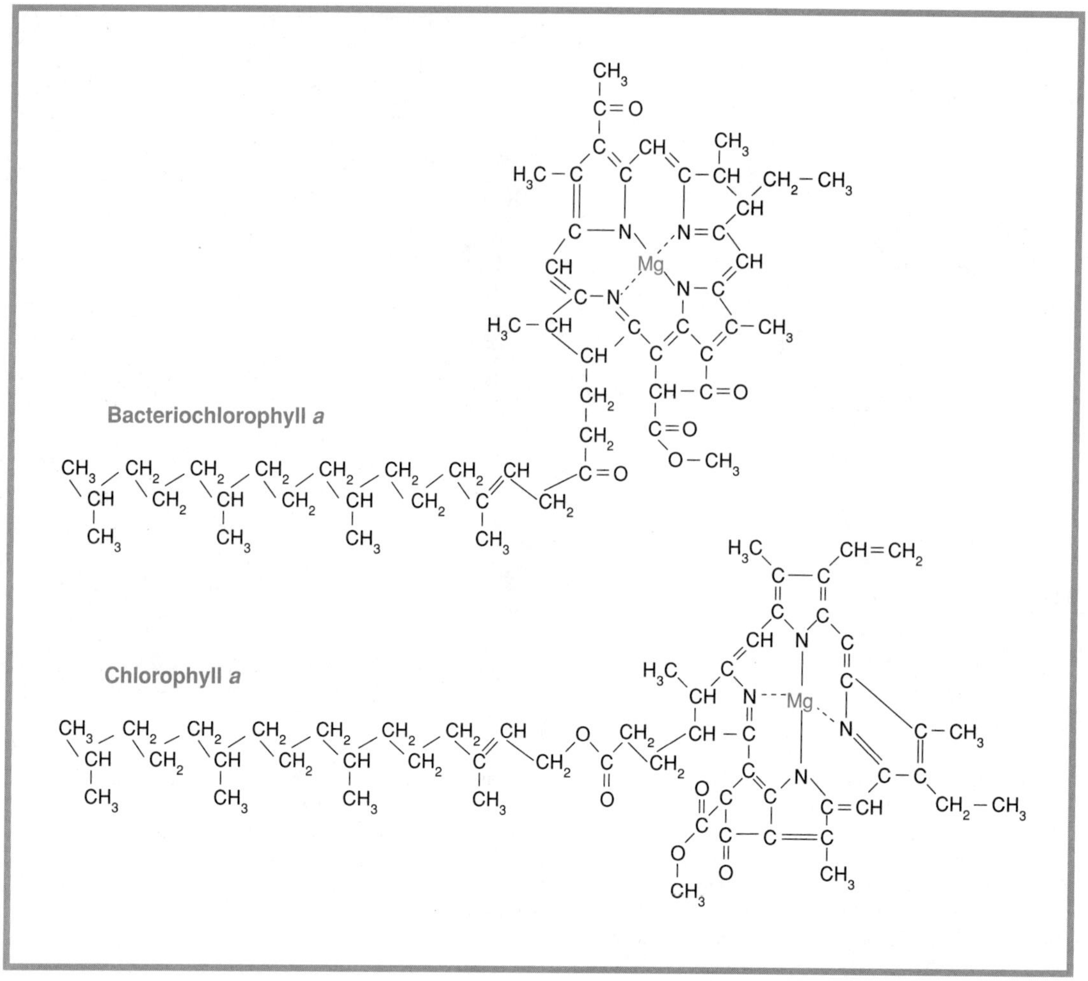

FIGURE 4.6
Comparison of the chemical structure of bacteriochlorophyll *a* and chlorophyll *a*.

TABLE 4.3 Examples of bacteriochlorophylls and their in vivo absorption maxima in the longer wavelengths that could be expected to penetrate anaerobic areas of a body of water

Molecule	Maximum Absorption (nm)
Bacteriochlorophyll *a*	850–910
Bacteriochlorophyll *b*	1020–1035
Bacteriochlorophyll *c*	745–755
Bacteriochlorophyll *d*	725–745
Bacteriochlorophyll *e*	710–725
Chlorophyll *a*	680–683

Source: Data from H. G. Trüper and N. Pfennig, Characterization and Identification of the Anoxygenic Phototrophic Bacteria, pp. 299–312, in *The Prokaryotes,* edited by M. P. Starr, H. Stolp, H. G. Trüper, A. Balows, and H. G. Schlegel. Copyright © 1981 Springer-Verlag, New York.

penetrate to the depths in which anaerobic organisms are found. Photons with wavelengths greater than about 1,100 nm do not have sufficient intrinsic energy to drive the photosynthetic reactions; therefore, they contribute only heat to the photosynthetic antenna system. Thus, all bacteriochlorophyll molecules have their absorption maxima clustered in the range of wavelengths between 680 and 1,100 nm.

Whenever the reaction center absorbs a photon directly or receives energy from some portion of the antenna system, it is stimulated to pass an electron to bacteriophaeophytin, the primary electron acceptor. In turn, bacteriophaeophytin passes the electron to a compound designated as Q (or ubiquinone). The electron

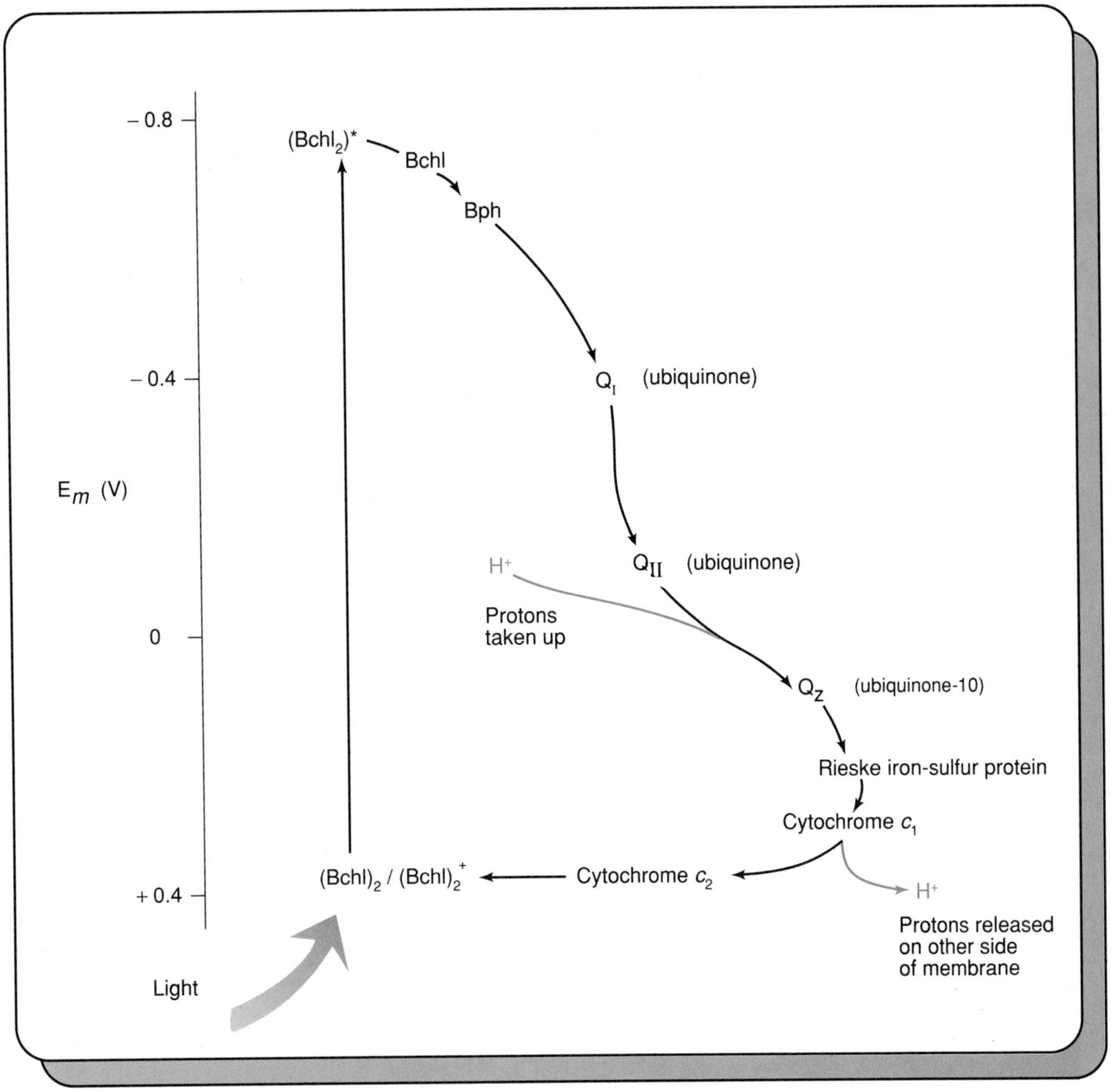

FIGURE 4.7

Photosynthetic system of the anaerobic bacterium *Rhodobacter sphaeroides* (analog of eukaryotic photosystem I). Arrows indicate the flow of electrons. The order of the steps within the brackets is uncertain, and the possibility of multiple pathways has not been eliminated. Bacteriophaeophytin (Bph) is a bacteriochlorophyll (Bchl) in which the magnesium atom has been replaced by two protons. In actual operation, the cycle waits until two electrons arrive at Q_{II} before proceeding. The scale along the left side of the diagram indicates the relative electrochemical potential of the various components; therefore, $(Bchl)_2^*$ has more intrinsic energy than $(Bchl)_2$.

then travels along a typical electron transport chain, reducing each molecule in its turn, until it returns to the reaction center from which it started (figure 4.7). Due to the electrochemical potentials and physical positions of the various electron acceptors within the cell membrane, movement of electrons results in a pumping action to generate the usual proton gradient, ΔpH. The potential energy inherent in the gradient then can be used by ATP synthase to make one ATP molecule for every electron that passes along the chain (figure 4.8).

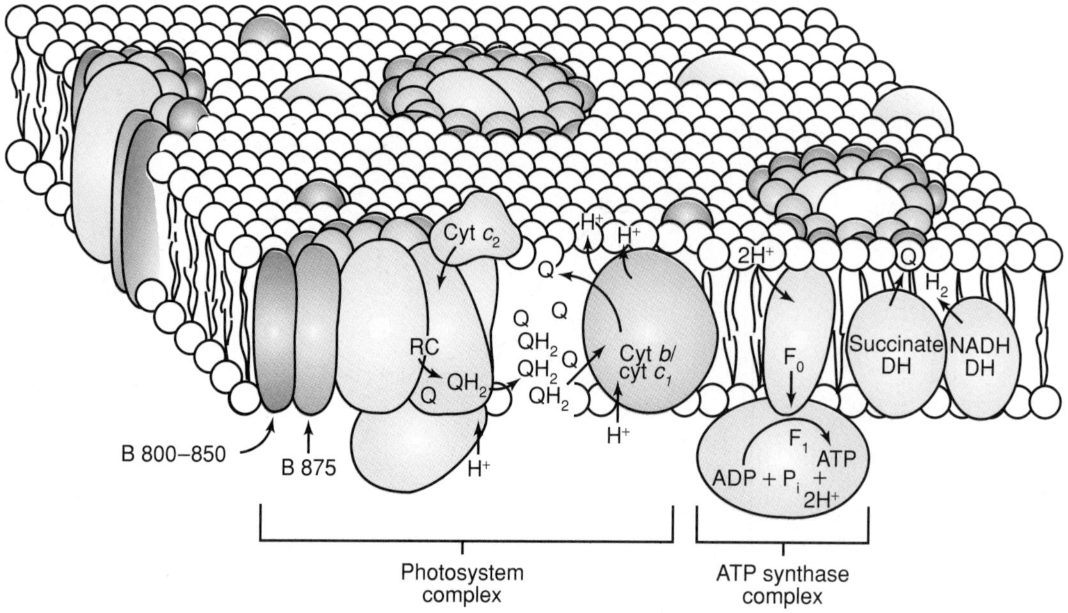

FIGURE 4.8

Relative position of the photosynthetic components within the intracytoplasmic membranes of *Rhodobacter sphaeroides*. Proteins labeled B800–850 and B875 are arrayed symmetrically around the reaction center (RC) and represent the light-harvesting complexes. Dehydrogenase enzymes are DH proteins in the diagram. Cytochromes are designated as cut. Note that ATP synthesis is coupled to photosynthesis only in the sense that the electrochemical gradient generated by photosynthetic electron transport is used to drive the reaction. ATP synthase is not a component of the photosystem itself (see also figure 4.3).

Because the energy to develop the proton gradient used to synthesize ATP came from light, the overall process often has been called photophosphorylation, even though the phosphorylation step is not linked directly to the actual photosynthetic pathway. This type of primary transport is called cyclic because the Bchl *a* molecule acts as both the initial electron donor and the final electron acceptor. The proton gradient generated by the transported electrons can be used for a variety of biosynthetic or secondary transport processes within the cell, including ATP synthesis.

There is a potential problem, however. In many cases, the only molecules available to a cell for its metabolic use are those that are too oxidized for direct participation in biosynthesis; for example, CO_2 is available instead of a carbohydrate, or N_2 instead of NH_4^+. Cells then have the choice of ceasing growth or finding some way to reduce these molecules to an appropriate valence. In order to carry out this reduction, it is necessary for a cell to have available some form of easily oxidized molecule that can be used as a reductant. If oxidoreduction reactions that break down a substrate such as glucose are energy-yielding processes (negative value for ΔE_m), then oxidoreduction reactions that use CO_2 to form glucose must be energy-requiring processes (positive value for ΔE_m).

The energetic electrons used to make the reduced C — H or N — H bonds are acquired from enzyme cofactors such as NADPH + H^+. In some anoxygenic photosynthesizers, H_2 gas can be reacted enzymatically with $NADP^+$ to yield NADPH + H^+ directly, allowing photosynthesis to continue as described. In most photosynthesizers, however, the synthesis of reducing molecules must come at the expense of ATP production by diversion of the electrons away from the electron transport chain (figure 4.9). If the electrons are diverted, photosynthesis becomes noncyclic.

In the noncyclic photosynthetic process, Bchl *a* still acts as an electron donor; however, the replenishing electron does not come from the cycle itself but rather from a primary electron donor, an organic molecule such as CH_4 or an inorganic molecule such as H_2S. Although all photosynthetic bacteria seem to be able to

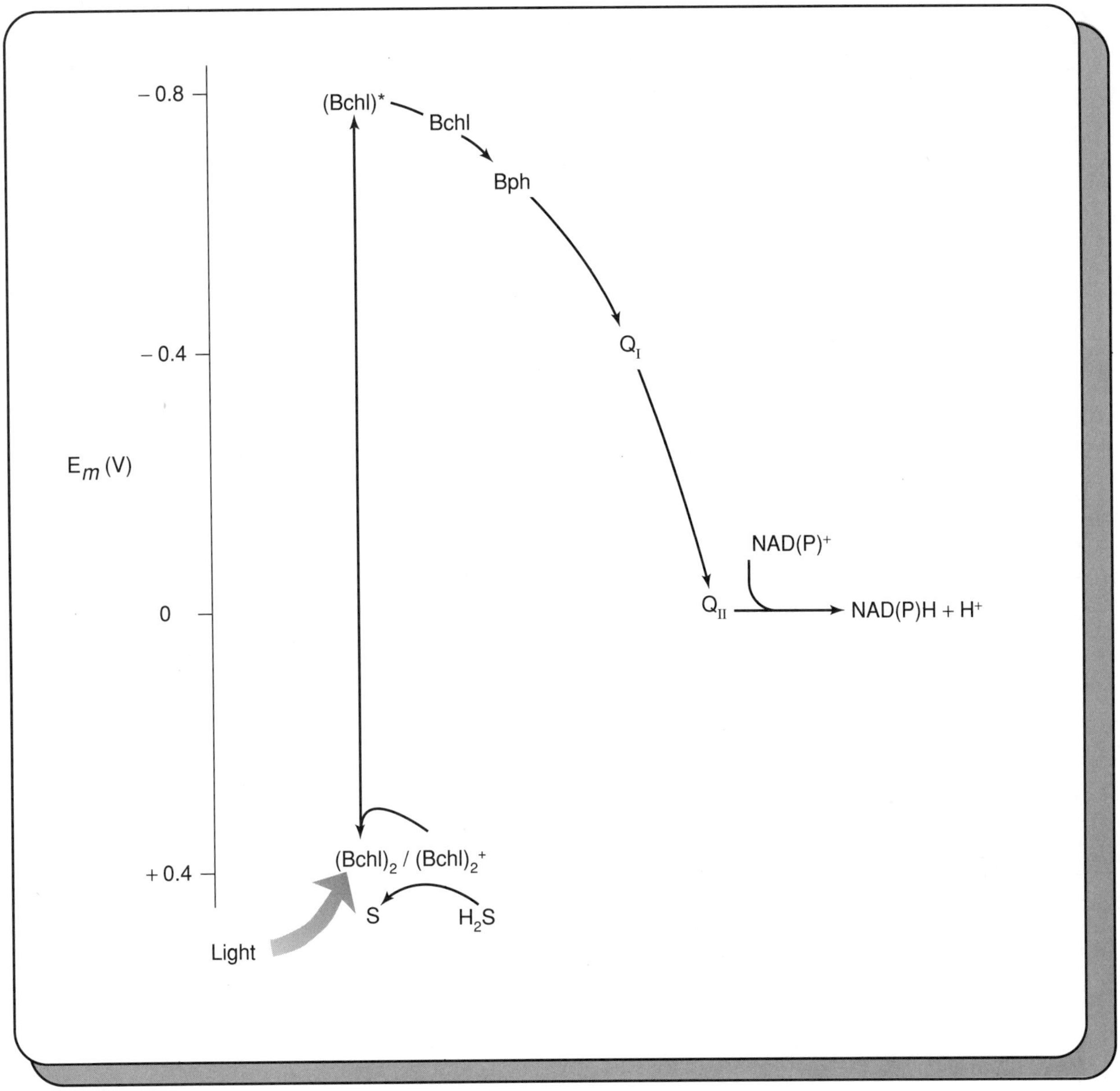

FIGURE 4.9

Noncyclic operation of bacterial photosynthesis. The abbreviations are the same as those used in figure 4.7. Both NAD^+ and $NADP^+$ are potential substrates for this reaction. The scale along the left side of the diagram indicates the relative electrochemical potential of the various components.

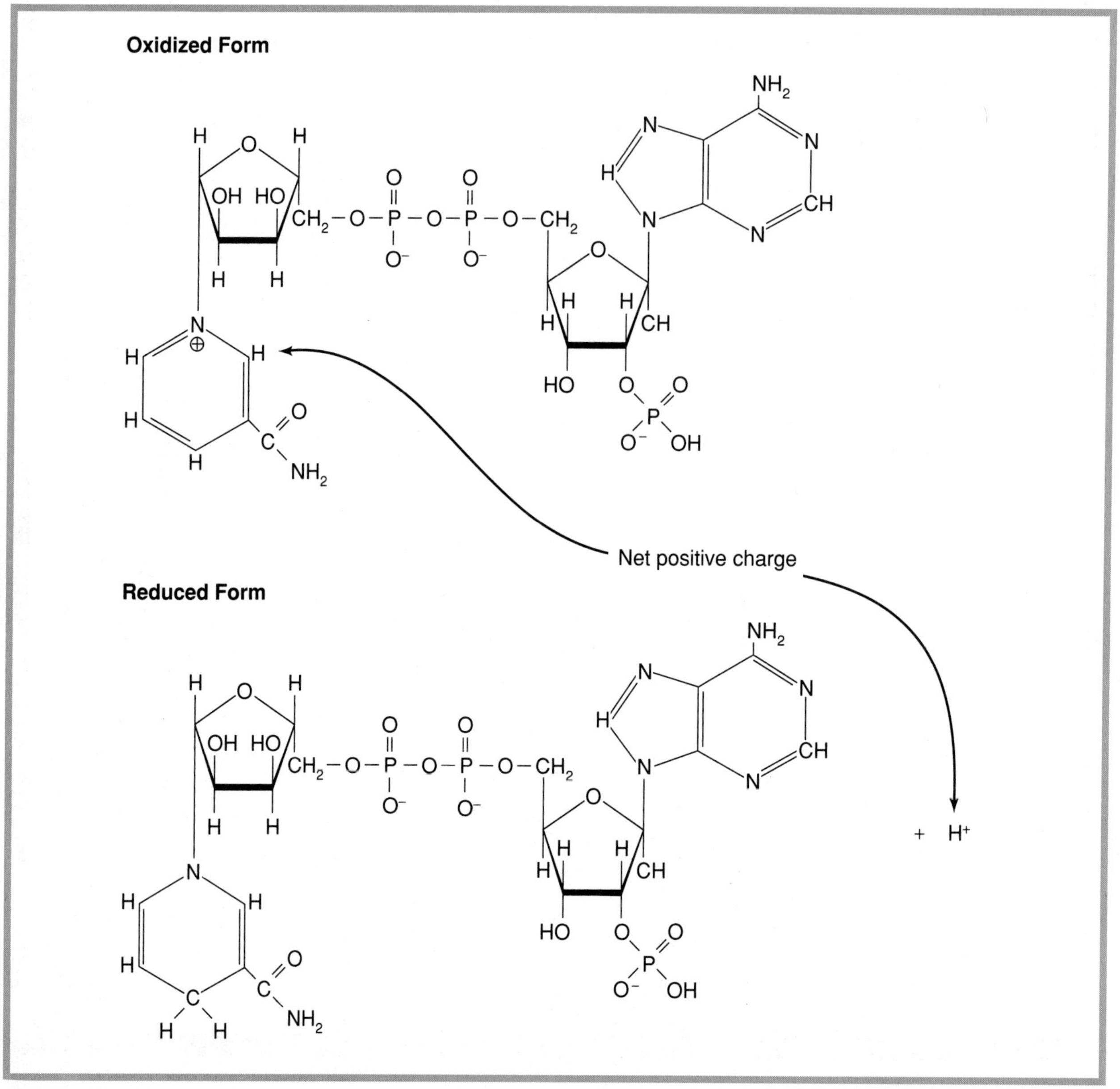

FIGURE 4.10

Reduction of the enzyme cofactor $NADP^+$ by the addition of an electron. The added electron breaks a double bond, causing a quaternary amine to be converted to a tertiary amine and eliminating a positive charge from the NADP molecule. Presence of the extra hydrogen ion maintains the electric neutrality of the system. The chemical behavior of NAD^+ and its conversion to NADH + H^+ is similar.

use sulfides in trace amounts, those like *Chromatium* or *Chlorobium* can tolerate quite high concentrations and are referred to as purple sulfur and green sulfur bacteria, respectively. The energetic electrons and the accompanying protons are transferred to the enzyme cofactor nicotinamide adenine dinucleotide phosphate ($NADP^+$) to form NADPH + H^+ (figure 4.10). NADPH + H^+ is used in biosynthetic reactions such as the Calvin cycle, which is discussed in the next chapter.

Oxygenic Photosynthesis

The oxygenic photosynthesizers include cyanobacteria (blue-green bacteria, blue-green algae) and green plants. These organisms synthesize oxygen as a direct

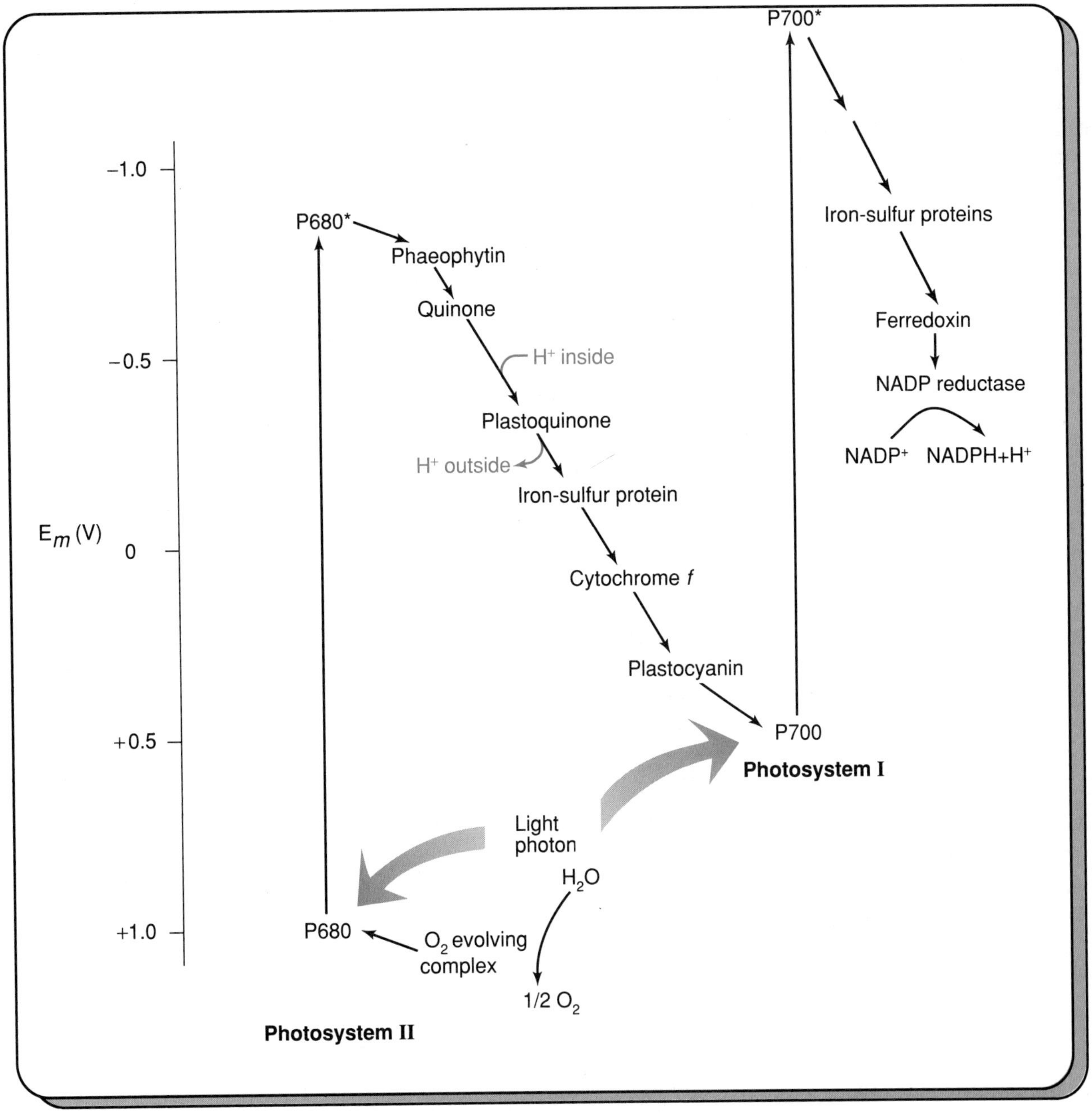

FIGURE 4.11

Oxygenic photosynthesis via photosystems I and II as observed in cyanobacteria. Both P680 and P700 contain chlorophyll *a*. Plastocyanin is a copper-containing protein that is sometimes replaced by a cytochrome *c* molecule. The scale along the left side of the diagram indicates the relative electrochemical potential of the various components. The difference between P680 and P680* or P700 and P700* is, therefore, one of intrinsic energy. Sequence of some intermediate electron carriers is still in doubt. Arrows to and from hydrogen ions represent points at which proton pumping activity occurs in the reaction pathways.

product of photosynthetic reactions and have a photosynthetic pathway analagous in function, but not in biochemistry, to the photosynthetic pathway observed in anoxygenic photosynthetic bacteria. This pathway is designated as photosystem I. A second photosystem, photosystem II, is operative only in oxygenic photosynthesizers and always functions in tandem with photosystem I. The paired systems are noncyclic in nature, the electron donor is water, and the overall reaction scheme usually is presented in the form of a sideways Z, as illustrated in figure 4.11. Note that the noncyclic nature of this combination of photosystems I and II means that it is possible to translocate protons in conjunction with the electron transport associated with photosystem II and provide reductant (NADPH + H^+ from photosystem I) at essentially the same time, if photons are introduced at the two separate reaction centers.

When cells do not require additional reducing molecules, it is possible for photosystem I to operate in a cyclic fashion, as in anoxygenic bacteria. When this happens, the electron emitted from the photosystem I reaction center is returned to the originating chlorophyll. Any electrons emitted from photosystem II during cyclic operation of photosystem I have no place to go and are fed into a fluorescent reaction that dissipates the excess energy rather than being passed through electron transport.

The photosynthetic apparatus of cyanobacteria is very similar to that of certain primitive eukaryotic algae, such as the red algae, in that both groups demonstrate the presence of unique accessory pigments called phycobiliproteins, which do not occur in other organisms. In addition, structures resembling fossil cyanobacteria can be isolated from rock strata not having any plant fossils. For these reasons, it has been suggested that cyanobacteria are organisms that share a common ancestor with eukaryotic photosynthesizers, an ancestor assumed to be the progenitor of the endosymbionts that produced the present-day chloroplasts.

Chemical Energy, the Secondary Energy Source

The general term **catabolism** describes any series of degradative reactions. Chemotrophic metabolism is, therefore, a collection of catabolic reactions. Production of energy in eukaryotic cells generally requires the presence of oxygen and mitochondria, although the yeast *Saccharomyces* and some protozoa are capable of fermentative growth. The same pathways, with some slight variations, are operative in many bacteria, although obviously not in subcellular organelles.

Glycolytic Reactions

Chemoorganotrophic metabolism is familiar to every biology student. It is exemplified by the aerobic utilization of glucose in the combined **Embden-Meyerhof-Parnas (EMP) pathway** and the **tricarboxylic acid (TCA) cycle (Krebs or citric acid cycle)** operating to generate primarily reduced enzyme cofactors that feed into an electron transport system similar to the one used by photosynthesizers. Note that the EMP pathway often is imprecisely called glycolysis, a term that can refer to any pathway that degrades glycogen to glucose and then onward to smaller molecules. The combined EMP and TCA cycles are presented diagrammatically in figure 4.12.

Metabolic differences between the TCA cycles of eukaryotes and prokaryotes are known. The prokaryotic isocitrate dehydrogenase enzyme that converts isocitric acid to 2–oxoglutaric acid (α-ketoglutaric acid) generally requires $NADP^+$ as a coenzyme rather than NAD^+ used by the enzyme found in mitochondria. Because NADPH + H^+ cannot feed directly into the electron transport system used to generate a proton gradient, it must undergo a transhydrogenase reaction with NAD^+ or else be used in biosynthesis (see figure 4.12 and chapter 5). In addition, the succinate thiokinase enzyme (that converts succinyl CoA to succinic acid with the direct formation of a nucleoside triphosphate) catalyzes the substrate level phosphorylation of GDP in mammals, but in bacteria frequently will also phosphorylate ADP.

It is often assumed that the EMP pathway and the TCA cycle operate in lock step, but this is not necessarily the case. In the short term, *Escherichia coli* cells apparently can derive all the energy needed for cell growth from the EMP pathway and the oxidation of some, but not all, of the acetyl CoA derived from pyruvate. Up to 17% of the carbon atoms in glucose can be exported from the cell as acetic acid during aerobic growth, resulting in degradation of all glucose in the medium and accumulation of acetate ions. Later, if the cells have not yet ceased growth for other reasons, they can take up the released ions again, oxidize them for energy, and use them for biosynthetic purposes (see chapter 5). Tempest and co-workers have suggested that such behavior indicates that bacteria put a premium on the rate of energy production rather than on the quantity of ATP that can be produced (see reference to this article in the suggestions for further reading). Although textbooks often emphasize maximum energy production from glucose, cells actually may be trying to maximize their rate of energy production and, thereby, maximize their rate of cell division.

Without respiration, the EMP pathway produces a net of two ATP molecules (one from each three-carbon unit) and two reduced enzyme cofactors, while the TCA

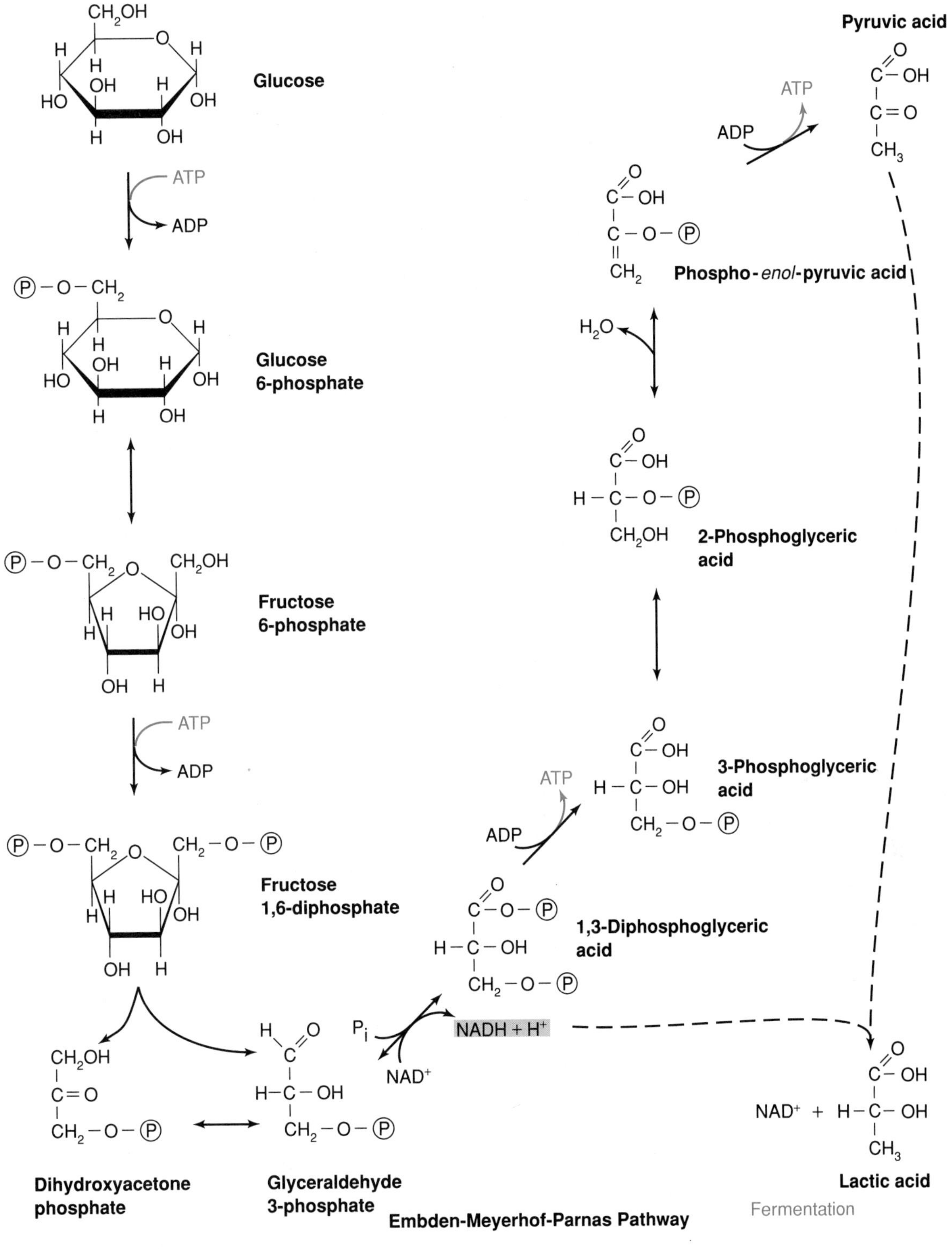

FIGURE 4.12

Continued on next page

Glucose utilization via the Embden-Meyerhof-Parnas (EMP) pathway and the tricarboxylic acid (TCA) or Krebs cycle in a prokaryotic cell such as *Escherichia coli.* The EMP pathway ends with pyruvate, and the TCA cycle begins with the condensation of acetyl coenzyme A and oxaloacetic acid. The reduced coenzymes ($NADH + H^+$ and $FADH_2$) are assumed to be reoxidized by the electron transport system (figure 4.4), except under fermentative conditions (see figures 4.15, 4.16, and 4.17). Circled Ps (Ⓟ) indicate bound phosphate groups, while P_i represents a phosphate ion. Coenzyme A is abbreviated as CoASH when it is uncombined and as —S—CoA when in the combined form.

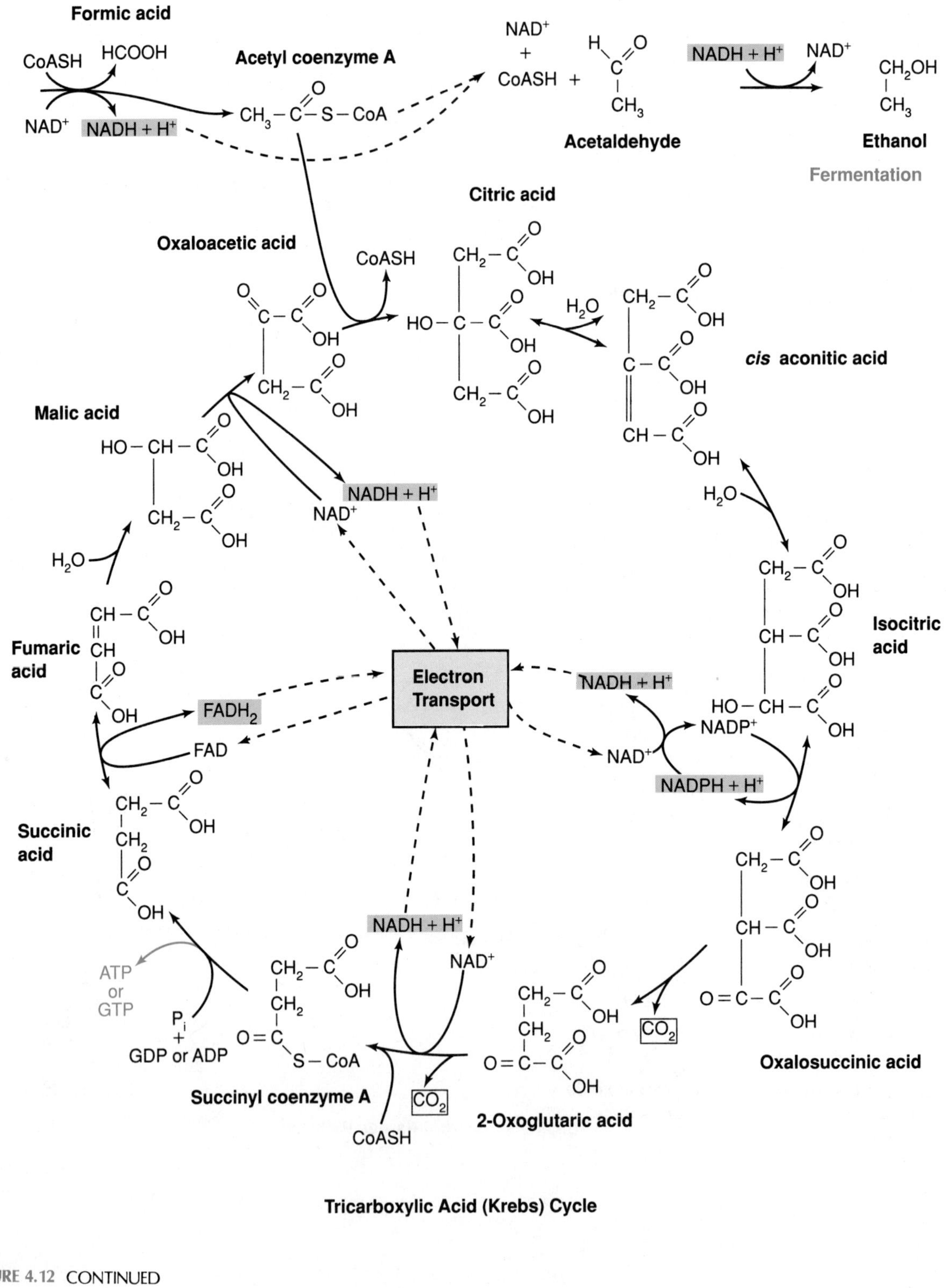

FIGURE 4.12 CONTINUED

cycle rapidly reduces all available cofactors, thereby preventing further catabolism. Thus, the extent of fermentative metabolism depends on the ability of the cell to reoxidize its cofactors without electron transport, and the TCA cycle normally does not operate during fermentation. Metabolic strategies for fermentation are discussed in the next subsection.

The energy balance for the reactions diagrammed in figure 4.12 can be calculated if it is assumed that each electron input in the form of NADH + H^+ will pump protons three separate times, potentially yielding up to three ATP molecules. By a similar analysis, electrons input as $FADH_2$ will pump protons two times (but see also figure 4.3). The NAD^+ necessary to keep the EMP pathway running under these conditions can be generated by one or more reactions indicated by the dotted lines in figure 4.12. In the presence of electron transport, however, the maximum theoretical yield can be as high as 38 ATP molecules, provided that all 18 protons pumped are later converted to ATP. In eukaryotes, the equivalent number is 36 ATP molecules because the equivalent of one ATP molecule is required to transport each three-carbon molecule from the cytoplasm into the mitochondria.

Despite its prominence, the EMP pathway is not the only means by which glucose can be catabolized (degraded) to smaller molecules with the release of some energy. Several other pathways of glucose degradation provide useful alternatives for some bacteria. The **hexose monophosphate (HMP) pathway** shown in figure 4.13 allows an organism to generate two ATP molecules per glucose-phosphate molecule degraded via glyceraldehyde 3–phosphate, but the HMP pathway bypasses the early steps in the EMP pathway. It serves to supply ribose phosphate for nucleic acid biosynthesis and NADPH + H^+ for reducing power that again can be used in biosynthesis. Erythrose 4-phosphate can be used as a precursor in the synthesis of aromatic amino acids (see chapter 5) or combined with xylulose 5-phosphate to yield fructose 6-phosphate and glyceraldehyde 3-phosphate that can be returned to the EMP pathway.

An advantage to the HMP pathway is that the sugars in question are produced from hexoses without a corresponding loss of a reduced carbon atom, as occurs in a simple decarboxylation reaction. While this pathway is present as an alternative in many bacteria and in some eukaryotes, such as *Saccharomyces,* it is the major pathway of hexose oxidation in *Acetobacter suboxydans.*

Pseudomonas and its related genera have yet another means of oxidizing hexoses, the **Entner-Doudoroff pathway** (figure 4.13). Like the HMP pathway, the Entner-Doudoroff pathway starts with glucose, but unlike HMP, after converting glucose to 5-phosphogluconate in the first step, it produces pyruvate rather than pentoses. The net yield is one ATP molecule per glucose molecule oxidized and one molecule of glyceraldehyde 3-phosphate that can be fed into the EMP pathway.

In *Pseudomonas aeruginosa,* glucose degradation via the Entner-Doudoroff pathway occurs in different cellular compartments. Within the periplasm (outside the cell membrane), glucose is oxidized to gluconic acid and then to 2-ketoglucose. In that form, it is then transported into the cytoplasm and phosphorylated by ATP. NADPH + H^+ next converts the molecule to 6-phosphogluconic acid, and the remainder of the pathway shown in figure 4.13 is followed.

One final mode of glucose oxidation is observed in a collection of **phosphoketolase pathways** found in genera such as *Leuconostoc* and *Bifidobacterium.* One such pathway, seen in *Leuconostoc mesenteroides,* is rendered in figure 4.14. As drawn, this is an example of a fermentation. In this particular case, the end products of glucose degradation are not fed into an electron transport system but rather are released into the surrounding medium. However, there is no intrinsic reason why a phosphoketolase pathway cannot be linked to the TCA cycle for reoxidation of NADH + H^+. In fact, according to three reports, *Thiobacillus* can use such a combination of pathways for respiratory growth.

Although this discussion has dealt with glucose, it is important to remember that sugars other than glucose can be channeled into the various glycolytic pathways, because most commonly encountered carbohydrates can be converted readily into glucose or fructose. Some examples are presented in table 4.4. Furthermore, amino acids, fats, and fatty acids can be fed into catabolic pathways by reversal of some biosynthetic reactions discussed in chapter 5.

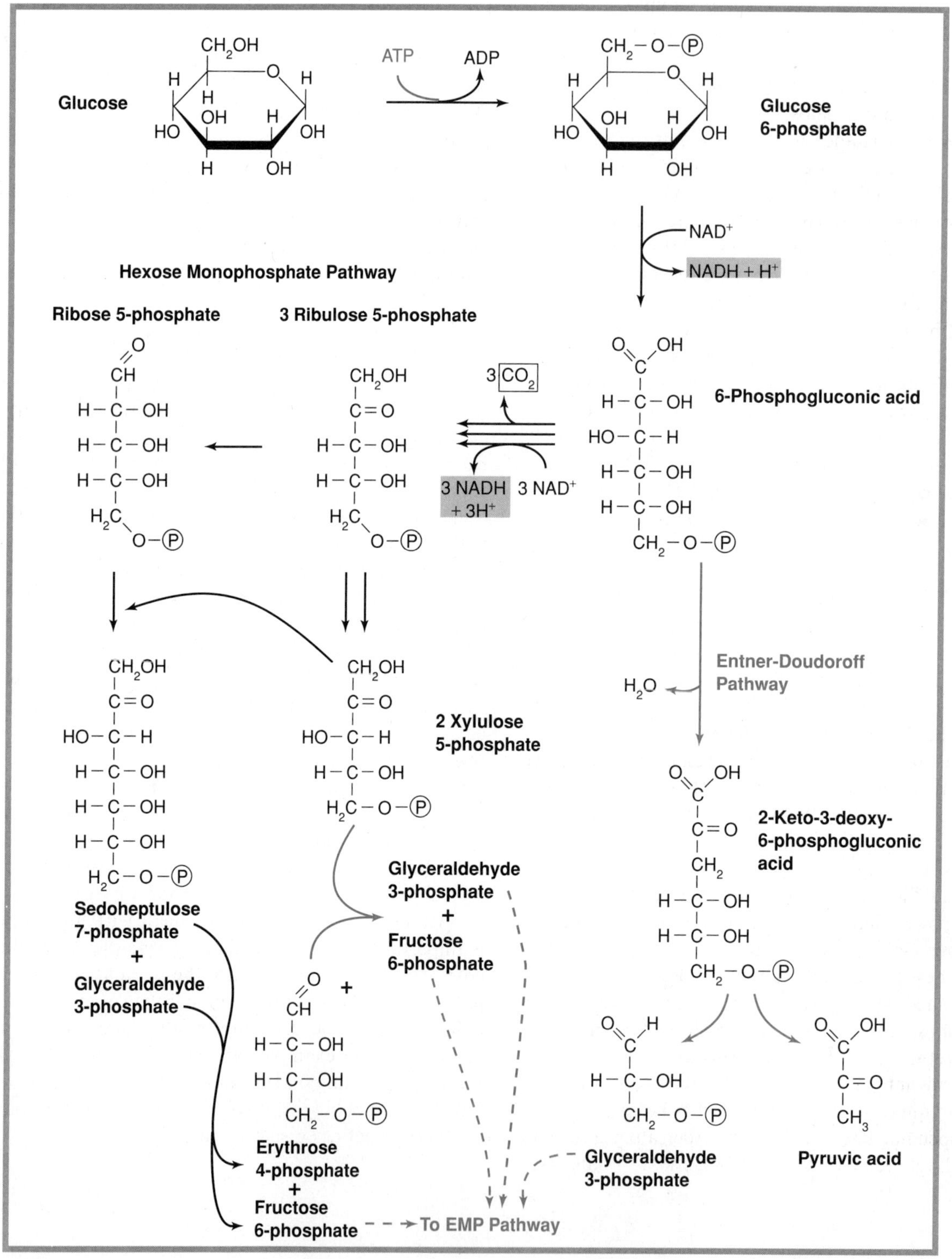

FIGURE 4.13

Glucose utilization via the hexose monophosphate (HMP) and Entner-Doudoroff pathways. Despite the common first step between these two pathways, the end products are quite different. Multiple parallel arrows indicate that the reaction must proceed more than once for the pathway to operate. Bound phosphate groups are indicated by Ⓟ. Structural formulas are not repeated for those compounds listed in figure 4.12.

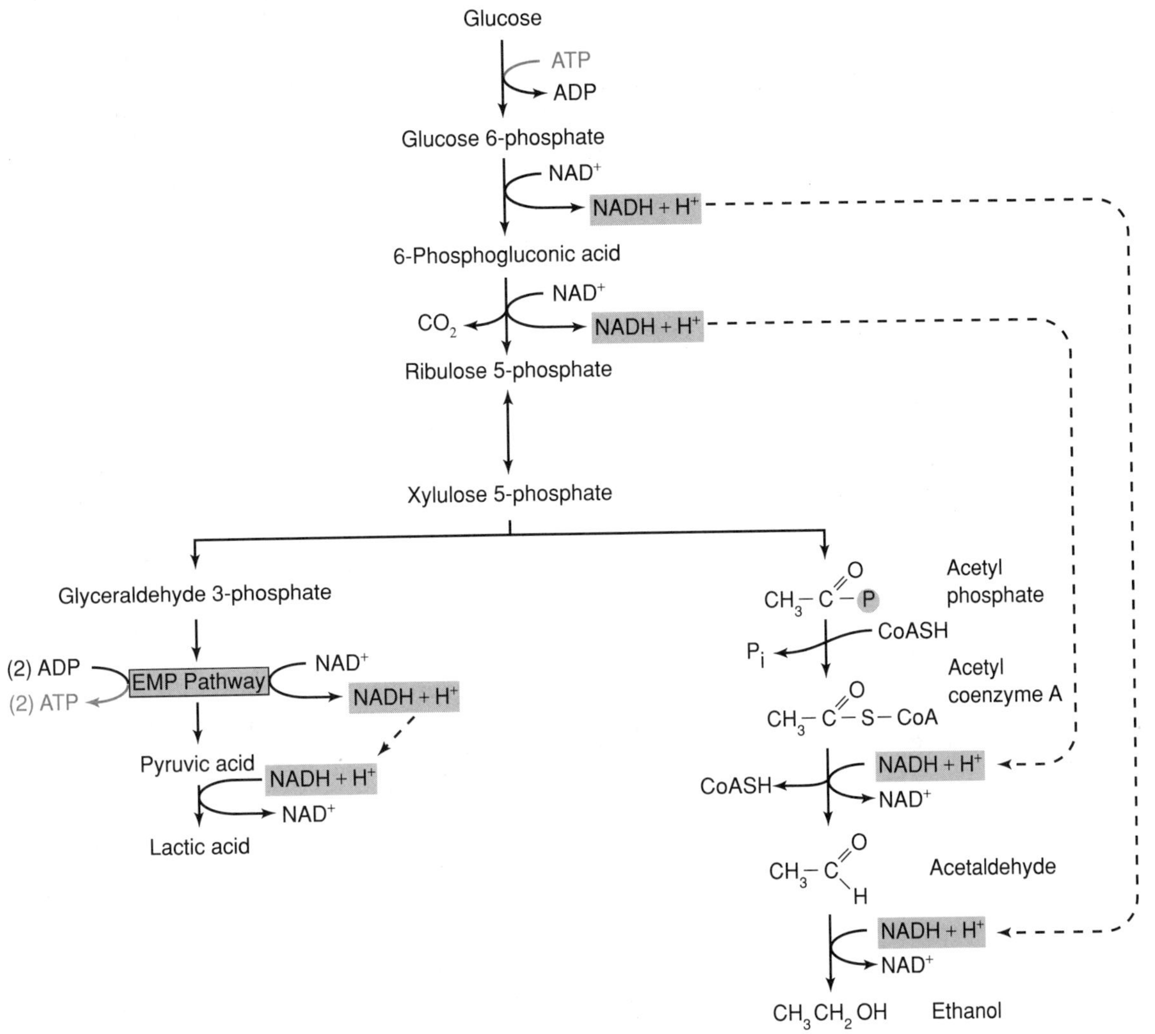

FIGURE 4.14

Leuconostoc mesenteroides utilization of glucose via a phosphoketolase pathway. Structural formulas not shown in this figure are given in figures 4.12 and 4.13. Note that all NAD molecules are reoxidized and one extra ATP molecule is produced. In the EMP pathway (figure 4.12), two extra ATP molecules would be produced.

TABLE 4.4 Some common carbohydrate conversions

Carbohydrate	Product
galactose	glucose via intermediate containing uridine diphosphate
glycogen	glucose 1-phosphate
lactose	glucose + galactose
maltose	two glucose molecules
mannose	fructose 5-phosphate
starch	glucose (variable number)
sucrose	glucose + fructose

It is not often emphasized, but some bacteria such as *Bacillus sphaericus* cannot use any sugars as sources of carbon and energy. Instead, molecules with fewer reduced bonds, such as amino acids, fatty acids, and bases from nucleic acids, must be used. Because *B. sphaericus* lacks the pyruvate dehydrogenase enzyme complex that links the EMP pathway to the TCA cycle, it grows on various compounds by feeding the molecules into reversed biosynthetic pathways (see chapter 5). The organism therefore is able to synthesize polysaccharides but not use exogenous sugars.

Fermentative Reactions

It might seem that the only way for cells growing fermentatively to generate a proton gradient would be to break down ATP and drive ATP synthase as a proton pump. However, in the early stages of growth when few waste products are present, protons can be pumped via reverse symport mechanisms in which movement of protons against a concentration gradient is coupled to movement of end products in the direction of their concentration gradient, as discussed in chapter 3. These protons then can be used by other chemiosmotic processes.

Reverse symport mechanisms can make a major contribution to cell metabolism under optimum conditions. For example, *Streptococcus cremoris* ferments glucose or lactose solely to lactic acid in a process called **homofermentation.** When the pH is near neutrality and the external concentration of lactic acid is negligible, two protons are transported for every lactic acid molecule excreted. Therefore, one molecule of glucose can yield two ATP molecules and four translocated protons.

Many anaerobic bacteria do not have the enzymes necessary to carry out a complete TCA cycle; therefore, they can never produce more than a few ATP molecules per glucose oxidized. This accounts for the slow growth rate and smaller mass of cells accumulated by anaerobes and facultative organisms growing anaerobically. Even cells with intact TCA cycles have a metabolic problem when growing fermentatively because they do not have enough NAD^+ and $NADP^+$ to operate the TCA cycle for very long, unless some process such as electron transport can regenerate it. When growing anaerobically in the absence of a suitable electron acceptor, bacteria like *Escherichia coli* will shut down their TCA cycles and exist on ATP molecules generated by substrate level phosphorylation and on the proton gradient generated by reverse symport. Some anaerobic organisms like *Desulfotomaculum acetoxidans* have an alternative pathway for the oxidation of acetyl coenzyme A that completely avoids the TCA cycle and instead breaks the acetyl moiety into carbon monoxide and ethanol.

The molecules released after partial degradation of glucose during anaerobic growth tend to be characteristic of a particular species or genus. A **homofermenter** is an organism that produces only one product, while an organism that produces two or more products is a **heterofermenter.** Many of the products are acidic and cause a dramatic reduction in the pH of the surrounding medium. For example, certain enteric bacteria such as *Escherichia* produce a mixture of acids (acetic, carbonic, formic, lactic) that can lower the pH of the medium below 5.5 and cause **methyl red,** a pH indicator, to change from yellow to red (mixed-acid fermentation; see figure 4.15). By contrast, other enteric bacteria such as *Enterobacter* catalyze an alternate reaction that yields 2,3-butanediol, a neutral end product, and the pH of the medium does not change markedly. Along the way, the intermediate compound acetylmethylcarbinol is synthesized also (figure 4.15). The presence of acetylmethylcarbinol (also called acetoin) can be detected by the **Voges-Proskauer test,** which usually is run in combination with the methyl red test and frequently used in water-quality assays. The combined test is known as the **MRVP test.**

Two biochemical pathways of considerable industrial importance are shown in figure 4.16. The production of propionic acid yields an additional ATP molecule for the producing cell. The acetone-butanol pathway, on the other hand, is once again concerned primarily with reoxidizing the $NADH + H^+$ produced in earlier enzymatic steps. Thus, many organisms degrade glucose in more or less the same way, but they can differ in their treatment of the pyruvic acid that is produced. Figure 4.17 provides a summary of fermentative reactions and emphasizes the central role of pyruvic acid in metabolism.

Lithotrophic Reactions

Chemolithotrophic bacteria are a heterogeneous group, although some aspects of their metabolism do lend themselves to generalizations. Originally, Sergei Winogradsky reported that chemolithotrophs were all autotrophic, but this has been proved incorrect. Nevertheless, all ammonia-oxidizing bacteria do require carbon dioxide as their major carbon source. The chemolithotrophs can use a wide variety of inorganic molecules as substrates for energy metabolism, two examples of which are presented in figure 4.18A–B.

The oxidation of ammonia shown in figure 4.18A is straightforward and combines reactions in the cytoplasm with reactions in the periplasm. The unusual feature seen in this instance is that the reaction donating the electrons is a periplasmic reaction instead of a cytoplasmic reaction. Nevertheless, the electron transport system pumps protons from the cytoplasm to the exterior to generate the usual ΔpH gradient. By

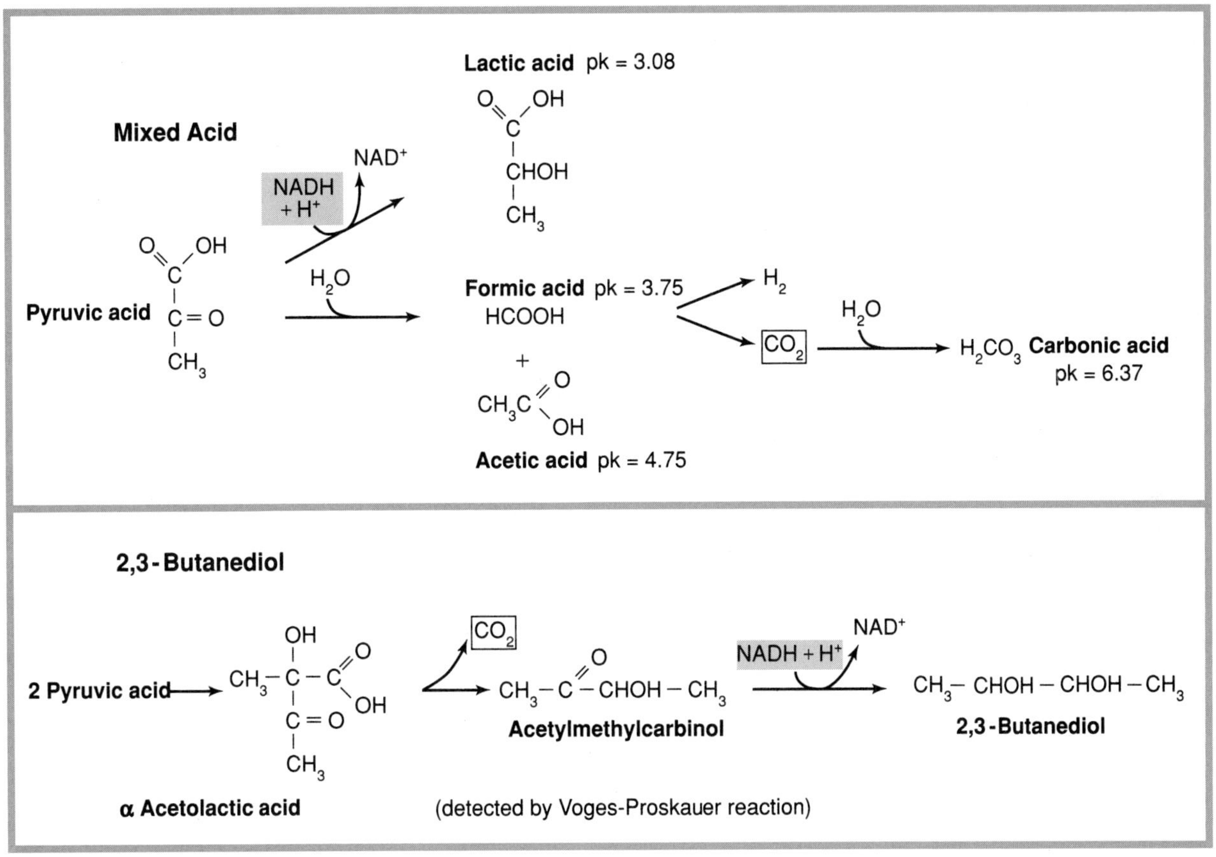

FIGURE 4.15

Biochemical pathways detected by the MRVP test. The mixed acid pathways drop the pH of the medium below 5.5 so that methyl red turns red. The 2,3–butanediol pathway yields essentially neutral products so that methyl red remains yellow. There is no convenient test for the presence of 2,3–butanediol, but, so long as the pathway is operative, the Voges-Proskauer test will demonstrate the presence of acetylmethylcarbinol (acetoin) in the surrounding medium.

contrast, the reactions for oxidation of thiosulfate occur entirely in the periplasm (Figure 4.18B). They generate large quantities of hydrogen ion that, owing to their physical position, automatically increase the pH gradient across the plasma membrane. These protons can be used by ATP synthase in the usual manner. The electron transport system still conducts the electrons to the cytoplasm for their final reaction. Note, however, that there is no proton pumping mechanism as is customarily seen in most respiratory organisms.

The principle of chemolithotrophic oxidoreduction reactions is the same as for reactions involving carbohydrates, namely, that one molecule must be oxidized while another is reduced; but in chemolithotrophic reactions, both molecules are inorganic. Bacteria that can grow using sulfur, nitrogen, iron, manganese, and stannous salts have been reported. The electrons generated by the oxidoreduction reactions are passed to more or less conventional electron transport chains that have been modified to cope with the specific inorganic molecule in question. Note that although the reactions shown in figure 4.18A–B have oxygen as the terminal electron acceptor, they need not do so. For example, *Thiobacillus denitrificans* can use nitrate as a terminal electron acceptor instead. Because electron transport is still occurring, this process is an example of anaerobic respiration.

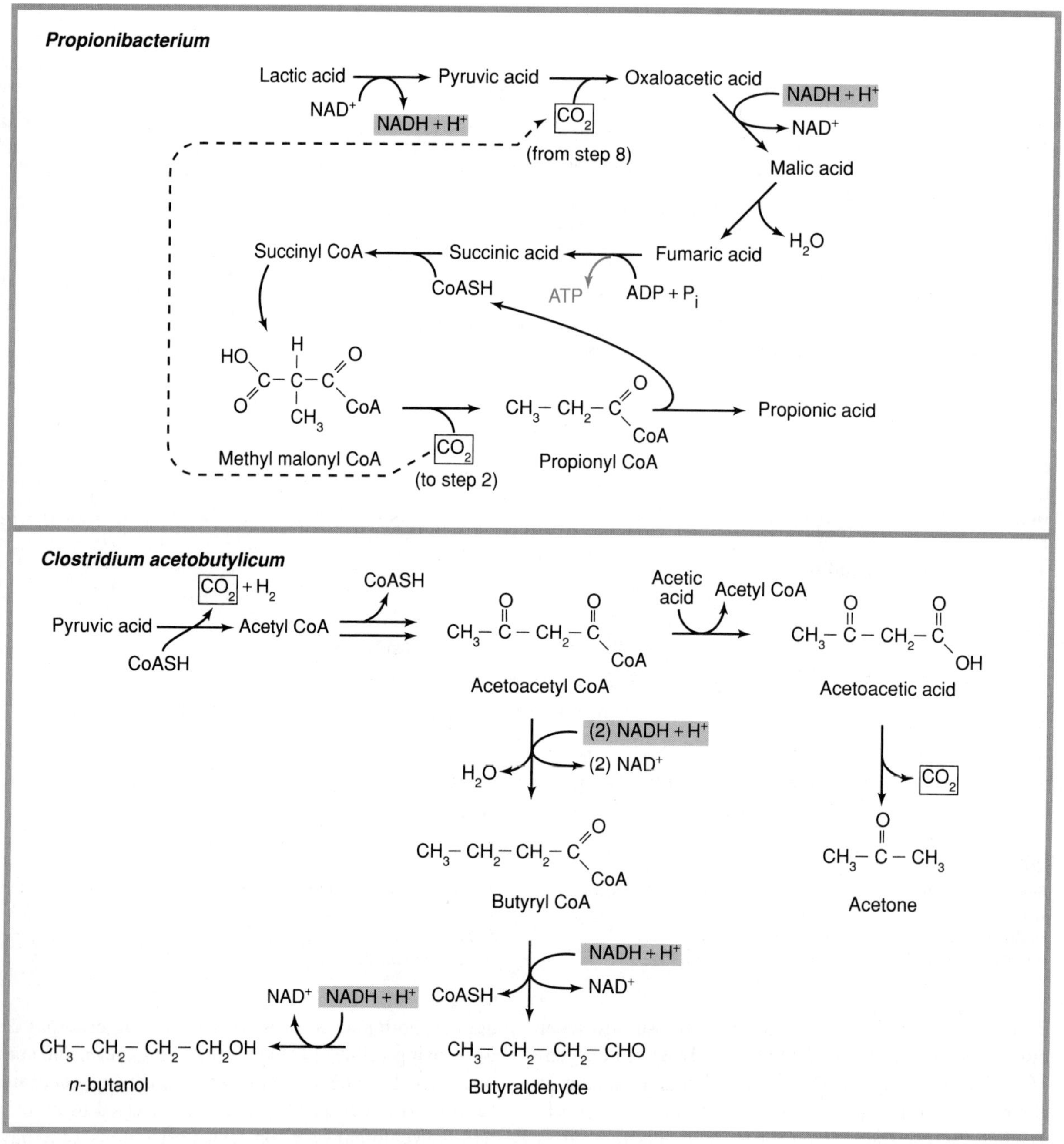

FIGURE 4.16
Some fermentative pathways of industrial importance.

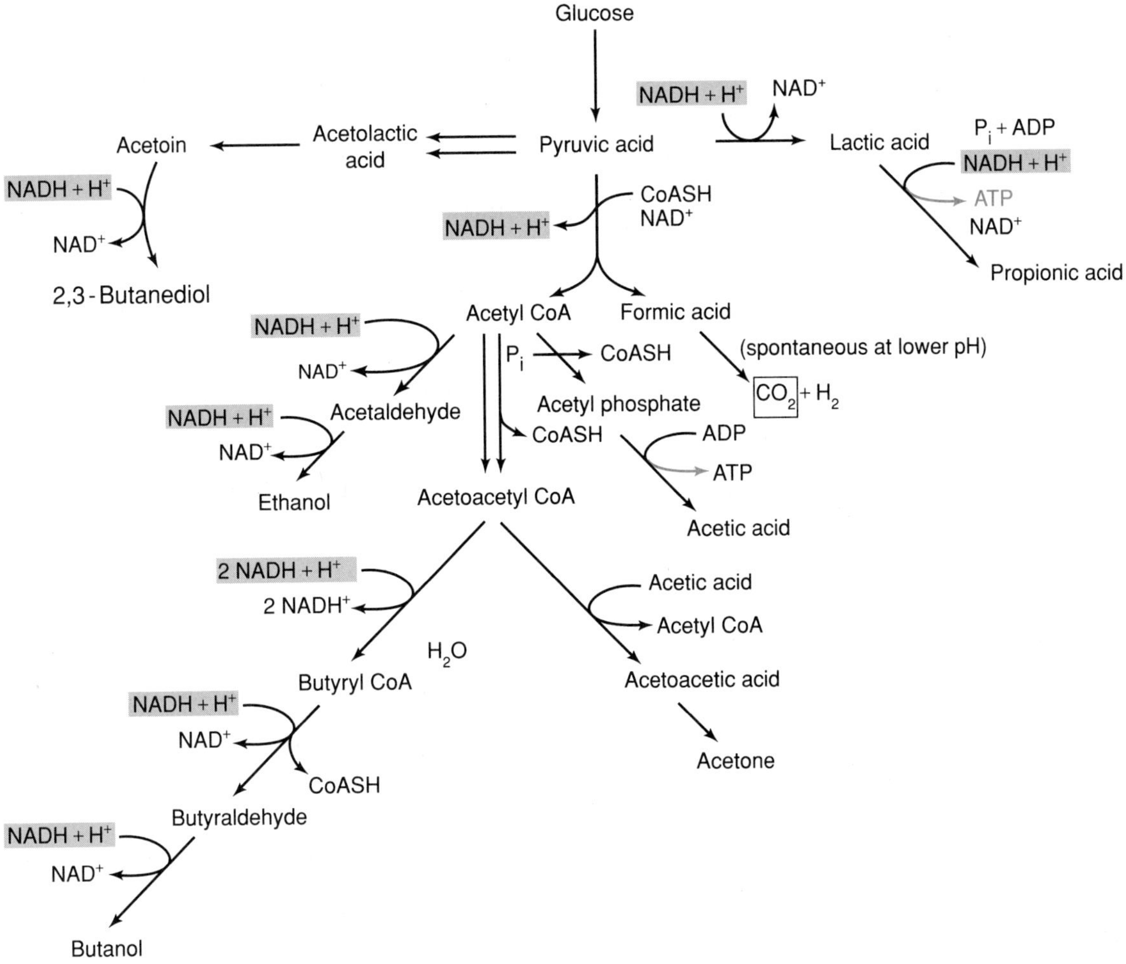

FIGURE 4.17
Alternative fates of pyruvic acid. Interactions of various glucose degradative pathways and their products are depicted schematically. No single organism is capable of carrying out all pathways indicated.

The methane-producing bacteria carry out oxidation of hydrogen with carbon dioxide as the terminal electron acceptor and methane as a waste product. Some, like *Methanosarcina barkeri,* also are able to oxidize methanol, methylamines, or acetate and are denoted as methylotrophs. Others, like *M. thermoautotrophicum,* are obligate chemolithotrophs. All methanogens are members of the archaebacterial group.

In terms of the discussion in the preceding section, chemolithotrophic bacteria must have unique ways of generating the reduced enzyme cofactors needed for biosynthesis. This conclusion follows from the observation that the electrons removed during oxidation reactions are passed directly to the electron transport system and not to an enzyme cofactor. Generation of NADH + H^+ must require that some of the proton gradient formed during metabolism be used to drive a special reverse electron transport whose function is to reduce the enzyme cofactors (figure 4.18A–B).

Stickland Reaction

Another set of combined oxidation-reduction reactions, this time involving organic molecules, has been identified in members of the genus *Clostridium* by L. H. Stickland. These reactions are fermentative (do not involve respiration) and produce ATP and reducing power

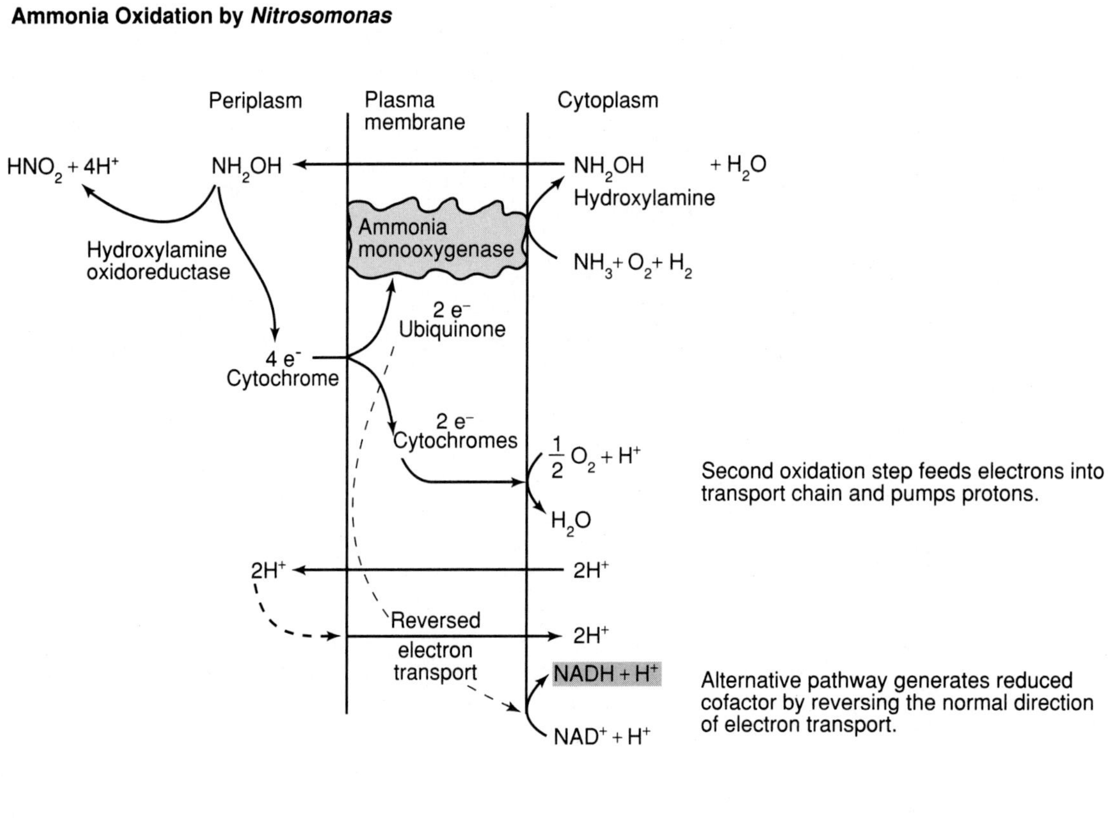

FIGURE 4.18

Two chemolithotrophic reactions. (**A**) Oxidation of ammonia by *Nitrosomonas*.

(A) From A. B. Hooper, "Biochemistry of the Nitrifying Lithoautotrophic Bacteria," *Autotrophic Bacteria*, edited by H. G. Schlegel and B. Bowien. Copyright © 1989 Science Tech Publishers, Madison, WI. Reprinted by permission.

from certain pairs of amino acids. One possible set of reactions is shown in figure 4.19. Other reactions also have been observed, depending on the species of *Clostridium* being studied. Potential electron donors are alanine, isoleucine, leucine, and valine; while potential electron acceptors are glycine, proline, and ornithine. The ATP yield depends on the actual pair of amino acids used.

Dual Energy Systems

A prokaryote need not maintain a typical phototrophic metabolism at all times. For example, most members of the genus *Rhodobacter* can grow as chemoheterotrophs in the dark under aerobic or reduced oxygen conditions. Cyanobacteria can grow anaerobically in the presence of H_2S, which inhibits photosystem II and serves as an electron donor for the reduction of carbon via NADPH produced by photosystem I. The oxybiontic facultative anaerobes like *Escherichia coli* can switch from respiratory growth to fermentative growth.

Walther Stoeckenius identified another example of dual energy metabolism that involves members of the archaebacterial, halophilic genus *Halobacterium*. *Halobacterium halobium*, the best-studied organism in the group, has the interesting property of producing a reddish, light-absorbing pigment that serves a protective function while the organism grows aerobically as a chemotroph in shallow, highly saline ponds. As the

Sulfur Oxidation by *Thiobacillus versutus*

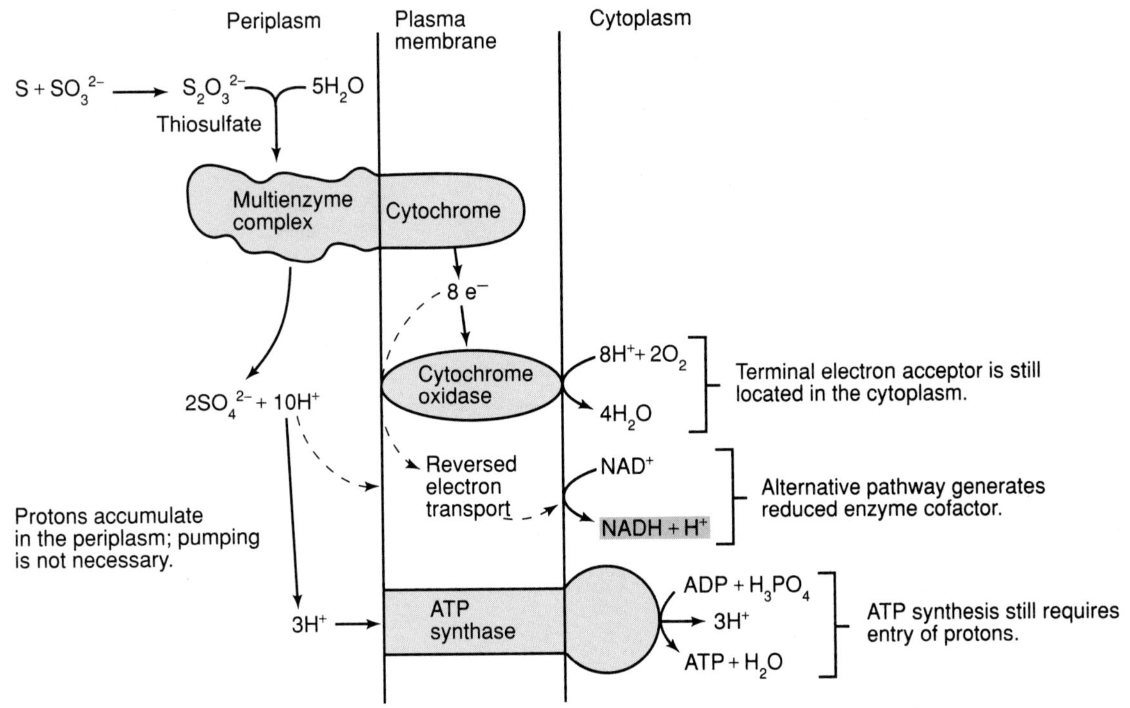

B

FIGURE 4.18 CONTINUED

(**B**) Oxidation of thiosulfate by *Thiobacillus versutus.*

(B) From D. P. Kelly, "Physiology and Biochemistry of Unicellular Sulfur Bacteria," *Autotrophic Bacteria,* edited by H. G. Schlegel and B. Bowien. Copyright © 1989 Science Tech Publishers, Madison, WI. Reprinted by permission.

salt concentration increases in these ponds, it becomes progressively more difficult for oxygen to dissolve in the water, and the environment becomes anaerobic. With anaerobiosis comes a color change in *H. halobium* due to the formation of a purple chromoprotein, **bacteriorhodopsin,** which is found in patches within its cell membrane. This pigment has a structure similar to the pigment rhodopsin in that it consists of the retinal derivative of vitamin A conjugated to a protein. Like true rhodopsin, bacteriorhodopsin also has been observed to bleach (oxidize) when struck by light.

The cyclic nature of the oxidized-reduced states of bacteriorhodopsin results in the purple membrane acting as a proton pump to develop a pH gradient across the cell membrane (figure 4.20). This gradient can be used to synthesize ATP in the usual fashion. Stoeckenius and his collaborators have suggested that the proton pumping activity is associated with a shift from the all-*trans* to a 13-*cis* configuration for the molecule. In the 13-*cis* state, a proton is lost to the surface of the membrane and later replaced during the conversion back to the all-*trans* form. Although there are obvious similarities to the photosynthetic processes discussed earlier in this chapter, from an energetic standpoint the *Halobacterium* reactions are quite different because, among other things, all activity necessary to generate a proton gradient is carried out by a single molecule.

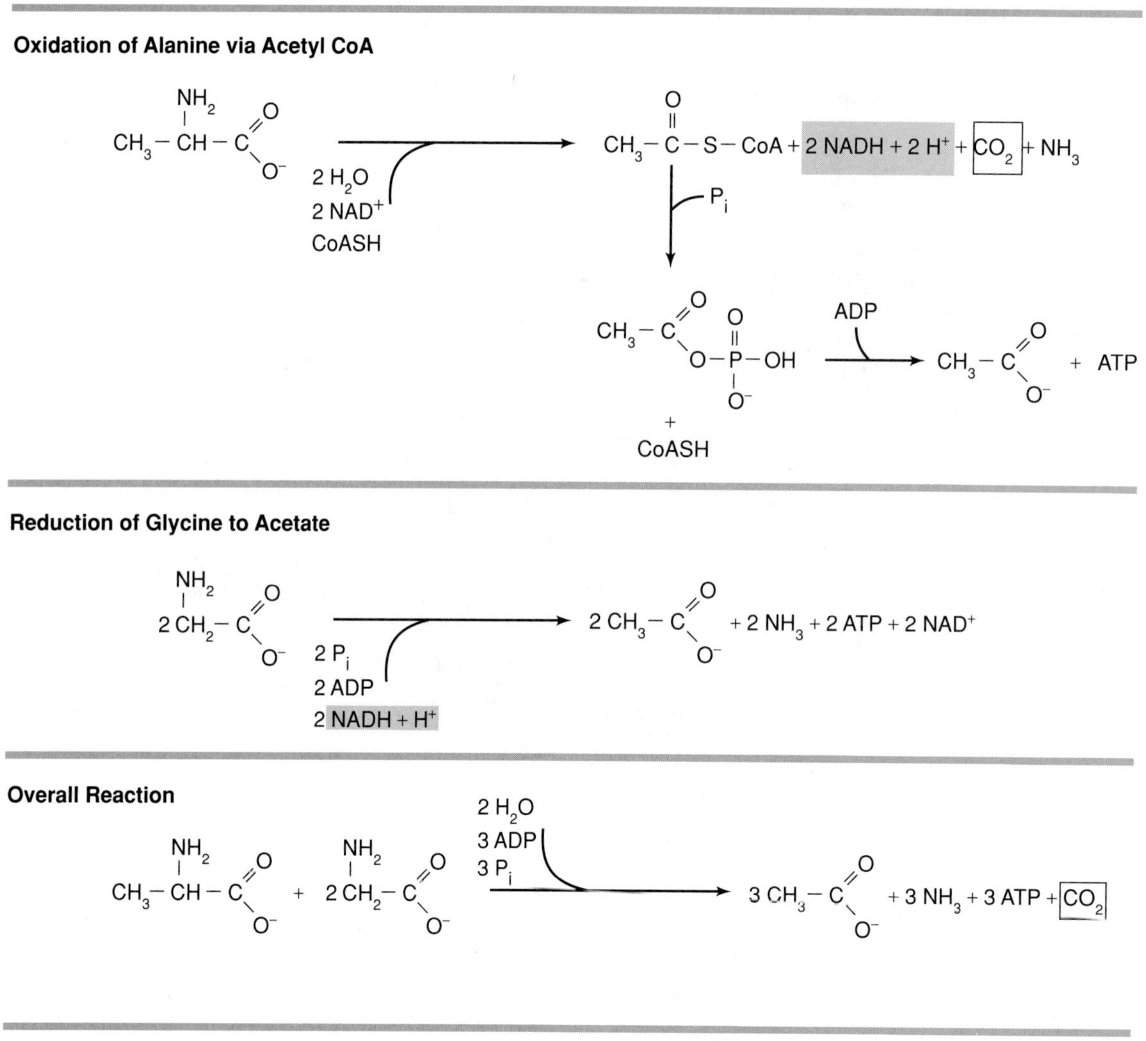

FIGURE 4.19

Example of oxidation-reduction patterns in the Stickland reaction as they occur in some clostridia.

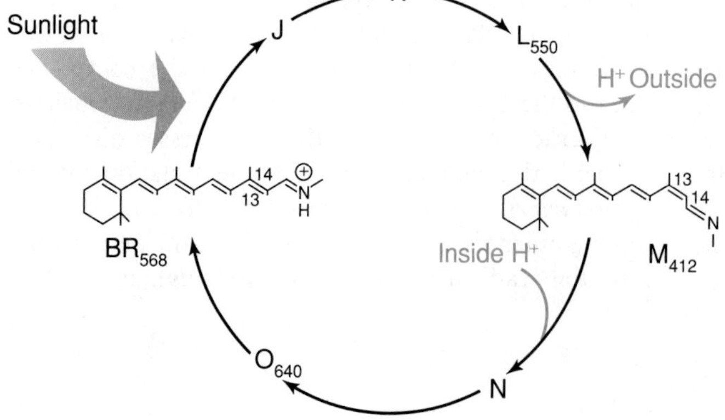

FIGURE 4.20

Development of a proton gradient by the purple membrane of *Halobacterium halobium*. Capital letters represent intermediate forms of bleached bacteriorhodopsin, and subscripts indicate the absorption maxima. The all-*trans* form is BR_{568} and the 13–*cis* configuration is designated as M_{412}.

From S. P. A. Fogor, et al., "Chromophore Structure in Bacteriorhodopsin's N Intermediate: Implications for the Proton-pumping Mechanism," in *Biochemistry* 27:7097–7101. Copyright © 1988 Plenum Publishing Corporation, New York, NY. Reprinted by permission.

Summary

Prokaryotic organisms can obtain the energy that they need for growth from a variety of sources. These sources can be categorized according to whether the organisms predominantly use light energy (phototrophs) or chemical energy (chemotrophs). Some bacteria can use both, although not simultaneously. Chemotrophs can be subdivided further, depending on whether they use inorganic chemicals (lithotrophs) or organic chemicals (organotrophs).

Photosynthesis can be oxygenic (releases oxygen) or anoxygenic (does not release oxygen) according to the bacterium and/or the conditions of growth. The major anoxygenic photosynthetic bacteria (gram negative) are subdivided by the color of their photosynthetic pigments into the purple and green bacteria. They carry out reactions considered equivalent to those of photosystem I in eukaryotes. The oxygenic photosynthetic bacteria (cyanobacteria) possess both photosystems I and II and seem to be closely related to certain eukaryotic algae and to chloroplasts of more complex eukaryotes. *Halobacterium* can carry out a very primitive form of photosynthesis that appears to bear little or no relationship to the conventional photosystems.

Chemotrophic bacteria can employ some of the same reactions as eukaryotes, but they also use others that are uniquely their own. Fermentative bacteria degrade various compounds to yield ATP without using the TCA cycle or an electron transport system. As a result, they derive relatively little energy from their substrates and tend to grow slowly. Chemotrophic bacteria that use electron transport are said to carry out respiration, which may be aerobic or anaerobic depending on the terminal electron acceptor.

The end products of metabolism tend to be characteristic of the producing organism. This is particularly true of organisms that grow fermentatively. Pyruvate is a common product of glucose catabolism, but it is not excreted as such. Rather, each organism produces other compounds from pyruvate whose production may reoxidize reduced coenzymes and/or generate more ATP molecules. These secondary pathways are frequently important to industry and also are useful in the classification of bacteria.

Questions for Review and Discussion

1. Name the various biochemical pathways that can be used to degrade glucose and describe their different features. What are the products of each pathway? What are the advantages to the cell of each?
2. How can chemiosmotic theory be used to explain the synthesis of ATP? What supplies the energy for ATP synthesis? If a cell has ATP, how might it use the ATP to generate a proton gradient?
3. What is electron transport? What functions does it perform for the cell?
4. What are the similarities and differences between energy production by chemotrophs and energy production by phototrophs?
5. How are the photosynthetic pathways in bacteria similar to those found in green plants? How are they different?
6. What are some alternative metabolic fates of pyruvic acid? What advantages do these conversions hold for the cell?
7. The presence or absence of oxygen is often a critical factor in culturing a microorganism. What are the different types of oxygen relations that have been observed in bacteria? Which of these types might be capable of carrying out respiration?

Suggestions for Further Reading

Anthony, C. 1989. *Bacterial energy transduction.* Academic Press, Ltd., London. A somewhat detailed discussion of most aspects of actual energy production. The introductory chapter is very good as are the thorough discussions of photosynthesis, electron transport, substrate transport, and chemolithotrophy.

Dawes, E. A. 1985. *Microbial energetics.* Chapman & Hall, New York. A concise summary of energy-yielding reactions in microorganisms.

Fulghum, R. S. 1983. Definition of anaerobic bacteria. *ASM News* 49:432–434.

Gottschalk, G. 1981. The anaerobic way of life of prokaryotes, pp. 1415–1424. *In* M. P. Starr, H. Stolp, H. G. Trüper, A. Balows, and H. G. Schlegel, (eds.) *The prokaryotes. A handbook on habitats, isolation, and identification of bacteria.* Springer-Verlag, New York. An excellent, brief review.

Gottschalk, G. 1986. *Bacterial metabolism,* 2d ed. Springer-Verlag KG, Heidelberg. A general survey of the major biochemical pathways found in bacteria.

Gray, M. W., and W. F. Doolittle. 1982. Has the endosymbiont hypothesis been proven? *Microbiological Reviews* 46:1–42. A good introduction to the endosymbiont theories.

Harashima, K., T. Shiha, and N. Murata (eds.) 1989. *Aerobic photosynthetic bacteria.* Springer-Verlag KG, Berlin. A monograph describing the aerobic, anoxygenic photosynthesizers.

Knaff, D. B. 1988. Reaction centers of photosynthetic bacteria. *Trends in Biochemical Sciences* 13:157–158. A brief review of new data on the physical locations of the various components.

Krieg, N. R., and P. S. Hoffman. 1986. Microaerophily and oxygen toxicity. *Annual Review of Microbiology* 40:107–130. A lengthy discussion of the biochemistry of microaerophiles with examples.

Nicholls, D. G. 1982. *Bioenergetics: An introduction to the chemiosmotic theory.* Academic Press Inc., New York. A well-written introduction for the nonspecialist.

Olson, J. M., J. G. Ormerod, J. Amesz, E. Stackebrandt, and H. G. Trüper (eds.) 1987. *Green photosynthetic bacteria.* Plenum Press, Inc., New York. Papers from a workshop on the subject. Some are quite detailed, others are a good review.

Prince, R. C. 1988. The proton pump of cytochrome oxidase. *Trends in Biochemical Sciences* 13:159–160. A brief review of models for a proton pump.

Stokes, G. B. 1988. Estimating the energy content of nutrients. *Trends in Biochemical Sciences* 13:422–424. Asks (and answers) the question of where all the energy released by oxidation of acetyl CoA was originally located in the molecule.

Stanier, R. Y., N. Pfennig, and H. G. Trüper. 1981. Introduction to the phototrophic prokaryotes, pp. 197–211. *In* M. P. Starr, H. Stolp, H. G. Trüper, A. Balows, and H. G. Schlegel, (eds.) *The prokaryotes. A handbook on habitats, isolation, and identification of bacteria.* Springer-Verlag, New York.

Tempest, D. W., O. M. Neijssl, M. J. Teixeira de Mattos. 1985. Regulation of carbon substrate metabolism in bacteria growing in chemostat culture, pp. 53–69. *In* I. S. Kulaev, E. A. Dawes, and D. W. Tempest, (eds.), *Environmental regulation of microbial metabolism.* Academic Press, Inc., Orlando, Fla. This paper suggests that bacteria emphasize rate over quantity of energy produced in determining which metabolic pathways to use.

Thauer, R. K. 1988. Citric-acid cycle, 50 years on. Modifications and an alternative pathway in anaerobic bacteria. *European Journal of Biochemistry* 176:497–508. The chemical details of anaerobic TCA metabolism.

chapter

5

Biosynthesis of Subunit Molecules and Enzyme Cofactors

chapter outline

■

GENERAL GOALS

Be able to describe the general structure of each type of polymer subunit as well as name specific examples.
Be able to show how the various energy pathways discussed in chapter 4 can be diverted for production of subunit molecules or enzyme cofactors and outline special reactions necessary for these functions.
Be able to outline the various reactions employed in nitrogen and carbon dioxide fixation.

Now that the ways in which bacteria can break down molecules in order to release energy for metabolism have been presented, it is time to consider how they use that energy in **anabolic reactions** (those reactions that make new molecules for the cell) to produce molecules needed for their own structure and function. Many of these molecules are quite large and hence are collectively termed macromolecules. Macromolecules include DNA, RNA, proteins, polysaccharides, etc. All these macromolecules are examples of **polymers,** high-molecular-weight-chains whose individual links are nearly identical subunits.

This chapter surveys the various small molecules used as subunits for macromolecules as well as the pathways by which many cells can produce these subunits from intermediate compounds of the energy pathways discussed in chapter 4. A related topic also considered in this chapter is the biosynthesis of some enzyme cofactors used in the same energy-yielding reactions. The ways in which bacteria take up their nitrogen and carbon atoms and convert them into usable forms as well as the mechanism by which cells make sugars from smaller molecules are discussed in addition.

Carbon Metabolism

Just as glucose plays a central role in the energy metabolism of chemoorganotrophs, so it also plays a central role in anabolic reactions, either directly or via the intermediate compounds derived from the EMP pathway or the TCA cycle. Therefore, bacterial cells supplied with organic molecules other than glucose or sugars readily convertible into glucose must be able to synthesize glucose from available carbon-containing molecules. There are several groups of reactions that contribute to this process.

Carbon Dioxide Utilization by Autotrophic Bacteria

The subunit molecules for the macromolecules discussed in chapter 6 are all built around carbon skeletons. For chemoheterotrophic organisms, the sources of these carbon atoms are the molecules being catabolized for energy. However, autotrophic organisms can obtain their carbon atoms directly from the atmosphere around them. Carbon (CO_2) fixation, the use of carbon dioxide as the major carbon source for growth, is very commonly found among photosynthetic organisms of all types but is not a necessary part of photosynthesis. For example, many purple and green bacteria are capable of growing either as autotrophs or organotrophs, while all *Halobacterium* species are strictly organotrophic. Additionally, nonphotosynthetic organisms such as *Nitrobacter* and *Nitrosomonas* are autotrophic.

The most familiar process by which carbon dioxide is reduced to a usable form is a cyclic one called the reductive pentose phosphate cycle or **Calvin cycle.** Melvin Calvin, who first described the reactions of this cycle, won a Nobel Prize for his efforts. These reactions are essentially identical in eukaryotes and certain prokaryotes, such as *Chromatium, Nitrobacter,* and *Thiobacillus,* and are diagrammed in figure 5.1. The process begins with condensation of ribulose bisphosphate and carbon dioxide. The network of subsequent reactions is most easily visualized as three parallel CO_2 fixation reactions. The reaction products interact to regenerate ribulose bisphosphate, and the final product of the three parallel reactions is one molecule of 3-phosphoglyceric acid. A brief comparison of figures 4.12 and 5.1 shows that the Calvin cycle has many other compounds in common with the EMP pathway; therefore, it would be theoretically possible to fix carbon dioxide and then immediately degrade the resulting compound via the EMP pathway and the TCA cycle. This type of activity would be a waste of energy for the organism because the ATP recovered by oxidation would not equal the amount of ATP that ATP synthase could have produced if NADPH had been used for electron transport and proton translocation instead. Interconnection between the Calvin and TCA cycles is prevented in eukaryotes as well as in such prokaryotes as *Chromatium vinosum* by the loss of the enzymes 2-oxoglutaric acid oxidase and malate dehydrogenase.

However, certain organisms such as *Rhodospirillum rubrum* are capable of growing anaerobically in the light and aerobically in the dark (that is, they are facultative phototrophs). They, therefore, need all TCA cycle enzymes for maximum growth efficiency and must depend on regulation of specific enzymes to prevent energy waste. At the present time, the method of this regulation is not known.

The green bacteria as well as those nonphotosynthetic anaerobic bacteria that also are autotrophs do not seem to employ the Calvin cycle for CO_2 fixation because they lack one or more of the enzymes necessary for the cycle to function. Instead, they make use of a process called **reductive carboxylation** (or the reductive TCA cycle) in which new organic acids are formed by the addition of carbon dioxide molecules to existing molecules. This is in contrast to the Calvin cycle that yields glyceraldehyde 3-phosphate exclusively. The

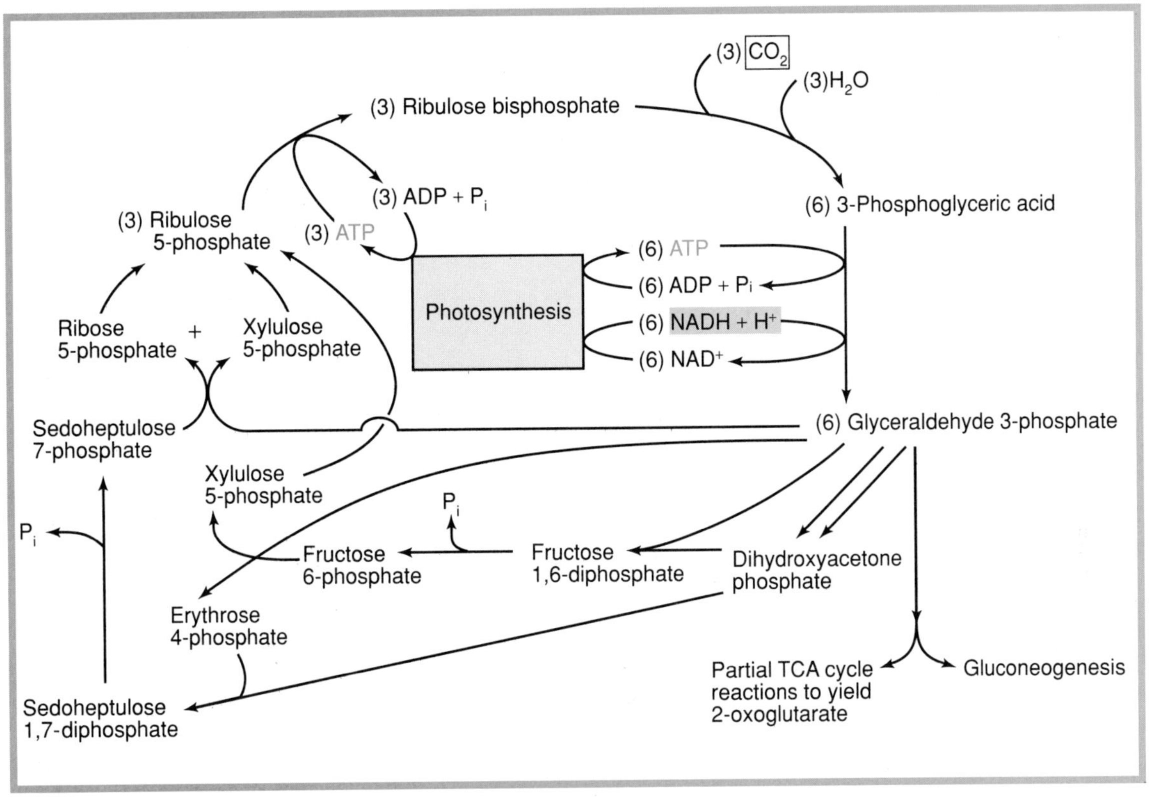

FIGURE 5.1

The Calvin cycle as it occurs in *Chromatium vinosum* strain D. Structural formulas for the compounds presented in this figure may be found in figures 4.12 and 4.13. P_i represents phosphate ion, and parenthetical numbers indicate the number of molecules that are produced per turn of the cycle. Note the ATP energy input and the NADH + H^+ reducing power input.

TABLE 5.1 Examples of reductive carboxylation reactions

Organism	Reaction
Methanobacterium thermoautotrophicum	Acetyl-coenzyme A + $CO_2 \xrightarrow{\text{reduced ferredoxin}}$ pyruvic acid + coenzyme A
Clostridium kluyveri	Succinyl coenzyme A + $CO_2 \xrightarrow{\text{reduced ferredoxin}}$ 2-oxoglutaric acid + coenzyme A
Chlorobium limicola	Phosphoenolpyruvic acid + H_2O + $CO_2 \rightarrow$ oxaloacetic acid + PO_4^{-3}

Note: Reduced ferredoxin, the hydrogen donor, is formed by a reaction with NADH or NADPH. All enzymes catalyzing these reactions also have been observed in bacteria that have an operative Calvin cycle. Their exact physiologic role is somewhat obscure.

biochemistry of the reductive carboxylation pathways is not well understood, but certain reactions that have been identified are presented in table 5.1. Note that the end products are more varied than in the Calvin cycle.

Methane Utilization

The problem facing methane oxidizers is exactly the reverse of that facing autotrophs. Autotrophs must deal with a molecule that is too oxidized to be usable in metabolism, and methane oxidizers must deal with a molecule that is too reduced. Because the process of carbon assimilation by a methanotroph includes several oxidoreduction reactions, energetic electrons are available to the cell as a result and can be fed into the electron transport system. Organisms can be obligate or facultative methanotrophs, the latter being able to use multicarbon substrates as well as methane.

The typical pathway for methane oxidation is shown in figure 5.2. The oxidation proceeds stepwise to alcohol, aldehyde, and carboxylic acid. Note that, unlike

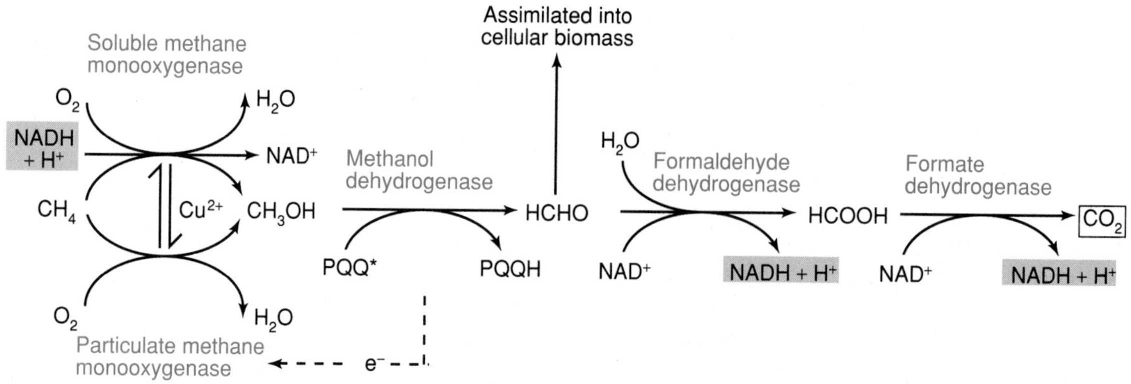

FIGURE 5.2

Methane oxidation. The enzyme methane monooxygenase has been observed in both soluble and particulate forms, depending on the organism studied. The source of electrons for the reduction portion of the oxidoreduction reaction can be either NADH + H^+ or, in the case of some particulate forms of the enzyme, a reduced cofactor from the oxidation pathway. PQQ is pyroquinoline quinone. A copper ion seems necessary for the conversion of the soluble form of the methane monooxygenase enzyme to the particulate form.

Adapted from C. Bedard and R. Knowles, "Physiology, Biochemistry, and Specific Inhibitors of CH_4, NH_4, and CO Oxidation by Methanotrophs and Nitrifiers," in *Microbiological Reviews* 53:68–84. Copyright © 1989 American Society for Microbiology, Washington, DC. Reprinted by permission.

the typical aerobic energy pathways discussed in the preceding chapter, oxygen is involved as an initial electron acceptor instead of as a terminal acceptor. An input of reduced cofactor is required by the soluble enzyme, but the cofactor is regenerated during the remainder of the oxidation steps. The particulate enzyme may receive electrons from the second step in the pathway. ATP can be generated if NADH + H^+ is fed into electron transport to drive proton translocation. Alternatively, the formaldehyde intermediate can be channeled into gluconeogenesis (see following subsection).

Glyoxylate Bypass

One interesting example of a biochemical problem that cells can face is how to produce their needed molecules when the available carbon is in the form of a small molecule such as acetic acid. This situation frequently arises in both gram-positive and gram-negative organisms that have been growing on glucose. When the supply of glucose is exhausted, a cell still can obtain energy from acetate molecules by transporting them back into itself, converting them to acetyl CoA (a process that requires the equivalent of two ATP molecules), and oxidizing them through the TCA cycle. The theoretical net yield of ATP from one acetate molecule is 10 ATP molecules, so the energy derived can be quite good. As is discussed later, however, the TCA cycle is used not only as a source of energy but also as a source of intermediates for various other biochemical pathways, such as those for amino acid and porphyrin biosynthesis. As such, the TCA cycle must be replenished with appropriate molecules to regenerate oxaloacetate and succinyl CoA needed for biosynthetic reactions. **Anaplerotic pathways** are those that act as alternatives to usual biochemical mechanisms and serve to replace molecules that have been diverted to other biochemical pathways. The anaplerotic pathway that serves to augment the TCA cycle is called the **glyoxylate bypass,** a diagram of which is shown in figure 5.3.

As can be seen from the diagram, the glyoxylate bypass skips two energy-yielding steps of the TCA cycle in order to produce succinic acid and malic acid. It is not surprising, therefore, that the glyoxylate bypass is only minimally active during glucose metabolism because it would compete for the same substrates being used for energy production. The basal level of glyoxylate bypass activity serves to replace those few TCA cycle intermediates diverted to amino acid biosynthesis.

Gluconeogenesis

The process of **gluconeogenesis** is one of assembling glucose molecules from simpler structures. Many reactions from processes such as the EMP pathway are reversible and, therefore, can contribute to the overall mechanism. However, certain steps, those that normally yield energy in the form of substrate level production of ATP, are not directly reversible, at least by

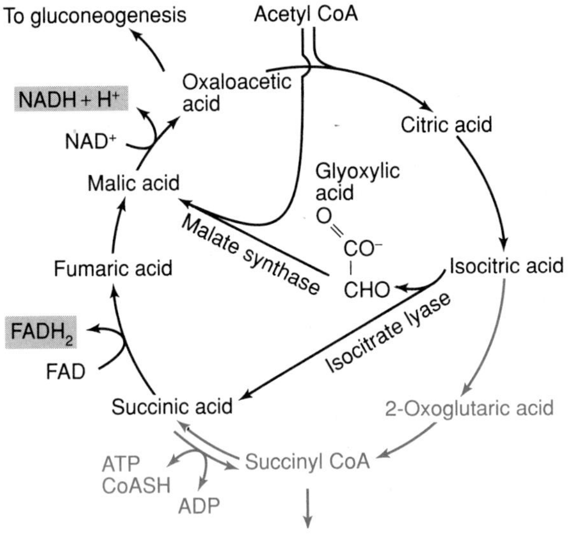

Glyoxylate bypass. Acetyl CoA is used as the source of carbon atoms to make succinic and malic acids. For reference purposes, the normal TCA cycle reactions not involved in the shunt are shown in color. Note that succinate can form either oxaloacetic acid, yielding FADH + H^+ in the process, or succinyl CoA, requiring energy in the process.

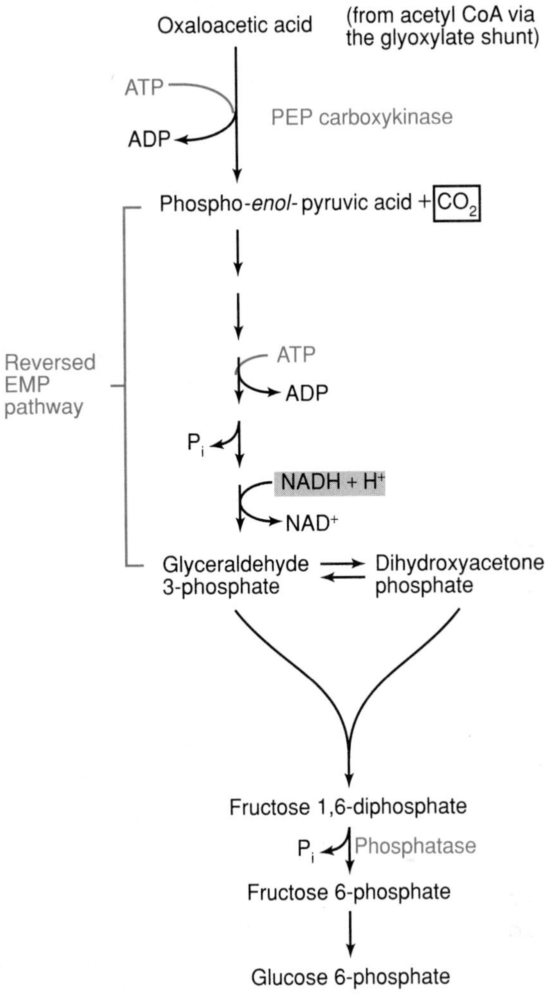

FIGURE 5.4

Pathway for gluconeogenesis. The ultimate source of the carbon atoms in this example is acetyl CoA. The energy from ATP molecules is needed to bypass or reverse those steps of the EMP pathway that normally yield energy, thus the resulting glucose molecules have the same energy content as any glucose molecules derived from other sources. Phospho-*enol*-pyruvic acid is designated as PEP.

the same enzymes. An example of a gluconeogenic pathway is shown in figure 5.4. In this example, oxaloacetic acid from the glyoxylate bypass is the starting point. Along the pathway, some intermediate compounds used in amino acid biosynthesis also are produced. Once glucose has been formed, all other necessary sugars can be made either by simple isomerization reactions or by reactions such as those of the pentose phosphate pathway discussed in chapter 4.

Nitrogen Metabolism

Production of Amino Acids

Most bacteria derive the nitrogen they need for growth from ammonium ions in the surrounding medium. Ammoniacal nitrogen is already at the correct oxidation level for cellular metabolism; therefore, all that is required is ammonium ion uptake and conversion into amino groups or other nitrogen derivatives. The actual reaction used depends on the external concentration of ammonium ion.

In the **reductive amination** reactions shown in figure 5.5A–B, either glutamic acid or 2-oxoglutaric acid is reduced and aminated to yield an L-amino acid. In figure 5.5B, the term *direct* means that ammonium ion is a substrate for this reaction rather than for a reaction that puts an amino group on an intermediate molecule. Note that both types of reactions require NADPH, which the cell obtains from reactions like those discussed in chapter 4. Because NADPH + H^+ also can pass electrons to the electron transport system to pump protons up to three times (figure 4.3) and because those protons can be used later to synthesize ATP, the reductive amination reactions require energy even if ATP does not appear directly in the reaction. With the two amino acids L-glutamic acid and L-glutamine available

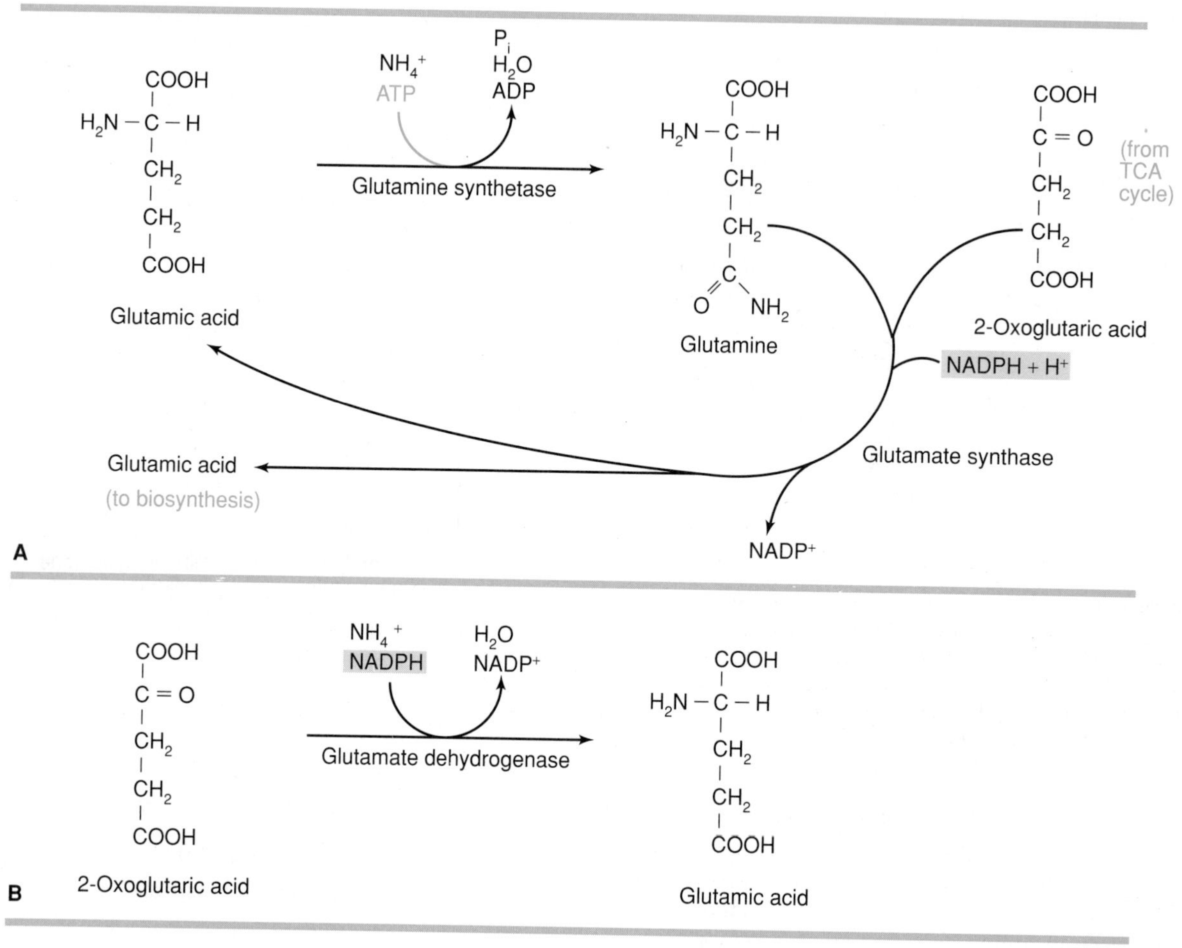

FIGURE 5.5

Conversion of ammonium ion (NH_4^+) to organic nitrogen. (**A**) Production of the amino acids glutamic acid and glutamine as it occurs at low ammonium ion concentrations. (**B**) At higher ammonium ion concentrations, it is possible to synthesize glutamic acid directly without the intervention of glutamine. Arrows indicate the reactions catalyzed by the specific enzymes listed next to each arrow. Inorganic phosphate is designated as P_i.

to act as donors of amino groups, cells are theoretically capable of producing all nitrogen-containing molecules that they need for their metabolism. However, remember that some bacteria require certain amino acids as growth factors, just as humans do, indicating one or more enzymatic deficiencies in their metabolic pathways.

An organism's primary method of synthesizing other amino acids is via a **transamination** reaction in which an amino group is moved from one molecule to another. For this type of reaction, glutamate is used as a donor of an amino group, and a 2-oxo acid is used as an acceptor. No energy is required for transamination because there is a simple exchange of functional groups. If an organism is supplied with an excess of one amino acid, it frequently can use some of the excess molecules as amino donors to make glutamic acid from 2-oxoglutaric acid and later use transamination to produce the amino acids that it does need.

The stereoisomerism of the original amino acid also is preserved during the transamination reaction so that L-glutamic acid yields only L-amino acids. This raises the obvious question of how cells manage to produce the D-amino acid molecules required for peptidoglycan structure. **Racemase,** an enzyme that interconverts the L and D forms of alanine, provides the answer. The D-alanine thus produced can be used as a donor to make D-glutamic acid by the usual transamination step.

The 20 or so amino acids are not all synthesized by separate pathways. Rather, the cells produce a few amino acids directly, usually by transamination, and then produce the remainder by enzymatic modification of the original amino acids. It is therefore customary to speak of amino acid "families," meaning groups of amino acids that are biochemically interrelated. Figure 5.6 displays these relationships.

The 2-oxo acids used in these transamination reactions are derived from the TCA cycle or from various glycolytic pathways; therefore, any organism that synthesizes all of its own amino acids at least must be capable of operating the pertinent portions of those biochemical pathways. There are, however, two families of amino acids that cannot be derived by simple transamination of intermediates from energy metabolism. These are histidine and the aromatic amino acids (phenylalanine, tyrosine, and tryptophan). The aromatic amino acids have a ring structure with carbon-carbon double bonds and are derived from a pathway that begins with erythrose 4-phosphate (from the oxidative pentose phosphate cycle) and PEP (from the EMP pathway). The histidine pathway begins with ribose 5-phosphate that can be derived from glucose (for example, in the hexose monophosphate pathway; figure 4.13).

Nitrogen Fixation

There are a few bacteria that synthesize their amino acids via reductive amination and transamination except that they do not obtain their nitrogen in the form of ammonium ion. Instead, they carry out nitrogen fixation, reducing nitrogen gas from the atmosphere around them to an appropriate oxidation level in the following reaction catalyzed by the enzyme nitrogenase:

Equation 5.1

$$N_2 + 3 \text{ reduced cofactor} + 3\ H^+ + {\sim}15\ ATP \rightarrow 2\ NH_3 + 3 \text{ oxidized cofactor} + {\sim}15\ ADP + {\sim}15\ P_i$$

The exact stoichiometry of this reaction is not known, but it is certain that the reaction consumes a large amount of cellular energy. The reduced enzyme cofactor varies with the organism involved, but in respiratory organisms it is reduced during the reoxidation of $NADPH + H^+$.

The nitrogenase enzyme is very sensitive to the presence of oxygen; therefore, nitrogen-fixing organisms have evolved appropriate mechanisms to protect the enzyme from atmospheric oxygen. Species of the gram-negative genus *Rhizobium* have a symbiotic relationship with certain plants to eliminate oxygen. Other organisms such as *Clostridium* grow only anaerobically and thus have no significant problem with oxygen sensitivity. Organisms like *Azotobacter* (also gram negative) that grow only aerobically seem to have no special adaptation to deal with oxygen, however. It is theorized that because the respiratory enzymes are located in the cell membrane while the nitrogenase enzyme resides in the cytoplasm, the cell can protect its nitrogen-fixing capability by maintaining a rate of aerobic electron transport that is sufficiently high to reduce all oxygen in the air into water molecules before it can diffuse throughout the cell.

Biosynthesis of Nitrogenous Bases

The nitrogenous bases, adenine, cytosine, guanine, thymine, and uracil, that comprise the nucleic acids are derived from certain amino acids and their precursors. When the appropriate ribosyl sugar is added, they are called nucleosides (adenosine, cytidine, guanosine, thymidine, and uridine). As successive phosphates are added, the molecules are designated as mono-, di-, or trinucleotides. Figure 5.7 shows how the purine and pyrimidine skeletons are produced. The purine pathway produces inosinic acid first, and then adenosine and guanosine are derived from it. The pyrimidine pathway produces uridine first, with thymidine and cytosine triphosphate being derived from uridine.

The sugars for the nucleosides are derived from the same phosphoribosyl pyrophosphate (PRPP) used in the synthesis of histidine and the aromatic amino acids. They are added at an early stage in the biosynthetic pathway so that the final products are mononucleotides. Nucleotides of higher order are produced by the addition of phosphate groups sequentially, with ATP used as a donor. ATP is then regenerated by any of the energy-transducing pathways.

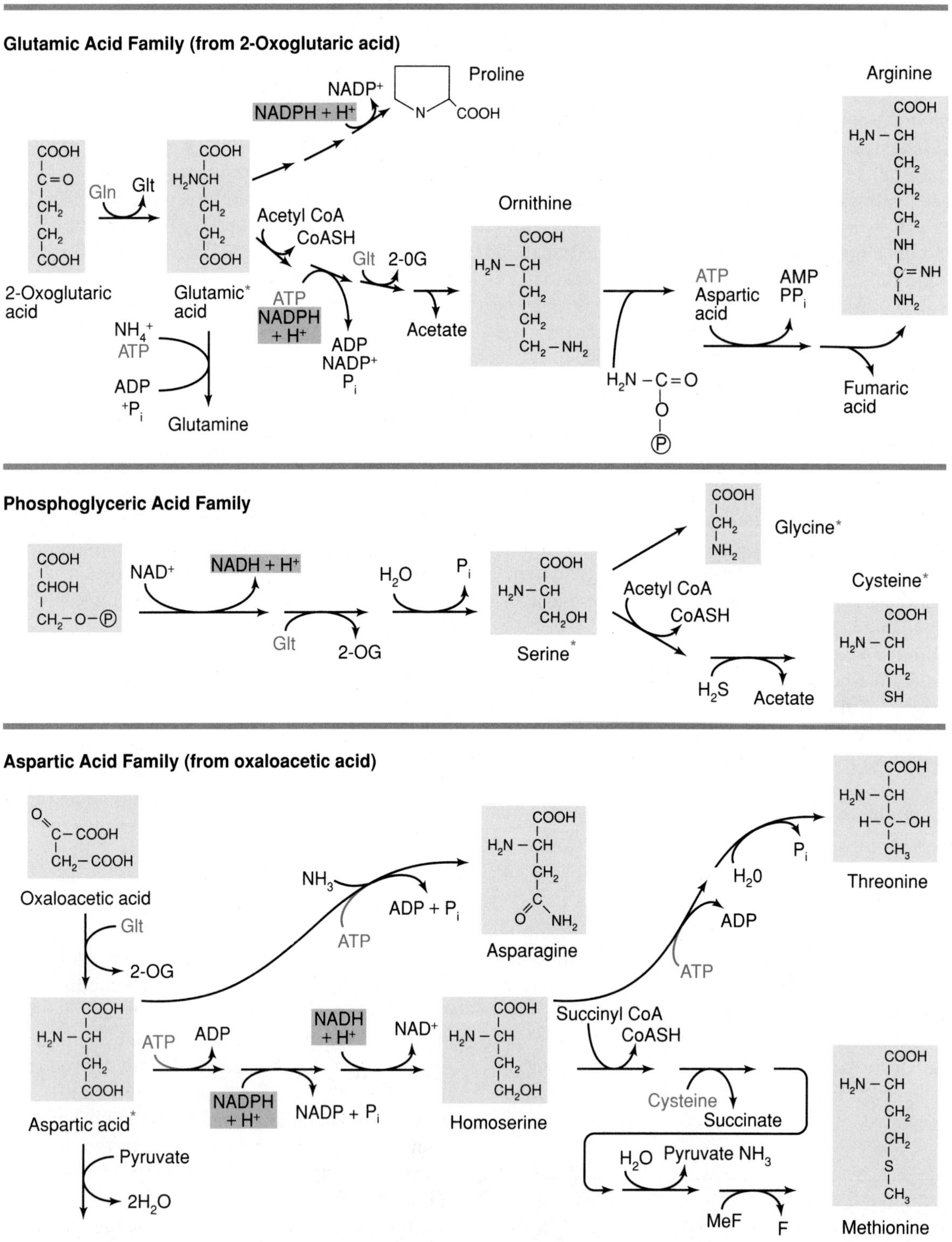

FIGURE 5.6

Continued on next page

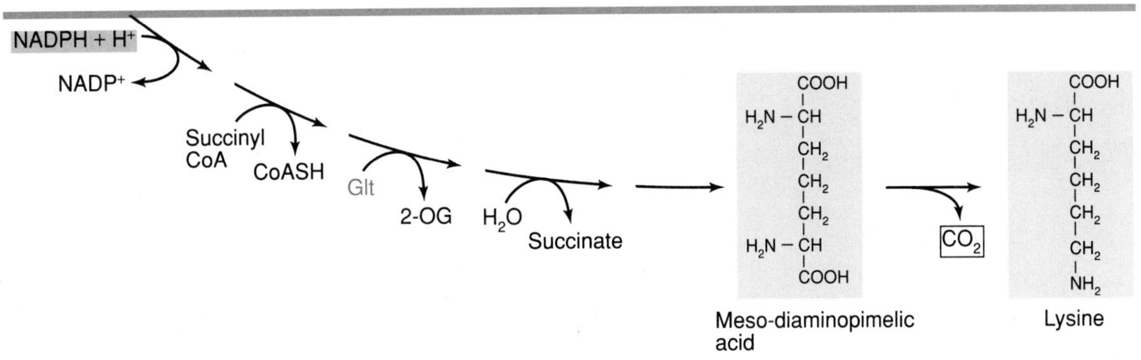

Histidine (from ribose L -pyrophosphate)

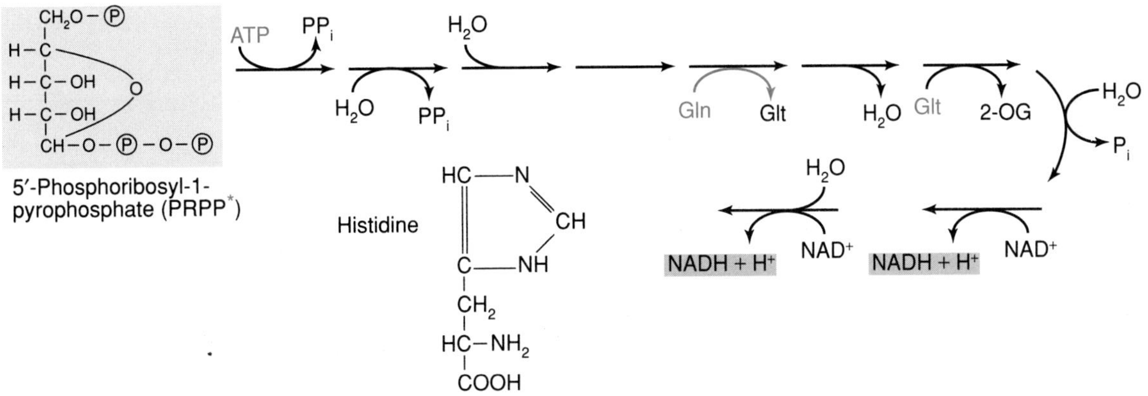

Alanine (from pyruvic acid)

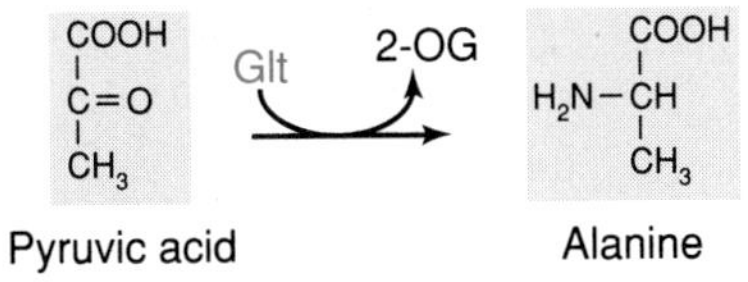

FIGURE 5.6 CONTINUED

Continued on next page

Amino acid biosynthetic groupings. The various amino acid biosynthetic pathways are shown in diagrammatic fashion. Each arrow indicates an individual enzymatic step. Parallel arrows of the isoleucine and valine pathways reflect the fact that the same enzymes are responsible for both syntheses. Compounds tagged with an asterisk are used in other biosynthetic pathways. Compounds rendered in color are synthesized in earlier amino acid pathways. Abbreviations: adenosine monophosphate (AMP); glutamine (Gln); glutamic acid (Glt); 2–oxoglutaric acid (2–OG); inorganic phosphate (P_i); inorganic pyrophosphate (PP_i); phosphoenolpyruvic acid (PEP); bound phosphate group (Ⓟ); phosphoribosyl pyrophosphate (PRPP); methyltetrahydrofolic acid (MeF). PEP and PRPP are derived from metabolic reactions involving glucose (see chapter 4).

Branched Chain Amino Acids

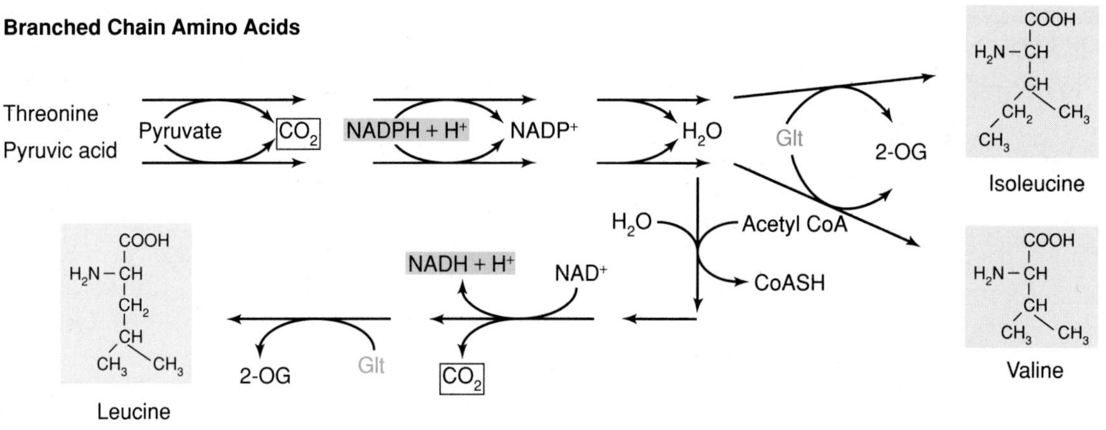

Aromatic Amino Acid Family (from phospho-*enol*-pyruvic acid)

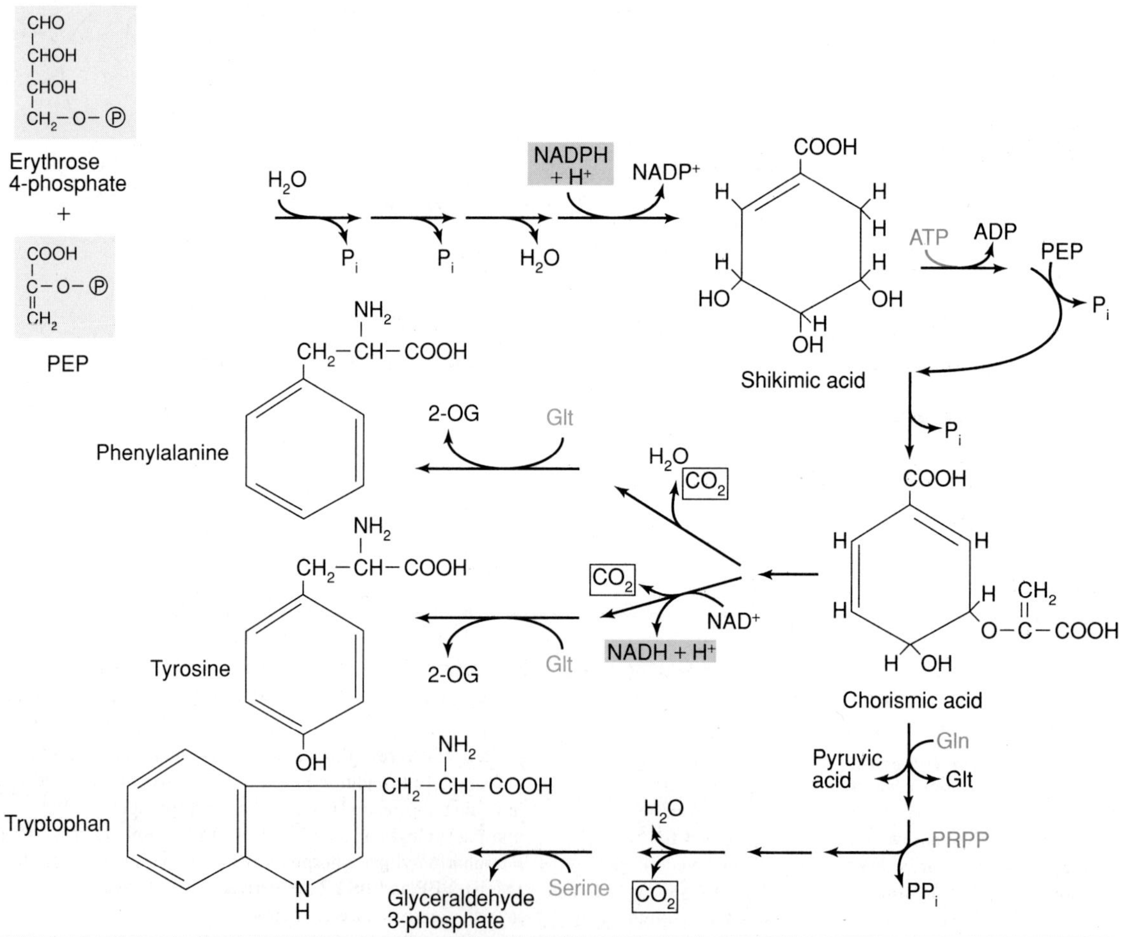

FIGURE 5.6 CONTINUED

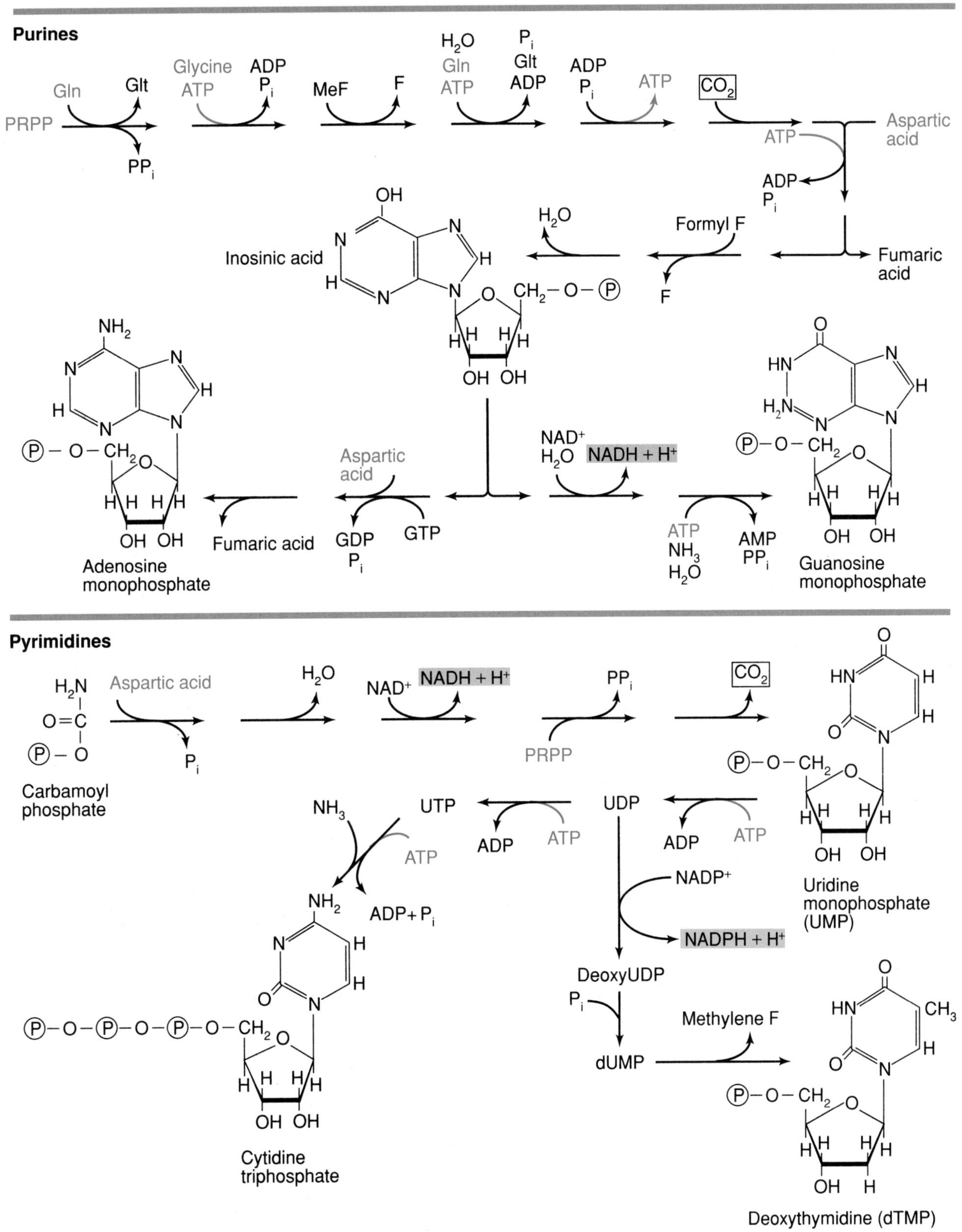

FIGURE 5.7

Biosynthesis of the nucleotides (nitrogen base + sugar + one or more phosphates). Separate pathways are used for the purines and pyrimidines, but phosphoribosyl pyrophosphate (PRPP) is used in both. The deoxyribonucleotides are produced by reducing the appropriate nucleotide diphosphate. Thymidine is made from deoxyuridine diphosphate (dUDP) by the addition of a methyl group. Compounds rendered in color come from the reactions shown in figure 5.6.

Lipid Production

Although a lipid bilayer is not exactly the same as a macromolecule, it is comprised of repeated subunits. These subunits are derived generally from various combinations of acetate moieties. In *Escherichia coli,* and presumably in other organisms as well, a special protein called the **acyl carrier protein** (ACP) is used to carry the intermediates in fatty acid biosynthesis. Initially, ACP reacts with acetyl CoA from pyruvate to yield acetyl ACP and with malonyl CoA, synthesized from acetyl CoA by carboxylation, to yield malonyl ACP. In a repeating series of reactions like those depicted in figure 5.8, two carbon units are added to make the 16- or 18-carbon fatty acid. Specific unsaturations also may be introduced during the synthesis, depending on the desired fluidity of the membrane.

In the case of phosphatidic acid and phospholipids, ACP again plays an important role, this time as a carrier of the entire fatty acid (figure 5.9). The glycerol phosphate backbone for the phosphatidic acid is derived from dihydroxyacetone phosphate in the EMP pathway. Various types of phospholipids can be produced depending on the final compound esterified to the phosphatidic acid. In the example in figure 5.9, phosphatidylserine is produced, but it equally could have been phosphatidylethanolamine. If a sugar is esterified in the third position, a **glycolipid** is produced.

The biosynthesis of ether lipids for archaebacteria is more problematic. The phytanyl chain is derived from acetyl CoA via mevalonic acid in a pathway analagous to that found in eukaryotes. However, the mechanism of formation of the ether linkage with glycerol is unknown, as is the mechanism by which the tetraether lipids are synthesized by condensation of the phytanyl chains.

Biosynthesis of Enzyme Cofactors and Heme

In terms of other types of molecules that are essential for cell metabolism, two significant groups remain to be considered. These are the enzyme cofactors involved in electron transport and biosynthetic reactions and the cytochromes that form part of the electron transport chain. Both sets of molecules have their origin in some compounds that have already been discussed.

The full names of the cofactors used as part of the TCA cycle and the EMP pathway are nicotinamide adenine dinucleotide (NAD^+) and flavin adenine dinucleotide (FAD^+). As the names imply, a major part of each of these molecules is ADP. Nicotinic acid is made from an intermediate in the tryptophan pathway, while flavin is derived from riboflavin that, in turn, can be synthesized from a guanosine derivative. The general mechanisms by which these cofactors are assembled are shown in figure 5.10. There is some uncertainty about the early steps in riboflavin biosynthesis.

Coenzyme A (CoA), the other major cofactor discussed in chapter 4, is derived from pantothenic acid. This acid is synthesized by combining 2–oxoisovaleric acid, an intermediate in the synthesis of the amino acid valine, and β-alanine, which can be made from the amino acid aspartic acid or from the fatty acid precursor malonic acid (figure 5.8). Reactions with ATP and with the amino acid cysteine complete the formation of coenzyme A.

The basic heme structure contains four porphyrin rings that are coordinately bonded to an iron atom in the center. Porphyrin rings are derived from succinyl CoA found in the TCA cycle and the amino acid glycine. The general pathway is depicted in figure 5.11. Individual types of cytochromes are produced by modifications of two porphyrin rings.

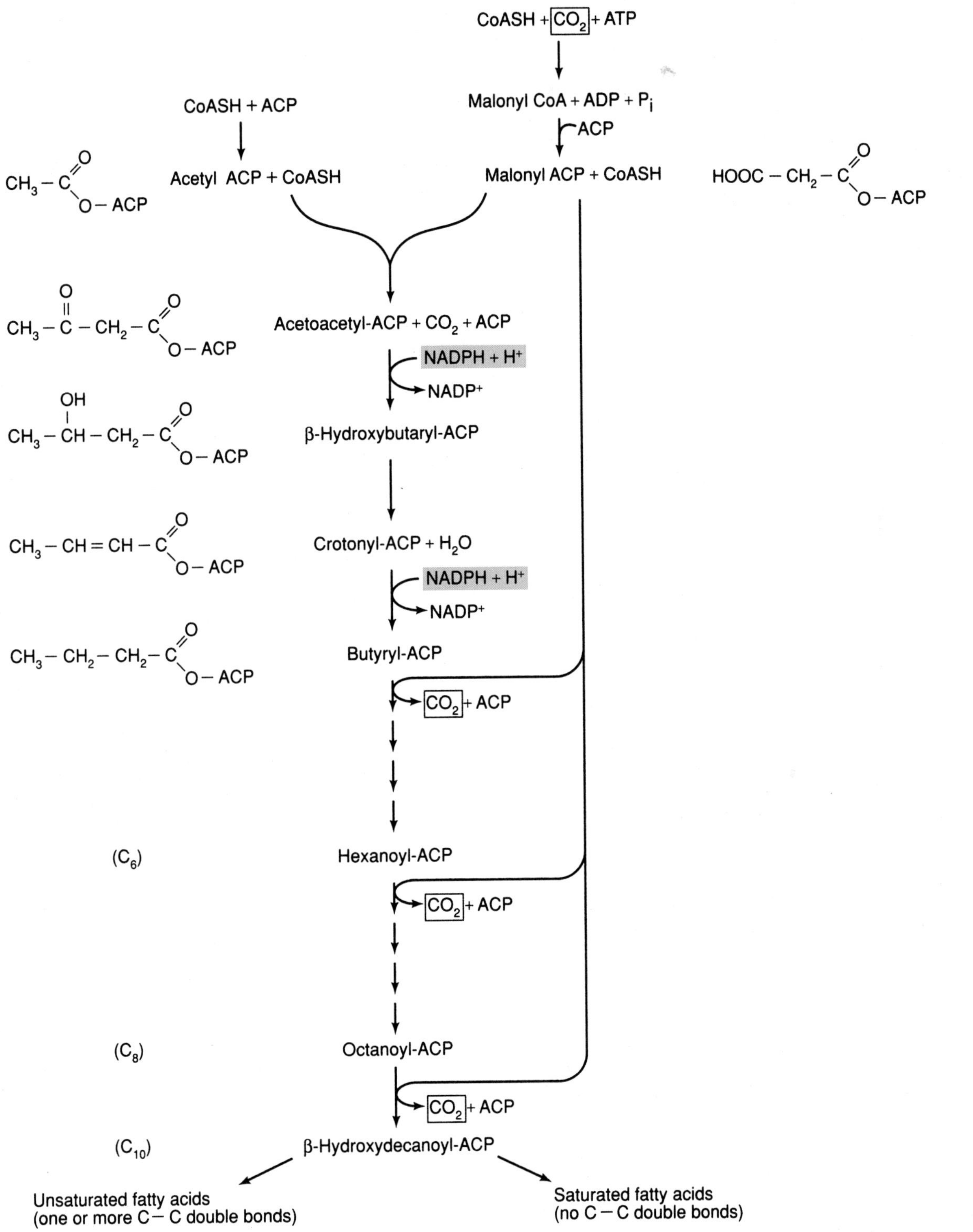

FIGURE 5.8
Biosynthesis of a fatty acid with acyl carrier protein (ACP). The same four reactions must be repeated for each two-carbon unit added. The degree of unsaturation, if any, is determined by the nature of the organism and by the temperature at which the cells are grown.

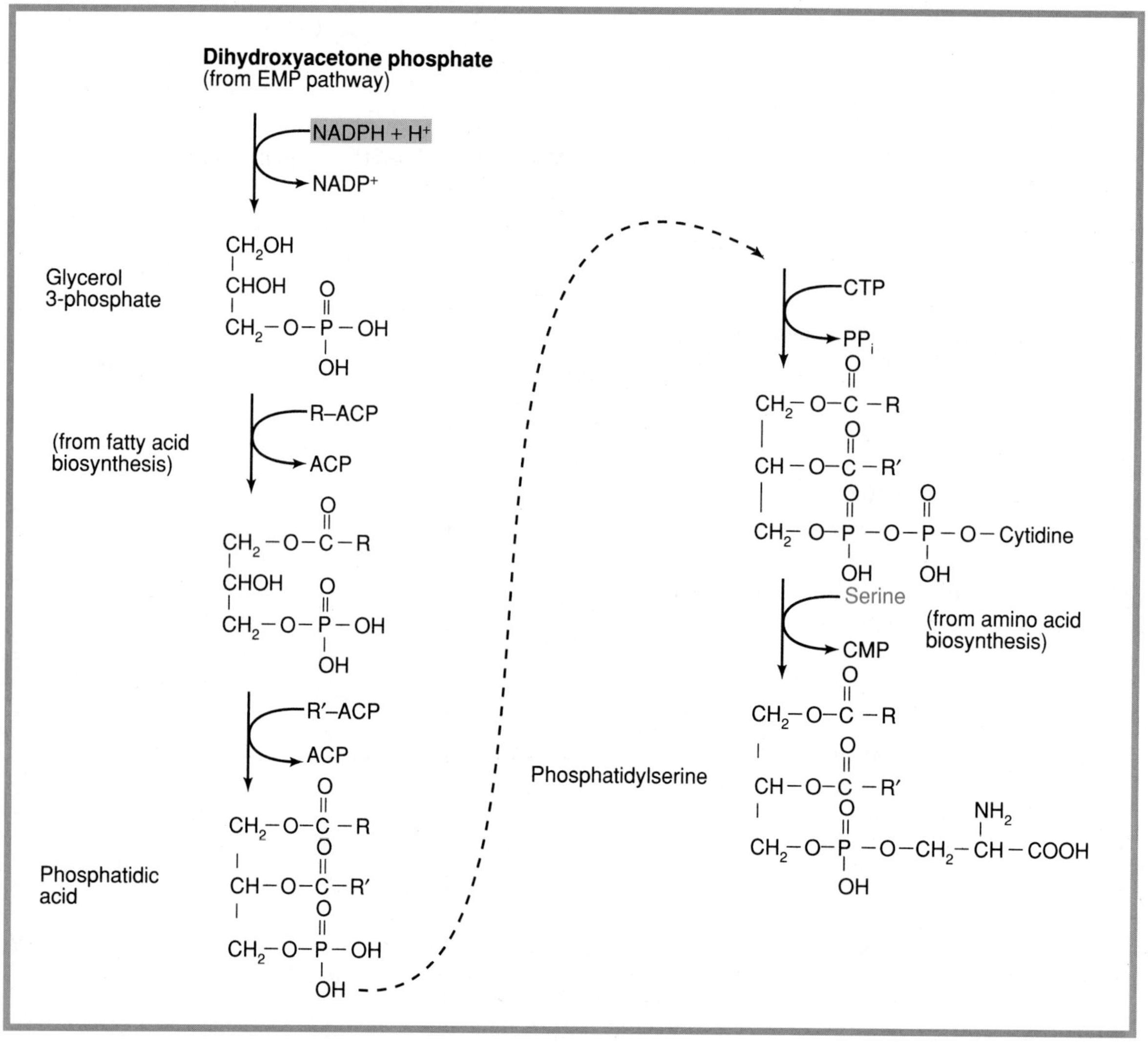

FIGURE 5.9

Biosynthesis of phosphatidic acid and phosphatidylserine, a phospholipid. It is possible to use a number of different fatty acids in this process, and therefore the fatty acid moiety is designated simply by R before it is esterified to glycerol phosphate. In the final molecule, the first carbon atom of the fatty acid is shown, and the remainder of the chain is indicated by R. Cytidine monophosphate is designated as CMP.

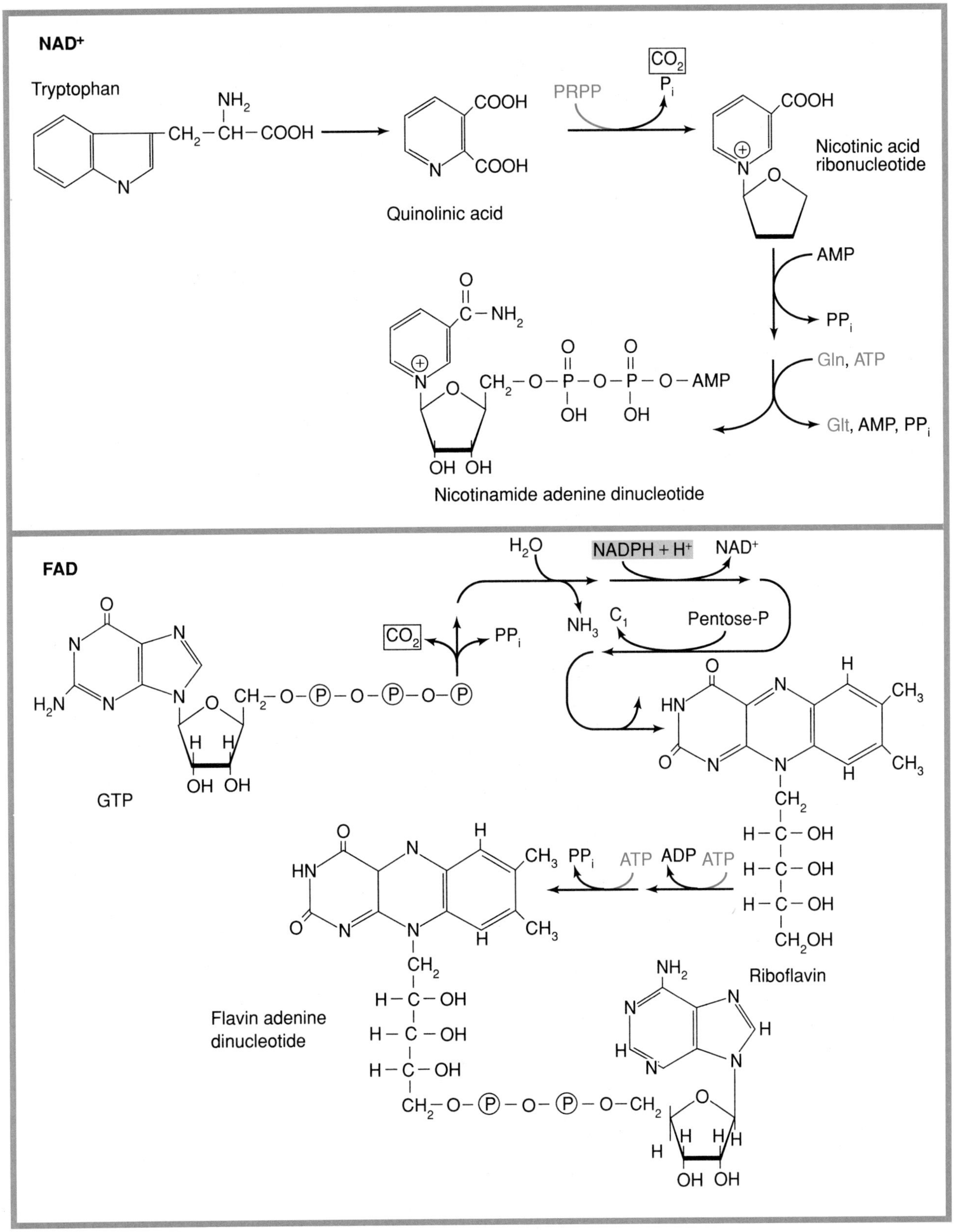

FIGURE 5.10

Biosynthesis of NAD+ and FAD. The early steps in the biosynthesis of the riboflavin intermediate for FAD are not certain, particularly the exact form of the guanosine nucleotide and the single carbon atom released.

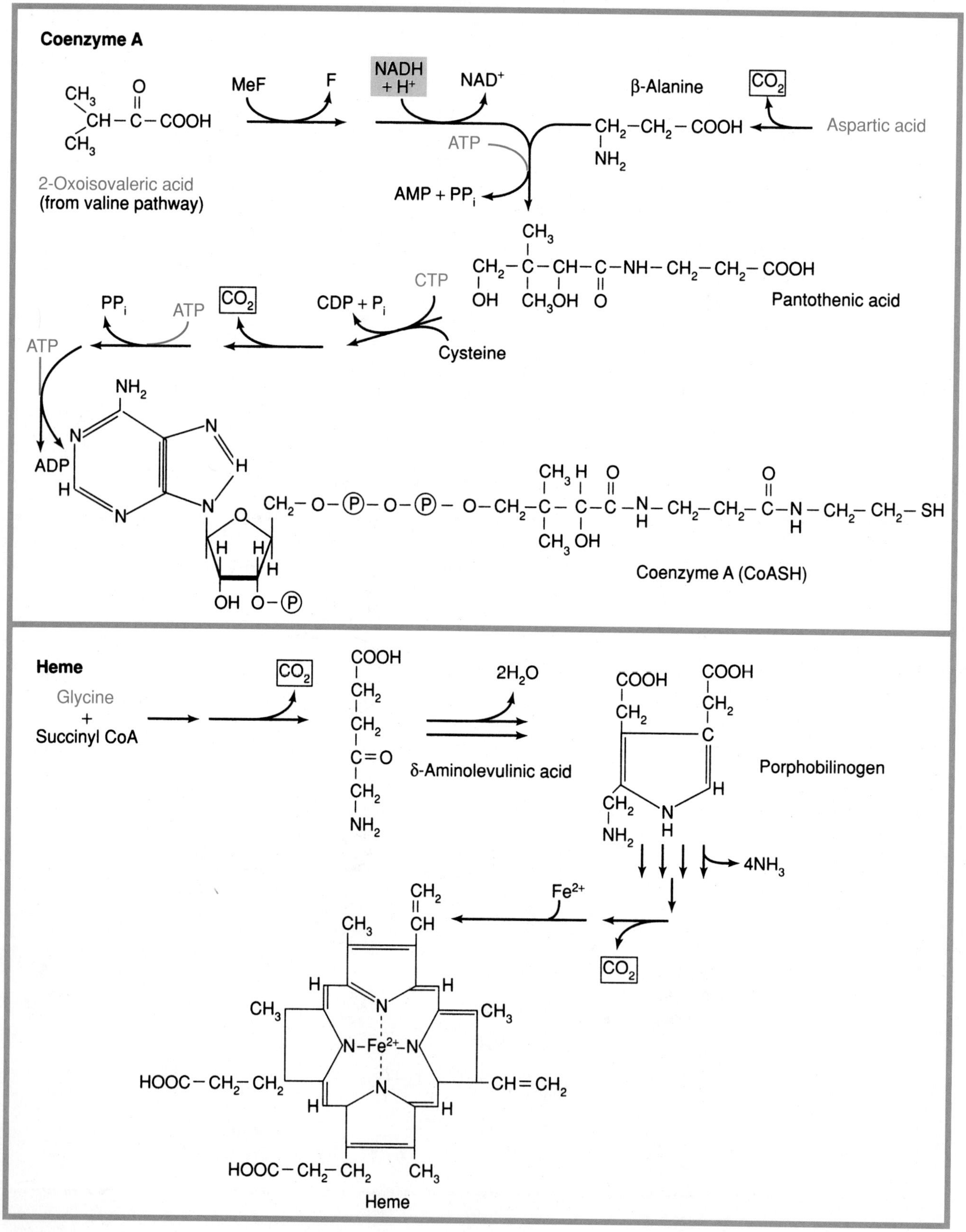

FIGURE 5.11

Biosynthesis of coenzyme A and heme. The individual heme molecules differ according to the chemical modifications introduced into two porphyrin rings of the final structure.

Summary

In addition to their energy-producing capacities, the TCA cycle and the EMP pathway are important to the cell because they also serve as the source of carbon skeletons for a wide variety of molecules necessary for normal cell metabolism. The addition of amino groups to several intermediate compounds that occur in these pathways results in the production of amino acids. The nitrogen atom for an amino group can be taken directly from an ammonium ion (reductive amination) or transferred from another organic molecule (transamination). Once the initial amino acids have been formed, they can serve as precursors for the synthesis of all other amino acids in the cell. The various amino acid biosynthetic pathways also act as starting points for the synthesis of purines, pyrimidines, heme, and enzyme cofactors. The anaplerotic pathway known as the glyoxylate bypass regenerates intermediate molecules of the TCA cycle that are used in biosynthesis.

Autotrophic organisms conduct many of the same biochemical reactions as heterotrophic organisms, but in the cases most studied, their carbon input comes from carbon dioxide that is fixed via the Calvin cycle. The ability of autotrophs to convert phosphoglyceraldehyde from the Calvin cycle into sugars necessary for cell structure exemplifies gluconeogenesis. This process is essential for any cell using molecules other than sugars as a carbon source. If the carbon source for a chemoautotroph is a particularly small molecule such as acetate, the glyoxylate bypass operates at a high level to replenish the TCA cycle intermediates diverted to gluconeogenesis.

Questions for Review and Discussion

1. Why is 2-oxoglutaric acid a critically important compound for biosynthesis? How can it be produced from glucose? From acetic acid?
2. What are the similarities and differences between the reactions of the Calvin cycle and the reactions of reductive carboxylation? Would you expect to find these reactions in a methane oxidizer? Why?
3. Outline three different ways in which a bacterial cell might be able to obtain the nitrogen needed for growth from its environment. Which method do you think would be the most common in nature? Why?
4. Would it make sense to call the amino acid aspartic acid the "head of the family"? Why? Give an example of another family member.
5. What are the similarities and differences between reductive amination reactions and transamination reactions?
6. List all molecules that are part of the initial reactions for the amino acid biosynthetic pathways. Now group them according to the degradative pathway from which they are derived. Based on these groupings, which pathways would you expect to find in most bacteria?
7. Why is the word *dinucleotide* a part of the full names of NAD and FAD even though they are not nucleic acid components? What is the nucleotide in question?

Suggestions for Further Reading

Bedard, C., and R. Knowles. 1989. Physiology, biochemistry, and specific inhibitors of CH_4, NH_4^+, and CO oxidation by methanotrophs and nitrifiers. *Microbiological Reviews* 53:68–84.

Gottschalk, G. 1986. *Bacterial metabolism,* 2d ed. Springer-Verlag, New York.

chapter

7

Biosynthesis of Macromolecules and Membranes

chapter outline

■

GENERAL GOALS

Be able to describe how each macromolecular synthetic process operates biochemically.
Be able to describe the cofactors necessary for the synthetic process.
Be able to describe the sources of energy for the polymerization reactions.

Macromolecules, polymers composed of nearly identical subunits, are essential to any cell. They contain its genetic information, carry out necessary metabolic reactions for it to live and reproduce, and help to protect it from the environment. The multitude of reactions necessary to link subunit molecules into cellular components necessary to accomplish these tasks compose the general subject of this chapter. Ways in which prokaryotes coordinate the synthesis of their DNA, RNA, and protein molecules, as well as how they assemble both the cell membrane and the cell wall are discussed. Although cell membranes are not macromolecules because their subunits are not truly polymerized, they represent a very large and important cell structure; therefore, they are discussed simultaneously for convenience.

DNA Replication

Structural Considerations

The fundamental structure of DNA proposed by James Watson and Francis Crick in 1953 is still considered valid today, although some modifications have been necessary. Inherent in this structure is the idea that DNA molecules consist of two antiparallel chains or strands in which the directionality of each chain is a product of the polarity of the sugar-phosphate bond (figure 6.1). Thus, in a linear, double-strand DNA molecule, each strand of the molecule has one 3′ and one 5′ end oriented oppositely. The width of the duplex is approximately constant because a purine is always paired with a pyrimidine (adenine:thymine, guanine:cytosine).

Although viral DNA may have single or double strands, cellular DNA is nearly always in the double-strand state. All prokaryotes that have been examined, as well as many single- and double-strand viruses, have their DNA molecules in the form of a circle. The general assumption is that a circular DNA molecule, because it has no free ends, is less susceptible to attack by **exonucleases,** enzymes that degrade DNA by attaching to an end and then moving progressively inward. Such enzymes are found commonly in the cytoplasm of prokaryotic cells.

In addition to the helicity predicted by Watson and Crick, most double-strand DNA molecules are also coiled so as to occupy less space within the cell (figure 6.2A). In eukaryotic cells, DNA is organized into true chromosomes in which DNA is wrapped around cylinders of histone proteins to form nucleosomes. Bacteria have been shown to use a different system in which superhelical turns are supplied by enzymes known as **topoisomerases** that break one or both strands of a DNA molecule, rotate the strands with respect to one another, and then reseal the break. The winding process can be either positive (leaves the helix more tightly coiled) or negative (leaves the helix more loosely coiled) and requires the energy of ATP. Only some extremely thermophilic archaebacteria have been reported as having positive supercoils. The DNA molecules now occupy a physically smaller space but are under tension. This condensing process allows a DNA molecule 1 mm in length to fit into an *Escherichia coli* cell that is 2×0.5 μm in size. If a **nick** (a break in a phosphodiester bond) is introduced into either strand, the superhelical twists will disappear spontaneously as the strands rotate relative to each other and "relax" the DNA molecule.

The DNA molecule in *E. coli* has been estimated to have about 43 supercoils, each of which behaves as a semi-independent domain stabilized by certain proteins that hold the loops in place. An electron micrograph of DNA released by gentle cell lysis is depicted in figure 6.2B. A nick introduced into one loop generally will not succeed in relaxing the adjacent loops.

There are some analogies between the chromosome structure of a prokaryote and a eukaryote. For example, both have regions of greater and lesser condensation. In prokaryotes this can be noted easily: some DNA from which RNA is being synthesized is found not within the fibrous nucleoid region (seen in figure 2.1B) but rather immediately adjacent to it where the ribosomes are located. The *E. coli* chromosome also has bound to it many histonelike proteins, even though a true nucleosome structure has not been demonstrated. Predominant among these histonelike proteins is one designated as HU.

Chemical and Enzymatic Considerations

The known structure of a DNA molecule mandates that the replication enzymes have certain characteristics. The **replication fork,** the region in which one double-strand DNA molecule is becoming two, is assumed to move as a unit along the molecule being replicated. If this is so, however, the enzymes doing the replication

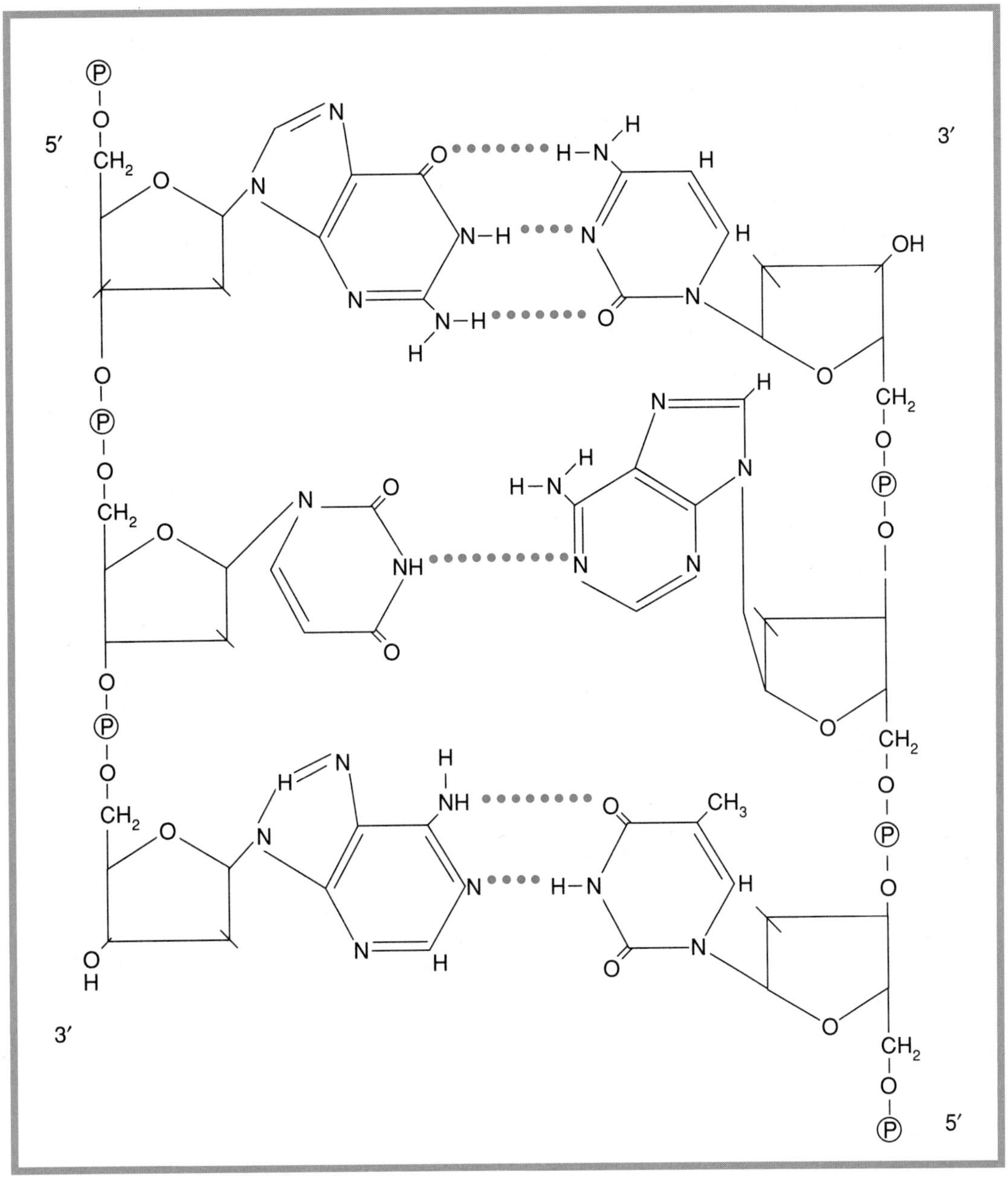

FIGURE 6.1

Chemical polarity in a DNA molecule. The two DNA strands are arranged in an antiparallel fashion so that each end of this linear molecule has one 3′-hydroxyl and one 5′-phosphate group. The base stacking arrangement maintains a constant width of 2 nm for the double helix. Phosphate groups are designated by Ⓟ.

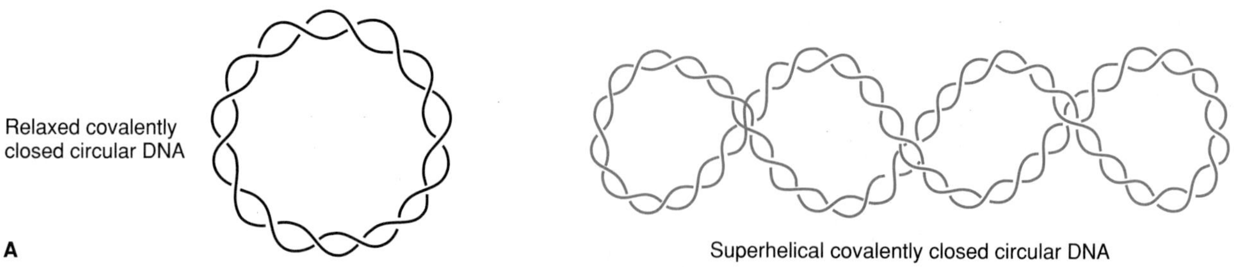

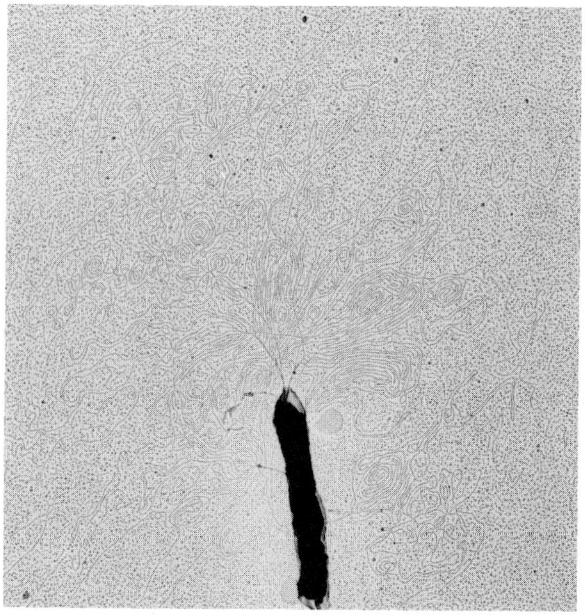

FIGURE 6.2

Superhelical turns in a DNA molecule. (**A**) Shown are two small circular DNA molecules. The molecule on the left is in the relaxed configuration, while the molecule on the right has had several superhelical turns added by the topoisomerase DNA gyrase. (**B**) *Escherichia coli* cell that was gently lysed, causing the nucleoid to spill out from the cell in a partially unwound state.

(A) From E. A. Birge, *Bacterial and Bacteriophage Genetics,* 2d edition. Copyright © 1988 Springer-Verlag, Heidelberg, Germany. Reprinted by permission. (B) Electron micrograph courtesy of Diane Apostolakos.

must deal with strands of opposite chemical polarity. Moreover, as the fork moves along the circular molecule, the superhelical turns must be removed and then later restored or else the unreplicated DNA will be wound ever tighter while the newly replicated DNA will be underwound.

At a minimum, replication requires the presence of one or more topoisomerases and one or more DNA **polymerases,** enzymes that join nucleoside triphosphates into chains of nucleic acid. The particular type of polymerase needed for this reaction is a DNA-dependent DNA polymerase, meaning that the enzyme uses a DNA molecule as a template to produce other DNA molecules. Beginning with the pioneering studies of Arthur Kornberg, several DNA polymerases have been identified in bacteria, more in fact than were originally thought to be necessary for the process. However, no polymerase is capable of synthesizing a new DNA strand along a 3′ to 5′ chemical gradient, neither is any capable of starting its own new strand of DNA (figure 6.3A–B). These observations created a problem in identifying which DNA polymerase actually is used during the replicational process.

It was discovered that several polymerases are involved in the process, but in unexpected ways. Some studies by Reiji Okazaki provided the key to this answer. Okazaki made newly synthesized DNA radioactive by using ^{32}P. He showed that much of the radioactive label was present as relatively short DNA fragments of more or less constant length and that only later were these fragments joined together to make a full-length molecule. If DNA is made in a discontinuous fashion, the problem of chemical polarity can be solved with two DNA polymerases that travel in physically opposite directions, as shown in figure 6.4. In point of fact, it seems likely that the strands of the replication fork can be twisted so as to allow both DNA polymerase molecules to move in the same physical direction (figure 6.4). The remaining problem of "priming" the reaction so that the DNA polymerase enzyme has something to extend can be solved if the second strand starts with a short RNA molecule that is readily synthesized de novo.

According to our present understanding of DNA replication, primarily as studied in *Escherichia coli,* the process begins at a specific location on the bacterial chromosome designated as *oriC,* the origin of chromosomal replication. On *oriC,* a complex of some 20 proteins forms a **primosome** or DNA-dependent RNA polymerase and makes a short piece of RNA complementary to one DNA strand of the double helix. This

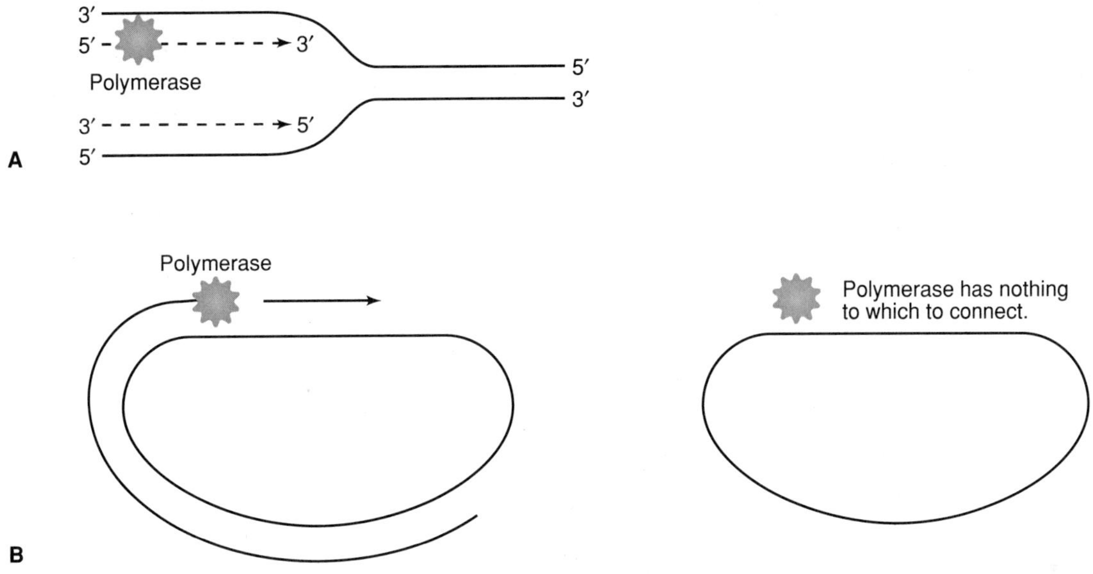

FIGURE 6.3

Potential problems in DNA replication. (**A**) The chemical polarity of the DNA strands means that two molecules of the same enzyme cannot travel along both DNA strands in the same direction. Dotted lines represent DNA to be synthesized. (**B**) All known DNA polymerases can extend the incomplete DNA strand in the structure on the left, but none of them is capable of starting a new strand from nothing, as in the structure on the right.

APPLICATIONS BOX

Making DNA

DNA replication occurs in vivo (within a living cell) and produces duplicate molecules having all the normal three-dimensional structures. By contrast, the term *DNA synthesis* often is used to refer to a process that occurs in vitro (within an artificial vessel like a test tube). DNA synthesis often implies that one or more aspects of the normal process has been omitted for purposes of simplicity.

The problems inherent in studying a multicomponent enzyme system are not trivial. Removal of a single, structurally significant component may completely destroy the enzymatic activity of the entire enzyme complex. On the other hand, there may be sufficient redundancy built into the various proteins to allow removal of a single protein with very little effect. Oftentimes, the ionic environment critically affects the way in which enzymes function. Because the local ionic environment within the cytoplasm is not really known, generally it is difficult to evaluate whether a particular experimenter's choice of experimental conditions is truly appropriate.

piece of RNA is always oriented so that it has a free 5′ end. It serves to allow a DNA polymerase III complex to bind and extend the RNA primer from its 3′ end. New deoxyribonucleotides are added at a rate greater than 1,000 bases per second, obeying the usual base-pairing rules. The DNA polymerase complex includes a 3′ $\longrightarrow$ 5′ exonuclease activity that is equivalent to a "backspace" on a computer; therefore, if a mistake is made, the enzyme reverses direction and removes the offending base. Another reversal of direction then resumes replication in the normal 5′ $\longrightarrow$ 3′ direction. Energy for the phosphodiester bonds is inherent in the deoxyribonucleoside triphosphates used as substrates and is, of course, derived directly or indirectly from ATP. The first DNA strand to be synthesized at a replication fork is called the **leading strand.**

Extension of the primer causes an unwinding of the DNA helix, making accessible the DNA strand with opposite chemical polarity. A primer RNA molecule, perhaps only four or five bases long, is inserted with opposite orientation to the original primer and across from a region of the original helix that has already been replicated. DNA synthesis now proceeds back toward the original start point. For obvious reasons, this new

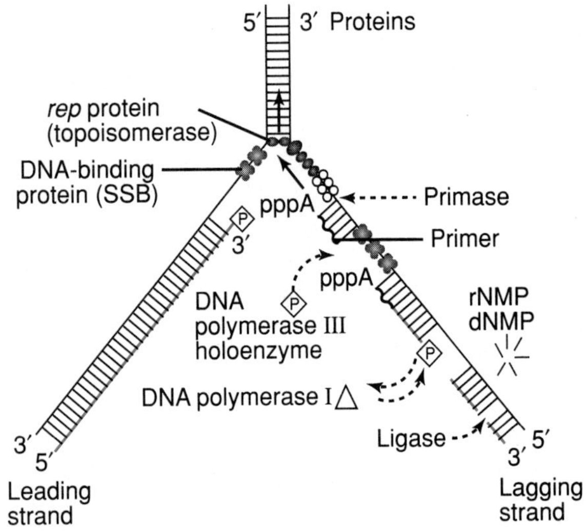

FIGURE 6.4

DNA replication as it occurs at a single replication fork. RNA primers are represented by small wavy lines. Newly synthesized DNA is rendered in color, and the original DNA is shown in black. The DNA polymerase III complex is shown as a diamond, while DNA polymerase I is represented by a triangle. rNMP is ribonucleoside monophosphate, and dNMP is deoxynucleotide monophosphate.

From A. Kornberg, "Enzyme Studies of Replication of the *Escherichia coli* Syndrome," in *Proteins Involved in DNA Replication,* edited by U. Hubscher and S. Spadari. Copyright © 1984 Plenum Publishing Corporation, New York, NY. Reprinted by permission.

strand is called the **lagging strand.** Clearly, the lagging strand must be synthesized discontinuously, which is not necessarily true for the leading strand. Technical difficulties involving the breaking of chemical bonds during normal DNA repair prevent experimenters from determining unambiguously whether the leading strand is synthesized in large segments or in one continuous process. As the initial Okazaki fragment of the lagging strand is synthesized past the *oriC* locus, it becomes the leading strand of a second replication fork and triggers bidirectional replication.

Completion of the actual DNA synthetic process requires the removal of RNA primers and the joining of DNA fragments, a process analagous to DNA repair. DNA polymerase I is the major enzyme involved in the removal process. It has both 3′ → 5′ and 5′ → 3′ exonuclease activities; therefore, the enzyme can remove bases from DNA while moving in either chemical direction. It also has the ability to lay down fresh DNA simultaneously while using the 5′ → 3′ exonuclease activity. The net effect of DNA polymerase I exonuclease and polymerase activities is that the enzyme binds to the 5′ end of the RNA attached to an Okazaki fragment, uses the adjacent DNA as a primer, and replaces the RNA bases with corresponding DNA bases (figure 6.4). In this process called **nick translation,** the position of a nick changes as a result of polymerase action, but nicks are neither removed nor added. The resulting all-DNA fragments are joined by the enzyme DNA ligase to make a single long strand of DNA.

Topoisomerase activity is required during all phases of DNA replication. In order for the replication fork to progress, **helicase** enzymes, such as DnaB, must locally unwind the DNA double helix. This unwinding puts extra tension on the remainder of the DNA molecule, which must be relaxed by other topoisomerases. Conversely, after a region of DNA has been replicated, it is necessary to introduce the normal supercoiled configuration of DNA with the aid of ATP and a topoisomerase like DNA gyrase. At this time, the way in which proteins act to stabilize the superhelical bacterial nucleoid is not well understood. The final product of DNA replication is two **semiconserved** molecules each having one old and one new strand.

Replication of single-strand DNA from viruses is different from replication of double-strand DNA. A discussion of that process is included with viral metabolism in chapter 9.

DNA Transcription into RNA

Structural Considerations

In **transcription,** a DNA-dependent RNA polymerase makes a single strand of RNA that is a copy of the base sequence of one of the two strands of a DNA double helix. RNA polymerase substitutes uracil for thymine and uses ribose-containing nucleotides instead of deoxyribose-containing ones. The size of the resulting RNA molecule is variable, ranging from a few hundred to a thousand or more bases. The same chemical polarity considerations outlined for DNA also apply to RNA; therefore, if one compares an RNA molecule to the DNA molecule from which it was transcribed, one strand is **complementary** (opposite polarity, purines across from pyrimidines) and the other is noncomplementary (same polarity, purines across from purines, pyrimidines across from pyrimidines).

Because nucleic acid synthesis operates on the principle of base pairing, the complementary DNA strand must be the **sense strand,** the one copied to make that particular RNA. For any given RNA molecule, there is only one DNA sense strand, but either DNA strand in the double helix can be used to synthesize RNA, at least along some portion of its length.

Chemical and Enzymatic Considerations

The RNA polymerase that carries out transcription is not a single protein but rather a protein complex. Several protein chains (two copies of α, one of β, and one of β') come together to form an RNA polymerase **apoenzyme,** an enzyme capable of synthesizing RNA although it does not have the normal specificity associated with the complex. The addition of any one of several other proteins called sigma factors (σ-factors), which act to confer DNA-binding specificity on the apoenzyme, results in the production of RNA polymerase **holoenzyme,** a complex that is fully functional in the biological sense (figure 6.5).

The existence of many different RNA transcripts for a given DNA molecule implies that DNA must contain numerous start and stop sites for RNA polymerase. Start sites are more commonly known as **promoters,** a term coined by Jacques Monod and Agnes Ullman. The holoenzyme binds to DNA at a promoter site in an oriented fashion, and recognition of any given promoter is a function of the σ-factor associated with the holoenzyme.

Actual RNA synthesis occurs in a very similar fashion to that of DNA. Nucleoside triphosphates are paired with the corresponding bases in the template strand and then polymerized at a rate of about four bases per second. Synthesis is continuous (nothing equivalent to Okazaki fragments) and again oriented in the $5' \longrightarrow 3'$ direction. This means that all polymerase enzymes are moving in the same direction on a given DNA strand, which presumably serves to minimize their interference with each other. The source of energy for the phosphodiester bonds is once again the nucleoside triphosphates themselves.

Normal termination of transcription occurs at specific signals within the DNA base sequence. Generally speaking, sequences of RNA that can pair intramolecularly to form a hairpin loop of the appropriate shape are likely candidates for termination signals. These can occur within protein-coding regions of the sequence or in spacer regions between coding segments. In order for termination to occur, it is necessary for RNA polymerase to pause briefly while the termination loop forms. Some termination signals require the presence of a specific protein called rho (ρ-factor) on the RNA polymerase holoenzyme before they can be recognized.

Eukaryotic cells carry out extensive RNA processing following transcription due to the dispersed arrangement of their genetic information. mRNA molecules must be spliced to link together **exons,** regions coding for parts of a specific protein, and to remove **introns,** noncoding regions. Moreover, a special "cap" of modified bases must be added to the 5′ end of an mRNA molecule before it can be translated, and, often, a tail of some 200 to 300 adenyl residues must be added to the 3′ end. In bacteria, only continuous DNA coding regions have been found; therefore, mRNA in prokaryotes does not require the splicing out of introns to give the proper code for a protein (this is not true for all bacterial viruses). Because prokaryotic mRNA does not need any processing, translation and transcription can be coupled, but both ribosomal RNA (rRNA) and transfer RNA (tRNA) require processing before use.

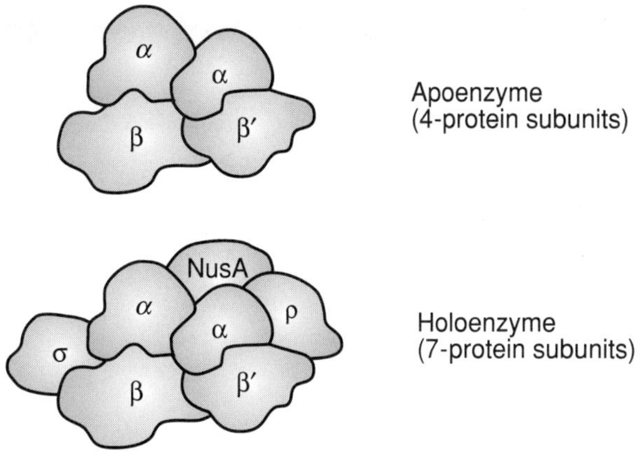

FIGURE 6.5

Formation of the RNA polymerase holoenzyme complex. The α, β, and β' proteins have the core enzyme activity. The extra proteins are responsible for the ability of the complex to recognize signals for start (σ) and stop (ρ and NusA).

In all prokaryotes, rRNA molecules are produced as a single long transcript that later must be cut apart to yield individual 16S, 23S, and 5S molecules. Sandwiched between the 16S and 23S species of RNA can be one or more discrete tRNA molecules that also must be cut out (figure 6.6). Occasionally, tRNA molecules are found at the ends of rRNA transcripts. In one instance, even an mRNA can be found within the rRNA transcript or a tRNA within an mRNA. The processing that supplies individual RNA molecules is carried out by several different ribonuclease molecules, RNase III being the best studied. This endonuclease cuts at specific sites within the large transcript to yield smaller molecules that can be processed further.

Even more processing is required for tRNA molecules. They are often transcribed in groups, ranging in size from three to five different tRNAs in *E. coli* to 20 in *Bacillus subtilis*. Each group must be cut apart to give individual molecules, which are then appropriately modified to make them functional. In the case of

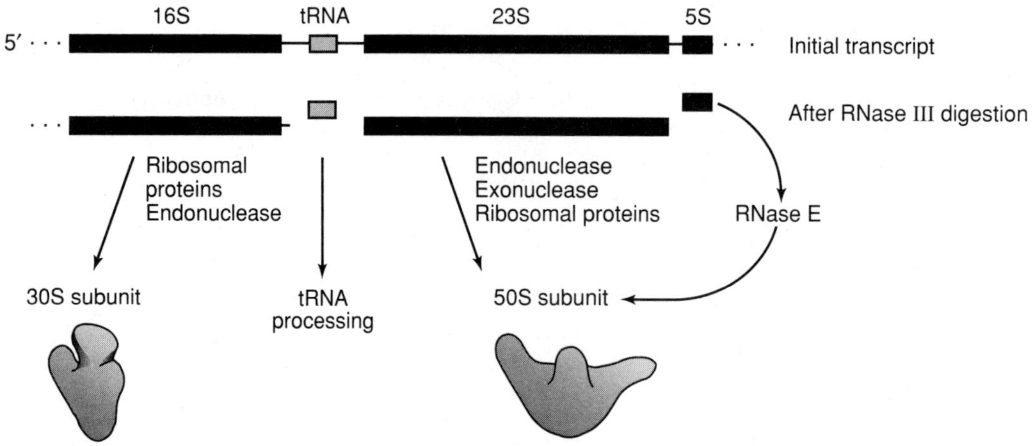

FIGURE 6.6

Processing of rRNA. First the transcript is cut into discrete pieces, and then nucleases trim the pieces to size. Ribosomal proteins spontaneously add to the processed RNA (black bars).

tRNA molecules produced by most bacteria, the modification consists of the addition of CCA bases, common to all tRNA molecules, to the 3′ end of each molecule. Other bases within the original transcript are chemically modified as well. For example, some uridine residues are converted to pseudouridine. Processed tRNA molecules are ready to carry out their cellular functions.

Protein Synthesis

Chemical Considerations

The synthesis of a protein requires that information inherent in the base sequence of DNA be converted into mRNA and then used to order a sequence of amino acids during the formation of a protein. **Translation** is the mechanism by which information contained in the base sequence of mRNA is converted into a chain of amino acids. This mechanism might seem to require the interaction of amino acids with an mRNA molecule, but Francis Crick pointed out that it is difficult to imagine how this might be accomplished with any precision. Instead, he proposed the concept of an adapter molecule to mediate between mRNA and the individual amino acids. We know these adapter molecules today as tRNA molecules. A tRNA molecule carries an amino acid covalently attached to its 3′ end and has in its middle an **anticodon,** a region that can form hydrogen bonds with a short sequence of bases on an mRNA molecule. The two- and three-dimensional structure of a typical tRNA molecule is illustrated in figure 6.7.

Each individual amino acid has at least one tRNA molecule specific to it, and some of the more commonly used amino acids may have four or five different species of tRNA specific to them. In such a case, however, members of the different tRNA species are not necessarily present in equal numbers. The specificity of amino acid attachment is governed by aminoacyl-tRNA synthetases, enzymes that catalyze the following reaction:

Equation 6.1

$$\text{Amino acid} + \text{ATP} + \text{tRNA} \xrightarrow{\text{synthetase}} \text{aminoacyl-tRNA} + \text{PP}_i + \text{AMP}$$

The reaction results in the hydrolysis of ATP to yield pyrophosphate (PP_i) rather than phosphate, which transfers a considerable amount of energy to the aminoacyl-tRNA produced. This reaction is therefore sometimes known as a charging or activating reaction.

Functional Considerations

The actual site of protein synthesis is the ribosome, a complex of RNA and protein described in chapter 2. Generally speaking, the small ribosomal subunit is concerned with translation of mRNA, while the large subunit is concerned with joining amino acids together and moving the complex along the mRNA molecule.

The genetic code used by prokaryotes is the same as that used by eukaryotes for cytoplasmic protein synthesis. Figure 6.8 provides a summary of this code. The

APPLICATIONS BOX

Molecular Biology of the Archaebacteria

The molecular biology of the archaebacteria is only now beginning to be understood. Already, however, it is becoming clear that archaebacteria constitute a heterogeneous grouping. For example, as mentioned earlier, the extremely thermophilic archaebacteria have DNA with positive supercoils instead of the negative supercoils found in all other organisms (including other archaebacteria). Another interesting observation is that certain archaebacterial tRNA transcripts contain introns (extra bases not found in mature tRNA molecules). These introns are located near the anticodon and must be removed during tRNA processing by a splicing mechanism similar to that used for mRNA and tRNA in eukaryotes (see diagram). The location of the introns is very similar to the position of introns found in eukaryotic tRNA transcripts. This finding, coupled with the original discoveries of similarities in rRNA sequence, gives added strength to the argument that archaebacteria may be more closely related to eukaryotes than to the rest of the prokaryotes. It also blurs what was considered to be a distinction between prokaryotes and eukaryotes: the presence or absence of RNA splicing. At this time, however, no mRNA splicing has been observed in any bacterial cell not infected with a virus so that trait can still serve to distinguish between the two groups.

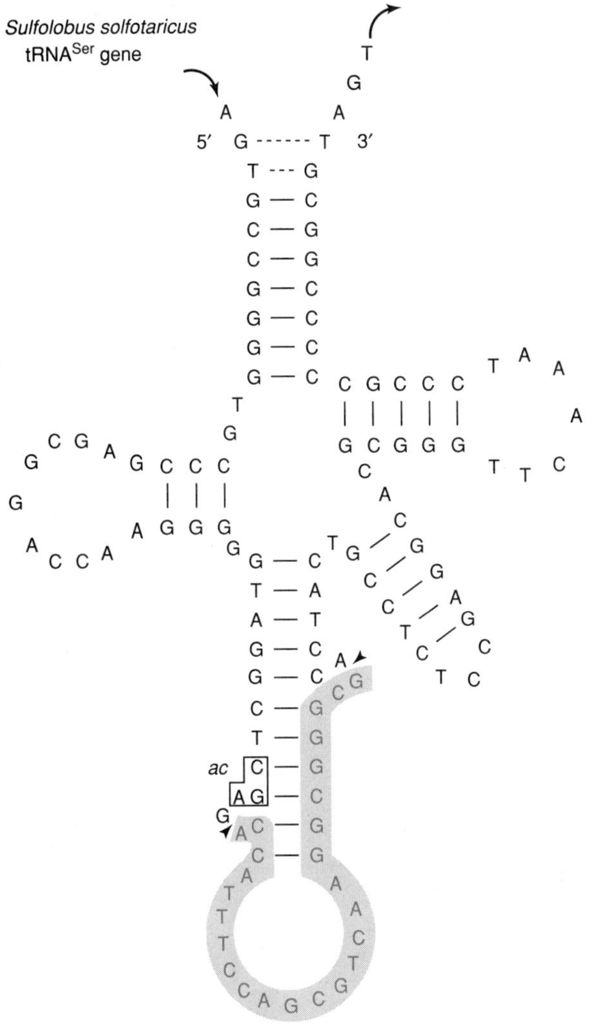

tRNA processing in archaebacteria. The sequence shown is the original transcription product. Shaded bases must be removed and other bases added at the 3′ end to give the mature tRNA molecule. The anticodon is boxed.

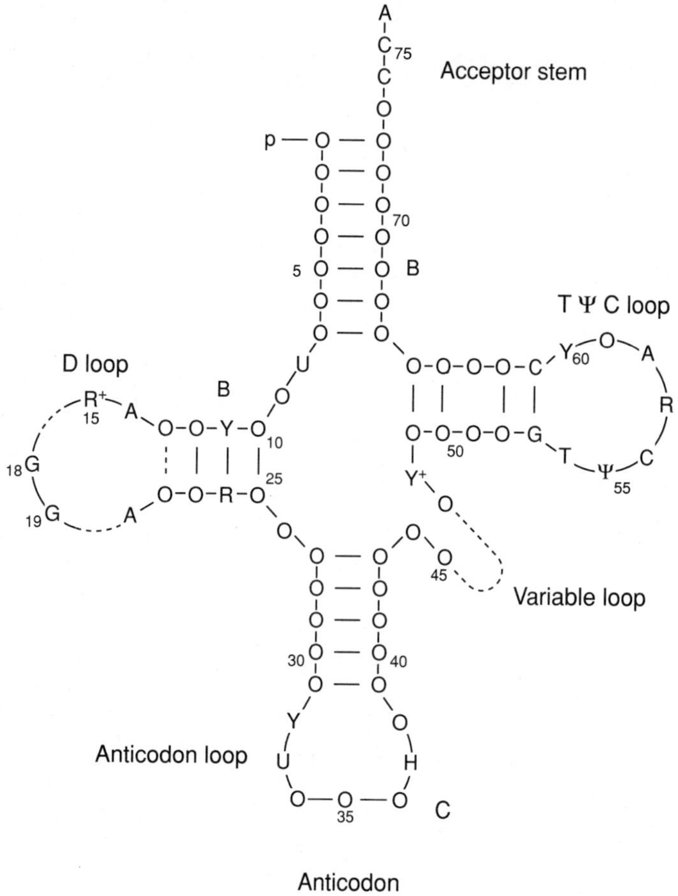

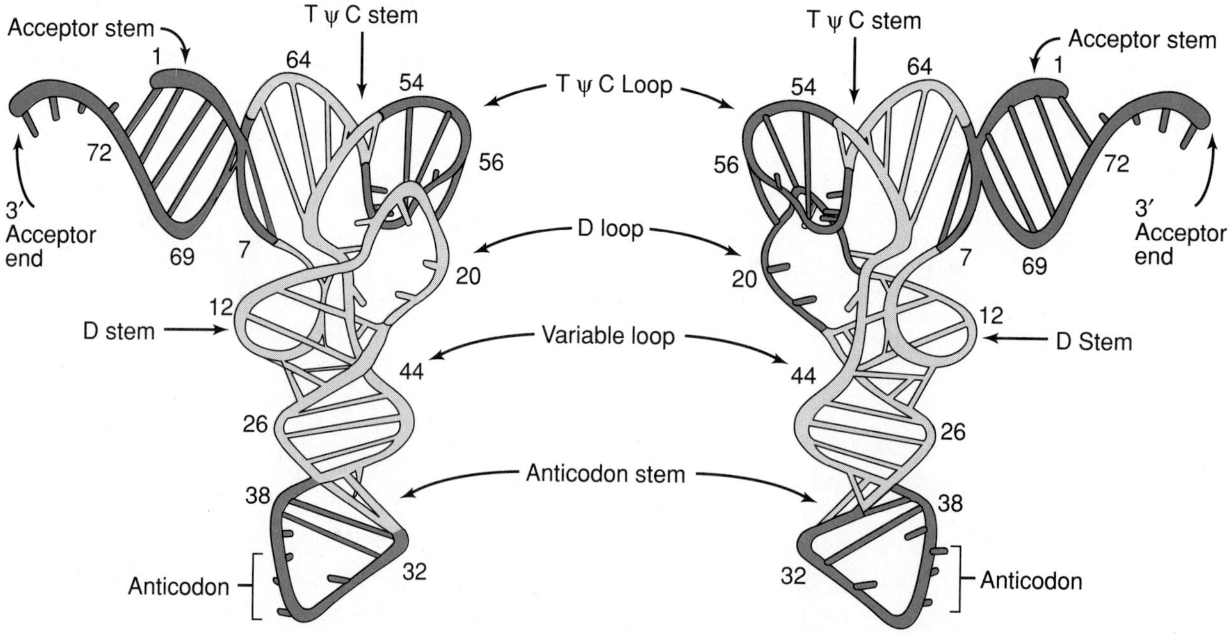

General structures of a tRNA molecule. Note the conventional cloverleaf representation of a typical molecule in the first diagram. Each of the three loops has a specific function. The D loop is responsible for binding the molecule to the ribosome. The areas designated by B are responsible for allowing only one kind of amino acid to be attached. The anticodon loop pairs with the codon on the mRNA molecule during translation, hence it is known as the anticodon loop. The second diagram depicts the same tRNA molecule folded into a three-dimensional L-shaped configuration.

First base \ Second base	U	A		C	G		A	T		G	C	
UA	UUU	AAA	Phe	UCU	AGA	Ser	UAU	ATA	Tyr	UGU	ACA	Cys
	UUC	AAG		UCC	AGG		UAC	ATG		UGC	ACG	
	UUA	AAT	Leu	UCA	AGT		UAA	ATT	Term	UGA	ACT	Term
	UUG	AAC		UCG	AGC		UAG	ATC		UGG	ACC	Trp
CG	CUU	GAA		CCU	GGA	Pro	CAU	GTA	His	CGU	GCA	Arg
	CUC	GAG		CCC	GGG		CAC	GTG		CGC	GCG	
	CUA	GAT		CCA	GGT		CAA	GTT	Gln	CGA	GCT	
	CUG	GAC		CCG	GGC		CAG	GTC		CGG	GCC	
AT	AUU	TAA	Ile	ACU	TGA	Thr	AAU	TTA	Asn	AGU	TCA	Ser
	AUC	TAG		ACC	TGG		AAC	TTG		AGC	TCG	
	AUA	TAT		ACA	TGT		AAA	TTT	Lys	AGA	TCT	Arg
	AUG	TAC	Met	ACG	TGC		AAG	TTC		AGG	TCC	
GC	GUU	CAA	Val	GCU	CGA	Ala	GAU	CTA	Asp	GGU	CCA	Gly
	GUC	CAG		GCC	CGG		GAC	CTG		GGC	CCG	
	GUA	CAT		GCA	CGT		GAA	CTT	Glt	GGA	CCT	
	GUG	CAC		GCG	CGC		GAG	CTC		GGG	CCC	

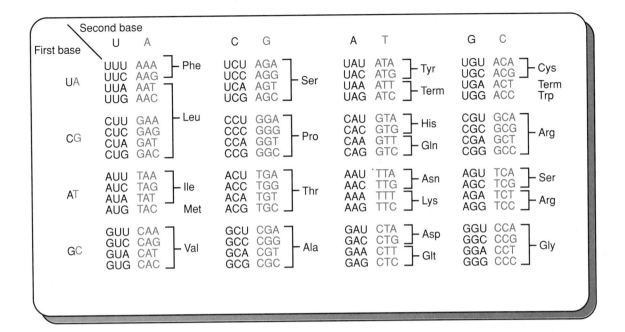

FIGURE 6.8
Genetic code. Each triplet of bases is a codon or unit of code. Abbreviations for the bases in the triplets are: adenine (A), cytosine (C), guanine (G), thymine (T), and uracil (U). The code used for mRNA is shown in black, and the corresponding code in the sense strand of a DNA molecule is shown in color. The first base is the 5′ end of the RNA codon but the 3′ end of the DNA codon. Abbreviations are: alanine (Ala), arginine (Arg), asparagine (Asn), aspartic acid (Asp), cysteine (Cys), histidine (His), glutamine (Gln), glutamic acid (Glt), glycine (Gly), isoleucine (Ile), leucine (Leu), lysine (Lys), methionine (Met), phenylalanine (Phe), proline (Pro), serine (Ser), terminator (Term), threonine (Thr), tryptophan (Trp), tyrosine (Tyr), and valine (Val).

signal for attachment of the ribosome to mRNA is a region called the Shine-Dalgarno box. Studies of DNA sequences indicate that the most common attachment sequence is 3′TAAGGA5′, which is basically the same as the 3′ end of 16S rRNA. The mRNA copied from such a DNA sequence will be complementary to the 3′ end of the 16S RNA molecule, therefore; and proper alignment of the ribosomes with the end of the mRNA will be automatic. The translational process actually starts at an **initiator codon,** usually methionine, and ends at one of the three terminator codons listed in figure 6.8.

Translation begins when the 30S ribosomal subunit binds to a Shine-Dalgarno sequence on mRNA in the presence of protein "initiation factors." The 50S ribosomal subunit then is added, the ribosome moves to the initiator codon, and the translational process is ready to begin. The first tRNA molecule used is a special one that carries an amino acid chemically modified or "blocked" on its amino group. It, therefore, can be bonded only to another amino acid through its carboxyl group. This tRNA molecule binds to the site on the ribosome where the growing peptide will be located (P site, figure 6.9A). Its anticodon loop (figure 6.7) complements the initiator codon of the mRNA, and the interaction is strengthened by the ribosome itself. A second tRNA molecule then arrives whose anticodon matches the second codon of mRNA (figure 6.9B). This second tRNA molecule binds first to activated elongation factor Tu (unstable at high temperature). The L-shaped, three-dimensional structure of tRNA molecules allows them to fit side by side into a space the width of a codon (1 nm). Elongation factor Tu is reactivated by a reaction with elongation factor Ts (temperature stable) and GTP.

Binding of tRNA molecules to the ribosome can occur whether or not they carry activated amino acids. If no amino acids are present, however, protein synthesis cannot proceed until the uncharged tRNA molecule is exchanged for a charged one. When two amino acids are adjacent to one another, the peptidyl transferase enzyme from the 50S ribosomal subunit, in conjunction with elongation factor EF-G, joins them

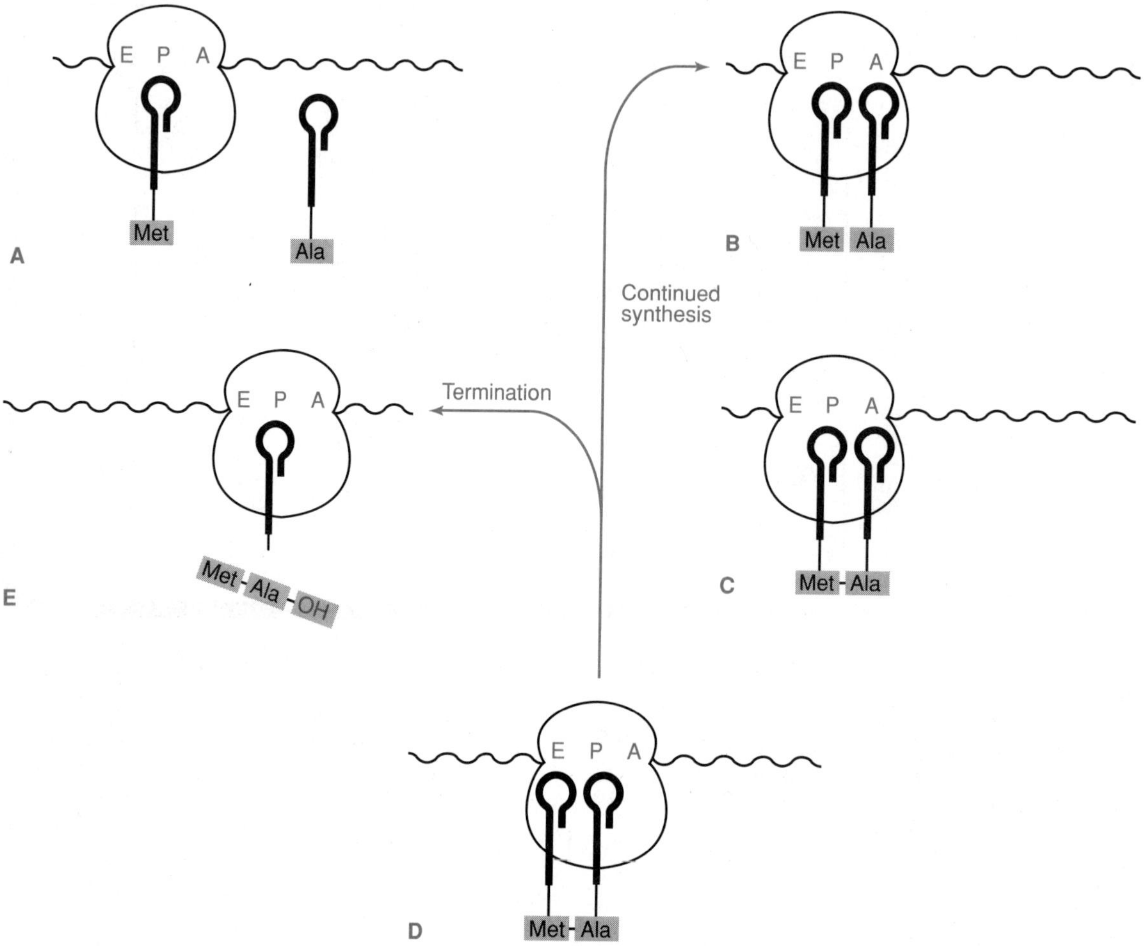

FIGURE 6.9

Role of the ribosome in translation. (**A**) During the initiation step, a tRNA molecule carrying a blocked amino acid binds directly to the P site. (**B**) The A site is then occupied by the tRNA carrying the anticodon corresponding to the next codon. (**C**) A peptide bond is formed, and the first tRNA molecule is released from the nascent chain. (**D**) Translocation moves the tRNA molecule attached to the nascent peptide chain into the P site and the discharged tRNA into the E site. The A site is ready to receive another tRNA molecule so that the cycle can repeat itself. When the A site is refilled, the E site is emptied. (**E**) When a stop signal is encountered, peptidyl transferase hydrolizes the bond connecting the tRNA molecule and the protein, and the separated molecules are released from the ribosome.

together as part of the growing peptide chain (figure 6.9C). At this time, the tRNA in the P site is moved to a third site designated as the E (exit) site, the tRNA formerly in the A site (together with its attached peptide chain) is moved into the P site, and the A site is left open (figure 6.9D). This process of **translocation** is powered by the hydrolysis of another GTP molecule. The three-site (EPA) model for the ribosome has been shown to be applicable to eubacteria and to archaebacteria but is still controversial with respect to eukaryotic organisms.

Because the tRNA molecules are assumed to still be paired with their respective codons, translocation causes the A site on the ribosome to move to the next codon on mRNA. The binding of a tRNA molecule to the A site then causes the release of the tRNA molecule from the E site. This cycle is repeated at the rate of about 17 amino acid residues per second, and translation continues until a terminator or stop codon is encountered (figure 6.9E). Then the completed protein is released from the complex as the peptidyl transferase enzyme adds a water molecule instead of an amino acid to the nascent chain. Release of the nascent peptide chain requires the presence of specific protein release factors.

Eukaryotic mRNA molecules are fundamentally monoinformational, but those from prokaryotes often contain information to synthesize more than one protein. If such information is present, the ribosome starts to make another protein. If not, the ribosome, with the aid of ribosomal release factor, has its tRNA molecules stripped off, is released from mRNA, and dissociates into subunits until needed again.

As soon as the 5′ end of mRNA has been cleared by the first ribosome, another takes its place. Thus, the mRNA molecule is covered with ribosomes at all times, the combination of mRNA and ribosomes being called a **polysome.** Untranslated mRNA is attacked generally by various endonucleases that cut bonds at multiple points, preventing ribosomes from reattaching and completing synthesis at a later time. Prokaryotic cells, therefore, unlike many eukaryotic cells, do not accumulate untranslated mRNA molecules.

The product of translation is a protein whose amino end must be unblocked. This can be accomplished by the removal of either the blocking group or the entire amino acid. In most species that have been examined, the blocking group is removed. Thus, about 70% of the proteins in *Escherichia coli* have methionine as their N-terminal amino acid.

Concerning the combined actions of replication, transcription, and translation, the predominant impression should be one of intense activity. Unlike eukaryotic cells that are compartmentalized by a nuclear membrane, prokaryotic cells have all three processes occurring at the same time and place. Transcription can occur whenever the replication complex is not physically present at a site. As soon as the first portion of an mRNA molecule has been synthesized, ribosomes will attach and begin translation; therefore, the first part of a protein may be translated before the last part of the mRNA has even been synthesized.

Synthesis of mRNA is driven obviously by the intrinsic hydrolysis of phosphate bonds of nucleotides. However, energy for the protein synthetic process also is derived from similar hydrolytic reactions. Energy for peptide bond formation comes from the ATP used to attach the amino acid to the tRNA molecule, and that used for the translocational step comes from the hydrolysis of two GTP molecules by the ribosome. Protein synthesis is an energy-expensive process and, as is discussed in the next chapter, highly regulated to avoid wasting energy.

Export of a Cellular Protein

All cells initiate synthesis of proteins in their cytoplasm, although some proteins actually function outside the plasma membrane. In gram-positive bacteria, these proteins usually are released into the surrounding medium, while in gram-negative bacteria, they are found commonly in the periplasmic areas of the peptidoglycan layer or in the outer membrane. Because such proteins are observed, the cell must have a mechanism for identifying them as export items and arranging for their specific exit from the cytoplasm. In other words, there cannot be a general protein pore in the cell membrane; otherwise all types of proteins would be found outside the cell.

A model currently used to explain protein-export observations in prokaryotes is a variation on the signal hypothesis first proposed for eukaryotic cells (figure 6.10). In this model, it is assumed that a protein to be exported is synthesized by ribosomes in the cytoplasm, but the protein has extra amino acids at the end first synthesized (amino terminus) that are not necessary for protein function. These extra amino acids always are arranged so as to form a short chain having a positively charged amino terminus, a hydrophobic central region, and polar groups at the point where the short chain joins with the main body of the protein. Taken together, the extra amino acids constitute a **signal peptide,** marking the protein as one designated for export. After protein synthesis has proceeded for some time and the signal sequence is complete, a "signal-recognition" protein binds to the growing protein and causes the ribosome to pause its synthetic activities.

The entire complex of growing peptide, ribosome, and signal-recognition protein travels to a membrane-associated protein called SecY, a **docking protein** that then binds the ribosome-protein complex. The hydrophobic signal sequence causes the physical interaction of the elongating peptide with the membrane. Either during protein synthesis or after synthesis completion, the entire protein is translocated to the other side of the cell membrane. The signal peptide also is removed

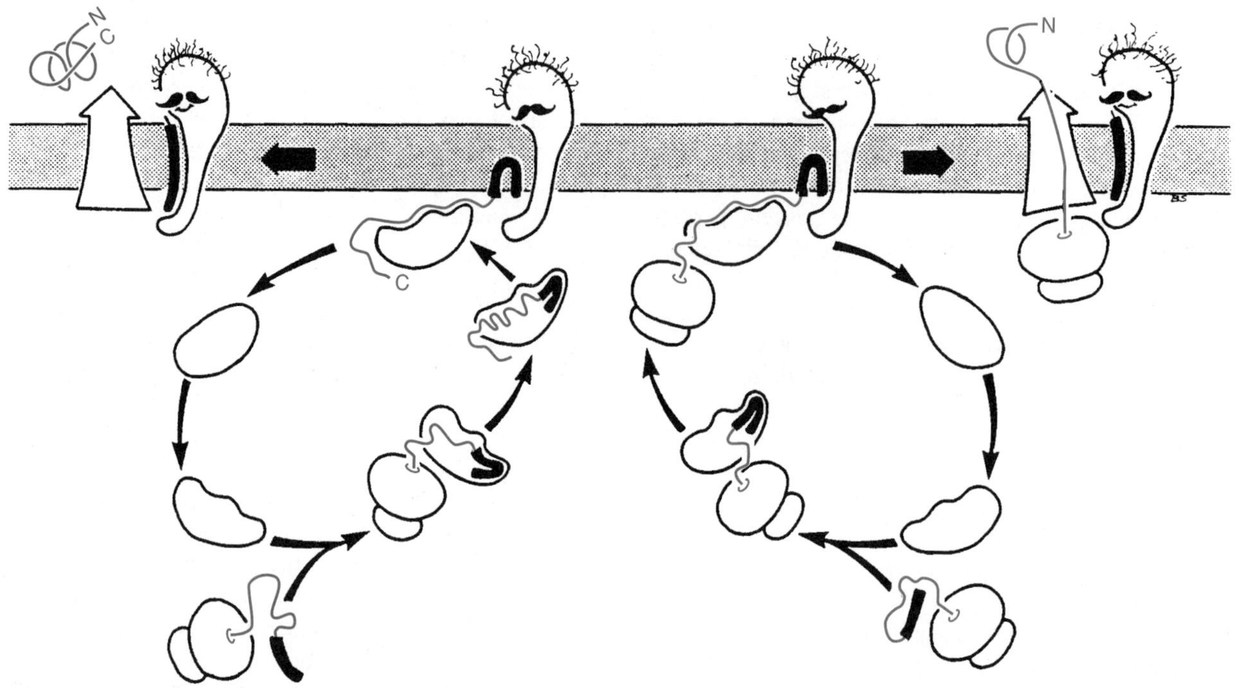

FIGURE 6.10

Models for protein export. The stacked disks with a central hole represent 70S ribosomes. The emerging filament (in color) is the nascent protein chain with its signal sequence indicated by a thicker line. As the protein chain elongates, it associates with a carrier protein that prevents excessive folding and assists in conveying it to the SecY (docking protein) located within the cell membrane. At the cell membrane, the protein is exported, and a leader peptidase (not shown) removes the leader peptide either during export or immediately thereafter. Once the protein has crossed the membrane, it can fold into its active configuration. In the left half of the diagram, synthesis of the protein chain has been completed before the leader peptide inserts into the membrane. In the right half of the diagram, synthesis of the protein chain is not completed until after the removal of the leader peptide.

Diagram courtesy of L. L. Randall and S. J. S. Hardy, Washington State University and University of York.

in many cases. The mechanism for the translocational step is still unknown, as are the mechanisms by which gram-negative cells manage to move some proteins to the surface of their outer membrane.

Polysaccharide Biosynthesis: Macromolecules without Templates

Protein and RNA molecules are synthesized by enzymes that are not specific to the amino acid or nucleotide sequence being synthesized. Rather, these enzymes derive their specificity from the template being used (DNA or RNA), and the same enzymes can produce any protein or RNA molecule. By way of contrast, enzymes that synthesize polysaccharides are sequence specific. In order to produce a radically different polysaccharide, an entirely new set of enzymes is required. The principles by which these different enzymes operate are fundamentally similar, however.

Energy for glycosidic bond formation comes from any of the ribonucleotides in the form of a sugar-nucleoside diphosphate derivative. For example, the following reaction yields an energetic glucose molecule:

Equation 6.2
Glucose + ATP $\rightarrow$ ADP-glucose + P_i

Similar reactions can occur with uridine triphosphate (UTP) and cytidine triphosphate (CTP). To make a **homopolymer** (polysaccharide with only one type of sugar) then requires only one enzyme to add repeating subunits. If the polysaccharide is a branched one, a separate branching enzyme is required to add on the polysaccharide arms (figure 6.11).

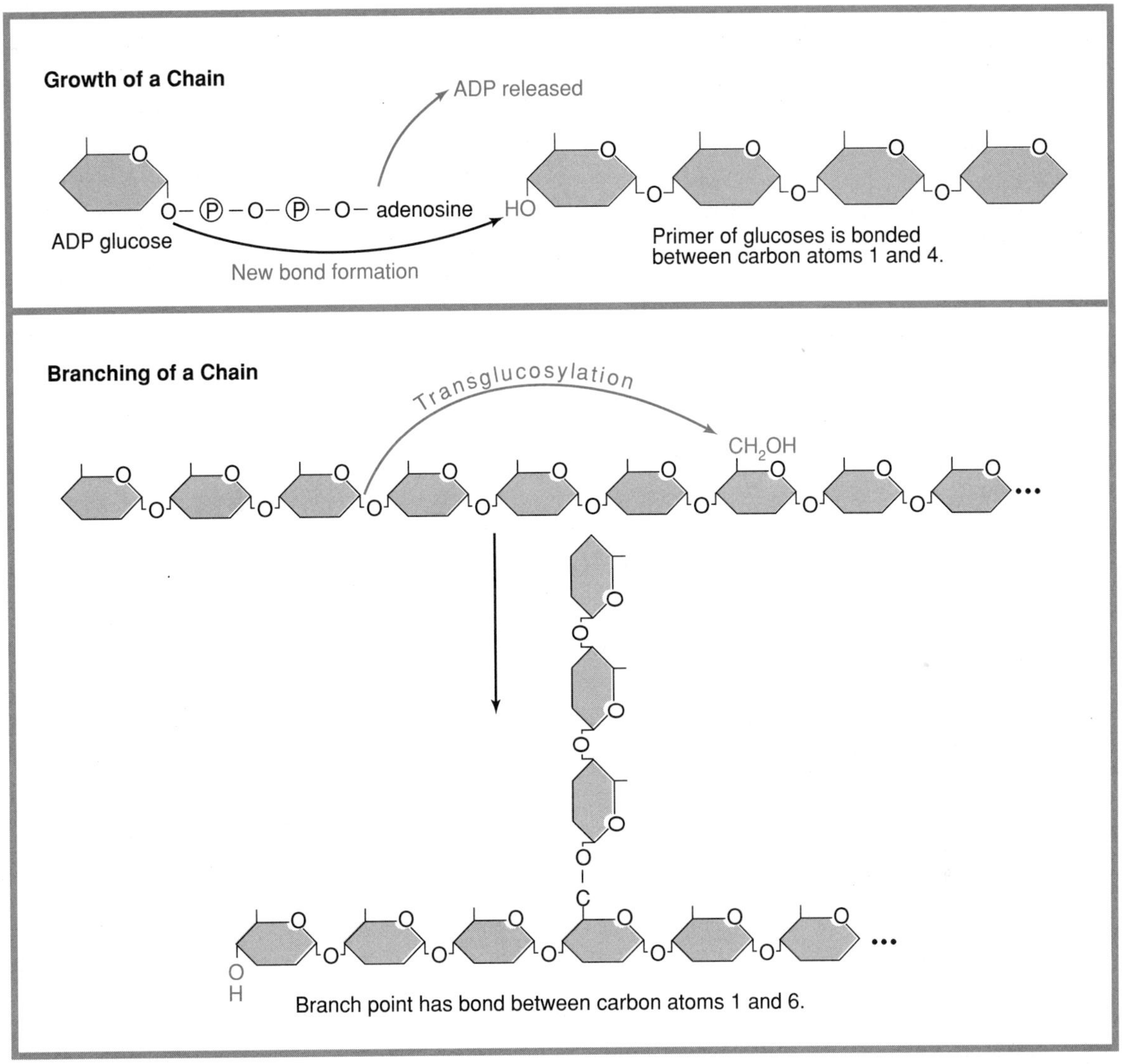

FIGURE 6.11

Synthesis of a branched homopolysaccharide. One enzyme is used to lengthen the chains and another to introduce the branch points that can be lengthened later.

More complex polysaccharides require additional enzymes that must act in the proper order, but the essentials of the system are the same. As an example, consider the synthesis of a glycerol teichoic acid used as a structural element in the gram-positive cell wall (figure 3.7B). The glycerophosphate moiety is activated by a reaction with CTP to yield CDP-glycerophosphate and inorganic pyrophosphate. UDP sugars are added by one enzyme, and glycerophosphates are polymerized by another enzyme complex.

Nearly every eubacterium has a cell wall that requires a minimum of one layer of peptidoglycan. Therefore, the most common type of polysaccharide biosynthesis is that used to construct cell walls. The general structure of peptidoglycan is described in chapter 3 and illustrated in figure 3.7A–D. Peptidoglycan is synthesized from UDP sugars to which the appropriate amino acid chains are added by specific enzymes (one for each amino acid), which is an example of nonribosomal protein synthesis. Questions have been raised, however, regarding whether the new material is added randomly throughout the existing cell wall or in specific "growth regions" and whether it augments all dimensions of the cell or only some.

Studies in which radioactively labeled molecules were added to the cell wall and their physical position determined by **autoradiography** (sensitization of a film emulsion by exposure to the decay products of radioactive atoms) have generated some controversy as to whether new material is inserted all over the surface of the wall or in specific growth zones. It is clear that the poles of a rod-shaped cell change little during growth and that a great deal of reshaping of the cell is necessary during formation of the septum in cell division. In most bacteria, somewhere between 30 and 50% of the material in the cell wall is turned over (excised from the cell wall) during each cell division cycle. In some organisms, the material is released, but in others it is reused.

The process of adding new material to the peptidoglycan layer is complicated mechanically in that the mechanism used must not weaken the structure materially to the point at which osmotic damage is possible. In gram-positive cells with their relatively thick peptidoglycan layer, the new material adds to the inside in essentially random orientation. Individual chains can be 30 to 50 disaccharides in length. Initially, they are only loosely cross-linked, but with time the percentage of cross-links increases. In mature *Gaffkya homari*, for example, 49% of the lysine is cross-linked to the alanine of another side chain. The situation for gram-negative bacteria is not understood as well because the peptidoglycan layer is so much thinner. Still, the apparent mechanism must be one of making new strands before breaking old ones.

Arthur Koch has proposed a model called the "surface stress" mechanism in which he attempts to set down rules for this activity and for the maintenance of different cell shapes. In its simplest form, this model shows that as the cytoplasm expands, the cell wall is stressed in a longitudinal direction. Initially, new material is cross-linked but unstressed, and then it is stretched as the cell expands causing the material at the surface to be under the greatest stress. Expansion occurs by the judicious cleavage of existing linkages, either glycosidic or peptide, so as to transfer the stress to the underlying layers (figure 6.12). The enzymes catalyzing these reactions are often referred to as **autolysins** because in excess they can cause the destruction of the wall of the producing cell.

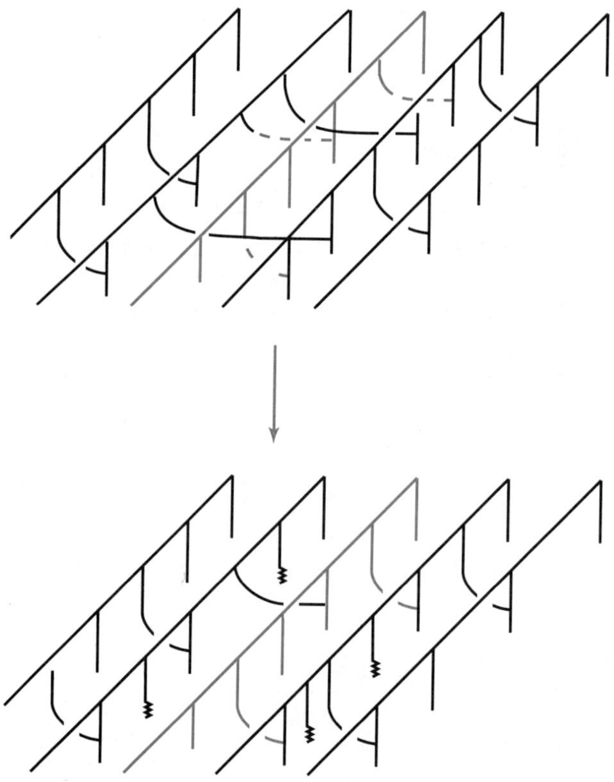

FIGURE 6.12

Insertion of new cell wall material. First the new chain (in color) is joined to the old material at several sites. Next, some linkages in the old material are broken, thereby allowing the mechanical tension to be passed to the new material.

The position of the new septum in cell division seems to be associated with zones of adhesion in the form of annular rings located at the potential division site. In a rapidly growing cell, there is a complete annular zone of adhesion located about the midline of the cell and two partial zones of adhesion located one-fourth and three-fourths of the way along the length of the cell. These latter two sites represent the presumptive septum for the next generation of cells.

Membrane Biosynthesis

Because membranes are held together by hydrophobic interactions that occur spontaneously, enzymes do not seem to be directly necessary for membrane assembly. Instead, it is assumed that as the new material is synthesized in the cytoplasm, it is assimilated into the existing membrane. Proteins are added to the membrane based on their signal peptides. The fluid mosaic model for biological membranes would accommodate the necessary rearrangements.

Membranes apparently increase in size due to the addition of new material more or less at random across the surface of the cell. This implies that if two adjacent cell structures are each bound to the cell membrane, with time they will move further apart. Although there is little direct evidence in favor of the hypothesis, it has long been assumed that segregation of bacterial nucleoids into daughter cells is accomplished by just such a process because nucleoids are known to be membrane bound during active DNA replication.

Summary

The mental image that a person should have when thinking about a rapidly growing bacterial cell is one of continuous activity at all biochemical levels. The DNA folded into the nucleoid is replicating with at least two forks traveling in opposite directions. A cell can use four or eight forks if the rate of division is high. RNA primers are being laid down for each of the Okazaki fragments and then quickly removed so that long continuous chains of DNA can be produced. Torsion of the DNA molecule by the replication forks is being removed by one type of topoisomerase, while another topoisomerase is supercoiling newly synthesized DNA.

Within the nucleoid itself, RNA chains are being made to support the protein synthetic machinery. In the normal course of events immediately after an mRNA molecule's 5′ end is clear of the RNA polymerase holoenzyme complex, a ribosome attaches and begins to translate it. As mRNA lengthens and the first ribosome proceeds along, other ribosomes attach so that mRNA is always fully coated with ribosomes. The proteins thus produced are released at the nucleoid surface and must diffuse in the cytoplasm to the point at which they are needed. A particular problem arises with respect to proteins that must function exterior to the plasma membrane. Such proteins are tagged for export by a signal peptide that usually is cleaved from the protein as it is translocated across the membrane.

Various enzymes produced by the cell are used for reactions of intermediary metabolism, as discussed in the last chapter, as well as to prepare the cell structurally for the next cell division. New membrane material is inserted randomly within the existing membrane, contributing among other things to the segregation of nucleoids into the daughter cells. New peptidoglycan is inserted at multiple sites within the existing framework, but primarily along the sides of a rod-shaped cell. The existing poles of the cell remain unaltered, and new poles for each daughter cell are produced by enzymes that can make a new septum.

Questions for Review and Discussion

1. What theoretical problems were resolved when DNA replication was discovered to proceed in a discontinuous fashion?
2. Compare and contrast the ways in which a prokaryotic cell synthesizes its cell membrane and its peptidoglycan cell wall layer.
3. Trace the flow of energy into each of the polymerization steps for each of the macromolecules discussed in this chapter.
4. What are topoisomerases? Why are they important in the DNA replication process?

Suggestions for Further Reading

Doyle, R. J., J. Chaloupka, and V. Vinter. 1988. Turnover of cell walls in microorganisms. *Microbiological Reviews* 52:554–567. A review of the processes that reshape the cell walls of various bacteria.

Drlica, K., and J. Rouvier-Yaniv. 1987. Histonelike proteins of bacteria. *Microbiological Reviews* 51:301–319. A review of all such proteins from eubacteria, archaebacteria, and chloroplasts.

Gottschalk, G. 1986. *Bacterial metabolism,* 2d ed. Springer-Verlag, New York. An excellent introduction to the subject.

Koch, A. L. 1988. Biophysics of bacterial walls viewed as stress-bearing fabric. *Microbiological Reviews* 52:337–353. A detailed explanation of the surface stress theory of cell wall biosynthesis.

Kornberg, A. 1988. DNA replication. *Journal of Biological Chemistry* 263:1–4. A brief but comprehensive review.

Mandelstam, J., K. McQuillen, and I. Dawes. 1982. *Biochemistry of bacterial growth,* 3d ed. John Wiley & Sons, New York. A very readable discussion of all aspects of bacterial growth.

Neidhardt, F. C., J. L. Ingraham, K. B. Low, B. Magasanik, M. Schaechter, and H. E. Umbarger. (eds.). 1988. *Escherichia coli* and *Salmonella typhimurium.* Cellular and molecular biology, vols. 1 and 2. American Society for Microbiology, Washington, D.C. A review of all aspects of the biology of the best-studied microorganisms. Within volume 1 are articles on DNA replication, transcription, mRNA translation, and most metabolic reactions discussed in this chapter.

Randall, L. L., and S. J. S. Hardy. 1989. Unity in function in the absence of consensus in sequence: Role of leader peptides in export. *Science* 243:1156–1159.

Schmid, M. B. 1988. Structure and function of the bacterial chromosome. *Trends in Biochemical Sciences* 13:131–135. This article summarizes information about nucleoids in vivo and in vitro. It includes a review of the properties of the major histonelike proteins.

chapter

7

Regulation of Cellular Functions and Differentiation

chapter outline

■

GENERAL GOALS

Be able to explain how allosteric proteins behave and how their properties are useful in cellular regulation.
Be able to describe how a cell can regulate the synthesis of any of its macromolecules and what impact this regulation has on cell metabolism.
Be able to explain some ways in which a unicellular organism can differentiate.

The biochemical reactions presented in the preceding chapters indicate how members of the kingdom *Monera* can interact with their environment. As the chemicals in their surroundings change, microorganisms have the ability to invoke new pathways to provide energy and/or carbon skeletons needed for growth. This ability of cells to change in order to meet altered conditions has some important implications for metabolism. Principal among these is that cellular proteins cannot endure forever: if protein degradation did not occur, there would be no short-term change in a cell's enzyme complement. The stability of a protein is reflected in its half-life, the length of time required for 50% of its molecules to be inactivated. Different proteins have different half-lives, with structural proteins, like ribosomal proteins, lasting longer than enzymes.

With the ability to carry out multiple biochemical pathways comes the necessity for controlling which pathways are active at any given time. There is no point in making enzymes to degrade glucose, lactose, and maltose simultaneously when any one of these sugars can be oxidized to yield sufficient energy for growth. Moreover, because macromolecular synthesis is very energy intensive, an organism that makes all possible proteins all the time is at a disadvantage when competing with an organism that is able to regulate its functions and, thereby, use less energy for the same amount of growth. To take a specific example, ribosomes used to synthesize cytoplasmic proteins are not available to synthesize ribosomal proteins, and cell growth is dependent on the production of new ribosomes, among other cytoplasmic components. Although two different energy-yielding pathways may be available to an organism, it might be advantageous for the organism to suppress one in favor of the other because of differing energy yields. An example of this type of activity is seen in the aerobic growth of *Escherichia coli* with the preferential degradation of glucose to acetate rather than to carbon dioxide exclusively via the TCA cycle.

Microorganisms, like multicellular organisms, strive for **homeostasis,** the maintenance of a constant internal environment in the face of changing environmental conditions. This chapter presents ways in which microorganisms, primarily bacteria, can modulate the function of enzymatic reactions and also control the timing and amount of macromolecular synthesis.

Regulation of Enzyme Function

Competitive Inhibition

Enzymes are protein molecules that have folded in such a way as to act as catalysts for a chemical reaction. This folding is usually represented as a lock-and-key arrangement in which the three-dimensional structure of a portion of the protein matches the shape of the substrate molecule. The portion of the protein that matches the substrate is called the **active site.** When the substrate is bound to the active site, rearrangement in the structure of the substrate occurs, and the product(s) of the reaction is released. This general process is depicted in figure 7.1.

APPLICATIONS BOX

Enzymes All around Us

Although enzymes are synthesized only by living organisms, they can have uses other than those originally intended. Indeed, proteolytic enzymes can be found in many kitchens and laundry rooms because proteases degrade proteins to short peptide chains or to amino acids. Their normal cellular function is one of digestion, and that same function can be applied in cooking. Many meat tenderizers are proteolytic enzymes that, in essence, tenderize by predigesting muscle protein partially so as to make it easier to chew.

The fabrics of which most clothing is constructed are either artificial or carbohydrate polymers. Only more expensive fabrics like wool or silk are based on protein-containing substances. Stains that accumulate on clothing, on the other hand, are often proteinaceous. The enzyme presoaks formerly used in the laundry were based on these differences. They contained proteases and amylases, enzymes that could break down a number of different types of molecules while not harming the carbohydrate or artificial polymer.

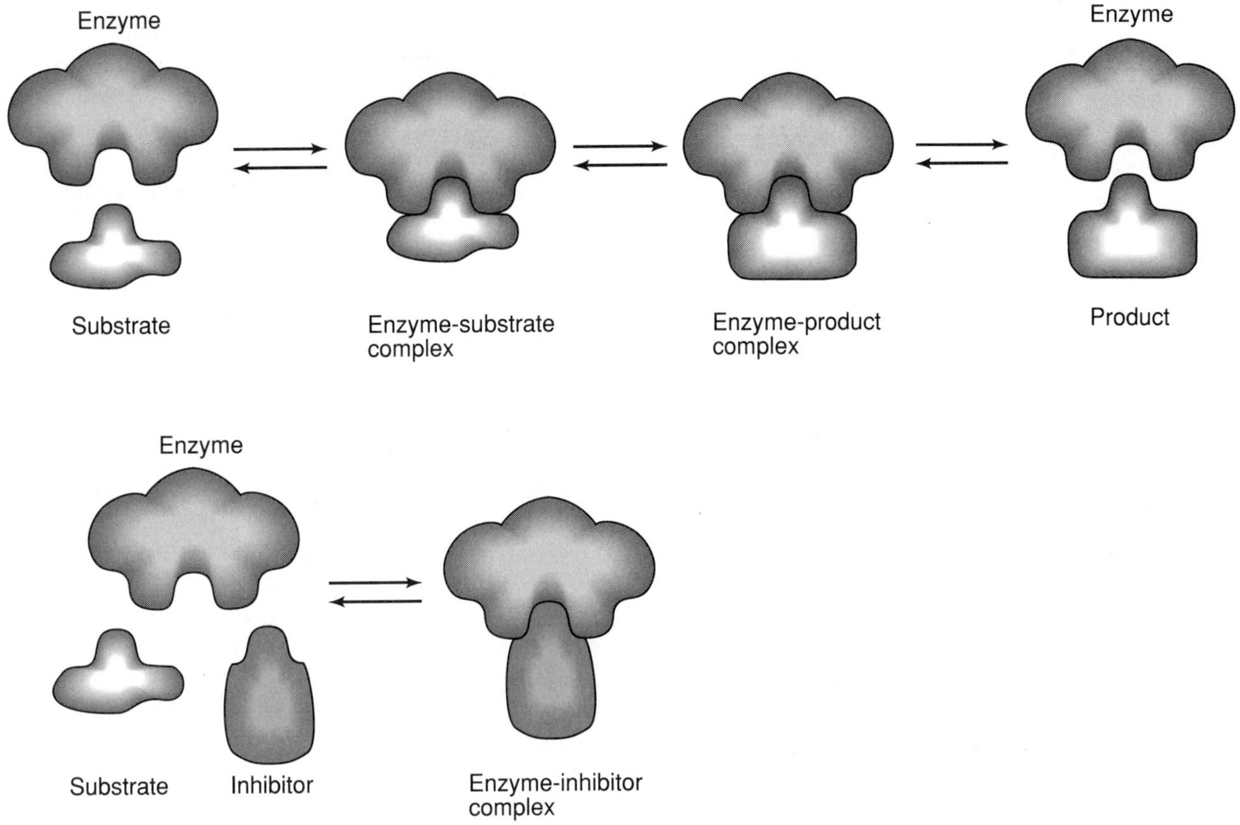

FIGURE 7.1

Interaction of enzyme and substrate. Binding of a substrate to the active site of an enzyme results in the formation of an enzyme-substrate complex. A change in the molecular configurations gives an enzyme-product complex that can then dissociate to yield the product and a free enzyme molecule. In competitive inhibition, the inhibitor (in color) may also bind to the active site and exclude the substrate.

It is often possible to find **substrate analogs,** molecules that more or less resemble the normal substrate in structure. When an enzyme is presented with a substrate analog, the outcome depends on the substrate specificity of the enzyme. If an enzyme is not particularly substrate specific (for example, it may be sufficient for the substrate to have a carboxyl group), the enzymatic reaction proceeds normally. If the enzyme is quite substrate specific, however, it may be that only one molecular type can actually be converted into product. Nevertheless, analogs may bind to the active site to form an enzyme-substrate analog complex (figure 7.1), but that complex cannot be converted to the enzyme-product complex.

Competitive inhibition results when significant concentrations of a substrate analog are present to prevent some or all enzyme molecules from binding to their normal substrate. Such an analog is then acting as an inhibitor of the enzyme. While the enzyme molecule is bound to the analog, it cannot catalyze its normal reaction and, therefore, is transiently inactivated. For situations in which competitive inhibition is occurring, the critical factors are the ratio of substrate to inhibitor and the relative strengths of their binding to the active site. In general, successful competitive inhibition requires that the concentration of the inhibitor be much higher than the concentration of the substrate. An example of a competitive inhibitor is shown in figure 7.2.

Allosteric Proteins

The original conception of allosteric proteins is derived from the work of Jacques Monod and Jean-Pierre Changeaux. They knew that complex macromolecules like proteins assume a steric (three-dimensional) configuration; however, they postulated that it should be

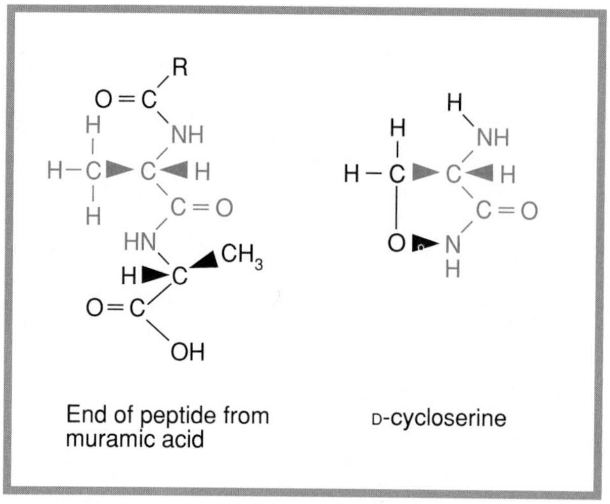

FIGURE 7.2

Example of competitive inhibition. D-cycloserine inhibits the transpeptidation reaction of cell wall biosynthesis (figure 3.9) because of its structural similarity to the peptide backbone of the D-alanine residues (in color) used to form the peptide cross bridges.

possible for some proteins to be **allosteric,** that is, to have more than one three-dimensional configuration. They further assumed that each steric configuration is in equilibrium with the other(s). At any given time, therefore, such a protein could exist in solution in multiple steric configurations although not necessarily in equal amounts (figure 7.3).

The regulatory importance of allosteric proteins can be understood if it is assumed that only one of the two (or more) allosteric configurations can serve as a functional enzyme. In this case, the amount of enzyme function observed in a particular protein preparation depends not only on the amount of allosteric protein present but also on the steric configuration of that protein. Thus, if the equilibrium between two allosteric states of an enzyme could be shifted, it would be possible to increase or decrease the amount of available enzyme activity without correspondingly changing the amount of protein actually present.

The shift in equilibrium can be accomplished by means of an **allosteric effector,** a small molecule that binds specifically to only one allosteric configuration of the protein at a (allosteric) site other than the active site. Because the different allosteric states are assumed to be in equilibrium, any molecule that binds to only one state will tend to shift the equilibrium toward that state. If the enzyme and the allosteric effector combine to make a functional enzyme, the effector also is called an activator; if the combination forms an inactive enzyme, the effector also is called an inhibitor.

Inhibition of this type is **noncompetitive** and fundamentally different from competitive inhibition in that the relative concentration of the substrate is irrelevant. If the allosteric effector is bound, enzyme activity will not be present regardless of the substrate level, because the effector is not in competition for the active site.

Feedback Inhibition

Allosteric proteins are involved at specific points in most multienzyme biochemical pathways presented in chapter 5. Cells must avoid the general problem of overproducing a metabolite in the cytoplasm. For example, *Escherichia coli* can both synthesize its own tryptophan and take it up from the environment. For energetic reasons, the preferred state is to obtain tryptophan from outside the cell. However, sometimes the metabolic needs of the cell can be met only by endogenous synthesis of tryptophan at either low or high levels.

As discussed later in this chapter, there are ways for the cell to regulate the amounts of enzymes produced. In the short run, however, it is sometimes necessary for the cell to regulate the actual functioning of entire pathways. This regulation can be accomplished by a mechanism known as **feedback (end product) inhibition.** In this process, the final product of the biochemical pathway serves its obvious biochemical function and acts as an allosteric effector for the first enzyme in the pathway. When the final product is present in excess of metabolic needs, it acts as a noncompetitive inhibitor of the first reaction in the pathway. By inhibiting this reaction, all subsequent reactions are deprived of their substrates; therefore, the potential problem of accumulation of unneeded biochemical intermediates is avoided.

Figure 7.4 illustrates the process of feedback inhibition as it occurs in the leucine biosynthetic pathway (compare with figure 5.6). In this example, there are a series of enzymatic reactions that convert the amino acid threonine into leucine. If the concentration of leucine begins to rise, leucine binds to the enzyme catalyzing the first reaction in the sequence and converts it into an inactive form. By this means, further accumulation of leucine is prevented.

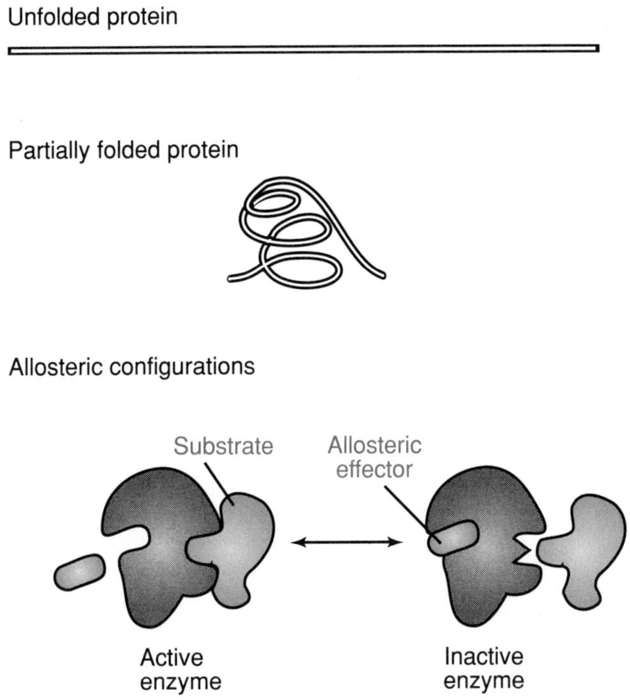

Allosteric protein. In this example, the protein can fold into two configurations. The binding of an allosteric effector can shift the equilibrium between the two allosteric states.

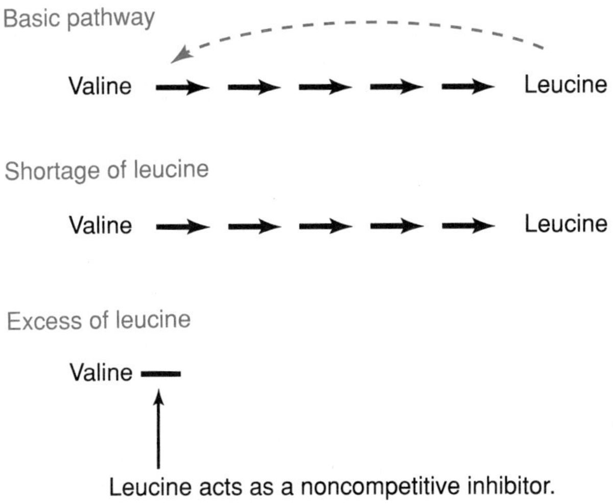

FIGURE 7.4
Feedback inhibition. There are multiple enzymatic steps leading from threonine to leucine. Accumulated leucine acts as an allosteric effector for the first enzyme in the pathway (dashed colored line), inhibiting its own synthesis thereby.

Regulation of Transcription

Transcription and translation are two cellular processes that consume a great deal of energy. Cells have evolved a large number of controls designed to ensure that relatively few unneeded macromolecules are produced. The primary way in which this is done, of course, is at the level of transcription because if there is no mRNA produced, no translation can occur.

The Operon Model

The concept of the operon originated in France, this time with Francois Jacob and Jacques Monod. The fundamental nature of this concept was recognized by the Nobel Prize committee in 1965. In essence, what Jacob and Monod postulated was the existence of regulatory units in the cell. Each such unit is designated as an **operon** and consists of one or more discrete functions expressed in a coordinate fashion. In molecular terms, each operon corresponds to the DNA coding for one mRNA molecule. In other words, it begins with a promoter, has an internal coding region, and ends with a transcriptional terminator. Unlike the situation in eukaryotes, individual prokaryotic mRNA molecules can code for multiple proteins, occasionally as many as ten or more. Each protein is translated separately from the bacterial mRNA code. In eukaryotes, multiple proteins from a single transcript are normally the result of posttranslational processing of a single large protein molecule into fragments of various sizes.

The regulatory elements for the operon all involve molecules binding directly to the DNA in the vicinity of the promoter. Hence, there are two components to each regulatory step, the bound molecule itself and the region of the DNA molecule to which it binds. Both are necessary for proper function. One group of regulatory elements can be thought of as activators or positive regulators. These function to turn on transcription of the operon, generally by increasing promoter function. The other group of regulators can be thought of as **repressors** or negative regulators. These function to keep transcription of the operon turned off by binding to the DNA near the point at which transcription begins and by blocking access of RNA polymerase to the promoter site.

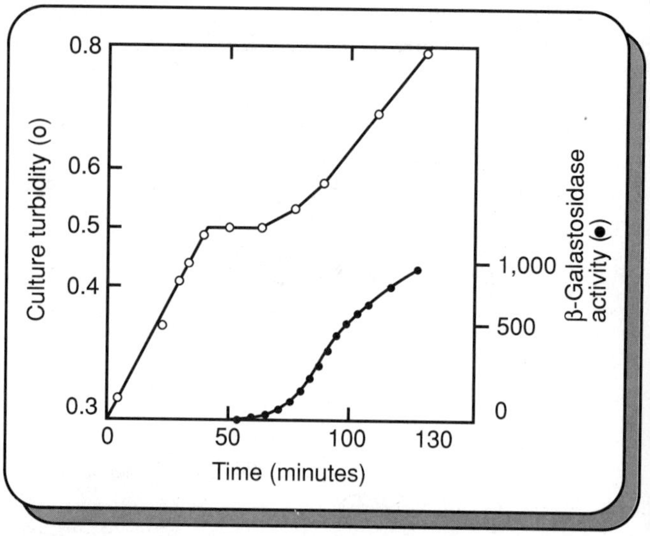

FIGURE 7.5

Diauxie. A culture of *E. coli* started on a medium containing both glucose and lactose as carbon sources exhibits two periods of growth, the early one in which only glucose is used and the later one in which only lactose is used.

From W. Epstein, S. Naono, and F. Gros, "Synthesis of Enzymes of the Lactose Operon During Diauxic Growth of *Escherichia coli*," in *Biochemical and Biophysical Research Communications* 24:588. Copyright © 1966 Academic Press, Inc., San Diego, CA. Reprinted by permission.

The Lactose Operon

Jacob and Monod did most of their work on the lactose operon of *Escherichia coli,* one of the regulatory elements still most studied in all bacteria. The lactose operon codes for three enzymes: β-galactosidase *(lacZ),* galactoside permease *(lacY),* and thiogalactoside transacetylase *(lacA)*. The permease, whose function is discussed in chapter 3, is responsible for sugar transport across the cell membrane. β-galactosidase degrades lactose to glucose and galactose, which can be metabolized by the pathways outlined in chapter 4. The transacetylase enzyme is thought to detoxify potentially harmful sugar derivatives that accidentally may enter the cell. The lactose operon provides a good introduction to modes of operon function because all the standard types of regulation (positive control, negative control, use of dual promoters, for example) can be observed with it.

When growing cultures of *E. coli* are examined for the presence of β-galactosidase, invariably the enzyme is found to be **inducible,** that is, produced in large quantity only when the substrate lactose is present in the medium. The presence of lactose in the medium is not the only factor, however. Monod demonstrated that *E. coli* displays a growth phenomenon called **diauxie** in which growth occurs in two separate stages when cells are cultured with a mixture of glucose and lactose as carbon sources (figure 7.5). Somehow during the first phase, the cells are able to use glucose preferentially and keep synthesis of β-galactosidase at a low level. The lag period that occurs in the middle of the diauxic curve represents the time cells require to synthesize the significant quantities of β-galactosidase needed to degrade lactose. Without substantial amounts of this enzyme, growth on lactose is not possible. Similar effects are seen with other sugar degradation pathways (maltose, arabinose, galactose) when glucose is present.

This predominance of glucose over other possible energy sources is the so-called **glucose effect** or **catabolite repression** assumed to be due to the concentration of glucose metabolites in the cytoplasm. Basically, in the presence of glucose or other substrate transported by the phosphotransferase system, cells do not attempt to use other carbon sources for energy. When glucose is exhausted, the cell activates other pathways to provide oxidizable molecules for metabolism. DNA sequence analysis and various genetic studies have provided an explanation for both the inducibility of the β-galactosidase enzyme and the basis for the glucose effect.

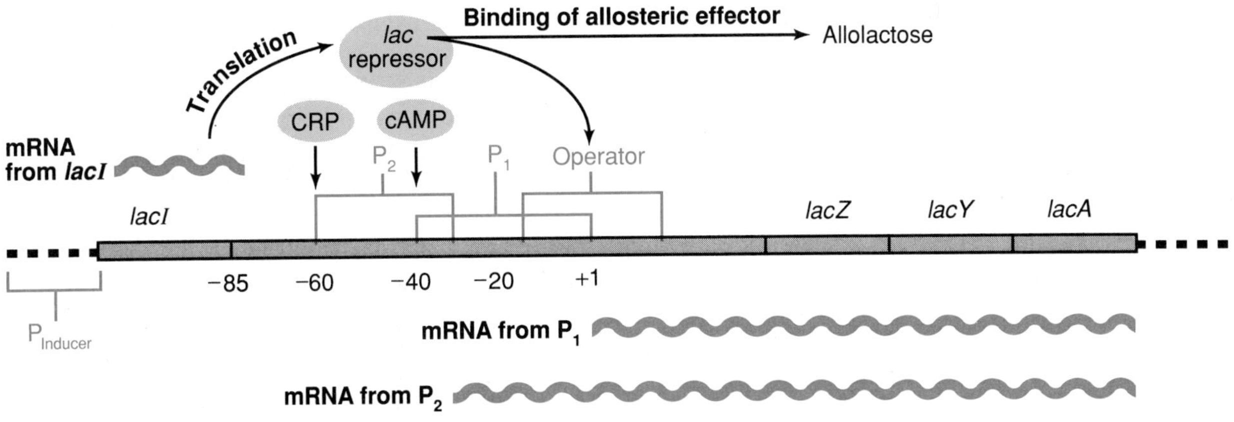

FIGURE 7.6

Lactose operon. The various structural elements of the lactose operon on DNA are shown together with the regulatory molecules that interact with the DNA (color). Base pair +1 is the start of the major mRNA species. Negative numbers refer to bases upstream to this position. Each wavy line represents an mRNA molecule. The diagram is not drawn to scale in that the right hand portion is greatly compressed.

In the following discussion, refer to figure 7.6 for the contemporary view of the structure of the lactose operon. The operon consists of sequences coding for β-galactosidase, galactoside permease, and thiogalactoside transacetylase as well as sequences upstream (negative base numbers in figure 7.6) from the transcribed region that act as regulators. Promoter P_1 can be quite efficient but is normally inactive. Its DNA sequence does not exactly match the standard promoter sequence, particularly in the region around base pair −35; therefore, RNA polymerase does not bind to it. In the presence of two activators, cyclic AMP (cAMP; figure 7.7) and catabolite repressor protein (CRP), P_1 becomes a strong promoter and transcription begins.

The concentration of cAMP varies inversely with the concentration of glucose in the cytoplasm because the activity of the enzyme adenylate cyclase is stimulated by the accumulation of the phosphorylated form of enzyme III^{glc}, the protein that supplies phosphoryl groups to the membrane protein used in the group translocation of glucose (figures 3.14 and 7.8). From this, an explanation for the diauxic growth curve is possible. When cells are presented with both glucose and lactose, adenylate cyclase activity is low; therefore, cAMP concentration is too low to activate P_1.

The other observation, that β-galactosidase is produced only in the presence of lactose, can be explained by the presence of a protein repressor in the cytoplasm. This *lac* repressor is encoded by an independent transcription region called *lacI* (inducibility) and is produced at a constant low rate. The *lac* repressor can be shown to bind to the lactose operon DNA in the region between bases 1 and +20 (figure 7.6). When it is bound in this position, the −10 element of promoter P_1 is not available for binding RNA polymerase, hence, no transcription is possible. If the *lacI* coding region is deleted, transcription of the *lac* operon occurs whenever cAMP is present at high levels.

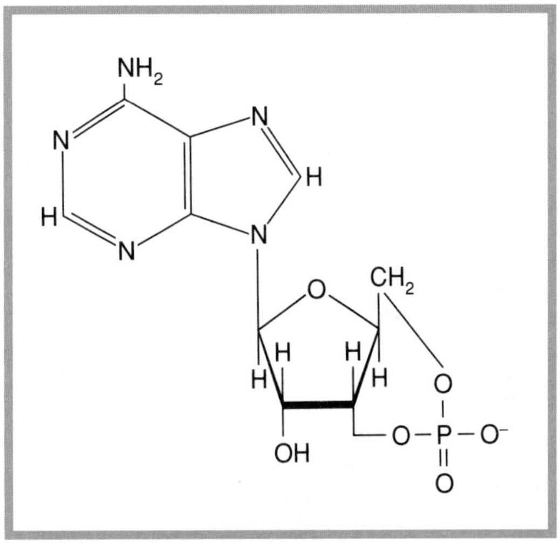

FIGURE 7.7

3′–5′ Cyclic AMP (cAMP). The enzyme adenylate cyclase synthesizes a cyclic form of adenosine monophosphate from ATP.

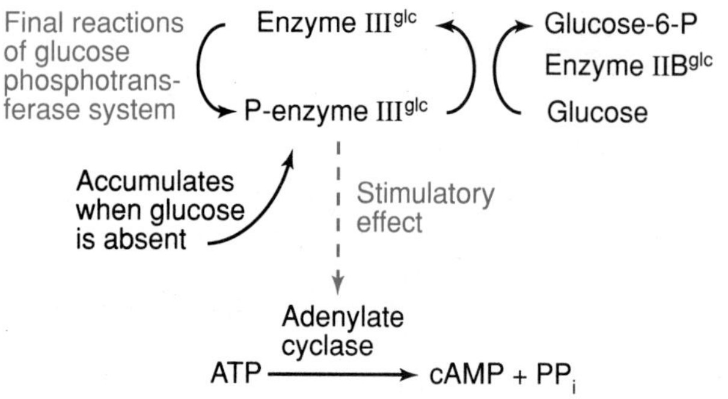

FIGURE 7.8

Stimulation of adenylate cyclase by intermediates of glucose transport. Accumulation of the phosphorylated form of enzyme III serves to activate the synthesis of cAMP.

The lactose repressor is a nonenzymatic, allosteric protein whose effector is allolactose, an occasional by-product of the reaction of β-galactosidase with lactose. When lactose enters the cell, the trace amounts of β-galactosidase that are present act to convert some of it to allolactose, which binds to the *lac* repressor and shifts the allosteric equilibrium to a state that does not bind to DNA. If P_1 is active, high-level transcription then can begin. Allolactose is the actual **inducer** of lactose operon transcription. When the lactose concentration drops, so does the concentration of allolactose, and the repressor returns to the allosteric state that binds to the DNA molecule and inactivates P_1. Thus, the lactose operon is transcribed only when the substrate is present. Mutations in the *lac* repressor can "freeze" the molecule in one of the two allosteric states. If repressor activity is present, the molecule is a "super repressor" and permanently bound to the operator.

It is possible to find **gratuitous inducers,** substrate analogs that are not degraded by β-galactosidase but are still inducers. Cells grown in the presence of compounds such as isopropyl thio-β-D-galactoside or thiomethyl galactoside transcribe the lactose operon at high rates indefinitely, so long as glucose is not present. If glucose is present, the levels of cAMP will be low, allowing little transcription to take place from P_1 regardless of the availability of the promoter.

Recently found within the *lac* operon regulatory region is a second promoter site, P_2. P_2 is a relatively inefficient promoter that overlaps P_1 but lies upstream from it. As shown in figure 7.6, the -10 region of P_2 is actually the -35 site of P_1. P_2 seems to function when catabolite repression is operative, to require no cofactors for its operation, and to maintain a low level of the various lactose operon enzymes in the cell. It provides the essential presence of enough permease and hydrolytic enzyme to ensure that the inducer molecules will enter the cytoplasm. Note that RNA polymerase must be capable of displacing bound repressor molecules in order for this mechanism to work. The presence of cAMP and CRP inhibits the functioning of P_2 while turning on P_1. This results in approximately a 1,000-fold stimulation in β-galactosidase production.

Elements of the regulatory system found in the lactose operon are duplicated elsewhere. This is particularly true of the use of cAMP to activate the operon. High concentrations of cAMP have been shown to activate a number of operons coding for catabolic functions. Whether the activated operons actually are transcribed then depends on the presence or absence of the appropriate inducers. Protein repressors also are associated with other operons, and so are protein activators as well. The two cases can be distinguished by ascertaining what happens when the base sequence coding for the regulatory protein is altered so that the protein is no longer made. Jacob and Monod found that such alterations of the *lac* repressor give **constitutive** expression of the enzymes (produced at all times). If the protein had been an activator instead of a repressor, a similar alteration would have prevented expression of the operon. As a general rule, activators are seen for operons used frequently, and repressors for operons used infrequently.

Attenuation in the Tryptophan Operon

The tryptophan operon has many elements in common with the lactose operon, including the presence of a protein repressor; however the tryptophan operon does not depend on cAMP concentration for its regulation. It has a different mode of regulation known as **attenuation,** which has been studied extensively by Charles Yanofsky and his collaborators. The concept of attenuation is easy to understand, although the molecular details are somewhat complex. Basically, not all mRNA molecules that begin transcription are allowed to be elongated to full length. Instead, up to 85% are terminated shortly after they begin and before much energy has been used up.

The attenuation process occurs in the **leader region** of an mRNA molecule. This is the portion of mRNA measured from the point at which transcription starts to the point at which translation begins, and its length can be quite variable. The tryptophan mRNA has a leader region of 162 bases that have been completely

APPLICATIONS BOX

Overproduction of Amino Acids

An important segment of industrial microbiology is devoted to the production of amino acids. Some products are used for further syntheses, but many are also used as food supplements. These supplements literally can make the difference between life and death for poor, agrarian populations that are trying to grow certain crops as their sole source of protein. Many plant proteins are incomplete; that is, they do not contain the correct proportions of amino acids humans need for proper metabolism. As a result, such populations are subjected to certain deficiency diseases unless a mixture of plant proteins is used to provide all the essential amino acids. One way to avoid these diseases is to supplement locally produced flour with the critically deficient amino acid(s).

The economics of the situation is that these amino acids must be produced as cheaply as possible because those areas in which they are most needed are inhabited by people who can least afford to pay for them. Microbial geneticists often work to alter producer strains of bacteria so that they will overproduce the desired amino acid. The genetic changes include removal of feedback mechanisms, inactivation of repressor proteins, elimination of attenuation, and constitutive production of activators. After appropriate mutations are in place, the yield of a particular product may be as great as 8 to 10% of the dry weight of the culture.

sequenced. Within this region is a small **open reading frame** (ORF) between bases 27 and 68. An ORF is a sequence that could be translated into a protein because it is bounded by a ribosomal binding site and a termination codon. This theoretical protein or leader peptide would have 14 amino acids, including two adjacent tryptophan residues. Because tryptophan is normally a rare amino acid, consecutive tryptophan residues are a highly unusual event. Although the leader peptide could not be identified originally, its theoretical existence led to the prediction that attenuation was somehow associated with translation of the leader sequence. In subsequent experiments, a leader peptide with the predicted amino acid sequence has been identified.

Experiments using controlled alterations in the leader sequence have given rise to the model for attenuation presented in figure 7.9. In the model, translation is assumed to follow closely on transcription such that any given RNA polymerase complex is physically adjacent to a series of trailing ribosomes that are translating mRNA more or less as fast as it is produced. Note in figure 7.9 that it is possible to fold the leader RNA into several different configurations held together by hydrogen bonds creating local regions of double strandedness. If no ribosome is present, these are the structures that might be produced after an mRNA is synthesized.

If a ribosome is present, it will prevent any base pairing in its immediate vicinity. The permitted structures in the RNA will then depend on the exact position of the ribosome, and here the strategic position of the coding sequence for the leader peptide becomes apparent. If the leader peptide is produced, the ribosome is bound to region 2, and only regions 3 and 4 can pair to form a stem and loop. This type of small loop has been associated with the normal 3′ ends of mRNA molecules and is assumed to mark the site for termination of mRNA synthesis (attenuation). The situation is different, however, when tryptophan is in short supply. In this case, the ribosome will have to pause in region 1 while it waits for a charged $tRNA^{trp}$ to arrive. A space is then left between the paused ribosome and the advancing RNA polymerase. This space would allow regions 2 and 3 to pair, forming a stem and loop structure that is broader than the 3–4 loop; therefore, the loop is not a termination signal, and the RNA polymerase molecule would continue synthesizing mRNA. Prompt release of the ribosome from mRNA after synthesis of the leader peptide would allow occasional formation of the 2–3 antiterminator loop and result in the observed basal level synthesis of the enzymes.

This model allows for sensitive control of the rate of production of the tryptophan biosynthetic enzymes in that the rate of synthesis of a peptide rich in tryptophan determines whether or not the molecule of mRNA currently being synthesized will be completed. It also allows for a higher basal level of enzymes without the need for two separate promoters, as in the case of the lactose operon. Similar attenuation mechanisms have been found for a number of other amino acid biosynthetic pathways in *E. coli* as well as in the *Bacillus subtilis* tryptophan operon.

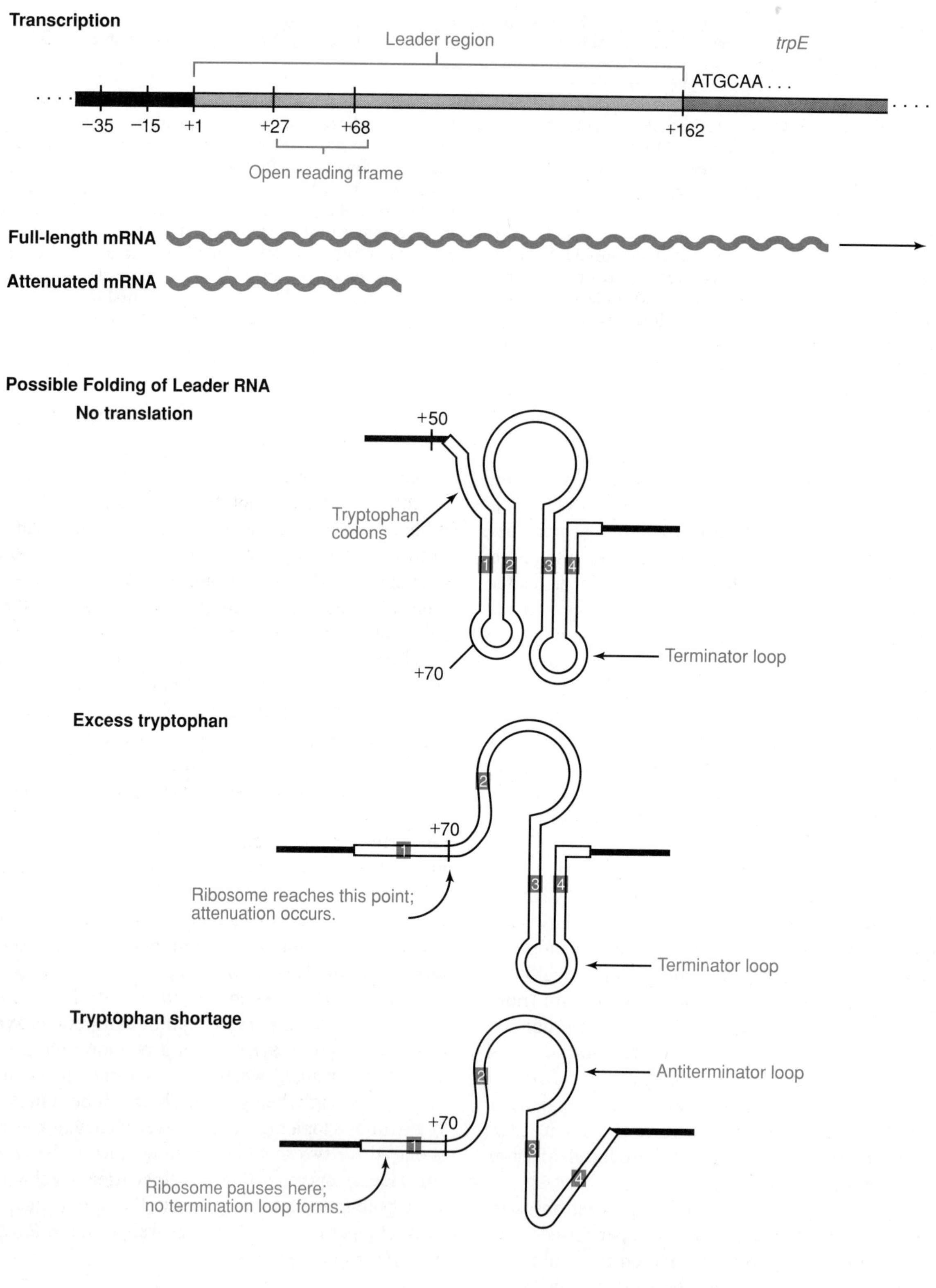

FIGURE 7.9

Attenuation in the tryptophan operon of *E. coli*. Only a portion of the operon is shown, including specifically the transcription and translation start sites and the leader region. Various ways in which the leader RNA can be folded are depicted also. It is assumed that certain stem and loop structures represent termination signals and, when present, cause RNA polymerase to stop transcription at that point.

Sigma Factors as Regulators

Because sigma factors are used to determine the promoter specificity of the RNA polymerase holoenzyme complex, a change in sigma factors within the complex can eliminate or stimulate the synthesis of whole classes of proteins. An excellent example of this type of activity is the sporulation process of *B. subtilis*, described in the following section. Similar adaptive strategies are used by some viruses and by such familiar organisms as *E. coli*. In the latter case, raising the temperature of a culture causes production of a new sigma factor (σ^{39}, molecular weight 39,000) that replaces the normal σ^{70} and then triggers synthesis of new "heat shock" proteins thought to have a protective value for the cell. Normality is restored eventually because σ^{39} also stimulates transcription of the gene coding for σ^{70} so that σ^{70} will regain control of all RNA polymerase complexes.

Regulation of Translation

It is common for eukaryotic organisms to produce significant amounts of mRNA that will be translated only at some future time. For example, unspliced mRNA, even if translated, will not yield a functional protein. Another example is an egg cell preprogrammed with mRNA that will be used only if and when the egg is fertilized. This process requires the egg cell to have a very stringent mechanism for holding mRNA in an undegraded and untranslated state. Phenomena similar to these have not been observed in bacteria themselves, but different sorts of translational regulation have been found to occur.

It has long been known that there is a polarity of position on the mRNA molecule. Proteins encoded near the 5′ end of the molecule tend to be produced in higher quantities than those encoded at the 3′ end. This observation is understandable if one assumes that with every protein translated, there is a finite probability that the ribosome will dissociate from the mRNA molecule and discontinue producing the remainder of the proteins encoded in it. Frequently, translation is considered to be **coupled** in the sense that it is necessary to translate the code for the first protein in order to get to the code for the second, and so on. This coupling means that most initiator codons for proteins do not include ribosomal binding sites.

A different sort of phenomenon has been observed in the case of ribosomal proteins, some of which play a very important role in regulation of their own operons. When there is an excess of the first protein synthesized (such as does occur when assembly of new ribosomes is slow or stopped), the protein binds to the mRNA molecule coding for itself and prevents the ribosome from passing through the blockade (figure 7.10). A protein of this type is considered to be **autogenously regulated** in that it controls its own rate of transcription.

Differentiation and Developmental Stages in Unicellular Microorganisms

Regulatory processes can be used to affect more complex cellular activities than simple biosynthetic pathways. Although it is not immediately evident, even unicellular organisms can have various developmental stages in their life cycle, and they may actually differentiate portions of their cells into specific structures. The following subsections present a few examples of these activities and introduce some new ways in which microorganisms regulate their genetic activity.

Sporulation in *Saccharomyces cerevisiae*

The yeast *Saccharomyces cerevisiae* has a complex life cycle in which it spends part of the time as a **haploid** organism, having only one copy of each chromosome, and the remainder of the time as a **diploid** organism, having two copies of each chromosome (figure 7.11). Haploid cells result from meiosis occurring in a diploid cell, and diploid cells result from individual fusions of two haploid cells. Fusion will not occur between just any two haploid cells, however. They must be of different mating types, one designated as **a** and the other as **α**. The resulting heterozygous diploid cell can divide to produce additional diploid cells. Later, when subjected to nutritional deprivation, this diploid cell is capable of undergoing meiosis (reduction division) to yield four haploid spores in an **ascus** or sac. When nutrients are again available, the spores germinate to give new haploid cells that can reproduce themselves.

An organism that requires two different haploid cell types for mating to form a diploid cell is said to be **heterothallic,** while one that can mate any two haploid cells is said to be **homothallic.** The basic description for the

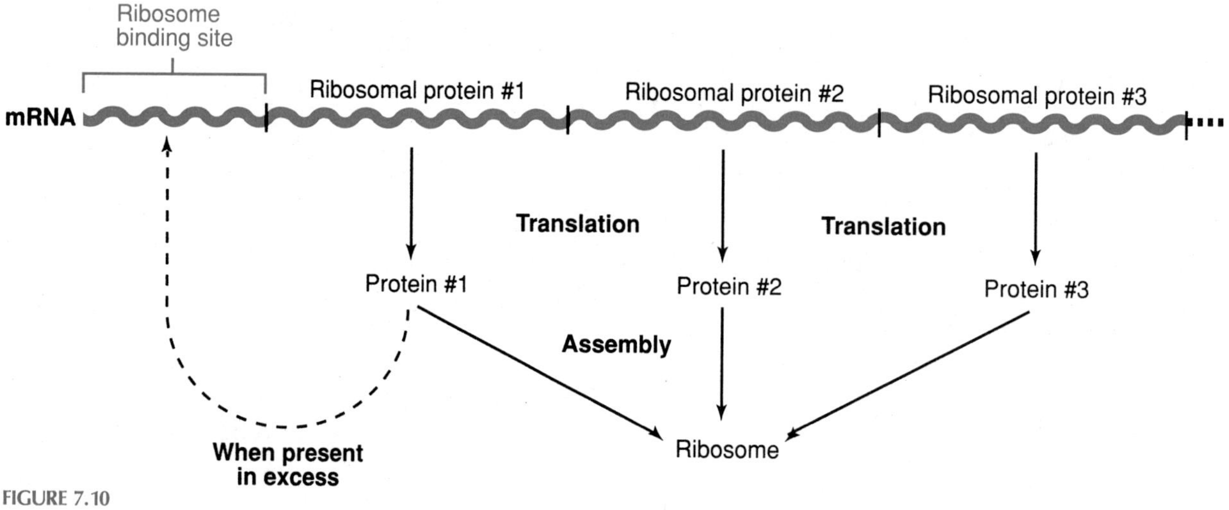

FIGURE 7.10

Regulation of translation. Certain ribosomal proteins, when present in excess, bind to their own mRNA and prevent further translation.

life cycle of *S. cerevisiae* is, therefore, one of heterothallism. However, it has been observed that some pure cultures of one mating type can change so that they have both types of cells present. In other words, the potential for both mating types is present in every cell, but only one mating type can be expressed at a time. Strains of *S. cerevisiae* that can change their mating type are homothallic, and the gene that controls this ability is designated as *HO*.

Expression of the mating types is determined by the *MAT* locus that can be in the form of *MAT***a** or *MATα*. In *MATα* cells, there are two specific proteins produced, α1 and α2, that have complementary regulatory functions. The α1 protein serves as an activator to turn on the functions necessary for the α mating type, and the α2 protein serves to repress all **a** mating type functions. In *MAT***a** cells, there are corresponding **a**1 and **a**2 proteins synthesized. The function of **a**2 is not known. The **a** mating type is expressed because of a lack of the α2 repressor and the α mating type is not expressed due to a lack of activator. In heterozygous diploid cells, the combination of α2 and **a**1 serves to shut down all mating type expression and trigger sporulation.

It can be shown that the information for both mating types is present at separate loci in the chromosome of all cells but in inactive forms. An additional copy of information for one mating type is present at the *MAT* site, and it is this information that is expressed. When *HO* is functional, it is possible for new mating type information to be copied from one of the inactive loci (either *HMLα* or *HML***a**) to the *MAT* locus (figure 7.12). This phenomenon is an example of **transposition,** a reaction in which a DNA sequence is moved from one site to another location on the same DNA molecule or to a different DNA molecule.

In fundamental terms, regulation of mating type in *S. cerevisiae* is accomplished when genetic information from a site that is not transcribed is moved to a site that is. Additional activator and repressor proteins are needed to prevent accidental mutations from creating promoter sites near the normally inactive loci.

Endospore Formation in *Bacillus subtilis*

Bacterial endospore formation is a cellular differentiation process related to cell division in some important senses but distinct from normal cellular function. The mechanism by which endospore formation is regulated is different from that seen in *S. cerevisiae*. From an ultrastructural point of view, the process can be easily described and is subdivided into seven morphologic stages identified by roman numerals (figure 7.13). A vegetative cell is considered to be in stage 0. Stage I, whose biochemical identification is uncertain, usually is represented as a cell with a chromosome filament. During stage II, the division septum forms off center. By stage III, the larger portion of the dividing cell (mother cell) engulfs the smaller portion (forespore). The membrane from the mother cell extends back over

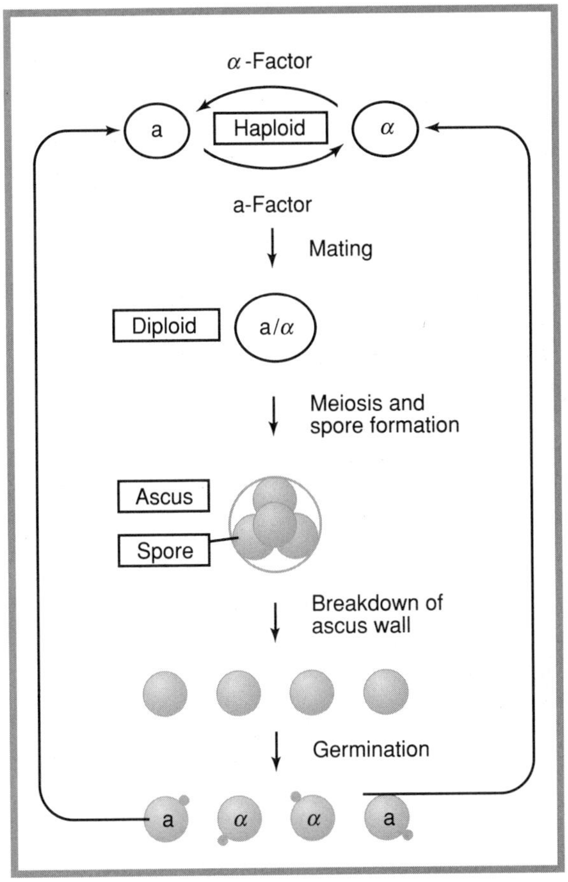

FIGURE 7.11

Life cycle of *Saccharomyces cerevisiae*.

From I. Herskowitz, "Life Cycle of the Budding Yeast *Saccharomyces cerevisiae*," in *Microbiological Reviews* 52:537. Copyright © 1988 American Society for Microbiology, Washington, DC. Reprinted by permission.

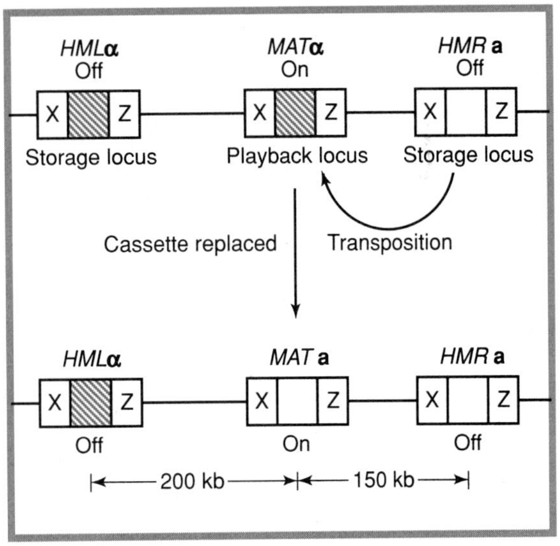

FIGURE 7.12

Cassette mechanism for mating type interconversion. The top line depicts the arrangement of clumps or cassettes of information for mating type found on chromosome III of an α cell. Only the cassette at *MAT* is expressed. Switching to **a** occurs by removal of the α cassette and replacement with the **a** cassette. The product of the *HO* gene initiates this transposition process.

From I. Herskowitz, "Life Cycle of the Budding Yeast *Saccharomyces cerevisiae*," in *Microbiological Reviews* 52:543. Copyright © 1988 the American Society for Microbiology, Washington, DC. Reprinted by permission.

the developing forespore so that the presumptive spore actually is encased within two layers of unit membrane.

The spore cortex, a modified form of peptidoglycan, is laid down by the outermost unit membrane during stage IV. Surrounding the cortex is a spore coat composed of proteins, some of which are similar to keratin. It is synthesized during stage V. Maturation and lysis occur during stages VI and VII. Dipicolinic acid is produced within the **sporangium** (remains of the cytoplasm of the mother cell) and deposited within the spore core. At this same time, the spore develops its typical heat resistance. In figure 7.13, notice how the sporangium begins to degenerate as the spore is completed.

The finished spore is very stable and does not respond readily to its environment. Spore germination can occur shortly after sporulation, but only if the spore is activated artificially by heating or by other special treatment. In the normal state, as the spore ages it gradually undergoes a spontaneous activation process so that it can respond to appropriate nutrients in the environment. When the critical nutrients are present, the spore germinates and starts a new cell. The nature of these nutrients varies with the type of spore, but typically they are amino acids.

During the sporulation process, a whole class of RNA consisting of a large number of different molecules, the sporulation-specific mRNA, is being activated. It can be shown that this wholesale change is due to alterations in the structure of the RNA polymerase holoenzyme, specifically to the sigma factor. Conceptually, the process is straightforward. The trigger for sporulation results in the production of a new sigma factor that causes transcription of a new set of genes, including yet another sigma factor, etc. Such a process automatically provides functions in the correct sequence.

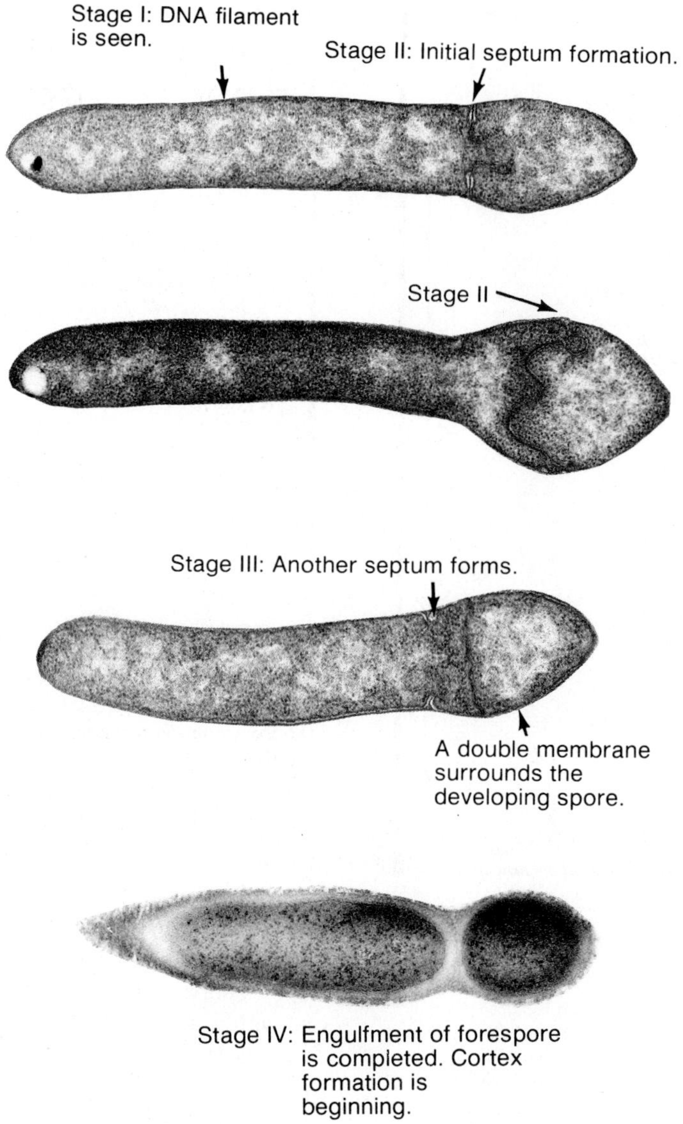

FIGURE 7.13

The stages of endospore formation in *Bacillus* sp.

Electron micrographs courtesy of Dr. Joseph Madden.

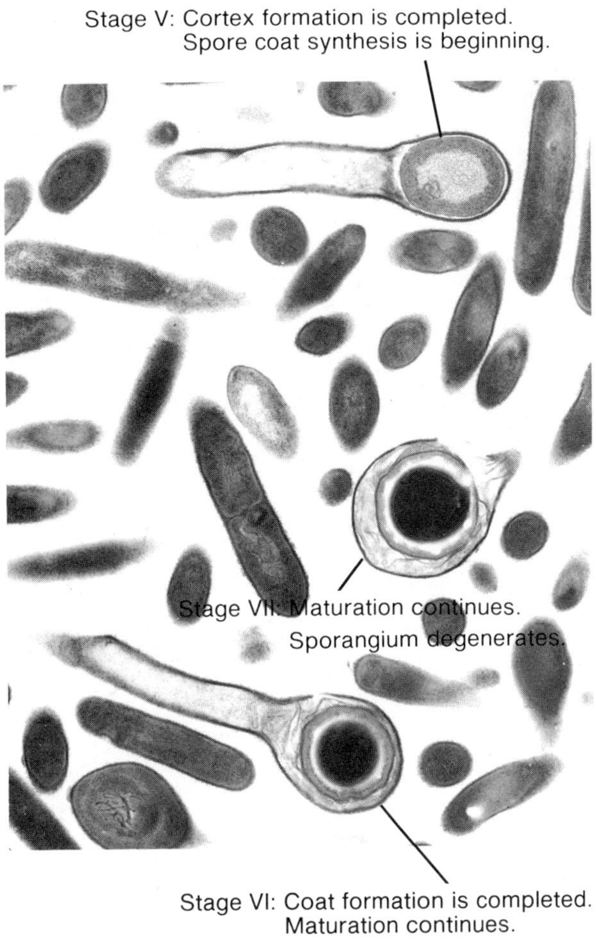

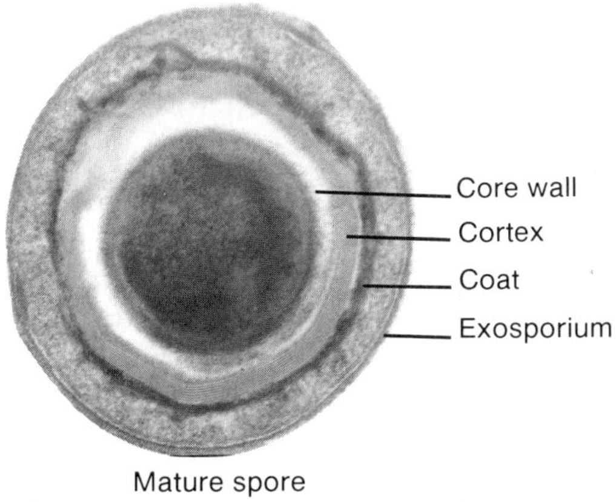

FIGURE 7.13 CONTINUED

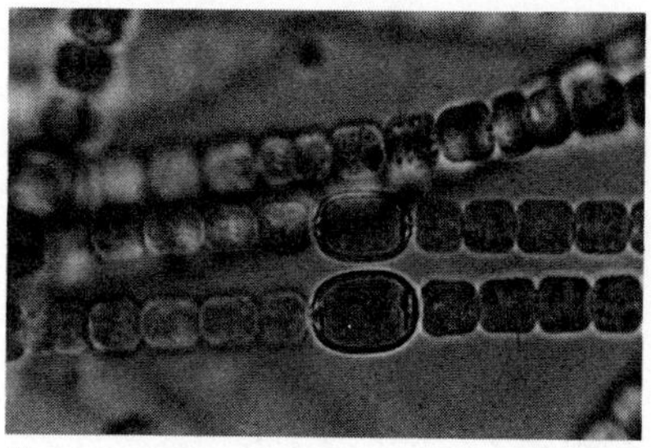
A

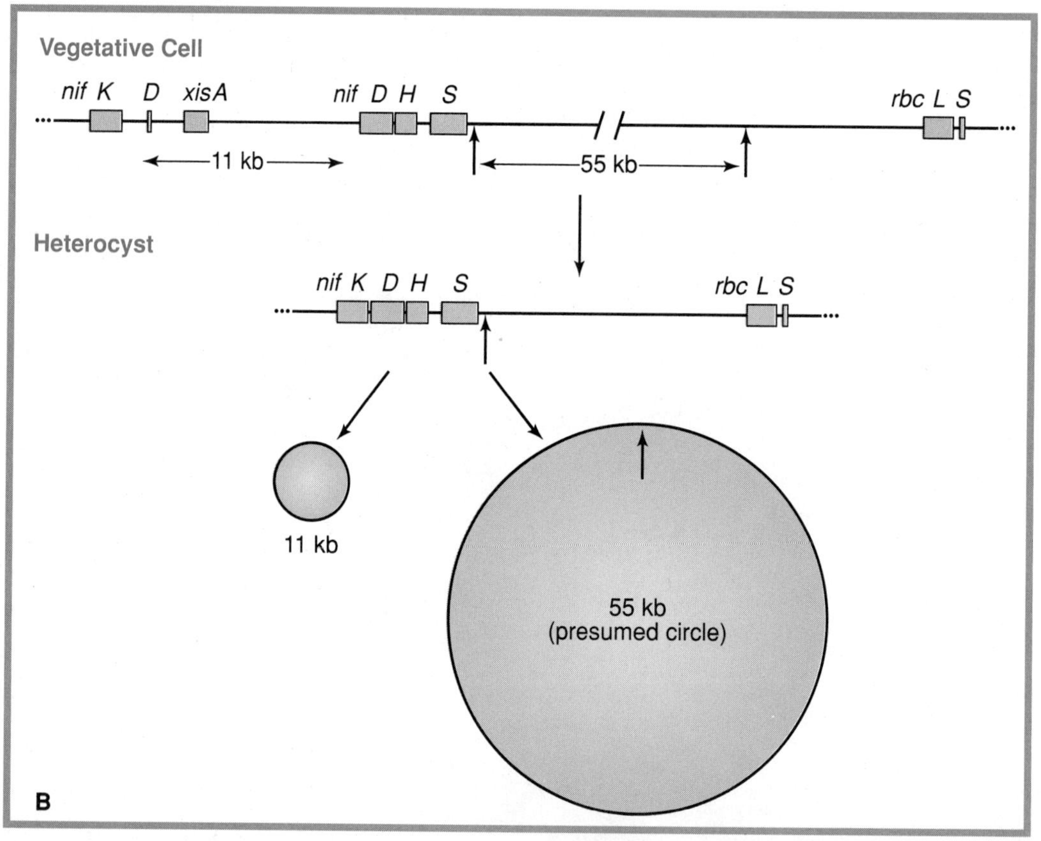

FIGURE 7.14

(**A**) *Anabaena* filaments showing occasional heterocysts.
(**B**) Chromosomal deletions seen in *Anabaena* heterocysts.

(A) Reprinted from Sze, P. 1986. *A biology of the algae*, p. 24. Wm. C. Brown Publishers, Dubuque, Ia. (B) From J. W. Golden, C. D. Carrasco, M. E. Mulligan, G. J. Schneider, and R. Haselkorn, "Deletion of a 55-Kilobase-pair DNA Element From the Chromosome During Heterocyst Differentiation of *Anabaena* sp. Strain PCC7120," *Journal of Bacteriology* 170:5034–5041. Copyright © 1988 American Society for Microbiology. Reprinted by permission.

Heterocyst Formation in *Anabaena*

Anabaena is a genus of the cyanobacteria whose members are capable of nitrogen fixation as well as photosynthesis. The organism generally appears as long filaments (figure 7.14A). Under conditions of nitrogen starvation, approximately every tenth cell within the filament differentiates to yield a **heterocyst,** a thick-walled structure that can fix nitrogen although it has given up most other cellular functions. The ability to break down nutrients via the Calvin cycle and photosynthetic enzymes is lost through synthesis of appropriate repressors, and the production of the required enzymes for nitrogen fixation (*nif* genes) is through induction.

A few *nif* genes require some DNA rearrangements before they are ready for transcription. The *nifD* gene must have 11 kilobase pairs (kb) excised from within its coding sequence, and the *nifS* gene must have 55 kb excised (figure 7.14B). The mechanisms by which these excisions are accomplished are not yet understood, but the 55-kb excision can occur under microaerophilic conditions, while the 11-kb excision

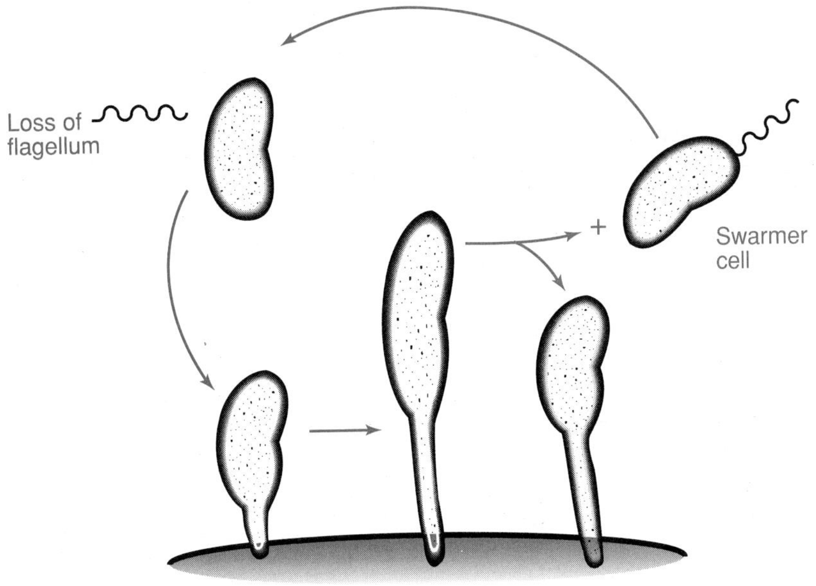

FIGURE 7.15

Life cycle of the stalked bacterium *Caulobacter crescentus*.

cannot. They must involve somewhat different processes, therefore. Similar DNA rearranging reactions can be seen in connection with antibody production.

Morphogenesis in *Caulobacter crescentus*

Not all microbial differentiation involves a change in the entire cell, although that is certainly the most common mode. There are a few organisms whose basic cell shape is asymmetrical and, therefore, must undergo some special processes at each cell division. The best studied of these organisms is the stalked bacterium *Caulobacter crescentus*. The molecular mechanisms of regulation have not yet been elucidated, but many details of the cell cycle itself are known.

The changing morphology of *C. crescentus* during its life cycle is depicted in figure 7.15. If the plane of cell division were along the line of symmetry, then each daughter cell would need to produce only its mirror image. Because the cell septum forms in a transverse fashion, however, one daughter cell has a stalk and the other does not. The daughter cell that receives a stalk continues to grow and later divides to give another set of stalked and unstalked daughter cells (figure 7.16). The daughter cell without a stalk synthesizes a set of pili and a polar flagellum prior to cell division. When the unstalked daughter cell is released at the conclusion of cell division, it becomes a motile swarmer cell. Typical chemotactic receptors also are present. Eventually, the swarmer cell loses its flagellum and pili and develops a stalk. Cell division is then possible.

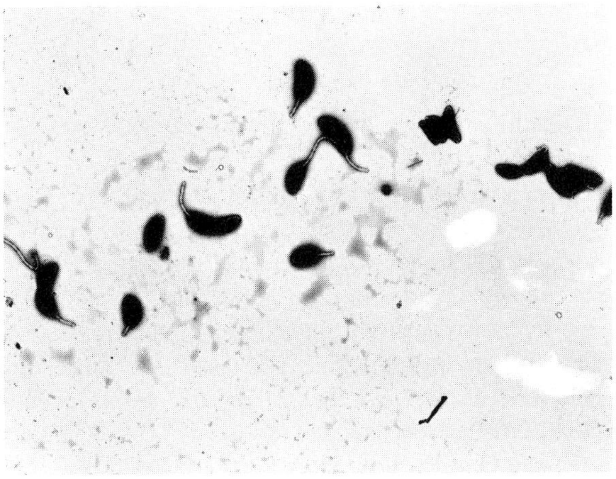

FIGURE 7.16

Different forms of *Caulobacter crescentus*. When a large cell, like the one shown dividing, completes its fission, it releases another stalked cell and a swarmer cell. The flagellum on the swarmer cell is not visible in this micrograph.

Electron micrograph courtesy of Dr. J. M. Schmidt, Department of Microbiology, Arizona State University.

What is most intriguing about the cell cycle of *C. crescentus* is that differentiation occurs automatically as the division cycle proceeds. Each specific event occurs at a particular percentage of the complete cycle, regardless of the growth rate. For example, the flagella are assembled at 80% of the total time between divisions, and the pili are assembled at 10% of the next division cycle. Unlike sporulation or heterocyst formation, there is no external trigger for the differentiation events. Instead, the act of cell division itself seems to trigger the process. In the case of **conditional** division mutants, those expressed only under particular circumstances such as high temperature, pili are assembled only after the daughter cells have separated. At the present time, the mechanisms used to coordinate these events remain unknown. One fairly simple model suggests that the initial events are triggered by a specific cell mass. Until that mass is attained by synthesis of new cytoplasm, nothing else happens. As soon as the critical mass is exceeded, an irreversible sequence of events is triggered, resulting in cell division. Temporal sequencing may result from mechanisms like those used for sporulation in *Bacillus*.

Summary

Bacteria have a variety of ways in which they can regulate their synthesis of mRNA and protein molecules. These include repressors to turn off mRNA transcription by binding to operators and activators to turn on transcription by binding to promoters. The unit of transcription controlled by a given promoter and operator is the operon. The act of initiating the transcriptional process does not guarantee that a full-length mRNA molecule will be produced because the process of attenuation (shortening of transcripts) has been shown to occur in many instances. Some proteins, once made, can bind to their own mRNA molecule to regulate the amount of translation that takes place (autogenous regulation). Codon usage is also important in determining how much protein can be made from a given mRNA molecule.

Enzyme activity itself is subject to regulatory processes. The most commonly encountered types are competitive and noncompetitive inhibition (feedback inhibition). Competitive inhibition involves two or more substances trying to bind to the same active site on the enzyme molecule, while noncompetitive inhibition involves the interconversion of allosteric states of the enzyme by an effector molecule that binds away from the active site.

Some microbial cells, prokaryotic as well as eukaryotic, can carry out true differentiation. In some cases, the entire cell is changed as in sporulation or heterocyst formation. These processes seem to be regulated by alterations in RNA polymerase specificity and by literal DNA rearrangements. One particularly striking system is the production of motile swarmer cells by *C. crescentus* that later differentiate to produce a stalk in place of a flagellum and pili. Molecular mechanisms regulating this series of events are not yet known.

Questions for Review and Discussion

1. Describe the differences between transcriptional and translational control. What kinds of regulatory molecules (if any) are involved in these controls?
2. Explain how the lactose operon is regulated; indicate whether each control element is acting in a negative or positive fashion. What are the sensors that determine whether or not the operon should be activated?
3. Try to design an operon that performs the same functions as the lactose operon but uses only positive controls.
4. What is an allosteric protein? How are allosteric proteins involved in the various regulatory pathways that you have studied? Give some specific examples.
5. Why might it be beneficial to a cell to have both transcriptional regulation and feedback inhibition of the same pathway?

Suggestions for Further Reading

Ely, B., and L. Shapiro. 1984. Regulation of cell differentiation in *Caulobacter crescentus*, pp. 1–26. *In* R. Losick and L. Shapiro (eds.), *Microbial development*. Cold Spring Harbor Laboratory, Cold Spring Harbor, N.Y. An extensive but understandable review of the developmental events in *C. crescentus*.

Herskowitz, I. 1988. Life cycle of the budding yeast *Saccharomyces cerevisiae*. *Microbiological Reviews* 52:536–553. A more detailed discussion of the differentiation of this organism and other aspects of its life cycle.

Miller, J. H., and W. S. Reznikoff (eds.). 1978. *The operon*. Cold Spring Harbor Laboratory, Cold Spring Harbor, N.Y. Probably the most comprehensive discussion of the subject.

Vold, B. S. 1985. Structure and organization of genes for transfer RNA in *Bacillus subtilis*. *Microbiological Reviews* 49:71–80.

Yanofsky, C. 1982. Attenuation in the control of tryptophan operon expression, pp. 17–24. *In* M. Grunberg-Manago and B. Safer (eds.), *Interaction of translational and transcriptional controls in the regulation of gene expression*. Elsevier Biomedical Press, New York. A general review of Yanofsky's work on attenuation.

chapter

8

Bacterial Growth: Biological and Environmental Considerations

chapter outline

Quantifying Bacterial Growth

Estimating the Number of Bacteria in a Culture
Bacterial Growth Curves
Growth in a Chemostat

Effect of Environmental Parameters on Bacterial Growth

Temperature Requirements
pH and Salt Requirements
Other Important Environmental Parameters

Selective and Differential Media
The Bacterial Cell Division Process
Summary
Questions for Review and Discussion
Suggestions for Further Reading

GENERAL GOAL

Be able to describe bacterial growth in standard terms and explain how that growth can be regulated internally and externally.

The result of all the biochemical pathways and processes discussed in the preceding chapters is growth and reproduction of an organism. In multicellular organisms, growth can be a separate phenomenon from reproduction, which is not possible in unicellular microorganisms. As soon as a unicellular organism achieves a characteristic size, it divides and in so doing creates two individuals from only one. This chapter summarizes various aspects of bacterial growth. It also considers how that growth can be quantified and can be influenced by changes in the environment.

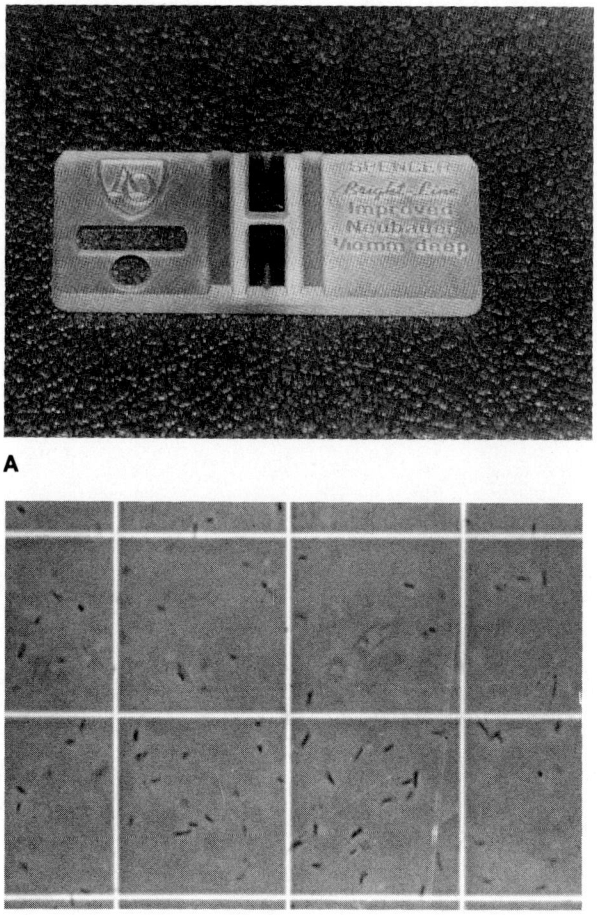

FIGURE 8.1

(**A**) Neubauer counting chamber. A glass coverslip placed over the center area creates two chambers 0.1 mm deep. Within each chamber are ruled lines to facilitate counting. (**B**) The counting chamber as seen through a microscope. Cells are *Saccharomyces cerevisiae*.

Quantifying Bacterial Growth

Estimating the Number of Bacteria in a Culture

Each bacterial cell grows for some period of time (**generation time** or interdivision time) and then divides to form two daughter cells of approximately equal size in the process of **binary fission.** In this sense, bacterial cells resemble unicellular, eukaryotic microorganisms. The obvious difference between the two groups is that eukaryotes divide by mitosis, whereas bacteria do not. As each new bacterial cell repeats the process of binary fission, it gives rise to a line of cells that theoretically is immortal because there does not seem to be any aging process in bacteria. The **growth rate** for a culture dividing by binary fission can be expressed as the generation time (minutes or hours) or the instantaneous growth rate constant μ, calculated as

Equation 8.1

$$\mu = \frac{\ln 2}{g}$$

where g is the generation time and ln the natural logarithm. The constant μ has the advantage of being applicable to any component of cell growth, such as protein synthesis, increase in cytoplasm, etc.

The consequences of growth by binary fission easily can be described mathematically. The equation has the general form

Equation 8.2

$$N_T = N_0(2^{T/g})$$

where N_0 is the number of cells initially present, N_T the number of cells present at time T, and g the generation time expressed in the same units as T. The quantity T/g represents the number of generations that have elapsed.

In order to use this equation, it is necessary to measure accurately the number of cells present in the culture. There are several ways in which this measurement can be taken. In theory, the number of cells present in an aliquot of any culture can be counted with the aid of a light microscope and a special counting chamber, such as the Neubauer counting chamber, that holds a very precise volume of liquid (figure 8.1A–B). There are two disadvantages of this method. First, both live and dead (incapable of further division) cells are counted. Second, if the culture is relatively dilute, there

A

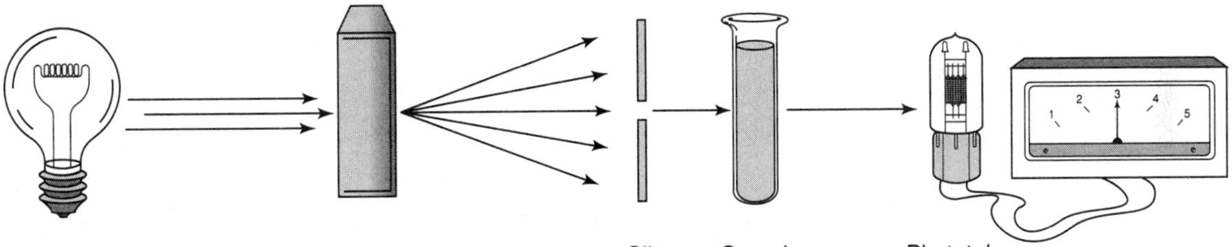

FIGURE 8.2

(**A**) Spectrophotometer with digital readout used to measure culture absorbance. (**B**) Diagram of the general principles of spectrophotometer operation. Light rays are passed through a monochromator to separate rays of different wavelengths. A slit selects light of a particular wavelength. The selected light then passes through the sample and strikes a photodetector. A meter reads the output from the phototube, the output being proportional to the intensity of light striking the phototube. Comparisons are made normally between a container of uninoculated medium and one in which bacteria have been growing.

(A) © Runk/Schoenberger/Grant Heilman

will be too few cells in a microscopic field to give any real confidence in the estimate of cell number.

Live cells can be counted if the culture is appropriately and carefully diluted and a measured sample placed onto solid medium. Because each viable cell put onto an agar plate will give rise to a colony, a count of the number of colonies will reveal the original number of viable cells present in the sample of diluted culture. The experimenter, however, will have to wait anywhere from one day to several weeks before counting in order to allow time for the organisms to grow. The dilution method cannot be used when speed is important, therefore. Moreover, the accuracy of the method depends on how carefully the original dilutions were performed.

Sometimes use is made of Tyndall's observation that the turbidity resulting from bacteria in a suspension will cause a beam of light to be reduced in intensity because the light is scattered at acute angles as it is reflected off the cells. A spectrophotometer or colorimeter (figure 8.2) is an instrument that measures the amount of light passing directly through a solution and the light **absorbance** of the culture (also called turbidimetry). The amount of absorbance is proportional to the number of cells present per milliliter below a certain upper limit that varies from instrument to instrument. This is a very quick and easy method that works well for cultures having cell densities from about 10^7 to 5×10^8 cells/ml. All particulate matter has similar effects on the instrument, even dead cells, and the instrument must be recalibrated if cell sizes are much larger or smaller than average.

A more recent development is the automatic particle counter that passes a solution through a very thin tube containing two parallel electrodes (figure 8.3). Particles present in solution disturb an electric field set up between the electrodes; therefore, they are counted by the electronic circuitry. The instrument can be set to discriminate between different size particles based on the amount of disturbance in the electric field. This type of counter can deal quickly with cultures of almost any cell density because, as is the case with a colorimeter, more dense solutions are simply diluted before being passed through the instrument. Once again, however, the count only indicates the number of particles present and not the number of living cells.

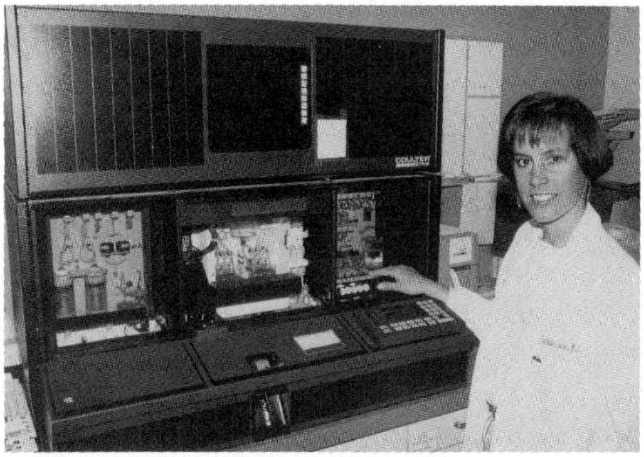

FIGURE 8.3

Coulter counter, an instrument that can be used to count particles (cells) in a solution.

Courtesy of Robin Scharferd, St. Luke's Medical Center, Phoenix, Arizona

Bacterial Growth Curves

If equation 8.2 is analyzed graphically, the resulting curve has a constantly changing slope because of the exponential nature of the equation. A typical graph might look like the one rendered in figure 8.4A. Predicting the cell density in a culture at some later time becomes a burdensome task because it is difficult to extrapolate a parabola.

However, equation 8.2 can be modified into a much more useful form if the logarithm of both sides is taken to give

Equation 8.3

$$\log N_T = \log N_0 + (T/g)\log 2$$

This is a linear expression that can be extrapolated easily. Because of the predictive possibilities of an extrapolated growth curve, the curve normally is presented in a graph of the logarithm of the cell number as a function of time and is multiphasic (figure 8.4B).

The initial or lag phase represents the period of time during which cells are adjusting their metabolism to

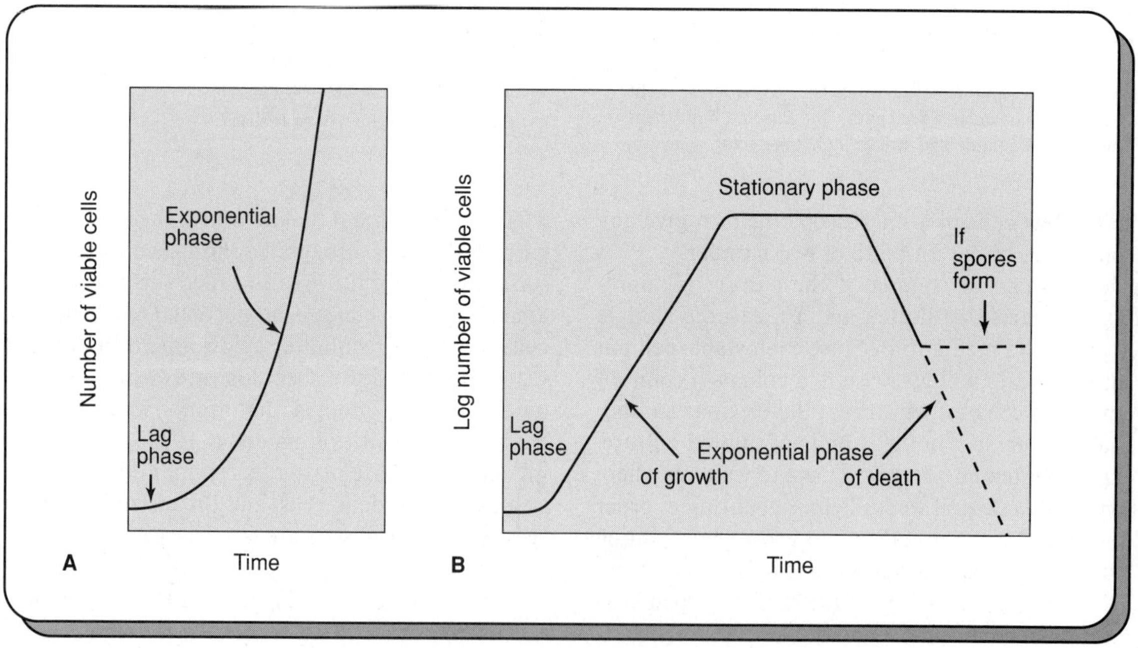

FIGURE 8.4

(**A**) Bacterial growth curve predicted by equation 8.2. (**B**) Bacterial growth curve predicted by equation 8.3. Log and exponential growth phases are the same as those shown in (**A**).

prepare for a new cycle of growth. Synthesis of RNA and necessary enzymes is occurring, and DNA replication begins as the phase draws to a close. If the cells used to inoculate this culture were growing actively already, the lag phase can be short or nonexistent, but nongrowing cells or cells that have been downshifted can stay in lag phase for hours. During the second or exponential phase of growth, the cells are growing and dividing at the maximum rate possible for that medium and those incubation conditions. As a result, equation 8.3 is applicable, and the graph indicates a straight line. Most experiments in bacterial physiology are carried out with "log phase" cells, that is, with cells harvested during the exponential phase of growth. The absorbance of an exponential phase culture is directly proportional to the dry weight of the cells present, a relationship that is not necessarily true during other growth phases.

The typical bacterial culture is grown in a closed container such as a tube or a flask with a cap. The culture can be aerobic or anaerobic, depending on the nature of the gas filling the container, and there is no limit to the amount of gas that can be present. The system is a closed one because no nutrients are added after cells are inoculated and no waste products are removed. Consequently, the cells cannot grow indefinitely and must stop eventually due to nutrient shortage and/or accumulation of inhibitory waste products. These inhibitory products can include hydrogen ions (low pH) or organic acids like lactic or acetic.

In terms of the bacterial growth curve, transition from a growing culture into a nongrowing one occurs during a third or stationary phase of growth. During this period of time, the rate of production of new cells exactly equals the rate of death of old cells, and the total number of cells remains constant. It is important to note, however, that cell metabolism does continue during this stationary phase and that some cells are still dividing.

The final portion of the bacterial growth curve is the exponential death phase, occasionally referred to as the decline phase, during which the number of viable cells in the culture undergoes a rapid drop. If nothing is done to reverse the trend, that is, if no new nutrients are added and no waste products removed, all cells in the culture may eventually die. Another possible outcome is the production of endospores that can, of course, survive for many years and serve as inocula for a new culture. Examples of both outcomes are depicted in the graph in figure 8.4B.

Growth in a Chemostat

Not all bacterial cultures are grown in closed vessels. Open systems (continuous cultures) that are useful for certain types of experiments can be devised. The most common type of open culture system is the **chemostat,** shown in figure 8.5. It consists of a culture vessel of variable shape that has provisions for the addition of new medium and for the overflow of excess liquid into a collection vessel. The chemostat conducts this process **aseptically,** that is, without contaminating the fresh medium.

The chemostat operation is deceptively simple. Fresh medium is added to the culture at some measured rate, which causes an overflow of liquid into the waste container. The overflow removes some cells and some waste products, but relatively little newly added

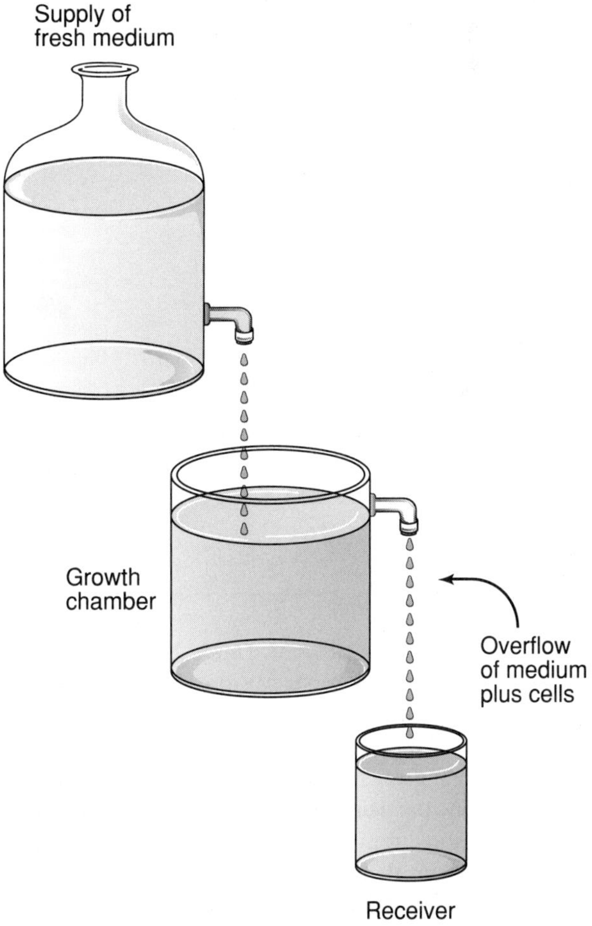

FIGURE 8.5

Bacterial chemostat. The reservoir holds fresh medium, and the receiver takes the overflow liquid. In the bottom of the culture chamber is a small stirring bar that can be turned by a magnet.

APPLICATIONS BOX

Bacterial Growth Curves

It is often difficult to get excited about the subject of bacterial growth curves, yet they describe events that always are occurring in our everyday life. For example, sourdough bread derives its name from the taste of the various acids secreted by bacteria growing in the flour mixture. Cooks can maintain their sourdough cultures for long periods of time provided that they remove some of the old material and add fresh flour (new nutrients) to the culture. If you buy a commercial package to start your sourdough culture but fail to feed it regularly, all the bacteria will eventually die. Any attempts to prepare sourdough bread from this culture will be disastrous.

In another vein, an outbreak of disease in a community often follows the same general pattern as a bacterial growth curve. The graph in this applications box provides an example of how a countywide outbreak of measles can be assessed. Initially, there were only a few cases in the community, but as time passed, logarithmic increase in the total number of reported cases was observed.

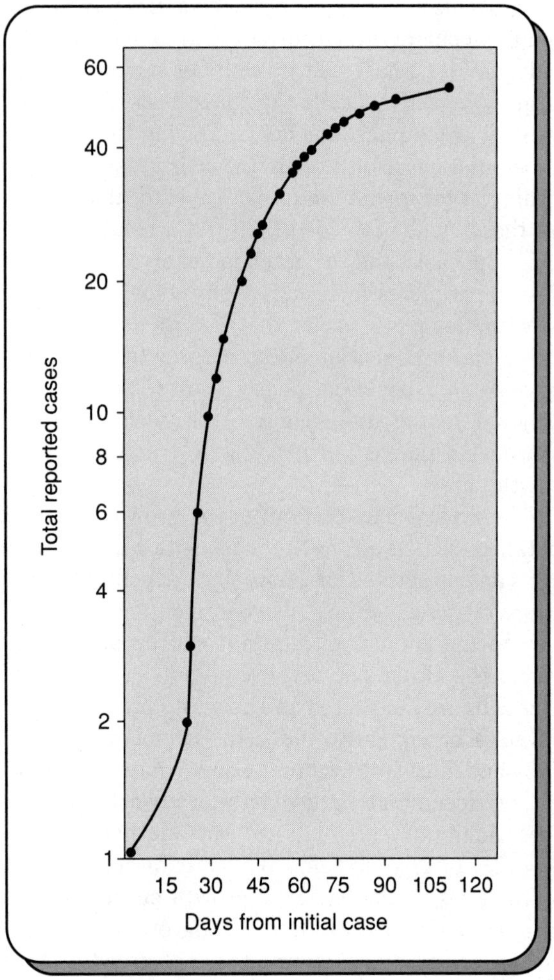

Growth curve of a disease. Cases of measles observed in a recent outbreak are plotted in the same manner as bacterial growth curves.

Source: Data from *Morbidity and Mortality Weekly Report* 35:99, 1986.

nutrients if mixing is efficient. Under these conditions, it is possible to produce a steady-state culture in which all cells are in the same phase of growth all the time. This occurs when the rate of addition of fresh medium is such that the cells in the culture are constantly deficient for one particular nutrient, the **limiting nutrient.**

When more limiting nutrient is added, the cells take it up rapidly but then must wait for the next portion of that nutrient before continuing their metabolism. Cells in a chemostat culture are, therefore, not quite the same as cells growing exponentially in a batch culture. Chemostat cells are always undergoing nutrient deprivation, while cells growing in the exponential phase in a closed culture are not. If the addition of fresh nutrient is slow but continuous, growth is correspondingly slow but continuous. If the addition of fresh nutrient is rapid, growth is rapid as well; however, the maximum growth potential of the cells imposes an absolute upper limit on the rate of addition of fresh medium. If the medium is added too fast, the rate of cell division cannot keep pace, and the culture will be lost by "washing out" of the chemostat.

A **turbidostat** is a variation on the chemostat that exploits the possibility of washout to achieve its effects. The medium used in a turbidostat is not a limiting medium but rather a complex one. Instead of cell growth being controlled by the rate of addition of a limiting nutrient, cell density is regulated by a controlled approach to washout conditions. At a flow rate for new medium that is slightly below the washout rate, the culture density is at a minimum. As the flow rate is reduced, the cells have more time for growth before being lost to the overflow stream, and the culture density increases. Thus, a characteristic culture turbidity is associated with the input rate of a particular medium. If a mechanical medium-dispensing system is used with a **turbidimeter,** an instrument that measures light scattering by cells, it is possible to grow a culture at a constant cell density without the use of a limiting nutrient.

Chemostats are useful particularly for experiments in which population dynamics are studied. Mixtures of organisms can be prepared and inoculated into a chemostat. Over time, the experimenter then can observe whether the organisms grow consistently in the ratio in which they were prepared initially or whether one organism becomes predominant. In the case of organisms like *Paramecium* that eat bacteria or other microorganisms, it is possible to study predator-prey relationships in a chemostat. Certain types of evolutionary experiments also can be carried out with a chemostat, such as the ability of a population of cells to adapt to the presence of low levels of a toxic chemical.

Effect of Environmental Parameters on Bacterial Growth

Temperature Requirements

Bacteria are much more versatile than plants and animals in their ability to grow at temperature extremes. In terms of the range of common experience, some forms of bacteria can be found growing in the polar seas at subzero (Celsius) temperatures and others can readily be observed growing in the near boiling water of thermal springs, such as those in Yellowstone National Park. In the laboratory, special equipment is used to control growth temperatures accurately (figure 8.6).

The temperature at which an organism grows is a function of at least two cellular components, cellular enzymes and lipids of the cell membrane. At lower temperatures, enzymes are less efficient catalysts; therefore, reactions are slower, resulting in slower growth of the organism. Lipids tend to become more solid and less fluid at lower temperatures, which has important consequences for membrane function. At higher temperatures, there is risk of too much membrane fluidity and/or possible protein unfolding and consequent loss of enzyme activity.

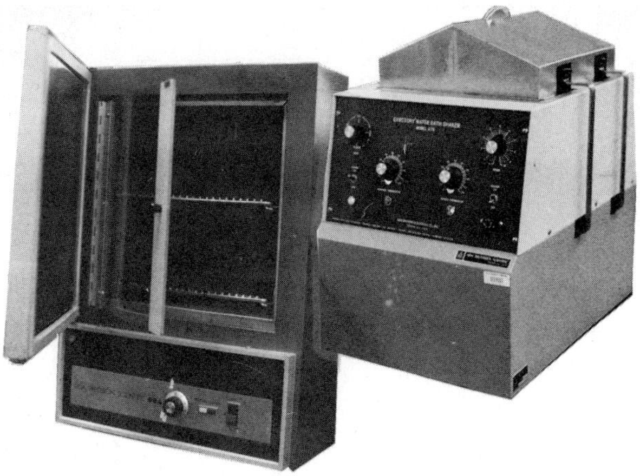

FIGURE 8.6

Microbial cultures are often grown at constant temperature in special equipment. On the left is an incubator that uses warm air to heat cultures. The inner glass door assists in preventing heat loss. More elaborate incubators also have cooling capabilities. On the right is a water bath shaker that can be set to either heat or cool water. Because water has a greater capacity to absorb heat than air does, it acts as a better heat sink, and water incubators (or shakers) can control temperatures more precisely. The shaking function allows for better aeration of cultures.

When graphs of growth rate as a function of temperature are prepared for various bacteria, there is a broad range of temperatures over which the rate of growth of the organism is relatively constant, with rather abrupt transitions at either end of the normal growth range (figure 8.7). Organisms that grow preferentially at high temperatures are said to be **thermophiles,** those growing only at low temperatures are **psychrophiles,** and those growing at moderate temperatures are **mesophiles.** In practice, a fourth group of organisms, sometimes designated as **psychrotrophs,** falls on the boundary between mesophiles and psychrophiles because the organisms tolerate cool temperatures while still having a relatively high optimum growth temperature.

There is always some discussion regarding the exact range of temperatures that should be assigned to each class of organisms. One generally accepted system classifies psychrophiles as those bacteria that will grow at $<0°$ C, but that are capable of growth at temperatures as high as 20° C. Mesophiles grow at temperatures ranging from 10 to 45° C, but thermophiles grow only at temperatures $>40°$ C (table 8.1). Note that the temperature ranges overlap, requiring that each bacterium be tested at several different temperatures. In all cases, the optimum or preferred temperature for growth is near the upper end of the range.

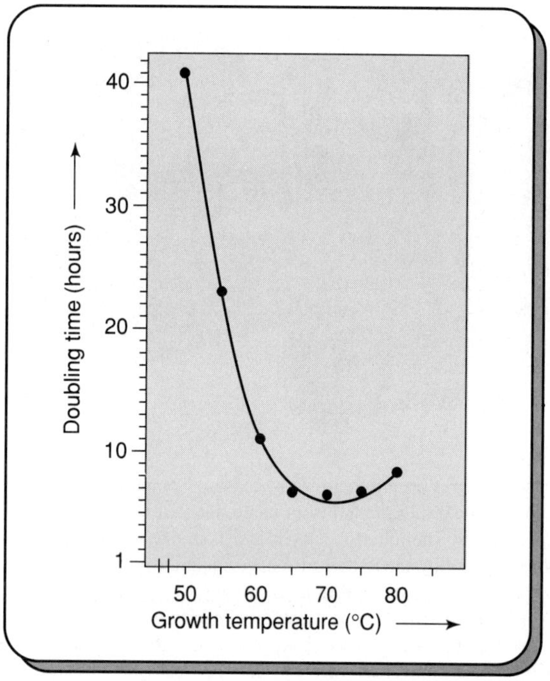

FIGURE 8.7

Doubling times for *Metallosphaera sedula* at various temperatures as calculated from the slopes of the exponential phases for individual growth curves.

From G. Huber, C. Spinnler, A. Gambacorta, K. O. Stetter, "*Metallosphaera sedula* gen. and sp. nov. Represents a New Genus of Aerobic, Metal-mobilizing, Thermoacidophilic Archaebacteria" in *Systematic and Applied Microbiology* 12:38–47. Copyright © 1989 Gustav Fischer Verlag, Stuttgart, Germany. Reprinted by permission.

All temperature classifications should be considered somewhat tentative because new bacteria are still being discovered. For example, only bacteria can grow (not merely survive) at temperatures greater than 75° C, and it has long been assumed that their upper limit was essentially 100° C. However, water samples taken from deep in the ocean where pressures can reach many hundreds of atmospheres indicate that chemolithotrophic bacteria can survive in the vicinity of sulfur emissions from volcanic vents. In the laboratory, the maximum permissible temperature for growth of some hyperthermophilic bacteria has been shown to be as high as 108° C.

pH and Salt Requirements

The availability of water itself has been shown to create a definite problem for cell growth in certain circumstances. The difficulty is a consequence of the semipermeable membrane surrounding all cells. According to a fundamental thermodynamic principle, if two solutions are touching and differ in their solute concentrations, there will be a net movement of molecules to equalize the solute and solvent concentrations in each solution. If, however, a semipermeable membrane prevents movement of solute molecules, solvent molecules will still tend to cross the membrane, increasing their number on one side of the membrane and thereby reducing the solute concentration. This process results in a net flow of water into or out of the cell cytoplasm.

TABLE 8.1 Classification of bacteria by preferred growth temperature

Type of Organism	**Customary Defining Temperatures**	**Proposed New Defining Temperatures (*T*)[1]**		
		T_{min}	T_{opt}	T_{max}[2]
Psychrophile	Grows from −5 to +20° C	<261 K <−12° C	<285 K 12° C	<290 K <17° C
Psychrotroph		261 to 269 K −12 to −4° C	285 to 305 K 12 to 32° C	290 to 315 K 17 to 42° C
Mesophile	Grows from 10 to 45° C	270 to 289 K −3 to 16° C	300 to 315 K 27 to 42° C	305 to 330 K 32 to 57° C
Thermophile	Grows only above 40° C	≥290 K ≥17° C	>315 K >42° C	>330 K >57° C

[1]Based on the work of Ratkowsky, D. A., R. K. Lowry, T. A. Meekin, A. N. Stokes, and R. E. Chandler. 1983. Model for bacterial culture growth throughout the entire biokinetic temperature range. *Journal of Bacteriology* 154:1222–1226.

[2]Minimum (*min*), optimum (*opt*), and maximum (*max*) temperatures are calculated according to the method of Ratkowsky et al. Temperatures are calculated in Kelvin (K) as well as in Celsius (C).

APPLICATIONS BOX

Temperature Relations

An attempt has been made to define psychrophile, mesophile, and thermophile on a more quantitative basis with the following equation:

Equation 8.4

$$k_r = b(T - T_{min})\{1 - exp[c(T - T_{max})]\}$$

where k_r is the growth rate constant (reciprocal of the time needed to reach 35% of maximum growth beginning with a standard number of cells); T is the absolute temperature; b is the regression coefficient for suboptimal temperatures; c is the regression coefficient for supraoptimal temperatures; and T_{max} is the highest temperature at which growth occurs. The hypothesis is that the significant characteristic should be T_{min}, the hypothetical low temperature at which the growth rate becomes zero. T_{min} and T_{max} are calculated by extrapolation, as shown in the graph in this box, and a comparison of the different defining temperatures is given in table 8.1.

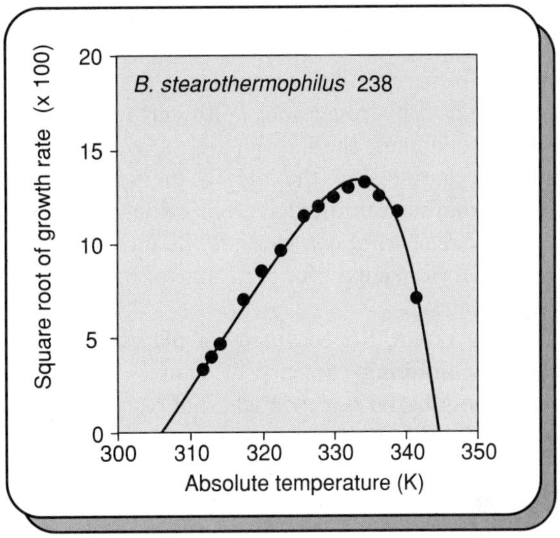

Prediction of maximum and minimum growth temperatures for *Bacillus stearothermophilus*. The graph has been prepared using equation 8.4. Extrapolation to zero growth rate gives the theoretical maximum and theoretical minimum temperatures. Dark circles indicate actual data points.

From D. A. Ratkowsky, "Model for Bacterial Culture Growth Rate Throughout the Entire Biokinetic Temperature Range," in *Journal of Bacteriology* 154:1222–1226. Copyright © 1983 by the American Society for Microbiology, Washington, DC. Reprinted by permission.

Unless a cell membrane is readily expandable, the different amounts of water result in the development of an actual pressure differential (usually positive but occasionally negative for cells in high salt medium) with respect to the interior of the cell. The absolute value of this pressure is the **osmotic pressure,** and it is equal to the amount of external pressure that would need to be applied to prevent any net movement of water.

An example of osmosis can be seen when cells are put into a very dilute medium, that is, one of low osmotic strength. Water tends to flow into a typical cell causing it to swell and possibly burst, as can easily be seen when red blood cells are put into distilled water. By way of contrast, a very concentrated medium of high ionic strength tends to draw water out of a cell, literally dehydrating the cell even though it is surrounded by water. If red blood cells are put into such a medium, they shrivel somewhat like raisins.

The term **water activity** (a_w) is often used to express the actual availability of water molecules to a cell. A high a_w value means that most water molecules are available to the cell (pure water has an a_w equal to 1.0), while a low value means that most water molecules are in tight complexes with solute molecules, rendering them unavailable for other purposes. Bacteria are highly variable in their water requirements. Organisms like *Pseudomonas cepacia* will grow in a distilled water medium using trace chemical contaminants as a source of nutrients. Others like *Halobacterium halobium* require large amounts of salt in the medium in order to grow; therefore, they are said to be **halophiles.** Extreme halophiles like members of the genus *Halobacterium* will not grow in media containing less than 15% NaCl, but they will manage to grow in solutions saturated with NaCl. Such conditions are found in the Great Salt Lake of Utah and in the Dead Sea of Israel. More moderate halophiles can tolerate salt concentrations up to 15%.

Because bacteria are surrounded by a semipermeable membrane similar to that seen in eukaryotic cells, they are subject to the same sorts of osmotic constraints. The concentration of substances within a "typical" bacterial cell (*Bacillus megaterium,* for example) is such that the osmotic pressure of the cell relative to a commonly used medium is on the order of 16 atmospheres. That such a cell does not burst is a tribute to the strength of the cell wall surrounding it. Extreme halophiles have an internal salt concentration as great as 3 M KCl and 1 M NaCl, whereas a more conventional organism might have a concentration one-tenth of that. In dilute solutions, therefore, extreme halophiles would be subjected to even greater osmotic pressures. Their lack of growth in dilute medium can be attributed to their requirement for high exterior salt concentrations to relieve some osmotic pressure on the cell.

Regardless of the salt concentration outside a cell, the concentration inside the cell tends to be quite constant. Nowhere is this more evident than with the concentration of hydrogen ion. With certain physicochemical techniques, the interior pH of a cell can be estimated regardless of the pH of the surrounding medium. From such studies has come evidence that the bacterium *Escherichia coli* can hold its interior pH at 7.3 (± 0.05) over an exterior pH range of 6.5 to 8.0 (a 30-fold change).

To some extent, the constancy of pH can be credited to the buffering capacity of many cytoplasmic components. A **buffer** is a chemical that resists changes in pH by releasing hydrogen ions as the pH rises or by combining with them as the pH falls. Particularly useful as buffering agents are carboxyl and amino groups; therefore, free amino acids in the cytoplasm aid in buffering pH changes.

Other Important Environmental Parameters

Other environmental factors can become important when attempting to grow bacteria isolated from special environments. For example, photosynthetic bacteria must be supplied with light of an appropriate wavelength. Bacteria isolated from the depths of the ocean are frequently **baryophilic** in that they will grow only under high pressure, generally from 300 to 500 atmospheres. Other bacteria may be able to grow only in the presence of a second, unrelated organism that provides some essential factor in their nutrition (syntrophy). Often this phenomenon occurs with organisms isolated from lakes or ponds and creates considerable frustration for microscopists. The organisms are clearly present visually, but they will not grow in pure culture and so cannot be further studied. Until a microscopist understands the syntrophic nature of such organisms, it will not be possible for him or her to grow the dependent organism in pure culture.

This brief sample of unusual aspects of bacterial culture should serve as a reminder of just how variable and omnipresent bacteria are. It is easy to mistakenly think that all bacteria are essentially similar and can readily be grown in the laboratory, whereas the exact opposite is more probable.

Selective and Differential Media

Culture media can be used in a way that accomplishes more than just growth of a bacterium. For example, a medium can be used for the purpose of **selection.** A selective medium is one that allows only a certain organism or group of organisms to grow while preventing growth of all other organisms. Selection can be based on any trait possessed by bacteria. For instance, if no nitrogenous compounds are added to a medium, it becomes selective for nitrogen-fixing organisms. A medium that contains carbon only in the form of toluene is selective for heterotrophic organisms that can break down the toluene compound.

Other types of media are considered to be **differential** in the sense that colonies growing on them display some sort of readily visible reaction indicative of one aspect of their metabolism. An agar plate (agar in a Petri dish) can contain a variety of carbon compounds including casein. Casein is not soluble in water; therefore, it appears as particulate matter throughout the medium just as it does in milk. If the organisms growing on the agar can make the proteases necessary to break down casein, casein particles will become soluble and a clear area will develop around the colony (figure 8.8). Because a variety of carbon sources have been provided, many bacteria will be able to grow regardless of whether they can use the casein present.

The construction of differential media is not quite as simple as it appears because of the interactions between biochemical pathways, but the principle is perfectly sound and a wide variety of differential media is available. Examples of differential, selective, and combined media are listed in table 8.2.

Occasionally, a situation arises in which an experimenter might want to search for organisms displaying particular characteristics that cannot be selected immediately. Although it is certainly possible to search through many thousands of colonies to find one rare organism, it is more practical to try to increase the relative size of the population of the desired organism by means of an **enrichment medium.** On such a medium, all or nearly all organisms will grow to some extent, but the rarer organisms should grow better and/or faster than the unwanted organisms because of the particular combination of nutrients provided. An increase in the proportion of desired organisms within the culture should result. Table 8.3 presents some examples of enrichment media.

In many instances, what should be a selective medium is in fact only an enrichment medium. Con-

APPLICATIONS BOX 1

Controlling the Culture Environment

Microbiologists must be able to control all parameters discussed in this chapter in order to optimize the growth of organisms in laboratories or factories. The problem becomes particularly acute as the volume of culture increases. The photograph in this box shows a fermenter (apparatus for growing cultures of microorganisms under carefully controlled conditions) together with its accompanying circuitry to control physical parameters.

Inside the fermenter is a large metal impeller with variable speeds that keeps the culture well stirred. The many pipes permit the introduction of whatever gas or gases are required by the organism. By maintaining a positive gas pressure in the fermenter, the experimenter can grow either aerobic or anaerobic cultures. The apparatus also permits automatic monitoring of various environmental parameters such as pH. An automatic dispenser allows for close control of the pH of the medium. The temperature of the culture is controlled by chilled or heated water that is circulated through a closed system of pipes within the fermenter vessel.

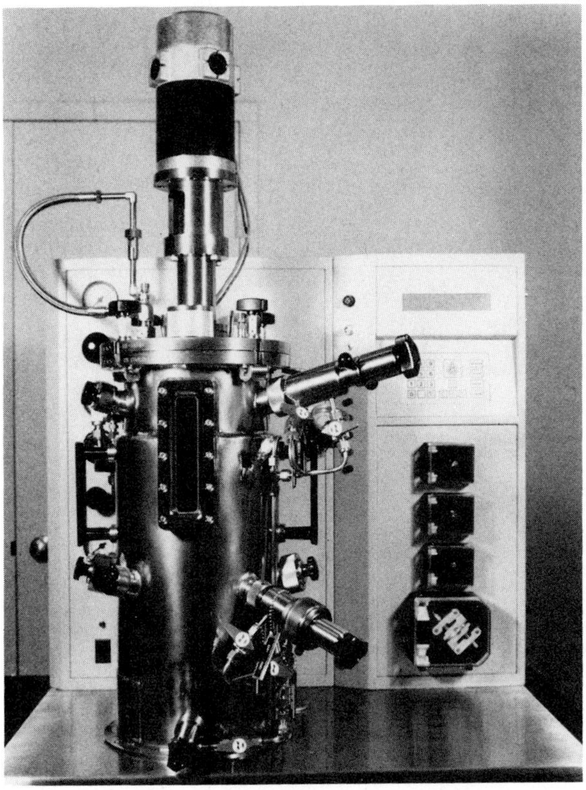

Fermenter used for batch culturing of microorganisms on a small scale. This particular instrument features computer control of the major physiological parameters.

Photograph courtesy of New Brunswick Scientific.

sider the specific case in which an experimenter wants to know if there are any photosynthetic bacteria in a sample of pond water that contains a multitude of organisms. By taking some pond water, mixing it with a nutrient solution in the laboratory, and allowing it to stand in front of a lamp, the experimenter can establish an enrichment culture. If the nutrient solution does not include any carbon-containing compounds, heterotrophs will have to try to get their carbon molecules from the low concentration of suitable chemicals in the pond sample or stop growing. Only autotrophs can grow extensively under these conditions. Moreover, only phototrophic autotrophs can grow in an unrestricted fashion in this culture, and they will be enriched relative to the other organisms in the sample. In practice, however, an environmental sample is likely to contain significant quantities of organic material; therefore, an attempted selection for phototrophs is really more of an enrichment. Repeated culturing of small aliquots of newly grown cells on the same medium will result in increased selection with each passage. Eventually, only photosynthetic autotrophs will remain.

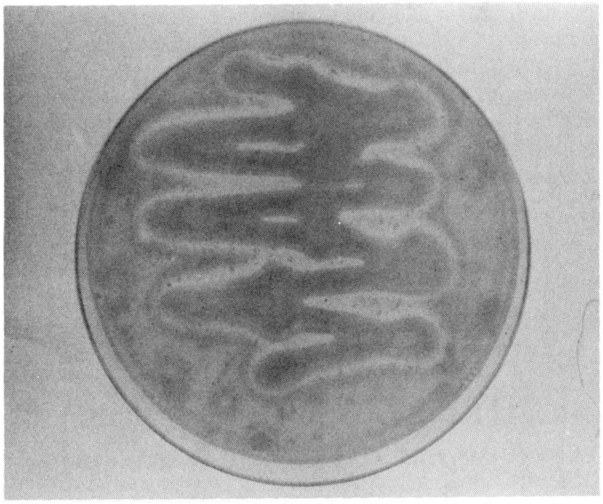

FIGURE 8.8

Differential reaction on an agar medium. The medium contains an excess of powdered milk that gives it a cloudy appearance due to the presence of casein (milk protein). The clear space seen surrounding the bacterial streak results from the production of proteases, enzymes that degrade casein to soluble amino acids.

TABLE 8.2 Examples of differential and selective media

Medium	Characteristic Properties
Differential	
Phenol red glucose broth	Turns yellow when the organism converts glucose to an acid and lowers the pH.
Lead acetate agar	Forms a black precipitate (PbS) if the organism produces hydrogen sulfide gas.
Selective	
Pseudomonas selective agar	Contains nalidixic acid, a chemical that prevents the growth of most other organisms by inhibiting DNA gyrase.
Mycoplasma agar	Contains penicillin, an antibiotic that inhibits transpeptidation and allows only bacteria lacking a cell wall to grow.
Selective and Differential	
MacConkey agar	Contains bile salts that allow only organisms adapted to living in the intestine to grow. Those bacteria that also can use lactose will form pink colonies. Those that cannot will form yellowish colonies. The color change is the result of a drop in pH.
Mannitol salt agar	Contains 7.5% sodium chloride and selects for moderate halophiles, usually for members of the genus *Staphylococcus*. Presumptive pathogens have yellow zones around the colonies due to the breakdown products of mannitol.

TABLE 8.3 Examples of enrichment media

Medium	Characteristic Properties
1% peptone in pond water	Enrich for certain aquatic chemoheterotrophs
Ammonium sulfate in basal salts with no organic carbon source	Enrich for ammonium-oxidizing autotrophs
Bacteria from a pond or lake on a basal salts agar with 60% hydrogen, 30% air, 10% carbon dioxide atmosphere and no organic compounds	Enrich for hydrogen-oxidizing autotrophs

Note: Basal salts medium contains only the necessary inorganic chemicals for growth. The environmental samples may contain variable quantitites of substances that can be used for growth by a wide variety of bacteria; therefore, these media, when used to enrich such samples, are not as selective as they might first appear.

The Bacterial Cell Division Process

Although there are several basic cell shapes, most of the work on bacterial growth mechanisms has been done with rod-shaped bacteria that divide by binary fission. This discussion focuses on that group, therefore. Microscopic observations of growing rods have indicated that cell width is essentially unchanged throughout the division cycle and that cell growth is accomplished by gradual lengthening. Hence, in a homogeneous culture of cells growing exponentially under a given set of conditions, a mother cell that is ready to divide will be approximately twice as long as two progeny or daughter cells that have just begun their independent growth. This proportionality is true regardless of the length of time between cell divisions as long as the cells are in a steady-state relationship (that is, metabolism is unchanged with time).

In order to maintain a balanced growth relationship within the cells of a culture, the components of each cell also must increase at an appropriate rate. The cell membrane must lengthen by the addition of new material, and as the time for division approaches, it must invaginate to form a septum separating the two daughter cells. The cell wall must lengthen at a corresponding rate and have some chemical bonds broken so as to create new corners for the eventual daughter cells. The actual site of division within the cell is bounded by two periseptal annuli, ring-shaped zones in which the cell wall and the plasma membrane are closely adherent (figure 8.9). After an annulus is used for cell division, it moves near the newer pole of the daughter cell and is altered chemically so that it is no longer functional. This alteration makes the cell wall material near the poles of a cell very stable. Before the next cell division can occur, a new pair of periseptal annuli must be synthesized, one on each side of the cell midline between the attachment points for the two DNA molecules.

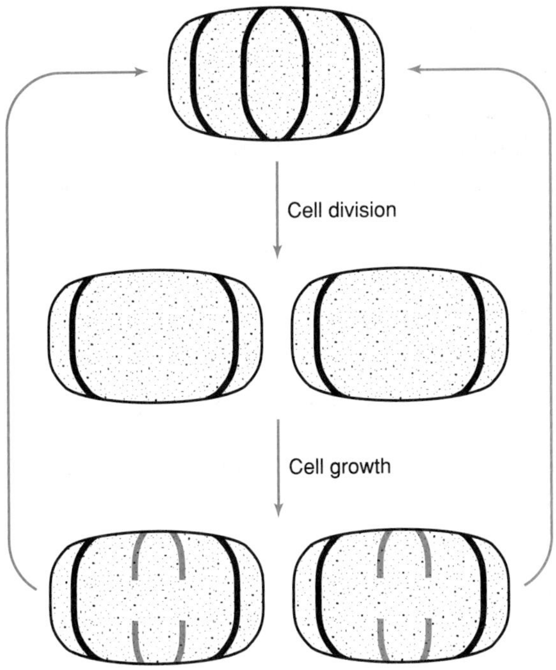

FIGURE 8.9

Periseptal annuli are essential for cell division. A newly arisen cell has two annuli, one located at each pole. Prior to the next cell division, another pair of annuli (in color) must be laid down near the center of the cell. Only the new annuli are involved in cell division.

A **steady state** implies that the number of basic cytoplasmic components per unit volume remains the same throughout a cell's growth cycle. In particular, the ribosomes must increase at the same rate as the cytoplasm itself does so that the protein synthetic capability of the cell per unit of cytoplasm is constant. In rapidly growing cells, more than 70% of the cell mass can be composed of ribosomes. Thus, ribosomes are assembled at all times during the cell cycle. If cells are moved to a richer medium (a "shift up" experiment), any increase in growth rate must be preceded by a period of increased ribosomal synthesis. Conversely, during a "shift down" experiment from a rich medium to a poor medium, ribosomal synthesis must be curtailed.

DNA molecules, on the other hand, basically are required in a ratio of only one per cell. A newly arisen daughter cell may have only one chromosome, but during cell division, a mother cell must have two chromosomes or else one daughter cell will have no DNA and, therefore, be inviable. Although ribosomes and other cytoplasmic components are made continuously, DNA replication is triggered only once for each anticipated cell division, regardless of how often cell division occurs. In other words, DNA replication must be initiated only when the cell is nearing the time for division (high ratio of cytoplasm to DNA). Once replication is triggered, it proceeds to completion, and the cell must divide or the proportionality between DNA and cell size will be lost.

It is customary to distinguish between the "C period" during which a bacterial chromosome replicates, and the "D period" or "division" interval between the end of the replication and the actual formation of two daughter cells. At very slow growth rates, there is a variable I or "interreplicational" period in which no DNA synthesis or cell division is observed after cell division is complete. From a biochemical point of view, events occurring during the C and D periods should require a roughly constant amount of time for a given temperature. For *Escherichia coli* growing at 37°C, C and D values are about 40 minutes and 20 minutes, respectively. Cells vary their growth rate by changing the rate at which cytoplasmic components are synthesized, but the rate at which any individual replication complex travels along a DNA molecule is essentially invariant. Similarly, the process of septum formation occurs at a standard rate. Until the interval between cell divisions is less than the sum of the C and D periods, a cell has no difficulty in accommodating all necessary biochemical events. When cell division occurs more frequently, however, the cell clearly has a problem.

All bacteria studied thus far already employ bidirectional replication in which two replication complexes begin at the same point on a DNA molecule and proceed in opposite directions, cutting the time needed to replicate the entire DNA molecule in half. Additional gains in time can be accomplished only if cells begin a new round of replication before the last round has been completed (figure 8.10). Thus, if a culture has cells dividing every 40 minutes, a new round of replication must be initiated 20 minutes before the next cell division. It will not be completed by the time of the next cell division but 20 minutes before the subsequent division. The total time required will be $40 + 20 = 60$ minutes, thereby allowing the standard time for a D period. The presence of incomplete genomes in rapidly growing cells compels geneticists and physiologists to talk of **genome equivalents,** the amount of DNA equivalent to one complete double-strand DNA molecule, rather than complete chromosomes.

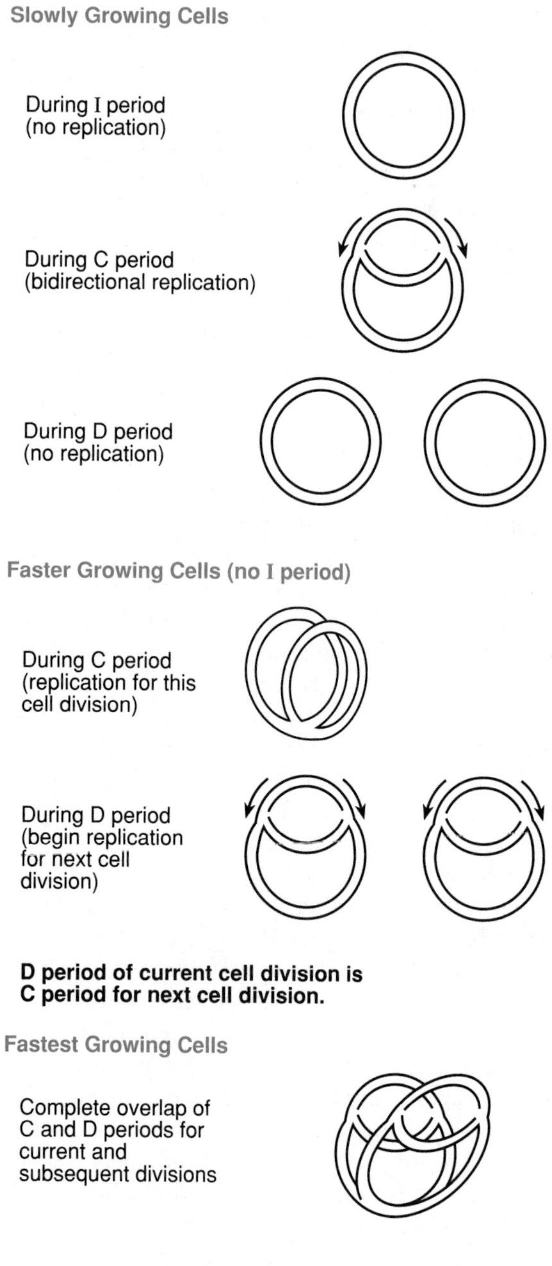

FIGURE 8.10

Multiple replication forks are necessary in rapidly dividing cells.

The cell division strategies employed by bacteria would not work for cells with the genetic complexity of eukaryotic cells. One problem concerns the size of DNA molecules needed to code all cellular functions. Bacteria have only a single origin of replication and depend on multiple initiations from that origin to allow cell division to occur more frequently than the time represented by the sum of the C and D periods. Multiple initiations lead to an excess of DNA from the origin region and, if used with a DNA molecule of a size typically found in a eukaryotic cell, would result in a mass of DNA equal to the entire weight of a bacterial cell. Instead, eukaryotic cells employ multiple, bidirectional replicational origins per chromosome.

Multiple replicational origins, however, necessitate a different means of segregation than that employed by bacteria. Bacteria segregate their DNA by attaching the single origin region to the membrane and moving DNA along as the membrane grows. With multiple origins of replication, DNA segregation of this nature is not possible because the DNA molecules would gradually be stretched to the breaking point by their multiple attachments. It has been suggested that mitosis, with its network of microtubules that serve to move the chromosomes, has evolved so as to allow the existence of DNA molecules with multiple origins of replication.

Summary

The physical conditions appropriate for bacterial growth can vary widely. Thermophilic organisms grow best at high temperatures, psychrophilic organisms grow best at low temperatures, and mesophilic organisms fall in the middle range. Halophiles grow only in high-salt media.

Because most bacteria reproduce by binary fission, the equation describing their growth is an exponential one. Semilogarithmic plots of the logarithm of cell number versus time produce more usable growth curves. Logarithmic curves usually designate four phases of cell growth: lag phase (preparation for growth), exponential phase of growth (period of maximum reproduction), stationary phase (transitional period in which death rate equals rate of reproduction), and exponential phase of decline (culture dies or sporulates). The basic cell cycle is subdivided into three periods: interreplication (I), chromosomal replication (C), and cell division (D). The I period is of variable length, but the duration of the C and D periods normally is constant.

Measurement of cell number in a culture can be accomplished by direct microscopic counts, by colony counts on an agar plate, by changes in an electric field induced by the passage of individual cells, or by turbidimetry. Regardless of the method used, none is without drawbacks. Quick methods may not permit the

APPLICATIONS BOX 1

Synchronized Cell Divisions

Microbiologists have long sought to have synchronized cultures in which all cells divide simultaneously so that they can study the events of a cell cycle. Various methods of synchronization have been tried including: (1) starvation of cells for a required nutrient that can be restored abruptly to initiate the experiment; (2) shift up and shift down experiments; and (3) permanent attachment of bacteria to a nitrocellulose membrane. In the third case, daughter cells are not bound and, therefore, can be washed away to yield a population of cells that have all recently completed a division.

The major problem with all synchronization experiments has been that any synchronization achieved is not maintained for a significant period of time. Normally, loss of synchronized division is visible by the second cell division. Instead of a truly synchronized population of cells, what is achieved is a population in which the growth rate is changing constantly with time.

investigator to distinguish between live and dead cells or between cells and dust particles. Other methods may be more accurate but substantially slower to use.

Bacterial cultures can be maintained in closed (nothing enters or leaves once the culture is inoculated) or open systems (nutrients are added, cells and waste products are removed). Closed culture systems are much easier to use, but continuous cultures have their place in physiological and evolutionary studies. Care must be taken in a continuous culture to control the rate at which the limiting nutrient is added because all cells literally can be washed away in a flood of nutrient solution.

Questions for Review and Discussion

1. How can bacteria manage to divide as often as once every 10 minutes when chromosomal replication requires some 40 minutes to complete?
2. How can the ingredients in a medium be altered to make the medium differential? Selective? What constitutes an enrichment medium?
3. What is the difference between a closed culture and an open one? What is the effect of nutrient limitation on each type of culture?
4. What sorts of culture conditions must be controlled in order to induce proper bacterial growth? What does the suffix *philic* signify? How is it used to describe some properties of bacterial cultures?
5. Why is the parameter T_{min} described as the "hypothetical" minimum growth temperature?

Suggestions for Further Reading

Baross, J. A., and J. W. Deming. 1983. Growth of "black smoker" bacteria at temperatures of at least 250°C. *Nature* 303:423–426.

Cavalier-Smith, T. 1981. The origin and early evolution of the eukaryotic cell, pp. 33–84. *In* M. J. Carlile, J. F. Collins, and B. E. B. Moseley (eds.), *Molecular and cellular aspects of microbial evolution.* Cambridge University Press, Cambridge. A comparative discussion of the main similarities and differences between prokaryotes and eukaryotes. The text also includes a detailed analysis of why the prokaryotic cell cycle does not work for a eukaryotic cell.

Cooper, S. 1991. *Bacterial growth and division: Biochemistry and regulation of prokaryotic and eukaryotic division cycles.* Academic Press, Inc., San Diego. An extremely interesting (and opinionated) text. Much of the material presented in this chapter has been influenced strongly by this source. It is an excellent reference for additional information as well.

Gerhardt, P., R. G. E. Murray, R. N. Costilow, E. W. Nester, W. A. Wood, N. R. Krieg, and G. B. Phillips. 1981. *Manual of methods for general bacteriology.* American Society for Microbiology, Washington, D.C. An extensive laboratory manual covering all aspects of bacteriology and intended as a reference tool.

Helmstetter, C. E. 1987. Timing of synthetic activities in the cell cycle, pp. 1594–1605. *In* F. Neidhardt, J. L. Ingraham, K. B. Low, B. Magasanik, M. Schaechter, and H. E. Umbarger (eds.), *Escherichia coli and Salmonella typhimurium: Cellular and molecular biology.* American Society for Microbiology, Washington, D.C. Considers the factors that influence the duration and timing of the I, D, and C periods of the cell cycle.

Yayanos, A. A., and A. S. Dietz. 1983. Death of a Hadal deep sea bacterium after decompression. *Science* 220:497–498.

Zillig, W., I. Holz, D. Janekovic, H.-P. Klenk, E. Imsel, J. Trent, S. Wunderl, V. H. Forjaz, R. Continhuo, and T. Ferreira. 1990. A hyperthermophilic sulfur-reducing archaebacterium that ferments peptides. *Journal of Bacteriology* 172:3959–3965.

chapter

9

Viruses and Related Entities

chapter outline

■

GENERAL GOALS

Be able to describe the types of experiments used to study viruses and to understand the relevance of the information these experiments can provide.
Be able to explain possible viral life cycles and different types of nucleic acid replication.
Be able to discuss some insights into the workings of molecular biology that viruses have provided.

Viruses are ubiquitous in the biological world, infecting all types of organisms regardless of size. This chapter first presents the bacterial viruses. Some have been studied very intensively and serve as a good introduction to the techniques used in classic virology. Then selected viruses that infect eukaryotes are surveyed and compared with bacterial viruses to demonstrate some similarities and differences.

Bacterial Viruses

All viruses are acellular, obligate intracellular parasites possessing only one type of nucleic acid encased within a protein coat, but beyond that fundamental similarity there is enormous heterogeneity. Despite the fact that bacteria are small, they host a multitude of different viruses, far too many for a thorough discussion of each type. Instead, three viruses that infect *Escherichia coli* and have been studied intensively are discussed in this section. It is important to remember, however, that most bacteria seem to have associated viruses that can become active under appropriate conditions.

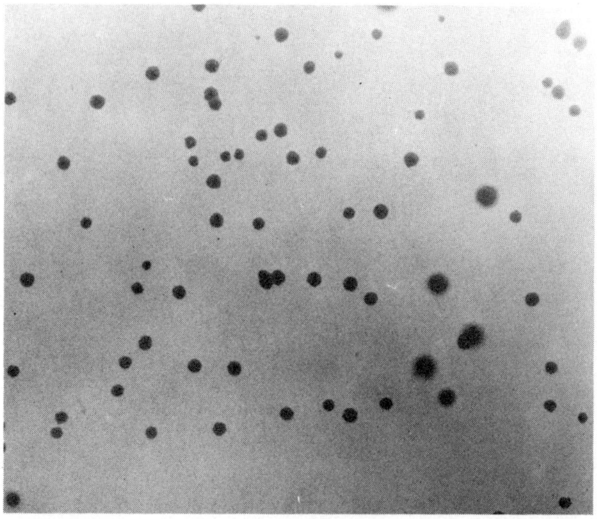

FIGURE 9.1

Viral plaques in a bacterial lawn. Each plaque or hole interrupting the continuous growth of the bacteria is the result of infection of a progenitor cell by a single virus particle. The progeny released from the subsequent cycles of infection have destroyed sufficient cells to make a macroscopically visible hole.

An alternative name for a bacterial virus is a **bacteriophage,** a term coined by the French scientist Felix D'Herelle when he observed that a virus seemed to "eat" a hole in an otherwise confluent **lawn** or layer of bacterial cells on an agar surface (figure 9.1). In common usage, the term is often shortened to "phage" and given either a French or American pronunciation. Each hole or **plaque** present in the lawn represents an area of bacterial cell destruction or growth inhibition resulting from a viral infection. A plaque bears the same relationship to the original population of viral particles as does a well-isolated bacterial colony to the cells in a culture. That is, each plaque is the result of the multiplication of viruses arising from a single infectious center. A **plaque-forming unit** gives rise to a single plaque. It can be a single viral particle that later infects a cell or a host cell already infected by one or more viral particles. The term is analagous to the bacteriologists' colony-forming unit; therefore, in a plaque assay, the number of plaques produced can be used to measure the number of viral particles and/or infected cells present in the original population.

Bacteriophage T4, a Lytic Virus

Bacteriophage T4 is one member of a larger group of "T phages." Max Delbrück and his colleagues did exhaustive studies of this virus and gathered a huge body of information that has made phage T4 one of the two best-understood bacteriophages. It is a large virus having both a "head" (capsid) and a "tail" region (figure 9.2A–B). Morphologically and genetically phage T4 is similar to phages T2 and T6; therefore, these viruses are often called the T-even group. The **capsid** or head of a **virion** or viral particle is comprised of a protein coat surrounding a large double-strand DNA molecule; the tail is composed entirely of protein. At the end of the tail is a base plate to which tail fibers are attached on its lower surface. DNA and associated proteins located within the capsid are sometimes called the **nucleocapsid** of a virion. Plaques produced by T4 are mostly devoid of cells, indicating that it is a **lytic virus** that has the ability to destroy its host cell by inducing cell lysis at the end of its infectious cycle. The few colonies that do appear within a plaque can be shown to contain only cells resistant to T4 infection.

Classically, two types of experiments are used to study viral infections. The first is the **one-step growth experiment.** In this experiment, cells are infected under conditions in which attachment is permitted only for a limited period of time, and then infection/reinfection is prevented. Preventive measures include dilution of the culture to a point beyond which collision between virions and cells becomes improbable (usually $<10^7$/ml) and addition of special molecules that bind to the tail fibers and block infection. Samples of the culture are taken at various times and plated to deter-

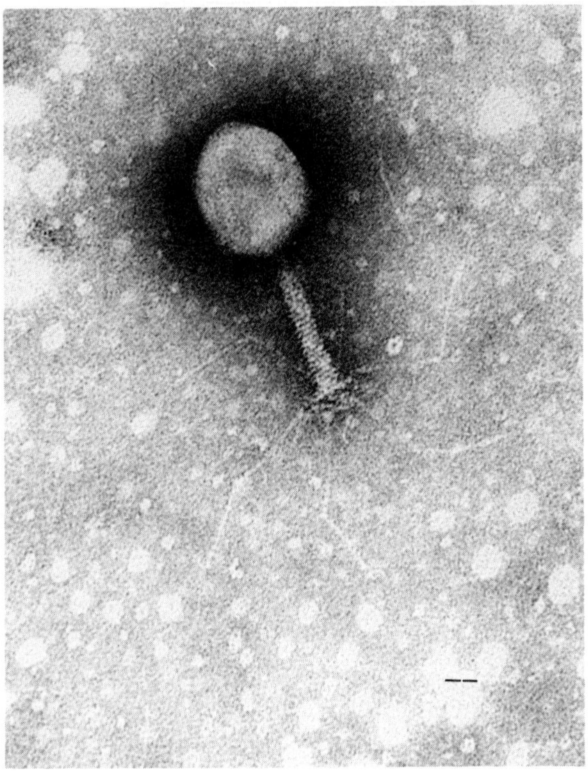

A

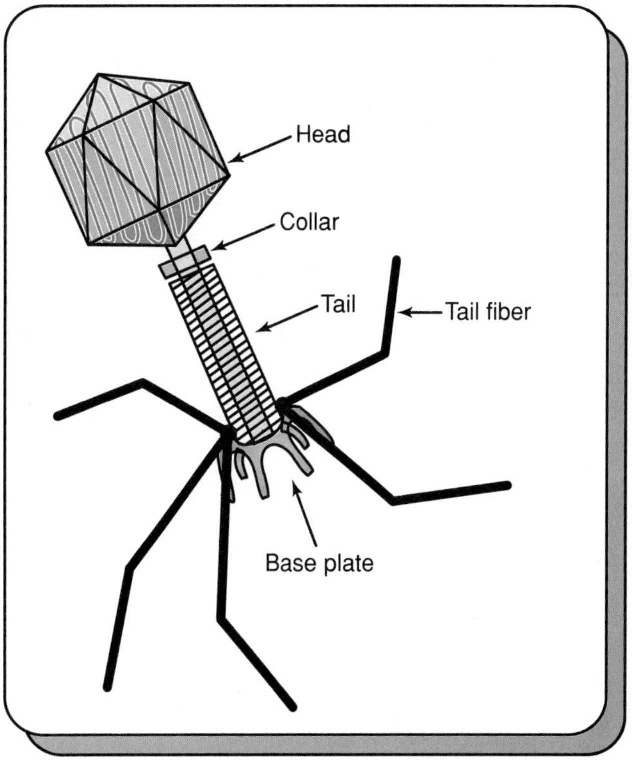

B

FIGURE 9.2

Bacteriophage T4. (**A**) Electron micrograph of the virus. (**B**) Schematic drawing detailing the various structures shown in the micrograph.

(A) Reprinted from Birge, E. A. 1988. *Bacterial and bacteriophage genetics: An introduction,* 2nd ed. Springer-Verlag, New York. Used by permission.

mine how many plaques will be formed using extra **indicator bacteria** that serve as hosts. A typical triphasic graph prepared from data obtained in a one-step growth experiment is presented in figure 9.3. The **latent phase** represents the interval during which the viruses are making new DNA and viral particles within the infected cells. Each infected cell thus represents a single infectious center until it lyses. No matter how many viruses are present within an infected cell, they can give rise only to one plaque because they are all released at exactly the same tiny spot on the agar surface.

During the **rise phase,** however, the cells of the culture have begun to lyse or break open. A mixture of free virions and infected cells results, and the number of individual plaque-forming units increases. The increase continues until all infected cells have lysed, leaving only free virions and uninfected cells present in the culture. Lysis of all infected cells establishes a stable state in the culture called the **plateau phase.** In the case of dilute cultures, constancy in the number of plaque-forming units will persist until the number of cells in

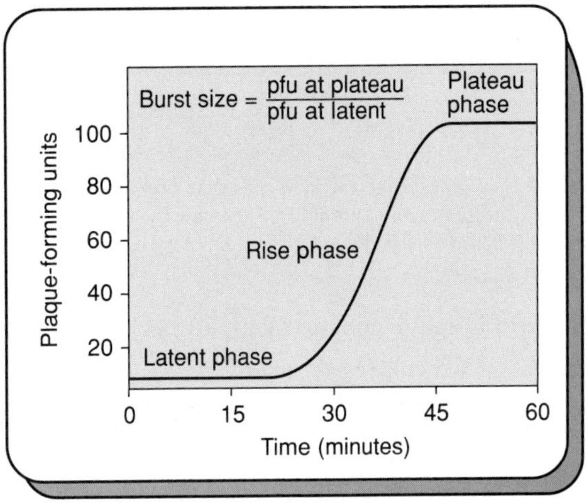

FIGURE 9.3

One-step growth curve for phage T4. Plaque-forming units (Pfu) may be either infected bacteria or free virions. The various phases of the growth cycle are labeled. Note that the average burst size is obtained from the ratio of the number of Pfu in the plateau phase to the number of Pfu in the latent phase.

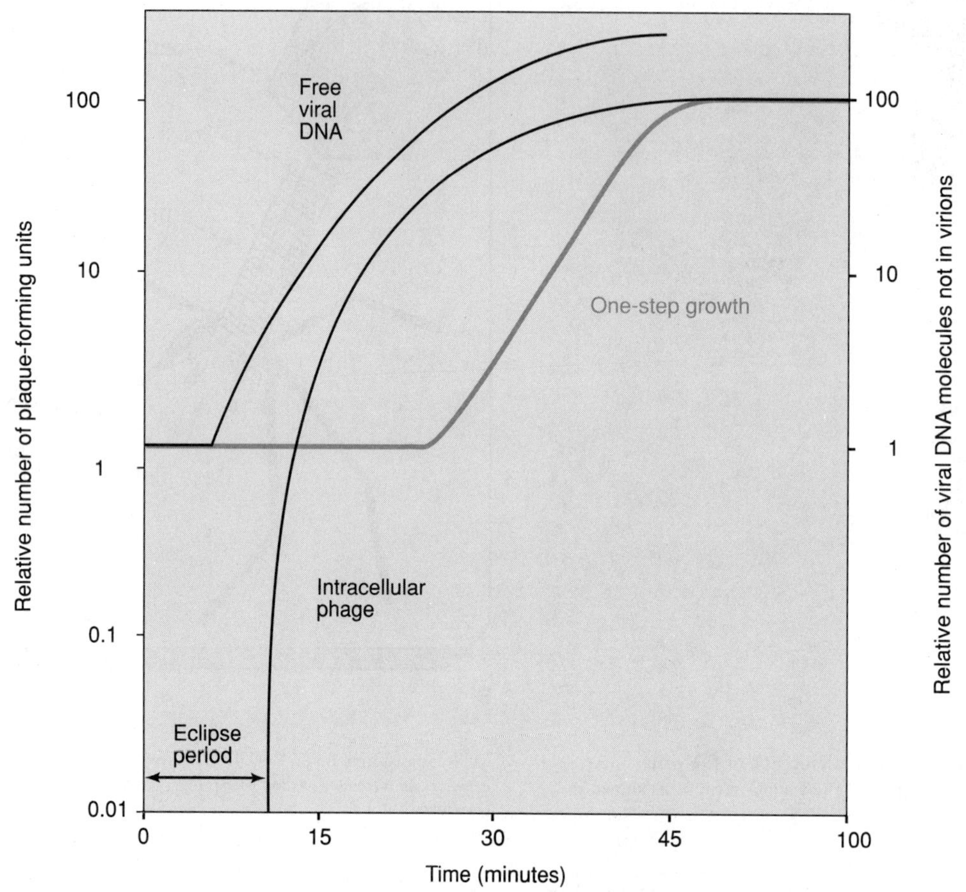

FIGURE 9.4

Premature lysis experiment. Infected cells were broken open before they were expected to do so normally. Subsequently, the number of functional virions and the amount of viral DNA were determined.

the culture is again sufficiently high to make collision with a virion probable.

The ratio between the number of plaque-forming units present in the plateau phase and the number of plaque-forming units present during the latent phase represents the **burst size,** the average number of virions released by one infected cell. It is important to remember that the burst size is merely an average. The actual number of virions released by a single cell can vary from as few as 1 particle to as many as 90. The T4 burst size has been estimated to be approximately 50.

The second type of classic experiment is the **premature lysis experiment** developed by A. H. Doermann. Infected bacterial cells are lysed artificially before the infecting phage would normally initiate lysis, and the released contents are examined for number of infectious phages, amount of viral DNA present, etc. A typical set of graphs from a premature lysis experiment is shown in figure 9.4. One important feature to notice is the **eclipse period** during which the infected cell does not contain any infectious virus. In addition, once infectious phage particles do appear within the cytoplasm, their number increases at an arithmetic rate rather than at a geometric one (compare with the bacterial growth curves of chapter 8). Arithmetic growth implies that each viral particle is assembled separately as though on an assembly line in a factory and is not the result of cleavage of old viruses. This kind of growth reaffirms the acellular nature of viruses.

Consider one infectious cycle for phage T4. When T4 is not associated with a cell, it is metabolically inactive. After bacterial cells are added to a suspension of virions, viral particles can bind to the surface of the cells. This is an intrinsic activity of viral tail fibers by which they specifically adsorb to certain structures on the surface of a cell. In the case of phage T4, the viral tail fibers actually attach to either of two sites, a disaccharide in the core fragment of one of the polysaccharides attached to the lipid A moiety in the outer membrane (figure 3.11) or the outer membrane protein C (OmpC). Cells lacking both these structures are resistant to phage T4, while cells lacking only one of the receptors show a reduced efficiency as hosts. Once the appropriate surface structures have been bound by the tail fibers, electron micrographic studies have shown that the fibers bend and allow the base plate of the virus to touch the surface of the cell. One component of the base plate acts like a lysozyme, having the ability to degrade the glycan chains that comprise the peptidoglycan layer and, thereby, facilitating contact of the phage with the cell's inner membrane.

Arthur Hershey and Martha Chase labeled capsid proteins with radioactive sulfur (^{35}S) and DNA of virions with radioactive phosphorus (^{32}P). Their experiments showed that after attachment of the virus, only the DNA of the virion actually enters the bacterial cell. This observation was one of the early proofs that DNA is the genetic material of a virus. It also implied that there is some sort of specific mechanism to remove the DNA from the phage particle. Apparently, when the base plate touches the cell surface, several things occur more or less simultaneously. A plug in the center of the base plate is released, opening a channel through the hollow tail core and into the head. The tail sheath contracts like a spring, and it is assumed that this contraction provides some energy to move part of the DNA molecule through the tail channel. There have been suggestions that phage T4 behaves like a syringe in moving the bulk of its DNA into the host cell.

However, this is incorrect because, unlike a syringe, the total volume occupied by the T4 head and tail remains unchanged. Instead, phage DNA is ejected rather than injected, and the bacterium cooperates in its own infection. This is inferred from the fact that phage infection is unsuccessful if the electrochemical gradient ($\Delta\psi$) across the cell membrane is destroyed. The DNA molecule is assumed to be transported in some fashion across the cell membrane and into the cytoplasm. This phenomenon also has been shown true for other viruses, including phage lambda (discussed in the following subsection), as well as for naked DNA molecules (see chapter 10).

The observation that only viral DNA enters the host cell leads to an interpretation of the eclipse phase. The eclipse must correspond to the time immediately after infection when only viral DNA is present in the cells. The corollary to this conclusion is that, under experimental conditions, naked viral DNA is not infectious.

Once inside the cell, viral DNA usurps control of many cellular functions. One protein product of initial transcription and translation acts as a sigma factor to modify RNA polymerase of the host cell so that it preferentially recognizes viral promoters instead of bacterial promoters. Other viral proteins produce additional modifications in the RNA polymerase holoenzyme complex. Certain species of tRNA molecules are present normally in relatively low concentrations, presumably because *E. coli* rarely uses those codons; but T4 DNA codes for additional copies of these tRNA molecules so that there will be plenty available during viral protein translation. Other viral proteins are involved in destroying DNA of the host cell. This destruction provides the virus with extra material for synthesizing its own DNA and does not hinder its ability to reproduce because the virus codes for many of its own functions. The enzyme responsible for degrading DNA of the host cell does not attack viral DNA because that DNA contains a modified base (5-hydroxymethyl cytosine) often with one or more glucose molecules attached (figure 9.5). The modified cytosine prevents the degrading enzymes from recognizing the viral DNA molecule as a substrate.

After the necessary proteins and DNA have been synthesized, new viral particles can be assembled. The actual assembly process is most interesting in that it seems to closely resemble the formation of crystals from a supersaturated chemical solution. When proteins of the viral head and tail have accumulated in sufficient quantity, the head and tail self-assemble in an orderly fashion similar to the ways in which eukaryotic microtubules and bacterial flagella do. The empty heads then are filled with DNA and its associated proteins. At this point, the tails are attached and a functional virus results, marking the end of the eclipse period. It is thought that one end of viral DNA protrudes into the tail core. This end is the one ejected by tail contraction.

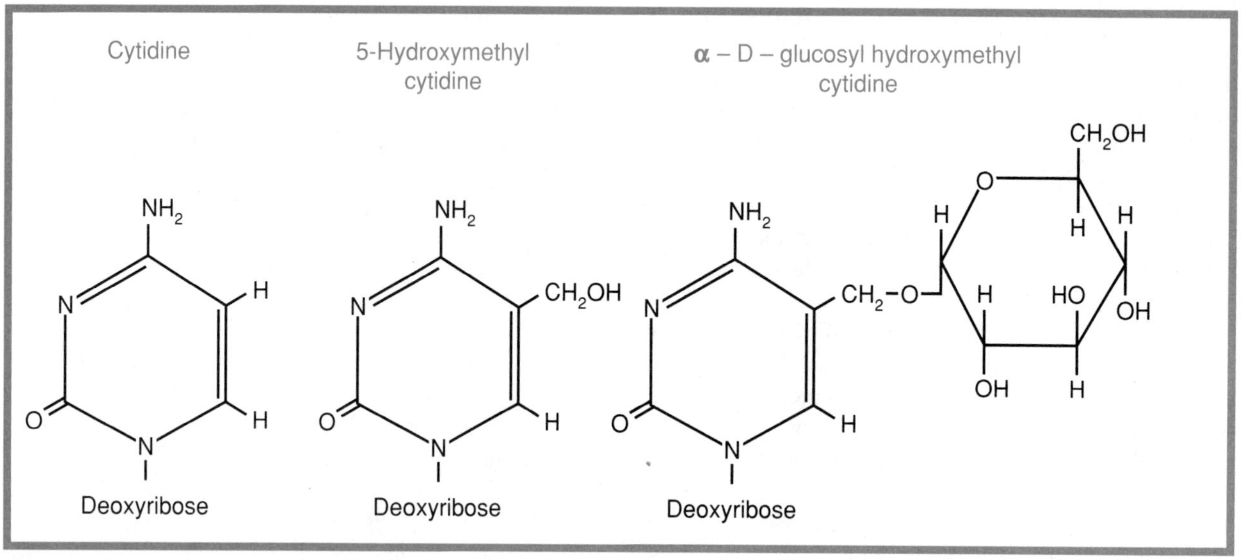

FIGURE 9.5

DNA modifications found in phage T4. Instead of cytidine, the DNA contains 5–hydroxymethyl cytidine. The extra hydroxyl group allows one or more glucose molecules to be attached to the DNA. The amount of glucosylation varies among the T-even phages.

APPLICATIONS BOX

Lytic Viruses in Industry

When bacteria are grown in a large commercial vat, it may be necessary to provide very large quantities of air to obtain the best energy yields per substrate molecule oxidized during aerobic growth. Unless this air is sterilized, there is always the possibility of lytic bacteriophages being introduced into the system to infect the culture.

It is true that a few bacteriophages distributed throughout 5,000 liters of culture are not going to have a major effect. In one unfortunate instance, however, a large culture facility was designed so that the air intake and air exhaust systems for the bacterial growth vessel rested side by side. When a virus was introduced via the unsterile air intake system, it infected cells, caused their lysis, and eventually was recycled as droplets in the exhaust air. This contaminated exhaust was sucked back into the air intake system, resulting in the complete destruction of the culture by viral activity before the industrial process could be completed.

Similar problems can arise during milk fermentations at the time when specially prepared bacteria are added to the milk medium. If viruses have been growing on the bacteria naturally found in milk, they also can infect the bacteria added, destroy them, and prevent formation of product (cheese or yogurt, for example). Lytic viruses are not a strictly academic phenomenon, therefore. They also can have a significant impact on industrial microbiology.

Lysis of the bacterial cell is accomplished by the combined action of at least two viral proteins. A lysozyme attacks the glycosidic linkages in the peptidoglycan structure of the cell wall causing it to lose its rigidity and allowing the cell to rupture from osmotic forces. The mode of action of the other protein is unknown. The virus controls the timing of cell lysis by regulating the rate of transcription of the appropriate genes.

Bacteriophage Lambda, a Temperate Virus

Not all viruses produce solely lytic infections. Many are capable of triggering either a lytic response or a **temperate response** upon infecting a cell. A temperate response is one in which a virus establishes a stable relationship with its host cell such that transcription of the lytic functions of viral DNA is repressed, enabling the host cell to survive the infection. The repressed viral DNA is a **prophage** that is inherited by all daughter

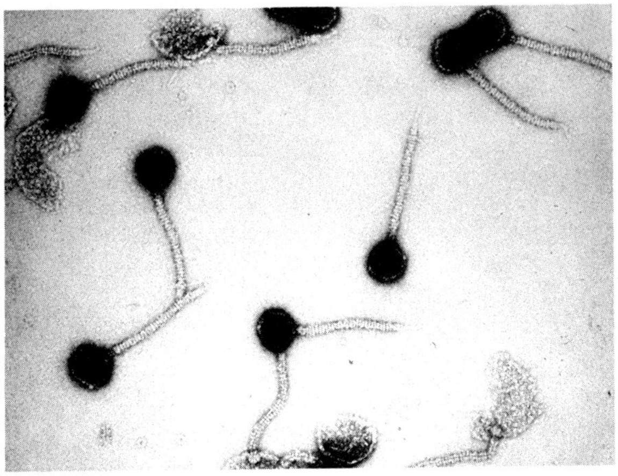

FIGURE 9.6

Phage lambda (λ). Compare this electron micrograph with that in figure 9.2A. Note the smaller size and decreased complexity.

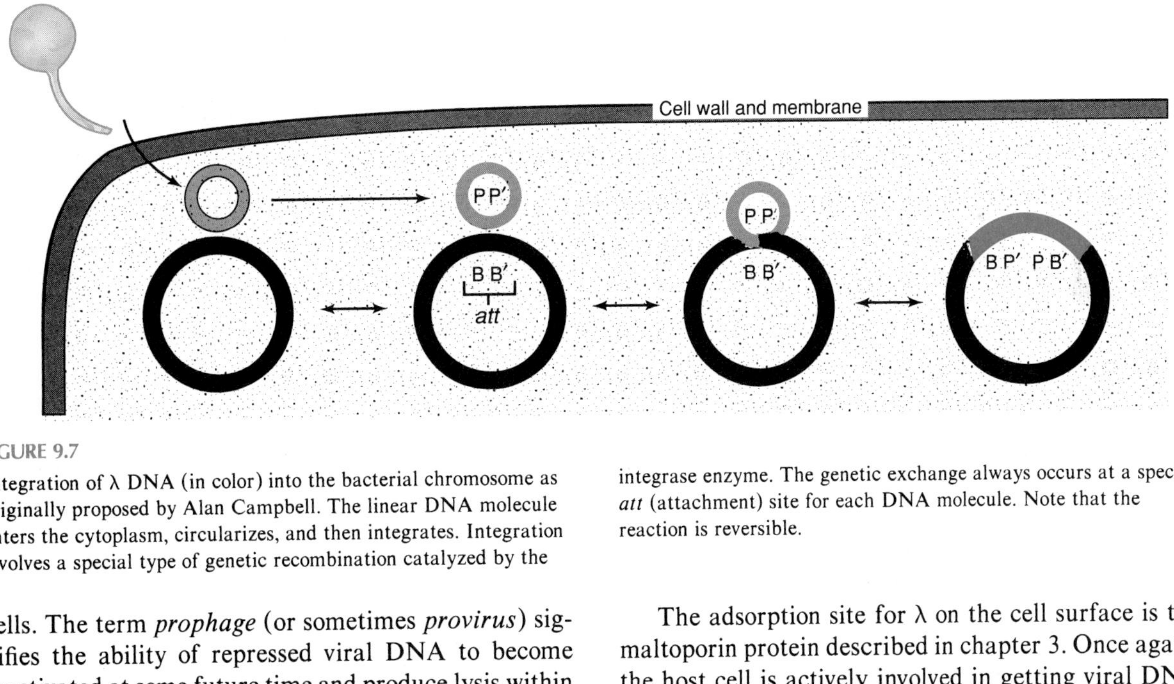

FIGURE 9.7

Integration of λ DNA (in color) into the bacterial chromosome as originally proposed by Alan Campbell. The linear DNA molecule enters the cytoplasm, circularizes, and then integrates. Integration involves a special type of genetic recombination catalyzed by the integrase enzyme. The genetic exchange always occurs at a specific *att* (attachment) site for each DNA molecule. Note that the reaction is reversible.

cells. The term *prophage* (or sometimes *provirus*) signifies the ability of repressed viral DNA to become reactivated at some future time and produce lysis within host cells. A cell carrying a prophage is called a **lysogen.**

The best-studied temperate bacteriophage uses *E. coli* as a host and is denoted by the Greek letter lambda (λ). Like phage T4, phage λ has a head and a tail, but its structure is much less complex (figure 9.6). The tail is not contractile, and there is only a single tail fiber.

The adsorption site for λ on the cell surface is the maltoporin protein described in chapter 3. Once again, the host cell is actively involved in getting viral DNA transported into its cytoplasm. The linear DNA molecule circularizes by joining single-strand regions at either end of the molecule (figure 9.7). When λ DNA is in the cytoplasm and begins transcription, there are two possible outcomes. In a lytic response, a more or less typical lytic infection can result. Unlike phage T4, phage λ does not degrade the host DNA nor does it

APPLICATIONS BOX

Lysogenic Conversion

Every prophage must have at least a small region of active transcription so that necessary repressor molecules can be provided. In some cases, other proteins can be included in the transcript. If these proteins confer an observable trait on the host cell, it is easy to distinguish a lysogen from a nonlysogen by simple examination of the properties of a pure culture. In some instances, the proteins produced make a bacterial cell better able to cause a disease, as is the case for *Corynebacterium diphtheriae* whose presence in the throat is significant only if it carries phage β, which codes for a poisonous protein. If this occurs, the lysogenic cell has undergone **lysogenic conversion** into a more harmful state.

provide its own tRNA molecules. The RNA polymerase complex also is less extensively modified in that no new sigma factor is produced, although changes in the termination properties of the complex do occur. The burst size is about the same as that for T4, and the mechanism of cell lysis involves the production of a lysozyme as well.

In a temperate response, the major λ protein synthesized is a specific λ repressor molecule that prevents transcription of the DNA regions coding for proteins necessary for viral DNA replication, virion protein production, and cell lysis, many of which are initially active. During the transitional period required to shut down lytic functions and stabilize the virus, a special enzyme called integrase is produced. Integrase catalyzes a **site-specific recombination** event between λ DNA and the bacterial chromosome (figure 9.7). The event is site specific because there is only one spot on each DNA molecule (the attachment site) at which the DNA strand exchange can occur under normal conditions. The λ DNA is thus **integrated** (becomes a physical part of the bacterial chromosome) and inherited just like any other trait. It no longer requires its own replication functions because the bacterial system replicates the prophage along with the rest of the bacterial chromosome.

This physical combination of phage and bacterial DNA is the most commonly observed form of lysogeny. A minor variation is seen in phages that have more than one potential site for integration into bacterial DNA. In such instances, there is normally only one prophage present per bacterial chromosome, but it can be integrated at any one of the available sites. The rarer form of lysogen is one in which the semiquiescent viral DNA remains as a separate molecule, independent of the bacterial chromosome. It replicates itself at the same rate as the bacterial chromosome but does not synthesize mRNA for any lytic function. Viral DNA that remains a separate molecule is a true plasmidlike lysogen (plasmids are discussed in chapter 10).

The molecular mechanisms by which phage λ regulates the lysis-lysogeny decision after an infection are much too complex to explain in detail in an introductory text. The basic elements of regulation are easy to understand, however. As shown in figure 9.8, the alternative transcriptional states of λ DNA each result in the production of a repressor protein, λ repressor from the temperate pathway and *cro* from the lytic pathway. Both repressors are capable of shutting down the opposing metabolic pathway. If the lytic pathway is shut down, a lysogen results. If the temperate pathway is shut down, phage production is followed by cell lysis. In a normal cell, the temperate pathway predominates somewhat more than half the time. Once again, host cell proteins play an important role in the process.

Because a lysogen remains a lysogen due to the presence of the λ repressor protein, it follows that if another λ virion tries to infect the same cell, the new λ DNA will be repressed before it has time to express any of its own functions. A lysogen is thus immune to **superinfection,** that is, a second infection by an identical virus. This observation has important consequences with respect to plaques produced by temperate phages. As can be seen in figure 9.1, plaques produced by lytic phages have clear centers. Plaques produced by temperate phages, on the other hand, have turbid centers. Turbidity results from lysogenic cells that continue to grow in the presence of the virus due to their superinfection immunity. Generally, lysogens do not manage to grow quite so dense as uninfected cells; therefore, plaques remain visible.

Lysogens are not necessarily stable indefinitely. A culture of lysogenic cells always has associated with it some viral particles of the type that produced the lysogen in the first place. In fact, it is impossible to completely purify a culture of lysogenic cells to the point at which no free virions are present, which is how lysogens are identified in the first place. The presence of free phage particles results from the spontaneous **induction** of prophage. As in the case of the regulatory

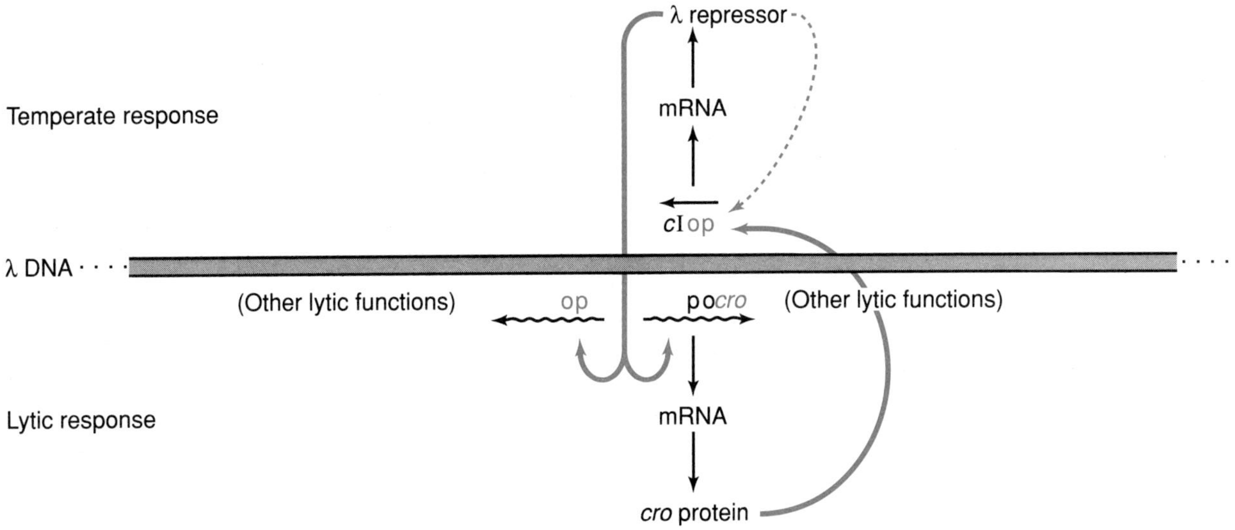

FIGURE 9.8

Regulation of the lysis-lysogeny decision in phage λ. The upper portion of the figure depicts the pattern of transcription, repression, and activation during lysogeny. The lower portion illustrates the main points of the pattern during lytic infection. Each pathway produces a repressor specific to the other; that is, when the lysogeny pathway is dominant, it represses the lytic functions and *vice versa*. Arrows indicate the direction of transcription; solid lines indicate repression; dotted lines indicate self-activation; operator site (o); promoter site (p).

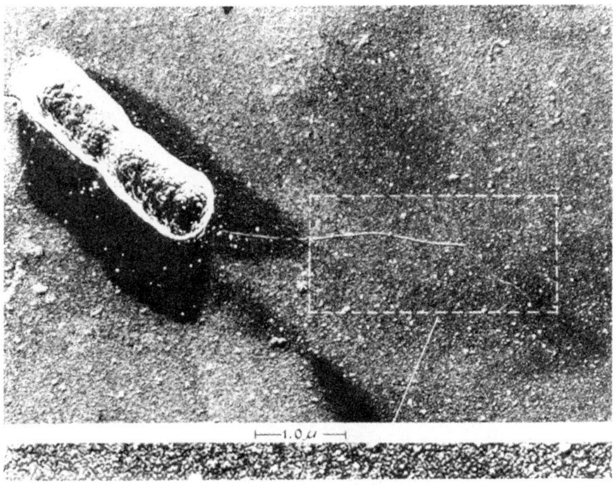

FIGURE 9.9

The helical phage M13. This virus attaches to the tips of F pili, and then the pili retract and pull the virus to the surface of the cell.

Dan S. Ray, Department of Biology, UCLA.

mechanisms for the *lac* operon discussed in the preceding chapter, induction results from an inactivation of the phage repressor protein (λ repressor in this case). All temperate viruses undergo spontaneous induction at a low but measurable rate ($\sim 10^{-6}$/cell division). Some temperate viruses, including λ, also are inducible externally in the sense that if the host cell DNA is damaged by some treatment such as ultraviolet irradiation, the repressor protein is destroyed and phage DNA reactivates to initiate a lytic infection.

Bacteriophage M13, a "Leaky" Virus

Some viruses cannot be categorized neatly as lytic or temperate. An example is M13, a small virus that specifically infects *E. coli* cells carrying an F plasmid (see chapter 10) by using as a binding site the filamentous F pili found on the cell surface and encoded by the plasmid. Unlike phages T4 and λ, M13 is rod shaped with protein subunits arranged in the form of a helix around its DNA molecule (figure 9.9). The DNA molecule is also unusual in that it is a single strand and circular. Bacterial cells normally do not make single-strand DNA; therefore, some special processes must come into play.

When viral DNA arrives in the cytoplasm, it first must synthesize a complementary strand, creating a double-strand **replicative form** (RF). The resulting circular DNA molecule then replicates by an unusual DNA synthetic process called a **rolling circle** (figure 9.10). In rolling circle replication, a newly synthesized DNA strand is used to displace an old strand. The old strand can then serve as a template for the synthesis of a complementary strand or be packaged into a virion. The route taken by any given DNA molecule depends

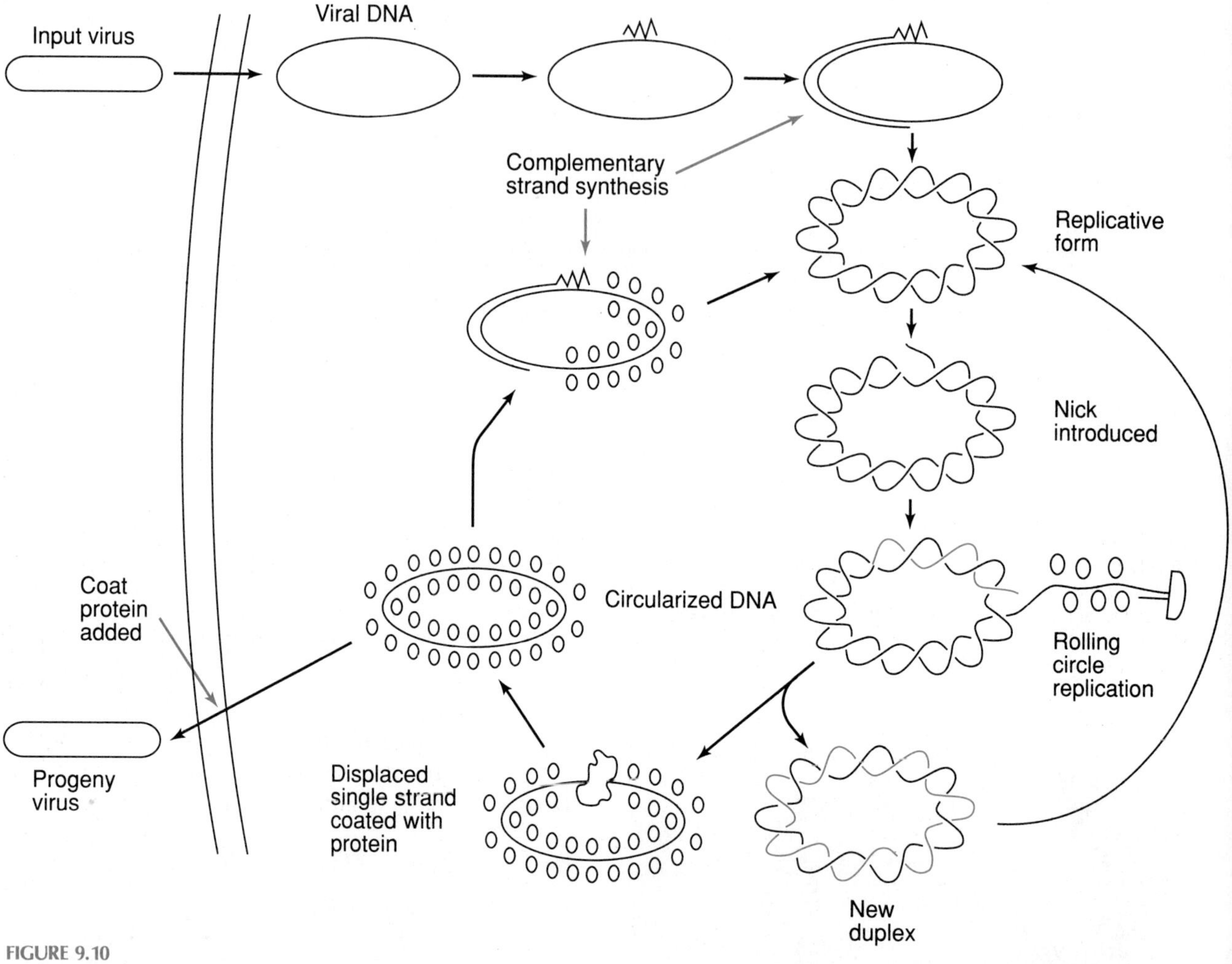

FIGURE 9.10

Rolling circle DNA replication. The circular, double-strand molecule is first nicked (breakage of a single phosphodiester bond). Then synthesis of only one strand is initiated with the existing strand used as a primer. The new strand (in color) displaces the old one that can be used later to make more virus or to synthesize its own complementary strand to make another RF molecule.

From E. A. Birge, *Bacterial and Bacteriophage Genetics*, 2d edition.

on the concentration of phage structural proteins in the cell at the time of synthesis. If the concentration is high, DNA will be packaged to make more virions.

Assembly of the infectious virion, however, does not follow the pattern seen in λ or T4. Instead, the displaced DNA strand first is coated by a special phage protein that specifically binds to single-strand DNA and then is extruded through the cell membrane, where coat proteins displace the single-strand DNA binding protein. Thus, the intact virion is found only on the exterior of the cell. Due to the presence of signal peptides (discussed in chapter 3), coat proteins locate themselves within the cell membrane to await DNA molecules. Because of the manner in which virions are released, the host cell is not destroyed; it merely has some of its metabolism diverted into the production of phage particles. The infection is, therefore, not a lytic one. Neither is it a temperate infection because M13 DNA is not repressed and does not form a stable relationship with the host cell. The infection perhaps is best described as "leaky" because progeny virions are released slowly. Plaques are formed only because infected cells grow more slowly than uninfected ones, not because of cell destruction.

RNA Phages

Not all bacterial viruses contain DNA as their genetic material. There are a large number of single-strand RNA viruses that infect bacteria as well as a few containing double-strand RNA molecules. The basic

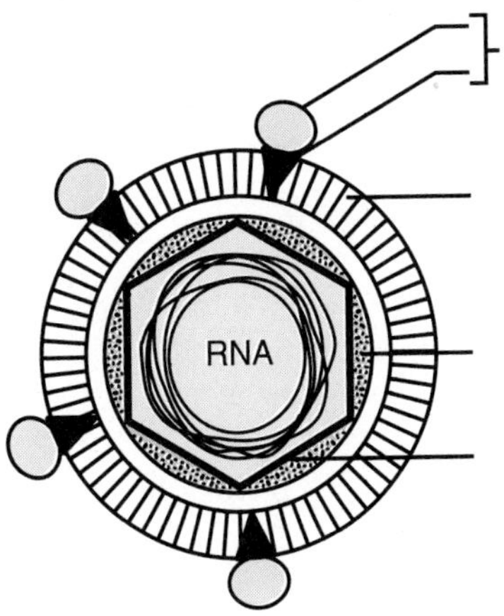

FIGURE 9.11

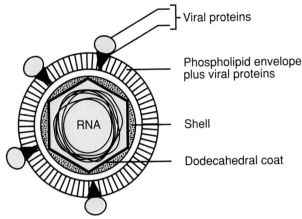

Diagram of the structure of $\phi 6$. A dodecahedron composed of five proteins surrounds the inner core of three double-strand RNA molecules. The outer membranous layer contains host lipids and additional viral proteins.

From L. Mindich and D. H. Bamford, "Lipid-Containing Bacteriophages," in *The Bacteriophages*, Vol. 2, edited by R. Calendar. Copyright © 1988 Plenum Publishing Corporation, New York, NY. Reprinted by permission.

problem facing any RNA phage is the lack of an RNA-dependent RNA polymerase in a normal bacterial cell. Some RNA viruses provide their own polymerase. Others combine host proteins and viral proteins to make one. A particularly interesting example of an RNA phage is $\phi 6$, a virus that has double-strand RNA as its genetic material and a lipid **envelope** around its capsid. Phage $\phi 6$ was isolated from a plant-pathogenic *Pseudomonas phaseolicola,* but it also will infect *Ps. syringae* and *Ps. pseudoalcaligenes* if these hosts are provided. Within the $\phi 6$ nucleocapsid, three discrete double-strand RNA molecules have been found. They are designated as S(mall), M(edium), and L(arge).

Structurally, $\phi 6$ is a multilayered virus consisting of an RNA core, a protein coat, and a phospholipid outer layer (figure 9.11). It attaches itself to the cell via a cell fimbria that brings it in touch with the cell's outer membrane. The viral membrane and the bacterial outer membrane fuse, peptidoglycan is digested by an intrinsic viral enzyme, and the virus is delivered into the periplasm. The cell membrane encases the nucleocapsid forming a vesicle that is released into the cytoplasm. The nucleocapsid includes four proteins necessary to provide RNA-dependent RNA polymerase activity. Transcription proceeds on all RNA molecules, but only the L class is translated initially.

The viral proteins and RNA assemble in the usual way. During this time, phage $\phi 6$ proteins are added to the cell membrane. The membrane invaginates to form a free vesicle used later to create the membrane surrounding the virus. Eventually, a lytic enzyme destroys the host cell, releasing progeny virions.

Other Bacterial Viruses

Many other viruses infect bacterial cells, although not all bacteria have been shown to be infected by viruses. This may reflect a deficiency in our knowledge, however, rather than the actual state of affairs. At the other extreme, over 250 tailed viruses are known to infect enteric bacteria. Clearly, it is not possible to consider all or even most bacterial viruses in this text. References to more detailed compendia are provided in the suggestions for further reading at the end of this chapter.

Plant and Animal Viruses

The diversity of viruses that can infect plant and animal cells is even greater than that observed in the bacteria. Although the basic pattern of viral infection is the same, there are some differences in technique used for observing eukaryotes infected with viruses. It is not always practical to study the biology of viruses by infecting entire eukaryotic organisms. Instead, cells are often taken from susceptible eukaryotic organisms to be maintained as cell cultures and used as substitute hosts. The procedures used for growing cell cultures have elements in common with those used for growing bacteria, but there are some important differences.

Experimenters produce eukaryotic cell cultures by removing a piece of tissue from an organism, breaking the adhesion between cells with enzymes such as trypsin, and then transferring the cells into a nutrient solution contained in a plastic culture dish or bottle (figure 9.12A). When provided with appropriate salts and specialized nutrients, such as fetal calf serum for animal cells or plant growth hormones for plant cells, cells will grow and divide to cover the surface of the container. Therefore, this process is often referred to as **tissue culture.** Incubation for animal cells includes providing a 5 to 10% CO_2 atmosphere to aid in pH maintenance. Unlike bacterial cells, normal animal cells (except for those that produce blood cells) grow only when actually touching a solid support. They are not free floating within a medium like most bacterial cells.

When all cells are touching other cells or the walls of their container, **contact inhibition** occurs and the cells stop dividing (figure 9.12B). This is the normal response of a **primary cell culture.** Cells will start growing again if diluted twofold and transferred into fresh medium; however, this process can be repeated only so

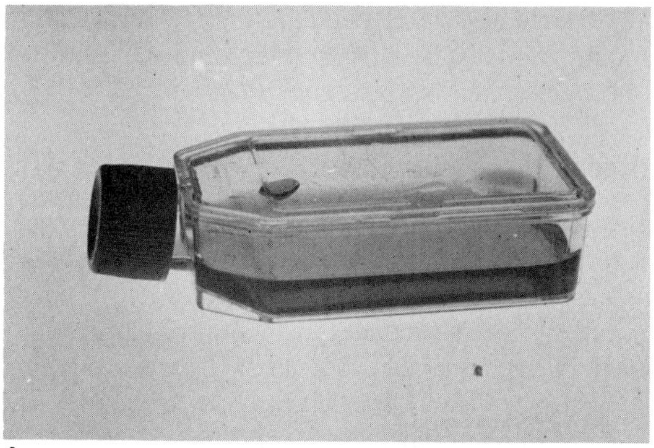

A

FIGURE 9.12

Cell cultures. (**A**) Typical cell culture bottle. (**B**) Hamster embryo cells maintained in cell culture. Note the typical shape of the cells and the monolayer they have produced. (**C**) Hamster embryo cells transformed with bovine papilloma virus. Compared with the preceding photograph, they show alteration in cell shape and arrangement. (**D**) Vesicular stomatitis viral plaques in fibroblast cell culture. The dish in the center received no extra treatment. Dishes A through G received increasing amounts of interferon, which has the effect of reducing the number and size of the plaques. Interferon induces the expression of certain cellular genes that function to inhibit the translation of viral mRNA without affecting the translation of most cellular mRNAs.

(B–C) Reprinted from Morgan, D. M., and W. Meinke, 1980. Isolation of clones of hamster embryo cells transformed by bovine papilloma virus. *Current Microbiology* 3:247–251. (D) Photograph courtesy of Dr. Charles E. Samuel, Department of Biological Sciences, University of California, Santa Barbara.

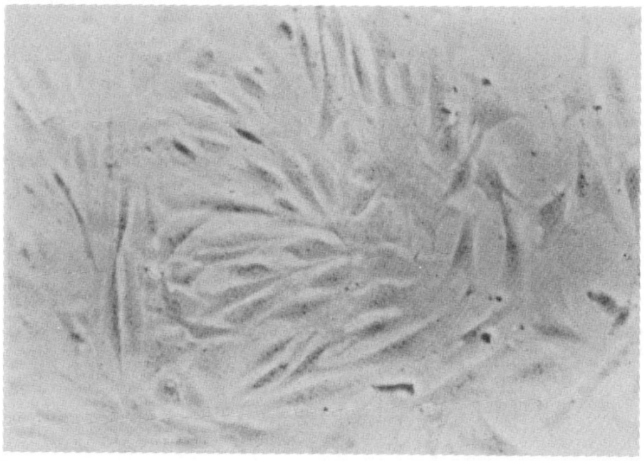

B

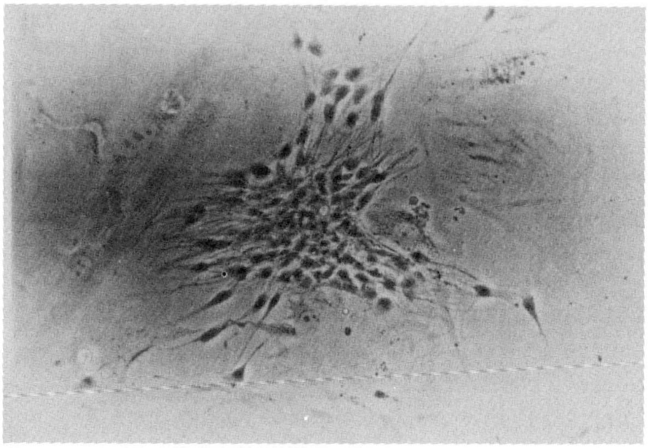

C

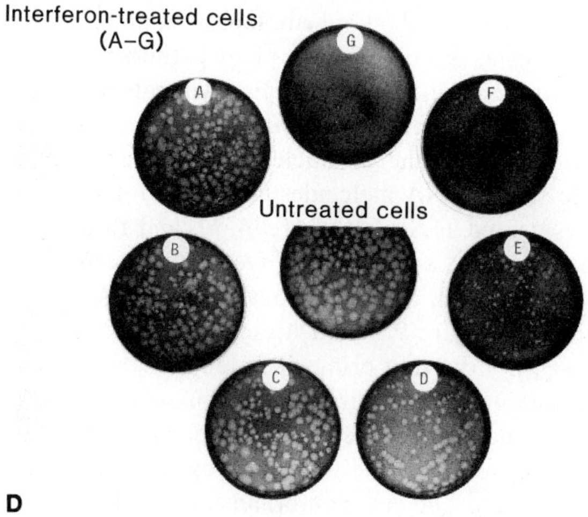

D

many times because the growth of animal cells ceases after about 50 cell divisions. Apparently, either the culture systems used today are lacking one or more components essential for long-term survival of these cells, or the cells have a build-in senescent (aging) mechanism.

By the time animal cell cultures cease to grow after being transferred to fresh medium, selection for certain cell types has occurred. Some cells in each culture continue to grow after numerous transfers, but these are found to have accumulated chromosomal abnormalities. They are not grossly different from cells of the primary cell culture, but they are designated as a **diploid cell strain** to distinguish them from cells of a primary culture. Eventually, even diploid cell strains will lose their ability to grow, but once again a few cells will survive. These survivors constitute an **established cell line** and are **heteroploid** (that is, grossly abnormal with many chromosomal rearrangements and/or deletions). They exhibit a great deal of heterogeneity within a culture. Sometimes, cell lines lose contact inhibition (figure 9.12C). They will grow in a cell culture indefinitely as long as the medium is replenished, making them effectively immortal. These cells are said to have undergone **transformation.** Transformed cells lose much of their oriented growth pattern and no longer require a solid support; therefore, they can be grown more like bacterial cultures.

It is possible to obtain transformed cells directly from a eukaryotic organism if cells are taken from tumor tissue instead of normal tissue. In fact, the most famous of all animal cell lines is the HeLa cell line derived from human cervical cancer cells removed decades ago. These cells are tenacious growers and have caused substantial problems in tissue culture laboratories by taking over other types of cell cultures being maintained in the same laboratory. Tumor cells of plants, such as those induced by the Ti plasmid (see chapter 10), grow independent of the presence of plant growth hormone. It is assumed that the process of transformation observed in the progression from primary cell culture to cell line reflects, in some fashion, the induction of a tumor, but the mechanism(s) of this process remains uncertain.

With cell cultures of the appropriate type, animal and plant viruses can be grown in the laboratory. Cell destruction or inhibition results in the production of a plaque similar to those observed with bacterial viruses (figure 9.12D). It is possible, therefore, to carry out the same types of experiments on animal and plant viruses that have been described for bacterial viruses.

An alternative method of growing some animal viruses is to use embryonated eggs. The egg is opened carefully, inoculated with a virus, and then sealed with tape. Many viruses, including rabies, influenza, and polio virus, grow well under these conditions. Depending on the specific virus used, the embryo itself or its amnionic membrane can be inoculated.

Descriptions of specific viruses and viruslike elements that infect eukaryotic cells follow. The emphasis of the discussion is on major similarities and differences between plant and animal viruses and bacteriophages.

Adenovirus, a Double-Strand DNA Virus

Adenoviruses are a family (*Adenoviridae*) of double-strand DNA viruses that have been shown to cause respiratory diseases in humans and other animals. In certain animals, some human adenoviruses can cause cell transformation leading to the production of tumors, although there is no positive evidence that these adenoviruses cause tumors in their natural human hosts. A few DNA sequences of adenoviruses have been identified in tumor cells, but their role in the **oncogenic** (cancer-causing) process is not well understood.

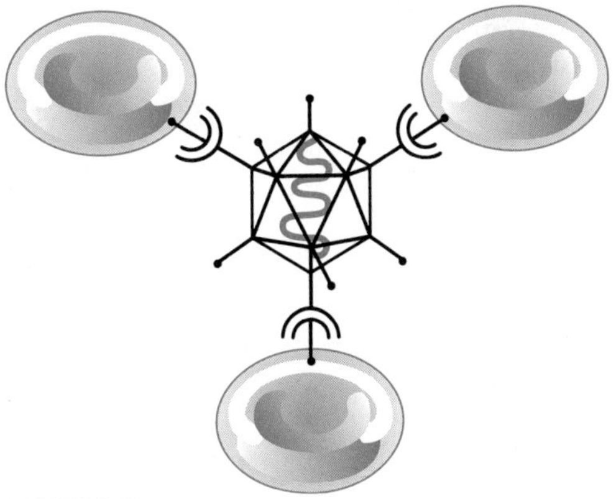

FIGURE 9.13

Hemagglutination. The multiple red-blood-cell-binding sites on the virion surface allow viruses to form large clumps of red cells that are visible as a precipitate.

The virus has an **icosahedral** shape (a regular polyhedron that has 20 faces and 12 corners) but no tail or tail fibers. There is, however, one (sometimes two) fibers that project from each corner. These fibers can cause **hemagglutination** (clumping of red blood cells) because of their ability to bind to the surface of red blood cells and thus allow a virion to hold several red cells together at the same time (figure 9.13). The lattice of cells and viruses created thereby becomes so large that it is visible to the unaided eye and sinks to the bottom of the cell suspension. The corner fibers also are somewhat toxic to host cells and can inhibit cell division.

Adenoviral infection of host cells occurs by a slightly different process than that seen in bacteriophages. The virus binds to specific receptors on the cell surface as do bacteriophages. However, instead of adenoviral DNA protruding from the capsid, the entire virion enters the cell either by penetrating the cell membrane directly or by **endocytosis,** a generalized phagocytic process in which the virion is enclosed in a membrane vesicle. This process was known formerly as viropexis. Following endocytosis, the virion still must penetrate the vesicle membrane layer somehow to reach the cytoplasm.

Once inside the cell, the adenovirus begins an **uncoating** process in which the protein coat surrounding the virion is removed, releasing the nucleocapsid that then migrates to the cell nucleus where transcription begins. The linear DNA molecule consists of approximately 35,000 base pairs (bp). The mRNA produced

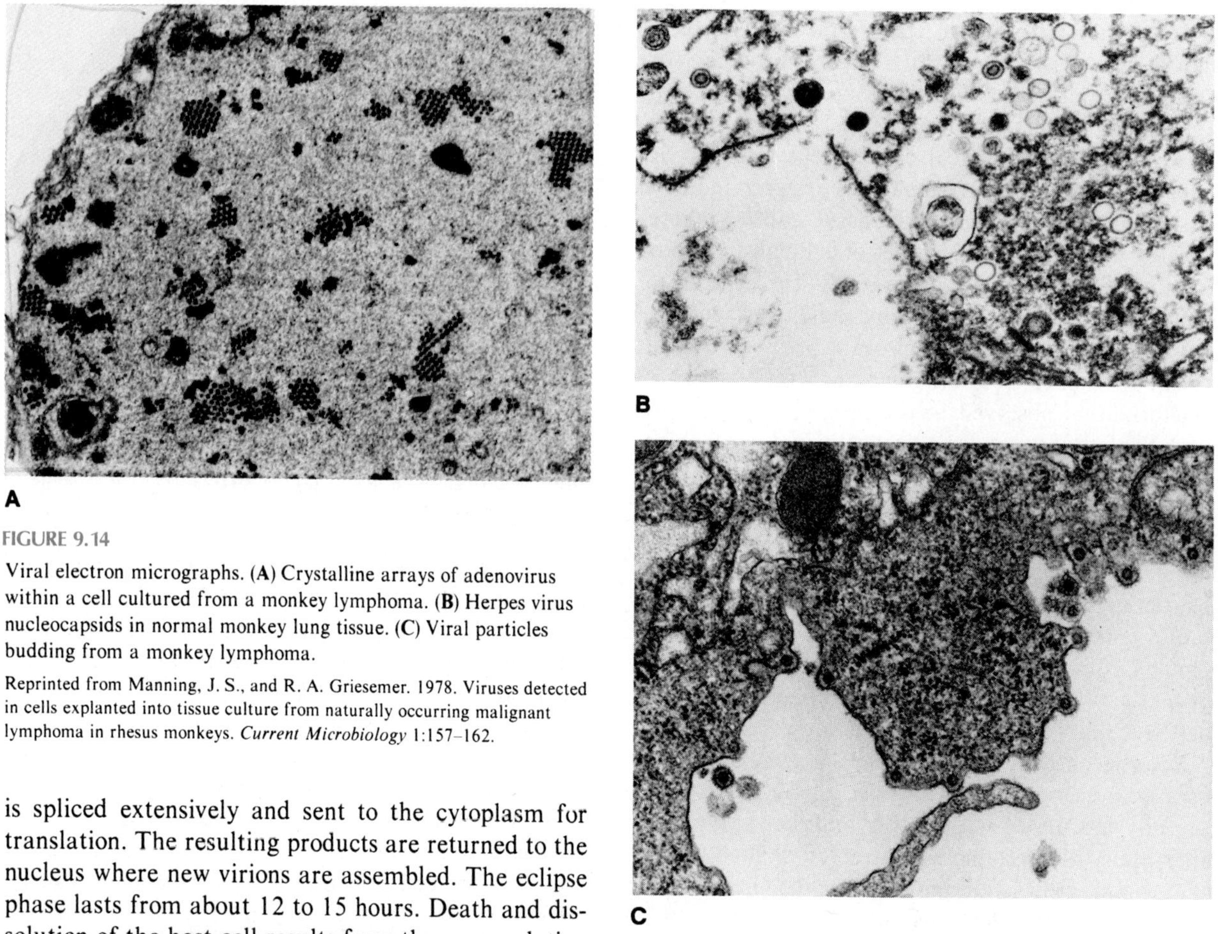

FIGURE 9.14

Viral electron micrographs. (**A**) Crystalline arrays of adenovirus within a cell cultured from a monkey lymphoma. (**B**) Herpes virus nucleocapsids in normal monkey lung tissue. (**C**) Viral particles budding from a monkey lymphoma.

Reprinted from Manning, J. S., and R. A. Griesemer. 1978. Viruses detected in cells explanted into tissue culture from naturally occurring malignant lymphoma in rhesus monkeys. *Current Microbiology* 1:157–162.

is spliced extensively and sent to the cytoplasm for translation. The resulting products are returned to the nucleus where new virions are assembled. The eclipse phase lasts from about 12 to 15 hours. Death and dissolution of the host cell results from the accumulation of viral particles over a period of several days. Virions are released only after cell lysis has occurred. Meanwhile, viral particles can be seen forming crystalline arrays within the host cell (figure 9.14A).

Herpes Simplex Virus, a Double-Strand DNA Virus

The herpes simplex viruses (family *Herpesviridae*) cause persistent infections of mucous membranes in humans. The virion itself is composed of an icosahedral capsid surrounded by an envelope derived from the nuclear membrane of the previous host. The space between the capsid and the envelope is termed the tegument (figure 9.14B). Presence of the nonrigid envelope gives the mature virion a pleomorphic shape when viewed with the electron microscope. The virus binds to fibroblast growth factor that, in turn, binds to specific receptors on the host cell surface. It then enters the cell along with the growth factor. The capsid moves to a pore in the nuclear membrane where the linear, 155,000 bp viral DNA is released into the cell nucleus. The resulting eclipse phase lasts from five to six hours.

When herpes virus infects a cell, it inhibits DNA, RNA, and protein synthetic capabilities of its host. The virus replicates its own DNA within the cell nucleus by a modified rolling circle mechanism. Assembly of new nucleocapsids takes place in or near the nucleus, and they reach the cytoplasm by budding through the nuclear membrane that now contains specific viral glycoproteins. It is the inner leaflet of this membrane that forms the viral envelope. The host cell does not lyse; instead, viral particles are released via the endoplasmic reticulum or reverse endocytosis, such as that depicted in figure 9.14C.

Herpes simplex infections are familiar to most people as "cold sores," small ulcerative lesions that occur within the oral cavity for a period of time and then spontaneously disappear only to reoccur at some later date. The basis for viral latency is not yet understood, although it presumably bears some relationship to phage lysogeny. Viral DNA apparently persists within ganglia cells of the peripheral nervous system, probably as a linear, nonintegrated molecule. Within the past decade, an epidemic of genital herpes, infection of the mucous membranes of the genitalia by herpes simplex, has developed. The pattern of latency and reactivation is seen also in genital herpes.

Vaccinia Virus, a Double-Strand DNA Virus

Vaccinia virus is a member of the family *Poxviridae* and seems to be a variant form of the virus responsible for smallpox. For many years, vaccinia virus has been used as the immunizing agent against smallpox. More recently, it has been used as a vector in genetic engineering experiments.

The virus is somewhat ovoid or elongate, measuring from 250 to 390 nm by 200 to 260 nm. Its DNA is contained in a central nucleoid that is surrounded by lipoproteins and enclosed in the usual capsid. Following attachment and endocytosis, the virus is uncoated in a two-step process. First the DNA and associated proteins are released, and then the free viral DNA is liberated some 30 to 60 minutes later. Transcription is accomplished with a viral DNA-dependent RNA polymerase that is part of the viral core. Transcription begins before uncoating is complete, and some viral proteins are required to complete the uncoating process. Unlike other DNA viruses, all activities including replication and assembly take place in the cytoplasm. There is no nuclear involvement.

DNA replication is more complicated than usual because the ends of the viral DNA strands are cross-linked with short polynucleotides. Bidirectional replication is initiated internally at one end of the DNA molecule. In order for replication to be completed, an endonuclease must cleave the cross-links, and new cross-links must be introduced into the daughter molecules.

A summary of the life cycles of adenovirus, herpes simplex, and vaccinia virus is diagrammed in figure 9.15.

Cassava Latent Virus, a Single-Strand DNA Virus

The Cassava latent plant virus is a typical member of the family *Geminiviridae* that consists of paired, seemingly identical quasi-icosahedra each containing a circular, 2,700 bp single-strand DNA molecule. Both DNA molecules are required for a successful infection. The virus is carried from plant to plant in the hemolymph of whiteflies, but it does not seem to enter the seeds of the Cassava plant. Replication and viral assembly occur in the nucleus. The pattern is similar to that seen with phage M13 in that a second DNA strand is synthesized and further replication is thought to occur by the rolling circle mechanism.

Reovirus, a Double-Strand RNA Virus

Members of the family *Reoviridae* are parasites of vertebrates, insects, plants, and oysters. They have a number of unusual features, not the least of which is their genetic material. Instead of DNA, the mature virion contains 10 distinct linear molecules of double-strand RNA, making it similar to phage $\phi 6$. Each individual RNA molecule codes for a different protein, except for one having two different promoters that can code for two distinct proteins. The route of infection involves endocytosis followed by uncoating in a lysosomal vesicle within the cytoplasm. Uncoating, however, does not completely degrade the virion; rather it results in the production of a subviral particle, an association of specific internal proteins and viral RNA molecules.

Internal proteins provide the necessary enzymes to produce a capped mRNA molecule suitable for translation by the cell. They include a double-strand RNA-dependent RNA polymerase and various capping enzymes. Production of mRNA molecules (also called **plus-strand RNA** in viruses) operates in much the same way as transcription of a double-strand DNA molecule; therefore, the process is said to be conservative, that is, parental viral RNA molecules are left intact (figure 9.16). As the infection proceeds, viral proteins alter the cell's translation machinery so that it becomes specific for uncapped mRNA. Late in the infection process, subviral particles produce only uncapped mRNA; therefore, all proteins being synthesized by the infected cell are viral in origin.

Translation of viral mRNA results in the production of virus-specific proteins necessary for the assembly of new particles. Initially, a particle consists of a protein coat with one of each type of plus-strand RNA (untranslated mRNA) inside it. Associated with these new provirions is RNA replicase that produces minus-strand RNA by copying the existing plus-strand RNA, thereby completing the conservative replication of the double-strand RNA. At the end of the viral infection, the host cell lyses, releasing progeny virions to begin new infections.

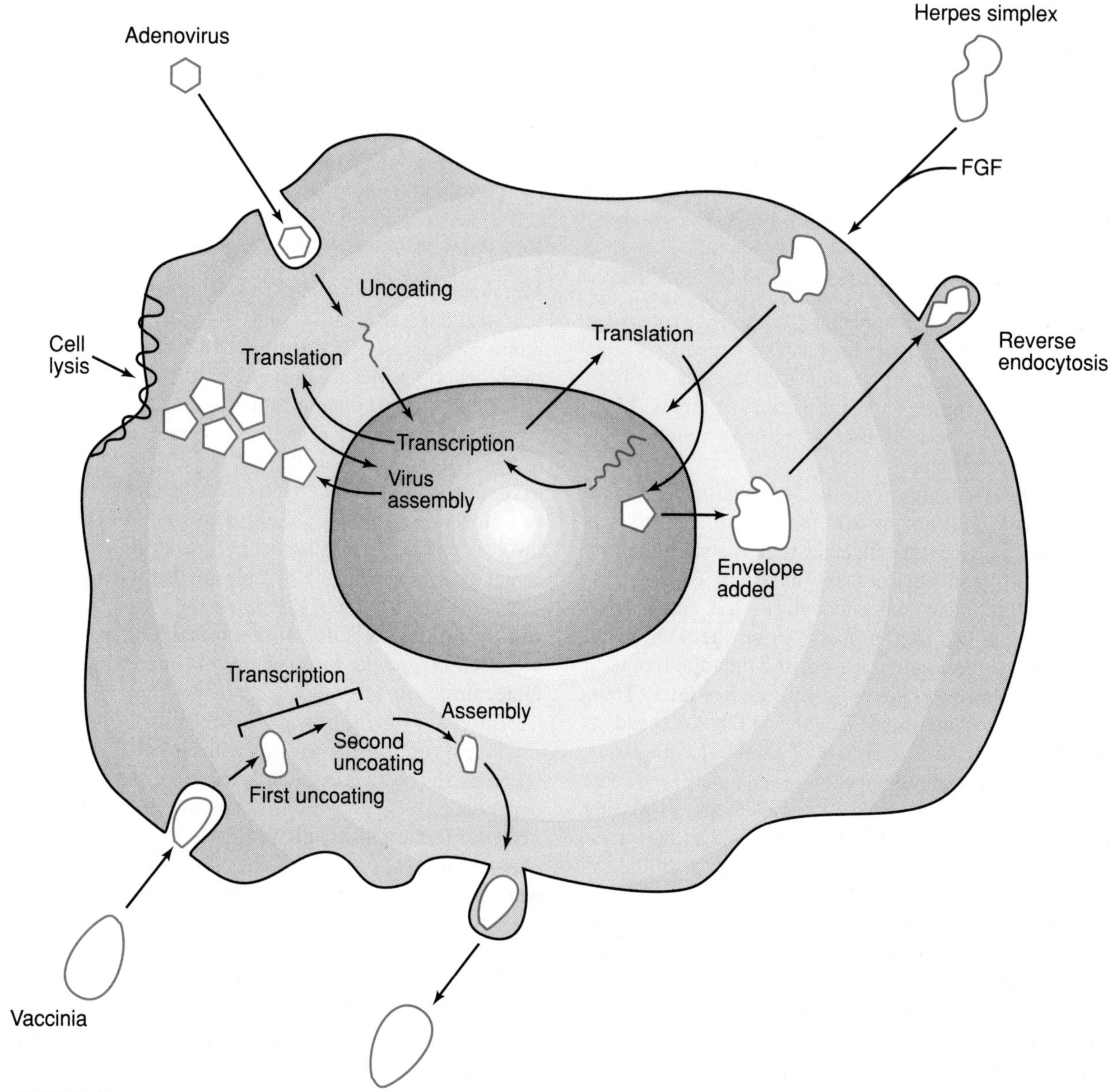

FIGURE 9.15

Summary of the life cycles of adenovirus, herpes simplex, and vaccinia virus. The circle in the middle represents the cell nucleus.

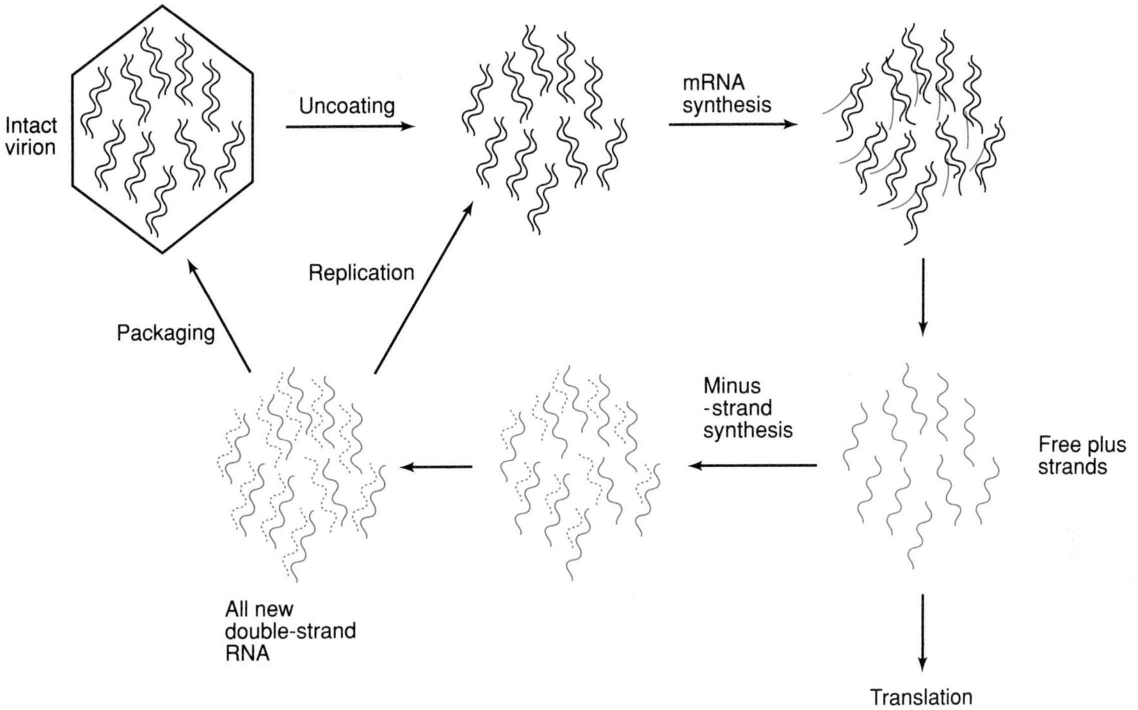

FIGURE 9.16

Conservative replication of reovirus. After uncoating, the subviral particle provides enzymes to synthesize plus-strand RNA. These molecules serve both as mRNA and as template for the synthesis of minus-strand RNA that will be incorporated into mature virions. Because the parental RNA molecules remain unchanged throughout this process, it is termed conservative replication.

Tobacco Mosaic Virus, a Single-Strand RNA Virus

Tobacco mosaic virus (TMV) is a small rod, 300 by 9 nm. Dimitri Ivanovski and Martinus Beijerinck described the virus independent of each other in the late nineteenth century. As such, TMV was the first virus discovered. It causes the formation of yellow spots (mosaics) on the surface of susceptible plant leaves (figure 1.6). Later scientists have shown the TMV genome to be a single strand of RNA that is the plus strand (viral RNA can serve as its own mRNA). Replication of the virus is by means of a minus-strand RNA intermediate (figure 9.17). The virus codes for an RNA-dependent RNA polymerase, a replicase enzyme that copies its mRNA into a nontranslated molecule and then copies that molecule back into plus strands for translation or packaging. Packaging occurs spontaneously within the cytoplasm of the cell to yield crystalline arrays of virus visible with the electron microscope.

Polio Virus, a Single-Strand RNA Virus

Polio virus, the best-studied representative of the *Picornaviridae* family, is very small and contains a single strand of plus RNA. The viral particle has a diameter of 26 nm and contains 7441 bases of RNA. It adsorbs to specific cell receptors, and the virions penetrate into the cytoplasm where uncoating occurs. The 5′ end of the virus's RNA is joined covalently to a short chain of amino acids, the VPg peptide, and the 3′ end has about 60 adenine residues. RNA that is to be translated must have its VPg protein removed, and this cleavage is observed in about half the infecting molecules. There is only one open reading frame, 6627 bases in length, that codes for a polyprotein. This protein can be cleaved at multiple sites by proteolytic enzymes to yield 11 or 12 different peptides with various functions.

RNA replication occurs in the same manner as for TMV but in a membrane-bound complex. Replicative intermediates in which multiple plus strands are associated with a single minus strand have been observed.

APPLICATIONS BOX

Useful Mosaics

Not all viruses cause extensive damage to their hosts. In the plant kingdom, mosaic viruses have turned out to be quite valuable, at least in one instance. Tulips infected with a mosaic virus produce flowers of unusual patterns that are considered to be more valuable commercially than flowers from uninfected plants.

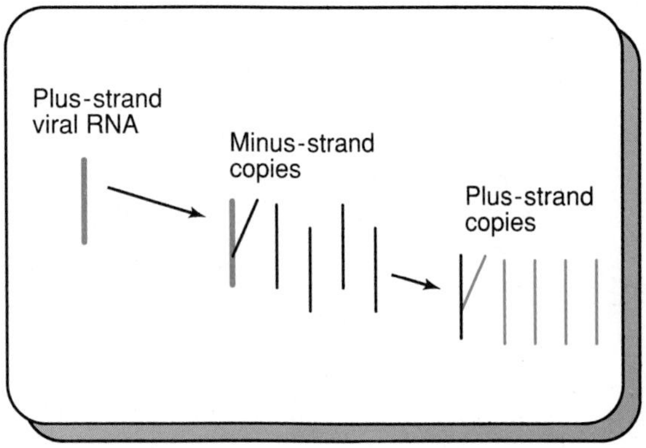

FIGURE 9.17

Replication of TMV RNA. The plus-strand RNA (in color) is copied to yield minus strands, and they in turn are copied to yield more plus strands.

Initially, all plus strands synthesized become mRNA (no linkage to VPg). Once sufficient proteins have accumulated, newly replicated RNA is assembled rapidly into viral particles.

There is some leakage of viral particles from the cell, but the majority of virions remain inside until cell lysis occurs. The infectious cycle takes eight or more hours.

Influenza Virus, a Single-Strand RNA Virus

The influenza virus is an enveloped virus in the family *Orthomyxoviridae* and contains negative-strand RNA. Presence of negative-strand RNA necessitates the inclusion of an RNA replicase molecule in the virion itself. A, B, and C are the three types of influenza virus known, and each differs slightly in genome size and base composition. Types A and B have eight discrete RNA segments in their genome while type C has only seven.

The virus enters the cell by fusing its envelope with the cell membrane. The integral RNA replicase synthesizes mRNA molecules (with polyadenylated tails) and templates (untailed). The templates are used later to synthesize more viral RNA segments. How duplication of segments is prevented during viral assembly is not known. Virions gradually bud out of infected cells over a period of hours, but unlike the case with phage M13, host cells eventually die.

Proteins of the influenza A viral coat are variable and undergo both major and minor changes. The minor changes are apparently due to simple mutation of viral RNA while it is replicating (RNA replication is intrinsically less accurate than DNA replication). The major changes result from recombination between different influenza A viruses infecting the same cell. Major changes in coat proteins often result in a broader range of susceptible hosts because the body's defense mechanisms are not prepared to cope with the new type of virus.

Retroviruses, RNA Viruses with DNA Intermediates

Retroviruses are spherical enveloped viruses originally identified as animal tumor viruses that contain a single-strand RNA molecule. The mode of replication of retroviruses, first worked out in the laboratories of Nobel laureates Howard Temin and David Baltimore, violates the conventional doctrine of molecular biology in which the flow of genetic information always proceeds from DNA to RNA to protein. In these viruses, the genome is a plus strand of RNA that codes, among other things, for a special multifunctional enzyme called **reverse transcriptase.** In biochemical terms, reverse transcriptase is an RNA-dependent DNA polymerase that catalyzes **retroreplication.** The enzyme makes a DNA copy of viral RNA and then, using its ribonuclease H activity, degrades the RNA strand while making a second DNA strand (figure 9.18). This duplex DNA can then migrate to the nucleus where it is integrated by reverse transcriptase into host chromosomes to establish a **provirus** (analagous to a prophage). When the time comes to produce RNA molecules for incorporation into new virions, conventional transcription of the DNA intermediate by host cell RNA polymerase produces the requisite molecules as well as the mRNA molecules needed for synthesis of viral proteins.

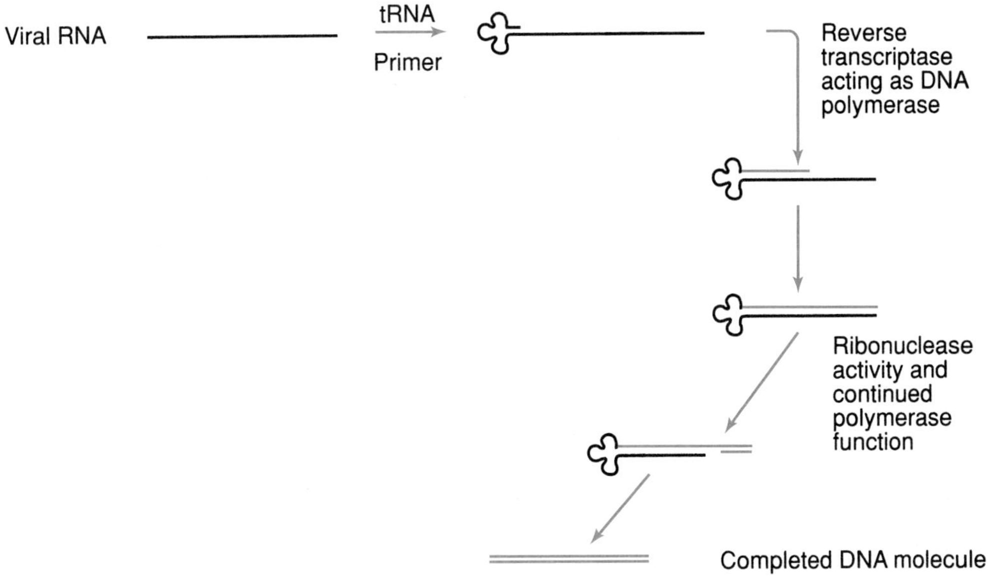

FIGURE 9.18

Reverse transcriptase activity. The enzyme initially binds to a tRNA molecule paired with the end of the viral RNA strand serving as a primer for the reaction. A single strand of complementary DNA (in color) is laid down, and then the enzyme loops back on itself, degrading the original viral RNA and replacing it with DNA. The final result is a double-strand DNA molecule containing all the viral genetic information.

Some retroviruses code for a transforming protein that rapidly transforms the host cell and induces a tumor in the host animal. Such viruses are the fastest-acting carcinogens known. Their tumor-inducing ability usually reflects the presence of an *onc* gene that can cause cell transformation. The most notable of this type is the Rous sarcoma virus (RSV) of chickens whose *onc* gene is also known as *src*. Other retroviruses, however, do not carry an *onc* gene; therefore, they do not cause rapid transformation of cells. Instead, tumors may never appear or appear only late in the life of the organism. Human T-lymphotrophic viruses (HTLV), all of which infect specific cells of the immune system, are examples of retroviruses without *onc* genes. One particular member of this family, human immunodeficiency virus type 1 (HIV-1), has been prominent in the medical news for the past several years as the causative agent of acquired immune deficiency syndrome (AIDS). Figure 9.19 summarizes the life cycle of the HIV-1 virus.

Viroids and Virusoids, Subviral Nucleic Acids

Viroids and virusoids are unusual viruslike entities found only in plants. Viroids are circular, single-strand, self-replicating RNA molecules of low molecular weight that have no protein coat; nevertheless, they can cause such infectious diseases as potato spindle tuber or tomato apical stunt. Generally, the mode of transmission is unknown. Possibly the viroid comes in direct contact with a plant or is transmitted via some insect; however, viroids are known to cause disease primarily in cultivated plants. The RNA molecule folds on itself to form a rodlike secondary structure. Replication of this RNA molecule requires host cell enzymes, and probably occurs via a rolling circle mechanism. Translation of viroid RNA is not required, neither is the presence of any true additional virus necessary for viroid function.

Virusoids are slightly more complex. They are viroids or viroidlike molecules located inside the capsid of a true, multisegmented RNA virus that still contains viral RNA. Virusoids can be linear or circular. They require the presence of the virus in which they are found normally because virusoids are incapable of independent replication. In some cases, virus-virusoid dependence is mutual, and the virus can replicate properly only in the presence of the virusoid.

Prions, Infectious Proteins

Prions have been associated with spongiform encephalopathic diseases, such as scrapie, Kuru, and Creutzfeld-Jakob disease, originally thought to be caused by "slow" or "unconventional viruses." In the case of scrapie, a disease of sheep and goats, Stanley

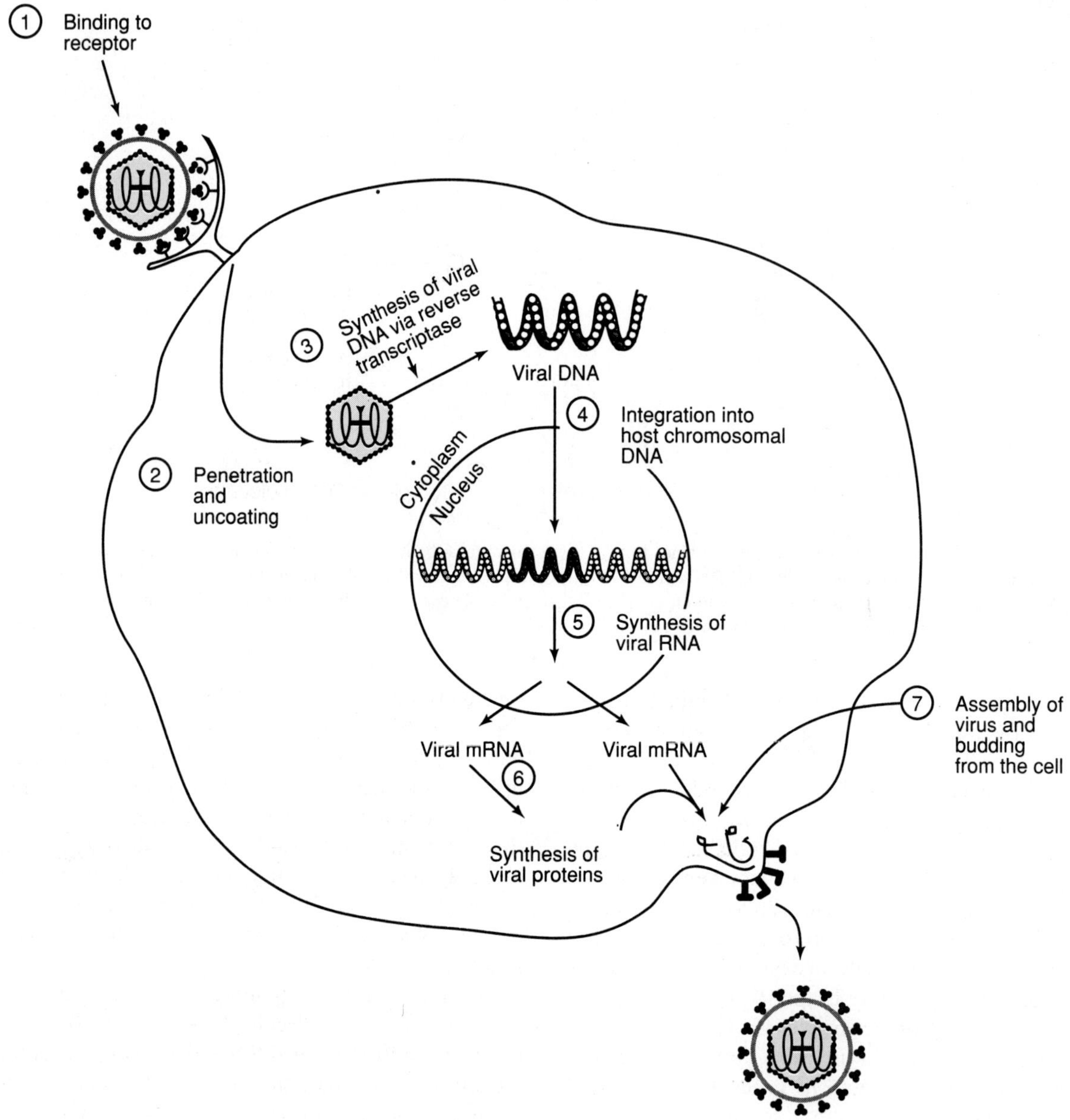

FIGURE 9.19

Life cycle of the AIDS virus (HIV-1), an example of a retrovirus.

From Fan/Conner/Villarreal: *The Biology of AIDS,* copyright © 1989 Boston: Jones and Bartlett Publishers, Fig. 4.5, p. 22, Life Cycle of the AIDS Virus. Reprinted by permission.

Prusiner and his collaborators have shown that an infectious preparation contains no nucleic acid, only protein. The major protein present is designated as PrP, but involvement of minor proteins in the infectious process has not been ruled out. Recent studies have shown that the gene for PrP is present normally in a repressed form in cells. Presence of the extra PrP causes increased production of the protein with concomitant detrimental effects on the central nervous system of the host. There also is the potential for infecting other hosts as well if they are exposed to infected tissues.

Summary

All types of cells, prokaryotic as well as eukaryotic, seem to be capable of hosting viruses. A virus is an acellular, almost crystalline structure consisting of a single type of nucleic acid enclosed within a protein coat or capsid that can multiply and reproduce itself when introduced into an appropriate host cell. The capsid may be surrounded by a lipid envelope derived from a membrane of the previous host cell. Viruses have all possible types of nucleic acids, single- or double-strand RNA or DNA. The extent of their dependency on host cell functions generally varies inversely with the size of the viral genome. Surprising new modes of nucleic acid replication have been identified in certain viruses, such as rolling circles, conservative RNA replication, and retroreplication.

Susceptible host cells are used to study viruses. In the case of prokaryotes, normal bacteria suffice. For animal and plant viruses, however, special cell cultures are often needed. Animal cell cultures are classified as primary cultures, diploid cell lines, and established cell lines. The latter are equivalent to tumor cells. Common experimental methods for studying viruses include one-step growth and premature lysis experiments. Table 9.1 summarizes the properties of the viruses that have been discussed in this chapter.

Subviral entities also have been identified. Viroids are very small, untranslated, circular RNA molecules that are independently infectious. Virusoids are viroid-like molecules found inside normal RNA viruses. Both viroids and virusoids have been observed only in higher plants. Prions are infectious proteins that act to deregulate their own production from normal genes.

Questions for Review and Discussion

1. What unusual modes of nucleic acid replication have been observed in viruses? Why are some viruses not able to use the host cell apparatus and copy their nucleic acid in the same way that a cell does?
2. What is the difference between a primary cell culture, a diploid cell strain, and an established cell line? What are the advantages and disadvantages of working with each type of culture?
3. What is a tumor? Give an example of a virus that can cause tumors.
4. How can the intracellular growth phase of a virus be studied? How might you tell whether a particular protein is one produced normally by a cell or one that results from a viral infection?
5. What different kinds of subviral particles are known? Is it likely that they represent degenerate viruses? Why?

Suggestions for Further Reading

Birge, E. A. 1988. *Bacterial and bacteriophage genetics: An introduction,* 2d ed. Springer-Verlag, New York. A general introduction to bacteriophages.

Casjens, S. (ed.). 1985. *Virus structure and assembly.* Jones and Bartlett, Boston. An extensive discussion of bacterial and animal viruses.

Cooper, J. I., and F. O. MacCallum. 1983. *Viruses and the environment.* Chapman & Hall, New York. Virology from an environmental standpoint. Includes some population biology.

Davies, J. W., R. Hull, K. F. Chater, T. H. N. Ellis, G. P. Lomonossoff, and H. W. Woolhouse (eds.). 1986. *Virus replication and genome interactions.* The Company of Biologists Ltd., Cambridge. Monograph about replication patterns of many of the viruses discussed in this chapter.

Dulbecco, R., and H. S. Ginsberg. 1988. *Virology,* 2d ed. J. B. Lippincott, Philadelphia. These are the virology chapters (44–65) from a much larger medical microbiology text. They provide a more extensive survey of the animal viruses discussed in this chapter.

Gallo, R. C., and F. Wong-Staal (eds.). 1990. *Retrovirus biology and human disease.* Marcel Dekker, Inc., New York. Various chapters survey the biology of human and animal retroviruses.

Maramorosch, K., and J. J. Mckelvey, Jr. (eds.). 1985. *Subviral pathogens of plants and animals: Viroids and prions.* Academic Press, Inc., Orlando, Fla. Presents both introductory and detailed articles.

Ptashne, M. 1986. *A genetic switch: Gene control in phage λ.* Cell Press, Cambridge, Mass. The definitive discussion about regulation in this bacterial virus.

Van Regenmortel, M. H. V., and H. Fraenkel-Conrat. 1985. *The plant viruses,* vol. 2. Plenum Publishing Corp., New York. The first seven chapters of this book discuss in detail the life cycle of the tobacco mosaic virus.

TABLE 9.1	Properties of bacterial, plant, and animal viruses					
Family	**Virus**	**Type of Nucleic Acid**[1]	**Envelope**	**Host**	**Type of Infection**	**Characteristics and Functions**
Myoviridae	T4	2-DNA	No	*E. coli*	Lytic	———
Styloviridae	λ	2-DNA	No	*E. coli*	Temperate	———
Inoviridae	M13	1-DNA	No	*E. coli*	Leaky	Cells survive; virus transported across membrane
Cystoviridae	$\phi 6$	2-RNA	Yes	*Pseudomonas phaseolicola*	Lytic	———
Tobamovirus	Tobacco mosaic	1-RNA	No	Tobacco	Lytic	———
Adenoviridae	Adenovirus	2-DNA	No	Epitheleal tissues	———	Virions released when cells die
Herpesviridae	Herpes simplex	2-DNA	Yes	Mucous membranes	Leaky	May set up latent infection
Poxviridae	Vaccinia	2-DNA	Yes	Humans	———	Modified smallpox virus
Reoviridae	Reovirus	2-RNA	No	Respiratory and intestinal epithelium	Lytic	Conservative replication
Orthomyxoviridae	Influenza	1-RNA	Yes	Respiratory epithelium	Leaky	A negative-strand virus
Retroviridae	HIV-1	1-RNA	Yes	T cells	Temperate	Retrovirus; uses a DNA replication intermediate
Picornaviridae	Polio virus	1-RNA	No	Intestinal epithelium	Lytic	Makes a polyprotein
Geminiviridae	Cassava latent virus	1-DNA	No	Tobacco	———	———

1. The nucleic acid within the virion is shown as the number of strands followed by the type of nucleic acid (e.g., 1-RNA designates single-strand RNA; 2-DNA designates double-strand DNA).

chapter

10

Microbial Genetics

chapter outline

■

GENERAL GOALS

Be able to explain the similarities and differences among various types of genetic transfer processes seen in prokaryotes and eukaryotes and give examples of each.
Be able to describe the impact of these genetic processes on microbial populations and on human activities.
Be able to synthesize the information in this chapter and chapter 9 so as to summarize the similarities and differences between plasmids and viruses and the ways in which they contribute to genetic transfer.
Be able to explain how genetic transformation and transposons can affect microbial populations.

The small size of microorganisms led early investigators to conclude that they were unlikely to have any significant genetic phenomena associated with them. Although we now know that such is emphatically not the case, it is true that the characteristics studied by microbial geneticists are unlike those studied by classical geneticists of that time. Instead of looking at traits such as the height or color of pea plants, microbial geneticists study mutations in biochemical pathways, such as those discussed in chapters 4 and 5. Nevertheless, the concept of the **phenotype** or catalog of presently active genes in an organism is still valid, as is the concept of the **genotype** or catalog of all genes present in an organism. Because numerous microorganisms can be grown in a very small space, microbial geneticists have been able to study rare genotypic changes comparatively easily. In this chapter, certain genetic processes occurring in eukaryotes are compared with those found in prokaryotes.

Mutations and Mutagenesis

A **mutation** is most easily defined as any change in the base sequence of the nucleic acid that is the hereditary material of a cell or virus. Anything that induces a mutation is denoted a **mutagen.** From a logical point of view, there are only a limited number of possible types of mutations. These include base substitutions, deletions, insertions, and inversions.

Base substitution mutations are the simplest to visualize in terms of both cause and effect. They arise when the normal base-pairing rules are violated during replication or repair. One method of inducing a base substitution is to use a base analog like 5-bromouracil that can replace thymine. Occasionally it will pair with guanine instead of adenine (figure 10.1A–B). A mispairing of this sort that results in the substitution of a purine for a purine or a pyrimidine for a pyrimidine is called a **transition,** whereas the substitution of a purine for a pyrimidine or vice versa is designated a **transversion.**

Base substitutions also can arise during repair of damaged DNA (stimulated by an altered base, mismatched base, etc.). Excision repair is the mechanism that has been studied most extensively. It involves an enzyme, sometimes called correndonuclease (correction endonuclease), that makes a nick in the DNA strand at a point adjacent to the damaged base. DNA polymerase I then binds to the site of the nick and, using both its exonuclease and polymerase activities, removes the damaged DNA and replaces it with normal DNA (figure 10.2A). This process is analagous to the removal of RNA primers from Okazaki fragments during DNA replication (figure 6.4). The average amount of DNA replaced is about 20 bases. Excision repair is essentially error free and does not introduce mutations into the repaired DNA.

When DNA damage is heavy, the error-free repair process is too slow to keep up; therefore, most organisms invoke a very fast but not nearly so accurate system referred to as error-prone or **SOS repair** (figure 10.2B). The system derives its name SOS because it is used only in emergencies. Accumulating DNA damage products released by excision repair enzymes act to stimulate the cleavage of a repressor molecule, thereby activating the synthesis of some new proteins in the cell. These proteins affect a variety of metabolic activities including cell division. The process also forces the replacement of large portions of the helix (about 100 bases). During replacement, numerous mistakes in base pairing are made, and the mutation rate increases substantially (figure 10.2B).

The effect of a base substitution can be determined from a listing of the genetic code (see figure 6.7). If the new codon is synonymous with the old, the same amino acid is inserted into the protein during translation, and the mutation is a silent one. If a different amino acid is inserted, the effect on the protein depends on the chemical nature of the two amino acids in question. If the substitution changes an amino acid codon into a stop signal (TGA, TAA, or TAG), the protein will be shortened. Once again, the effect is uncertain except for the general rule that drastically shortened proteins rarely retain enzyme activity. If the substitution changes a stop signal into an amino acid codon, the protein will be longer. Increased protein length does not necessarily affect its function.

With insertion and deletion mutations, bases are simply added to or removed from the nucleic acid. These mutations are induced commonly by intercalating agents. Chemicals like 5-aminoacridine can take the physical space of a base but lack the means to be joined into the sugar-phosphate backbone of the nucleic acid. When bases are added or removed in multiples of three, insertion or deletion mutations result in the addition or subtraction of one or more amino acids (figure 10.1B). Any number of bases not divisible by three results in a frameshift mutation in which the amino acid sequence downstream from the site of the mutation is gibberish, completely destroying protein function.

Inversion mutations result when a piece of nucleic acid is removed, rotated 180°, and reinserted into the

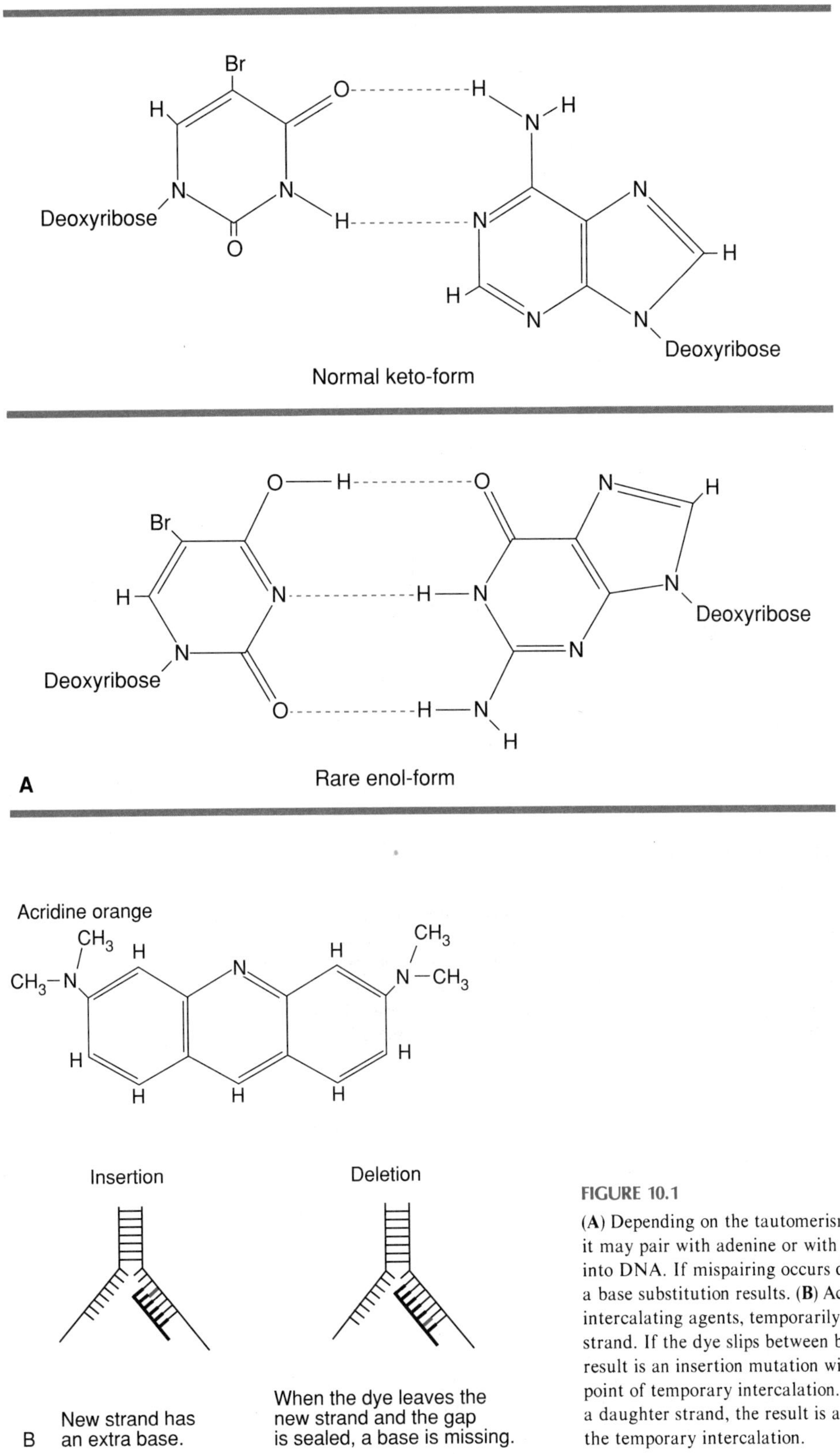

FIGURE 10.1

(**A**) Depending on the tautomerism of the molecule 5–bromouracil, it may pair with adenine or with guanine when it is incorporated into DNA. If mispairing occurs during DNA replication or repair, a base substitution results. (**B**) Acridine dyes may function as intercalating agents, temporarily slipping between bases of a DNA strand. If the dye slips between bases in the template strand, the result is an insertion mutation with the extra base opposite the point of temporary intercalation. If the dye intercalates itself into a daughter strand, the result is a deletion of one base at the site of the temporary intercalation.

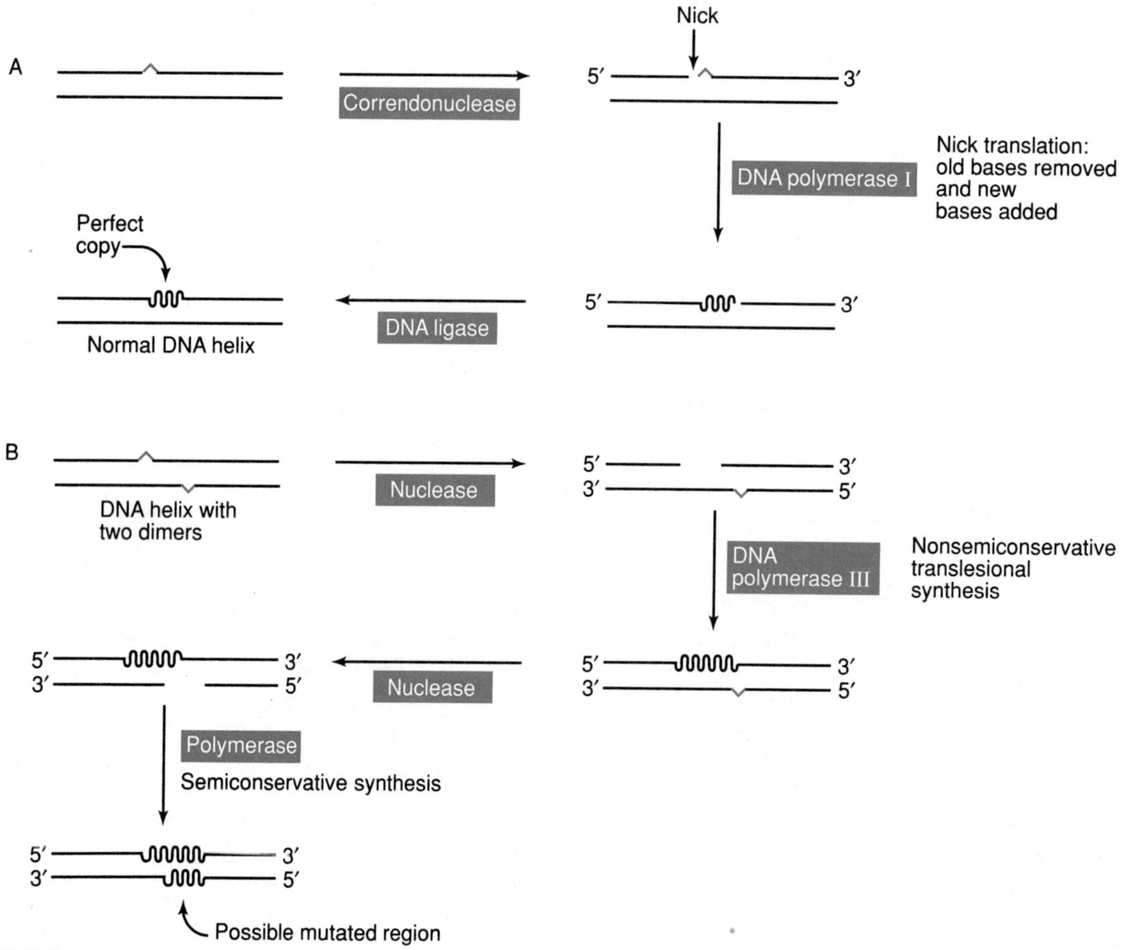

FIGURE 10.2

Error-free and error-prone DNA repair. The caret (^) indicates a damaged DNA base that DNA polymerase cannot replicate. (**A**) In the excision process, DNA polymerase I removes the damaged base from a DNA strand and replaces a short section of that DNA strand. Mistakes in base pairing are not usually made. (**B**) SOS repair occurs when the damaged areas are more numerous. In this case, excision repair processes cannot function because the damaged sites are too close together, and no functional template exists for the strand resynthesis (as diagrammed in A). Repair can occur if DNA polymerase III is forced to go past the lesion, but there is no certainty as to what bases will be inserted in the damaged region (wavy line). When excision repair occurs on the second strand, a mutation is likely to occur (parallel wavy lines).

nucleic acid. Normally, they are associated only with certain specific base sequences, and the mechanisms by which they are generated are complex.

Genetic Processes Seen in Selected Eukaryotic Microorganisms

Multicellular eukaryotic organisms are fundamentally diploid (or higher ploidy in the case of some plants). Normally, each cell has two copies of each chromosome, except for the gametes produced by meiosis that have only the haploid number of chromosomes. Fertilization consists of fusion of the gametes followed by restoration of the diploid number of chromosomes and growth of a new organism. This general type of process is seen among unicellular eukaryotes as well, but with some variations.

Mating Behavior in *Saccharomyces cerevisiae*

The life cycle of the unicellular fungus *Saccharomyces cervisiae* is depicted in figure 7.11. Its cells can be either diploid (34 chromosomes) or haploid (17 chromosomes). Each of these states is stable, meaning that a haploid cell can give rise to more haploid cells, or a diploid cell can give rise to more diploid cells. As discussed in chapter 7, however, when the correct mating types occur (figure 7.12), haploid cells can fuse to form a diploid zygote. The resulting diploid cell can undergo the usual proliferation process but, under conditions of severe nitrogen and carbon limitation, meiosis is initiated followed by sporulation. When spores germinate, new haploid cells are produced. This type of life cycle is called heterothallic because two different mating types are required for fertilization. If haploid cells of the same type can fuse, then the process is called homothallism.

Conjugation among Ciliates

Protozoa classified as ciliates display an interesting sexual variation, a true sexual exchange known as conjugation. They also exhibit nuclear dimorphism, an unusual approach to cellular function. Each cell contains a large macronucleus and a smaller micronucleus. Although the macronucleus is derived from the micronucleus, it is greatly modified. Certain DNA sequences (variable among different cells) are lost and others are significantly amplified. All cellular functions are controlled by the macronucleus; therefore, individual phenotypic variations result from changes in the macronucleus.

Tetrahymena thermophila is one of the best-studied ciliates. Its cells must go through a considerable period of growth before they are sexually mature, as many as 40 to 60 cell divisions from the last sexual episode. Conjugative behavior itself, however, is not triggered until starvation conditions occur and two different mating types are present, similar to that seen with *Saccharomyces*. Conjugation begins when cells of two different mating types touch and actually fuse their cell membranes to the extent that some cytoplasm and its inclusions are exchanged. At that time, the diploid micronucleus in each cell undergoes a typical meiotic cycle yielding four identical haploid nuclear products (figure 10.3). Three products eventually decay but the fourth divides mitotically to yield two haploid pronuclei. One of these remains in the cell in which it was formed as a stationary pronucleus, and the other becomes a migratory pronucleus that is transferred to the conjugative partner. The two cells separate, and the resident stationary pronucleus fuses with the newly arrived migratory pronucleus to form a diploid zygote. Note that each exconjugant cell now has the same genotype.

Mitotic divisions yield nuclei that follow two separate paths. One path, followed by the posterior nucleus, leads to the formation of a new micronucleus that remains inactive until the next conjugative episode. The other path, followed by the anterior nucleus, leads to the formation of a new macronucleus that replaces the existing macronucleus. Once again, certain sequences are lost and others are amplified; therefore, the genetically identical offspring of the conjugation may display different phenotypes nevertheless.

Extrachromosomal and Insertion Elements in Prokaryotic Microorganisms

Although they are haploid, bacterial cells are not necessarily limited to just one replicating DNA molecule. Much of the diversity found within the kingdom *Monera* can be attributed to DNA molecules that are in some sense autonomous (independent of the normal bacterial chromosome). By moving from one cell to another, these **extrachromosomal elements** can provide a host cell with entirely new properties. Only in the past few decades have scientists begun to realize just how pervasive these special molecules are and how much they contribute to the genetic processes observed in bacteria.

Extrachromosomal elements are structurally distinct pieces of DNA that can be present within a cell but do not necessarily occur in all cells of a population. Such an element usually is thought of as being dispensable to the cell in the long run because it is capable of severing its relationship with the bacterial chromosome. Loss of an extrachromosomal element can alter some properties of a cell, but its presence is not intrinsically necessary for survival of the species.

Extrachromosomal elements in bacteria are referred to as **plasmids,** DNA molecules capable of an independent existence within the cell. Plasmids are circular DNA molecules that can replicate their own DNA independent of the bacterial chromosome, although they may require some of the same enzymes for that replication (figure 10.4). Information encoded in plasmid DNA can be fully active in a metabolic sense; therefore, plasmids can contribute to the phenotype of their host cells. While a virus is present within a cell's cytoplasm, its function is similar to that of a plasmid, replicating its nucleic acid and contributing to the

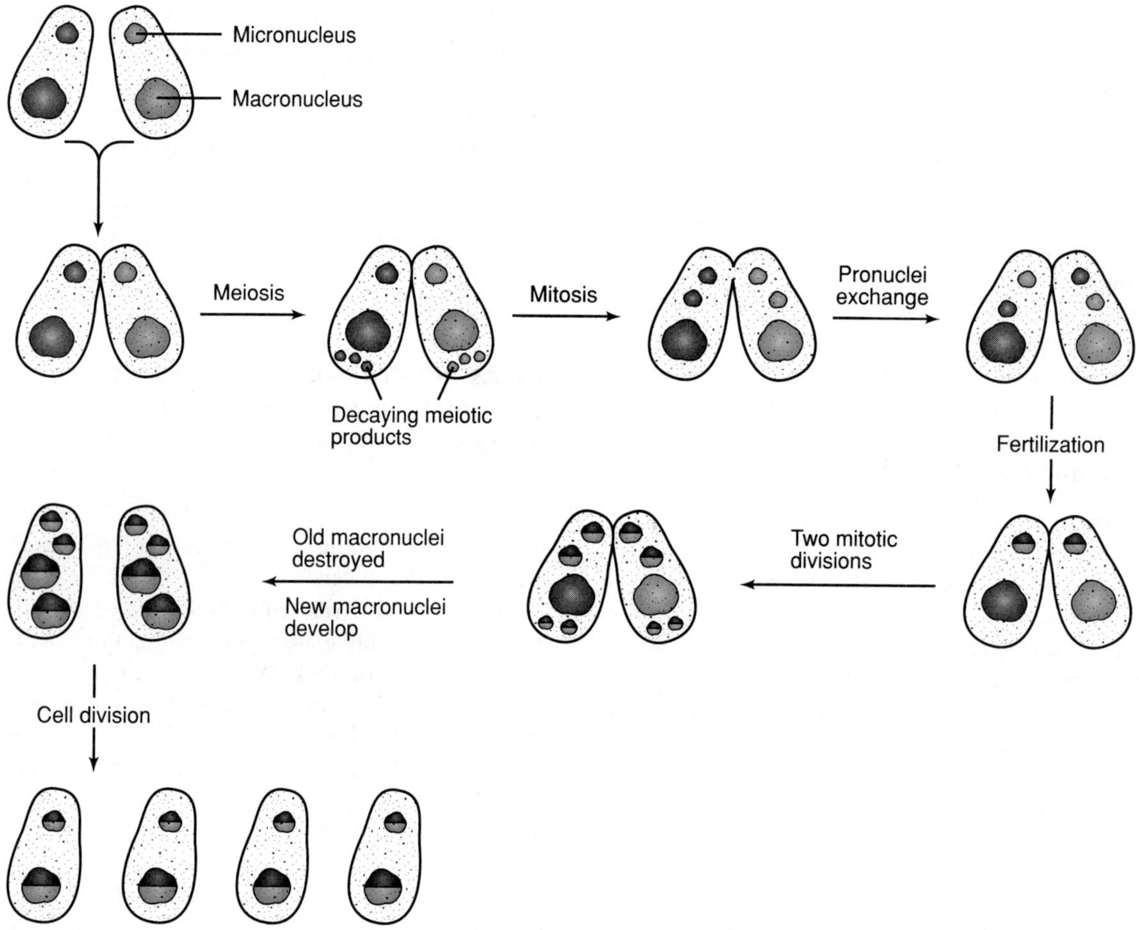

FIGURE 10.3

Conjugation in *Tetrahymena thermophila*. When the cells are suitably mature and of different mating type, they touch and fuse their membranes. The diploid micronuclei go through typical meiosis. Only one meiotic product survives as a pronucleus. It duplicates itself and then exchanges one copy for a similar pronucleus from the other cell. The existing macronuclei begin to break down as the micronuclei fuse to give a new diploid nucleus. After another mitotic division, new micro- and macronuclei are formed.

Source: Original version of the figure was published in the journal, *Genetics*, Genetics Society of America.

phenotype of its host cell. Viruses represent a specialized subset of plasmids in that their normal life cycle includes a period of independent existence outside a host cell.

Transponsons are a group of highly specialized DNA elements with DNA sequences having well-defined boundaries. These sequences possess the ability to synthesize proteins that specifically catalyze the movement of the original sequence or its copy to another site either on the same or on a different DNA molecule. Transposons come in a variety of sizes and exhibit widely variable phenotypic effects, but they do not seem to be capable of independent existence like plasmids. Because transposons are always associated with another DNA molecule and have no independent existence, they are referred to as insertion elements.

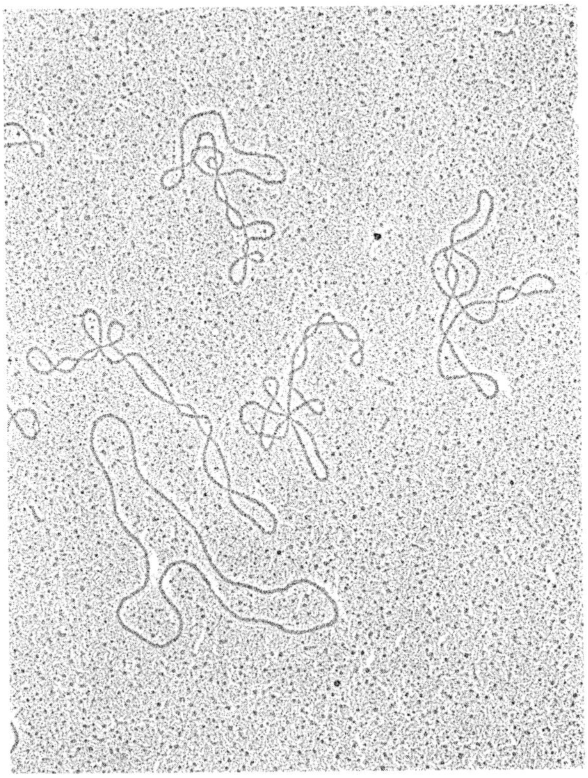

FIGURE 10.4

Plasmid DNA molecules spread on a protein film, stained and coated with heavy metals, and then examined with a transmission electron microscope.

Transduction: Transfer of Bacterial DNA Mediated by a Virus

The term **transduction** refers to a process of genetic transfer in which bacterial DNA is transported from one cell (the donor) to another (the recipient) encased within the capsid of a virus. The process is a simple extension of the viral assembly principles discussed in chapter 9 and first seen in the *Salmonella* genus by Norton Zinder and Joshua Lederberg. Since that time, transduction has been observed in many other genera with a wide variety of bacteriophages and can occur in either a generalized or specialized form.

Generalized Transduction

Generalized transduction is a property of certain lytic viruses, all of which have two things in common. First, during a normal infection, the phage does not produce enzymes that break down the host DNA molecule.

TABLE 10.1 Examples of generalized transducing phages and their hosts

Host	Transducing Phage
Escherichia coli	P1
Salmonella typhimurium	P22
Bacillus subtilis	PBS1, SPP1
Bradyrhizobium japonicum	M–1, 11
Pseudomonas aeruginosa	F116, G101, E79*tv-2*

Second, when a virion carrying transducing DNA attaches to a new host cell, the mere presence of capsid proteins does not cause cell death.

The actual production of transducing virions is thought to proceed by **wrapping choice** packaging. In this model, it is assumed that the enzymes responsible for stuffing DNA into phage heads are not infallible. Occasionally, they will process bacterial DNA rather than viral DNA. This type of error results in a phage capsid that contains only bacterial DNA, in accord with what is observed. Providing that the bacterial DNA sequences recognized by the viral enzymes are selected more or less at random, any bacterial sequence has an approximately equal probability of being packaged into a virion, and thus this type of transduction is truly generalized. The size of the transduced bacterial DNA fragment is determined by the amount of space available in the capsid. Some examples of generalized transducing phages that seem to operate in this manner are presented in table 10.1.

When a transducing particle attaches to a host cell, its DNA is ejected normally and transported into the cell. No infection can result, however, because only bacterial DNA is present. The linear transducing DNA generally cannot replicate itself and, therefore, must recombine with the resident DNA if it is to be maintained. Recombination between a linear molecule and a circular molecule requires a substitution process in which the original recipient DNA sequences are lost. Only in this way can the circularity of the bacterial chromosome be preserved.

Specialized Transduction

Specialized transduction is accomplished by temperate phages whose prophage DNA actually is integrated into the bacterial chromosome (or sometimes into an appropriate plasmid). When examined, specialized transducing phage particles are found to carry a single large DNA molecule that is predominantly viral but has some covalently linked bacterial sequences.

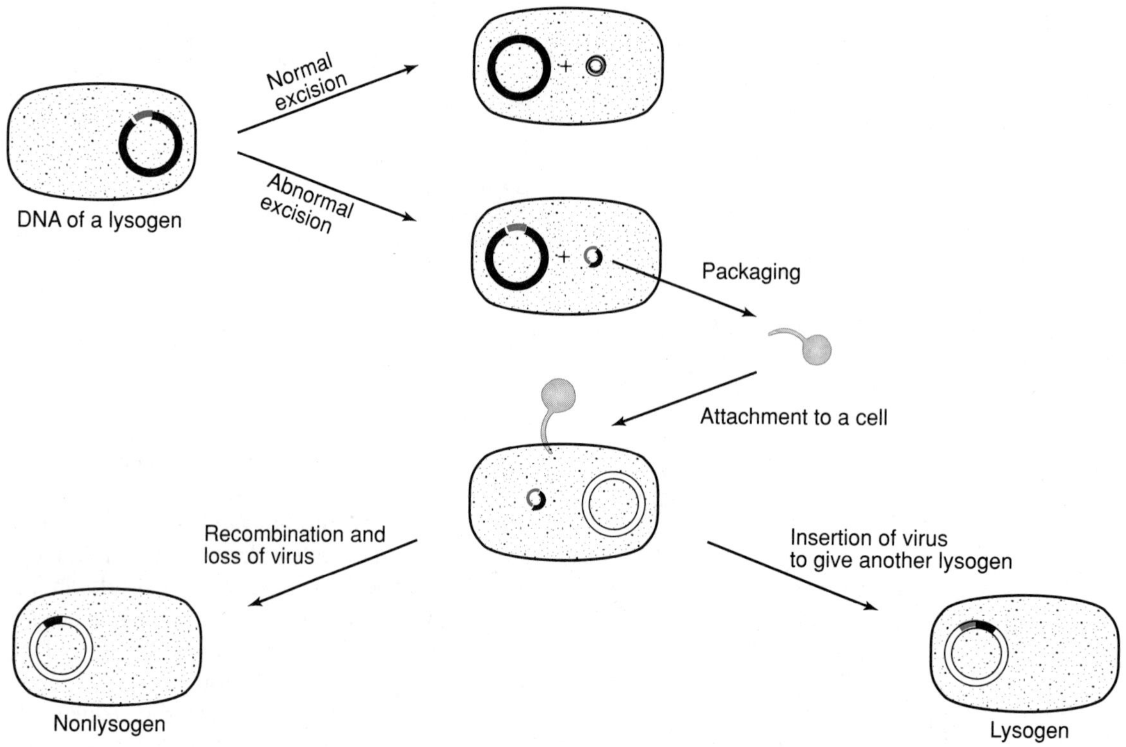

FIGURE 10.5

Specialized transduction. The process begins with a lysogenic cell. After induction of prophage, excision occurs. In the normal process, phage DNA is removed precisely. In the transduction process, removal occurs off center so that some phage DNA is left behind and some bacterial DNA is carried along. When the resulting phage attaches to a new host cell, the bacterial-viral DNA conglomerate is delivered into the cytoplasm. The conglomerate may integrate itself to give a new lysogen, or homologous bacterial DNA sequences may recombine with each other. Homologous sequences are those that are nearly identical in base sequence. In the latter case, the DNA conglomerate is eventually lost.

When the prophage is induced and excision ensues, it is possible for a mistake to occur in the way in which the prophage DNA excises itself from the host chromosome. If the excision is positioned off center, some prophage DNA is left behind and some bacterial DNA is acquired (figure 10.5). There is a size constraint operative in this instance, however. The viral capsid can hold only so much DNA; therefore, a temperate prophage is excised as roughly the same size molecule, even if the excision takes place off center. In the specific instance of the λ prophage, DNA molecules ranging in size from 78 to 105% of normal can be packaged efficiently, provided that one normal DNA end is present within the circular molecule. All other molecules cannot be packaged properly and are not found in intact virions.

The standard size of the transducing DNA implies that whenever a transducing particle is formed, it has lost some viral functions. The amount of missing DNA is proportional to the amount of bacterial DNA acquired; therefore, the degree of viral deficiency is quite variable. In some cases, a specialized transducing phage particle cannot even form a plaque unless another, normal virion (helper phage) also infects the cell in order to provide the missing functions. On the other hand, the pieces of bacterial DNA acquired may be so small that essentially no viral information is lost.

When a transducing particle attaches to a new cell and the DNA molecule is transported into the cytoplasm, there can be two possible outcomes. Sometimes there is simply recombination between the resident bacterial chromosome and the bacterial DNA fragment located on the viral DNA, as is the case for generalized transduction. Recombination results in the substitution of some of the donor sequences for some of the recipient DNA. The recombinant viral DNA molecule then fails to replicate or insert itself and is diluted from the culture as a consequence of cell growth and division. At other times, viral lysogenic functions are still available and the transducing phage DNA inserts itself to form a prophage. The result is a newly lysogenic cell that now carries a duplication for a small region of its DNA, namely, the region of bacterial DNA carried by the virus (figure 10.5). If this lysogen is induced later, it will produce very large numbers of transducing particles identical to the original infecting virion.

Plasmids and Conjugal DNA Transfer

The F Plasmid

The F or fertility plasmid was the original plasmid discovered. It was observed first in *Escherichia coli* and called the F factor (fertility factor) following its discovery. In its normal state, the F plasmid does not confer any unusual metabolic properties on its host cell. Instead, it is best known for its self-transfer capabilities. A cell carrying an F plasmid is said to be F^+, and a cell lacking an F plasmid is said to be F^-. If a mixture of F^+ and F^- cells is allowed to stand for about an hour, most F^- cells are observed to acquire the F plasmid and become F^+, while the original F^+ cells remain F^+. If contact between the two cell types is prevented by an interposed filter that allows the passage of liquid but not cells, no transfer of F plasmids occurs. From these observations, it is concluded that DNA is transferred directly from the donor cell to the recipient cell only when the cells are in direct contact. Such transference is called conjugation, and the F plasmid is said to be conjugative.

The F plasmid codes for special cell structures called F pili or sex pili that appear on the surface of the host cell (figure 10.6A–B). These pili mediate the formation of **mating aggregates** (clumps of F^+ and F^- cells). It has been suggested that F pili identify F^- cells by their lack of F pili and then bind to the surface of those cells. The pilus-mediated attachment then can be used to overcome the normal electrostatic repulsion of the charges on the cell surfaces. By retracting the pili back into the producing cell, a donor cell can cause the two cell types to touch physically. The assumption is that portions of the cell membranes are fused to allow transfer of DNA from the donor cell to the recipient after initial contact is made, although wholesale exchange of cytoplasm has not been demonstrated.

Joshua Lederberg and Edward Tatum, the Nobel laureates who first studied the *E. coli* conjugative system, observed that although the F plasmid was transferred to F^- recipient cells very efficiently, the ge-

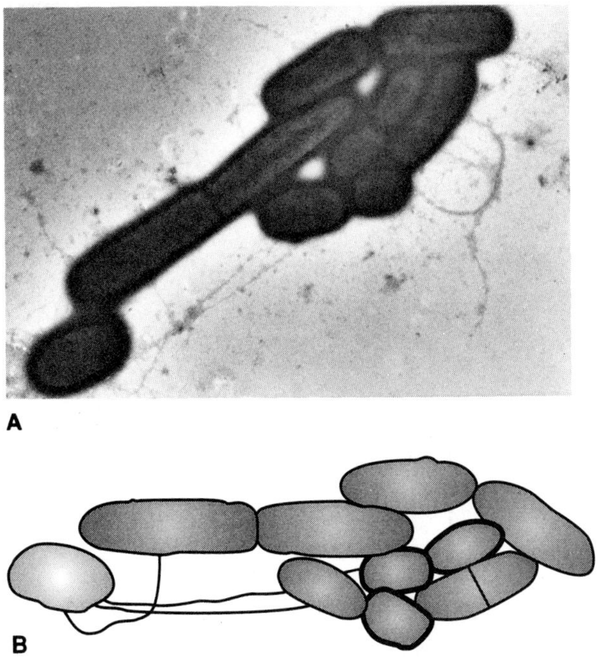

FIGURE 10.6

F pili. (**A**) The long thin structures in this micrograph of negatively stained *E. coli* are F pili. Note that they are longer, thinner, and more flexible than the normal fimbriae on a cell. The longer cells are Hfr cells, and the shorter are F^-. (**B**) In the cartoon, the thick-walled cells are the F^-; the unshaded cell could not be classified.

(B) From M. Achtman, G. Morelli, and D. S. Schwuchow, "Cell-cell Interaction in Conjugating *E. coli*: Role of F. pili and fate of mating aggregates," in *Journal of Bacteriology* 135: 1053–1061. Copyright © 1978 American Society for Microbiology. Reprinted by permission.

netic traits coded by a bacterial chromosome itself were inherited only rarely (typically a probability of 10^{-7}/recipient cell). Other scientists soon showed that variants occurred within the population of F^+ donor cells that seemed to be able to transfer genetic markers (genes) from a bacterial chromosome at a substantially higher frequency (10^{-2}/recipient cell), but rarely if ever were they able to transfer the F plasmid itself. The new type of cell was called an **Hfr cell** because it seemed to cause a high frequency of recombination.

Our present understanding of the relationship between F^+ and Hfr cells is summarized in figure 10.7. It has been found that the F plasmid can function as an **episome,** meaning that it can exist either in an autonomous state (F^+) or in an integrated state like the λ prophage (Hfr). The model proposed by Allan Campbell for the integration of phage λ (chapter 9) is also applicable to the F plasmid. As in the case of phage λ, the integration reaction is actually an equilibrium one. Regardless of its physical state, an F plasmid is

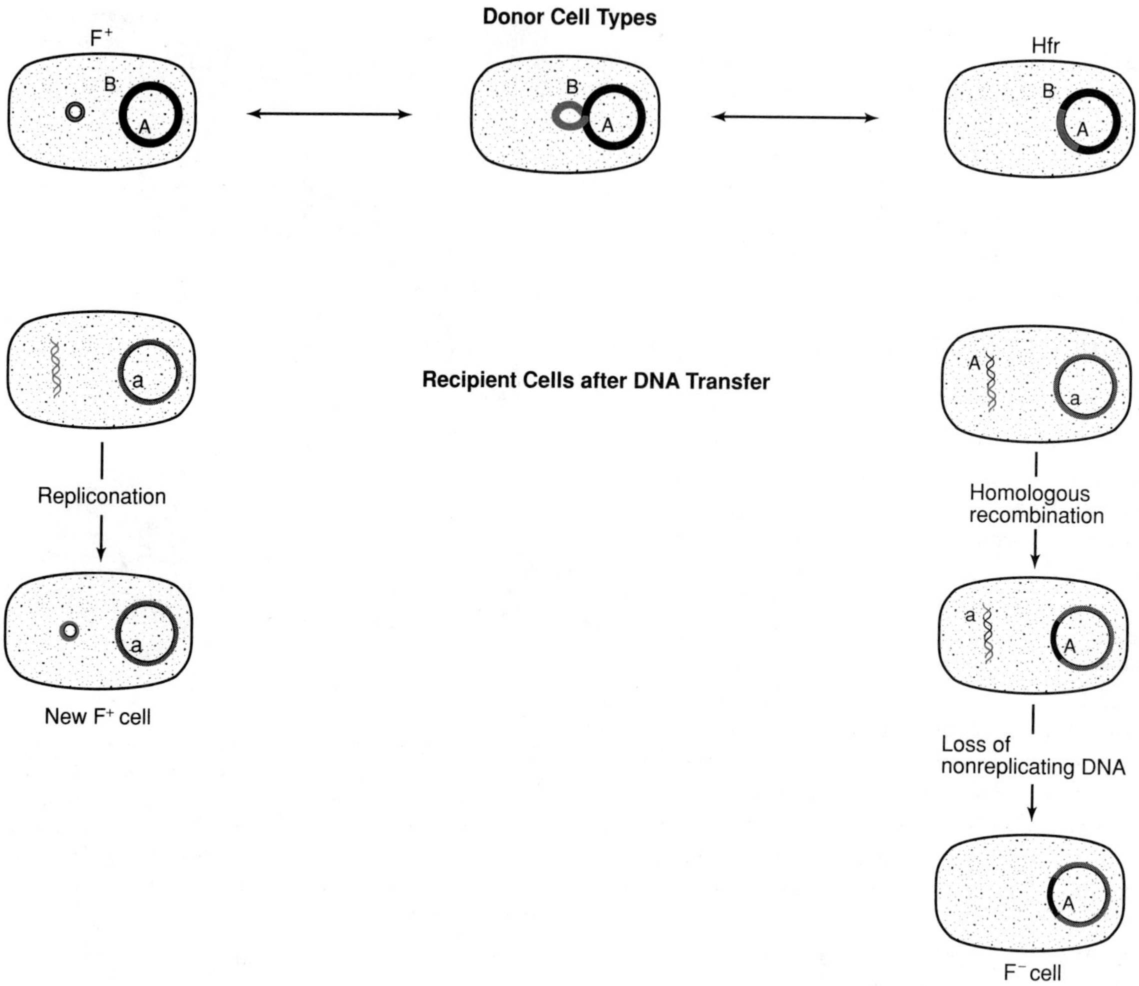

FIGURE 10.7

Formation of Hfr cells from F^+ cells and behavior of Hfr cells as donors. F plasmid DNA is depicted in color. In an F^+ cell, the plasmid is autonomous, while in an Hfr cell, it is integrated. The mechanism of integration is similar to that for the integration of phage λ. Note that the DNA transferred by an Hfr cell must recombine in order for subsequent generations of bacteria to inherit it, whereas DNA from an F^+ cell can circularize and resume replication. The recombination process again is homologous.

capable of self-transfer from the same origin of transfer site on its DNA. Because this origin of transfer is not located at one of the physical ends of the integrated plasmid, when transfer does occur, about 30% of the F plasmid is transferred followed by DNA sequences from the bacterial chromosome. The total amount of DNA transferred is quite variable but generally substantially shorter than the entire bacterial chromosome. Thus, a recipient cell mated to an Hfr cell receives primarily bacterial DNA rather than plasmid DNA. Note that the conjugative process is not functionally the same as that seen in members of *Tetrahymena* because DNA is not exchanged between the cells but passed unidirectionally.

No matter which type of donor cell is used, the transferred DNA is in the form of a single, linear strand generated by the rolling circle replication mechanism (see chapter 9). If the DNA includes the entire F plasmid, however, mechanisms are available to cause **repliconation** (recircularization of the DNA and resumption of replication). Repliconation cannot occur with the fragment of bacterial chromosome normally transferred by an Hfr cell. Therefore, the Hfr donor

DNA must recombine with the resident DNA in the recipient cell in order for donor genetic markers to be transmitted to progeny cells. The repliconation process is shown also in figure 10.7.

The linear transfer of donor DNA has turned out to be of extraordinary benefit to bacterial geneticists. Because a given culture of Hfr cells has its F plasmid integrated in only one site on the bacterial chromosome, transfer always begins with the same DNA sequence. Moreover, the linearity of the transferred molecule assures that genetic markers are always transferred in the same order in which they occur on the bacterial chromosome. Thus, the genetic map for *E. coli* theoretically can be read if one merely observes the order of transfer of the genetic markers by one or more Hfr cultures.

From this understanding of Hfr transfer has evolved the **interrupted mating experiment.** In such an experiment, donor and recipient cells are mixed and allowed to conjugate. At various times, samples are removed and mating aggregates disrupted by some sort of blender. The samples are plated on medium that selects against (prevents from growing) both donor and recipient cells but allows growth of a certain type of recombinant. As can be seen in figure 10.8, no recombinant bacteria are observed initially. As time elapses, the number of recombinants begins to increase steadily. **Time of entry** refers to the time at which **transconjugants** (recombinant cells) are first observed. A comparison of the times of entry for different markers transferred by the same Hfr donor makes it possible to estimate the distances between the markers. The amount of DNA transferred by an Hfr cell in 1 minute is used as the unit of length for the genetic map.

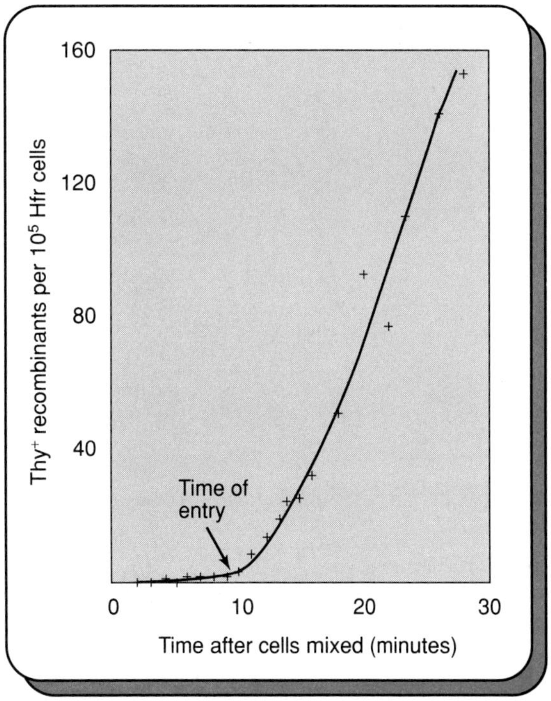

FIGURE 10.8

Interrupted mating curve. Samples were removed from a mating culture of Hfr and F^- cells. They were blended to separate the conjugating cells and then plated on selective medium. The arrow indicates the time of entry of the selected markers.

Figure 10.9 depicts a greatly simplified genetic map for *E. coli*. Although the map is 100 units long, in reality a single Hfr strain would require something on the order of two hours to transfer the entire bacterial chromosome because the speed of DNA transfer is not constant. Enough Hfr insertion sites are marked on the map to demonstrate that a multiplicity of Hfr strains is available to the geneticist.

According to the rules of the Campbell model, the F plasmid can integrate or excise itself. Most of the time, the excision reaction occurs very precisely, restoring the F plasmid to its original form. Sometimes, however, excision occurs outside the boundaries of the F plasmid (analagous to the formation of a specialized transducing phage), and some DNA from the bacterial chromosome is included in the newly enlarged F plasmid. In order to distinguish this type of plasmid from a normal F plasmid, a plasmid carrying DNA from the host cell is referred to as an F prime (F′). F primes behave like normal F plasmids in terms of their conjugation, and the bacterial markers that they carry can be inherited by the recipient cell without the need for recombination as long as the entire F prime plasmid is transferred.

Because there is no need to package plasmid DNA like a transducing phage, there is no real size limitation for an F prime. In a practical sense, however, the larger the plasmid DNA, the greater is the drain on the resources of a cell because of the extra replication activities required. Therefore, all other things being equal, natural selection tends to favor cells that have smaller F prime plasmids over those with larger ones, and deletions are often observed in very large F primes.

Although the F plasmid is most commonly studied in *E. coli,* it is capable of transferring itself to bacteria in other genera, specifically in *Salmonella* and *Shigella.* The ability of a plasmid to transfer itself to other

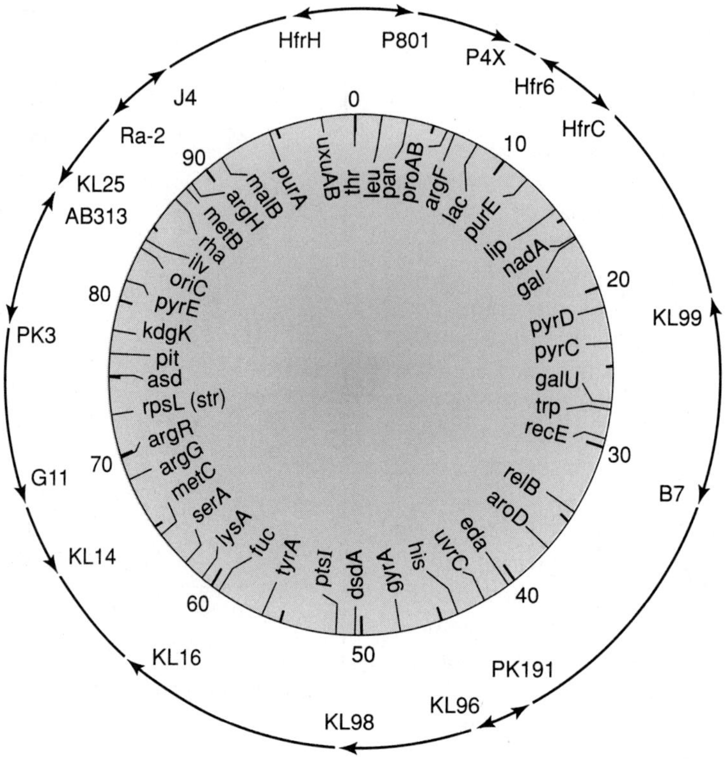

FIGURE 10.9

Simplified genetic map for *E. coli.* This circular rendition of the bacterial chromosome is shown together with some commonly used genetic markers (mutated genes). Also indicated are the points of origin and directions of transfer of some common Hfr strains.

From B. J. Bachmann and K. B. Low, "Linkage Map of *Escherichia coli,* K-12," Edition VI. *Microbiological Reviews* 44:1–56. Reprinted by permission.

genera and species reflects its **host range,** the number of different cell types with suitable receptors on their surface. Many other plasmids have a much broader host range than the F plasmid.

Other Fertility Plasmids

The F plasmid is not the only conjugative plasmid, merely the one studied most extensively. Many other plasmids and bacterial hosts are capable of conjugation. Some plasmids commonly used in genetic studies are listed in table 10.2. The mechanisms of conjugation involving plasmids other than F have not been so thoroughly studied. Nevertheless, it is clear that not all conjugation follows the pattern seen with *E. coli.* In the case of *Streptomyces,* for example, Hfr-like cells transfer in both directions from a single point of origin, while in *Pseudomonas* plasmid integration normally does not occur.

TABLE 10.2 Some conjugative plasmids and their typical hosts

Plasmid	Host
F plasmid	*Escherichia coli*
	Salmonella typhimurium
	Shigella dysenteriae
FP series	*Pseudomonas aeruginosa*
SCP1, SCP2	*Streptomyces coelicolor*
AD1	*Streptococcus faecalis*

Members of the genus *Bacillus* also seem to exhibit "conjugation" in the sense that they undergo genetic exchange requiring cell-to-cell contact. This process is attained usually via **filter mating** in which cells from two different strains are placed in close proximity when two cultures are mixed and then filtered to trap all the cells in a very dense, thin layer. The cells are later resuspended and checked to see if recombination of traits has occurred.

APPLICATIONS BOX 1

Bacteriocins and Other Killer Factors

Although some bacteria have bacteriocins, other microorganisms have been shown to possess equivalent systems. There is a killer factor present in some yeast cells and a kappa particle (bacterial endosymbiont) in *Paramecium* species that have effects similar to bacteriocins. Practical use can be made of these factors when large quantities of microorganisms are grown.

Contamination by extraneous microorganisms often can be a problem in large cultures. If the microorganism being cultured carries a killer factor, however, this desired organism has a chance of killing any organisms similar to it that might contaminate the culture. Such methods are of particular interest when naturally occurring microorganisms on a substrate need to be replaced by specific cultivated microorganisms. A patent application has been filed for a special strain of *Saccharomyces cerevisiae* used to ferment grape juice into wine. The new strain carries a killer factor that is supposed to eliminate endogenous *Saccharomyces* from the grape juice.

Bacteriocin-Producing Plasmids

Some bacteria produce **bacteriocins** that are toxic substances to certain other bacteria including closely related bacteria that are not bacteriocin producers. Bacteriocins are named according to the species of the producing organism. For example, *E. coli* can produce colicins, *Bacillus subtilis* can produce subtilisins, and *Bacillus megaterium* can produce megacins. Some of the originally identified bacteriocins have turned out to be defective bacterial viruses that retain their ability to kill nonlysogenic cells but have lost their ability to assemble an infectious virus. Others are proteins of variable size that inhibit specific cellular functions and are encoded by specific plasmids. In the latter case, the producing cells are not harmed because they also make an immunity protein that prevents the toxin from acting. Thus, regardless of the mode of action, all bacteriocin-producing cells can grow at the expense of their immediate, nonproducer neighbors.

Some bacteriocins studied extensively are the colicins. Different colicins are identified by capital letters after their name (for example, colicin E). If it is shown subsequently that more than one colicin is produced by a particular population of cells, an arabic number is added after the letter designation. For instance, the original colicin E is actually a group of proteins (colicins E1, E2, and E3), each having a unique mode of action.

Many, but not all, bacteriocin plasmids are conjugative. Some use the same conjugative apparatus as the F plasmid and are designated as F-like. Others use a different conjugative apparatus that was first seen in the colicin I plasmid, and hence these are called I-like plasmids. While the basic conjugative process is similar, the pili are different in structure from F pili. Furthermore, in a cell that carries both F and I plasmids, each plasmid preferentially uses its own conjugative system.

Other bacteriocin plasmids are nonconjugative, meaning that they cannot transfer themselves from one cell to another. Some, however, can take advantage of the conjugative apparatus prepared by another plasmid and use it to transfer themselves. The mechanisms by which the remaining nonconjugative plasmids spread through a bacterial population are uncertain, but possibly transduction or temporary insertion into a conjugative plasmid might be used.

R Plasmids

R plasmids were first identified because of their ability to confer antibiotic resistance to a host cell. In Japan in the 1950s, hospital isolates of *Shigella dysenteriae* were found to be simultaneously resistant to the antibiotics streptomycin, chloramphenicol, and tetracycline as well as to the drug sulfonamide. Because the probability of a bacterium acquiring a single resistance mutation is approximately 10^{-8}, the probability of one bacterium becoming simultaneously resistant to all four antimicrobial agents is about 10^{-32} (essentially zero). The investigators knew that something unusual had been found.

Many R plasmids are conjugative using the F or I systems. Still others are nonconjugative, as in the case of certain bacteriocin producers mentioned in the previous subsection. What all R plasmids have in common is their ability to make the host cell resistant to various environmental influences, such as antibiotics or heavy metals like mercury or arsenic. In the light of present-day knowledge, R plasmids constitute a very heterogeneous group. They are subdivided initially on the basis of their **incompatability** responses, that is, on their inability to coexist in a single cell with other specific plasmids. Two plasmids that are mutually exclusive (cannot coexist) are said to be members of the same incompatability group. Each incompatability group is designated by an arabic letter (for example, IncF members cannot coexist with the F plasmid).

The resistance encoded by an R plasmid is of an unusual type. When a cell mutates to become resistant to an antibiotic, it changes the genetic code for the cell structure normally affected by the antibiotic. For example, a bacterium acquires resistance to streptomycin by altering one protein that is part of its 30S ribosomal subunit. However, R-plasmid-mediated streptomycin resistance results from the production of an enzyme that modifies streptomycin chemically so that it can no longer function. A typical modification might be adenylylation (the addition of an adenyl group) or acetylation (the addition of an acetate moiety). The level of resistance of a cell, therefore, depends on the amount of inactivating enzymes present in its cytoplasm. Note the import of this observation because it is possible to overwhelm R-plasmid-encoded drug resistance by increasing the concentration of an antibiotic to the point at which the enzyme molecules cannot modify all of the antibiotic molecules arriving in the cytoplasm. Unfortunately, this often takes an antibiotic concentration outside the safe therapeutic level so that what works in a laboratory will not work in a clinical setting.

In many instances, the antibiotic-resistance genes of R plasmids are contained within transposons (see section on the genetics of transposons). The practical effect of this observation is that antibiotic-resistance genes that are part of transposons tend to be exchanged between plasmids or between chromosomes and plasmids as a group rather than as individuals. Therefore, an organism's resistance to one antibiotic often is correlated with its resistance to several chemically unrelated antibiotics. A further extension of the transposon concept is that R plasmids probably developed from preexisting conjugative plasmids that accidentally acquired antibiotic-resistance transposons. Later they were inadvertently enriched when the bacteria were treated with antibiotics. Cells with an R plasmid would have a selective advantage over cells without under those circumstances.

FIGURE 10.10
A gall on a plant produced by *Agrobacterium tumefaciens* carrying a Ti plasmid. This photo was taken about six weeks after the plant was infected.

Ti Plasmids

A group of plasmids with most unexpected properties are the tumor-inducing (Ti) plasmids occurring in members of the genus *Agrobacterium.* Ti plasmids enable these bacteria to cause tumors in plants. *Agrobacterium tumefaciens* provides a good example because it has been studied extensively. This bacterium without the plasmid is harmless to plants, but bacteria carrying these large, conjugative plasmids can cause some bizarre effects. If a plasmid-carrying *A. tumefaciens* is inoculated into a tomato or tobacco plant by puncture or abrasion, uncontrolled cell proliferation (a tumor) results (figure 10.10). After the bacterium's initial infection, its presence is not required for further development of the tumor; therefore, the bacterium must cause some sort of permanent alteration in plant cells.

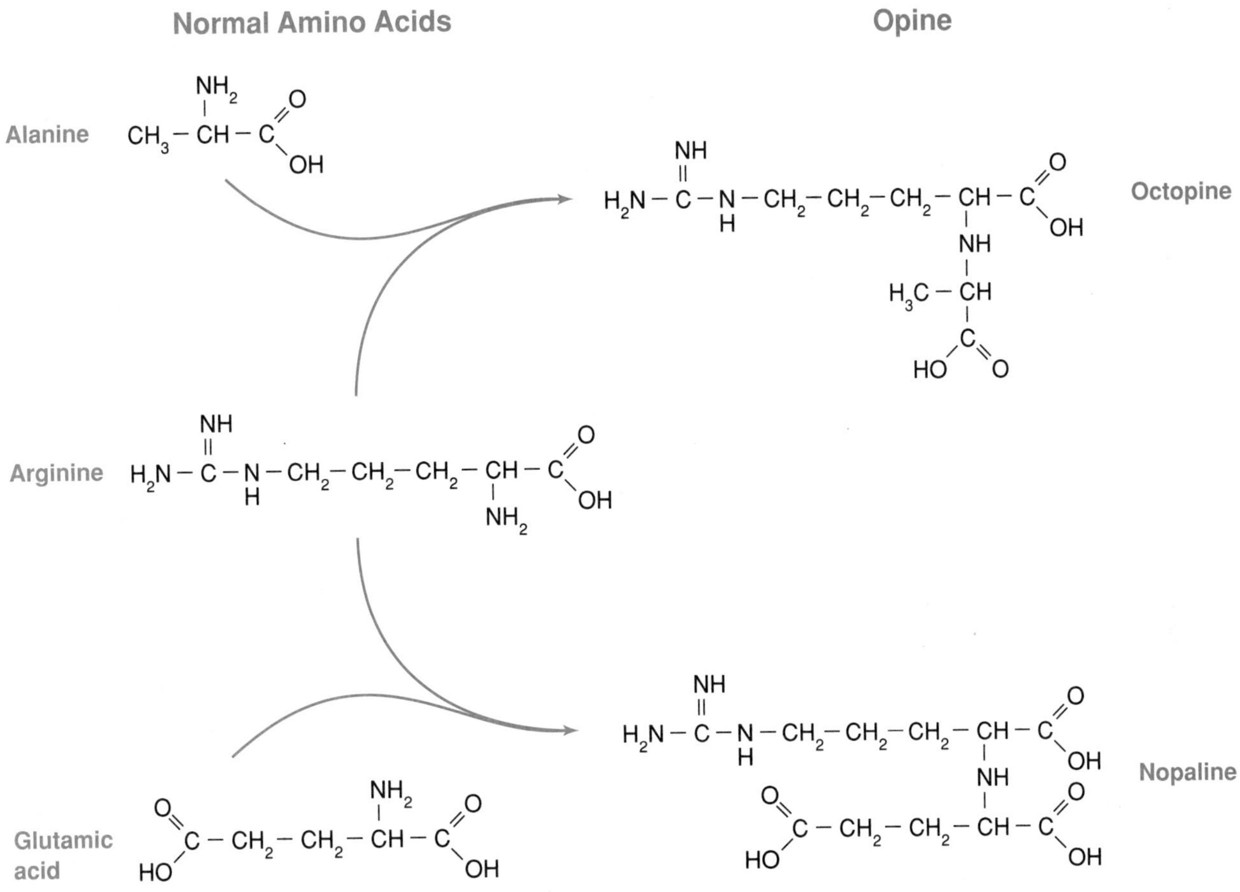

FIGURE 10.11

Some opines found in tumors induced by Ti plasmids together with the amino acids from which they are derived.

The ability of *A. tumefaciens* to attach to the plant is controlled by DNA sequences located on its chromosome, and molecular studies have shown that a portion of the Ti plasmid (T-DNA) is somehow transferred from the bacterium into the plant cell and integrated into one of the plant chromosomes. All cells carrying this DNA fragment become part of the developing tumor, while cells lacking the fragment are normal. This discovery marked the first time that transfer of DNA from a prokaryote to a eukaryote was observed to occur.

From the point of view of the bacterium, the relationship is a simple parasitic one. T-DNA, in addition to causing cell proliferation, also causes plant cells to produce unusual amino acids of the opine family (figure 10.11). The opines are absolutely useless to the plant, but the Ti plasmid codes for enzymes that allow the host bacterium to degrade these molecules for energy. Thus, in the normal situation, some agrobacteria can be found in intimate association with the cells of the tumor to derive nourishment from them. The degree of damage done to the plant varies, smaller plants being killed or grossly deformed, larger plants showing little or no damage.

Genetic Transformation: Transfer of Naked DNA

The first genetic transfer process to be reported in bacteria was **genetic transformation,** originally observed by Fredrick Griffith during his studies of the pathogenic organism now called *Streptococcus pneumoniae* (synonyms are pneumococcus and *Diplococcus pneumoniae*). Griffith observed that a mouse injected with *S. pneumoniae* cells having a capsule was killed rapidly (figure 10.12). A similar injection of cells that had mutated in the laboratory so as to lack a capsule had no effect, however. Simple mixtures of the two cell types were also lethal. An interesting result occurred when

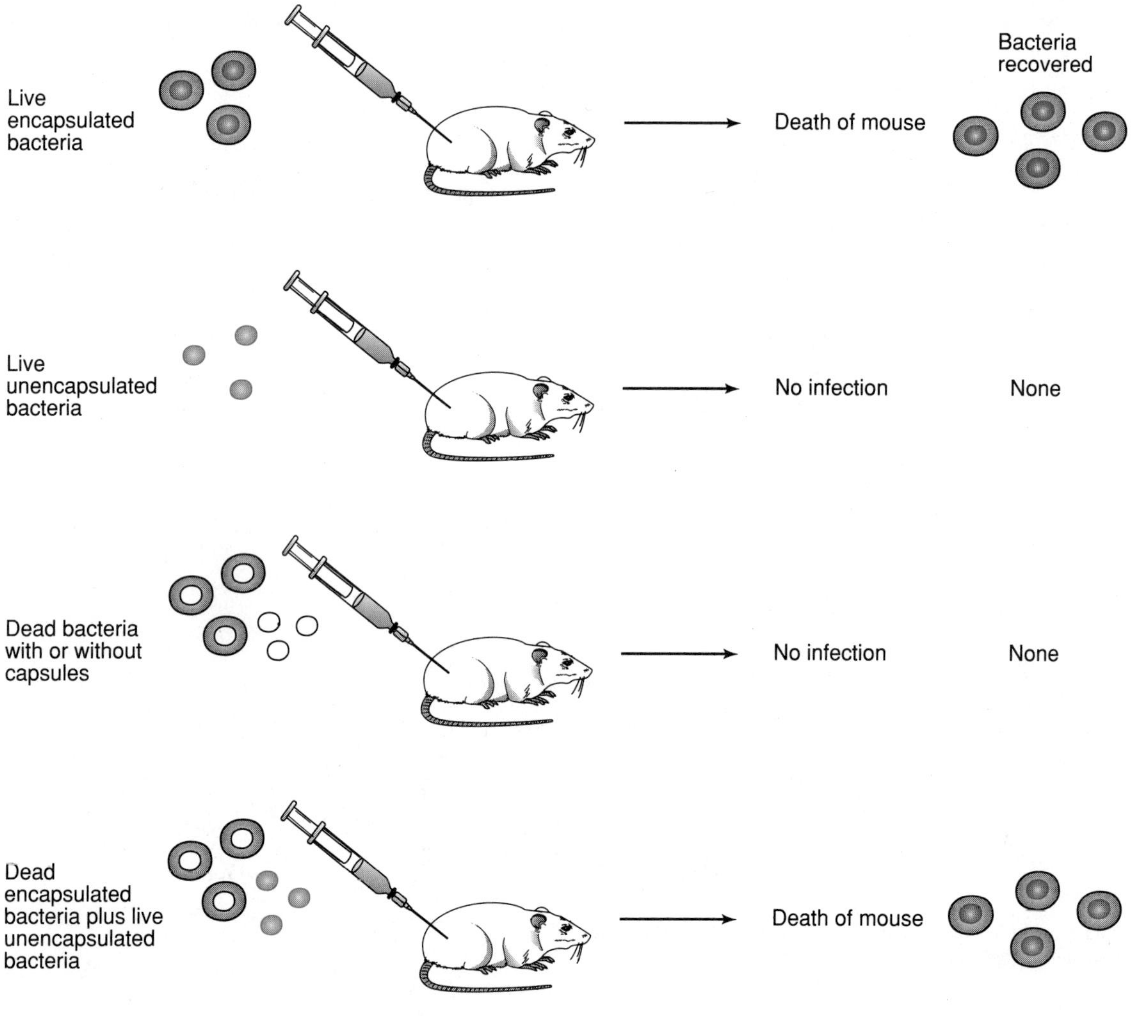

FIGURE 10.12

Diagram of Griffith's experiments demonstrating genetic transformation. These experiments demonstrate that capsules are protective for disease-causing bacteria. In addition, they illustrate how something released by dead, encapsulated bacteria can cause genetic transformation of live, unencapsulated bacteria, enabling the latter to acquire a capsule permanently.

Griffith injected heat-killed encapsulated bacteria mixed with live unencapsulated bacteria. The mice all died, and the bacteria isolated from them were all found to have a capsule and to produce encapsulated offspring. Griffith concluded that his bacteria had undergone some sort of genetic transformation, but he was never able to demonstrate what that meant in molecular terms.

Actual discovery of the true nature of the "transforming principle" was made by Oswald T. Avery, Colin McLeod, and Maclyn McCarty. Several years prior to the famous paper of Watson and Crick, Avery, McLeod, and McCarty published one concluding that the genetic material of the cell was DNA. Although the implications of their experiments were not fully appreciated at the time and they were never honored with a Nobel Prize, their work marked the beginnings of molecular biology.

Genetic transformation has been observed in a variety of bacteria, both gram positive and gram negative, but the major species studied have been *S. pneumoniae, Bacillus subtilis,* and *Haemophilus influenzae.* The general pattern of the process is the same in all cases, although some details may differ. Cells in a culture are not capable of acting as recipients of DNA at all times. Rather they go through periods of

APPLICATIONS BOX 1

Impact of Genetic Transfer on Natural Populations

Conjugation by Hfr cells allows the transfer of the largest pieces of DNA, and genetic transformation provides the smallest pieces. Nevertheless, it seems likely that generalized transduction, although transferring only moderately sized pieces of DNA, is the most widespread of the transfer processes. First, all genes can be transferred at roughly equal frequencies during the transduction process, but conjugation is limited to specific origins and directions of transfer. Second, conjugation requires that cells contact one another, while transduction involves contact only between a virus and a cell. Because generalized transducing phages are released in larger numbers than the producing cells, it is easier for at least one of them to encounter a suitable host.

competence during which DNA can be taken up; the remainder of the time, all DNA in the medium is ignored or actively degraded. Cells generally become competent as a result of stress, such as a shift from a rich medium to a minimal medium or from the transition from exponential phase into the stationary phase of growth.

When a cell is competent, one or more new proteins appear on its surface. These proteins serve as DNA-binding proteins for the cell. Competent gram-positive cells seem to bind DNA indiscriminately, while *Haemophilus* requires the presence of certain specific "recognition sequences" in any DNA before it is bound. Bound DNA is then taken into the cytoplasm across the cell wall and cell membrane (figure 10.13). As it is transported, bound DNA is shortened and, in the case of the gram positives, also made into a single-strand. In *Haemophilus,* the transported DNA is placed in a membranous vesicle from which it emerges as a single-strand molecule.

The newly arrived transforming DNA cannot replicate itself, even if it was derived originally from a plasmid, owing to difficulties in recircularization; therefore, in most cases, recombination is the requisite next step. A portion of a single strand of DNA is exchanged for some resident DNA, and the remainder of the DNA fragment, with its exchanged host DNA, is soon lost. Gram-positive cells will not recombine the transformed DNA unless it is homologous (similar) in sequence to the resident DNA. For example, *E. coli* DNA will not recombine with *S. pneumoniae* DNA. The only exception to the need for recombination is a case in which a self-replicating DNA molecule has been transformed. If the self-replicating DNA is a virus, the process is referred to as **transfection** because the result is a virus-infected cell. Transfection, like all forms of genetic transformation, requires competent cells; therefore, the process normally does not complicate other experiments such as premature lysis experiments that are conducted with exponential phase cells.

Transposons: Mobile Genetic Elements

One of the fundamental assumptions of geneticists has always been that the genome of an organism is essentially stable. Granted that mutations such as those discussed in chapter 7 are possible, but they do not occur with great frequency and their effects are highly localized. No wholesale rearrangements of DNA are known to result from simple changes of bases. On the other hand, Barbara McClintock, while working with corn in the 1950s, reported the existence of genetic elements that seemed to move from one chromosome to another. The pieces of the resulting puzzle did not begin to fall into place until bacterial geneticists in the 1970s showed that many bacteria contained **mobile genetic elements,** regions of DNA whose position within a bacterial chromosome did not appear to be fixed. The importance of McClintock's work was finally realized, and she received a Nobel Prize.

The mobile genetic elements appear to be of two sizes, one relatively small, on the order of 700 to 1,500 base pairs, and the other substantially larger, measuring in the thousands of base pairs. The smaller elements are called **insertion sequences** because they seem to have the ability to move from one place to another, inserting themselves more or less randomly within various regions of DNA. When they insert themselves, they generally disrupt the coding activities of the region because they usually have strong transcription terminator signals within them. They are designated by the arabic letters IS followed by a specific italicized number, for example, IS*1* or IS*50*.

The larger mobile genetic elements are known as transposons or transposable elements. These are regions of DNA that are bounded by IS elements and move as a unit. The central region of DNA (noninsertion sequence) can code for one or more functions that will be expressed by the host DNA molecule. The general structure of a typical transposon is shown in figure

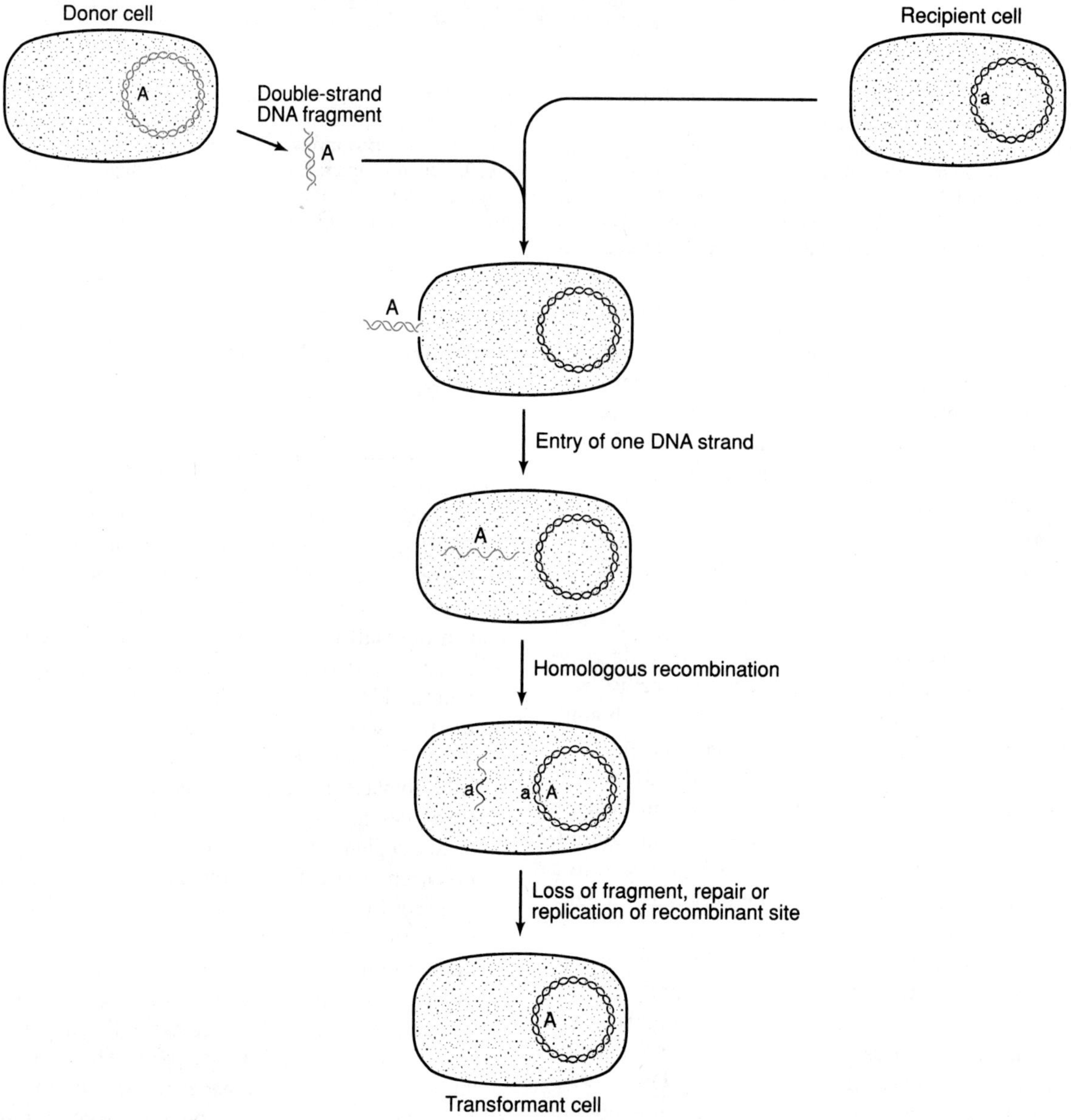

FIGURE 10.13

Genetic transformation. DNA is released by a cell as double-strand fragments. The fragments can attach to the surface of a competent cell and be transported across the cell wall and membrane. During the transport process, the DNA becomes a single strand. Upon arriving in the cytoplasm, the single-strand DNA can recombine with the homologous resident DNA to yield a transformant.

10.14. Many transposons replicate themselves during the process of transposition so that a cell beginning with one transposon may eventually have several copies of the same transposon. Others, such as Tn*5*, seem to be able to transpose themselves conservatively (no net increase in the number of transposons).

In addition to their ability to move about a bacterial chromosome, transposons also have the capacity to create inversions, deletions, and duplications of DNA in their immediate vicinity (see figure 10.14). These properties raise the question of why their ability to create genetic instability was not noticed earlier. The

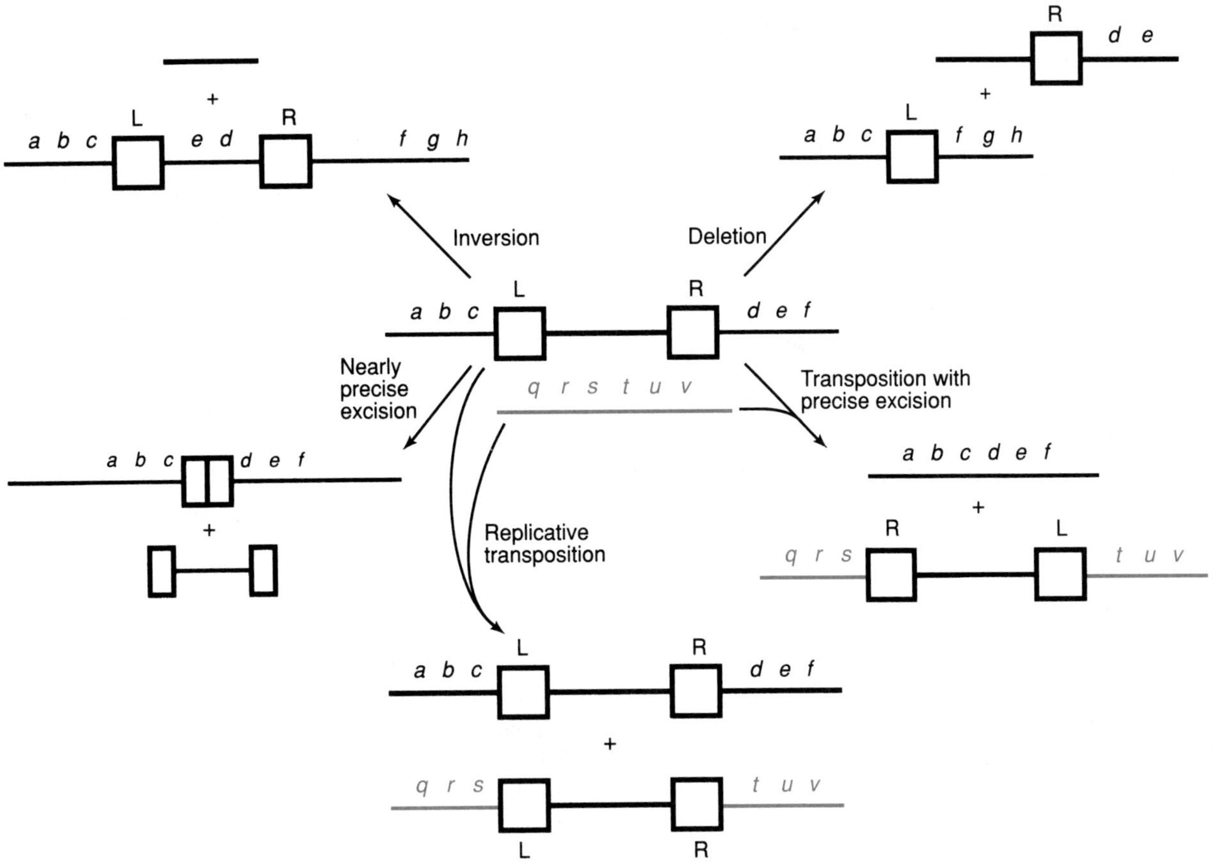

FIGURE 10.14

Transposons and transposition. The transposon is represented as a line flanked by two insertion sequences (boxes). Five possible transposition events are shown, only two of which leave a fully functional transposon(s). These are replicative transposition (bottom) and transposition with precise excision, a conservative process (lower right). The other three events are intramolecular and result in the removal of varying amounts of transposon. A specific transposon is not necessarily capable of carrying out all five events illustrated.

answer seems to be in the relatively low frequency with which transposition occurs, approximately 10^{-5}; therefore, unless the gene structure is checked very closely, no changes are seen. In point of fact, insertion elements were first noted because of the effects of their insertion on the expression of the galactose operon.

Insertion elements and transposons are significant in terms of a number of processes that were discussed earlier. For example, T-DNA that integrates into plant chromosomes has the structure of a transposon. Most antibiotic resistances encoded by R plasmids are also parts of transposons. In fact, some transposons may code for multiple antibiotic resistances. Perhaps most interesting of all is a double-strand DNA phage called Mu that infects *E. coli.* It is fundamentally a transposon with the ability to package itself. Mu DNA never exists in other than an integrated state, and it replicates by transposing itself to multiple sites on the bacterial chromosome. Packaging occurs directly from each individual site.

Hfr cells discussed earlier in this chapter also depend on insertion sequences for integration of F plasmids. The various points of F plasmid insertion for spontaneously arising Hfr cells are determined by the locations of insertion sequences designated as IS*2,* IS*3,* and $\gamma\delta$. The F plasmid carries corresponding insertion sequences, and integration is actually a simple recombination between homologous DNA regions.

There is a puzzling aspect to certain transposons found in gram-positive organisms. Transposons such as Tn*916* behave in a conjugative manner. They can transfer themselves from a donor cell to a recipient provided that cell contact is permitted, thus fulfilling the criterion for a conjugative process. On the other hand, there is no evidence of any plasmid existence for the transposon, no origin of replication, etc. The molecular mechanism for conjugative transposition remains a mystery, therefore.

Cell Fusion

Bacteria and other cells maintain their separation from the surrounding environment by their plasma membranes. Under certain conditions, it is possible to alter cellular integrity to permit the interaction of DNA molecules that do not occur normally. In order to manipulate bacteria in this fashion, it is necessary to remove the cell wall with some special treatment, such as with D-cycloserine (figure 7.2), to prevent synthesis of peptidoglycan. The resulting cells are known as **protoplasts** if they come from gram-positive cells and as **spheroplasts** if they are derived from gram-negative cells. Spheroplasts have remnants of cell wall remaining, whereas true protoplasts do not.

One commonly used technique in *Bacillus* genetics is protoplast fusion in which protoplasts from two different strains are mixed together in the presence of polyethyleneglycol. Their membranes fuse together to give a single, large diploid cell. The diploid state may or may not persist for some length of time. If it does persist, present indications are that it does so by inactivating one of the two bacterial chromosomes it has received.

The strength of the fusion technique lies in the fact that it is possible to fuse cells from radically different genetic backgrounds, even different genera, regardless of whether they would normally interact. Such fusions permit researchers to create combinations of traits that would never be found in nature.

Summary

Our knowledge of bacterial genetics has been enriched greatly by the information gained from experiments with plasmids and transposons. Plasmids come in many varieties and can contribute a number of different traits to the phenotype of a host cell, such as bacteriocin production, antibiotic resistance, and generation of tumors in plants. Many are also conjugative (self-transmissible). Conjugative plasmids can transfer only themselves, or they can interact with a bacterial chromosome to transfer comparatively large portions of the genome of the cell. Although most plasmids promote only interbacterial transfers, Ti plasmids can transfer prokaryotic DNA into eukaryotic cells.

Insertion elements are discrete regions of DNA that have the capability of moving from place to place on a DNA molecule or between two DNA molecules. Unlike the other genetic processes that have been discussed, there does not need to be any significant sequence homology between the site presently occupied by the insertion element and the target site. Transposons are regions of DNA bounded by insertion sequences that move as a unit. The DNA in the center of the transposon is not involved in the transposition process and can code for entirely different properties, such as for antibiotic resistance.

Transduction refers to the transfer of host DNA from a donor to a recipient cell packaged inside a viral capsid. Although the phenomenon theoretically can occur with most types of DNA viruses, it has been studied primarily in bacteria. Lytic phages give rise to generalized transducing particles, and temperate phages give rise to specialized transducing particles.

Genetic transformation involves the uptake of naked DNA molecules from the surrounding medium by competent cells. The DNA is then passed through the cell wall and membrane into the cytoplasm where it is available for recombination (or for replication if it is a plasmid). Although the process undoubtedly works in the laboratory, its significance in nature is uncertain.

Questions for Review and Discussion

1. What are some roles that plasmids can play in nature? How might they move from one cell to another? How do human activities encourage the spread of plasmids?
2. Why are certain viruses more like plasmids than others? In what ways are they similar and different?
3. Which bacterial genetic transfer process is likely to move the largest pieces of DNA? The smallest? Why?
4. Why do you suppose that the discovery of transposons has created so much excitement and controversy within the genetic community? Does it surprise you that entities like transposons can persist within a population of bacteria? Why?
5. Do you agree that transduction is the genetic transfer process that is most likely to have the biggest impact on natural populations? Why?

Suggestions for Further Reading

Berg, D. E., and M. M. Howe (eds.). 1989. *Mobile DNA*. American Society for Microbiology, Washington, D.C. An extensive compendium of transposable elements in prokaryotes and eukaryotes.

Birge, E. A. 1988. *Bacterial and bacteriophage genetics: An introduction,* 2d ed. Springer-Verlag, New York. Further information on the topics covered in this chapter.

Brock, T. D. 1990. *The emergence of bacterial genetics*. Cold Spring Harbor Laboratory, Cold Spring Harbor, N.Y. A complete tracing of all historical threads that have come together to create modern bacterial genetics.

McCarty, M. 1985. *The transforming principle: Discovering that genes are made of DNA*. W. W. Norton, New York. A reminiscence by one of the original workers in the field.

Orias, E. 1986. Ciliate conjugation, pp. 45–84. *In* J. G. Gall (ed.), *The molecular biology of ciliated protozoa.* Academic Press, Inc., Orlando, Fla. An extensive review of the conjugative phenomenon in which little prior knowledge of the ciliates is assumed.

chapter

11

DNA Technology and Genetic Engineering

chapter outline

■

GENERAL GOALS

Be able to describe how DNA molecules can be joined together artificially and placed into a living cell.
Be able to discuss some problems that face a scientist working on genetic experiments and others that face society as it attempts to cope with rapid technological changes.

Discoveries made during the past two decades have revolutionized the science of genetics as a whole and have dramatically altered bacterial genetics in particular. Their impact on the whole of biology is no less dramatic. More than ever, industry is looking to microbiology to provide new solutions to a wide variety of problems. With the genetic ideas presented in the preceding two chapters as a foundation, it is possible to develop a good understanding of these exciting developments.

Restriction and Modification Enzymes

The biological revolution that led to the new genetic technology began very quietly with the study of an unusual bacterial defense system known as restriction and modification. As is discussed in chapter 10, foreign DNA can arrive in a cell by a variety of means such as genetic transformation, transduction, transfection, conjugation, and even simple viral infection. However, if this DNA is not homologous to the host cell (that is, from the same or closely related species), it is not necessarily to the long-term advantage of the host species to allow it either to recombine or to persist as a self-replicating plasmid within the cytoplasm. Indiscriminate use of foreign DNA can destroy the species as a separate biological entity. Homology during recombination is a host cell's first line of defense; however, even if the foreign DNA fails to recombine, it may linger within the cell.

The means by which a host cell eliminates most foreign DNA actually involves two DNA site-specific enzyme activities. In each case, the enzyme recognizes the same base sequence on the DNA and binds to it. In the normal course of events within a cell, the first or **modification enzyme** adds a methyl group to a certain base within the recognition sequence during DNA replication (table 11.1).

This modification does not seem to affect the coding properties of the DNA. The second enzyme, a **restriction endonuclease,** will not bind to a modified DNA sequence, however, although it will bind readily to an unmodified one. Once bound, the restriction endonuclease acts to cut the unmodified DNA into pieces that are unable to replicate themselves and subsequently are lost from the cell. It is important to remember that although a cell can code for a modification enzyme without a corresponding restriction endonuclease, the reverse would be suicidal. These endonucleases can occur as separate proteins (types II or III) or as a combined large molecule (type I). The type of enzymes generally used in DNA technology are the separate variety.

The various restriction endonucleases are designated according to the first letters of the genus and species name of the producing organism followed by a roman numeral used to identify separate enzymes within the same organism. For example, *Anabaena subcylindrica* produces the enzymes *Asu*I, *Asu*II, and *Asu*III. If the information coding for the restriction and modification enzyme is located on a plasmid, that too can be noted. *Eco*RI is an enzyme produced by *E. coli* cells that carry the RY13 plasmid. In other cases, the final letter can indicate the strain (a variant within a species) that is the producing organism.

TABLE 11.1. Sample of restriction enzymes and their properties

Enzyme	Type	Source	Sequence Recognized[1]
*Asu*I	II	*Anabaena subcylindrica*	5′G/GNCC3′
*Asu*II	II		5′TT/CGAA3′
*Asu*III	II		5′GPu/CGPyC3′
*Bam*HI	II	*Bacillus amyloliquefaciens* H	5′G/GATCC3′
*Eco*B	I	*Escherichia coli*	5′TGA(N)$_8$TGCT3′
*Eco*PI	III	*E. coli* infected by phage P1	5′AGACC3′
*Eco*RI	II	*E. coli* carrying RY13 plasmid	5′G/AATTC3′
*Hpa*I	II	*Haemophilus parainfluenzae*	5′GTT/AAC3′
*Mbo*I	II	*Moraxella bovis*	5′/GATC3′
*Pst*I	II	*Providencia stuartii*	5′CTGCA/G3′

Note: Further details on restriction enzymes can be found in the catalogs of enzyme suppliers.

1. Abbreviations used to indicate sequences recognized are: adenine (A), cytosine (C), guanine (G), any base (N), thymine (T), purine (Pu), and pyrimidine (Py). A solidus (/) indicates the site of cutting by the enzyme if it occurs within the recognition sequence, and the underscored base is the one altered by the modification enzyme. *Eco*PI cleaves at a site located 24 to 26 base pairs from the 3′ end of the recognition sequence, while *Eco*B cleaves at a site located about 1,000 base pairs from the recognition sequence.

Although the binding of restriction endonucleases is specific, the cuts made by them are not necessarily so. A type I enzyme such as *Eco*B (table 11.1) cuts nonspecifically at some distance from the binding site (in this case about 1,000 base pairs). By contrast, a type II enzyme such as *Eco*RI both binds specifically and cuts specifically. It is this latter type of enzyme that is so interesting from the standpoint of DNA technology. A few type III enzymes such as *Eco*PI are known to have properties intermediate between types I and II. They bind to a specific DNA sequence but make their cuts a specific distance away from the binding site; therefore, the position of the cut can be predicted, but the sequence to be cut cannot.

In several different studies by Nobel Prize winners Werner Arber, Daniel Nathans, Hamilton O. Smith, and their various collaborators, it was shown that DNA fragments produced were of two types. They might have blunt ends (for example, *Hpa* I) or overlapping single-strand ends (*Eco*RI; see table 11.1). Of critical importance is the fact that all DNA fragments produced by a particular type II restriction endonuclease have absolutely identical ends regardless of the source of the DNA (figure 11.1).

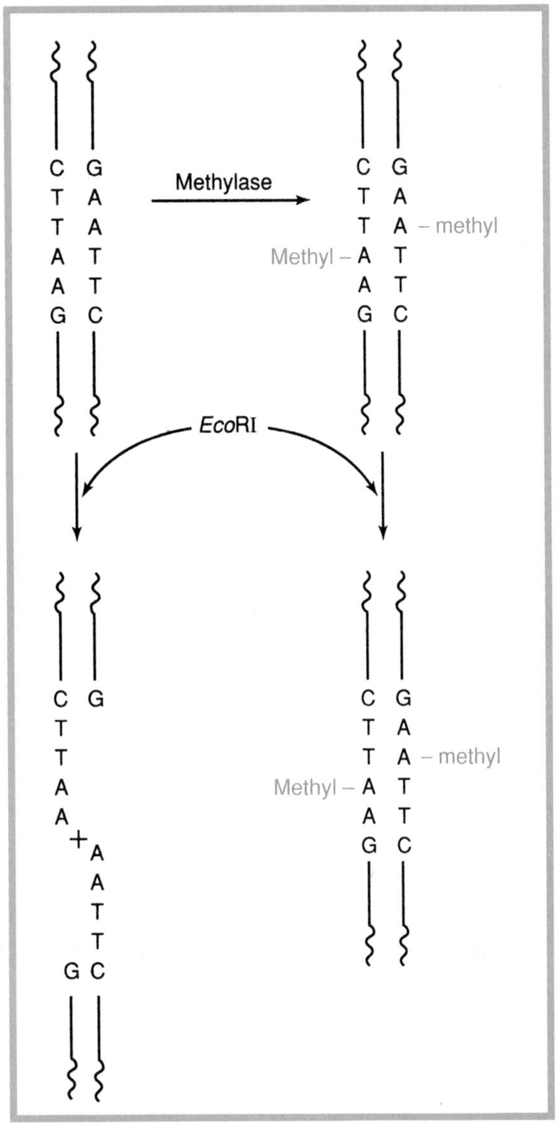

FIGURE 11.1

Modes of action of a typical pair of restriction and modification enzymes. Shown are portions of two DNA molecules that contain recognition sequences for the *Eco*RI restriction and modification system. Only the unmodified DNA is cut by the endonuclease.

DNA Splicing

Prokaryotic Cells as Hosts

The identical ends of restriction fragments make possible a real DNA splicing reaction (figure 11.2). DNA fragments can be brought together and the single-strand ends allowed to pair or reanneal. Random combinations of linear DNA molecules held together by hydrogen bonds result. As the molecules increase in length, however, they have a tendency to fold back on themselves so that they circularize. If these circular molecules have their nicks sealed by the same DNA ligase enzyme used for replication and if they have an appropriate origin of replication, new plasmids will have been formed.

There are several important points to note about DNA splicing reactions. Clearly, fragments produced by any restriction endonuclease can be used. If the fragments have blunt ends, however, highly concentrated DNA solutions are a necessity, and DNA ligase must be present during the mixing step in order to join fragments together as they collide. When fragments with cuts at specific base sequences do join, the recognition site for the restriction endonuclease is regenerated; that is, DNA fragments that have been spliced together can be cut apart later by the original enzyme to reform the original fragments. Joining of DNA fragments results both in random combinations of the fragments and in random orientation of the fragments within the growing molecules. If the fragments are

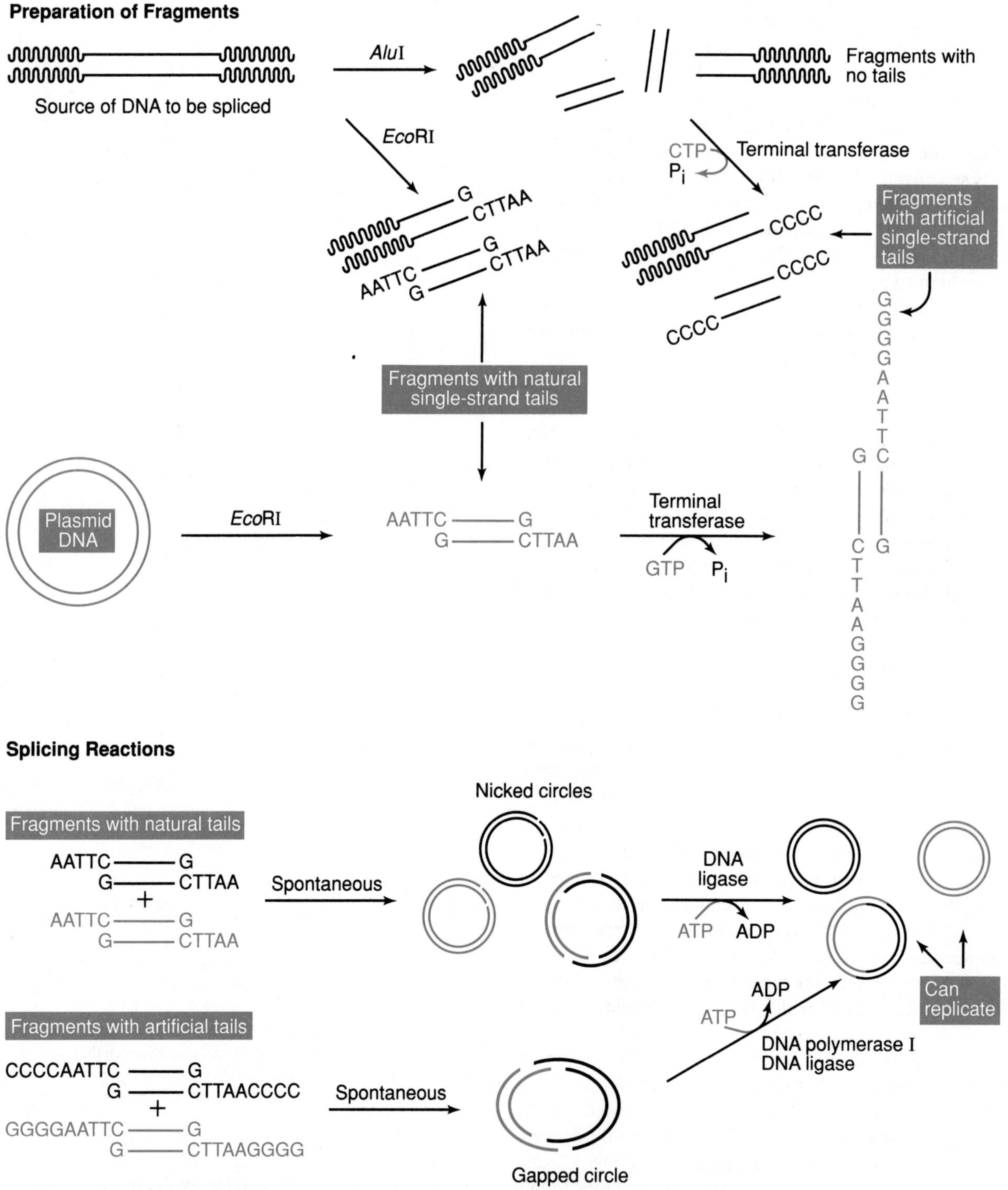

FIGURE 11.2
Splicing of DNA. DNA from a specific source is cut by a selected restriction endonuclease into fragments of various sizes. In these examples, *Eco*RI yields fragments with single-strand ends, while *Alu*I yields blunt-end fragments. Artificial tails can be added to either type of fragment with terminal transferase. When fragments with matching tails are mixed together, one or more fragments join together at their ends, or large fragments can self-close. In the case of artificial tails, the gaps remaining can be filled in with DNA polymerase. DNA ligase is used to restore the final phosphodiester bonds and give a continuous DNA molecule. The resulting molecule can self-replicate only if it contains a plasmid origin of replication.

APPLICATIONS BOX

Modification Enzymes and DNA Repair

There is a fundamental problem with extrareplicational DNA repair, DNA repair that must occur after a replication fork has passed by a specific point. As noted in chapter 7, DNA polymerase I can remove defective or incorrectly paired bases from DNA and then replace them by hydrogen bonding a new base into the gap that has been created. If the defect is a chemically damaged base or an RNA primer, the enzyme has no problem in selecting which base to remove. However, if the defect is one of mispairing, A with G or T with C, for example, a problem does arise. Which is the original base and which is the incorrectly paired base? If the replication fork has passed only recently through this area, it may be possible to tell. DNA modification is not instantaneous, and for a while newly replicated DNA is hemimethylated (the old strand has methyl groups and the new one does not). Removal of a base from an unmethylated DNA strand is most likely to remove the incorrectly paired base and restore the correct sequence, therefore.

generated by the use of two different restriction endonucleases, the ends may not be identical and only certain orientations of fragments will result in splicing (figure 11.3).

The product of a DNA splicing reaction is frequently referred to as "recombinant DNA" because it represents a novel combination of coding sequences. However, DNA splicing occurs only in a test tube (in vitro) and requires no real base sequence homology except in the sense that all restriction fragments produced by a particular type II enzyme have identical ends. On the other hand, genetic recombination occurs in a living cell (in vivo) and usually requires substantial sequence homology throughout the region. For the purposes of this text, therefore, the terms *recombination* and *recombinant* are reserved for the natural process discussed in chapter 10, and the in vitro process is referred to as DNA splicing or gene splicing. Genetic engineering can refer to organismal crosses either from true recombination or from DNA splicing. In the modern vernacular, however, the term *genetic engineering* usually refers to DNA splicing.

In order for spliced DNA to be maintained stably within a cell, it must be able to replicate. Replicational functions might be provided if the spliced DNA is made to recombine with the host cell genome so as to integrate the new DNA into the host replicon. Most frequently, however, some sort of plasmid is used as a **vector** (carrier of the DNA fragment) to provide replicational functions. Vectors can be any temperate virus, resistance plasmid, bacteriocin plasmid, etc. that is capable of replication within the desired host cell. The ideal vector also carries some sort of selectable marker (gene) so that its presence in a cell can be detected easily. One type of selectable marker commonly used is an antibiotic-resistance gene from an R plasmid.

Plasmids carrying spliced DNA normally can be introduced into cells by genetic transformation. In some cases, it is possible to carry out in vitro packaging of viral DNA into protein coats, and in such instances, fully infectious transducing viral particles are formed. Whatever the means used to introduce spliced DNA into a cell, once that cell has begun to replicate the DNA and undergo cell division, the entire clone of cells that it generates carries the spliced DNA. This spliced DNA is said to have been **cloned.**

Eukaryotic Cells as Hosts

The same kinds of splicing reactions used for prokaryotes can also be used for eukaryotes, but appropriate vectors must be employed. DNA replication is sufficiently different that plasmids able to replicate well in a bacterium are unable to do so in a eukaryotic cell unless something like a viral origin of replication is added. Yeasts, however, have several plasmids that normally occur in their cells and can serve as good vectors. For more complex organisms, the vectors of choice usually are either transposons or viruses. The most frequently used viruses are retroviruses, because of their integration into the host chromosome, and vaccinia virus, because it has many nonessential genes that can be deleted to make room for spliced DNA. The most common method of getting spliced DNA into a eukaryotic cell is genetic transformation, although transduction is also possible.

Sometimes it might be desirable to perform most of the genetic manipulations on a piece of DNA in a prokaryotic cell and then transfer it into a eukaryotic cell to study its expression. This feat can be accomplished with a **shuttle vector** that carries two origins of replication, one for the prokaryote and one for the eukaryote. A similar tactic is often used for prokaryotes: The basic cloning work is done in a well-studied host

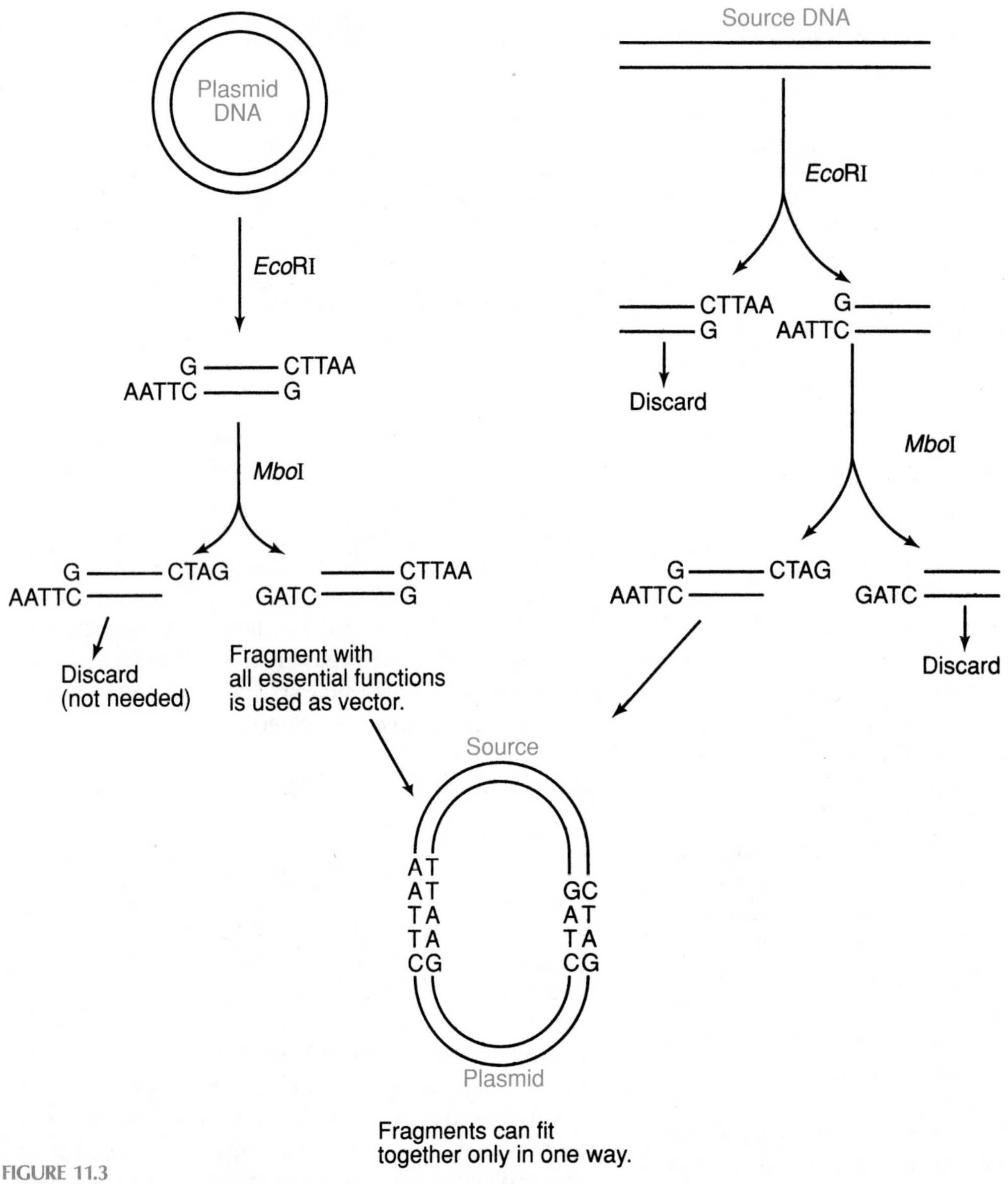

FIGURE 11.3

Controlling the orientation of a DNA insert. DNA fragments that have been generated by cutting with two different restriction enzymes will have nonidentical ends. If the vector DNA has been cut with the same two enzymes, the DNA insert then will be able to form a hydrogen bond only when it is in one specific orientation.

like *E. coli* and then the spliced DNA is transferred into a bacterium of interest to generate the finished product.

Expression of Cloned DNA

Successful **cloning** of a spliced DNA molecule does not, however, guarantee that the inserted DNA fragment will be expressed in a manner that will alter the phenotype of its host cell. For this reason, vectors usually carry one or more selectable genetic markers, as mentioned previously. In the case of a vector with two or more markers, it is frequently advantageous to splice the extra DNA into the middle of the coding sequence of one selectable marker so as to create an insertion mutation that can inactivate the coding sequence. Then,

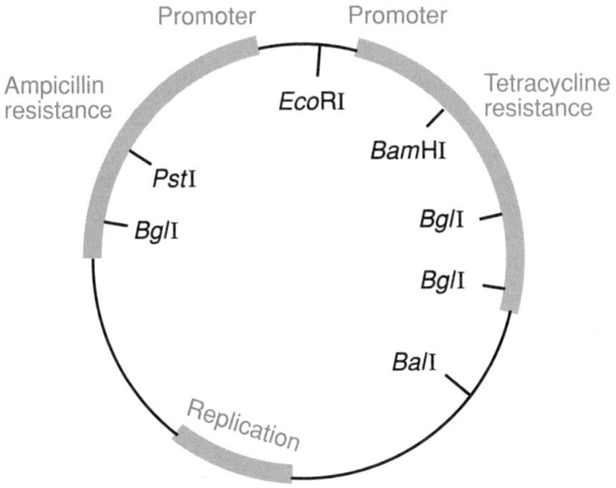

FIGURE 11.4
Simplified genetic and restriction map of pBR322. This plasmid was specially constructed for use as a cloning vehicle or vector. It is nonconjugative and carries single recognition sites for a number of important restriction endonucleases. The boxed areas indicate major plasmid functions.

the presence of one genetic marker and the absence of another can be taken as evidence that the cell is carrying spliced DNA.

For example, a commonly used vector for cloning within *E. coli* is the pBR322 plasmid derived from colicin E1. The small arabic "p" denotes that pBR322 is a plasmid, and the capital arabic letters represent the initials of the workers (Bolivar and Rodriguez) who constructed it. Plasmids generally are numbered for convenience of identification. The pBR322 vector confers enzymatic resistance to tetracycline and to ampicillin. A simplified genetic map for the plasmid is rendered in figure 11.4.

It is possible to clone into the tetracycline region, into the ampicillin region, or into the areas in between. However, DNA fragments inserted into the replicational region act to prevent the plasmid from replicating; therefore, it would be lost. Insertion of a DNA fragment coding for a protein into a site, such as *Pst*I, that is downstream from a promoter results in the formation of a **fusion polypeptide,** a protein having the amino terminus of the β-lactamase enzyme (an inactivator of ampicillin) attached to the amino end of the protein encoded on the cloned DNA.

Fusion polypeptides may or may not function as enzymes, depending on the exact nature of the fusion. In order for a spliced DNA fragment to express its function uniquely, therefore, it is generally necessary for the fusion polypeptide to be separated into discrete proteins or for transcription to occur in a manner that separates cloned DNA from vector DNA. Transcription of this nature means that a promoter site must be provided either on the spliced fragment itself or on the neighboring region of the vector (see figure 11.4). In addition, if a protein is to be produced, it is necessary for the appropriate ribosomal binding site to be present after the promoter. In the case of spliced eukaryotic DNA, there must be no need for RNA processing, because prokaryotes do not have enzymes to remove introns. Furthermore, the protein product must not require any processing other than that normally occurring in the host cell (for example, removal of signal peptides).

Guidelines for Safe Use of Spliced DNA

As the properties of restriction enzymes became clear, there was widespread debate within the scientific community over the extent to which the use of these enzymes presented a biological hazard. The general theme of the arguments raised revolved around how DNA splicing could be used to circumvent the so-called "species barrier" to genetic exchange. It was thought that once this barrier was broken, there might be wholesale rearrangements of genetic traits and inadvertent production of genetic "monsters" that could have serious effects on other members of the biosphere.

Clearly, such a concern was justified based on the information available at that time. The conversion of a disease-causing bacterium into a form resistant to the commonly used therapeutic antibiotics would be of grave concern. Similarly, any improvement in the ability of such a bacterium to cause disease would be undesirable indeed. Even more elaborate scenarios were envisioned in which combining segments of eukaryotic and prokaryotic DNA would lead to entirely unexpected results such as an *E. coli* cell that carried DNA capable of stimulating tumor production in animals.

Scientists involved in this research were so concerned that they met in Asilomar, California, in 1975 and declared a moratorium on DNA splicing research until such time as appropriate safety guidelines could be developed. Stringent guidelines were prepared and implemented in 1976 under the auspices of the National Institutes of Health and its new component, the Recombinant DNA Advisory Committee. In recent years, this committee has relaxed the guidelines significantly as it has become evident that early fears were

APPLICATIONS BOX

Protein Stability

Every cell has its own characteristic set of proteins. Within the cytoplasm are proteases that preferentially attack proteins not from the characteristic set. These abnormal proteins can result from mutations in the original gene or from transcription and translation of spliced DNA. The more closely similar the fusion polypeptides are in structure to the normal protein set, the longer they are likely to be maintained in the cytoplasm and to retain enzyme activity. Some enzymes like β-galactosidase are sufficiently stable and large so as to retain their active enzymatic configuration almost regardless of what is fused to them during gene splicing activities.

vastly overestimated. In part, this overestimation was due to a lack of specific knowledge regarding the considerable difficulties with which eukaryotic DNA can be expressed in prokaryotes, as indicated previously. We also now know that transfer of DNA from prokaryotic to eukaryotic cells is promoted routinely by the Ti plasmid (chapter 10). Additionally, careful examination of the older literature as well as contemporary experimentation has indicated that prokaryotes are capable of biochemical feats once thought to be the sole province of eukaryotic cells. For example, bacteria that spontaneously produce substances that mimic the activity of insulin, progesterone, and other animal hormones have been observed.

The guidelines for safe use of spliced DNA are adapted from the principles outlined for the containment of disease-causing microorganisms. These guidelines address both the biological and physical aspects of spliced DNA containment. Physical containment guidelines are taken directly from the Centers for Disease Control biosafety rules. They are easier to understand because the basic idea is to reduce the probability of an organism that carries spliced DNA escaping from the laboratory. All laboratory facilities are rated on a four point scale, a BL1 facility (biosafety level 1) being the least contained and a BL4 facility being the most contained. Previously, a BL1 facility was called P1, and BL4 was designated P4, etc.

A BL1 laboratory is a typical well-maintained microbiology laboratory in which lab coats are worn; eating, drinking, and smoking are prohibited; and handwashing facilities are available. In addition to these same features, a BL2 facility has limited access to approved personnel only. A BL3 facility is not only limited in access but also has a double door arrangement to aid in preventing any contamination from escaping. Much of the work in a BL3 facility must be performed in biological safety cabinets that filter the air before recirculating it to a room. The ultimate or BL4 facility is one in which self-contained anticontamination suits or glove boxes are used for all manipulations (figure 11.5A–B). Reduced air pressure relative to the external environment is maintained at all times; therefore, when a door is opened, air rushes into the building. All liquid, solid, and gaseous wastes are sterilized before they are allowed out of the building.

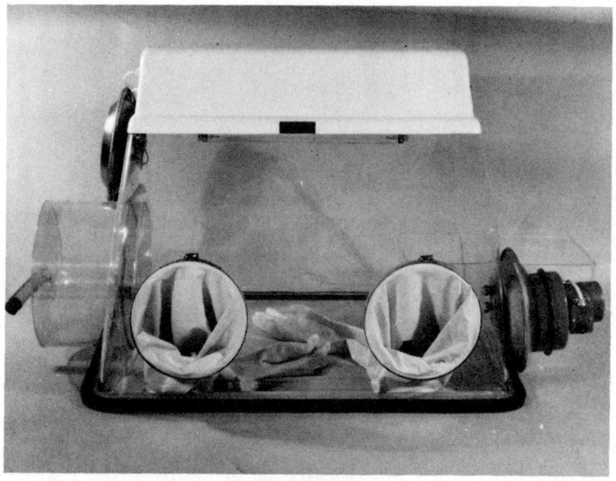

A

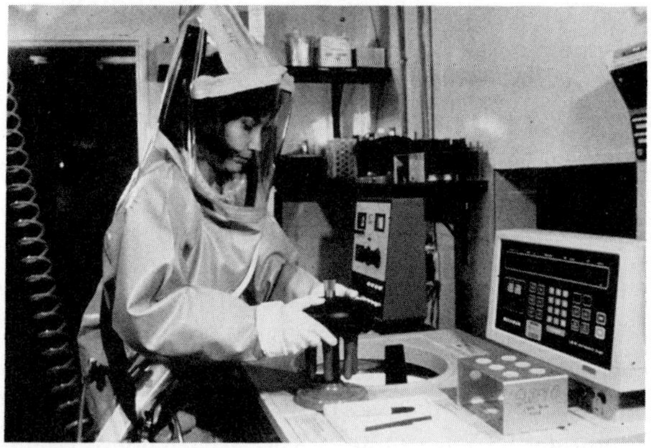

B

FIGURE 11.5

Equipment suitable for use with biohazardous materials. (**A**) A glove box provides a self-contained small working space. (**B**) In a full scale laboratory, anticontamination suits are used.

TABLE 11.2. Examples of the types of containment required to clone random pieces of DNA into *Escherichia coli*

Type of Donor DNA	Type of Containment	
	Physical	Biological
Escherichia coli	Exempt	Exempt
Salmonella	Exempt	Exempt (exchanges naturally with *E. coli*)
Bacillus	BL1 facility	Exempt
Drosophila	BL1 facility	Exempt
Rabies virus (laboratory strain)	BL2 facility	EK1 system
Rabies virus (wild-type strain)	BL3 facility	EK2 system

The problem of biological containment is analogous to physical containment, but the object is to prevent cloned DNA from leaving its host cell and escaping into an unwanted or undesirable host. If the vector used in the cloning is a conjugative plasmid like F or a normal virus like phage λ, the probability of the vector's escaping the host cell is high, provided that the host can survive outside the laboratory environment for a reasonable period of time. On the other hand, if a nonconjugative plasmid such as pBR322 is used, there is little chance of the vector escaping from the host no matter how long the cell survives outside the laboratory.

These considerations are taken into account in the creation of host-vector systems for cloning. If the host is highly viable, the vector is weakened and vice versa. With experiments perceived as potentially dangerous, certain levels of biological containment are required. For example, in cloning experiments involving *E. coli* as the host cell, an EK1 host-vector system employs a normal laboratory strain of the bacterium, while an EK2 host-vector system utilizes a specially weakened strain that cannot grow on naturally occurring amino acids. The weakened strain survives only in the research laboratory, therefore.

Biological and physical containment are complementary systems; therefore, an increase in the level of one allows for a decrease in the level of the other. Some specific cases are listed in table 11.2 to illustrate how biological and physical containment are used together in a laboratory setting.

Special Techniques for Testing DNA

Blotting Techniques

Many blotting techniques have been developed in which macromolecules are separated by **gel electrophoresis** first. Most often a gellike support composed of acrylamide (small pore size) or agarose (large pore size) is used. A sample of protein or nucleic acid is placed at one end of the gel to which an electric field is applied, forcing the charged molecules to migrate toward the appropriately charged electrode (figure 11.6 A–B). The rate of migration is affected by the size and shape of the molecules. The position of the macromolecules on the gel can be detected by the use of dyes such as Coomassie blue for protein or ethidium bromide for nucleic acids. A general dye, however, shows the position of all macromolecules, not just the one in which the experimenter is interested. This problem requires for its solution a method in which a protein or nucleic acid that has a specific sequence of amino acids or bases can be recognized.

After the molecules have been separated, they can be blotted or transferred sideways to a paper-thin matrix that has been laid on top of the gel (figure 11.7). This procedure gives a durable substrate that can be handled with less fear of breaking. Originally, the matrix material used for transferring nucleic acids was nitrocellulose, but more recently certain charged nylon derivatives have been employed. Once the macromolecules are attached firmly to the matrix, various techniques can be used to detect the position of the molecules of interest.

Essentially similar procedures can be used to trap nucleic acids or proteins within individual bacterial colonies or phage plaques. The blotting material is laid carefully on an agar plate so that cells or phages can adsorb to it. The matrix is treated to break open the cells or viral particles and then washed to remove unbound material.

For nucleic acids, the simplest sequence-specific detection method is to use a single-strand DNA **probe** that has been labeled in some fashion, usually with ^{32}P (figure 11.7). This probe can be made of DNA synthesized in the laboratory or produced from a piece of denatured DNA from a cell. The basic strategy is to denature the nucleic acids bound to the matrix and allow the probe to form hydrogen bonds with them.

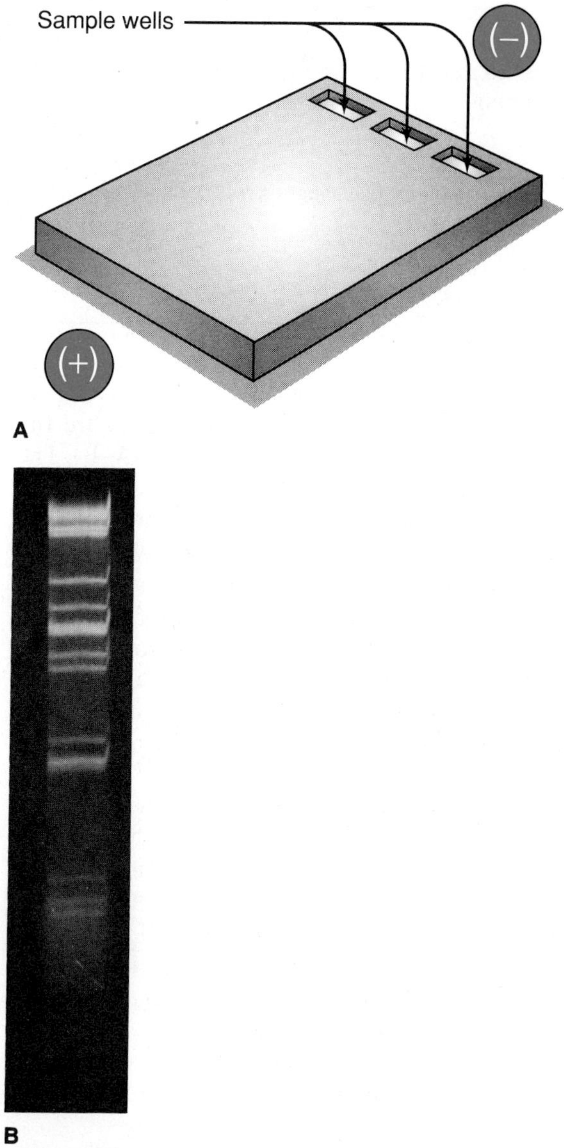

FIGURE 11.6

Gel electrophoresis. (A) A sample of DNA is applied to a small "well" or cavity in an 0.8% agarose gel, and an electric field is applied to the gel. The DNA migrates toward the anode, with smaller pieces of DNA traveling more rapidly through the gel than larger ones. (B) A DNA gel that has been stained with the dye ethidium bromide and then photographed using ultraviolet radiation. The dye fluoresces red in the presence of nucleic acid, and the fluorescence can be used to expose a photographic film.

After washing, the solid support is tested for the presence of radioactivity. If there is no homology between probe and bound nucleic acid, the radioactive label is lost. If there is perfect homology, the probe binds strongly, and the regions to which it has bound can be detected by autoradiography (a piece of photographic film is pressed against the blot, allowing the decay of radioactive atoms to expose the film). If there is imperfect homology, the result depends on the stringency of the incubation conditions. High pH, high temperature, or high concentrations of formamide all tend to break hydrogen bonds. Probing under any of these conditions requires better sequence homology than probing under less stringent conditions.

Some laboratory workers prefer to use nonradioactive methods of detecting DNA probes. In one such method, the biotin-avidin system is employed (figure 11.8). The vitamin biotin binds exceptionally strongly to avidin and forms a very stable complex. It is possible to attach biotin derivatives to nucleic acids in such a way that the nucleic acids' ability to form hydrogen bonds is not affected. When in a duplex, the biotin groups on the DNA are still accessible to avidin. The avidin protein itself is joined covalently to an enzyme, such as alkaline phosphatase. After the blotting process is completed, the substrate for alkaline phosphatase is washed across the surface of the solid support. Wherever the avidin has bound, a colored compound is formed. As with autoradiograms, the presence of a dark spot on the stained blot indicates binding of the probe.

The initial blotting technique was developed with single-strand DNA probes in the laboratory of Earl Southern and quickly became known as **Southern blotting** (figure 11.7). Of course, there is no reason why RNA cannot be used also. Because the opposite type of nucleic acid is used, this technique is known as **Northern blotting.**

In the same vein, there is now a Western blotting technique in which proteins are separated using gel electrophoresis and then transferred onto a matrix of diazotized cellulose. Various antibodies can then be used to detect the presence of specific proteins (see chapter 16).

Polymerase Chain Reaction

The polymerase chain reaction (PCR) is the long-awaited solution to the problem of detecting rare DNA sequences within a heterogeneous population of DNA. In order for the technique to work, it is necessary for the experimenter to have two DNA fragments that contain a short sequence of DNA known to flank or be part of the region of interest. In the case of cloned DNA, these fragments can simply be portions of the vector DNA itself that are located at each end of the cloning site. The two fragments must be from opposite strands.

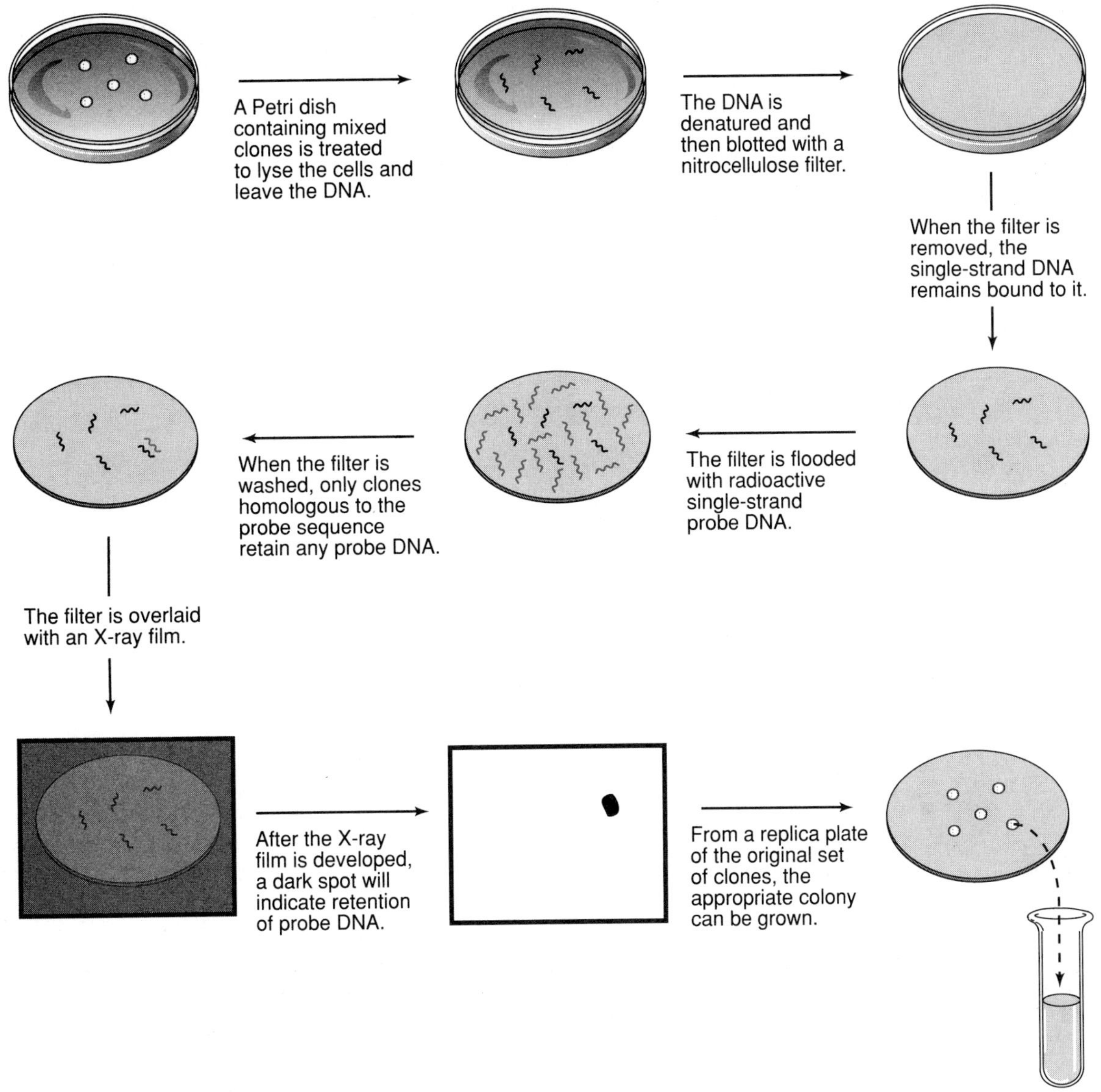

FIGURE 11.7

Southern blotting. Single-strand DNA is attached permanently to a filter made of nitrocellulose or certain charged nylon derivatives, following which a single-strand radioactive probe of known sequence is added. The radioactive probe remains bound to the filter only where significant base sequence homology is present. An autoradiogram can be prepared if the filter is overlaid with an X-ray film. Dark spots on the autoradiogram indicate DNA that is homologous to the probe.

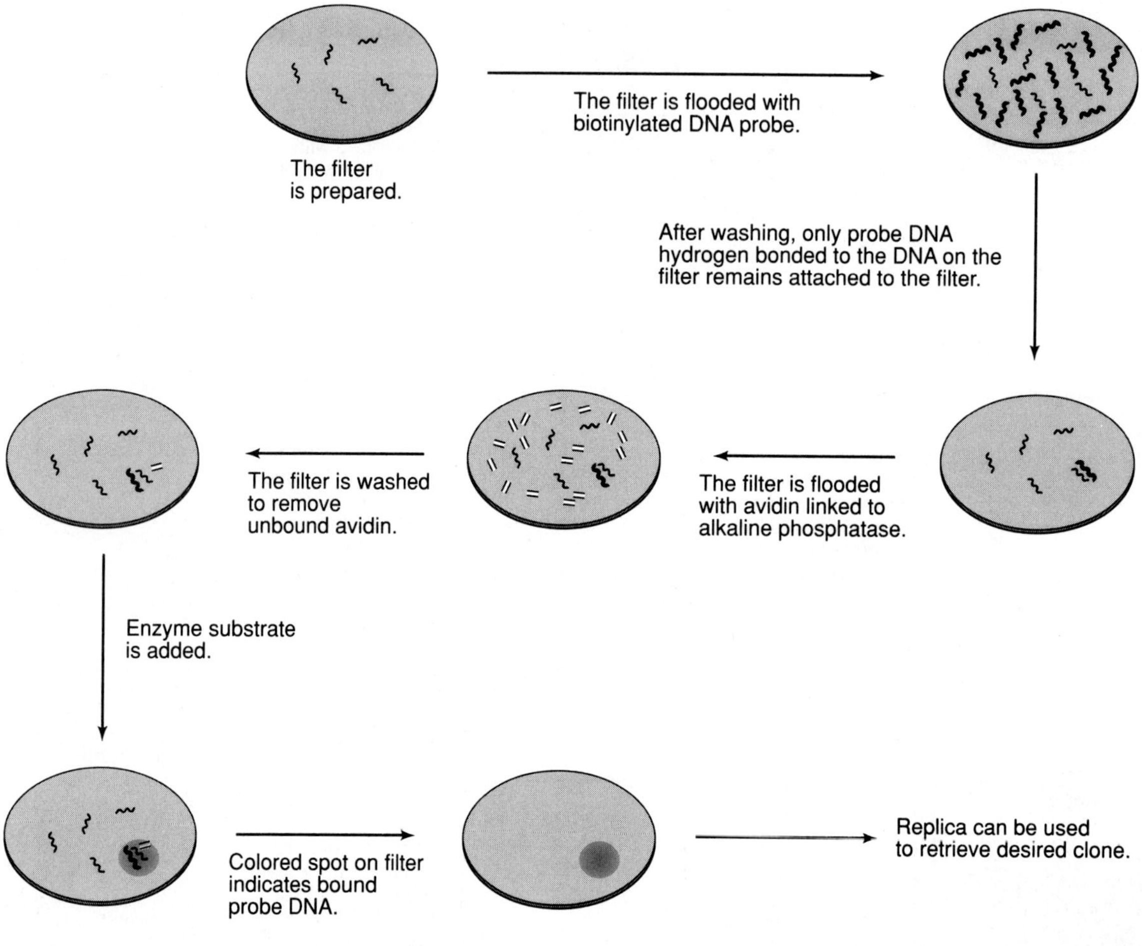

FIGURE 11.8

Nonradioactive method for detection of DNA. Blotting experiment is prepared as in figure 11.7, but the probe is biotinylated rather than made radioactive. When the probe is bound, therefore, there are biotin residues projecting away from the duplex. When the avidin protein binds to the biotin, it brings along covalently attached alkaline phosphatase molecules. If a substrate for the alkaline phosphatase is provided, a colored compound precipitates, indicating by its presence where biotinylated DNA can be found.

Use of these fragments is outlined in figure 11.9. The experimenter sets up an in vitro DNA replicational system using the fragments as primers. If the temperature is raised briefly, the DNA fragments can be denatured. DNA polymerase is added at this time so that a leading strand of DNA can be synthesized using an added DNA fragment as a primer. The corresponding lagging strand is not replicated in this system. Because the primers flank the region of interest, however, each DNA strand is copied. In the region in which there originally was one double-strand DNA molecule carrying the sequence of interest, there are now two. Reheating of the mixture denatures both molecules. New primers and enzymes then can be added, and the number of available molecules ought to double again. This chain reaction derives its name from the repeated cycles of DNA replication.

One problem with the PCR technique is that each time DNA fragments are denatured by heating, the activity of the DNA polymerase enzyme also is destroyed. A major breakthrough in using this technique came when DNA polymerase from the extreme ther-

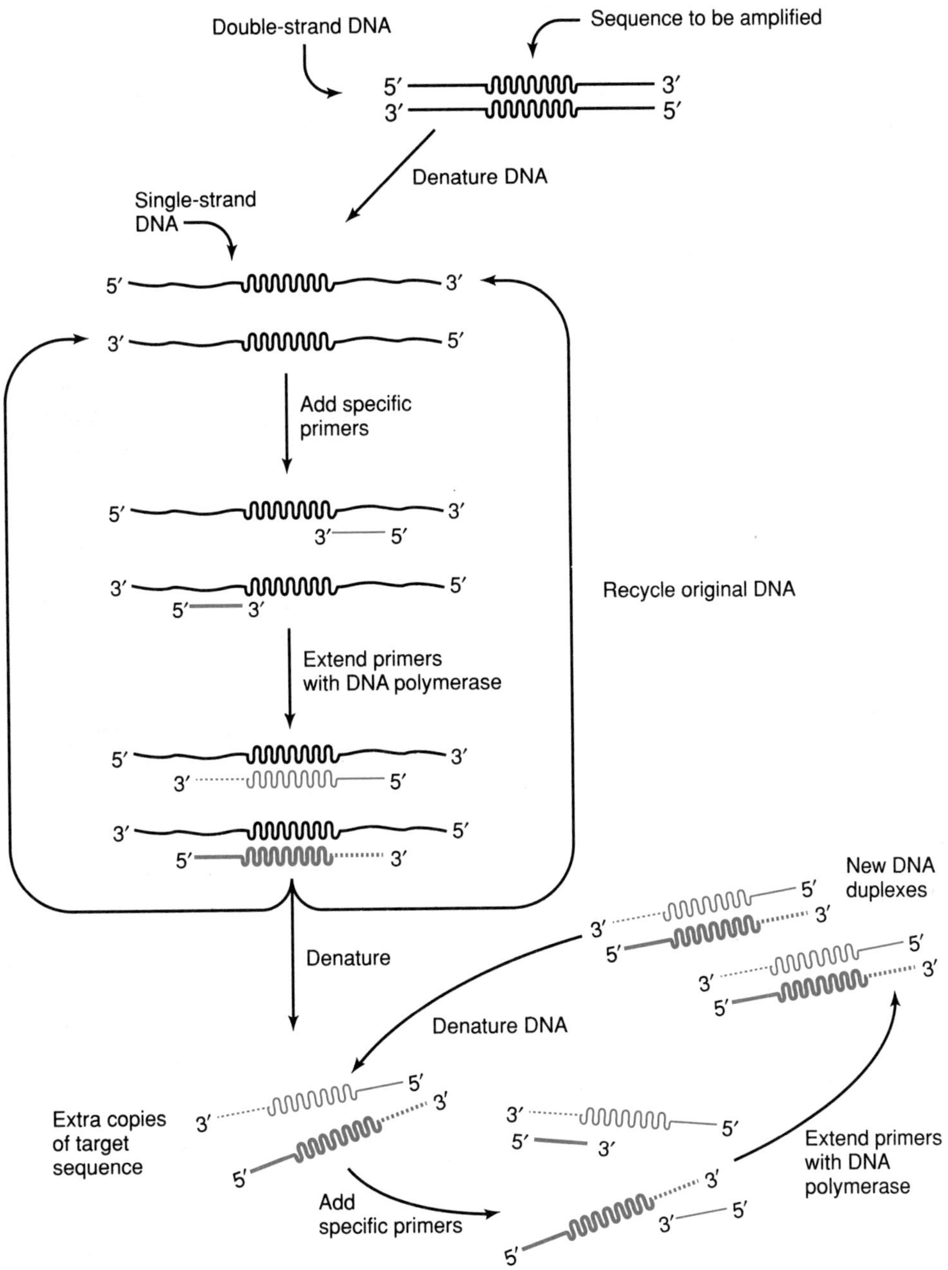

FIGURE 11.9

Polymerase chain reaction. DNA is denatured, and a primer of specific sequence is added. Once the primer is bound, it can be extended by DNA polymerase in the usual semiconservative manner. If all original DNA molecules have bound a primer, the number of sequences of interest has doubled. This process is then repeated many times, with the number of sequences approximately doubling each time.

mophile *Thermus aquaticus* was isolated. This enzyme, called *Taq* polymerase, is stable at temperatures that denature DNA; therefore, the enzyme need be added only at the beginning of the chain reaction. Automatic equipment is now available to take a small sample through a series of 20 to 30 cycles, each time amplifying the desired sequence approximately by a factor of two.

Practical Applications of Genetic Technology

The genetic technology described in the preceding sections has effects on all facets of microbiology. In subsequent chapters, the information presented in this chapter is enlarged upon and used to illustrate specific appications for the products of genetic engineering. The technology involved has already become a part of routine microbiology in taxonomic studies (presented in chapter 20) and in diagnostic laboratories.

From the point of view of the diagnostician, genetic engineering provides a major technological improvement, the ability to detect rare macromolecules within a heterogeneous population. For example, the standard test to confirm whether an individual is infected with the HIV-1 virus that causes AIDS is a Western blot that detects specific viral proteins within the sample of patient blood. Specimen identification of pathogenic bacteria often is performed with Southern blots and DNA probes specific to the particular genus or species.

The most recent advance is in the realm of PCR technology. In theory, an appropriate PCR should be able to amplify a single molecule of DNA to the point at which it can be detected visually on an electrophoretic gel. The only real problem is to locate DNA sequences that can be used as flanking primers for the synthetic reaction. Sequences chosen must be highly specific to the organism(s) of interest so that false positive reactions do not occur. The reaction itself requires part of one day to run, and DNA analysis adds only a few more days to the assay. For a slowly growing organism, therefore, identification may be possible in the amount of time that formerly was spent in waiting for a colony to appear on an agar plate. A theoretical limitation to the assay is that there must be at least one of the pathogens in question physically present in the laboratory sample.

Example of Successful Cloning: Interferon

The best way to understand the operation of DNA technology and genetic engineering is to follow the ways in which various techniques are applied in a practical situation. A number of different products are available today as a direct result of cloning. One of the most intriguing is interferon, a family of chemicals transiently produced by virus-infected animal cells. Interferon has a twofold effect: It acts directly to inhibit viral multiplication and causes the producing cell and its immediate neighbors to synthesize additional antiviral products. Thus, interferon assists animal cells in resisting viral infection and also may antagonize viral oncogenic effects.

Consider the cloning of human leukocyte interferon, a compound that has been available in small quantities for some time. Originally, 2 liters of blood were required as starting material to isolate 1 μg of interferon. Although some laboratory work had been carried out with the trace quantities of human leukocyte interferon isolated in this manner, no real clinical trials were possible prior to the advent of DNA splicing. At the time that this particular project began, the researchers did know that an interferon molecule has a molecular weight of about 17,500 to 21,000 and that certain **peptides** (short chains of amino acids) produced when interferon was cleaved with the enzyme trypsin could be isolated and their amino acid sequence determined.

To prepare a clone, the researchers had to find a source of the required DNA. The gene coding for interferon is not well characterized; therefore, it was not possible for them to cut up the appropriate region of a leukocyte chromosome directly. Moreover, it was reasonable for them to assume that the DNA sequence coding for interferon contains one or more introns that must be removed from the corresponding mRNA before it can be translated. Instead of using fresh leukocytes as a source of DNA, the researchers decided to use an established line of tissue culture cells. The principle characteristic of this particular cell line was that the cells produced comparatively large amounts of interferon when stimulated with a virus. About five hours after addition of the virus, when interferon mRNA synthesis was assumed to be maximal, the researchers extracted all of the RNA in the cells.

Before any DNA splicing could begin, the researchers needed to purify the interferon mRNA away from the rest of the cellular RNA. Because eukaryotic mRNA generally contains a "tail" of some 200 adenyl residues, mRNA purification is straightforward. As diagrammed in figure 11.10, when RNA passes through a glass tube (column) containing cellulose to which chains of single-strand polythymidylic acid [poly(T)] are bound, poly(A) tails of mRNA stick to the poly-(T) bound to the column, while the rest of the RNA passes on through. The researchers used this procedure

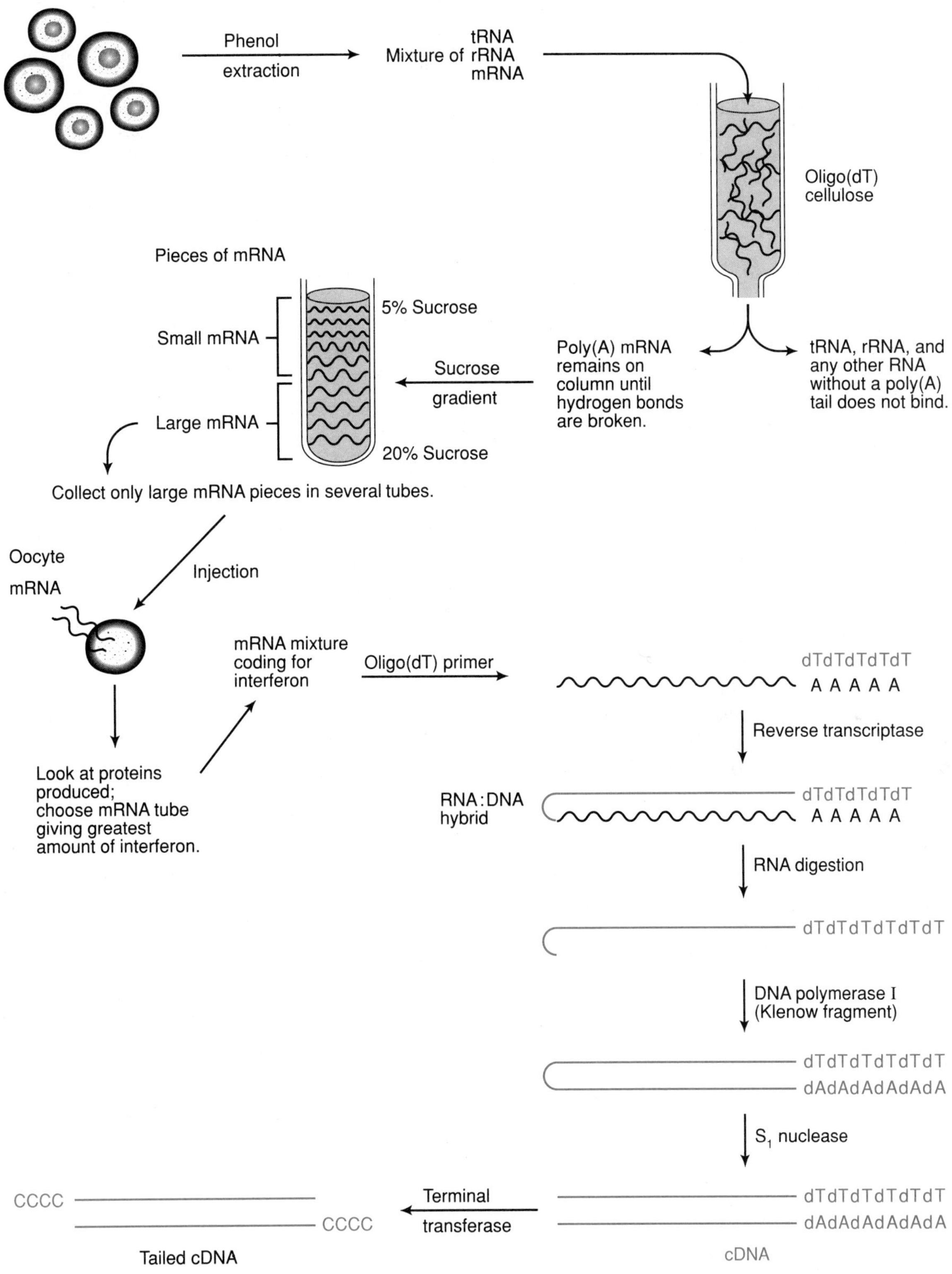

FIGURE 11.10

Preparation of mRNA from a eukaryotic cell and its use to make cDNA. Wavy lines represent RNA molecules and straight lines, DNA molecules. An oligo (dT) tail is a short chain of thymidine residues. This tail will pair with the poly (A) tail found on most eukaryotic mRNA molecules.

APPLICATIONS BOX

Who Owns Surgically Removed Tissue?

An interesting legal and ethical question has arisen with respect to the commercial preparation of valuable biochemicals derived from tumor cells. When patients undergo surgical procedures, they usually sign consent forms agreeing that the material removed can be used for experimental purposes. However, if the experiment with the tumor cells is a success and a potentially valuable product results, should the surgical patient share in the profits? The question has not yet been decided nor has a general law been developed to apply to all situations; however, at least one court has ruled in favor of a patient's being able to share in any profits.

to separate all poly(A)-containing mRNA from all other RNA molecules. Later, they changed the salt concentration of the fluid passing through the column, and hydrogen bonds holding the mRNA to the column broke, releasing the mRNA. The released RNA molecules could have been of almost any size; therefore, the researchers separated the RNA molecules into different size groupings by centrifuging the purified mRNA in a tube containing increasing concentrations of sucrose. Under these conditions, the rate of migration of molecules is proportional to size, and only the large molecules, those 500 to 1,500 bases that presumably represented intact mRNA, were purified further.

The researchers then needed to identify the precise mRNA molecules that coded for interferon. They allowed the RNA to be translated and examined the product. The researchers injected aliquots of RNA fractions taken from the sucrose gradient into *Xenopus laevis* oocytes. These cells will synthesize proteins encoded by any injected mRNA. Interferon is naturally secreted by cells; therefore, the fluid around the oocytes will tend to accumulate any interferon being synthesized. The researchers tested the tissue culture fluid surrounding the oocytes by adding it to fresh tissue culture cells infected with a virus and looking for inhibition of cytopathic effects. Duplicate samples of RNA shown to stimulate the production of large amounts of interferon were used in the next step.

The enzyme reverse transcriptase, produced by retroviruses (chapter 9), was used to copy RNA into DNA. A synthetic polynucleotide was bound to the mRNA and used as a primer to stimulate the synthesis of single strands of DNA complementary to every mRNA molecule in the sample. A proteolytic fragment of DNA polymerase I (the Klenow fragment) was used to make complete the DNA duplex by synthesizing a second strand. Although the Klenow fragment still synthesizes DNA, it has no 5′ to 3′ exonuclease activity and cannot accidentally degrade template sequences. The resulting double-strand DNA molecules were **complementary DNA** (cDNA) molecules that contained most or all the information found in the various mRNA molecules in the original fraction taken from the sucrose gradient. These cDNA molecules then were treated with terminal transferase to add short chains of poly(C) to the ends of the molecules. The pBR322 plasmid was cut open at the *Pst*I site (figure 11.4) and "tailed" with poly(G). Then the poly(C) tails of cDNA were allowed to anneal to the poly(G) tails of the pBR322 vector, the nicks were ligated, and *E. coli* cells were transformed (figure 11.2). Cells carrying cloned DNA were resistant to tetracycline (had received pBR322) but not resistant to ampicillin (carried an inserted cDNA in the ampicillin-resistance gene).

The next problem for the research team was to identify which DNA clone within the mixture of clones actually coded for the entire interferon molecule. Portions of the amino acid sequence of the desired protein already were known from the tryptic peptides, and the researchers could synthesize short oligonucleotides containing all possible combinations of bases coding for the appropriate amino acids to use as radioactive probes in Southern blots (table 11.3). Single-strand DNA from each of the 500 different clones was bound to nitrocellulose filters. Later, single-strand, radioactive probe DNA was washed onto each filter under conditions that would allow hydrogen bonds to form with the clone DNA. Formation of hydrogen bonds requires sequence homology, however. Only those probes partly or completely homologous to the clone DNA remained bound to the filter, therefore. The rest of the radioactive DNA washed off. Only those clones with sequence homology to interferon became radioactive, and their positions were detected by autoradiography.

TABLE 11.3 Preparation of DNA probes for a known sequence of amino acids

Amino Acid Sequence:	Methionine	Serine	Proline
Possible Codons:	TAC	TCA TCG	GGA GGC GGT GGG
DNA Probes Needed:	TACTCAGGA TACTCAGGC TACTCAGGT TACTCAGGG TACTCGGGA TACTCGGGC TACTCGGGT TACTCGGGG		

The clone chosen for further work contained 1,000 extra base pairs in the pBR322 vector. The researchers recovered the inserted DNA by recutting with *Pst*I (the tailing process had regenerated the recognition site) and then inserted it into an **expression vector,** a special plasmid designed to give high-level expression of a cloned fragment (figure 11.11). This particular plasmid contained the tryptophan promoter and ribosomal binding site, but not the attenuator region, and provided for good transcription and translation of the cloned DNA. *E. coli* cells carrying this clone produced about 480,000 units of activity per liter of culture. However, the protein produced was a preinterferon, a molecule carrying extra amino acids at the amino terminus that act as a signal peptide in the leukocyte. This signal peptide is not needed for therapeutic uses of interferon; therefore, it was necessary for the researchers to eliminate it.

They cut the cloned DNA with a specific restriction endonuclease that removed a fragment of DNA coding for all the signal peptide as well as for the first codon of the normal protein and part of the second codon. A chemically synthesized DNA fragment was then attached to restore codons one and two, and the new DNA was religated to the tryptophan DNA (figure 11.11). When this fragment was cloned, the *E. coli* cells produced about 2.5×10^8 units (roughly 650 μg) of biologically active interferon per liter of culture.

As a result of these and other cloning activities, interferon is now available for clinical trials on a mass production basis. Recently, it has been proposed as a treatment for hepatitis C viral infections. The success of this example has been repeated many times already. The procedure is a general one and can be used to clone a variety of hormones (see the reference to Habener's compendium in the suggestions for further reading at the end of this chapter). An entire new industry devoted to the production of economically significant products through the use of DNA splicing and its related technologies has developed. Further examples of applied genetic engineering are presented in chapters 14 and 15.

Problems Facing Society as a Result of Genetic Engineering

As with any new technology, genetic engineering has presented a number of difficult issues that need to be resolved over the next few years as we attempt to apply present-day knowledge and knowledge to be acquired in the foreseeable future. In this section, just a few of these issues are surveyed, but no solutions are outlined because there are no easy solutions. Instead, you are urged to keep these problems in mind and consider what are appropriate and inappropriate uses of DNA splicing technology.

One significant regulatory issue that has arisen concerns the use of bacteria in the environment. Should bacteria be genetically engineered for environmental use? There are many separate questions that require answers. Among them are:

1. Is it appropriate for a company to produce large quantities of a normal bacterium for release into an area in which it is not normally found? If the answer is no, some potentially important activities like the use of bacteria to degrade environmental compounds are eliminated.
2. If we assume that releasing naturally occurring bacteria in new habitats is acceptable, is it appropriate to do the same thing with genetically engineered bacteria? If the answer to that question is no, how are the situations different?
3. If genetically engineered bacteria are released into the environment, what sorts of risks is society taking in the process? What kinds of information should be provided about the bacterium before it can be released?

Questions like these are presently before the U.S. Environmental Protection Agency. Over the course of the next few years, answers to these and a number of other questions will have to be given. Presently, the general tendency is to treat bacteria like new chemicals

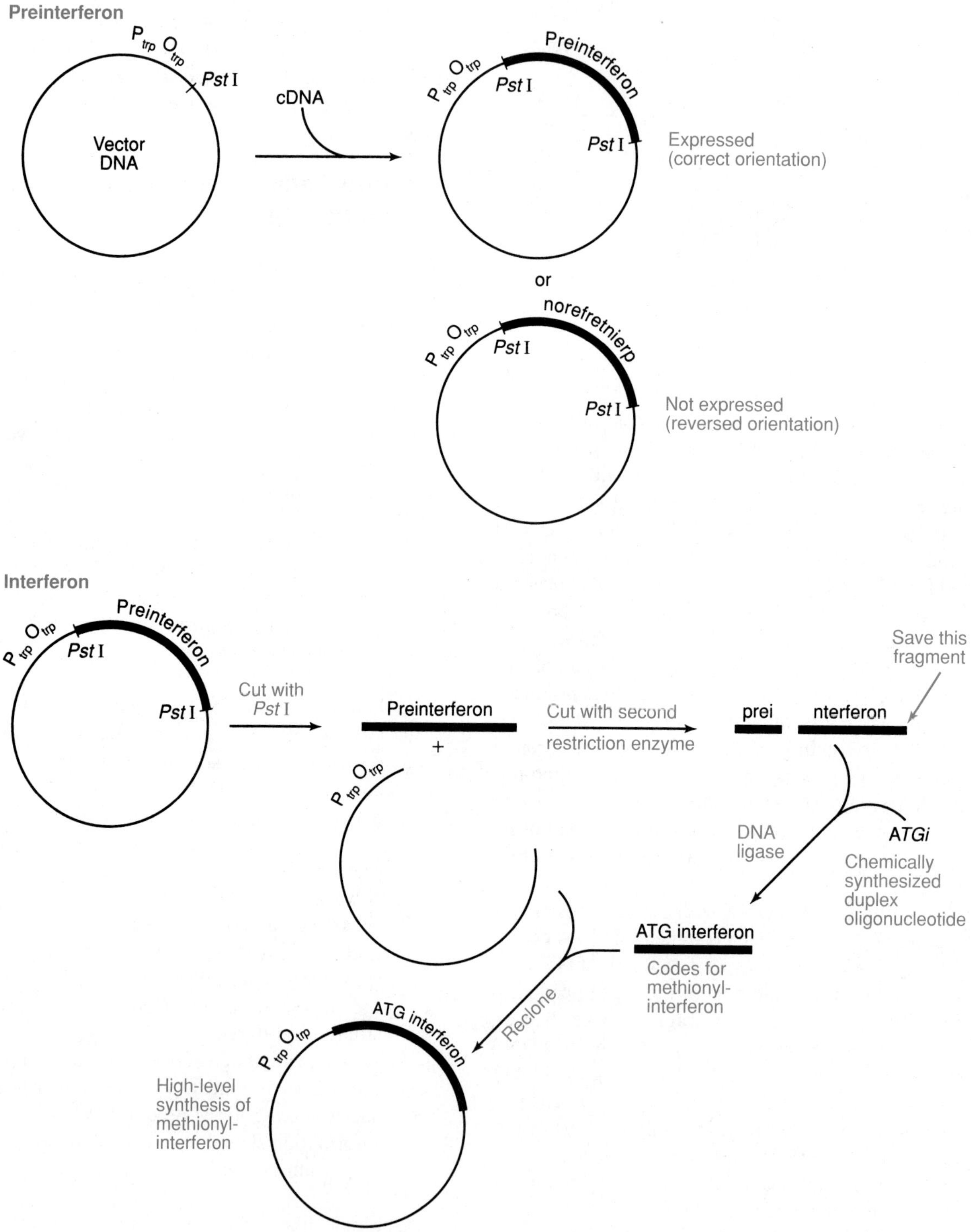

FIGURE 11.11

Plasmids for the expression of cloned interferon DNA. The *Pst*I cloning site is indicated as are the sites for the promoter (P_{trp}) and operator (O_{trp}) taken from the tryptophan operon.

and to rule that bacteria altered in any way other than by simple deletion mutations in the laboratory must be tested for safety before they are released. The regulatory process already has engendered complex lawsuits as companies with marketable products clash with those concerned about the potential effects of large numbers of novel bacteria displacing those present normally in the environment.

Turning to a different dilemma, consider the ethical questions that can be raised by the therapeutic uses of genetic engineering. Human interferon can be produced in large quantity. Is the use of this protein in therapy much like any other drug, or is it not? What safety and efficacy tests should be applied to biochemicals produced by DNA splicing? If the cloned interferon is the same as that found in the human body, no unknown molecules have been introduced if the recombinant interferon is administered. Currently, if the recombinant protein is 100% identical to the naturally occurring one, a manufacturer only has to prove its purity and safety. What if a modified protein can be shown to work better? How much testing should be required before that protein is allowed on the market? This is a very important question because manufacturers spend tens of millions of dollars to run the full battery of tests needed to bring a new drug onto the market.

Finally, consider what happens as tools become available to treat the causes of various genetic diseases. Suppose that an individual is ill due to the lack of a particular enzyme in her body. If vaccinia virus was used as a vector and DNA that coded for the missing enzyme was spliced into the virus, it might be possible to infect the sick individual with the recombinant virus. If the viral infection was successful, and if the spliced DNA integrated in the chromosome, and if the inserted DNA was expressed properly, the sick individual might be cured. This seems like a very desirable goal.

Suppose, however, that a person who needed the enzyme could not be assured that her body would produce it only in those tissues in which it was needed. Instead, all the cells in her body might start to make the enzyme. Is the treatment still appropriate? Suppose you had a way to introduce that virus into the germline cells so that all the offspring of that individual would inherit the virus and the spliced DNA. Is that sort of experiment appropriate? Should some sort of line be drawn for experiments on humans? If so, where? Who should draw such a line? What effects might we have on ourselves as a species if we start to do these kinds of experiments? These are important questions to which society must find answers over the next few years.

Summary

Discovery of restriction and modification enzymes has proved to have wide-ranging significance. Base sequence recognition provided by restriction endonucleases has made possible easy rearrangements of DNA to produce combinations of codes never before seen in nature. In general, DNA of interest is spliced into the middle of some form of plasmid vector that serves to provide the spliced DNA with replicational functions after it has been inserted into a cell (cloned). If appropriate transcriptional and translational signals have been provided either on the plasmid or on the cloned DNA itself, the cloned DNA is functional and will modify the phenotype of the host cell. This property can be used to provide large quantities of specific proteins or RNA molecules. Such products are on the market already, and more are announced every week. Society faces many difficult questions arising out of this new technology. We will have to find the answers quickly before the rate of new discoveries makes these questions moot.

Questions for Review and Discussion

1. Why has the technique of DNA splicing generated such a controversy about safety? What types of concerns have been expressed? What kinds of safety regulations are presently in place? To what extent do you think that these concerns are justified?
2. What are the different types of restriction enzymes? Can each type be used for genetic engineering, and if so how?
3. What is a vector for genetic engineering? What kinds of properties are essential for such a vector? What kinds of properties would be useful but not essential?
4. What kinds of products are being made presently from genetically engineered DNA? What kinds of products might be genetically engineered in the near future?

Suggestions for Further Reading

Barton, J. H. 1991. Patenting life. *Scientific American* 264 (3): 40–46. A discussion of legal issues affecting various biotechnology industries.

Domsch, K. H., A. J. Driesel, W. Goebel, W. Andersch, W. Lindenmaier, W. Lotz, H. Reber, and F. Schmidt. 1988. Considerations on release of gene-technologically engineered microorganisms into the environment. *FEMS Microbiology Ecology* 53:261–272. A working document for those considering how to regulate this type of activity. It is a good summary of the problems to be considered.

Goeddel, D. V., H. M. Shepard, E. Yelverton, D. Leung, R. Crea, A. Sloma, and S. Pestka. 1980. Synthesis of human fibroblast interferon by *E. coli. Nucleic Acids Research* 8:4057–4074.

Goeddel, D. V., E. Yelverton, A. Ullrich, H. L. Heyneker, G. Miozzari, W. Holmes, P. H. Seeburg, T. Dull, L. May, N. Stebbing, R. Crea, S. Maeda, R. McCandliss, A. Sloma, J. M. Tabor, M. Gross, P. C. Familletti, and S. Pestka. 1980. Human leukocyte interferon produced by *E. coli* is biologically active. *Nature* 287:411–416.

Habener, J. F. (ed.). 1987. *Molecular cloning of hormone genes.* Humana Press, Clifton, N.J. Brief discussions of the cloning of a wide variety of peptide hormones.

Klingmuller, W. (ed.). 1988. *Risk assessment for deliberate releases: The possible impact of genetically engineered microorganisms on the environment.* Springer-Verlag KG, Berlin.

Rodriguez, R., and D. T. Denhardt. 1988. *Vectors: A survey of molecular cloning vectors and their uses.* Butterworth Publishers, Boston. A detailed compendium.

William, J. G., and P. K. Patient. 1988. *Genetic engineering.* IRL Press, Oxford. A short treatise on the subject.

Winnaker, E.-L. 1987. *From genes to clones. Introduction to gene technology.* VCH Publishers, New York. An extensive text covering all facets of modern DNA technology.

part

3

THE INTERACTIONS OF MICROORGANISMS AND THE WORLD AROUND THEM

The first 11 chapters of the text deal primarily with the structures and functions of microorganisms. It also is important to consider the ways in which microorganisms interact with the other members of the various biological kingdoms and with the environment in general. Part III, encompassing chapters 12 through 19, is concerned with how the activities of microorganisms affect human society. Many times these activities are of great benefit, either directly or indirectly. Sometimes, however, microorganisms can harm humans, either by disrupting the economy of society or by using humans as parasitic hosts.

Part III addresses what is often called "applied microbiology" in all its myriad forms. The ways in which bacteria participate in the universal cycling of nutrients within the biosphere are discussed in chapter 12. Equally important, however, are the ways in which the presence or absence of microorganisms are used to provide commercially valuable products (chapters 13–15). Chapter 16 takes a more personal turn with a discussion of the immune system and how it acts to protect, and sometimes harm, the human body. In chapter 17, the nature of the interactions between host and parasite are considered. Epidemiology, the science in which immunology and host-parasite relationships are studied in conjunction, is outlined in chapter 18. Finally, some specific examples of diseases, the ways in which they are transmitted, and how they can be prevented are analyzed in chapter 19.

chapter

17

Microbial Ecology

chapter outline

Basic Ecological Principles

Special Features of Microbial Ecology
Recycling of Chemical Elements
Interactions of Microorganisms within an Ecosystem

Biogeochemical Cycling of Chemical Elements

Carbon and Oxygen
Nitrogen
Sulfur
Phosphorus
Iron Uptake
Pollution Problems Resulting from Interference with Bacterial Metabolism

Water Purification and Sewage Disposal

Natural Water Cycle
Sewage Treatment
Treatment of Drinking Water

Microbial Symbioses

Examples of Mutualistic Relationships
Examples of Commensal Relationships

Summary
Questions for Review and Discussion
Suggestions for Further Reading

GENERAL GOALS

Be able to outline the various mechanisms by which the chemical elements necessary for life can be converted from one oxidative state to others.

Be able to present examples of how organisms can interact to accomplish biochemical conversions that are impossible for a single organism.

Be able to discuss sewage and water treatment procedures appropriate for various situations.

Be able to give some examples of symbioses involving prokaryotes with other prokaryotes or eukaryotes and discuss how these relationships function to the mutual benefit of both partners.

When studying a particular organism or group of organisms, we easily can forget that organisms do not live in isolation but rather in a larger community. Their positions within the community, both physical and biological, are referred to as **niches.** Sometimes these niches lie in obvious places, like in a rotting log, or an animal intestine. At other times, niches are found in unexpected areas. Microorganisms can produce a biofilm hundreds of micrometers thick on the hull of a ship, reducing her maximum speed by as much as 20%. They also can be found inside turbines of electric power generating stations where they reduce efficiency and increase costs.

Some relationships of microorganisms to each other and to animals and plants are presented in this chapter. Emphasis is given to the biochemical interactions between various organisms and to the interdependencies they have developed. The ways in which nutrients are recycled and some important organisms that perform this task are considered. Finally, the closely related processes of sewage and water treatment are analyzed because they provide specific examples of how human society takes advantage of normal environmental processes.

Basic Ecological Principles

In thinking about our global ecosystem, bear in mind that the **biosphere** (portion of the planet containing living organisms) we inhabit is fundamentally a closed system in the sense discussed in chapter 8. Although the sun provides radiant energy that enters the earth's ecosystem from outside its atmosphere, no significant amounts of matter arrive on or leave from the surface of our planet. Various wastes produced from biological processes or from human industrial activities remain with us. If not degraded by some natural or artificial process, these wastes eventually may accumulate to levels that are toxic or otherwise hazardous.

Special Features of Microbial Ecology

The relatively small size of microorganisms and their unusual types of growth and cellular organization present some problems for microbial ecologists. Consider, for example, the number and quantity of organisms that abound within a given microbial ecosystem. One way to determine what organisms are present is to attempt to grow them in the laboratory. An immediate question that arises concerns what culture medium to use. No one medium is certain to grow all types of organisms. Furthermore, in terms of colony-forming units, both a single cell and a filament of cells will each yield one colony, yet the latter represents considerably more biomass within the ecosystem than the former. The most practical (and laborious) solution is to determine the microorganisms present by direct microscopic counts.

Even a direct count of microorganisms can be misleading, however, because there is no guarantee that the organisms observed with a microscope are actually growing and dividing. Some may be present by accident, passing through as it were but not growing because they lack one or more critical nutrient. Others may be present normally but in a dormant stage of their life cycle; therefore they do not impact the ecosystem at this time.

The physical nature of the ecosystem is also different from what might have been expected. In macroecology, it is possible to talk about large, relatively homogeneous areas like a pine forest although there may be regions of local heterogeneity. Similarly, there are local heterogeneities present within microbial ecology. Consider, for instance, the habitat of "soil bacteria." The soil itself is highly variable, its constitution depending on the nature of the underlying strata as well as on the erosive forces present. Over long periods of time, a vertical stratification is established that provides optimum growth conditions for different organisms at different levels. This has important implications when competitive studies between different organisms are conducted in the laboratory. One organism may clearly grow better than another in a laboratory, but they in fact are not competing at all in their natural environments if they are confined to different strata.

The microenvironment within a particular stratum of soil is also highly variable. Soil consists of individual particles and accompanying spaces and pores. The surface area of particles within soil can be enormous. Nedwell and Gray have estimated that in a cubic meter of clay composed of particles with a diameter of 2 μm, the total surface area available is about 3×10^6 m^2 (see the reference to Nedwell and Gray's discussion in the suggestions for further reading at the end of this chapter). If a flagellated bacterium is to move from one soil particle to another, there must be a column of water connecting the two. When water is not saturating the soil, a film of water molecules surrounds each particle. The surface tension of these molecules tends to trap bacteria on the particle surface.

The physiology of microorganisms inhabiting the soil also is affected by the nature of the soil. The amount of organic material present has an obvious impact. So too does the amount of water. If the soil is saturated with water, the system becomes anaerobic and microorganisms must metabolize nutrients accordingly. In sea sediments, for example, the concentration of oxygen is reported to be only 0.23 mM, while the concentration of sulfate is 20 mM. Therefore, about 50% of all reactions in which carbon is broken down in the sea tend to involve sulfate reduction.

It is also important to note that growth conditions in nature are very unlike those in the laboratory with respect to nutrient availability. Laboratory cultures frequently have interdivisional times ranging from 20 to 60 minutes. On the other hand, interdivisional times of microorganisms living in soil have been estimated to range from 26 to more than 1,000 hours. Similar conditions exist in the sea in which it has been estimated that dissolved organic carbon may only be available in the range of 50 fg (femtograms; 10^{-12} g) per bacterial cell. Microorganisms in nature are strongly limited in the amount of energy they can generate, therefore. The simple addition of glucose to a particular region can greatly increase biomass production.

Thus, microbial ecology involves the study of populations of organisms growing in microenvironments under semistarvation conditions. Unlike the metabolic reactions discussed in chapters 5 and 6, which are observed in cultured cells and organisms, the real world does not deal with pure cultures of organisms growing in isolation on rich media. Instead, the metabolic activities of one group of organisms have a direct impact on other groups in the same or physically linked ecosystem.

Recycling of Chemical Elements

In the normal course of events, all the chemical elements that comprise living things are recycled endlessly. The major elements of concern are the same ones required for microbiological media (chapter 3). Some of these elements are found in nature only in one oxidative state; hence, they present no real problems in terms of assimilation by biological processes. Phosphorus, for example, is always present as some sort of phosphate.

On the other hand, elements such as carbon, oxygen, nitrogen, and sulfur all can be found in nature in various combined states or as uncombined molecules. In many of these states, they are sequestered and unavailable to most organisms. Sequestration can result from the insoluble nature of some compounds or from the large amounts of energy investment required to change the oxidative state of an element into a form

TABLE 12.1. Sequestered forms of some essential elements

Element	Sequestered Form(s)
Carbon	Coal, oil, insoluble carbonates (such as limestone)
Iron	Insoluble ferric compounds, iron and steel fabrications, iron sulfides
Nitrogen	Nitrogen gas
Sulfur	Elemental sulfur, insoluble sulfates or sulfides
Phosphorus	Insoluble phosphates

Note: When in sequestered forms, elements are effectively removed from the biosphere; therefore, they contribute to the continued growth only of specialized organisms.

that can be used in the metabolism of a majority of organisms. Some examples of sequestered forms of essential elements are presented in table 12.1.

Sequestered forms of elements can, of course, be converted into molecules appropriate for living systems. The process of fixing nitrogen into a usable form is discussed in chapter 6. Coal can be burned and limestone treated with acid to release carbon dioxide into the atmosphere. Such processes occur continuously. Occasionally, they are conducted on such a large scale as to present ecological problems. An example concerns the potential **greenhouse effect** of carbon dioxide released by the burning of fossil fuels. From 1850 to 1986, the estimated amount of carbon (as carbon dioxide) in the atmosphere has increased from 560 to 760 metric tons. Increased growth of plants and autotrophic microorganisms tends to mitigate greenhouse effects but cannot completely compensate for them.

Interactions of Microorganisms within an Ecosystem

Before the roles played by various organisms in an ecosystem are considered, a distinction should be made between producer organisms and consumer organisms. A **producer** is an organism that, as a part of its life cycle, synthesizes carbon-containing molecules usable by consumer organisms. A **consumer** then is one that eats the producer or its products. Primary producers are either phototrophic or chemolithotrophic autotrophs. With respect to autotrophs, all chemoorganotrophs are consumers. Primary consumers eat autotrophs or their products, while secondary consumers eat primary consumers or their products.

The existence of a consumption chain is an important facet within an ecosystem. Because there is recycling of all elements, even secondary consumers must

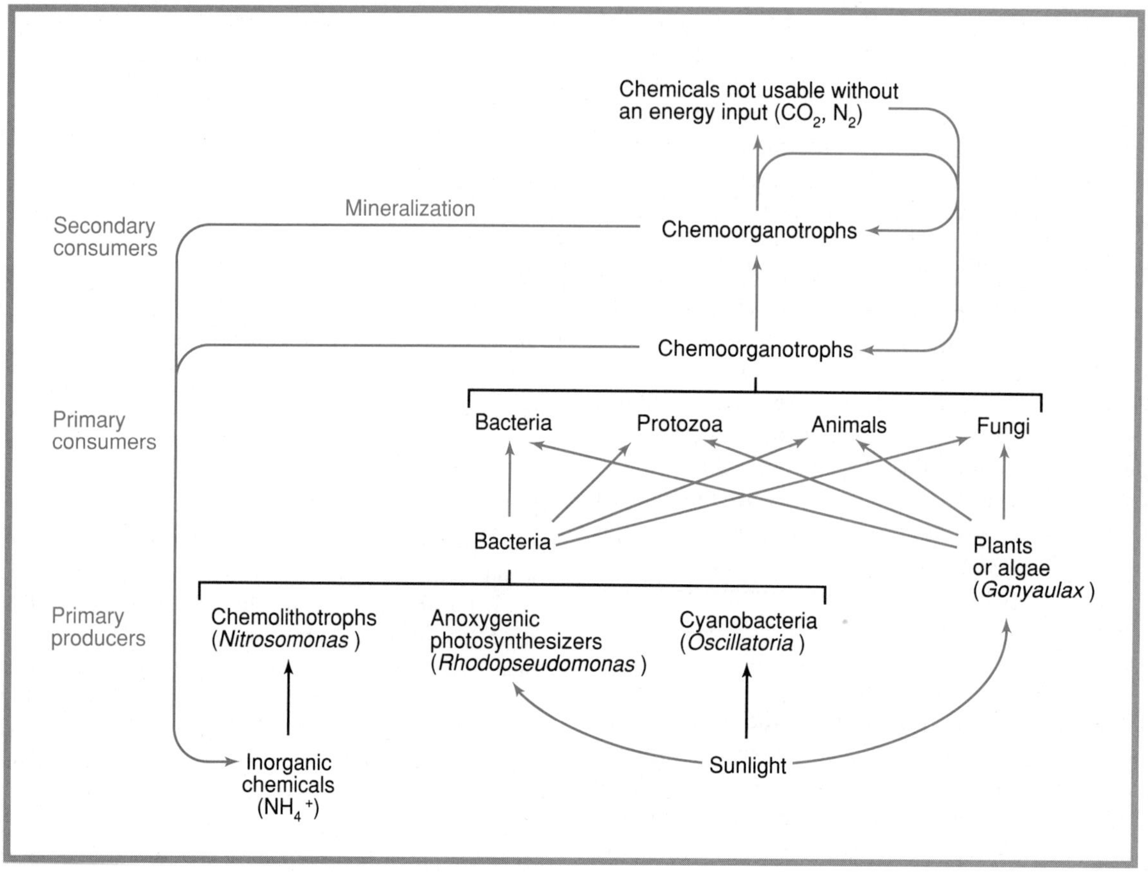

FIGURE 12.1

Interrelationships of various types of organisms within an ecosystem. Note that the primary producers may get their energy either as phototrophs or as chemolithotrophs. Colored arrows trace the flow of energy and chemical products within the system. Examples of specific organisms and chemicals are given in parentheses.

be producers with respect to some other organisms. The so-called **decomposers** or degradative organisms act to return chemical elements derived from larger organisms into forms that can be used by many organisms in the biosphere, specifically by primary producers (figure 12.1). Decomposers are, of course, fundamentally secondary consumers. A good example of secondary microbial consumers can be found in the dry savannas in which there is a reciprocal relationship between microbial biomass and plant growth rate. Microbial biomass accumulates during the dry season, and nutrients are released to the plants during the rainy season. The rot of a fallen tree or the decomposition of an animal carcass also illustrates the actions of secondary microbial consumers.

Biochemical reactions found in decomposers include many that are essentially the reverse of those used for biosynthesis. Assorted macromolecules, whose synthesis was described in chapter 6, are degraded to their various subunits; that is, proteins give rise to free amino acids, nucleic acids to mononucleotides, polysaccharides to individual sugars, etc. These individual subunits can be reused as is to make new proteins, nucleic acids, and polysaccharides, or they can be degraded further to provide energy and/or carbon skeletons for the synthesis of other types of molecules. Some examples of secondary consumer microorganisms specific for certain substrates can be found in table 12.2.

TABLE 12.2	Organisms that act to decompose particular types of molecules
Substrate	**Organism**
Cellulose	Slime molds *(Dictyostelium)*
	Fungi *(Trichoderma)*
	Bacteria *(Cytophaga)*
Crude oil	Bacteria *(Pseudomonas)*
Iron	Bacteria *(Thiobacillus)*
Lignin	Fungi *(Coriolus)*
Methane	Bacteria *(Methylococcus)*

Note: These organisms are representative samples and should not be taken as a comprehensive list.

TABLE 12.3	Examples of nitrogen-fixing bacteria
Free-Living Organism	**Symbiotic Organism**
Aerobic phototrophs	*Bradyrhizobium*
Anabaena	*Rhizobium*
Spirulina	*Frankia*
Anaerobic phototrophs	
Rhodospirillum	
Aerobic chemotrophs	
Azomonas	
Azotobacter	
Beijerinckia	
Derxia	
Klebsiella	
Anaerobic chemotrophs	
Clostridium	

Biogeochemical Cycling of Chemical Elements

Chemical elements can exist in different oxidative states within the biosphere. Cycles in which there are interconversions of different forms of the same element have both reactions that occur spontaneously and those that must be catalyzed by living organisms. These cycles are, therefore, referred to as biogeochemical cycles.

Carbon and Oxygen

The biological carbon cycle is comparatively simple (figure 12.2). Biochemically useful carbon is taken to occur in the form of carbon dioxide, methane, or cellular components. Those organisms that carry out decarboxylation reactions release carbon dioxide into the environment. Autotrophic bacteria, plants, and algae take carbon dioxide from the air and fix it in the form of carbohydrates. Methanogens use carbon dioxide as their terminal electron acceptor.

Free oxygen is used as the terminal electron acceptor for respiration by animals and many microorganisms. In this process, it is converted into water. Oxygenic photosynthesizers use water as their electron donor for photosystem II, releasing molecular oxygen as a by-product (figure 4.11). Note that a nonbiological process such as a lightning-caused forest fire also can use up oxygen and release carbon dioxide.

Nitrogen

Unlike carbon and oxygen, nitrogen can be found in a wide variety of oxidative states in nature. Figure 12.3 illustrates the interrelationships between these different valence states. Although nitrate (NO_3^-) is not chemically sequestered, it generally is not available to terrestrial organisms. Because of its extreme solubility in water, nitrate is leached rapidly out of upper layers of the soil. Ammonium ion (NH_4^+), in contrast, tends to form complexes with the negatively charged soil particles and remain in the root zone. Nitrate is, of course, commonly available to aquatic organisms, particularly in the form of fertilizer runoff. Atmospheric nitrogen (N_2) is the chemically sequestered form for this pathway and provides a convenient starting point for the discussion.

Various free-living and symbiotic bacteria expend much energy to fix nitrogen in the soil. A typical group of free-living nitrogen fixers is the genus *Azotobacter,* and the most famous group of symbiotic nitrogen fixers is the genus *Rhizobium.* A more extensive listing of nitrogen-fixing organisms can be found in table 12.3. Reduction of nitrogen also occurs due to environmental causes such as lightning bolts. Large quantities of nitrogen compounds are prepared every year for use as fertilizers by means of standard chemical processes, such as the Haber-Baush process, in which nitrogen gas reacts directly with hydrogen gas. The process requires high temperatures and pressures, making the resulting chemical fertilizer rather expensive. This is especially true with respect to agriculture in third world nations.

Microorganisms can assimilate their nitrogen either as free ammonium ion (see figure 5.5 A–B) or as combined nitrogen in the form of amino acids or nucleic acid bases. If an organism has an oversupply of nitrogen-containing molecules, it generally will **deaminate** the molecule (remove the nitrogen) by reversing the transamination reaction used to synthesize the amino acid and proceed to funnel the carbon skeleton

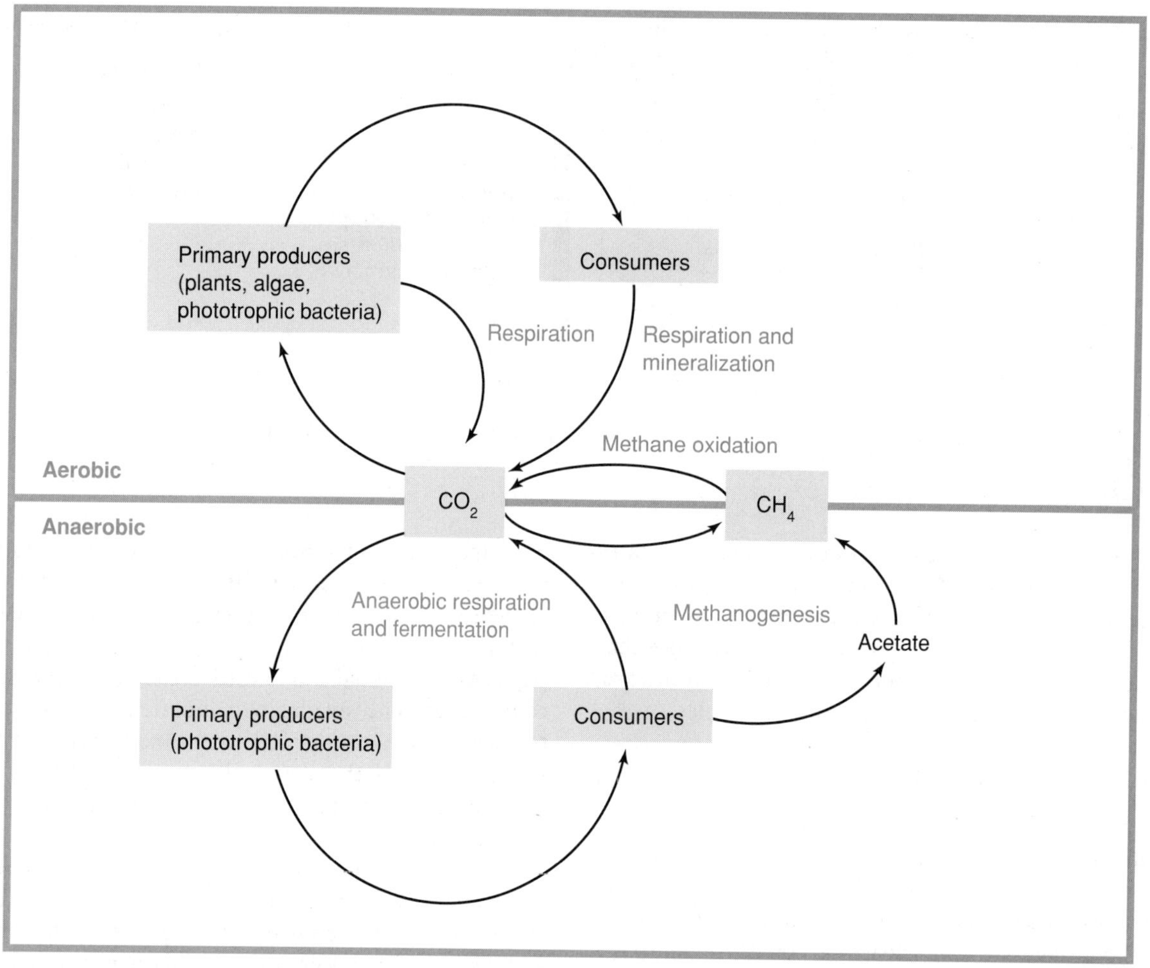

FIGURE 12.2

Biological reactions of the carbon cycle. Carbon found within organisms may be in the form of carbohydrates, lipids, nucleic acids, or any other organic molecules. Oxygenic photosynthesizers can reduce carbon at the expense of the oxygen in water. Methane production is always an anaerobic process, and its oxidation is always aerobic. The burning of methane or of many other organisms themselves is an example of a nonbiological process.

into the energy pathways. If an excess of amino nitrogen accumulates, as frequently occurs in the case of secondary consumers, it will be discarded into the surrounding medium as ammonium ion. The term **ammonification** often is used to describe the conversion of amino nitrogen into free ammonium ion.

Sergei Winogradsky, working in Paris, first demonstrated that some bacteria living in the soil are capable of deriving energy from the oxidation of ammonium ion to nitrite (NO_2^-). The best-known example is *Nitrosomonas*. The nitrite produced as a waste product of *Nitrosomonas* metabolism can serve as an energy source for *Nitrobacter,* whose metabolism oxidizes it to nitrate. This chemolithotrophic pathway is a good example of how the combined activities of two organisms can accomplish a biochemical conversion that is impossible for either organism alone.

A competing reaction reverses the nitrogen oxidative process. Many organisms such as *Pseudomonas* or *Escherichia* are capable of using nitrate as a terminal electron acceptor during anaerobic respiration. The final product of this nitrate reduction reaction depends on

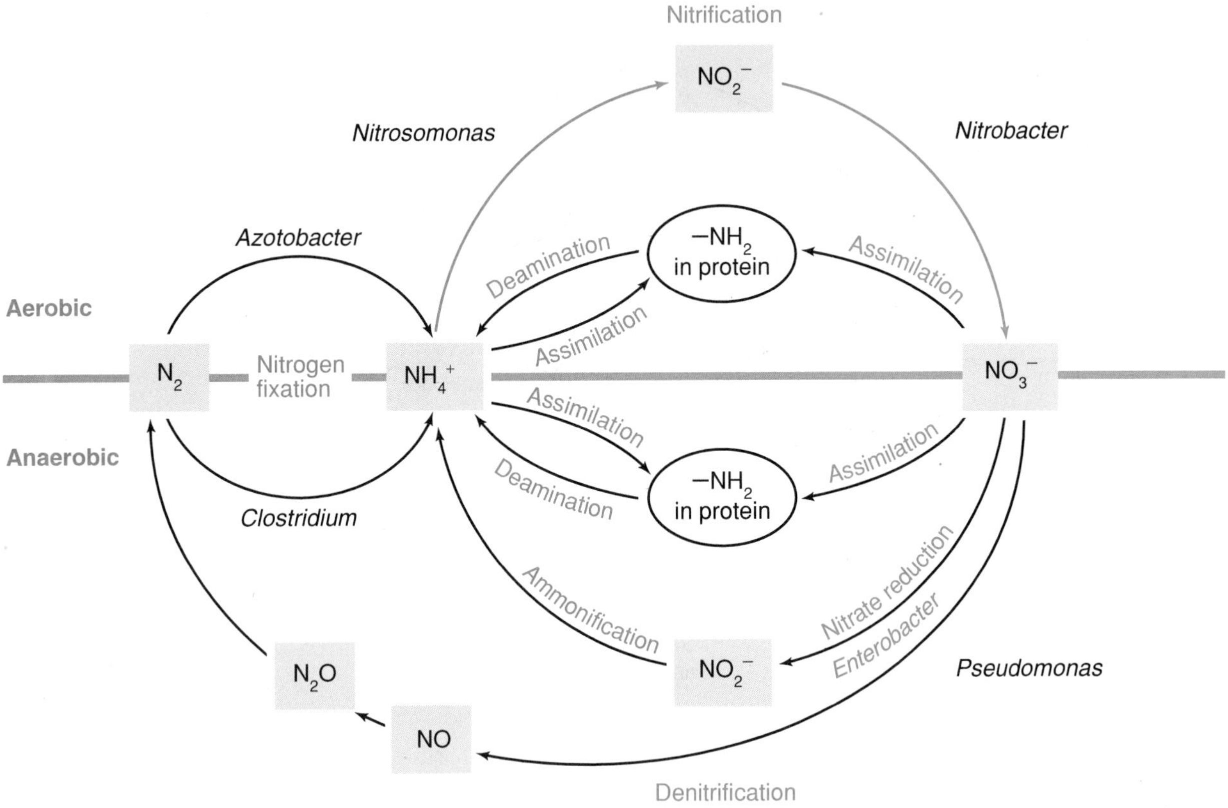

FIGURE 12.3

Biological reactions of the nitrogen cycle. The upper half of the diagram depicts aerobic reactions and the lower half, anaerobic reactions. Gaseous nitrogen (N_2) occupies a central location because it is the most abundant form of the element and thus acts as a reservoir for the biosphere. Amino nitrogen ($-NH_2$) and ammonium ion (NH_4^+) have the same valence and are essentially in equilibrium with one another, as indicated by the black arrows between them. The oxidative pathways of aerobic respiration are shown by colored arrows, and the reductive pathways including anaerobic respiration, are indicated by black arrows. Genera that carry out particular reactions are adjacent to the arrow indicating the direction of the reaction.

From H. Stolp, *Microbial Ecology: Organisms, Habitats, Activities.* Copyright © 1988 Cambridge University Press, New York, NY. Reprinted by permission.

APPLICATIONS BOX

Nitrogenous Fertilizers

Plants can take up ammonium ion directly, although they generally prefer to take up nitrate ion instead. The only plants that seem to have a preference for ammonium ion are those that like acidic soils, because nitrification proceeds best at neutral to slightly alkaline pH. Nitrate is an excellent fertilizer even though it is highly soluble in water. Natural deposits of nitrate are found only in very arid regions of the world, such as in the Chilean deserts where they are mined for use as fertilizer. Rapid nitrification can be disadvantageous for farmers because nitrates can be leached away from the rhizosphere (root zone) by heavy rains or extensive irrigation. Many farmers take advantage of the fact that ammonium ions tend to bind to soil particles and remain in the rhizosphere substantially longer than nitrate ions. They fertilize their fields with anhydrous ammonia that, in the presence of water, yields ammonium ion.

$$NH_3 + H_2O \rightarrow NH_4^+ + OH^-$$

The *Nitrosomonas* and *Nitrobacter* organisms found in the soil gradually convert the ammonium ion to nitrate for plant consumption.

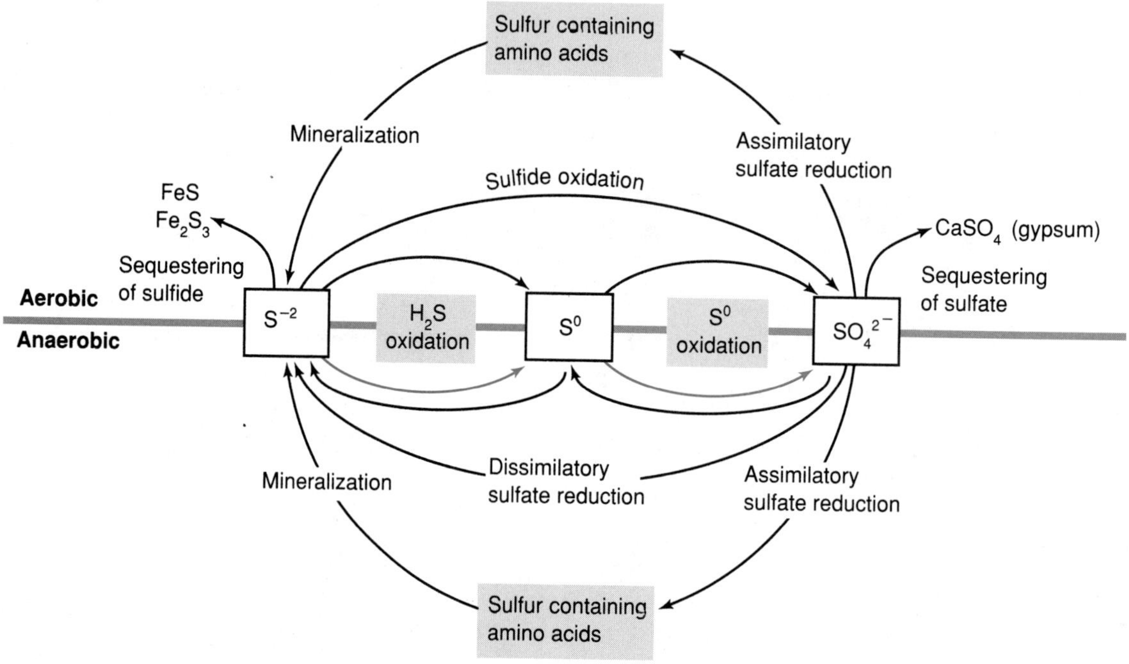

FIGURE 12.4

Sulfur cycle. Elemental sulfur (S^0) is a storage product. It is possible to sequester SO_4^{-2} or S^{-2} by the formation of insoluble precipitates. Note that anaerobic organisms can carry out both dissimilatory sulfate reduction and sulfide oxidation (although not at the same time), whereas aerobic organisms can only oxidize sulfide or sulfur.

the organism involved. *Escherichia* generally produces only nitrite, while *Pseudomonas* can cause **denitrification** by converting nitrite to various nitrogen oxides or to free nitrogen gas. If reduced nitrogen is taken into the cell and used for biosynthesis, the process is called **assimilatory nitrate reduction.** If reduced nitrogen is released to the environment, the process is called **dissimilatory nitrate reduction,** and the extent of the process is dependent on the amount of organic carbon available to be oxidized.

Large-scale production of nitrogen oxides can have important environmental implications because some of these compounds are found in the smog that blankets so many large cities. *Nitrosomonas europaea,* for example, converts about 0.3 to 10% of the ammonium ions that it oxidizes into nitric oxide (NO) or nitrous oxide (N_2O). It has been estimated that bacterial production of these two compounds can be as much as 5 to 10 $\times$ 10^6 tonnes/year on a global scale.

Sulfur

The sulfur cycle is quite similar to the nitrogen cycle (compare figures 12.3 and 12.4), but the organisms that carry out its reactions are not as well characterized as those for nitrogen metabolism. Beginning with the elemental form, sulfur can be oxidized by organisms like *Thiobacillus thiooxidans* in a reaction of the following type:

Equation 12.1

$$S + H_2O + 1\frac{1}{2}O_2 \rightarrow 2H^+ + SO_4^{2-}$$

Depending on the organism involved, the product of the reaction may be sulfate (SO_4^{2-}) or sulfite (SO_3^{2-}) ion. Gardeners often take advantage of this reaction when confronted with alkaline soils. The addition of inexpensive powdered sulfur stimulates the growth of thiobacilli and the production of acid in the soil.

Sulfate-reducing bacteria are all anaerobic, and their contributions to the environment often have been underrated. Their production of hydrogen sulfide (H_2S) is partly responsible for the disagreeable odors that emanate from stagnant pools of water heavily contaminated with organic matter. On a global scale, sulfate-reducing bacteria have been extremely active. Sulfur

APPLICATIONS BOX

Sulfur and Acid Rain

If sulfur-containing compounds are not removed from coal, the sulfites and sulfates produced during combustion can contribute to the problem of acid rain plaguing industrial nations worldwide. Industrial processes produce large amounts of sulfur oxides that, when dissolved in rainwater, drastically lower its pH. Permanent bodies of water fed by rainwater, primarily lakes, gradually acidify over the years as inorganic ions accumulate. In severe cases, the foliage of trees can suffer acidic burns. The heavily industrialized regions of the world have sustained the worst damage. These include the eastern United States and Canada, Scandinavia, and the German Black Forest.

deposits in Louisiana and Texas represent about 90% of the world's natural sulfur. John Postgate has pointed out that they are **biogenic** in origin (produced by living cells). It is thought that these deposits formed when huge inland seas gradually dried up millions of years ago.

The reactions of sulfur bacteria are of considerable economic importance. *Thiobacillus ferrooxidans* carries out an oxidation reaction that can be used to remove sulfur from coal before it is burned.

Equation 12.2

FeS_2 (iron pyrite) $+ 2Fe^{3+} \rightarrow 3Fe^{2+} + 2S$ (spontaneous)

Equation 12.3

$$2Fe^{2+} + 2H^+ + \frac{1}{2}O_2 \rightarrow 2Fe^{3+} + H_2O$$

(bacterial)

Reaction 12.1 can then oxidize sulfur to water-soluble sulfate that can be removed from the insoluble coal. Technologies are now available that allow continuous treatment and processing of coal via the reactions outlined in equations 12.1, 12.2, and 12.3. From 40 to 60% of the iron pyrites are removed. If chemoheterotrophs are included in the bioreactor to remove inhibitory carbon molecules, 50% coal and water slurries can be treated in batches.

The ability of the various members of *Desulfovibrio* to oxidize iron can lead to significant problems with iron pipes that are buried. Sulfide ion (S^{2-}) is itself corrosive. In addition, a solubilizing reaction results from the action of a bacterial enzyme on the surface of the iron; therefore, corrosion occurs much more rapidly than would be expected for simple rusting of iron pipes.

Phosphorus

Phosphates are present in all regions of the world, primarily in an insoluble form, however. Consequently, phosphate is oftentimes a limiting nutrient in environmental systems. Production of acids by various bacteria results in the solubilization of phosphate and its eventual use by various organisms. Ammonium phosphate [$(NH_4)_3 PO_4$], a commonly used chemical fertilizer, will supply large amounts of two essential elements.

Problems can arise when the level of phosphate in natural waters rises above the normal level. High levels can result from overfertilization of nearby cropland or from use of detergents and other household cleaning products containing phosphates. The presence of extra phosphates in bodies of water stimulates excessive growth of plants and algae. In particular, cyanobacteria, such as *Anabaena,* that are capable of nitrogen fixation are limited in their growth only by the levels of phosphate available to them. One early effect of phosphate pollution (in the absence of concomitant nitrogen pollution) is extensive growth of such organisms.

Lakes that are heavily fertilized by phosphates may have such extensive growth that water plants form a thick mat over their surfaces, choking the waterways (figure 12.5). The only real solution to the problem is to reduce the input of fertilizer into the water; however, harvesting of these plants does act as a stopgap measure. Otherwise, the accumulated organic matter can stimulate enough bacterial growth to render the subsurface water anaerobic (see subsection on sewage treatment).

FIGURE 12.5
Effects of pollution on an inland lake. Large quantities of cyanobacteria, algae, and plants accumulate in the water.
© Grant Heilman

Iron Uptake

Although iron is present only in trace amounts within a cell, it is critical to the functioning of several important molecules, such as the various cytochromes of the electron transport chain. Therefore, bacterial cells have specialized systems to concentrate low levels of iron to a point at which it can be used by cells. Usually, bacterial cells release low-molecular-weight compounds called **siderophores** that bind very strongly to iron. The complexes can then be transported back into the cytoplasm where the iron is stripped off. In some cases, siderophores are recycled, but in others they are degraded.

The type of molecule used as a siderophore varies. Most enteric bacteria use catechol derivatives, although clinical isolates of *E. coli* usually produce aerobactin, a hydroxamate. *Erwinia chrysanthemi* synthesizes chrysobactin, a dipeptide linked to 2, 3-dihydroxybenzoic acid.

Bacteria not living in or on other organisms may have trouble finding iron in the oxidative state they require for growth. Some reactions that can be used to change the oxidation level of iron are discussed with the sulfur cycle.

Pollution Problems Resulting from Interference with Bacterial Metabolism

In terms of the normal functioning of the biosphere, the biogeochemical recycling processes outlined and others too numerous to mention operate in a cooperative fashion to ensure that the chemical elements needed for continued life are made available. In addition, they serve to prevent the accumulation of undesirably large amounts of organic material. Indeed, under normal conditions, the question is not whether there is an organism that will grow using a particular organic molecule for energy but rather which of several possible organisms will do so. In general, those organisms that are most successful at maintaining their physical position in an area of high-nutrient concentration will prevail. Fimbriae (common pili) or stalks are often used to anchor cells.

A general underpinning of all human society is an abiding faith in the recycling of all matter. If something is discarded and left exposed to the elements, something else will degrade it into forms that can be reused by living organisms. The degradative organisms in question are usually microorganisms. This fact has given rise to a strong belief in the "infallibility" of microorganisms; that is, microorganisms always are capable of breaking down any and all organic molecules, ranging from wastes disposed in bodies of water to those covered with dirt. Historically, this belief has been the rationale used by governments and private individuals for dumping of raw, untreated sewage into lakes and streams, a centuries-old procedure that is rarely used in the United States today.

When human populations were smaller and less technological, the dumping system worked quite well. With increasing population density, however, came increasing amounts of organic waste material that needed to be decomposed. Waste material fueled the growth of large numbers of decomposer organisms that created visually unappealing, smelly conditions. Moreover, the high probability that disease-causing organisms were contained within waste material created justifiable concerns about the health of people living in the immediate vicinity of bodies of water and land used for dumping. These concerns led to the development of present-day sanitation systems in which waste material is trucked and/or piped away to some distant disposal

APPLICATIONS BOX

Biodegradation and Plastics

Wire insulation is certainly not the only form of plastic used at the present time. Some other plastics really are not intended to last for a long time although they too contain halogens. Garbage bags sold at your local supermarket should be tough and durable, but the recycling process cannot begin if they can never be decomposed. The first plastics used for disposable bags, for example, were too stable and did not break down even when they were exposed to the weather for many months.

In response to the developing problem of nonbiodegradable plastics, plastics manufacturing industries have created new formulations that still maintain the strength in plastics but make them susceptible to the action of sunlight. In one case, ultraviolet radiation from the sun causes deliberately weak links in the polymeric chain of a particular plastic bag to break, shredding the bag and releasing its contents. More recently, starch polymers are being intermixed with plastic resins. The starch is degraded easily by environmental organisms. Both these phenomena are observed when materials are stored in plastic bags kept in exposed locations. Major problems still exist with the disposal of plastics in the oceans, however.

facility. For many years, the fundamental assumption on which disposal was based was that as long as few people lived in the vicinity of the disposal site, the waste material was being properly discarded. No thought was given to what would happen when people did move in or to the actual fate of discarded chemicals.

Most municipalities use **landfill sites** consisting of holes or depressions in the ground that are gradually filled in with trash and garbage. Sometimes, these sites can be dirt-covered mounds. The solid waste generated by cities and towns are disposed of at these sites. At first glance, landfill might seem like the ideal way to promote decomposition of organic matter. Modern landfill sites are not operated in that manner, however. Waste material is packed so tightly that the core is basically anaerobic and lacking in water. Growth of microorganisms is difficult under these conditions, and materials do not break down very rapidly in fact. Most tests to detect whether a particular substance, like a plastic bag, can be degraded by microorganisms do not mimic landfill conditions; therefore, they are not relevant to actual practice.

As landfill sites are filling rapidly, different approaches to reduce the volume of waste material are being tested. An obvious but nonmicrobiological tactic is recycling. Some states have embarked on a major effort to limit the amount of yard waste (leaves, mulch, grass clippings) disposed of in a landfill because such wastes can constitute as much as 50% of residential trash during the summer months. Instead, these states are urging residents to turn to the old technology of composting. Leaves, grass clippings, certain types of kitchen waste (vegetable or fruit peelings) are mixed in a container about 0.04 m^3 so as to achieve an optimum carbon-to-nitrogen ratio of 30:1 (roughly two parts grass clippings to one part fallen leaves). This size container allows air to penetrate most of the way into the compost, yet the pile still has sufficient volume to retain the heat generated by microbial metabolism. A good compost pile has a mild, pleasant odor with internal temperatures from 60 to 70° C. Decomposers can convert the wastes into material suitable for use as a soil conditioner in about three to four months.

The rise of technology, and most particularly of organic chemistry, has created new problems in waste disposal. The synthetic organic molecules created have become essential for the normal operation of a technological civilization. Imagine for a moment what would happen if all the plastic insulation on the various wires in a home or place of business were subject to decay processes. They would gradually disappear with time! These plastics, however, have been designed specifically to resist the action of decomposer organisms. They do so primarily because the halogen substituents on their organic molecules are toxic to bacteria attempting to metabolize them.

Plastics are not the only chemicals that have caused environmental problems. Sometimes, the optimum performance derived from a chemical is not environmentally sound. Consider the action of an insecticide. An ideal insecticide is toxic to insects but not to humans or animals. In addition, it has persistence, remaining active on environmental surfaces for long periods of time. Because of this persistence, repeat applications are needed only infrequently and costs are reduced. Another way to describe persistent molecules is to say that they are recalcitrant.

The insecticide DDT (dichlorodiphenyltrichloroethane) is just such a chemical. Problems arose when it became apparent that decomposer organisms were not keeping pace with the rate at which DDT was being added to the environment. Within a few decades of its introduction, traces of DDT were found worldwide, including in such inaccessible places as the Antarctic. As they accumulated in the environment, DDT molecules were absorbed into the fat layers of various animals, with particularly disastrous results for certain birds. Their eggs had such fragile shells that they tended to break during incubation, resulting in precipitous declines in the population. The United States has banned the use of DDT except in emergency situations. Since the ban, the domestic situation has greatly improved, although DDT is still widely used in third world countries.

DDT presented a twofold problem. First, the insecticide is not sufficiently **biodegradable,** that is, it cannot be broken down rapidly by the normal metabolic processes of decomposer organisms. Second, some of its initial breakdown products, like dichlorodiphenylethane (DDE), are even more toxic than DDT. Similar problems have arisen with many other chemicals, polychlorinated biphenyls (PCBs), trichloroethylene, and crude oil, for example, that make newspaper headlines on a regular basis. These chlorinated and aromatic chemicals have produced a staggering problem of hazardous waste disposal, the magnitude of which is only now becoming clear. In order to remedy certain aspects of hazardous waste management, microbiological solutions offer attractive possibilities (see chapter 15).

Water Purification and Sewage Disposal

Natural Water Cycle

Water comprises some 70% of the surface area of our planet, but in many regions of the world today, water shortages are a reality or becoming so rapidly. The amount of available water is not always the problem as much as the purity of the available water. An abundance of seawater, for example, is of no use to the conventional farmer. Good-quality water must fulfill three criteria:

1. The water must have a sufficiently low mineral content to present no hazard to land animals and plants (in other words, it must be freshwater).
2. The water must not contain appreciable concentrations of toxic materials such as nitrates, arsenic, chlorinated hydrocarbons, and organophosphate pesticides.
3. The water must not contain any disease-causing organisms.

For centuries, settlements were established by the side of a freely flowing stream, or wells were dug in order to meet these criteria. Both methods took advantage of the natural circulation of water that ensures its purity (figure 12.6). Water is lost by evaporation from all sources (oceans, lakes, rivers, artificial pools, etc.). Evaporation is fundamentally a purification process, just as a chemist uses distillation to purify a compound. Water vapor gradually condenses to form clouds from which rain falls. Traditionally, rainwater was assumed to be both pure and mild. In these industrial times, however, this is not necessarily an accurate assumption as our recent problems with acid rain have demonstrated.

Not all water that falls on earth evaporates. Significant amounts of surface water sink into the soil and move through the pores of many layers of rock. Such water is denoted as **groundwater,** an important resource in many parts of the world. Unless soil formations contain chemicals, such as nitrates, that present a significant hazard in themselves, sufficient filtration produces good drinking water. The mineral content may be higher than that of surface water, but the groundwater is still acceptable. Organic chemicals and hazardous microorganisms are removed by the combined filtering and adsorbing action of the soil particles as well as by bacterial metabolism. In regions in which there is a high concentration of organic material, bacterial growth is mainly by fermentation. Within the acquifers themselves, bacterial growth is mainly by anaerobic respiration. Such growth will tend to reduce nitrate levels.

Groundwater accumulates in **aquifers** (underground pools and rivers) that can be tapped by drilling. A suitable well penetrates through a sufficient depth of soil to ensure proper filtration action (figure 12.7). It has a watertight casing that surrounds the actual well bore to prevent surface water from contaminating the groundwater. As long as the rate of removal of water from the well is balanced by the seepage of groundwater, groundwater represents a limitless, self-renewing resource. Unfortunately in recent years, the rate of groundwater withdrawals (100 billion gallons/day in the United States) has drastically exceeded the replenishing rate, and regions of the Great Plains are threatened with a lack of sufficient water to maintain present levels of agriculture.

Another problem that has arisen to threaten groundwater as a resource is the careless dumping of organic wastes. There has been significant contamination of groundwater supplies in various areas. Such

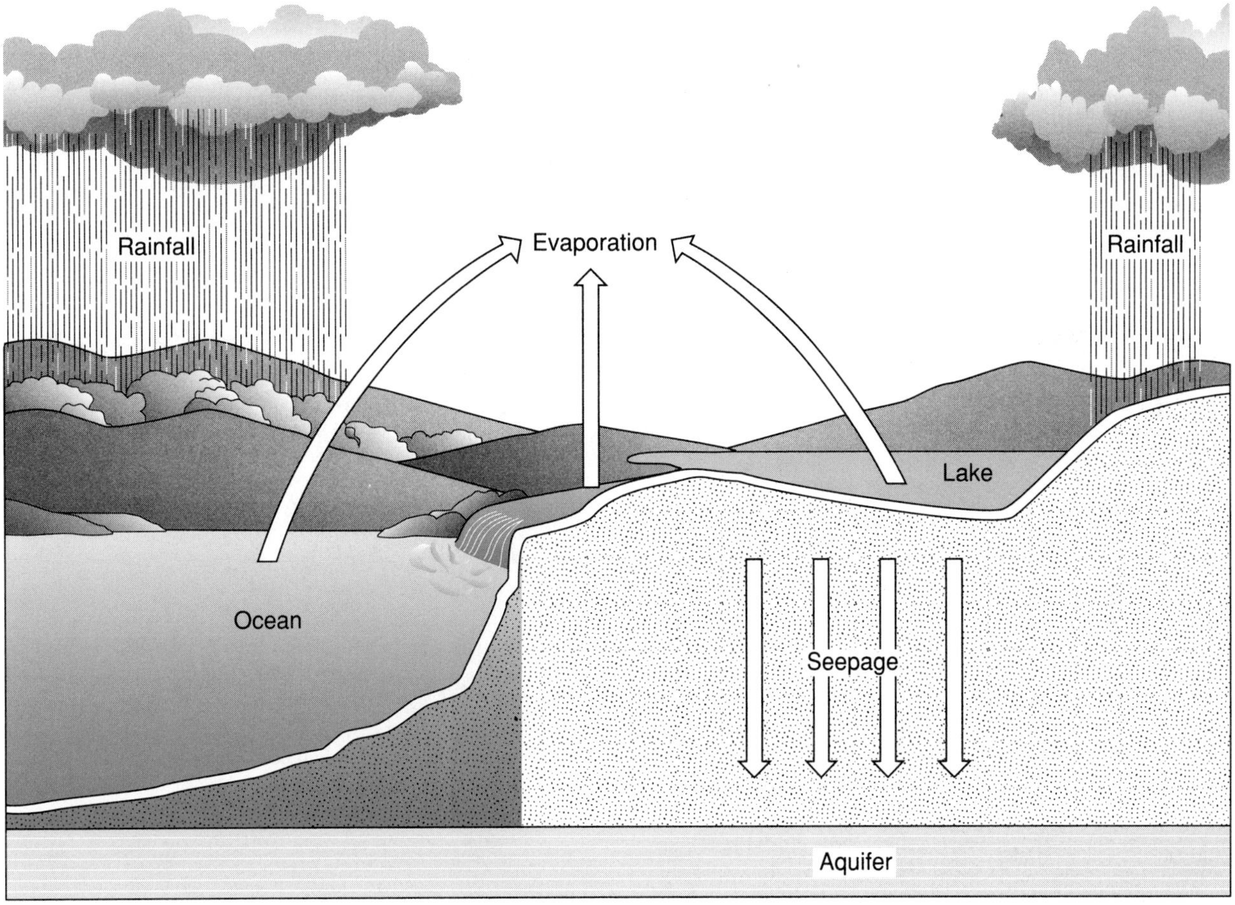

FIGURE 12.6

Natural water cycle. Evaporative loss of water from lakes, streams, and oceans (upward arrows) results in an automatic purification of water. When water falls on the land as rain or snow (downward arrows), it acquires only those impurities present in the air or on the land at that time. For this reason, mountain streams traditionally are considered to have very pure water. Groundwater is surface water that has seeped into the soil or through pores in rock surfaces and has collected in underground pools. The filtering action of the soil provides a general purifying effect, provided that the water sinks deep enough.

contamination can result from landfill sites in which water leaches through the accumulated waste before entering a local aquifer. Contamination also can arise when industries dispose of chemicals by pumping them into deep holes in the ground (dry wells). Fertilizers from large-scale agriculture are another source of contamination. The Office of Technology Assessment estimates that 260,000 tons of pesticides and 42 million tons of fertilizer are applied each year. Unless all these chemicals are digested by microorganisms or removed during harvesting of plants, they will eventually reach the ground- or surface water. To take a specific case, it has been estimated that 260,000 of the 400,000 tons of fertilizer applied in Denmark each year reaches the Baltic Sea. The growth of microorganisms eventually may help to repurify the water by decomposing the organic material, but this will not occur overnight. Meanwhile, the water quality deteriorates (color plate 6).

Sewage Treatment

Various municipalities have developed a partial solution to the problems of water contamination by treating various wastes before they can enter the water cycle. This may seem like a common-sense approach, but it is a relatively expensive proposition. Because of the cost, many municipalities, large and small, have not adopted complete treatment procedures, preferring instead to dump their wastes into a local body of water and assume

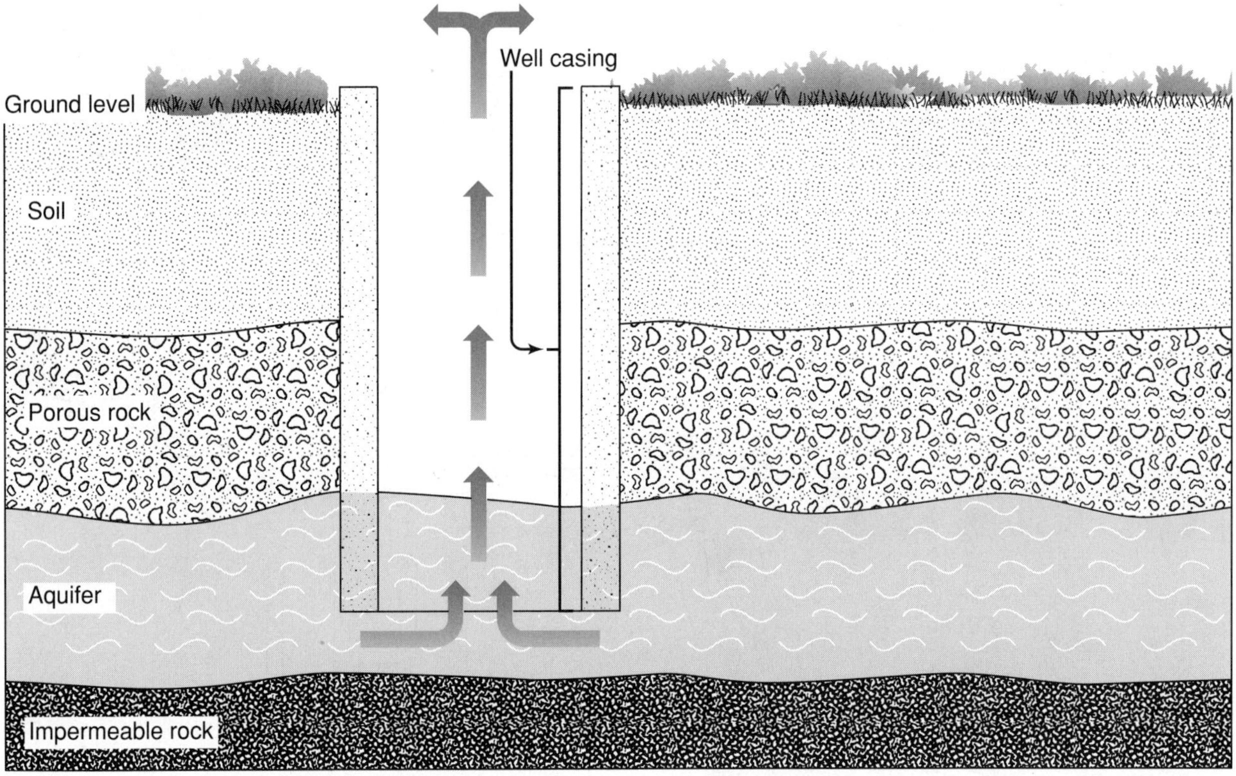

FIGURE 12.7

Typical well. The well is drilled into strata of rock in which water is known to collect. The well casing starts above ground level and extends down to the bottom of the area drilled in order to prevent surface water from entering the filtered water supply.

that the outgoing current will carry the problem elsewhere. For example, in 1983 the Centers for Disease Control reported that ">188,000,000 gallons of raw sewage are discharged daily into the Hudson and East Rivers of New York City." A sewage treatment plant capable of treating the remaining half of sewage from Manhattan was not completed until 1986. Another study in 1976 estimated that Hawaii was discharging 8×10^{10} enteroviral particles daily into the surrounding ocean. One priority of the Environmental Protection Agency is to promote the construction of appropriate sewage treatment facilities in all cities and towns that need them, although in recent years funding for this program has been cut back.

Good sewage treatment should be capable of producing water suitable for normal freshwater needs. All disease-causing microorganisms and hazardous chemicals should be removed and the level of biodegradable materials within the water should be at acceptable levels. One way to estimate the load of biodegradable materials (usually carbon-containing molecules) in water is to determine the **biochemical oxygen demand** (BOD). The BOD is the amount of oxygen necessary to support aerobic respiration of all organisms (chemoorganotrophs or chemolithotrophs) that can grow using the nutrients contained in the water. Obviously, as the amount of oxidizable organic material or nitrogen compounds increases, so does the BOD. In practice, BOD is measured from diluted samples of water incubated in sealed containers for a standard period of time, usually for 5 days (BOD_5).

Consider the simple case in which some water is heavily contaminated with glucose. If this water is poured into a stream, many bacteria will start to grow and use the glucose as a carbon and energy source. Some of these bacteria will be strict aerobes and some will be oxybionts (facultative anaerobes). Any anaerobes are of no concern. Because maximum energy is obtained via aerobic respiration, the oxybionts will use oxygen whenever possible, switching to a fermentative metabolism only when oxygen is in short supply. If the rate at which oxygen diffuses into the body of water is less than the rate at which it can be used by the aerobes and oxybionts, that portion of the water away from the actual air-water interface will become anaerobic. Clearly, the BOD is too high for the local stream conditions, and various consequences such as the death of most fish and the growth of anaerobic bacteria with their concomitant odoriferous by-products will ensue. If the excess of organic molecules is not too great, sometimes artificial aeration of the body of water with a pump may be sufficient to match the oxygen supply to the BOD and solve the problem. In the open ocean, colloidal carbon aggregates on bubbles of foam promote more rapid microbial growth.

On the other hand, if the organic molecules are at a lower level from the beginning, natural diffusion of oxygen into the water will permit both fish and bacteria to have all the oxygen they can use (BOD is less than the available oxygen). The remaining nutrients will be degraded gradually in accord with the various nutrient cycles outlined previously. An important goal in sewage treatment is to lower the oxidizable organic material and nitrogenous compounds (reduction in BOD).

A typical municipal sewage treatment plant is shown schematically in figure 12.8. Disposal pipes running from all houses and other liquid waste generators in a community are connected to larger pipes that eventually drain into the main treatment facility. The amount of liquid delivered can be quite substantial. A single large metropolitan facility can process 120 to 200 million liters (30 to 50 million gallons) of wastewater per day. The design of the system is such that the inflow of sewage cannot be stopped. If it were, waste would back up into homes. Consequently, if treatment is impossible, raw, untreated sewage must be dumped, usually into the nearest body of water. Current practice is to design plants with backup receiving tanks to avoid dumping of raw sewage.

The first step in the treatment process is the separation of liquid and solid materials. Really large solids, such as tree limbs and other types of debris, can be removed by coarse filters. The remaining suspension is then piped into settling tanks for some period of time. Various flocculent chemicals, such as aluminum or ferric hydroxide, can be added to hasten settling of solid materials. At the conclusion of the settling process, the liquid is separated from the **sludge** (settled solid material) and each is processed separately. Removal of solids from the sewage constitutes primary treatment and reduces the organic material in the water (as measured by BOD) by more than 40%.

Secondary treatment is intended to reduce the remaining organic material in the wastewater to acceptable levels and to eliminate any potentially harmful microorganisms. The two goals can be achieved simultaneously. The liquid is piped to an **aerobic digester,** such as an activated sludge tank or a trickling filter (figure 12.9 A–B). The trickling filter is a device that traditionally looks very much like an elaborate sprinkler system. In reality, however, it is a "biological fixed-film reactor." Surrounding the sprinkler system is a container approximately 2 m deep that is filled with angular rocks or other support material underlain with slightly larger rocks. A slow spray of the liquid over the rocks will encourage microorganisms to grow in a thin film and use all or nearly all the nutrients that may be present. In the activated sludge process, sewage is pumped into moderate-sized tanks that are vigorously aerated. Some of the solid material accumulated during the process (activated sludge) is used to inoculate the next batch of sewage. In either system of aerobic digestion, a variety of organisms, especially photosynthesizers, can use up phosphates and nitrogenous compounds. Aerobic bacteria like *Pseudomonas* degrade a wide range of organic compounds. A final filtration serves to remove the cells and other solid material from the treated sewage. When the trickling filter begins to clog from accumulated solid material, the solids are scraped into a container and added to the sludge of primary treatment.

The water that emerges from an aerobic digester has gone through a process that mimics the purification of surface water into groundwater, but the time scale has been accelerated greatly. If all has gone according to plan, the water is ready for discharge. Interestingly, if known disease-causing microorganisms, including viruses, are tracked through a sewage treatment plant, it is found that the combination of primary and secondary treatment also eliminates them along with the excess nutrients. Apparently, most are removed by flocculation ($>$ 97% of bacteria and viruses checked), and the remainder do not maintain themselves well in the presence of the flora that normally grow independently on the aerobic digester.

One way in which bacterial populations might be reduced is by predation from protozoa like *Paramecium* that grow on the aerobic digester and feed on bacteria. Large numbers of bacteria (about 1 to 10 million/ml) stimulate the feeding behavior of protozoan populations. During heavy predation, bacterial species, such as disease-causing organisms, normally present as a

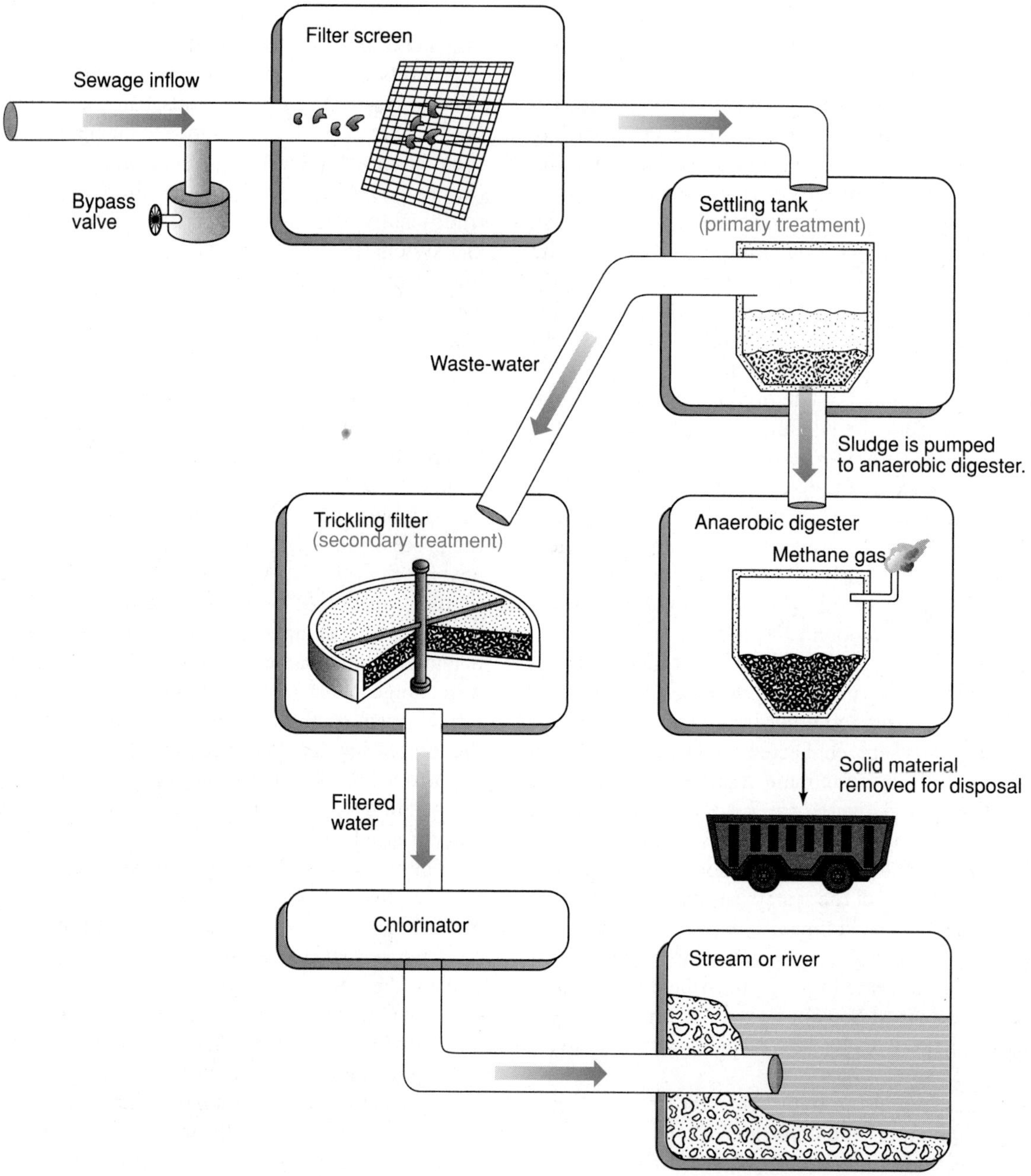

FIGURE 12.8

Schematic diagram of typical sewage treatment plant.

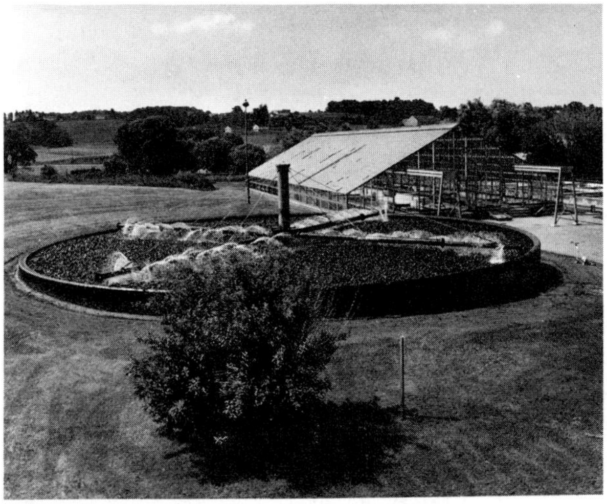

A

B

FIGURE 12.9

Aerobic sewage digesters (secondary sewage treatment).
(**A**) Trickling filter with sludge drying pits in background.
(**B**) Activated sludge tank.

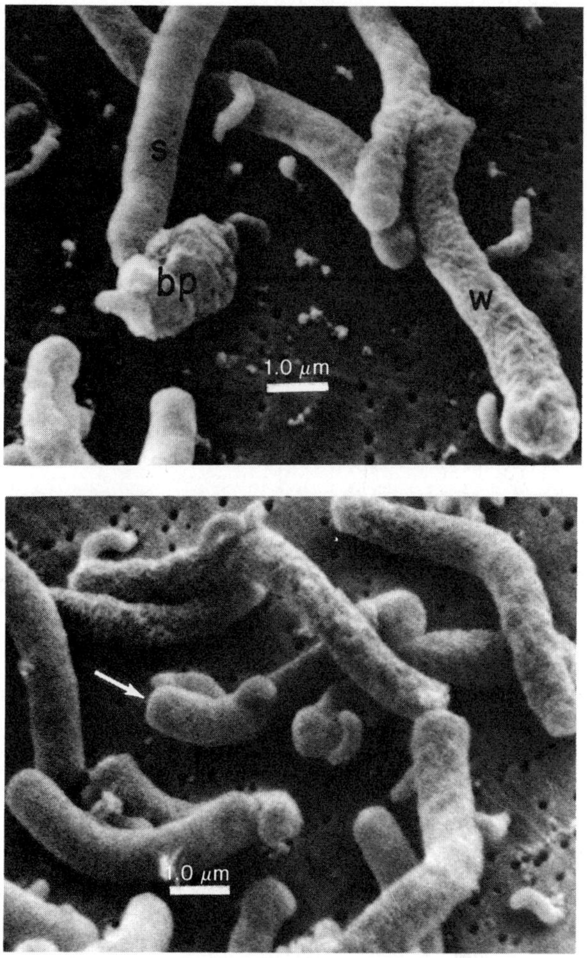

FIGURE 12.10

Bdellovibrio bacteriovorus (arrow) attacking *Spirillum serpens.*

Reprinted from Snellen, J. S., and M. P. Starr. 1976. Alterations in the cell wall of *Spirillum serpens* VHL early in its association with *Bdellovibrio bacteriovorus* 109D. *Archives of Microbiology* 108:55–64.

small percentage of the population can be eliminated totally. *Bdellovibrio* is a prokaryote that preys on other prokaryotes (figure 12.10). It too can be expected to reduce the population size of rare organisms.

Circumstances that can make the effluent from secondary treatment unacceptable in the environment are usually peculiar to specific situations. For instance, if local industries have a process whereby significant quantities of arsenic are released, the wastewater may have enough arsenic in it to violate water-quality standards. Nothing in primary or secondary treatment is designed to handle such problems. Some form of tertiary treatment must be used to solve them. Tertiary treatments are designed to cope with specific problems and they vary in the technology used; however, they are all expensive. For this reason, municipalities may require waste generators to pretreat their own sewage rather than establish municipal tertiary treatment facilities.

As a final safeguard, chlorine or ozone gas is added to the effluent. These strong oxidizers are disinfectants (destroy infectious microorganisms) that can inactivate any harmful bacteria or viruses remaining (see chapter 13). Ozone imparts no extra taste to the water, but it is substantially more expensive to use than chlorine. The organic material that remains is at a low level; therefore, the BOD of the stream or lake receiving the discharge will not be significantly affected.

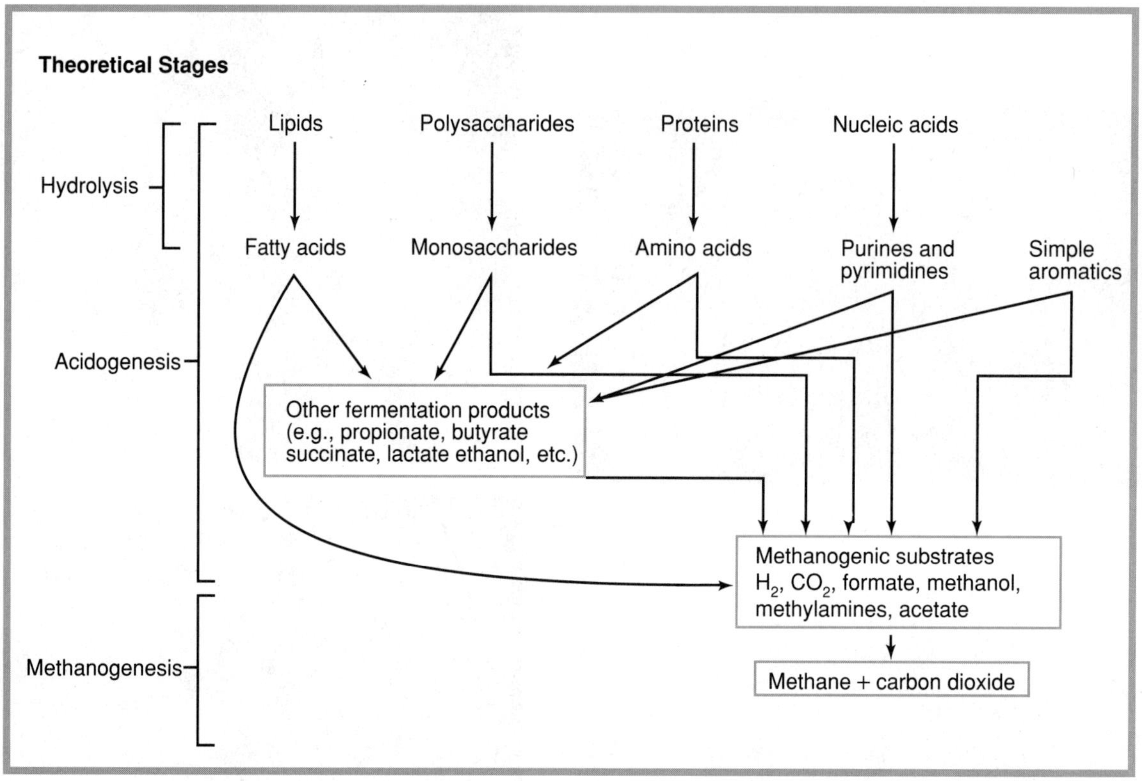

FIGURE 12.11

Patterns of carbon flow in anaerobic digestion.

From Holland, K. T., Knapp, J. S., and Shoesmith, J. G. (1987). *Anaerobic Bacteria,* Blackie, Glasgow and London. Reprinted by permission.

Effluent from a sewage treatment plant need not be returned to the local body of water because it can be used for other purposes. Some alternatives include irrigation of golf courses, reuse as drinking water, or use as cooling water for a nuclear electricity generating facility.

The sludge and accumulated solids remaining from the primary and secondary treatment phases can be assumed to have numerous harmful microorganisms and a high-nutrient level. They are treated by a process of anaerobic digestion. Treatment takes place in large tanks and can be administered in batches or with a continuous flow system. The basics of the process are the same in either case and are highly cooperative, requiring the interaction of many organisms. Strictly anaerobic organisms such as *Clostridium* are encouraged to grow and break down the complex nutrients into simpler forms (figure 12.11). Using these breakdown products, methanogenic bacteria, for example, members of *Methanobacterium,* will produce significant quantities of methane gas. Up to 50% of the total carbon can be converted to methane in the 10 to 30 days required for digestion. If the facility has been set up to take advantage of this production, methane gas can be used to provide power for most of the pumps and other equipment necessary to the operation of a sewage treatment plant. For example, a sewage treatment facility operated by one rum manufacturer in Puerto Rico is reported to produce about 2 million cubic feet of methane gas per day. This is enough to supply 40% of the power needed by the entire plant.

Anaerobic digestion does not eliminate all the solid material, just simplifies it chemically. The resulting stabilized sludge makes a good fertilizer and is sometimes sold for that purpose. Any danger from disease-causing microorganisms can be eliminated if the sludge is heated during or after processing to kill these microorganisms. Some heat is generated by the metabo-

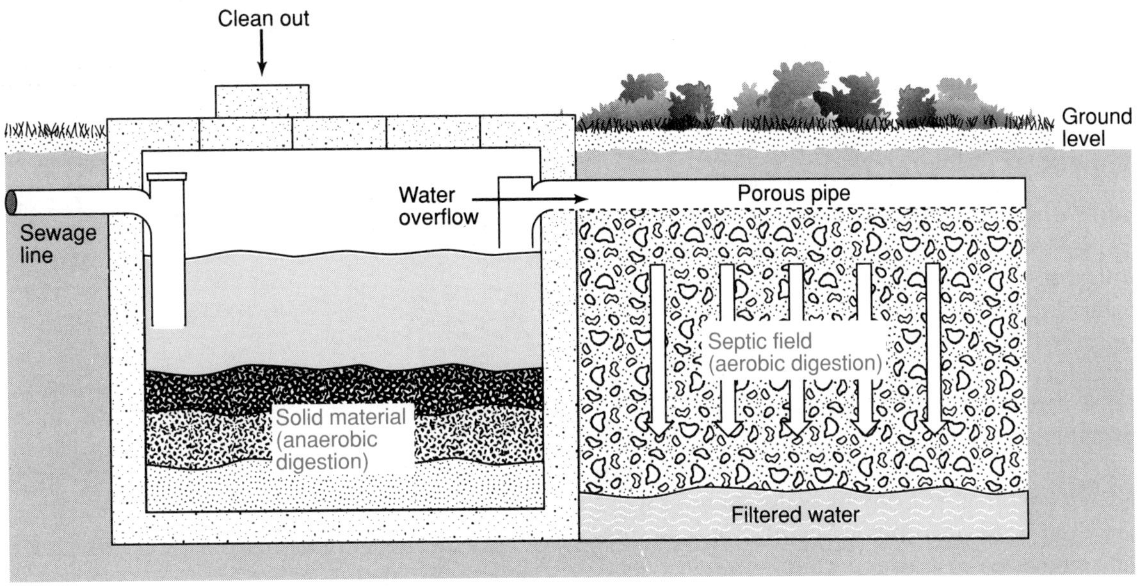

FIGURE 12.12
Schematic diagram of a septic tank. This system performs most functions of a sewage treatment plant. The tank acts as the site for primary treatment, with aerobic digestion occurring in the septic field. Solid wastes are pumped out on a regular basis and taken elsewhere for appropriate disposal.

lism of anaerobic bacteria in the digester, and additional heat can be supplied exogenously. There is, however, the possibility of accumulated heavy metals and other toxic compounds to consider.

Because sewage treatment is almost entirely accomplished by microorganisms, it is quite sensitive to environmental changes. When heavy-duty laundry detergents were first introduced, no thought was given to their disposal. It turned out that the detergent molecules could not be broken down by normal organisms used in sewage treatment; therefore, detergents were left in the effluent from treatment plants. As they entered the various waterways, they continued to suds, making many rivers and streams most unappetizing. Public clamor resulted in the passage of laws requiring all detergents to be biodegradable, a fact that is still noted on their labels.

In a similar vein, many detergents work better in the presence of sodium phosphate. At one time, the amount of phosphates added to wastewater by household detergents exceeded the capacity of secondary treatment facilities to remove these phosphates and maintain normal water flow rates simultaneously. Consequently, enhanced growth of algae in lakes and streams resulted. Eventually, detergents on the market were produced with no or low amounts of phosphate. Once again, a note was added to detergent labels indicating the amount of phosphate present.

Other problems can arise from the presence of toxic and/or carcinogenic compounds in raw sewage. Chemicals like toluene and benzene that are lipid solvents are obvious troublemakers. Chapter 13 introduces large families of compounds that inhibit bacterial growth. An appreciable concentration of any toxic molecule can retard the sewage treatment process; therefore, the Environmental Protection Agency and many municipalities have stringent rules governing the disposal of chemicals into a public sewer.

When large-scale sewage treatment facilities cannot be justified, an appropriate solution to waste disposal is a septic tank (figure 12.12). Septic tanks are specially fabricated with no bottom seams, and they come in various sizes. Most are used to serve single dwellings. Water and solids are piped into the tank in which the solids settle to the bottom. The water with its dissolved wastes overflows from the tank into porous pipes laid under the surface of the ground. It then trickles through the ground, stimulating the growth of plants and microorganisms. As long as the inflow of water is not too great and the ground remains porous, oxidation will occur, and the BOD of the wastewater will be reduced to reasonable levels before it runs off into a body of water or reaches the groundwater. The solid material remaining in the tank is continuously digested by anaerobic bacteria and eventually pumped out for disposal at a landfill site or sewage treatment plant.

Treatment of Drinking Water

In most cases, preparation of drinking water is much simpler than sewage treatment because it is primarily a physical process. The source of water can be a lake, a stream, or a well. At this time, water generally is not taken directly from a sewage treatment facility. In practice, however, cities whose water intakes are situated downstream from other sewage treatment plants are essentially recycling water generated by these facilities. Fortunately, most bacteria whose presence is considered abnormal in drinking water do not survive well in open water. A combination of protozoan predation and nutrient deprivation causes a rapid decline in their population.

Drinking water from open bodies of water is clarified by a flocculation and filtration process that, like aerobic digestion, is intended to mimic the natural filtration of groundwater (figure 12.13 A–B). After filtration, the water is pressurized for distribution into water mains. Pressurization can be produced directly by pumps or indirectly by storage of water in elevated tanks. As water enters the mains, chlorine gas is added to kill any microorganisms that may be present. The object is to maintain a residual chlorine level of 3 to 5 ppm (parts per million) out to the end of the main. This provides a safety margin in the event of a cracked pipe. If the water is particularly muddy, has a large amount of organic matter present, or is infested with certain protozoan parasites like *Giardia,* it may be necessary to increase the chlorine concentration. The pH of the water should be near neutrality because highly basic solutions antagonize the chlorine.

Chlorination is both a benefit and a hazard. The benefits are obvious, the hazards less so. Water these days is very likely to contain trace amounts of organic material, particularly after being used in a city. If each cycle of wastewater or drinking water treatment includes chlorination, the level of chlorinated hydrocarbons, such as chloroform, will build up gradually in the water. In the case of a river like the Mississippi, from which the water is reused many times, it can be shown that the chlorinated hydrocarbon level correlates with the water's movement along the river. New Orleans receives water relatively high in chlorinated hydrocarbons while St. Louis receives water with lower concentrations. Because chlorinated hydrocarbons are thought to increase the risk of cancer, the potential for long-term hazardous effects clearly is present. Under consideration are proposals to require that drinking water be filtered through charcoal to adsorb organic molecules.

A

B

FIGURE 12.13

Filtration at a water treatment plant. (**A**) Municipal facility serving about 75,000 customers. (**B**) Flocculation basins.

Photographs courtesy of Harry A. Meyer, Water Production Supervisor, Tempe, Arizona.

Other treatments can be used to kill microorganisms in drinking water (see chapter 13), but none are as effective and cheap as chlorination. Ozone is the most popular treatment system next to chlorination. Although it leaves no aftertaste, it is substantially more expensive. It is a favorite of bottled water manufacturers.

Water from private wells usually is not treated, except possibly for simple filtration to remove excessive amounts of iron or other minerals. The general assumption is that the aquifer has safe water and that the well casing is intact and will prevent any contamination. Nevertheless, it is always a good idea to have private wells tested occasionally to verify that all is working as it should. This is especially true in areas in which dry wells have been used for waste disposal or in areas in which leaching of landfill sites is known to occur. Campers can now purchase a small filter apparatus that can remove all microorganisms from water; however, they will not necessarily be protected against chemical contamination.

Microbial Symbioses

Symbiosis refers to two (or more) organisms living together. The relationship can be characterized further according to the nature of their interactions. An endosymbiont is a microorganism that lives within the cytoplasm of a eukaryote. If both members derive benefit from the relationship, it is said to be a **mutualism.** On the other hand, a **parasitism** is a relationship in which one member derives benefit from the relationship while the other suffers harm. Technically speaking, there may be such a thing as a **neutralism,** although true neutrality is difficult to establish experimentally. **Commensal** organisms are those that live in close physical association. Both may benefit from the association, neither may benefit, or only one member of the commensalism may benefit. The focus of the remainder of this chapter is on commensalism and mutualism. Topics related to parasitism are discussed in chapter 17.

Examples of Mutualistic Relationships

Chapter 2 introduces one very common mutualism, lichen symbiosis (color plate 5A–B). This composite organism uses its fungal component to provide anchorage and protection, while the algal or cyanobacterial component provides food for both itself and the fungus via photosynthesis. Because neither organism can live by itself in the same harsh environment as the composite, the relationship is clearly mutualistic.

The symbiotic relationship between members of the genera *Rhizobium* and *Bradyrhizobium* and various legumes (peas, beans, alfalfa, for example) also warrants attention. Legumes having small lumps or nodules on their roots have been observed to grow substantially better than legumes that do not (figure 12.14). The nodulation process begins when rhizobia found growing in the soil invade a plant as a rootlet comes near. The bacteria stimulate the plant cells to establish an **infection thread,** a tube through which the bacteria can gradually move further into the plant.

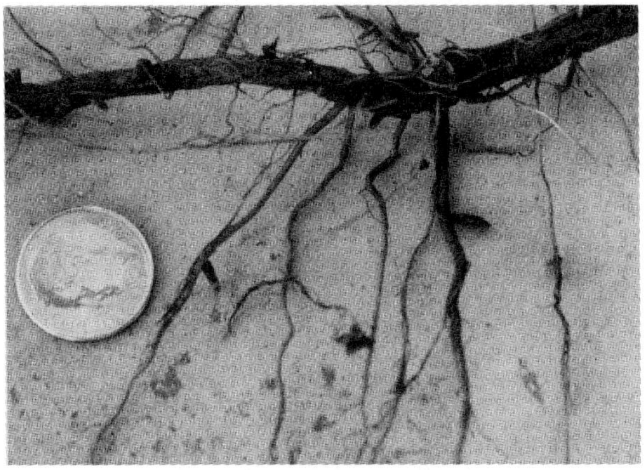

FIGURE 12.14
Root nodules on a pea plant. The dime indicates their relative size.

Eventually, the bacteria produce enzymes that degrade the walls of the infection thread, allowing them to reach the plant cells. They invade tetraploid cells (those with four copies of each chromosome) and take up residence as endosymbionts. In the process, they differentiate into **bacteroids,** prokaryotic cells that have an altered appearance and cannot live independently in their present form (figure 12.15). The bacteroid, however, retains the nitrogenase enzyme from the bacterial cell from which it came. Bacteroids stimulate tetraploid cells of the legume to proliferate and form the small nodules visible on the plant's roots. Within these nodules, nitrogen fixation occurs using food stores provided by the plant and enzymes provided by the bacteroids. Anaerobic conditions are maintained easily within the cytoplasm of the nonphotosynthetic plant cells. In fact, the plant cells contain a special form of hemoglobin called **leghemoglobin** whose apparent function is to transport bound oxygen to the bacteroids for metabolic purposes without adversely affecting the nitrogenase enzyme. It is thought that the plant cells contribute the globin portion of the molecule while the bacteroids add the heme.

Rhizobial symbiosis is the best-studied example of nitrogen fixation by mutualism, but it is certainly not the only one. Nonleguminous shrubs and trees of many types can set up similar associations with members of the genus *Frankia.* These organisms are so difficult to grow in the laboratory that little is known about their interactions with plant cells.

Other kinds of plant-microorganism interactions are possible. **Mycorrhizae** are mutualistic associations between various fungi and roots of certain trees. These associations have been shown to enhance the ability of

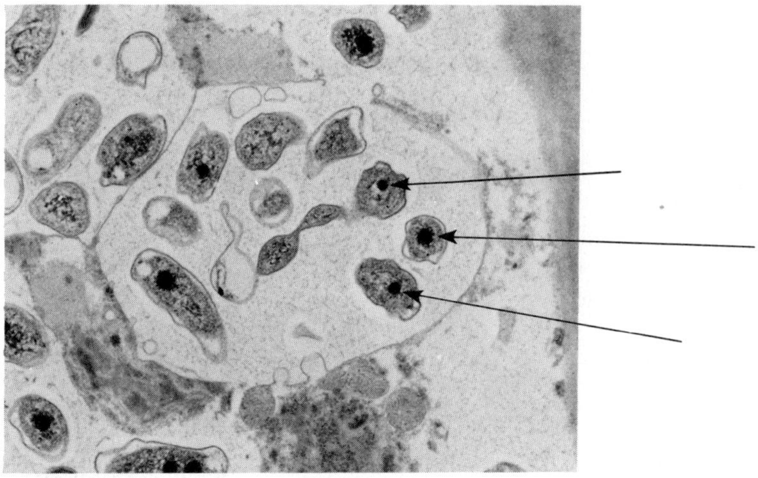

FIGURE 12.15

Bacteroids inside a legume cell. Arrows point to some bacteroids. These thin sections of soybean root nodules were prepared by Dr. John Burke, presently of the United States Department of Agriculture, Lubbock, Texas.

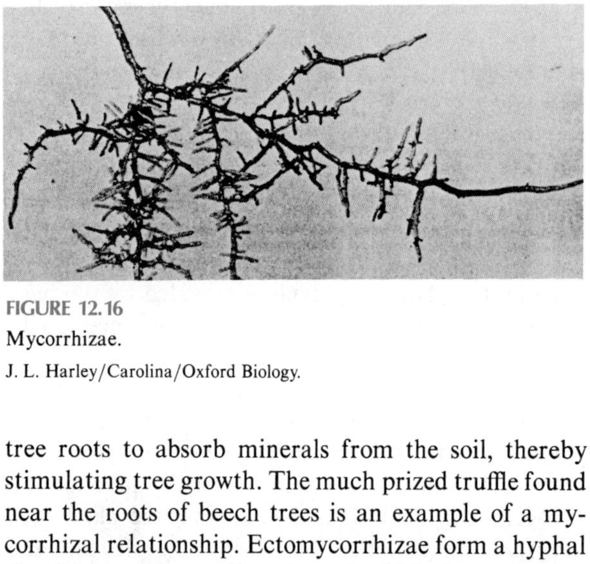

FIGURE 12.16

Mycorrhizae.

J. L. Harley/Carolina/Oxford Biology.

tree roots to absorb minerals from the soil, thereby stimulating tree growth. The much prized truffle found near the roots of beech trees is an example of a mycorrhizal relationship. Ectomycorrhizae form a hyphal sheath around a root fiber through which the root cortex is invaded by the mycelium (figure 12.16). With endomycorrhizae, on the other hand, the fungus penetrates into the root (inter- or intracellularly) as well as spreading into the surrounding soil (figure 2.14).

Examples of Commensal Relationships

Associations between animals and microorganisms have proved to be of considerable significance. Take for example the case of the common termite. This organism is a much feared pest because its ability to eat wood and, over a period of time, completely ruin a house. Yet from the point of view of physiology, the termite is absolutely incapable of deriving significant nourishment from the wood it eats. Instead, it depends on a number of commensal organisms, such as flagellated protozoa like *Trichonympha sphaerica* and bacteria like *Micromonospora propionici,* to break down the cellulose and lignin in the wood fibers and convert it into usable sugars.

On a larger scale, a similar phenomenon can be seen with respect to the ruminants, cud-chewing animals such as cows, camels, and deer. Their digestive tracts are modified so that a **rumen** or special pouch receives ingested food before it enters the stomach (figure 12.17). When cattle graze, they ingest large quantities of vegetation, generally of rather poor quality. After the material is swallowed, it moves through the esophagus and enters the rumen. Within this pouch, numerous anaerobic bacteria ($\sim 10^{10}$/ml) and some protozoans ($\sim 10^{6}$/ml) act to break down ingested plant fibers and use their components to grow in a manner analogous to the sewage anaerobic digester. The ruminant facilitates this process by chewing its cud (regurgitated material from the rumen) so that the microorganisms are thoroughly mixed with the food.

When swallowed again, the food may be returned to the rumen or passed on to the omasum. Once it enters the omasum, true digestion begins from the point of view of the animal. Digestive enzymes finish degrading

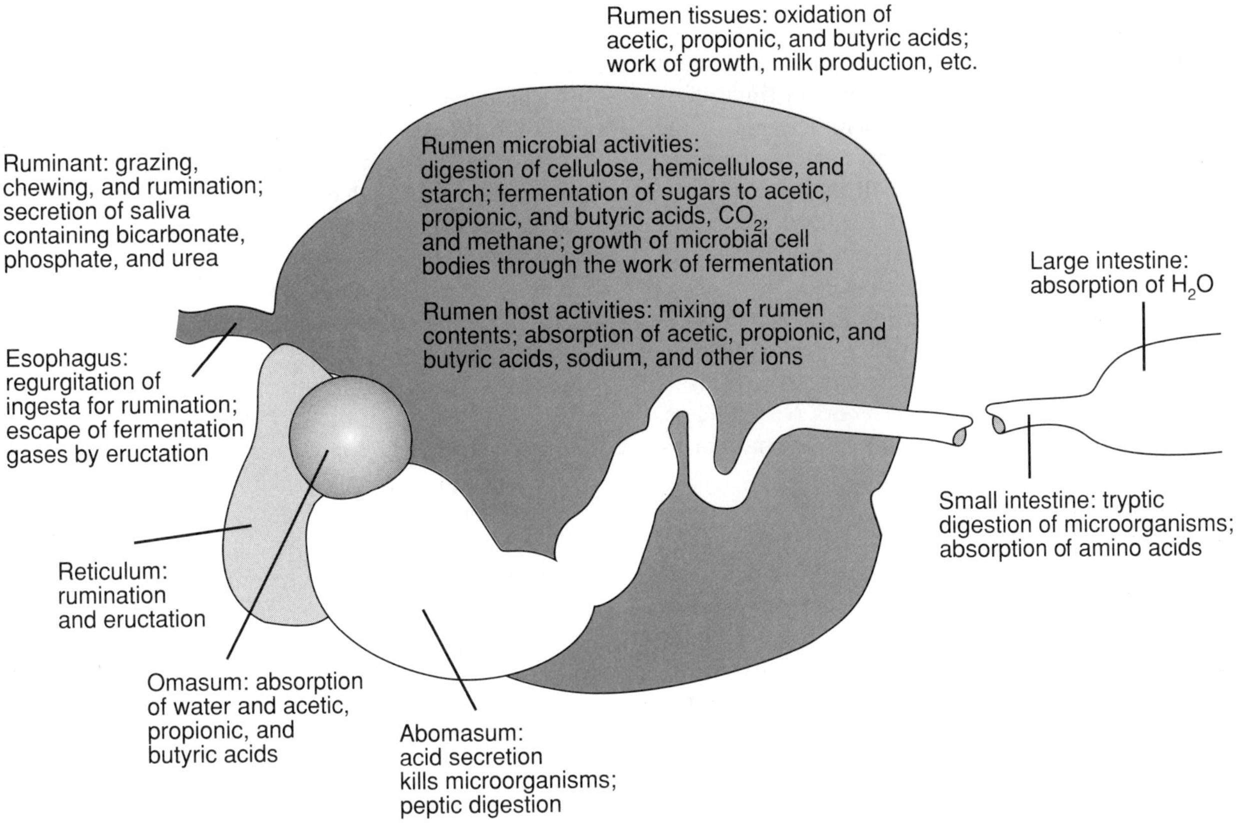

FIGURE 12.17

Ruminant anatomy. Instead of a simple stomach, a ruminant has a complex set of structures for digestion and bacterial culture. The specific activities associated with each area are listed. Eructation is belching.

From R. E. Hungate, "Anaerobic Biotransformations," *Bacteria in Nature*, Vol. 1, pp. 39–95, edited by E. R. Leadbetter and J. S. Poindexter.

the plant material and also decompose the bacteria, releasing their proteins and vitamins. These nutrients, in addition to fatty acids (by-products of fermentation) absorbed from the rumen, nourish the animal and account for its ability to survive in regions in which protein-containing foods are scarce. The rumen is a very efficient converter of biomass. More than 80% of ingested cellulose is converted to fatty acids, 10% to microbial biomass (amino acids, etc.), and only 5% to waste in the form of heat.

Bacterial genera found in the rumen include many specialists such as *Ruminococcus* and also some more familiar genera such as *Streptococcus, Bacteroides,* and *Methanobacterium.* With respect to *Methanobacterium,* it has been estimated that bacteria from this genus, present in the guts of termites as well as ruminants, are significant contributors to the 4×10^9 metric tons of biogenic gas normally present in the atmosphere. In fact, a bloating disease of cows occurs if a cow ingests material that foams enough to prevent the escape of gases produced during fermentation. Enough pressure can accumulate to rupture the rumen and cause the death of the cow.

Various adaptations are necessary in order for rumen symbiosis to persist. Unlike human saliva, a cow's saliva contains no digestive enzymes, for these might interfere with the activities of the bacteria. Behavioral patterns, such as the consumption of food regurgitated by adults, have also developed to ensure that a young ruminant is provided with suitable microorganisms for its digestive tract.

The human body offers some excellent niches for commensal bacterial habitation. Skin serves as a host to many different bacteria. *Staphylococcus epidermidis* is generally present on the surface, and anaerobes like members of *Propionibacterium* are found inside the pores. Within the oral cavity, members of

Streptococcus, Lactobacillus, Corynebacterium, Haemophilus, Neisseria, and several other genera are found. The lower portion of the intestinal tract offers several areas of excellent habitation. The pH of the human stomach is too acid for most bacteria, however, and the upper portion of the small intestine also has few inhabitants, primarily due to the flow rate of partially digested food and to the abrupt pH changes. In the ileum of the small intestine, bacteria begin to accumulate, and within the large intestine they are numerous. There may be as many as 10^{11} bacteria/g of feces in a normal individual. Most of these organisms are strict anaerobes like *Bacteroides*. Nevertheless, many facultative organisms are present, the most prevalent of which is *Escherichia coli*.

In addition to commensal bacteria preventing potentially harmful organisms from colonizing the human body, there is some indication of actual symbiosis between mammals and their intestinal bacteria. Most of the evidence comes from work done with **gnotobiotes** or "germfree" animals. These are animals delivered by cesarian section and maintained in sterile incubators for the rest of their lives. Based on nutritional studies, it appears that certain vitamins (K, perhaps B_1) can be supplied by intestinal bacteria.

Many other examples of microbial symbioses are known, and more are certain to be discovered. This is a very active research area in microbiology, and many important problems remain to be solved. The reference list at the end of this chapter offers some suggestions for further reading in microbial symbioses.

Summary

Microorganisms do not live in isolation. They are part of the global biosphere, and as such they interact with each other and with all other organisms. Sometimes, the relationships between organisms are relatively permanent, as is the case in plant root nodules, rumen symbiosis, and intestinal bacteria. Other times, the relationships are temporary, as when bacteria serve as prey for various protozoa. Transitory interactions have also been suggested, but by their very nature these types of relationships are difficult to verify experimentally.

The habitats of microorganisms can be exceedingly diverse owing to their small size. Bacteria are found bound to soil particles or freely floating in water. They can be single cells, small colonies, or long filaments. The best way to determine what organisms are present is by direct microscopic examination.

Microorganisms fill many roles in the constant recycling of nutrients. Cyanobacteria use energy from the sun to produce some of the free oxygen necessary for animal life. *Rhizobium* and *Azotobacter* are examples of organisms that can fix nitrogen. Other organisms figure prominently as decomposers. Fungi, funguslike bacteria (actinomycetes), clostridia, and numerous other bacteria are involved in the breakdown of macromolecules and in their conversion to simpler substances that can be used in energy and biosynthetic metabolism.

Humans derive significant benefit from these microbial activities either directly or indirectly. Disposal of wastes generated by modern civilization absolutely requires naturally occurring microbial activities. Solid wastes are buried in landfill sites in which soil organisms are free to attack organic molecules. With time, these molecules are decomposed and completely recycled into new protoplasm. Liquid wastes formerly were disposed of by simple dumping into local bodies of water. As pollution levels have risen, so have pressures for adequate sewage treatment. All sewage treatment, whether on a municipal or individual level, allows plants and microorganisms to grow and use nutrients contained in the wastewater. When this is carried out properly, effluent from a sewage treatment plant is suitable for immediate reuse.

Preparation of drinking water for delivery to consumers is one of the few processes not based on microbial action. Instead, it is primarily a filtration process coupled with a final chlorination step to ensure that no harmful organisms are present. Water purification in nature is much slower and is based either on evaporation or filtration through layers of rock and soil to aquifers located as much as hundreds of meters underground.

Questions for Review and Discussion

1. Describe some different types of symbiotic relationships that are possible and give an example of each. What specialized adaptations (if any) are present in symbionts to facilitate their interaction?
2. Give some examples of economically important symbioses. How do humans benefit from these interactions?
3. Trace the cycling of either sulfur or nitrogen through the biosphere. What kinds of organisms might catalyze the various steps of the cycle?

4. What natural processes act to remove pollutants from our environment? Why do you think that pollution seems more abundant today than a decade ago? How might microbiologists be able to contribute to the solution of pollution problems?
5. How does the sewage treatment process manage to eliminate disease-causing organisms and viruses? What safeguard is used generally to ensure that all microorganisms are dead before the water is released?
6. What is biochemical oxygen demand (BOD)? Why is it important to those people who manage our waterways and recreational areas? What are some methods used to reduce BOD? What types of activities can increase BOD?

Suggestions for Further Reading

Ahmadjian, V., and S. Paracer. 1986. *Symbiosis. An introduction to biological associations.* University Press of New England, Hanover, N.H.

Beevers, L., and R. H. Hageman. 1983. Uptake and reduction of nitrate: Bacteria and higher plants, pp. 352–375. *In* A. Läuchli and R. L. Bieleski (eds.), *Inorganic plant nutrition.* Springer-Verlag KG, Berlin.

Brock, T. D. 1987. The study of microorganisms in situ: Progress and problems, pp. 1–17. *In* M. Fletcher, T. R. G. Gray, and J. G. Jones (eds.), *Ecology of microbial communities.* Cambridge University Press, Cambridge. An interesting and personal analysis of what is wrong (and right) about research in microbial ecology today.

Forster, C. F. 1985. *Biotechnology and wastewater treatment.* Cambridge University Press, Cambridge. Intended for the civil engineer, this book details the appropriate design and operation of sewage treatment facilities.

Ghiorse, W. C., and J. T. Wilson. 1988. Microbial ecology of the terrestrial subsurface. *Advances in Applied Microbiology* 33:107–172. Tables at the back of this article supply the natural decay rates for a wide variety of chemical compounds.

Kepkay, P. E., and B. D. Johnson. 1989. Coagulation on bubbles allows microbial respiration of oceanic dissolved organic carbon. *Nature* 338:63–65.

Leadbetter, E. R., and J. S. Poindexter (eds.). 1985. *Bacteria in nature.* Plenum Publishing Corp., New York. A very nice collection of short essays written by noted workers in their fields and intended as introductions to the various subject areas.

Lidstrom, M. E. 1989. Molecular approaches to problems in biogeochemical cycling. *Antonie van Leeuwenhoek Journal of Microbiology* 55:7–14. A summary of how to identify bacteria in field samples and correlate their presence with ecological activities.

Lipschultz, F., O. C. Azfirious, S. C. Wofsy, M. B. McElroy, F. W. Valois, and S. W. Watson. 1981. Production of NO and N_2O by soil nitrifying bacteria. *Nature* 294:641–643.

Morita, R. Y. 1988. Bioavailability of energy and its relationship to growth and starvation survival in nature. *Canadian Journal of Microbiology* 34:436–441.

Nedwell, D. B., and T. R. G. Gray. 1987. Soils and sediments as matrices for microbial growth, pp. 21–54. *In* M. Fletcher, T. R. G. Gray, and J. G. Jones (eds.), *Ecology of microbial communities.* Cambridge University Press, Cambridge. A more detailed discussion of the points raised at the beginning of this chapter.

Paul, E. A., and F. R. Clark. 1989. *Soil microbiology and biochemistry.* Academic Press, Inc., San Diego. A short text with good discussions about nutrient cycles and detailed descriptions of nitrogen-fixing symbioses.

Reineke, W., and H. Knackmuss. 1988. Microbial degradation of haloaromatics. *Annual Review of Microbiology* 42:263–287. A summary of the biochemistry of these processes and how to isolate appropriate strains from nature.

Scheuerman, P.R., J. P. Schmidt, and M. Alexander. 1988. Factors affecting the survival and growth of bacteria introduced into lake water. *Archives of Microbiology* 150:320–325.

Senior, E. (ed.). 1990. *Microbiology of landfill sites.* CRC Press, Inc., Boca Raton Fla. A short collection of articles on specific aspects of commercial scale landfill operation. One chapter deals with the possibility of methane gas production for commercial use.

Singh, J. S., A. S. Raghubanshi, R. S. Singh, and S. C. Srivastava. 1989. Microbial biomass acts as a source of plant nutrients in dry tropical forest and savanna. *Nature* 338:499–500.

Sprent, J. I., and P. Sprent. 1990. *Nitrogen fixing organisms. Pure and applied aspects.* Chapman & Hall, New York. Appendices include some information on biochemistry and genetics.

Stolp, H. 1988. *Microbial ecology: Organisms, habitats, activities.* Cambridge University Press, New York. A concise but thorough introduction to microbial ecology.

chapter

12

Controlling Microbial Growth

chapter outline

■

GENERAL GOAL

Be able to describe how various antimicrobial treatments work and in which situations they should and should not be used.

■ In the previous chapter, possible interactions between microorganisms and their environment or between microorganisms and the rest of the biological world were highlighted. Normal functioning of the biosphere depends on these interactions. Nevertheless, there are instances in which microorganisms are not wanted. This chapter introduces the huge industry devoted to the various aspects of controlling the growth of microorganisms. Ways in which microorganisms can be eliminated or inhibited in growth in specific circumstances or places are discussed. Included in that discussion are antibiotic use and production.

Antimicrobial Treatments

After much discussion has been devoted to the promotion of microbial growth, it may seem odd to consider how that growth can be prevented. Nevertheless, this is an extremely important activity. Much of the pharmaceutical industry is based on the ability to produce drugs and solutions that are free from microbial contamination. The long-term storage of foods is possible only when some sort of protective treatment is used. Even an everyday item like mascara requires a preservative to prevent harmful bacteria from growing in it. Many of the commonly used methods of antimicrobial treatment are introduced in this section.

There first needs to be an understanding of the notion of sterility. When something is **sterile,** it lacks viable cells of any sort, prokaryotic or eukaryotic. In addition, any spores, viruses, viroids, virusoids or prions that may be present are inactivated completely. Note that cells still may be present, but they are incapable of growth at any future time. This point occasionally has led to some confusion in quality-control laboratories. A microscopic examination can reveal the presence of normal-appearing bacteria in a sterile solution, but attempts to culture the bacteria have proved unsuccessful. Although the solution is sterile, it obviously had been contaminated previously.

The key element in the definition of sterility is that of "failure to grow." Accordingly, the appropriate way to test sterility is to attempt to culture cells or viruses from the sterilized item in question. This plan, however, produces some difficulties. Microorganisms have a wide variety of habitats, and culture conditions suitable for one organism can be totally unsuitable for another. Therefore, what set of conditions is appropriate to use in testing sterility?

Tests for noncellular entities obviously require that appropriate cells be supplied as hosts because many viruses are very host specific. Similarly, media suitable for bacteria might not be suitable for the growth of a protozoan, yet anything that is sterile should be lacking these items as well. Microbiologists take advantage of the fact that certain bacterial endospores are the most durable structures in the entire biological kingdom, as far as we know. Frequently, small containers of particularly stable bacterial spores, such as those from *Bacillus stearothermophilus,* are introduced into a system before a sterilization process is carried out. If subsequent testing demonstrates that the spores have been killed, it is reasonable to assume that all other biological entities also have been killed or inactivated. For critical testing, however, it is prudent for more than one sample of the control organism to be used, and the various samples should be located within different regions of the material to be sterilized.

The effectiveness of a treatment that kills bacteria can easily be demonstrated by means of a graph, such as that presented in figure 13.1. The curves represent the effect of increasing doses of singlet oxygen ($^1O_2$) on cell viability. Note that the percentage of surviving cells is expressed as a logarithm. This allows one graph to illustrate the killing of a very large number of cells. As in the case of the bacterial growth curve, the use of logarithms has converted an exponential function into a linear one. The linearity of the graph indicates that each equivalent dose of singlet oxygen kills a constant percentage of the cell population.

Because even the smallest dose of singlet oxygen results in cell death for this streptococcal species, the general interpretation is that all doses of singlet oxygen are potentially lethal to a few cells at least, provided that the culture is sufficiently dense. Each additional dose results in greater cell death. This is known as **single-hit kinetics** because the cells are behaving as though they are being struck by "particles," any one of which will kill a cell when it hits. The number of these "particles" is then proportional to the dose rate. Multiple-hit kinetics, found in the *Salmonella* strains tested, occur when each cell must be struck by more than one "particle" for any killing to occur. The killing curve is initially flat until the cells build up the required number of hits, and then it falls off rapidly.

Extrapolation of linear graphs obtained in these experiments leads to a seeming contradiction. Eventually, the graph predicts that only a fraction of a cell remains, which is clearly a physical impossibility. The appropriate interpretation of a prediction of 0.1 cell remaining is that there is only 1 chance in 10 that a sur-

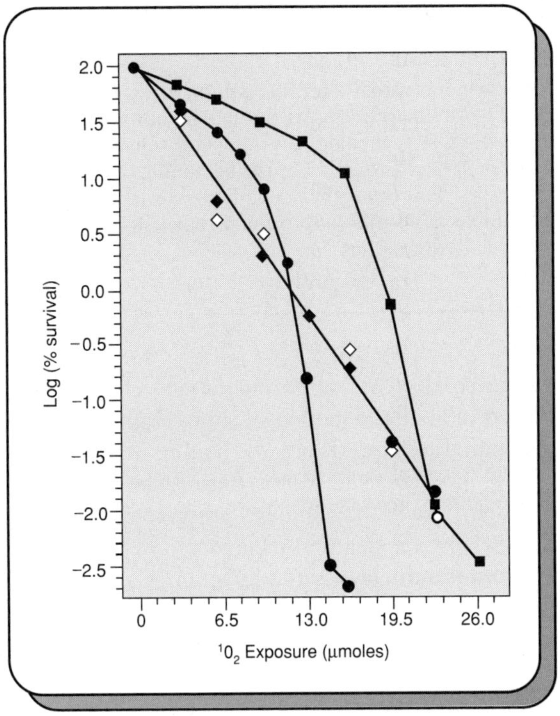

FIGURE 13.1

Bacterial killing curves. Bacterial cultures were exposed to various quantities of pure exogenous singlet oxygen ($^1O_2$). For each dose, a sample of the culture was taken and plated on nutrient medium to determine the number of cells still viable. Closed boxes designate *Salmonella typhimurium,* and closed circles represent a deep rough mutant of *S. typhimurium* that lacks most of the polysaccharides on its outer membrane. Open diamonds designate *Streptococcus lactis* and closed diamonds *Streptococcus faecalis.* Only the streptococcal curves show kinetics of the first order.

From T. A. Dahl, W. R. Midden, and P. E. Hartman, "Comparison of Killing of Gram-negative and Gram-positive Bacteria by Pure Singlet Oxygen," in *Journal of Bacteriology* 171:2188–2194. Copyright © 1989 American Society for Microbiology. Reprinted by permission.

viving cell remains. This observation implies that a manufacturer of sterile products can reduce the probability of an organism's surviving the sterilizing treatment to any appropriate value by increasing the dose of the sterilant.

The graphs in figure 13.1 do not describe exactly how the dose of singlet oxygen was administered to the cells. Rather, they express the total number of micromoles to which the cells were exposed. This amount of singlet oxygen can be presented to the culture for long or short times. The ultimate effect of singlet oxygen treatment depends both on the concentration used and on the time of exposure, an interrelationship often called the **time-dose-rate relationship** in antimicrobial treatments. The same net effect can be achieved in more than one way.

Generally speaking, for a linear killing curve such as that seen with *Streptococcus* in figure 13.1, an increase in the amount of chemical used will allow a proportional reduction in the time of treatment necessary to obtain the same result. Sometimes it is convenient to treat a culture slowly for a long period of time, while in other cases, it can be more advantageous to provide a high dose rate for a short period of time. In industry, a decision as to the appropriate dose rate is based on the cost of providing a given dosage compared to the cost of holding the material in the treatment vessel for the necessary length of time.

It is customary to speak of the **death point** or **death time** of a particular treatment. For example, in order to kill all cells in a particular culture, one can hold the time constant and vary the dose rate or keep the dose rate constant and vary the time. Either method results in sterilization. These terms are often used in connection with the application of heat. Thus, the thermal death time refers to the time (in minutes) necessary to sterilize the culture when a particular temperature is applied, and the thermal death point refers to the temperature necessary to sterilize the culture in 10 minutes.

Treatments such as singlet oxygen are said to be **bactericidal** because the treatment actually kills cells. These are obviously the treatments of choice for sterilization. In some circumstances, however, sterilization is not required. It is sufficient to use a **bacteriostatic agent** that merely prevents further growth of cells but does not kill them. If the bacteriostatic agent is removed, the cells will resume their growth with no apparent damage. However, if a culture is held in an inhibited state for long enough by a bacteriostatic agent, cell death will occur, although such a time is frequently a matter of days or weeks. It is important, therefore, to make the distinction between bacteriostatic and bactericidal agents when referring to a particular treatment. The suffixes *cidal* and *static* can be applied to all types of agents. Thus, a substance can be virucidal (kills viruses), fungistatic (inhibits fungi), or germicidal (kills disease-causing microorganisms).

Methods of Sterilization

Heat Sterilization

The use of heat to kill microorganisms is as old as the science of microbiology itself. Pasteur, Koch, Tyndall, and Spallanzani all used boiling to kill some or all the

APPLICATIONS BOX

Thermophilic Organisms

Extremely thermophilic organisms do not seem to follow the usual rules with respect to the instability of macromolecules at high temperatures. It is comparatively trivial to obtain bacteria that can grow at temperatures only a few degrees below that of boiling water by sampling the nearest volcanic hot spring. Even more interesting are the superthermophiles that are found in the depths of the ocean growing at temperatures several degrees above boiling. The use of the DNA polymerase from *Thermus aquaticus* for PCR testing is discussed in chapter 11. Its thermal stability has greatly simplified PCR technology by allowing scientists to add enzyme only once instead of at the beginning of each cycle. It is possible that other thermophilic bacteria have particularly stable molecules within them that will also become commercially important.

organisms in their infusion mixtures. Heat sterilization is an excellent and readily available method that causes **denaturation** or unfolding of the tertiary and quaternary structures of protein molecules so that they are no longer enzymatically active. Even when cooled again, the vast majority of proteins cannot refold themselves to regain this activity. Consider the change that occurs in the protein of an egg white when an egg is boiled. A protein denatures at a characteristic temperature that generally tends to be lower for proteins isolated from psychrophilic organisms than for proteins isolated from thermophilic organisms. Heat also can disrupt cell membranes and cause DNA duplexes to separate into single strands that may or may not return to their original pairing upon cooling.

Conceptually, heat can be applied wet or dry. Dry heat is provided by an oven, and wet heat by steam generated from boiling or some other similar process. In general, wet heat is the more effective in that a lower temperature can be applied for a shorter period of time to achieve the same killing levels. Hot water allows the various molecules to unfold more easily than dry heat does. In fact, dry sterilization is in large measure an oxidative process.

Although any uncertainties about sterility probably can be avoided if very high temperatures are used, it is important to remember that in the case of microbiological media many of the components are themselves sensitive to heat. Too much heat may sterilize the medium but render it unfit for use. In the discussion that follows, therefore, emphasis is placed on the minimum conditions necessary for sterility.

Tyndall's experiments demonstrating the existence of bacterial endospores also showed that boiling of an infusion for a few hours is not necessarily sufficient to inactivate all entities in it. The principles of the time-dose-rate relationship suggest that it might be advantageous to apply more heat in the form of a higher temperature so that the same amount of killing can be achieved in a shorter period of time. Higher temperatures than those produced by boiling water require pressures greater than atmospheric. The device used for this purpose in the home is a pressure cooker; in the laboratory, it is a steam autoclave.

A steam **autoclave** is fundamentally a pressurized vessel within a pressurized vessel (figure 13.2A–B). The space between the two vessel walls is known as the jacket, while the space within the innermost pressurized vessel is called the chamber. Moist heat is provided in the form of low-pressure steam that can be admitted either to the jacket or to the jacket and the chamber. A drain within the chamber floor serves to remove condensed steam as water and also allows any air trapped within the chamber to be expelled. Steam itself cannot pass through the drain. If the air is not completely removed, its compression provides a back pressure that prevents some steam from entering the chamber. Thus, any back pressure will result in a lower temperature within the chamber and may prevent sterilization.

Although the time and the temperature of steam sterilization can vary, about 15 pounds per square inch of steam pressure maintained for 15 minutes results in the sterilization of a small load, such as a few well-separated 1-liter flasks of media. Larger loads require more time or greater pressure. It is very important to bear in mind that it is not the pressure that kills the bacteria, it is the temperature. Therefore, it is probably more appropriate to remember that a temperature of 121°C (250°F) for 15 minutes will sterilize an object small enough to reach thermal equilibrium promptly. All commercial autoclaves are designed with built-in thermometers to monitor the interior temperature.

Any sort of oven can be used for sterilization, provided that it can achieve the desired temperatures. A reasonable balance between time and temperature is to

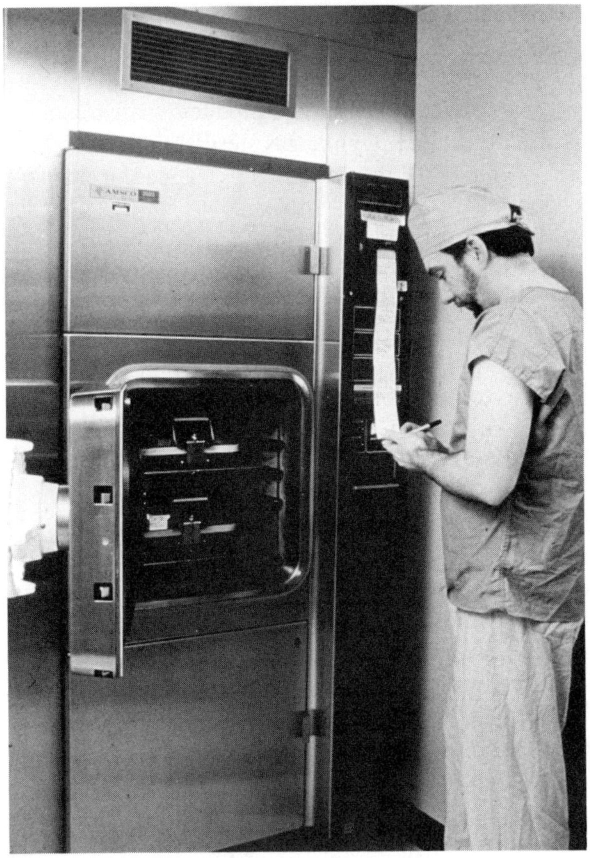

A

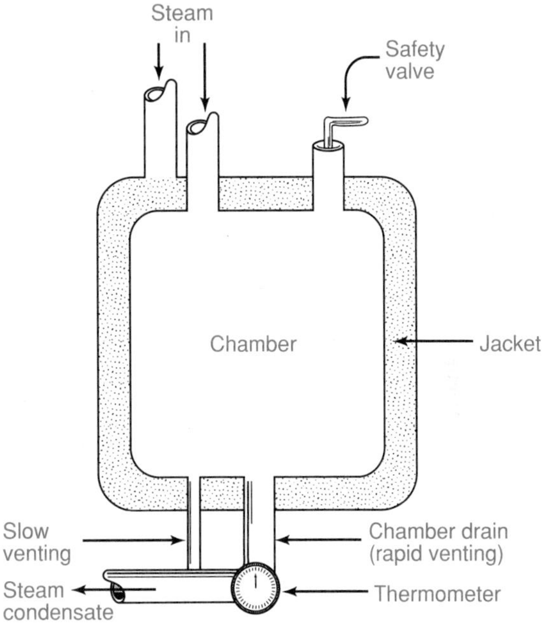

B

FIGURE 13.2

Autoclave. (**A**) Photograph depicts a typical autoclave that is computer operated. (**B**) Diagram illustrates its various components.

(A) © SIU/Visuals Unlimited

hold the load at 160°C (310°F) for 2 hours. This is an excellent and inexpensive method of sterilizing sturdy materials, such as metal and glass, and a few heat-resistant plastics, such as polymethylpentene or phenolic resins. Some additional plastics (for example, polypropylene) are sufficiently heat resistant to be sterilized in an autoclave even if they cannot be sterilized in an oven. The majority of plastics, especially polystyrene, cannot be sterilized by either method.

Chemical Sterilization

One way in which heat-labile materials can be sterilized is by treatment with an appropriate chemical agent. These agents are highly reactive molecules that modify most of the molecules in a cell to destroy their biological activity. As such, they also are hazardous to unprotected laboratory workers. The commonly used agents are one of two gases, ethylene oxide or propylene oxide (figure 13.3). These chemicals are thought to alkylate (attach ethoxy or propoxy groups) various reactive portions of molecules such as sulfhydryl groups, carboxyl groups, and even labile hydrogens.

Ethylene oxide and propylene oxide are used in special gas autoclaves in which the temperature and pressure of the gas can be controlled carefully. The temperature of a gas autoclave is always sufficiently low (about 50°C) that all plastics can be sterilized successfully. A complete cycle may require more than 5 hours, however. New federal regulations lowering the permissible concentration of these gases in the atmosphere of a workplace may significantly reduce the amount of gas sterilization conducted by industry.

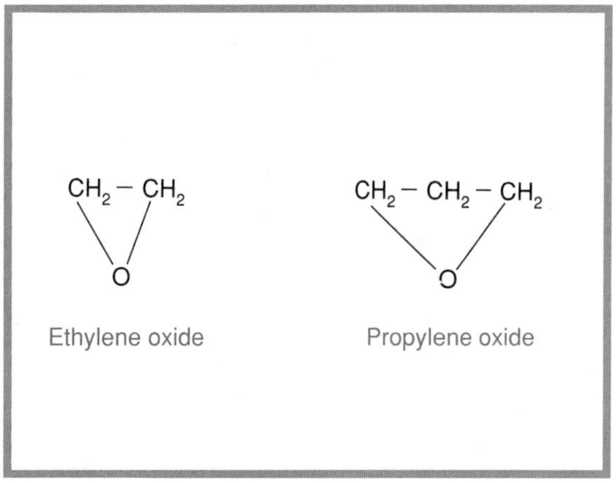

FIGURE 13.3
Common gaseous sterilants include ethylene oxide and propylene oxide.

Radiation Sterilization

The use of ultraviolet (UV) radiation has long been a favorite method for the sterilization of large surface areas, as in a laboratory. UV radiation has no real penetrating power and is strongly absorbed by the components present in many complex media. Where it can reach a cell, however, it penetrates the cytoplasm and is absorbed primarily by DNA molecules. The energy of UV radiation causes pyrimidine bases (cytosine and thymine) in DNA to undergo several types of chemical changes. Two adjacent pyrimidines may dimerize (fuse together) to give an intrastrand cyclobutane ring or a pyrimidine-(6→4)-pyrimidone (figure 13.4). The presence of pyrimidine dimers in a DNA strand prevents the replicational process from proceeding past that point. A gap is left that must later be repaired. If two such gaps should overlap, the DNA molecule will fall apart and the cell will die (figure 13.5). The probability that such an overlap will occur is proportional to the dose of UV.

One promising new technology for sterilization is that of X irradiation. Gamma-ray emissions of ^{60}Co are of sufficient energy that when they interact with DNA molecules in a cell, the phosphodiester backbone is literally broken. When enough breaks accumulate, the DNA is unable to replicate or repair those phosphodiester bonds, and transcription cannot occur. Consequently, the cell dies.

Because X irradiation is energetic enough to penetrate any reasonable size packaging, it is possible for a manufacturer to prepackage plasticware and then sterilize it by radiation treatment. Because of the high-energy particles emitted by ^{60}Co and the low-mass numbers of the elements in plastics, any radioactive isotopes that manage to form during the irradiation are quite short lived. The finished product is safe to handle and use, therefore. Unlike gas sterilization, there is no risk of harmful residues being trapped inside the product.

Even bacteriological media can be sterilized with X irradiation. Some of the most fastidious organisms can, however, have trouble growing on the medium unless the enzyme catalase is present. This presumably indicates that radiation induces the production of some reactive oxygen molecules.

The effects of radiation on DNA molecules can be reversed by various repair mechanisms (figure 13.6A–B). One type of repair, called **photoreactivation** or light repair, is specifically for pyrimidine dimers. Visible light, especially the blue wavelengths, can interact with a photolyase enzyme and provide the energy to separate a pyrimidine dimer into its component nucleotides, a simple reversal of the original process. Alternatively, there are two nonspecific dark repair processes (no visible light required), excision repair and SOS repair, that are discussed in chapter 10. They can remove a portion of the damaged strand and replace it with normal bases using an unaltered DNA as a template.

Filter Sterilization

Once the existence of microorganisms and their role as infectious agents were accepted, the possibility of sterilization by filtration was considered. The early workers turned to filters of sintered glass whose pores were fine enough to retain most bacteria. They were not totally successful because viruses are too small and bacteria without cell walls (for example, *Mycoplasma)* are too fluid in shape to be retained by such filters. For this reason, viruses were originally described as "filtrable."

It is often difficult to clean reusable microporous filters adequately. When the technology became available to produce disposable microporous filters, it was quickly adopted. Today most filtration is performed with membrane filters made of materials such as nitrocellulose, cellulose acetate, regenerated cellulose, polycarbonate, and polyvinyl chloride. Pore sizes can be precisely controlled so that filters with a uniform pore size as small as 0.1 μm are possible. Filters of this type will retain many viruses and most mycoplasmas. Some examples are depicted in figure 13.7.

Filtration is an excellent method to sterilize both liquids and gases. Filters can be manufactured to transmit liquids but not gases or vice versa so that the

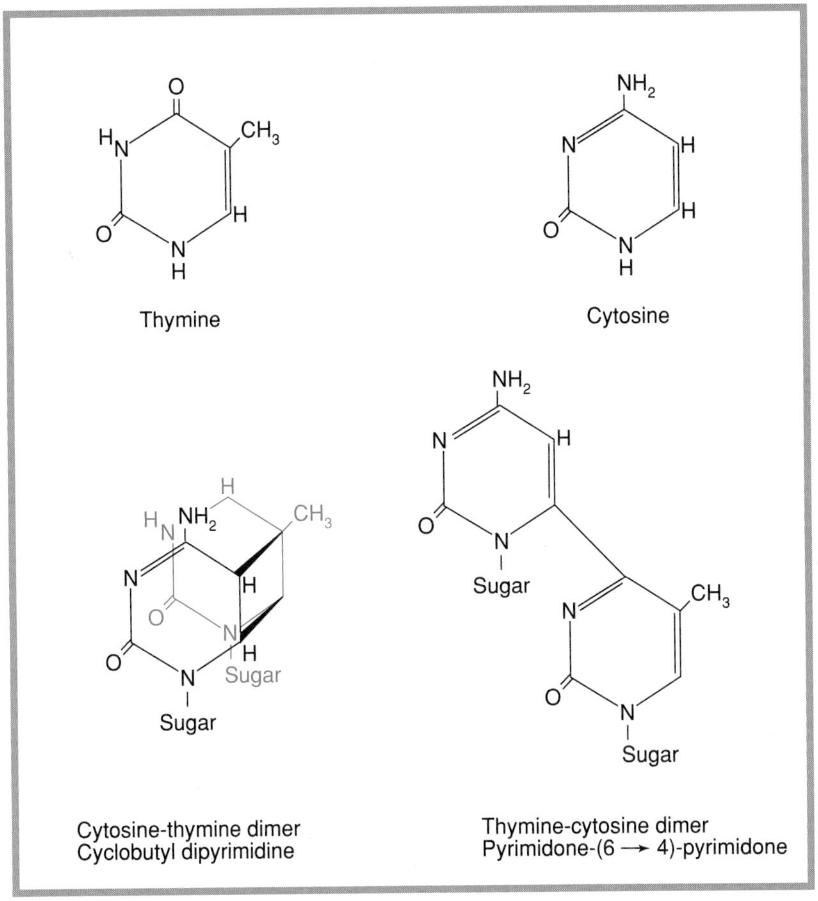

FIGURE 13.4
Effects of UV radiation on pyrimidine bases. Two types of structures are formed, cyclobutane dimers and pyrimidine-(6 → 4)-pyrimidones.

sterilizing operation also can have a purifying effect. This feature can be used as a safety device for vacuum systems (liquids cannot pass through the filter) and for intravenous syringes for injections (air embolisms cannot develop). An example of a sterilizing filter is shown in figure 13.8.

Chemicals That Inhibit Microbial Growth

Any deviation from the criteria outlined in chapter 2 for the preparation of microbiological media should be inhibitory for one or more types of organisms. This is particularly true for many bacteria that are likely to spoil foods. The specific methods used for controlling microbial growth in certain food products are discussed in chapter 14. For now, it is sufficient to know that variations in pH, osmotic strength, and/or temperature are traditional methods used for food preservation.

Sometimes, the best way to prevent microbial growth is to kill or inhibit the desired organisms by means of some more or less specific bacteriostatic or bactericidal chemical agent. A variety of descriptive terms have been used in connection with antibacterial chemicals; however, many of them are more a product of Madison Avenue than of science. Among the terms capable of some realistic definition is **antiseptic,** an agent that prevents **sepsis** or microbial infection caused by vegetative (growing) cells. Usually, the antiseptic

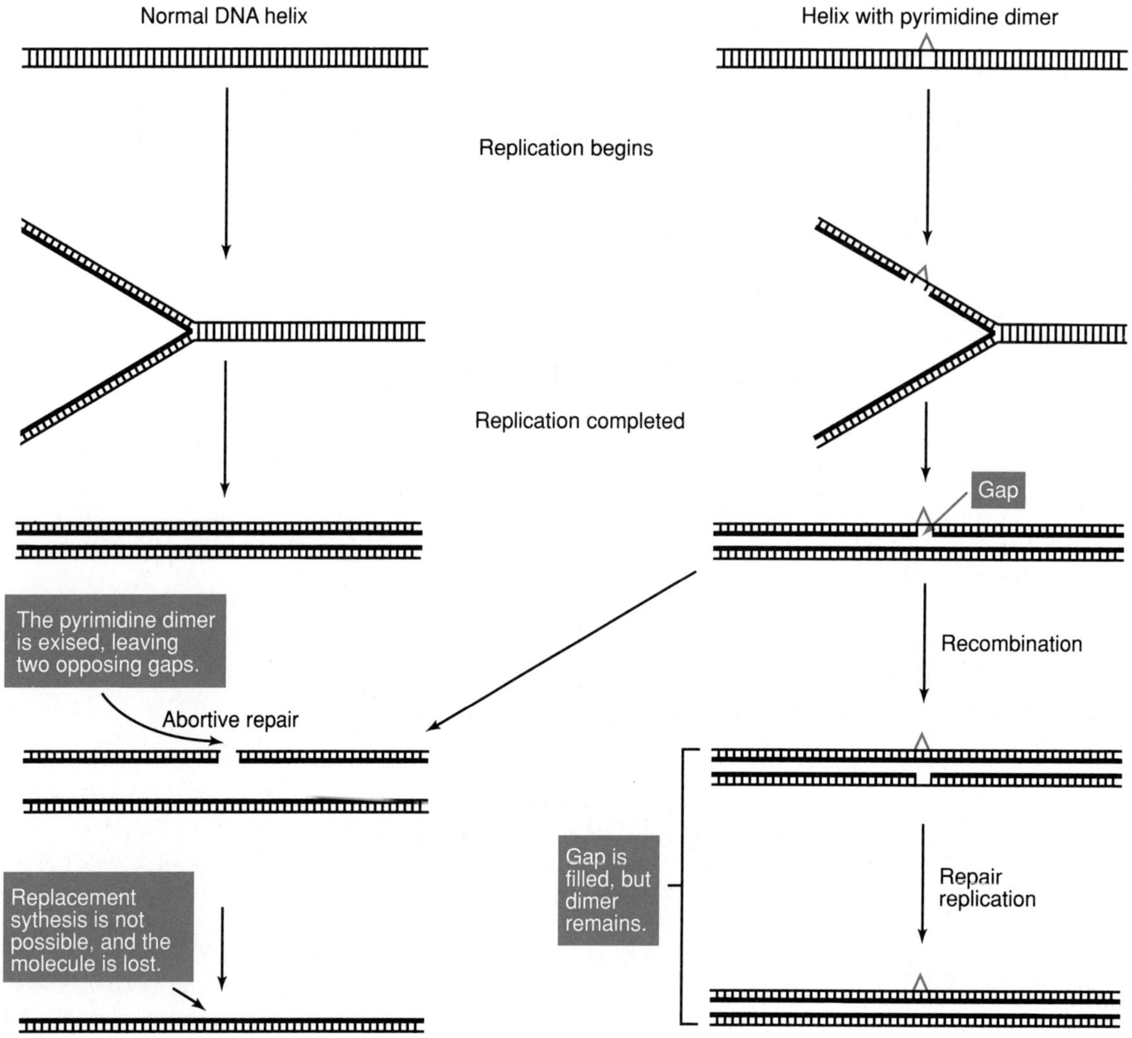

FIGURE 13.5

DNA damage can prevent DNA replication. Presence of pyrimidine dimers causes the DNA polymerase to stop before each dimer and to resume its synthetic activities some distance away. Gaps in the newly replicated DNA strand result. If the companion DNA molecule has normal DNA in the corresponding region, recombination can be used to fill in each gap; however, the dimer still exists and must be removed at some later time. If gapped regions overlap, the molecule will be unable to repair itself and will be lost. Unless an extra copy of that DNA molecule is available within the cell, cell death will result.

does not affect endospores or cysts. Such an agent is used externally to treat surface wounds. A **disinfectant** is a chemical that attacks the same spectrum of infectious agents as does an antiseptic, but it is intended for use on environmental surfaces. A **sanitizer** is a chemical that reduces the number of potentially harmful bacteria on a surface to acceptable levels, but it does not necessarily kill them all. A disinfectant is a harsher chemical than an antiseptic or sanitizer and is not appropriate for use on living tissue. The actual distinctions among the three terms are not sharp, however, and may depend on the standards applied at the time. A good case in point is the introduction of antiseptic surgery, in which disinfectants were used on skin wounds.

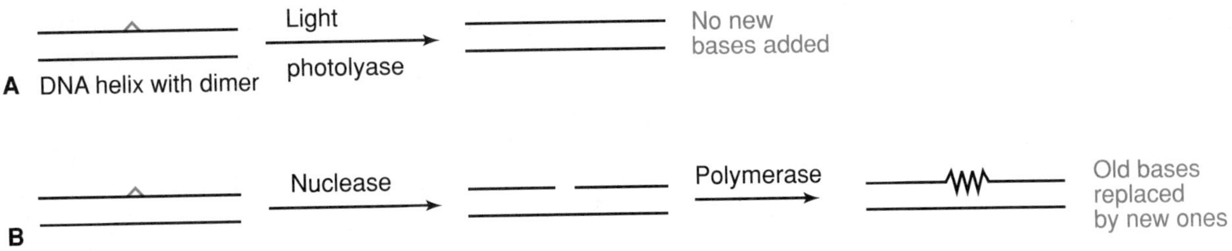

FIGURE 13.6

DNA repair mechanisms. (**A**) Photoreactivation or light repair. The chemical reaction is reversed with energy derived from visible light. (**B**) Dark repair. A nuclease removes the damaged region, and DNA polymerase replaces it with normal bases.

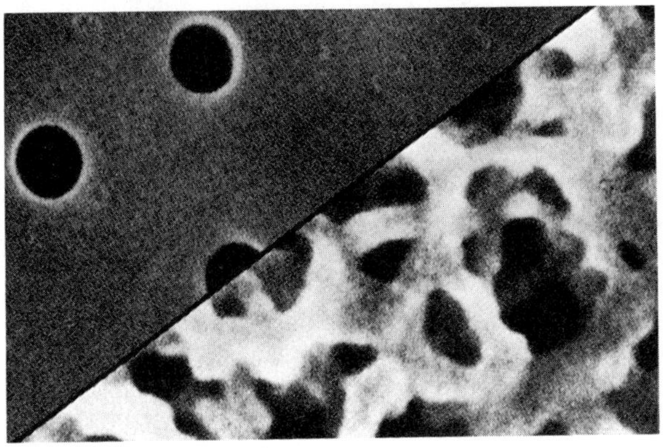

FIGURE 13.7

Electron micrographs of microporous filters. Iron bombardment was used to create the pores of the regenerated cellulose filter on the left. The filter on the right is a conventional nitrocellulose one composed of a thick mat of overlapping strands whose effective pore size is determined by the spacing between strands.

© Reproduced with permission from Fisher Scientific Lab Reporter Vol. 28, p. 3.

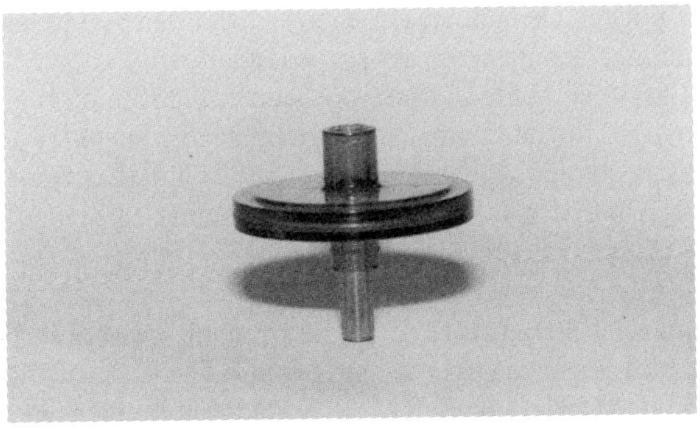

FIGURE 13.8

This disposable filter fits on the end of a syringe and can be used to sterilize a solution.

Effect of Phenolic Compounds

Joseph Lister (later Baron) of England was greatly impressed by Pasteur's early work illustrating that microorganisms were present in air. He accepted Pasteur's assertion that these organisms could be the causative agents of disease and formulated a method of antiseptic surgery to deal with them in 1867. Lister's method of antisepsis consisted of spraying the patient, the surgeon, and the surrounding area with a fine mist of phenol (carbolic acid) during an operation.

This method worked fairly well in reducing postsurgical infections, but it was rather hard on all the individuals concerned. Phenol is a poison that can be absorbed through the skin and can cause chemical burns as well. It acts to destroy cells of all types by dissolving lipids and precipitating and denaturing proteins. Today, it is considered as a disinfectant and not as an antiseptic. As other methods of antisepsis became available, the practice of using phenol was discontinued.

Nevertheless, phenol is still one standard used to evaluate water-miscible phenolic compounds for their antibacterial activity. The amount of phenol needed to kill a culture of test organisms in 10 minutes (but not in 5 minutes) is defined as the baseline. Dilutions of the compound to be tested are added to a similar culture, and the greatest dilution that meets the same killing criterion is determined. The **phenol coefficient** represents the ratio of the dilution of the compound to be tested to the dilution of phenol that will accomplish the same task. A coefficient greater than 1 means that the new chemical is more bactericidal than phenol. Examples of such chemicals are cresol (methylphenol) and hexachlorophene (formerly used in antiseptic soaps).

TABLE 13.1	Antimicrobial agents and their phenol coefficients
Chemical	**Phenol Coefficient**[1]
phenol	1.0
2-methylphenol	2.3
p-chlorophenol	4.3
p-bromophenol	5.0
4-*n*-propylphenol	16.3
n-heptyl orthochlorophenol	375
4-*n*-heptylphenol	625

Source: Data from R. F. Prindle, "Phenolic Compounds," pp. 197–224, in *Disinfection, Sterilization, and Preservation,* edited by S. S. Block. Copyright © 1983 Lea & Febiger, Philadelphia.

1. Phenol coefficients were determined at 37°C using *Staphylococcus aureus.*

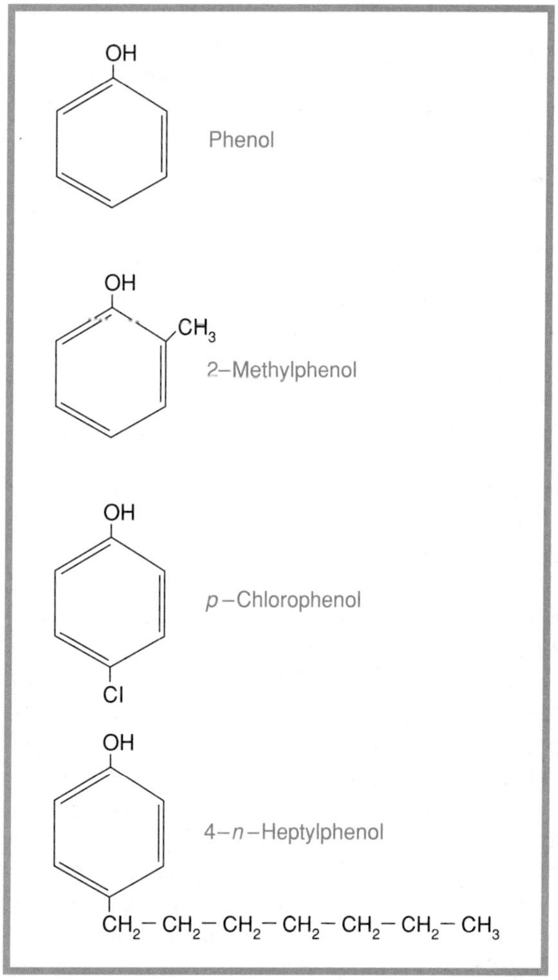

FIGURE 13.9

Examples of some phenolic compounds with antimicrobial activity.

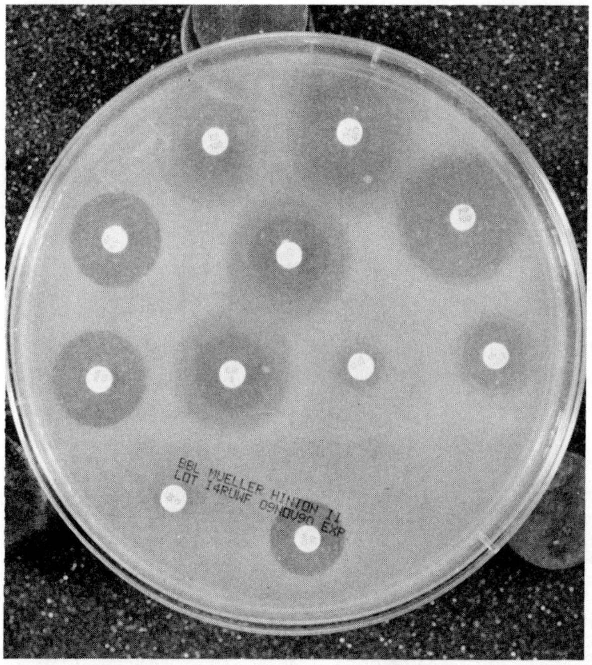

FIGURE 13.10

Zones of inhibition produced by antimicrobial agents. *Bacillus subtilis* cells were spread over the surface of an agar plate, following which various commercially prepared paper disks containing antimicrobial agents were placed on the agar.

Some examples of antibacterial agents and their phenol coefficients are listed in table 13.1. Chemical structures of some of these antibacterial agents are diagrammed in figure 13.9.

When chemically dissimilar antimicrobial compounds are compared, what is calculated generally is a germicidal coefficient. This coefficient is analogous to the phenol coefficient, except that the compound to which the comparisons are made is arbitrarily selected.

Another contemporary measure of antimicrobial activity is the **minimum inhibitory concentration** (MIC). This is defined as the amount of substance that must be applied to the surface of an agar plate to prevent the growth of a lawn of bacteria in its immediate vicinity. The area of no growth is referred to as the zone of inhibition (figure 13.10). The stronger the action of the antimicrobial agent, the larger is the zone of inhibition. Kirby and Bauer quantified this method and prepared tables showing the minimum size of the zones of inhibition necessary for an organism to be considered as sensitive to a particular antimicrobial agent. These minimum sizes vary with the concentration of the agent and its molecular weight.

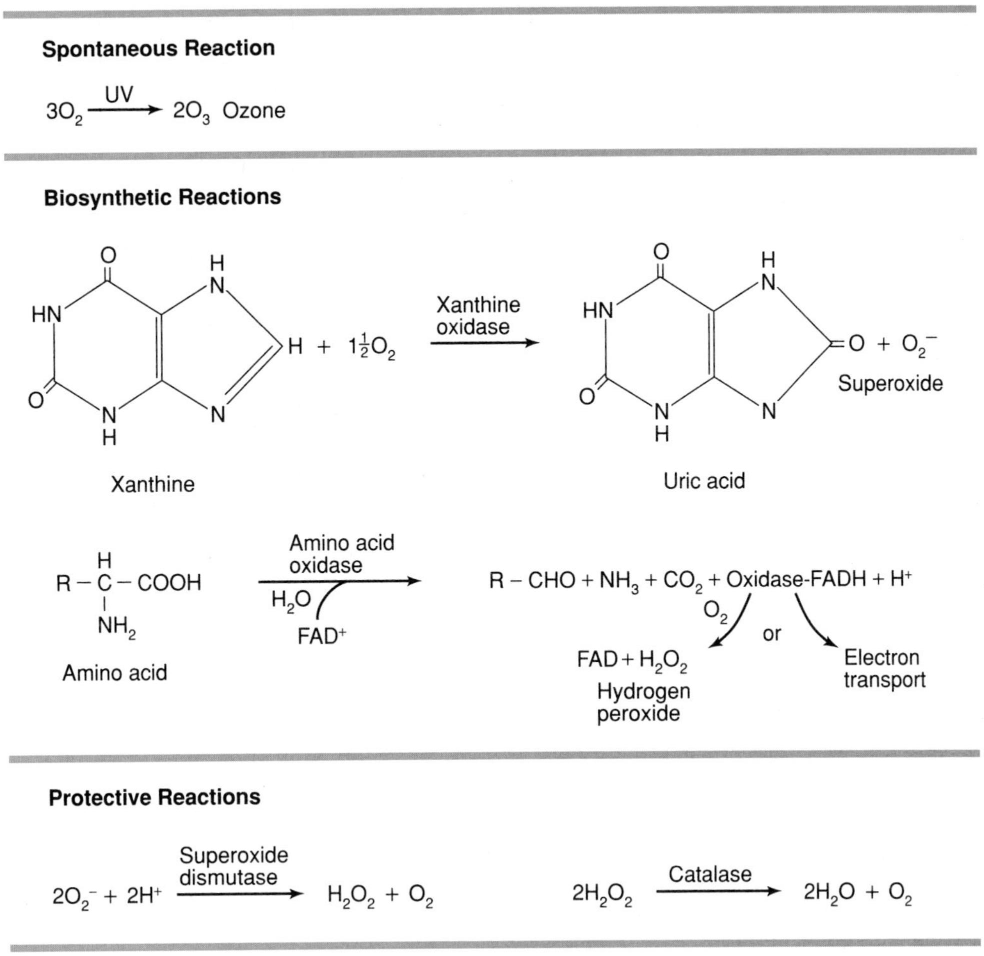

FIGURE 13.11

Reactions generating reactive oxygen molecules. Superoxide is a stronger oxidizer than is hydrogen peroxide. Superoxide dismutase will convert superoxide to hydrogen peroxide, and catalase will convert hydrogen peroxide to water and oxygen.

Effect of Oxygen

Molecular oxygen is essentially a two-edged sword with respect to the bacterial kingdom (and eukaryotic organisms as well). Although facultative anaerobes can grow in its absence or presence, strictly aerobic organisms require oxygen and strictly anaerobic organisms cannot grow in its presence. Why some organisms require molecular oxygen can be illustrated on straightforward biochemical grounds. Certain organisms lack the ability to reoxidize various coenzymes unless electron transport is operative, and molecular oxygen is needed as a terminal electron acceptor in electron transport (figure 4.3). The inability of an organism to tolerate oxygen, however, has a more subtle biochemical base.

Oxygen in the air is generally thought to occur as the diatomic molecule O_2. Actually, there are some variant forms of oxygen that can be present in trace amounts (see figure 13.11). Ozone (O_3) is produced by UV radiation from the sun striking the upper layers of the atmosphere. It remains trapped in the atmosphere and serves as a barrier, absorbing additional UV radiation. Ozone also is a product of air pollution, electric discharges, and artificial UV sources; therefore, ground-level ozone concentrations occasionally can be quite high. Regardless of its source, ozone is a strong oxidizer and will oxidize organic molecules such as odor-producing compounds and cellular components. Portable ultraviolet lamps are sometimes sold as air-freshening units. Oxidation of cellular components can result in cell death. Ultraviolet-producing fluorescent tubes that are often sold as "germicidal lamps" can kill bacteria in two ways. Organisms exposed to their radiation are killed by the disruption of normal DNA replication due to the production of pyrimidine dimers as previously discussed. Organisms protected from the radiation of these fluorescent tubes nevertheless may be killed by the oxidizing effects of ozone produced by the UV radiation from the lamps striking oxygen molecules in the air.

Oxygen metabolism itself also can be hazardous to a cell. Within a cell, there are a number of oxidase enzymes that carry out oxidation reactions in which molecular oxygen is a participant. In some cases, the end product of these oxidation reactions may be a form of oxygen with an extra electron(s). For example, as shown in figure 13.11, the enzyme xanthine oxidase generates superoxide (O_2^-), and the enzyme D-amino acid oxidase generates hydrogen peroxide (H_2O_2). Both superoxide and peroxide moieties can be very harmful to molecules in the cytoplasm due to their strong oxidizing capabilities; therefore, organisms that grow in the presence of oxygen also produce special enzymes that can detoxify the reactive oxygens. In the case of superoxide, the enzyme in question is superoxide dismutase, which catalyzes a conversion of superoxide to hydrogen peroxide (figure 13.11). Hydrogen peroxide, whether produced by superoxide dismutase or by another oxygenase, then can be broken down to water by the enzyme catalase.

Based on these considerations, it might be assumed that anaerobic bacteria would not have either superoxide dismutase or catalase. Nevertheless, careful examination of anaerobes has shown that some do indeed possess at least trace amounts of these enzymes. When the quantity of enzymes present during normal growth of aerobes and anaerobes is compared, however, it becomes clear that strict anaerobes do not maintain their enzymes at levels that can give significant protection against the reactive forms of oxygen.

From the terminology listed in table 4.1, obligately anaerobic bacteria (facultative anaerobes with strictly fermentative metabolism) must also be prepared to deal with reactive oxygen moieties. Many of them produce various peroxidases that reduce peroxides using NADH + H^+ to help cope with potentially troublesome molecules.

Organisms exposed to sunlight in the presence of oxygen also can accumulate a variant form of oxygen. In this case, the spin of one of the electrons is reversed, generating a molecule of singlet oxygen. If this oxygen molecule is not de-energized, it will pass the extra energy to other molecules and destabilize them, enabling them to undergo degradation reactions that normally occur only rarely. In microorganisms, carotenoid pigments normally are used to react with singlet oxygen and channel the extra energy into harmless outlets. Thus, microorganisms growing in sunlit areas tend to be pigmented, while those growing elsewhere tend to be nonpigmented. The importance of these pigments can be observed in *Rhodotorula mucilaginosa,* for example. If carotenoid pigment production is blocked in this red yeast, the organism becomes progressively more sensitive to singlet oxygen.

The effects of superoxide, singlet oxygen, and hydrogen peroxide on bacteria take on a wider significance in terms of the ability of bacteria to cause disease. As a mechanism of defense at times, body cells will engulf bacterial invaders in vacuoles in their cytoplasm and bombard these invaders with singlet oxygen, hydrogen peroxide, or superoxide. The degree of bacterial resistance to this treatment obviously depends on the amount of inactivating enzymes and pigments the bacteria produce.

Effect of Alcohols

Various shorter alcohols often have been used as antiseptics because they are mild lipid solvents that also can denature proteins. Interestingly, ethanol works best in a slightly diluted solution. At full strength, it tends to dehydrate the cell before cell proteins can be denatured; therefore, the cell may be able to survive for comparatively long periods of time, certainly longer than following a typical application of the alcohol. In a slightly dilute solution (70 to 90%), however, ethanol is a very good inactivator of bacterial cells and some viruses (primarily enveloped viruses), although it does not affect bacterial endospores. In other words, alcohols have an antiseptic effect, but they are not sterilants. At substantially lower concentrations, they actually can be used as carbon and energy sources by some organisms.

The most commonly used alcohols are ethanol and isopropanol in an approximately 70% solution, although the antimicrobial activity of isopropanol actually increases with concentration up to about 99%.

TABLE 13.2 Examples of commonly used disinfectants and their exposure times

Agent	Concentration	Contact Time[1] (min)	Inactivation Effect[2]			
			Bacteria	Endospores	Viruses	
					enveloped	plain
Ethanol	85%	10	+	−	+	±
Isopropanol	85%	10	+	−	+	±
Phenolics	2%	10	+	−	+	±
Chlorine	3%	10/30	+	+	+	+
Iodophor	2%	10/30	+	+	+	+
Quaternary ammonium compound	2%	10	+	−	+	−
Ethylene oxide	.45 g/liter	60/60	+	+	+	+

Source: Adapted from *NIH Guidelines for Recombinant DNA Research.* 1976. U.S. Government Printing Office, Washington, DC.

1. Exposure (contact) times are recommended by the Centers for Disease Control. The first number listed refers to the time an agent requires to inactivate lipid-containing viruses. The second number refers to the time an agent requires to affect a broad spectrum of microorganisms on environmental surfaces. If a second contact time is not listed, an agent is not considered to have a broad spectrum of action.
2. Has inactivating effect (+); has no effect (−); may or may not have inactivating effect (±).

Isopropanol is slightly more effective than ethanol against bacteria but slightly less virucidal. Necessary exposure times to kill typical bacteria vary, but some conservative estimates are available from the Centers for Disease Control, Atlanta, Georgia. A few examples are listed in table 13.2.

Effect of Heavy Metals

Paul Ehrlich, a collaborator of Robert Koch's, in attempting to find cures for the various diseases being studied in that laboratory, was the first to demonstrate that some metallic elements with high atomic numbers have antibacterial activities. He showed that arsenic-containing compounds such as arsphenamine were useful in the treatment of syphilis. Later bismuth- and mercury-containing compounds were added to the list. Ehrlich's work greatly advanced the treatment of disease, although the therapy fell into disuse as soon as other treatments became available owing to the highly toxic nature of heavy metals.

Today heavy metals are still used as antimicrobial agents but in less toxic forms. Both mercurichrome and merthiolate are mercury-containing compounds that have been widely used as antiseptics. Mercury is known to react with the sulfhydryl group of cysteine and may thereby cause proteins to precipitate. Its action is antagonized by nonprotein molecules that contain sulfhydryl groups. Various water-purifying membrane filters are now advertised as containing embedded silver particles intended to inhibit bacterial growth within the purification cartridge. Silver also precipitates proteins to produce its antibacterial effect, although not specifically through reactions with sulfur groups. Certain copper-containing compounds can be used as fungicides or algicides.

Effect of Halogens

Halogens can be used directly or as part of complex molecules. Chlorine gas, a strong oxidizing agent and excellent disinfectant, is the most commonly used halogen. It is conveniently provided in a solution of sodium hypochlorite (liquid chlorine bleach) that can be applied to desired surfaces. The active compound in the solution is thought to be hypochlorous acid, formed by the reaction of chlorine and water:

Equation 13.1

$$H_2O + Cl_2 \rightarrow HOCl + HCl$$

The amount of chlorine required for disinfection depends in large measure on the amount of organic material present. In order to be certain that all bacteria are killed, all organic material must be at least partially oxidized. Moreover, the oxidizing action of chlorine is antagonized by ammonium ions. Chlorine is used frequently to control microbial growth in swimming

APPLICATIONS BOX

Is Bar Soap Hazardous?

People are often concerned about sharing bars of soap because this might accidentally serve to spread disease. Certainly, some bars can get quite dirty from heavy use. An epidemiologic study was done to check the persistence of selected bacteria on the surface of bars of soap. Results indicated that there are not likely to be major problems. Apparently, the antibacterial action of soap coupled with its self-cleaning properties when rinsed with water prevents the accumulation of significant numbers of bacteria even after repeated use.

---

pools, treated wastewater, and drinking water. Bromine, which is more expensive but not inactivated so readily by sunlight, is replacing chlorine for some purposes.

Iodine has always been one of the traditional disinfectants. However, anyone who has ever used a **tincture** of iodine preparation (an alcohol solution) knows that it can inflict quite a bit of pain, particularly if it is applied to a wound in which the skin is broken. More recently, iodine-containing compounds called iodophors have been produced that retain the antibacterial activity of iodine but are much gentler on the skin. Iodophors are mixtures of iodine with a surfactant such as a nonionic detergent (see following subsection). These compounds are suitable as antiseptics or disinfectants. They precipitate proteins and/or iodinate organic molecules so that the amount of organic material present is again relevant. Despite their antibacterial activity, occasional contamination of iodophors by *Pseudomonas cepacia* has been noted. The ability of *P. cepacia* to grow in these solutions seems to be correlated with the presence of a glycocalyx.

Effect of Soaps and Detergents

Soaps (the product of alkali treatment of fats) and detergents (the product of the synthetic chemist) are often quite good antibacterial compounds. They can be categorized according to the nature of the charges on the molecule. Thus, one can have an anionic detergent (negatively charged), a cationic detergent (positively charged), or a nonionic detergent (no charge). Examples of each type are depicted in figure 13.12. Soaps disrupt cell membranes by separating the lipoproteins in the membrane and releasing autolytic (self-digesting) enzymes from the cell wall. They also can have an inhibitory effect due to the presence of various fatty acids. Cationic detergents based on nitrogen-containing molecules have particularly good antibacterial activity and are known as quaternary ammonium compounds ("quats"). Table 13.3 presents a summary of the antimicrobial chemicals discussed in this section.

Benzalkonium Chloride

$$CH_3-(CH_2)_{17}-\overset{\overset{CH_3}{|}}{\underset{\underset{CH_3}{|}}{N^{\oplus}}}-CH_2-C_6H_5 \quad Cl^{\ominus}$$

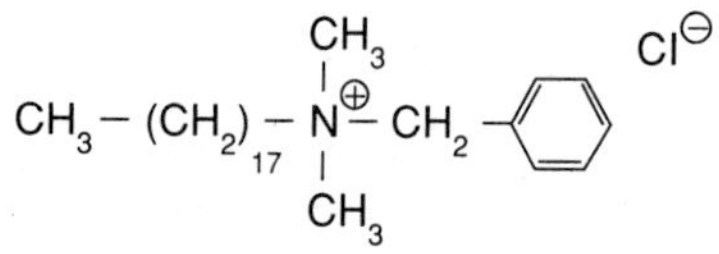

Triton X (polyethyleneglycol *p*-isooctylphenylether)

$$CH_3-\overset{\overset{CH_3}{|}}{\underset{\underset{CH_3}{|}}{C}}-CH_2-\overset{\overset{CH_3}{|}}{\underset{\underset{CH_3}{|}}{C}}-C_6H_4-O-(CH_2-CH_2-O)_x H$$

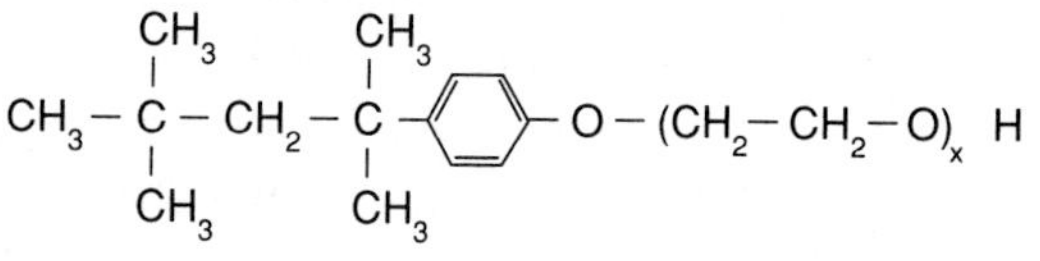

Sodium Dodecyl Sulfate

$$CH_3-(CH_2)_{10}-CH_2-O-\overset{\overset{O}{\|}}{\underset{\underset{O}{\|}}{S}}-O^{\ominus} \quad Na^{\oplus}$$

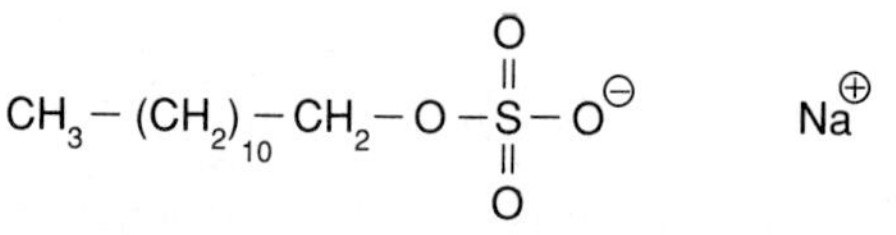

FIGURE 13.12

Examples of detergents. From top to bottom they are cationic, nonionic, and anionic.

Antimicrobials and Antibiotics

The disinfecting and antiseptic agents discussed previously are general in nature because they attack entire cell structures. They are not directed against a specific cell type (prokaryote or eukaryote) or against a specific cytoplasmic component. Neither are they suitable for internal use or consumption by humans and other animals. There are, however, various chemotherapeutic agents that can inhibit specific metabolic processes within a microbial cell while having no or only minimal effect on the metabolism of the host organism. This is particularly true with respect to prokaryotes living in

TABLE 13.3 Examples of chemicals that inhibit microbial growth			
Oxidizing Agents	**Chemical Modifiers**	**Protein Precipitants**	**Lipid Solvents**
Ozone	Iodine	Phenolics	Detergents
Chlorine	Chlorine	Mercury	Phenolics
Superoxide	Bromine	Silver	Alcohols
Hydrogen peroxide	Ethylene oxide	Alcohols	
Singlet oxygen			
Ethylene oxide			

association with eukaryotes. Some of these chemicals are designated as **antibiotics** because they have been produced by one living cell and are effective against other, nonproducing cells. Bacteriocins are examples of antibiotics, although they are not necessarily useful therapeutically. Other compounds synthesized in the laboratory have been shown to have antibacterial effects also. This latter group is referred to as synthetic antimicrobials. A molecule that is a chemically modified version of an antibiotic is considered a semisynthetic antibiotic.

Synthetic Antimicrobials

Ehrlich was not alone in searching for a "magic bullet" that would allow specific treatment of a disease. Many other scientists tried to identify compounds that would specifically inhibit or kill microorganisms without harming the eukaryotic organism they invaded. The first real success story came when sulfanilamide, a compound synthesized by German dye chemists, turned out to be an effective antibacterial agent.

As shown in figure 13.13, the structural formula of sulfanilamide is very similar to that of *p*-aminobenzoic acid (PABA), a compound required to synthesize the enzyme cofactor folic acid. Bacteria normally synthesize their own folic acid, but humans do not and must obtain it from their food. Sulfanilamide binds to the active site on the enzyme but does not allow the reaction to occur; therefore, it acts as a competitive inhibitor. It is only a bacteriostatic agent, however, because the inhibited reactions resume if the compound is removed from the medium. Sulfanilamide was the first of the family of "sulfa" drugs that are still used today.

Two other synthetic antimicrobials often used therapeutically are presented in figure 13.14. Their modes of action are diverse. Nalidixic acid is a molecule that acts to inhibit one subunit of the enzyme DNA gyrase. By so doing, it prevents the formation of supercoiled loops within replicating DNA molecules, and the replicational process soon comes to a halt. Trimethoprim is a competitive inhibitor of an enzyme used to synthesize an intermediate in thymine biosynthesis. In its presence, DNA replication stops unless thymine is provided exogenously. Both these agents are bacteriostatic and can be used to treat humans infected by bacteria sensitive to them.

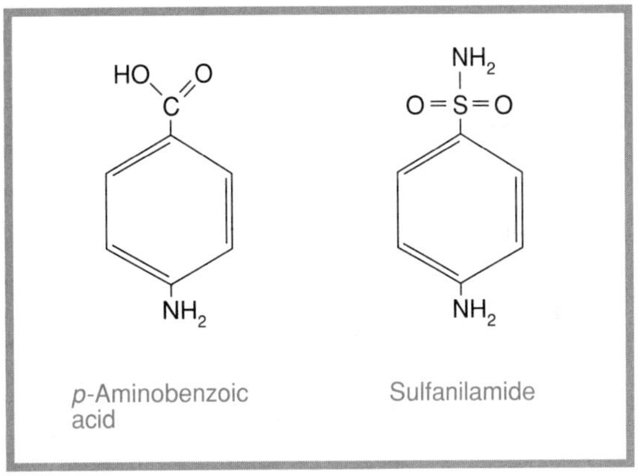

FIGURE 13.13
Sulfanilamide and *p*-aminobenzoic acid. These molecules have sufficient structural similarities so that sulfanilamide will act as a competitive inhibitor of metabolic reactions normally requiring *p*-aminobenzoic acid.

Representative Antibiotics

Alexander Fleming is credited as the first to observe an antibiotic. While in the process of discarding a Petri dish whose bacterial culture had been contaminated by the fungus *Penicillium notatum,* Fleming happened to notice that bacterial growth was inhibited in the immediate vicinity of the mycelia. From this observation, he theorized that the fungus produces some metabolite inhibiting bacterial growth. The metabolite was later isolated and called penicillin, the structure of which is illustrated in figure 13.15. Penicillin's great potential as a therapeutic agent was not fully realized until casualties incurred during the Second World War stimulated the development of a crash program to grow large enough cultures of the fungus from which significant quantities of the antibiotic could be prepared.

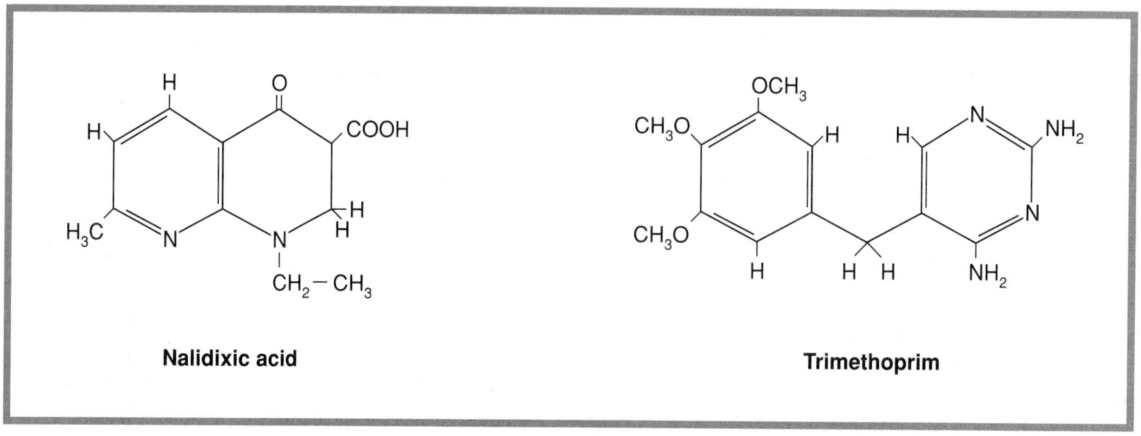

FIGURE 13.14
Some synthetic antimicrobial agents used therapeutically are nalidixic acid, an inhibitor of DNA gyrase activity, and trimethoprim, an inhibitor of dihydrofolate reductase used in thymine biosynthesis.

Penicillin acts to inhibit the peptidyl transferase enzyme that forms peptide cross bridges during peptidoglycan synthesis. Cells inhibited in peptidoglycan biosynthesis are also activated for peptidoglycan hydrolases (murein hydrolases) that break down existing cell wall. In this sense, penicillin is similar to D-cycloserine (figure 7.2). Both molecules prevent the formation of new cell wall, yet neither kills nor inhibits the cell directly. Instead, the cell becomes increasingly vulnerable to osmotic shock and, under normal circumstances, eventually lyses. The only time significant bacterial survival occurs is when a medium of high osmotic strength is used so that the osmotic pressure across the cell membrane is greatly reduced. Under those circumstances, gram-positive cells form **protoplasts,** round cells without any trace of peptidoglycan, while gram-negative cells form **spheroplasts,** round cells that still have remnants of peptidoglycan on their surface.

The distinction between protoplasts and spheroplasts is an important one because peptidoglycan can be synthesized only by adding *N*-acetylglucosamine and *N*-acetylmuramic acid subunits onto existing molecules. When removed from a penicillin solution, therefore, spheroplasts will regenerate normal gram-negative cell walls, while truly complete protoplasts can never regenerate their gram-positive wall except under special conditions that include very high concentrations of specific simple sugars. Protoplasts will continue to divide as long as the osmotic strength of the medium does not drop too low, however. The cells that they produce are very similar in form and structure to naturally occurring but rare **L forms** or L phase variants originally isolated at the Lister Institute in London. L form bacteria are prokaryotes with defective or missing cell walls that biochemically resemble members of several common genera. It is theorized that their origins are also similar, with the L forms resulting from exposure of bacteria to naturally occurring penicillin-like agents.

Obviously it is possible to confuse L forms with members of the genus *Mycoplasma,* but careful taxonomic studies show little relationship between the two groups. The only real similarity between L forms and mycoplasmas is the special strengthening adaptations in their cell membranes that allow them to grow in media of lower osmotic strength. A photomicrograph of L forms of *Haemophilus influenzae* is presented in figure 13.16.

Penicillin is unlike D-cycloserine in that it is not a substrate analog; therefore, it is not a competitive inhibitor of the transpeptidation enzyme. Nevertheless, the shape of the molecule is important, and the β-lactam ring shown in figure 13.15 is essential for the activity of the molecule. This ring can present a therapeutic problem because many plasmid DNA molecules and some bacterial chromosomes carry information coding for **β-lactamases,** enzymes that open the lactam ring and destroy the antibacterial activity of the molecule.

Because penicillin is sensitive to enzyme degradation and because it does not penetrate through the outer membrane of gram-negative cells very efficiently, nu-

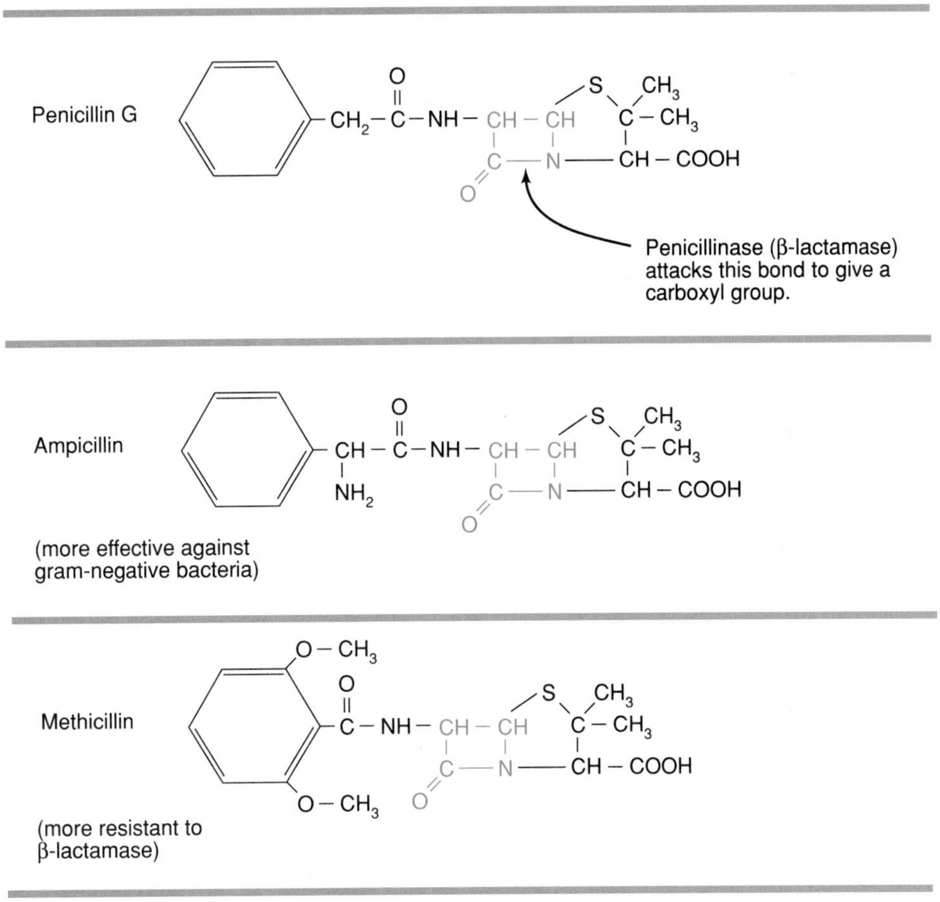

FIGURE 13.15
Structural formulas for penicillin and two penicillin derivatives. The β-lactam ring essential for the activity of these antibiotics is shown in color.

merous chemical derivatives (semisynthetic antimicrobials) have been prepared from penicillin that are either more effective against gram-negative cells and/or less sensitive to β-lactamases. Two such molecules are ampicillin and methicillin (see structural formulas in figure 13.15).

The discovery that therapeutically useful antimicrobial compounds are produced by microorganisms themselves set off an enormous revolution in the pharmaceutical industry. Selman Waksman showed that the streptomycete *Streptomyces griseus* (funguslike bacterium) produced an aminoglycoside compound with potent antibacterial activity that he called streptomycin. The structure of streptomycin can be seen in figure 13.17. Its mode of action is via the 30S ribosomal subunit, and its interference with translation causes cell death. Many other aminoglycoside antibiotics with similar modes of action are available today, including kanamycin, neomycin, and gentamicin. They all share the specific advantage of affecting prokaryotic ribosomes only. Like the penicillins, the streptomycins can be inactivated by enzymes encoded on R plasmids. In this case, the addition of an adenyl or phosphoryl group is sufficient to destroy antibiotic activity.

Members of the genus *Bacillus* also can produce antibiotics. Antibiotic synthesis typically occurs during the process of sporulation. It has been proposed that the antibiotic may represent a regulator or a waste product of some sporulation process. Examples of antibiotics of this type are bacitracin and polymyxin (figure 13.18). Both act to inhibit the formation of the prokaryotic cell wall.

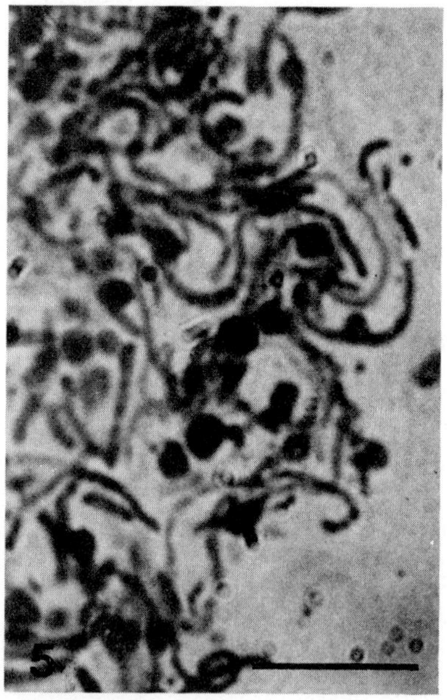

FIGURE 13.16

L forms of *Haemophilus influenzae.* The photomicrograph shows the periphery of a reverting colony. Note the large, round L forms as well as filamentous revetting ones.

Reprinted from Madoff, S. 1970. L forms of *Haemophilus influenzae. Current Microbiology* 2:43–46.

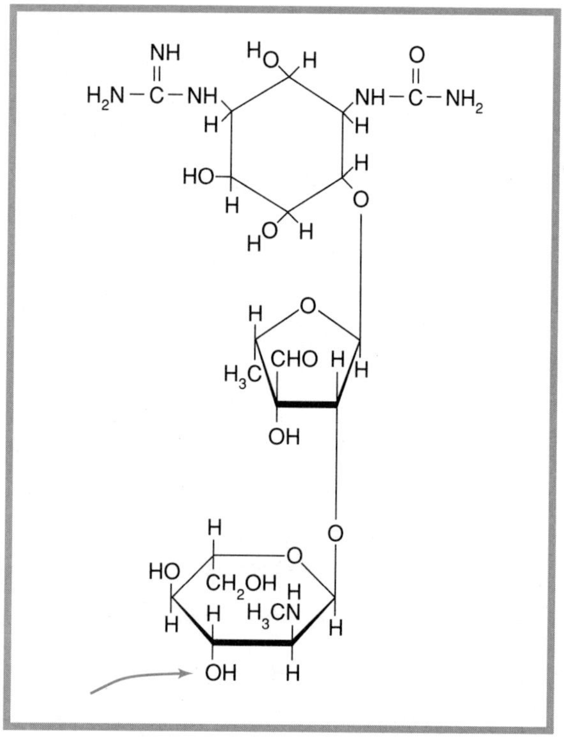

FIGURE 13.17

Streptomycin, an aminoglycoside antibiotic. The arrow indicates the site of enzymatic modification.

There are also antibiotics effective against eukaryotic organisms such as fungi and protozoa. As might be expected, these compounds tend to have more side effects because the metabolism and cell structures of host and parasite are basically similar. One such antibiotic that affects eukaryotic ribosomes is cycloheximide (figure 13.18) produced by *Streptomyces griseus.* Because the target microorganism is assumed to be growing faster and synthesizing more protein, cycloheximide should be more effective against it than against host cells. Another antibiotic effective against eukaryotic microorganisms is amphotericin B (figure 13.18), a polyene produced by *Streptomyces.* Its mode of action lies in its resemblance to cholesterol in that it strongly affects cells with sterols in their membranes. Thus, only a few mycoplasmas are sensitive to amphotericin B; however, it works well against fungi and also, unfortunately, against red blood cells.

Summary

The growth of bacteria can be controlled or prevented in a number of different ways. Sometimes the integrity of such products as freshly prepared media and food that must be stored for a period of time needs to be protected against bacterial growth. Sometimes environmental surfaces or wounds need to be disinfected or decontaminated. And finally, sometimes the growth of disease-causing microorganisms needs to be controlled in order to treat some kinds of illness.

Each of these situations presents special needs and requirements with respect to the treatments rendered. In certain instances, such as with environmental surfaces, very strong agents like chlorine gas, steam, or ethylene oxide can be used. For wound treatment, agents cannot be overly damaging to the skin. Frequently, iodophors, mercurials, or alcohols are used in treating skin wounds. For treatment of disease-causing microorganisms that have invaded the body, even more gentle agents are required: antibiotics, synthetic antimicrobials, or semisynthetic antibiotics.

Whatever the treatment given, it is important to recognize its limitations. Some chemicals, such as the alcohols, have no reliable effect on bacterial endospores. Others that carry out a chemical reaction with organic molecules are greatly susceptible to interference by extraneous organic matter. Some chemicals are

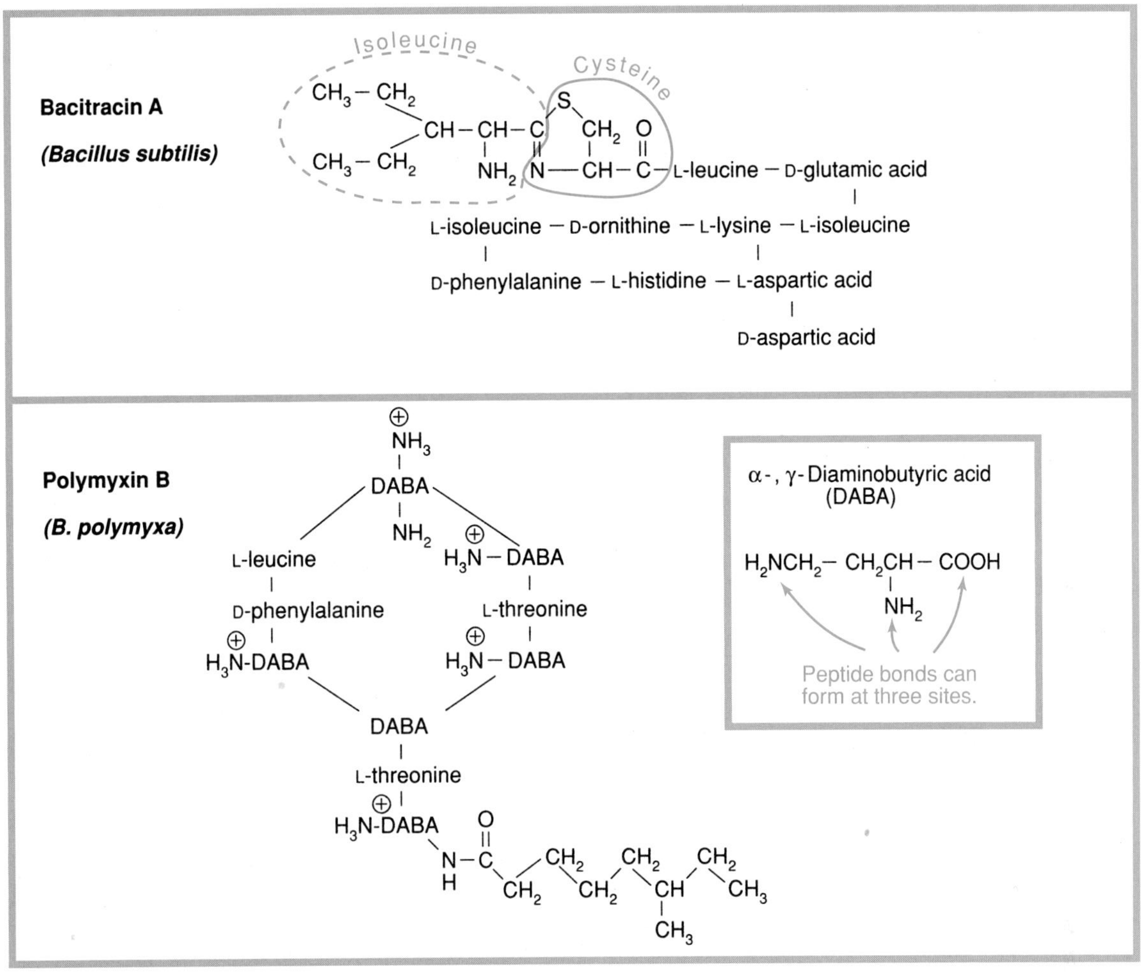

FIGURE 13.18
Examples of antibiotics and their producing organisms.

Continued on next page

bactericidal, but others that are bacteriostatic permit cell growth to resume if their use is discontinued. In the case of UV radiation, materials at the surface of a culture can absorb so much of the radiation that the cells beneath them will survive the treatment.

Regardless of the type of antibacterial or antiviral agent, a certain dose rate must be applied for a minimum period of time to ensure successful treatment. Unless these minimal requirements are carefully observed, treatment will fail. A bacteriostatic agent provided in suboptimum doses will not prevent further growth of cells. A bactericidal treatment provided in suboptimum doses will fail to sterilize the object. Recommendations for antimicrobial treatments are provided by agencies such as the Centers for Disease Control.

Questions for Review and Discussion

1. List some methods that can be used to sterilize culture media. What are the advantages and disadvantages of each? What additional methods might be used if the object to be sterilized does not contain liquid? Would your methods work with plastics? Why or why not?
2. Give an example of a disinfectant, an antiseptic, and a sanitizer. What is the action of each?
3. What are the similarities and differences between antibiotics and the compounds that you listed in question 2? From where do antibiotics come?

Cycloheximide

(Streptomyces griseus)

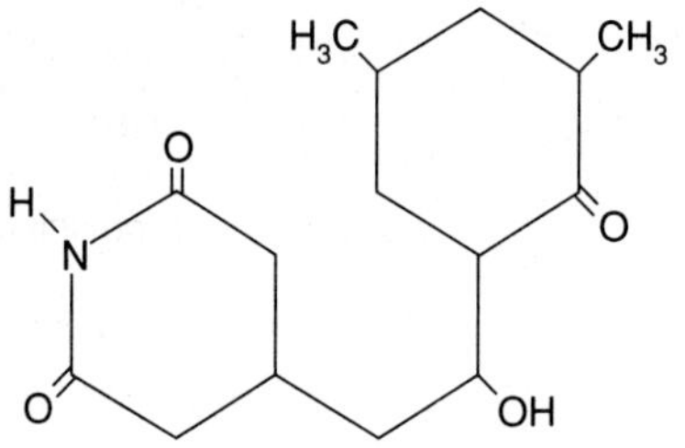

Amphotericin B

(S. nodosus)

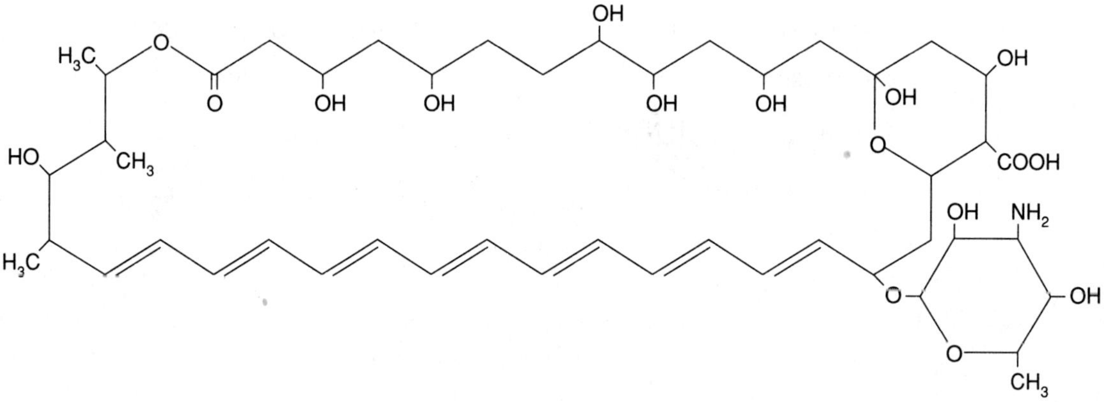

Rifamycin B

(S. mediterranei)

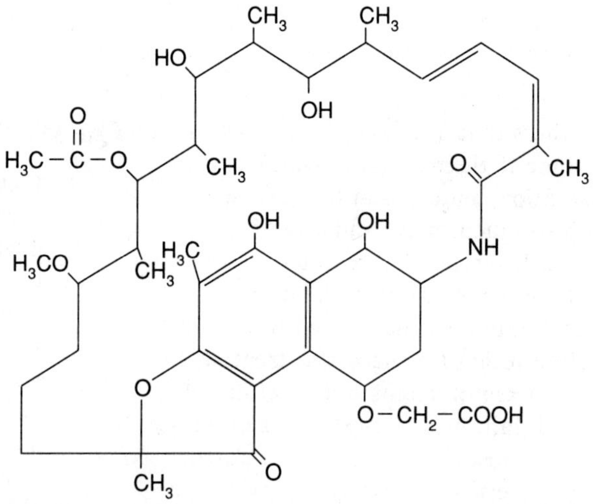

FIGURE 13.18 CONTINUED

4. What is a synthetic antimicrobial? How does it differ from an antibiotic? What is a semisynthetic antibiotic?
5. What is the time-dose-rate relationship for antimicrobial treatments? How can it be used to modify standard parameters of treatment?
6. What antimicrobial treatments will inactivate each of the following: enveloped virus, nonenveloped virus, bacterial cell, bacterial endospore, protozoan parasite, fungus? Which treatments are suitable for use on the surface of a living host? Inside a living host?

Suggestions for Further Reading

Block, S. S. (ed.). 1983. *Disinfection, sterilization, and preservation,* 3d ed. Lea & Febiger, Philadelphia. This extensive compendium provides 51 chapters on all aspects of antimicrobials.

Borich, P. M. 1973. *Chemical sterilization.* Dowden, Hutchinson & Ross, Inc., Stroudsburg. This is a member of the *Benchmark Papers in Microbiology* series and as such presents reprints of the classic papers in the field.

Brock, T. D. 1961. *Milestones in microbiology.* Prentice-Hall, Inc., Englewood Cliffs, N.J. This compendium includes original papers by Fleming and Lister.

Fridovich, I. 1982. The discovery of superoxide dismutases: A history, p. 1–9. *In* L. W. Oberley (ed.), *Superoxide dismutase,* vol. I. CRC Press Inc., Boca Raton, Fla. An entertaining review of the process of discovery in this branch of biochemistry.

Royal Society of Chemistry. 1980. *Oxygen and life.* Royal Society of Chemistry, London. A collection of lectures presented at the second BOC Priestley Conference at the University of Birmingham. It deals with all aspects of oxygen metabolism.

chapter

12

Food Microbiology

chapter outline

■

GENERAL GOALS

Be able to describe some types of organisms that can cause food spoilage and that can be harnessed sometimes to provide new sorts of foods.

Be able to describe the various methods used for food preservation and the reasons why they work.

Food microbiologists are concerned with two aspects of food: its preservation until use and its production in some cases. In our society, we assume that food can be grown in comparatively isolated regions and transported to urban areas for consumption. Processing plants at one or at most a few sites in the country usually supply all the markets for hundreds or thousands of kilometers around. This type of arrangement provides good economies of scale but requires that some method of preservation be used with nearly every food so that the growth of microorganisms does not spoil the food or present a health hazard for some reasonable time after it is packaged, shipped, and purchased by consumers. The more common preservation methods are discussed in the first part of this chapter. In the second half, examples of foods actually produced by microbial growth are considered.

Preservation of Foods

All our foods are derived directly or indirectly from living organisms and as such are subjected to the usual decay processes. Even so, it is important to remember that the internal tissues of any healthy plant or animal are of necessity sterile until cut and exposed to air. Thus, the problems faced by food processors revolve around the twin necessities of preventing growth of decay organisms and killing any disease-causing ones that can be present on the food surface or introduced during the food preparation process. Food intended for long-term storage also must be protected against oxidation by the air and decomposition by naturally occurring lysosomal enzymes that can degrade the molecular components of the food.

It is possible, of course, to sterilize all foods, thereby preventing spoilage for an indefinite period. Unfortunately, conventional heat sterilization ruins the flavor and/or texture of many foods and, therefore, does not present a real option in most cases. Instead, food processors use a combination of very old and very modern techniques in their preparation procedures. These techniques often result in a **commercially sterile** product in which no viable organisms can be detected or one in which so few are present that they pose no threat to long-term storage. The product is not truly sterile, however. Frequently, storage methods are designed to inhibit the growth of any surviving organisms so that they can cause no problems.

"Commercial sterilization" is generally defined with respect to the **decimal reduction time** denoted as D. This is the time in minutes required to reduce the number is the time in minutes required to reduce the number of viable cells or spores of a particular organism tenfold when the cells or spores are exposed to a specific bactericidal treatment. Table 14.1 presents a few typical D values for selected organisms. If a "commercially sterile" product has been exposed to a 12D dose, this dose is sufficient to reduce the number of viable cells or spores of a particular organism by a factor of 10^{12}. A related parameter is Z, the change in dose (for example, the change in temperature measured in degrees) required to change the thermal resistance of a particular organism tenfold. D and Z values are gradually replacing the death point and death time values discussed in chapter 13.

Changing Temperatures to Preserve Foods

The cooking of foods is thought to date back to prehistoric times. Cooking tenderizes many foods but, more importantly from a microbiological point of view, it also kills many bacteria. Boiling foods in water obviously kills most vegetative cells, but methods in which even lower temperatures are used also have a significant effect. A beef roast, for example, generally is cooked to a minimum temperature of 60°C (104°F) in the middle. This temperature is sufficient to kill all typically harmful bacteria that may be present within the meat due to processing. In addition, many bacteria that normally break down meat proteins are killed. The act of cooking, therefore, renders the meat safe to eat and retards its spoilage. Cooking serves to inactivate degradative enzymes as well. Such enzymes are part of the autolytic process and do break down proteins and polysaccharides in their own cells, resulting in bad flavors and loss of texture in the food. Conventional cooking positively does not sterilize a product, and spoilage does occur given sufficient time.

In a similar vein, milk produced by a healthy cow is a sterile product that acquires many different kinds of bacteria as it is handled and processed. If untreated (raw), it does not keep well and may spread various diseases. Milk can be rendered safe from harmful bacteria and have an extended shelf life if it is pasteurized, a process developed by Louis Pasteur. **Pasteurization** consists of heating milk to a temperature that will kill the most heat-resistant disease-causing organisms that might be present (*Mycobacterium tuberculosis* and *Coxiella burnetti*). In bulk pasteurization, milk is placed in a large vat at 62.8°C for 30 minutes; whereas in high-temperature short-time (HTST) pasteurization, milk is pumped through a pipe and heated to 71.7°C for 15 seconds. The number of other bacteria

TABLE 14.1 Some decimal reduction times for heat killing of selected microorganisms

Organism	Suspending Medium	Temperature (°C)	D Value (min)
Vegetative Cells			
Staphylococcus aureus	Custard;	60	7.7–7.8
	chicken a la king	60	5.2–5.4
Salmonella manhattan	Custard;	60	2.4
	chicken a la king	60	0.4
Spores			
Clostridium botulinum			
Type A	Phosphate buffer, pH 7.0	110.0	4.4
		121	0.2–0.4
Type E	Phosphate buffer, pH 7.0	70.0	29.3–37.5
		80.0	0.4–3.3

Source: Data from G. J. Banwart, *Basic Food Microbiology,* 2d edition. Copyright © 1989 Van Nostrand Reinhold, New York.

TABLE 14.2 Maximum bacterial counts permissible in various grades of milk

Grade	Bacterial Count/ml
Grade A raw milk for pasteurization	300,000
Grade A pasteurized milk and products	20,000
Grade A cultured milk and milk products	10 coliforms[1]
Certified raw milk (for drinking)	10,000 plus 10 coliforms

Note: A healthy animal produces sterile milk; therefore, bacteria are introduced during handling and processing. The relatively high bacterial counts reflect the ease with which bacteria can grow in milk even when precautions are taken.

1. Coliforms are cells morphologically and biochemically similar to *Escherichia coli*.

APPLICATIONS BOX

European Milk

Consumer preference often can dictate which of several technologies will be most commonly employed by industry. Nowhere is this more obvious than in the type of milk produced for western Europe. Traditionally, the home refrigerator in Europe is much smaller than in the United States. As a result, storage space for bulky items such as milk is extremely limited. Consequently, most milk sold in large supermarkets is processed by UHT technology. The product is packaged in 1-liter containers so that it fits comfortably into a small refrigerator after the container is opened. Because of commercial sterilization, unopened milk packages can be stored at room temperature for up to four months. Only after the package has been opened is refrigeration necessary. In most areas of the United States, the UHT process is used only with specialty items like whipping cream.

is greatly reduced as well, a fringe benefit of pasteurization that retards milk spoilage during storage. Typical bacteria counts for raw milk and pasteurized milk are listed in table 14.2.

The most recent development in pasteurization is the ultrahigh-temperature (UHT) process in which temperatures between 140 and 150°C are used only for a few seconds. Technically, this feat is accomplished by atomizing the milk through a cloud of superheated steam.

Many foods that eventually will be cooked and others that need no cooking can be stored well at cool temperatures for comparatively long periods of time. If temperatures are reduced sufficiently, mesophilic organisms will be unable to grow, and those psychrotrophic organisms that can continue to grow will be able to do so only at a slow rate. Thus, refrigeration provided by underground caves, cool springs, and areas filled with stored ice have been used for centuries to prolong the storage life of various food products.

The advent of the mechanical refrigerator offered an expanded range of storage possibilities and made possible a dietary diversity unknown 40 years ago. The conventional kitchen refrigerator operates at a temperature of 4°C, well below the minimum temperature for growth of mesophilic organisms. This cooler temperature also retards the rates of reaction by lysosomal enzymes released from cells cut during harvesting or processing.

Although simple refrigeration is useful for extending the shelf life of various foods, growers and processors realized that actual freezing offers even more advantages. Psychrophilic bacteria and fungi that grow slowly at refrigerator temperatures do not grow at all at $-20°C$. For this reason, frozen foods can be stored for much longer periods of time than foods that are merely refrigerated. In fact, the storage time for some foods is on the order of a year or more. Many types of meat freeze well for six to eight months with no significant change in quality. In the case of vegetables, a problem can arise during storage due to the presence of autolytic enzymes, as discussed at the beginning of this section. For this reason, vegetables are usually **blanched** (briefly immersed in boiling water) to denature degradative enzymes before freezing. Blanching does not actually cook the vegetable, however.

Commercially frozen foods usually are prepared at substantially lower temperatures than those provided by home refrigerators. Often, liquid nitrogen is used as a refrigerant. The nearly instantaneous freezing greatly enhances the quality of a product.

Despite improvement in quality, cold temperatures by themselves do not kill bacteria and other microorganisms, although formation of ice crystals can damage some cells. Therefore, no frozen foods can be assumed sterile, even after they are frozen for a year or more. As soon as an item has thawed, some microorganisms already present will resume their growth cycle. Hence, decay processes occur quite readily in thawed food.

Moreover, disease-causing organisms in humans that are generally mesophilic in nature are not killed by freezing either. If frozen or cooked food is allowed to stand for long periods of time at room temperature, microorganisms can grow and reproduce to a point at which they represent a significant danger to the consumer. The most important rule in processing food is to keep a product hot (thermophilic temperatures) or cold (psychrophilic temperatures), but never keep it at room temperature, even for a few hours. To do so is to invite the multiplication of potentially harmful bacteria that can cause food poisoning.

Adding Chemicals to Preserve Foods

Various methods other than cooking and freezing are available to preserve foods. These generally involve the addition of some chemical(s) that more or less specifically inhibits microbial growth. The ancient custom of salting meat is a good example. An increased salt level raises the osmotic strength of the liquid in the food, thereby reducing water activity. A concentration from 18 to 25% NaCl ($a_w < 0.9$) prevents the growth of most spoilage bacteria and yeasts. Many potentially harmful bacteria, such as enteric bacteria, cannot tolerate even more modest salt concentrations. There are, of course, moderate halophiles that actually grow better under these conditions, but they are not frequent enough to pose any real problem. Spoilage fungi can grow in a_w values as low as 0.8 and therefore remain a potential problem.

A similar effect is seen with sugar syrups often used to preserve fruits, although higher concentrations of sugar must be used as opposed to salt. Canned peaches, for instance, can be maintained in a 50 to 60% sucrose syrup. Although sucrose is a potential nutrient for microbial growth, water activity in the syrup is so low that only a few fungi can manage to live on it. Thus, sugar acts as an excellent preservative and also protects fruit tissue against oxidation.

Many meats, such as hams and turkeys, are smoked. The heat generated from the smoking process has a drying effect on the surface of the meat that acts to preserve it and simultaneously provides partial internal cooking (remember that tissues not having been cut open remain internally sterile). The smoke itself is the source of a complex mixture of formaldehyde and aromatic compounds, some of which are related to the cresols and phenolics described in chapter 13. As these deposit on the meat, they inhibit and/or kill the microorganisms present.

In today's modern food processing, an aromatic compound or an organic acid often is added deliberately to products for its antimicrobial effect. Two commonly used preservatives are calcium propionate and sodium benzoate (figure 14.1). So-called "natural foods" are prepared without such additives and have correspondingly shorter shelf lives and higher prices.

Canning to Preserve Foods

The term *canning* is something of a misnomer because the "can" in question can be made of either metal or glass. The essence of the process is the same in both cases. Food is placed into a container that can be sealed hermetically after cooking; therefore, no gases or organisms are able to enter once the "can" is sealed. This process has several advantages. The oxygen in the container is displaced completely by steam, leaving an anaerobic environment to minimize oxidation of the

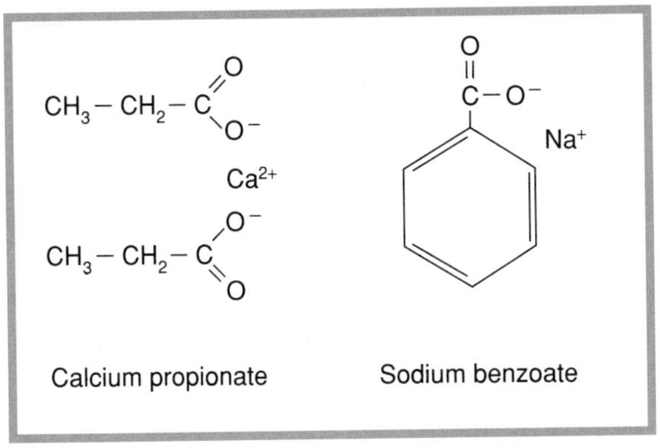

FIGURE 14.1

Two chemical preservatives used in foods: calcium propionate and sodium benzoate.

product during storage. Aerobic decomposers are unable to grow due to the lack of oxygen; however, anaerobes and facultative anaerobes will still be able to grow unless other measures are taken.

Generally speaking, the simplest solution to prevent growth of anaerobes is to sterilize the can and its contents in the cooking process. The resulting product can be stored at room temperature for long periods of time without significant degradation of the contents because all autolytic enzymes will have been denatured by sterilization. In one instance, some vegetables that had been in a can for 40 years were analyzed for their vitamin content and found acceptable by today's criteria. Because canning is a sterilization process, it is vital that all precautions regarding appropriate time and dose of heat be followed in order to have a safe product. This is particularly true for home-canned foods that are much more likely to present problems for public health officials.

There are a few cases in which canned foods do not require sterilization. These are foods having a high concentration of natural acids, so high in fact that their pH is low enough to prevent growth of nearly all microorganisms. The old-fashioned tomato is an example; however, modern varieties of tomatoes often have been bred for low acidity and, as a result, must be canned like any other vegetable.

Gamma-Ray Sterilization to Preserve Foods

Canned foods have been available for years, but they have certain drawbacks. As noted previously, many foods do not take well to heat sterilization and, therefore, cannot be canned. The weight of the container itself is significant and adds to the transportation costs of shipping food. These disadvantages can be avoided by a method of sterilization that does not require heat and can be used on foods packaged in plastic containers. This situation is analogous to the one faced by manufacturers of polystyrene labware, in which a similar solution is employed.

Gamma irradiation is a process that eliminates these disadvantages because gamma rays are an excellent sterilant that has almost no effect on the texture and flavor of the food irradiated. The radiation dose used is several million rads (500 rads is a lethal dose for humans). Foods prepared in this way are not radioactive and can be stored without refrigeration for long periods of time because of the denaturing effect of radiation on autolytic enzymes. Irradiated foods, however, are not very popular because the general public distrusts radioactive processes. The only real commercial success story with the gamma-irradiation process has been in the preservation of bacon for several years. However, radiation treatments given in low dosages (radurization) to seafood can achieve the equivalent of pasteurization and extend its shelf life two- to sixfold.

Drying to Preserve Foods

Dried foods in which $> 90\%$ of the water has been removed obviously have very low water activity and can be kept safely in that state. The precise degree of drying necessary for preservation depends in part on the nature of the food product. Whole milk powder needs to be dryer than does dehydrated fruit in order to prevent the growth of fungi. Once again, it is important to remember that drying is used to avoid thorough cooking of a product; therefore, dried foods cannot be assumed as sterile. Dried foods, however, are protected from significant enzyme degradation because there is no water to allow enzyme molecules to seek new substrates.

A large number of fruits can be preserved successfully in dry form, some of which are known by different names. Dried grapes yield raisins and dried plums yield prunes, for example. Many other fruits and vegetables can be obtained in dry form in a local grocery store.

In the case of meat, drying usually is carried out after cooking and also can be associated either with smoking or with addition of various preservative chemicals. Dried meat in the form of "jerky" is readily available. The basic techniques of drying were known to the ancient Indians, but modern shortcuts are now available.

Light weight is the big advantage of dried food. The most extreme example of dried products is **lyophilized** (frozen and then dried under vacuum or freeze-dried)

TABLE 14.3 Methods of food preservation

Method	Provides Sterility	Kills Mesophiles	Storage Time
Cooking	No (generally)	Yes (generally)	Few days
Refrigeration	No	No	Few days
Freezing	No	No	Weeks to months
Canning	Yes	Yes	Years
Gamma irradiation	Yes	Yes	Years
Drying	No	No	Months

prepared foods. These are commonly used by astronauts, backpackers, and others for whom minimal weight is worth a premium price. Lyophilization is an excellent method of preservation for both foods and microorganisms; therefore, the usual precautions must be taken when a product is rehydrated.

The various methods of food preservation are summarized in table 14.3.

Food Fermentations

There are some cases in which growth of microorganisms results in the creation of a new food. The origins of most of these processes have their roots in prehistoric times. Nevertheless, it is easy to see how the original discoveries could have been made by accident, for the processes used today are essentially ones in which controlled spoilage occurs. Every society today, as in centuries earlier, has its own fermented foods and beverages. The details of each process may vary, but the essence remains the same.

The basic substrate in these products is some food that contains reasonable quantities of a fermentable carbohydrate, usually glucose or a molecule readily convertible into glucose. The microorganism(s) grows fermentatively on that carbohydrate, producing pyruvic acid first and then one or more of the various possible derivatives of that molecule (see figure 4.17). Generally, among these products is at least one acid that drops the pH of the solution, providing a preservative effect on the substrate. The flavors characteristic of the food product are provided by the waste products of the microorganisms and, therefore, require that the food be held at moderate temperatures for weeks to months to allow time for microbial growth. This process is known as aging or ripening.

The fermented food itself is frequently more nutritious for humans than the original product from which it developed. The contents of the microbial cells can augment the vitamin content of the food, for example. During their growth, organisms also can act to partially degrade complex polysaccharides like celluloses, releasing nutrients that can be trapped inside aggregates of these cellular macromolecules and, incidentally, rendering the macromolecules digestible by the human intestinal tract.

Fermented Vegetables

There are many examples of fermented foods that can be discussed in this section, but the humble pickle provides a good beginning. Typically, young cucumbers (called pickling cucumbers) are placed in a brine (salt solution) with various spices and flavorings and covered to prevent access of oxygen. If allowed to stand for several days to a week, various bacteria normally present on cucumbers begin to grow. First, varieties of *Streptococcus* grow followed by the more acidophilic *Lactobacillus*, *Leuconostoc*, and *Pediococcus* species. Initially, growth of enteric and other undesirable bacteria is retarded by the brine, but later when the pH of the brine drops to about 3 to 4, their growth is completely precluded. Fermentation can continue for 2 to 6 weeks before the final product is ready for consumption.

Most picklers, whether commercial or home, prefer to leave less to chance than does the scenario just outlined; therefore, they will add sufficient amounts of vinegar to the brine to drop the pH significantly. Acetic acid immediately inhibits the growth of most spoilage bacteria and gives a firmer product. It does not shorten the time of fermentation significantly, however, because the pH of the brine will drop to the same level within the first few days under natural conditions. Some other types of fermented fruits and vegetables available in a supermarket are listed in table 14.4.

The fermentation procedure for olives is unlike most processes required for the other vegetables and fruits listed in table 14.4. As they grow on the trees, olives are inedible due to the presence of an extremely bitter compound in the seed. To prepare them for eating, they are first soaked in a sodium hydroxide solution for about

APPLICATIONS BOX

Fermented Beans

Both coffee and cacao beans cannot be used immediately after they are picked from the plant. Coffee beans have a mucilaginous envelope that must be removed before they can be roasted. Various organisms can be used to digest the mucilage including *Erwinia dissolvens* in Hawaii and the Congo, *Leuconostoc* and *Lactobacillus* species in Mexico, and species of *Saccharomyces* in India. Fermentation contributes nothing to the flavor but merely expedites processing.

In the case of cacao beans from which cocoa powder is made, the opposite is true. Fermentation is essential to flavor development and seems to proceed in two stages that are characterized by different groups of microorganisms. First, ethanol is produced and then oxidized to acetic acid. Only when the fermented beans are roasted does the typical flavor and aroma of chocolate develop.

TABLE 14.4 Examples of fermented vegetables and fruits

Fruit or Vegetable	Fermented Product
Cucumber	Pickle
Cabbage	Sauerkraut
Olive	Olive
Soy protein	Tofu (essentially similar to cheese)
Soybeans, wheat	Shoyu (soy sauce)

8 hours to extract the bitterness from them. Then they are brined and fermented in the usual manner. Depending on the conditions used in the fermentation process, olives can be black or green.

Fermented Dairy Products

Milk often is advertised as "nature's most nearly perfect food," which applies to bacteria as much as to humans. An enormous variety of bacteria can grow in milk, some disease-causing and some not. Under normal conditions, milk is stable in the refrigerator only for a few days. When it begins to spoil, it turns "sour" because the growing bacteria produce acids. If souring continues, casein (milk protein) eventually is precipitated by the accumulated acids, and a semisolid lump forms. The solid portion of the milk is now called **curds,** and the liquid is called **whey.**

From this basic natural process, a multitude of different dairy products can be produced. Like fermented vegetables, fermented dairy products are cross-cultural; therefore, only a few are discussed in this section. Table 14.5 presents some examples of fermented milk products.

At least two different kinds of buttermilk can be made (table 14.5). Originally, the liquid product remaining after butter was removed from churned fermented milk was called buttermilk. Today, cultured buttermilk is produced by simple fermentation of skim milk (all butterfat removed) for about 14 hours. The acidic taste comes from lactic acid produced during fermentation, and the buttery flavor results from diacetyl produced in the reaction shown in figure 14.2. Note that *Streptococcus lactis* also can synthesize acetaldehyde, a product that contributes to an off-flavor in milk, but *Leuconostoc* converts it to acetate or ethanol. A second type of buttermilk, known as bulgarian buttermilk, requires *Lactobacillus bulgaricus* for fermentation. In this case, the typical flavor results from synthesis of acetaldehyde, primarily by the following reaction:

Equation 14.1

threonine → glycine + acetaldehyde

One fermented dairy product familiar to millions of people, especially to those who are dieting, is yogurt. A mixed culture of *Streptococcus thermophilus* and *Lactobacillus bulgaricus* added to warmed milk produces yogurt. These organisms are both thermophilic; therefore, fermentation is carried out at temperatures of about 45 to 50°C. After some 8 hours, the resulting product has a creamy texture due to the precipitation of casein by lactic acid. The various organisms also contribute significant flavoring compounds, including acetaldehyde principally, with lesser amounts of diacetyl and acetoin and trace amounts of acetone and butan-2-one. Sometimes, a thicker product is desired. If this is the case, powdered dry milk can be added to raise the protein concentration of the milk and thus form more precipitate.

TABLE 14.5 Examples of liquid and semisolid fermented milk

Product	Fermentative Organism(s)
Cultured buttermilk	*Streptococcus cremoris* or *lactis; Leuconostoc cremoris*
Bulgarian buttermilk	*Lactobacillus bulgaricus*
Yogurt	*Lactobacillus bulgaricus; Streptococcus thermophilus*
Kefir	*Saccharomyces* species; *Candida* species; *Lactobacillus* species; *Leuconostoc* species
Acidophilus milk	*Lactobacillus acidophilus*

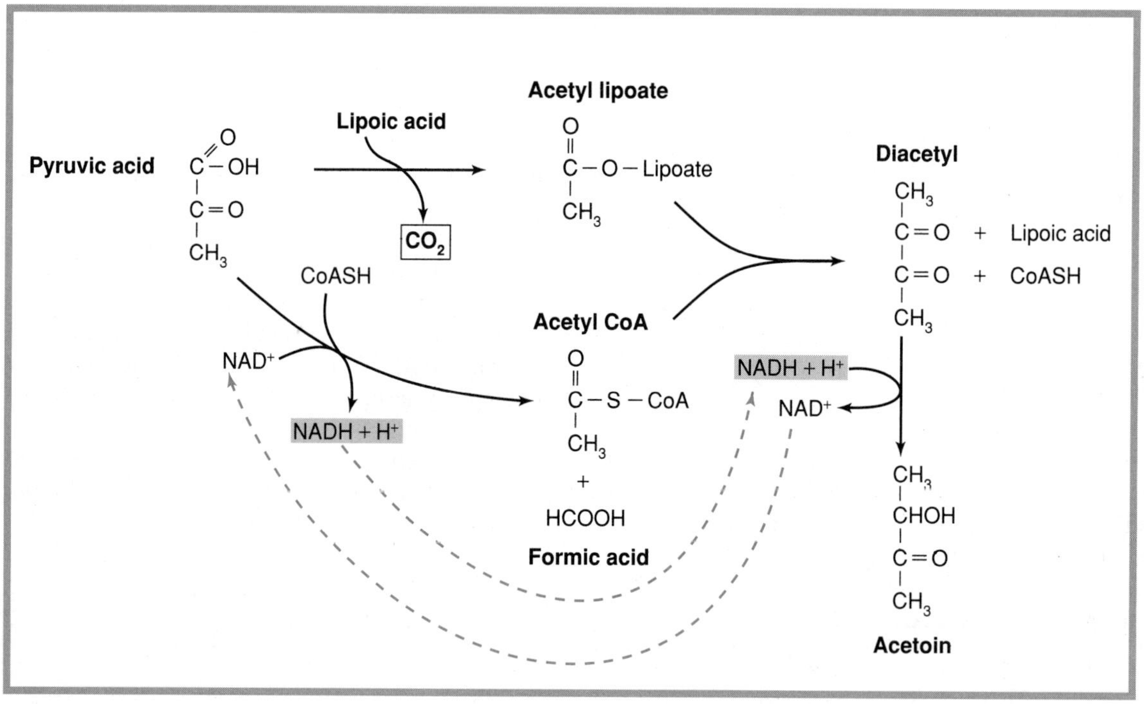

FIGURE 14.2

Production of diacetyl from pyruvic acid. An excess of $NADH + H^+$ will allow the conversion of diacetyl to acetoin with concomitant loss of flavor. Coenzyme A is abbreviated as CoASH when it is uncombined and as -S-CoA in the combined form (see chapter 4).

A most unusual fermented milk originating in the Caucasus of the USSR and now consumed there and in Scandinavia is kefir, an alcoholic fermented milk. The fermentation process is poorly defined in the sense that there are many regional variations leading to substantial differences in the organisms isolated from the product. Table 14.5 lists the bacterial and fungal genera that have been found in kefir. The product is a beerlike liquid containing small grains of carbohydrate in addition to the associated microorganisms. It is thought that the carbohydrate is synthesized by some of the lactobacilli in kefir, although there is some doubt because fermentation generally is not carried out under aseptic (exclusion of unwanted organisms) conditions. The major flavoring compounds are the same as those in yogurt, lactic acid, diacetyl, and acetoin, in addition to ethanol and carbon dioxide.

If curds formed by casein during fermentation are collected and dried, they can be used in the production of cheese. The major steps in the cheese-making pro-

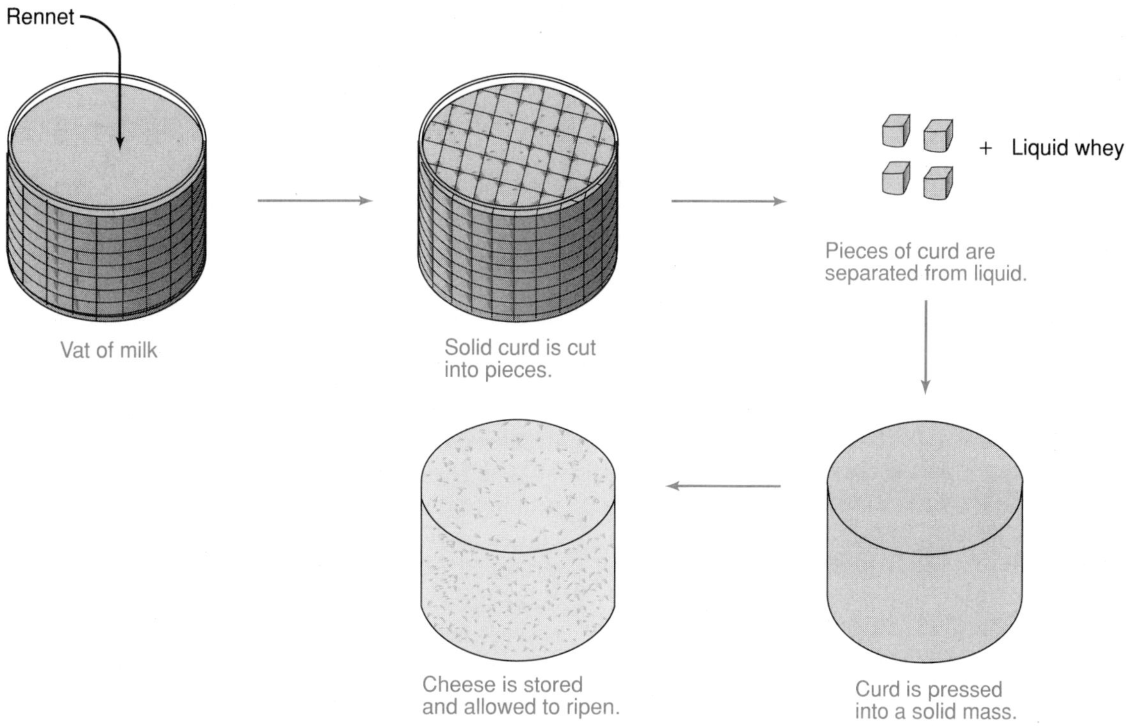

FIGURE 14.3

Five basic steps in the making of cheese.

TABLE 14.6 Some cheeses and the organisms responsible for their production

Cheese	Organism(s)
Soft cheeses	
Cottage cheese	*Streptococcus diacetylactis; Leuconostoc* species
Mozzarella	*Streptococcus thermophilus; Lactobacillus bulgaricus*
Brie	*Streptococcus cremoris; Streptococcus lactis*
Hard cheeses	
Gouda	*Streptococcus cremoris*
Roquefort	*Streptococcus cremoris; Penicillium roquefortii*
Cheddar	*Streptococcus lactis; Streptococcus diacetylactis; Leuconostoc* species
Swiss	*Propionibacterium shermanii; Lactobacillus helveticus; Lactobacillus lactis; Lactobacillus bulgaricus; Streptococcus thermophilus*

cess are shown in figure 14.3, and some of the more common cheeses together with the organisms used to prepare the curd are summarized in table 14.6. Most natural cheeses, however, do not begin by fermentation but by an enzymatic reaction involving rennin, an enzyme obtained from calf stomachs and certain fungi. This reaction causes casein to coagulate and form large clumps of curds. If the milk is strained through cheese cloth, the whey is removed and what remains is cottage cheese.

Mozzarella and cottage cheeses are soft cheeses served directly after initial fermentation. Both contain lactic acid, but cottage cheese also obtains some flavor from diacetyl production. It has not been as tightly compressed as mozzarella; therefore, it has substantially more water.

Some soft cheeses are ripened when various fungi grow on the surface of compressed curds. In the case of Brie, the surface organisms are assorted yeasts in addition to *Penicillium caseicolum*. They contribute fatty acids and aromatic hydrocarbons as flavoring compounds.

Small lumps of curd can be pressed firmly into a mold to squeeze out even more whey, reducing the water concentration below 50%. The curd is then ripened by allowing microorganisms to grow in or on it. Depending on storage conditions and on organisms used in this process, various types of firm cheeses can result.

Harder cheeses are produced by ripening from within the curd, while semihard cheeses use surface growth in addition. Because organisms are mixed into milk prior to the addition of rennin, they can be found throughout the curd, even inside sizable lumps. The ripening process can require months to years. Each cheese flavor depends on the organism(s) added to the milk and on the characteristic compounds synthesized. Many cheeses take their names from the locales in which they were produced originally; for example, Edam and Gouda were first produced in the Netherlands. They probably derive their unique qualities from unusual local strains of bacteria.

In the case of swiss cheeses, the characteristic holes or "eyes" are caused by bubbles of carbon dioxide produced by *Propionibacterium shermanii* using the metabolic pathway outlined in figure 4.16. Because propionibacteria grow slowly, swiss cheeses require a particularly long time to develop properly. The chief flavoring compounds are amino acids, peptides, butyric and acetic acids. Cheddar cheeses are ripened by lactobacilli and pediococci that contribute amino and fatty acids as flavoring compounds.

As stated previously, organisms growing on the surface of the curd also aid in ripening a number of semihard cheeses. Obviously, these cheeses must be produced as substantially smaller balls of curd in order for the flavoring compounds to be distributed throughout the cheese. Gouda is a cheese ripened primarily by starter organisms but some propionibacteria may accumulate on the surface of the cheese. Perhaps the best known semihard cheese is Roquefort, which takes its name from the region in France where it is produced. In the original process, the cheese ripened by aging in caves located outside the town in which the fungus *Penicillium roquefortii* is prevalent. Growth of the fungus in the cheese in addition to some yeasts and micrococci on its surface imparts the characteristic flavor and blue streaking of the cheese.

Obviously, anyone can produce the same type of cheese as the people of Roquefort do given the availability of the proper microorganisms. In fact, similar cheeses are made in many parts of the world; however, the name *Roquefort* is a trademark for that particular type of cheese made in France, and other producers are not allowed to use the name. Instead, their products usually are sold under the generic name of blue cheese.

TABLE 14.7 Sources of flavoring compounds in cheeses

Substrate	Metabolite
Lactose	Lactic acid; propionic acid
Citric acid	Carbon dioxide; diacetyl
Lipids	Fatty acids; ketones
Proteins	Amino acids; peptides
Amino acids	Ammonia; phenylethanol

The cheeses discussed thus far have all been natural in that they obtain their flavor from the growth of one or more microorganisms using some of the substrates listed in table 14.7. Many manufacturers also make "processed cheeses" in which the ripening step is shortened or omitted. The desired flavors are introduced by the addition of appropriate organic chemicals that mimic the end products of microbial metabolism. Because the lengthy ripening step is omitted, processed cheeses usually are cheaper.

Beverage Fermentations

One thing that seems to be common to peoples all over the world is the production of alcoholic beverages from one or more of the local crops. This activity is exemplified in Western cultures by the production of wine and beer.

Wine Fermentations

Wines can be produced from any fruit juice or liquid extract high in carbohydrates. Most wines are made from grape juice, but some also come from peaches, apricots, dandelions, and other fruits. In the case of conventional wines, the growers pick grapes when a certain concentration of sugar is present within the fruit. Up to the concentration at which the ethanol becomes inhibitory, the yield of ethanol from fermentation is proportional to the amount of carbohydrates present. When inhibition occurs at lower ethanol concentrations, the finished wine becomes progressively sweeter because some carbohydrate remains after fermentation. "Dry" wines are those that have little or no sugar remaining from the fermentation process.

Basically, fermentation results from the growth of yeasts (*Saccharomyces cerevisiae*) on juice stored in large vats approximately at room temperature (figure 14.4). Sucrose is converted to glucose and fructose that are then fed into the EMP pathway to be degraded to pyruvate. In turn, pyruvate is converted into ethanol and carbon dioxide is released (figure 4.17). In order to maximize the yield of ethanol, a vintner releases the carbon dioxide from the fermentation vats. According to the law of mass action, removal of one product of the chemical reaction tends to drive the reaction toward complete digestion of the sugar. The practical limitation to fermentation of fruit juice is the ethanol tolerance of the yeast, which may be as high as 14 to 18%. The entire process can last from 4 to 10 days.

Modern wines are produced with characteristic strains of yeast that are closely guarded secrets of each winery. Subtle differences in the flavor of wines result from extra substances produced by these yeasts. For this reason, vintners generally treat their grapes with sulfur dioxide to kill resident yeasts before adding their own cultures. Some resident bacteria remain after treatment, however, and organisms such as *Leuconostoc* carry out a malolactic fermentation process shown in figure 14.5. The effect of this reaction pathway is to reduce the acidity of the wine and enhance certain of its flavors.

Additional flavors are added to wine as it ages in oak barrels. The alcohol literally dissolves flavoring substances from the wood. Hence, the length of time that a wine is left in a barrel is quite important and varies inversely with the number of times a barrel has been used previously.

When appropriate flavors have developed, a wine is ready to be bottled. Although traditional cork seals are being replaced with plastic ones, the bottling process is very similar in principle to that in Pasteur's day. There is always a risk of some organisms being introduced that can grow using ethanol as a source of carbon and energy. If such an event occurs, the acid that they produce sours the wine. Pasteur recommended his pasteurization process to French winemakers in order to kill such organisms without drastically altering the flavor of the wine. The cork seal then prevents any new organisms from entering the wine as long as the cork remains wet. Storage of wine bottles on their sides fulfills this criterion.

Modern wine manufacturers often omit the pasteurization step entirely by using the membrane filters described in chapter 13. In one step, therefore, a manufacturer can both clarify the wine and sterilize it. If bottles have been sterilized also, there is no reason to expect the wine to be altered microbiologically during shipment.

A

B

FIGURE 14.4

(**A**) Fermentation vats used for the production of wine. (**B**) Large storage tank in which wine can be aged.

(A) © Christina Dittmann/Rainbow. (B) © 1984 David J. Cross/BPS.

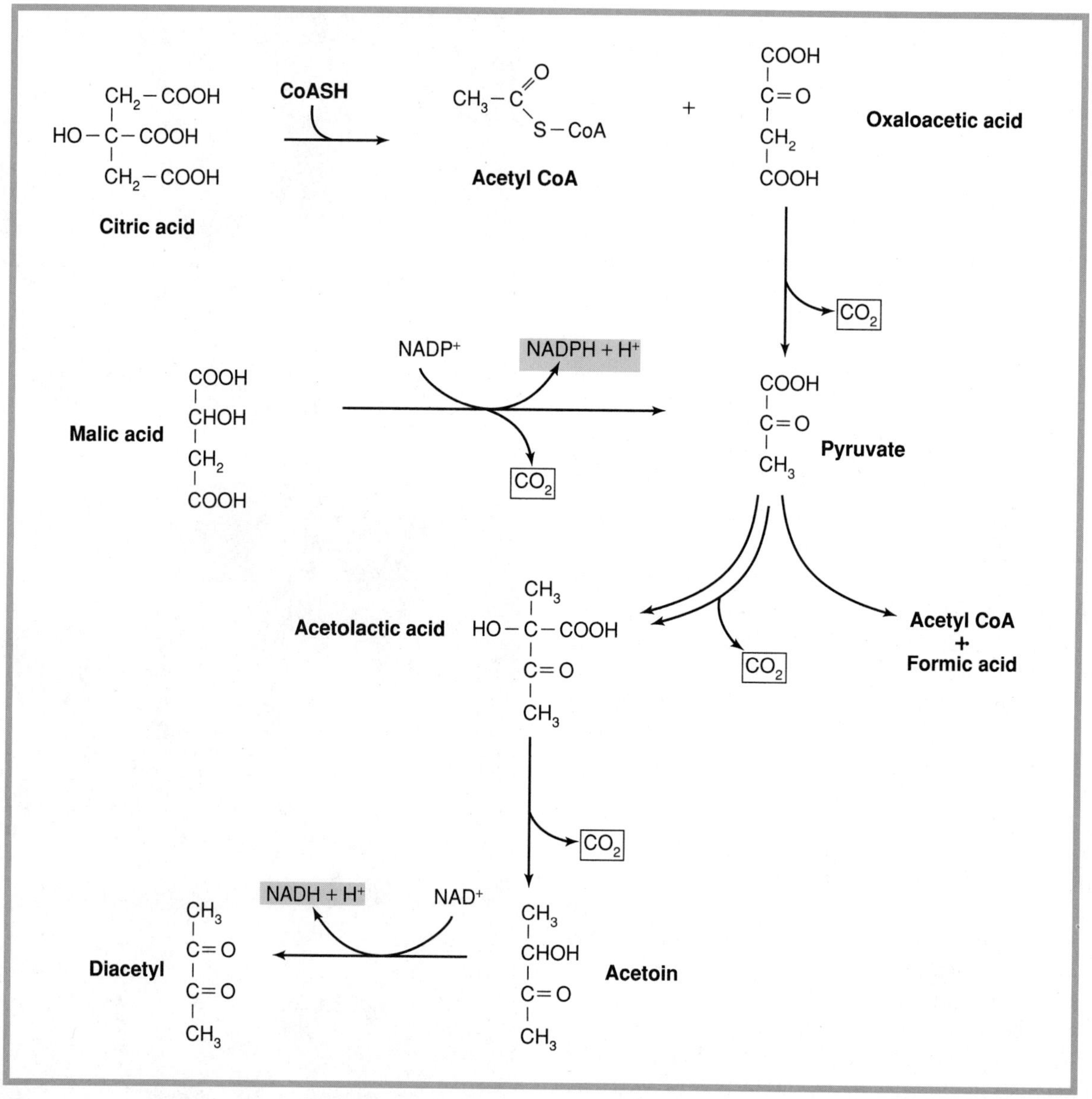

FIGURE 14.5

Malolactic fermentation carried out by *Leuconostoc mesenteroides.*

This process was proposed by Shimazu, Y., M. Uehara, and M. Watanabe. 1985. Transformation of citric acid to acetic acid, acetoin and diacetyl by wine making lactic acid bacteria. *Agricultural and Biological Chemistry* 49:2147–2157.

The alcohol produced by fermentation can have various fates. Many wines are sold as is or after varying periods of aging. Sometimes a wine is converted into a sparkling (carbonated) beverage, such as champagne. A champagne manufacturer who carries out fermentation in the traditional manner (in the bottle) takes unaged wine, adds more sugar and a special strain of alcohol-tolerant yeast, and bottles the wine using a temporary closure. Over the course of a year or more, the yeast converts some of the sugar to carbon dioxide and a little more ethanol. The change in ethanol concentration is negligible, but carbon dioxide pressures of

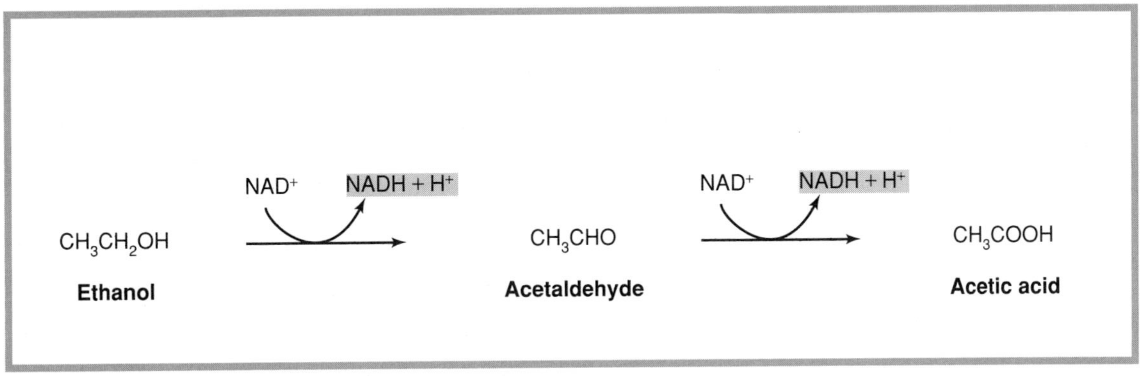

FIGURE 14.6
Oxidation of ethanol to acetic acid in *Acetobacter*. Each reduced NAD can feed into electron transport to drive proton translocation.

up to 7 atmospheres can develop. For this reason, sparkling wines have their corks wired in place so that they cannot accidentally be dislodged. In addition, they carry hazard warnings on their labels. A more inexpensive way to prepare champagne is the Charmat bulk process in which the second fermentation is carried out in large tanks and the champagne transferred to bottles later.

Vinegar Production

Vinegar manufacturers deliberately produce and market soured wine. They treat diluted wine under aerobic conditions with *Acetobacter aceti*. The oxidation reaction in which *Acetobacter* converts ethanol to acetic acid is depicted in figure 14.6. Under federal law, vinegar sold for human consumption in the United States must be produced by fermentation, not by chemical synthesis. It may be sold as wine vinegar, still containing the flavoring compounds from the wine, or as the distilled product, which is pure, dilute acetic acid.

Beer Fermentations

Beer is also produced by fermentation, but the process is distinct from that used for wine making. The fundamental ingredients are modified plant extracts rather than plant juices. First comes barley malt, barley seeds that have been germinated briefly to activate their enzymes. These enzymes break down the starch stored in the seeds into simpler sugars that can be used by yeasts. When the seeds are dried, the product is called barley malt. Without this step, fermentation is impossible. Other cereal grains, such as rice or corn, also can be added to provide extra starch. Enzymes from the malt serve to convert these starches to usable forms.

A brewer takes malt and any added cereals, mixes them with water, and cooks them in a large vessel (figure 14.7). The resultant mixture is called a mash. Temperature of the mash is increased gradually to allow the enzymes from the malt to carry out their appropriate functions before they are inactivated by high temperature (about 77°C). At the conclusion of the process, the solid material (a waste product) is removed and sold as animal feed. The liquid portion of the mash, now known as wort, is then piped into a brewing kettle.

In the brewing kettle, flowers of the hop plant are added to provide the characteristic bitter flavor and aroma of beer. The mixture is boiled and then the wort removed to the fermentation vessel. *Saccharomyces cerevisiae* is the organism of fermentation. Actual fermentation takes place at relatively cool temperatures (3 to 14°C) and requires more than a week to reach completion. The product of this stage contains about 5% alcohol and is mildly carbonated because some carbon dioxide of fermentation is retained. Final carbonation requires the addition of carbon dioxide gas from the fermentation vessels or a brief second fermentation analogous to the process used to produce champagne.

The major solid material in the beer is removed, and the beer is lagered (stored in tanks) for some period of time. More particulate matter settles out during storage, and a final filtration removes the last traces of sediment. Beer is not sterile; therefore, it must be kept cold and used promptly, pasteurized as wine is, or filter sterilized.

APPLICATIONS BOX

Low-Carbohydrate Beers

The so-called light beers contain fewer calories not because of a reduced alcohol concentration but because of a reduction in complex carbohydrates. Simple sugars are fermented to ethanol by *Saccharomyces,* but more complex sugars may not have been completely degraded by malting enzymes. The remaining carbohydrates usually are branched polysaccharides whose catabolism requires special debranching enzymes not normally found in yeast. Various microbial enzymes can be added to the malt or to the first fermentation to enhance the degradation of complex carbohydrates to simple carbohydrates. These simple sugars can be converted to ethanol by *Saccharomyces;* therefore, unless precautions are taken, the concentration of ethanol can rise to unacceptable levels. One other problem can arise when carbohydrates are broken down enzymatically. If the enzymes survive the pasteurization process, they will continue to degrade carbohydrates and can, in fact, render beer sweeter than normal by converting branched molecules that are normally tasteless into simple sugars with a pronounced sweet taste.

A

B

FIGURE 14.7

(A) Mash tank in a large brewery. The wort is prepared in the tank and then transferred into brew kettles **(B)**. After several hours of boiling, the brew is filtered, cooled, and placed in a fermentation vessel similar to that shown in figure 14.4.

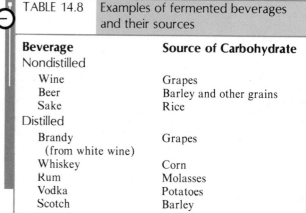

TABLE 14.8 Examples of fermented beverages and their sources

Beverage	Source of Carbohydrate
Nondistilled	
Wine	Grapes
Beer	Barley and other grains
Sake	Rice
Distilled	
Brandy (from white wine)	Grapes
Whiskey	Corn
Rum	Molasses
Vodka	Potatoes
Scotch	Barley

Distillation of Alcoholic Fermentations

There are other alcoholic beverages, such as brandy, whiskey, and rum, that are based on the distillation of various fermentations. Each beverage differs in the local plant product fermented and in the parameters of distillation itself (see table 14.8). After fermentation is completed, the product is distilled to concentrate the alcohol. The manner in which distillation is conducted determines the nature of the final product.

Brandy results from the distillation of a specially fermented dry white wine. In addition to ethanol, **fusel oils,** alcohols with longer chains of carbon (C_8 to C_{12}), have also been synthesized during fermentation and can be distilled into the product. The range of alcohols between ethanol and fusel oils, primarily isoamyl, *n*-butyl, and *n*-propyl, are significant flavoring compounds in brandy. Accumulated ethyl esters of fatty acids contribute other flavors. These and similar important flavoring compounds present in low concentration are called **congeners.**

A good brandy has very little in the way of fusel oils or acetaldehye, both of which detract from its flavor. This necessitates some changes in the fermentation process itself. The temperature must be kept below 23°C to minimize fusel oil production. Sulfur dioxide must not be added to the grapes otherwise the following reaction is possible:

Equation 14.2

$$CH_3CHO + SO_2 + H_2O \rightarrow CH_3CHOHSO^-_3 + H^+$$

The aldehyde-sulfurous-acid complex would distill with the alcohol and provide an off-flavor. When sulfur dioxide is not used to pretreat grapes, special care must be taken to choose quality fruit for fermentation.

The other distilled beverages listed in table 14.8 are based on mash fermentation like that used for beer. The differences among them depend on the nature of carbohydrates added and on the way in which distillation is carried out. Whiskeys, for example, are distilled from grain (corn or rye) so that some of the flavoring substances are carried over into the final product. On the other hand, vodka is practically pure ethanol distilled from potatoes.

The use of ethanol as a gasoline additive (gasohol) has been a recent development in the alcohol fermentation industry. Although still somewhat controversial because food is converted into fuel, the process has had a significant impact on both the corn producers and the fermentation industry. In 1985 it was estimated that about 220 million bushels of corn (about 2% of the crop) were fermented into ethanol for use as gasohol.

Microorganisms as Foods

Leavened Breads

Nearly every householder has had occasion to use yeast in baking at some time or another. The packaged yeast available in the stores is another strain of *Saccharomyces cerevisiae,* the same species used in the manufacture of wine and beer. In this case, however, the purpose of fermentation is to generate carbon dioxide instead of ethanol. Dough rises because the pressure of carbon dioxide gas being generated within it causes the dough to expand as the yeast ferments the extra sugar added during dough preparation. The ethanol that is coproduced with the carbon dioxide is driven off during baking.

Single-Cell Protein

Although a worldwide food shortage is debatable in light of crop surpluses and fallow farmland in the United States, there is, no doubt, a shortage of affordable food for developing nations. Similarly, farmers who raise animals for slaughter would like to have low-price feed available when their animals cannot forage for themselves. One proposed solution for both needs is the **single-cell protein,** a food product made from unicellular microorganisms.

The concept is fundamentally sound. For example, cyanobacteria from the genera *Spirulina* or *Anabaena* could be raised in solar ponds, shallow, water-filled depressions that have been designed to maximize a pond's exposure to sunlight. At intervals, cyanobacterial cells could be collected and sold as a food product or food supplement. A potential difficulty with this procedure is the possibility of bacterial contamination of the pond. Indoors, a substance such as molasses that is high in carbohydrate but low in protein could be used as a substrate for the growth of yeast. This product would serve as a more complete food.

The problems with single-cell protein are twofold. First, it is difficult to produce microbial protein inexpensively enough to be affordable to those who need it. In many third world countries, an annual income of $50 is not unusual. Second, the protein is accompanied by a large amount of nucleic acids that, when broken down, yield uric acid. Excess uric acid in the human body can lead to gout and various other problems. If the microbial cells are treated to remove the nucleic acids, the cost will increase, further aggravating the problem of affordability.

Food Spoilage

Until now, the emphasis in this chapter has been on the positive aspects of microbial growth on foods. Not all microorganisms do good things to the food on which they grow, however. Food spoilage in the strict sense of the term refers to detrimental changes in texture, taste, and appearance caused by microorganisms. Various disease-causing organisms considered in chapters 18 and 19 are not included in this discussion.

One way in which food can be spoiled is by degradation of macromolecules to the point at which the food has lost its ordinary texture. Proteases or cellulases may be responsible for this type of activity. Other problems can arise from the growth of organisms that degrade macromolecules to compounds having dis-

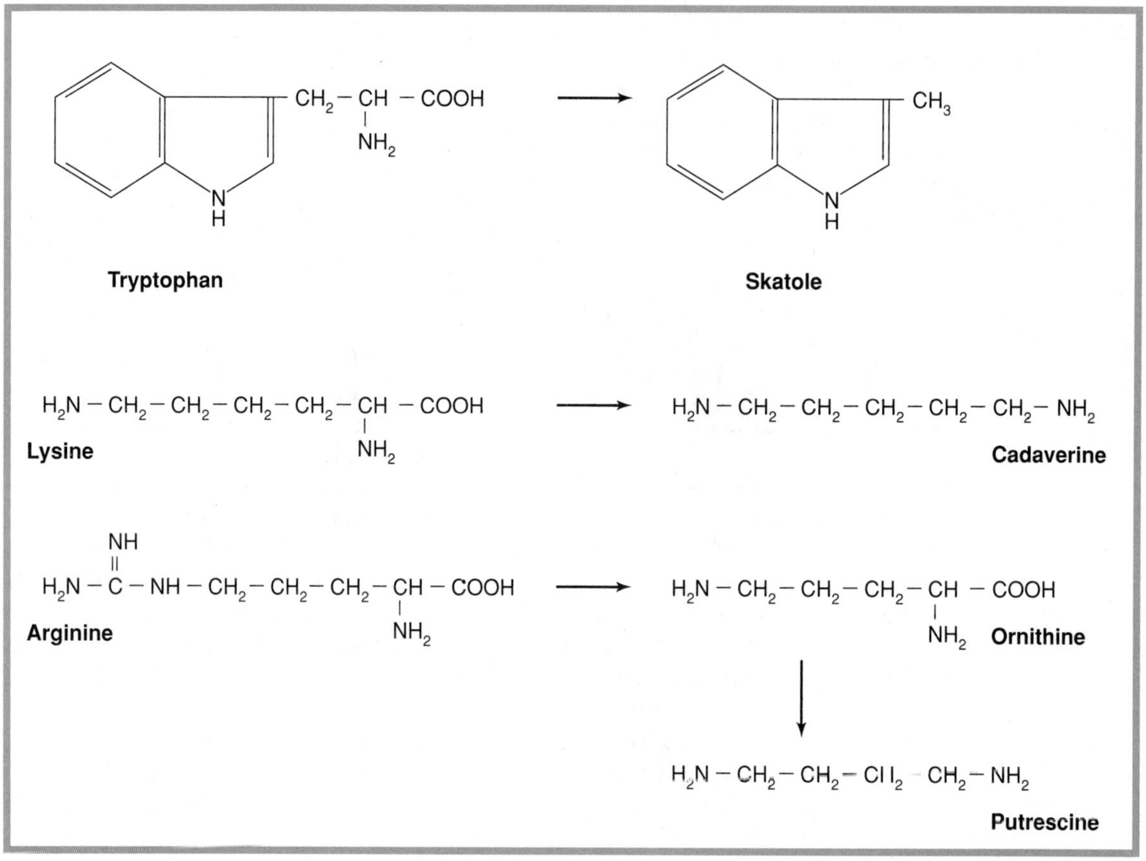

FIGURE 14.8

Skatole, an intermediate in the degradation of tryptophan, is one of the chemicals responsible for the typical odor of feces. As such, it contributes to off-flavors in foods but is not likely to present a problem unless extensive protein degradation occurs. Other compounds contributing to off-flavors include the polyamines putrescine and cadaverine, both derivatives from diamino acids.

agreeable flavors. Lipid degradation can give rise to organic acids that inhibit the growth of necessary organisms for flavor development. The breakdown products of amino acids, such as the aptly named skatole derived from tryptophan (figure 14.8), are frequently disagreeable. Certain peptides can impart a very bitter flavor to a food when present in any quantity.

Even desirable flavoring compounds are not always appropriate. As noted earlier, diacetyl is required in certain milk products, yet if it is present in excess, the flavor is quite unappetizing. It is the balance of flavors that is important in food fermentations. The presence of a "wrong" organism or the imbalance of a mixture of organisms can completely ruin a fermented food product. For these reasons, commercial food fermentations are carried out under aseptic conditions with specially prepared innocula. Naturally occurring microorganisms do not occur in sufficiently reproducible mixtures for the modern food processor.

The types of microorganisms involved in food spoilage depend on the nature of the food product. The most important factors that probably contribute to spoilage are water content and pH of a product. Vegetables have an average water content of 88%, and fruits are not far behind at 85%. Nevertheless, bacteria rarely spoil fruits because their pH is generally too low to support bacterial growth. Fruits are most susceptible to fungi, including yeasts. On the other hand, vegetables are primary targets of bacteria and are most often attacked by species of *Erwinia,* a ubiquitous genus occurring on plants.

Meats, of course, are particularly susceptible to spoilage by all types of microorganisms because they have a relatively neutral pH coupled with a high-nutrient density. The organisms involved vary, depending on the manner of slaughter or harvesting. Possible contaminating microorganisms include enteric bacteria and *Clostridium perfringens* from the intestinal tract; pseudomonads (resembling members of the genus *Pseudomonas*) from the skin of an animal; and a wide variety of fungi and yeasts including *Rhizopus, Mucor, Candida,* and *Rhodotorula* organisms. Ground beef is spoiled usually by bacteria, whereas cuts of meats, like steaks, can accumulate surface growth caused by bacteria or fungi. If storage conditions are high in humidity, bacteria predominate. Otherwise, spoilage is due to fungi because they can tolerate lower moisture levels.

Summary

The food processing industry in the United States is an enormous one, and microbiologists are involved on two levels. Food spoilage and preservation are one microbiological concern. Various bacteria and fungi act as decomposers and can grow on a food product as soon as it is harvested (and sometimes before). This activity does not present a major concern if the product is eaten promptly, but adverse microbial growth presents a significant barrier to long-term food storage and to central processing facilities.

Food can be preserved for comparatively long periods of time if it is subjected to appropriate treatments. Thorough cooking inactivates most bacteria present. Sterilization is frequently not an option because the great heat involved alters food flavor and texture in ways that consumers find unacceptable. Sterilization is used for canned and irradiated foods, however, both of which will keep for years. Some foods are considered to be "commercially sterile" in that all significant disease-causing and spoilage organisms are absent and shelf life is extended. Some foods can be cured with salt or sugar, a method of preservation that increases the osmotic strength of a product, thereby lowering water activity and inhibiting microbial growth. Of course, drying of foods can lower their water content also. A lower pH deters food spoilage organisms as well. Aromatic compounds added in the smoking process or as purified chemicals also are used for their preservative effects.

The other level on which food microbiologists work is in the actual production of certain types of foods. Generally, foods are fermented using the carbohydrates present in them as an energy source. Fermentation products lower the pH of the food (preservative effect) and also provide its characteristic flavor. Typical flavoring compounds are diacetyl, lactic acid, acetaldehyde, fatty acids, and amino acids. Many dairy products such as yogurt, buttermilk, cheese, and kefir can be produced in this manner. Vegetables and fruits also can be fermented for long-term storage. Products prepared from them include sauerkraut, tofu, and olives. Numerous alcoholic beverages can be made, all of which are based on wine or beer fermentation processes. The alcohol and/or flavoring compounds can be concentrated by distillation to create new flavor combinations or alcohol for commercial use.

Questions for Review and Discussion

1. Give an example of a food prepared by fermentation. What organism(s) carries out the fermentation process, and what chemical products of fermentation impart the characteristic flavor of this food?
2. Describe the general process for manufacturing cheese. What is the difference between a ripened and an unripened cheese? Between a hard and a soft cheese?
3. Describe the general process of alcohol production by fermentation. What are the differences and similarities between wine making and beer making? What kinds of products can be produced as a result of distillation of a fermentation mixture?
4. Describe three different methods of food preservation and tell why they are effective. Which of your examples would provide storage for the longest time?

Suggestions for Further Reading

Banwart, G. J. 1989. *Basic food microbiology.* Van Nostrand Reinhold, New York. An extensive and up-to-date textbook.

Davies, F. L., and B. A. Law. 1984. *Advances in the microbiology and biochemistry of cheese and fermented milk.* Elsevier/North Holland Publishing Co., New York. This collection of review papers includes two good articles on the subject of flavor development.

Mountney, G. J., and W. A. Gould. 1988. *Practical food microbiology and technology.* Van Nostrand Reinhold, New York. A comparatively short, general survey text.

Shimazu, Y., M. Uehara, and M. Watanabe. 1985. Transformation of citric acid to acetic acid, acetoin and diacetyl by wine making lactic acid bacteria. *Agricultural and Biological Chemistry* 49:2147–2157.

chapter

14

Applied and Industrial Microbiology

chapter outline

GENERAL GOALS

Be able to describe several different types of industrial processes in which microorganisms are used, explain the biochemical reactions important in these processes, and specify the genus and species names of the organisms involved.

Be able to explain how genetic engineers can contribute to applied microbiology and how they hope to take advantage of certain types of symbiotic relationships to improve plant cultures.

Be able to discuss short- and long-term methods for culture preservation.

In this chapter, some important uses of microorganisms in economically significant ways are summarized from the information presented in chapters 4 through 14. Among the topics considered are applied genetic engineering and biochemistry, including some unusual ways in which microorganisms can be used to contribute to new technologies. New methods to enhance traditional food fermentation processes are considered as well. The final topic of this chapter concerns the preservation of microbial cultures for future use.

Applied Genetic Engineering

Solutions to Some Environmental Problems

Chapter 12 presents some environmental problems that have been caused by technology. In many cases, these problems result from industrial chemicals that are not properly or promptly returned to their normal nutrient cycles. The chemicals can be toxic to microorganisms that normally would degrade them, or they can have a structure that is very refractory to attack by normal enzymes. In either case, if microbial degradation is to occur, special microorganisms will have to be provided.

A solution to the problem sometimes can be found in the context of what is called **directed evolution.** In this process, selective and enrichment techniques are employed to effect comparatively rapid changes in the phenotype of a culture in order to render it capable of new biochemical activities. Take for instance the situation in which a carbon-containing molecule (for example, Dinoseb, an herbicide) has accumulated in the environment. The carbon skeleton represents a potential source of energy for a microorganism, if the organism has the proper enzymes to strip off side groups and degrade the skeleton into appropriate smaller molecules.

In the normal course of events, Dinoseb accumulates in soil and groundwater for many years until one or more organisms happen by chance to develop the abilities to attack it. This process of evolution can be imitated in the laboratory with a series of culture tubes. The first tube is inoculated with organisms that have been isolated from forest soil and shown to possess a weak ability to degrade the herbicide. The medium used contains a low level of a metabolizable organic molecule, such as glucose, and a trace of herbicide. After the bacteria in the first tube reach stationary phase, an inoculum is transferred into a second tube that contains less glucose and slightly more herbicide. This culture is allowed to enter stationary phase, after which time the entire process is repeated.

The hope is that if the culture is transferred many times and the Dinoseb concentration raised slowly enough, a mutant organism that can use the herbicide as a significant carbon source will arise spontaneously. At first, the mutant probably will not be very efficient, but as the culture is transferred to fresh medium and provided with more Dinoseb at regular intervals, additional mutations will arise to provide increased efficiency for Dinoseb catabolism. Cells carrying these additional mutations will have increasing **selective advantage.** As the new mutants grow a bit faster, they will become proportionately more common in the culture than the original mutant (figure 15.1). If the process is continued over thousands of generations, a nearly pure culture of the desired mutant type can be obtained. The evolution of the culture has been directed by the culture medium and by the incubation conditions.

Directed evolution works fairly well when only a single enzyme is needed or when groups of organisms can provide all necessary enzyme activities. Sometimes, however, it is clear from biochemical principles that degradation of a particular molecule requires more than one enzyme. Perhaps there is a benzene ring to be broken as well as chlorine atoms to be removed. In such cases, an experimenter may combine directed evolution with other techniques, such as one in which plasmids coding for some necessary enzymes are added. If no naturally occurring plasmid can provide all the requisite enzymes, DNA splicing techniques (described in chapter 11) can be used to construct one. By means of experiments like these, Ananda Chakrabarty and his co-workers were able to prepare a mutant form of *Pseudomonas* sp. that is useful particularly for degrading crude oil like that found near shipwrecks.

In order to make the mutant *Pseudomonas* bacterium suitable for commercial use, the company concerned attempted to patent it. After numerous legal battles, the Supreme Court ruled that artificially prepared bacteria indeed could be patented, thereby opening the doors to commercial exploitation of laboratory constructs. The work on degradation of environmental pollutants continues, with particular emphasis being placed on herbicides and phenolics.

Solutions to Some Industrial Problems

Manufacturers of foods and chemicals sometimes find themselves faced with a shortage of a particular compound or enzyme necessary for synthesis of their product. At most companies involved in this business,

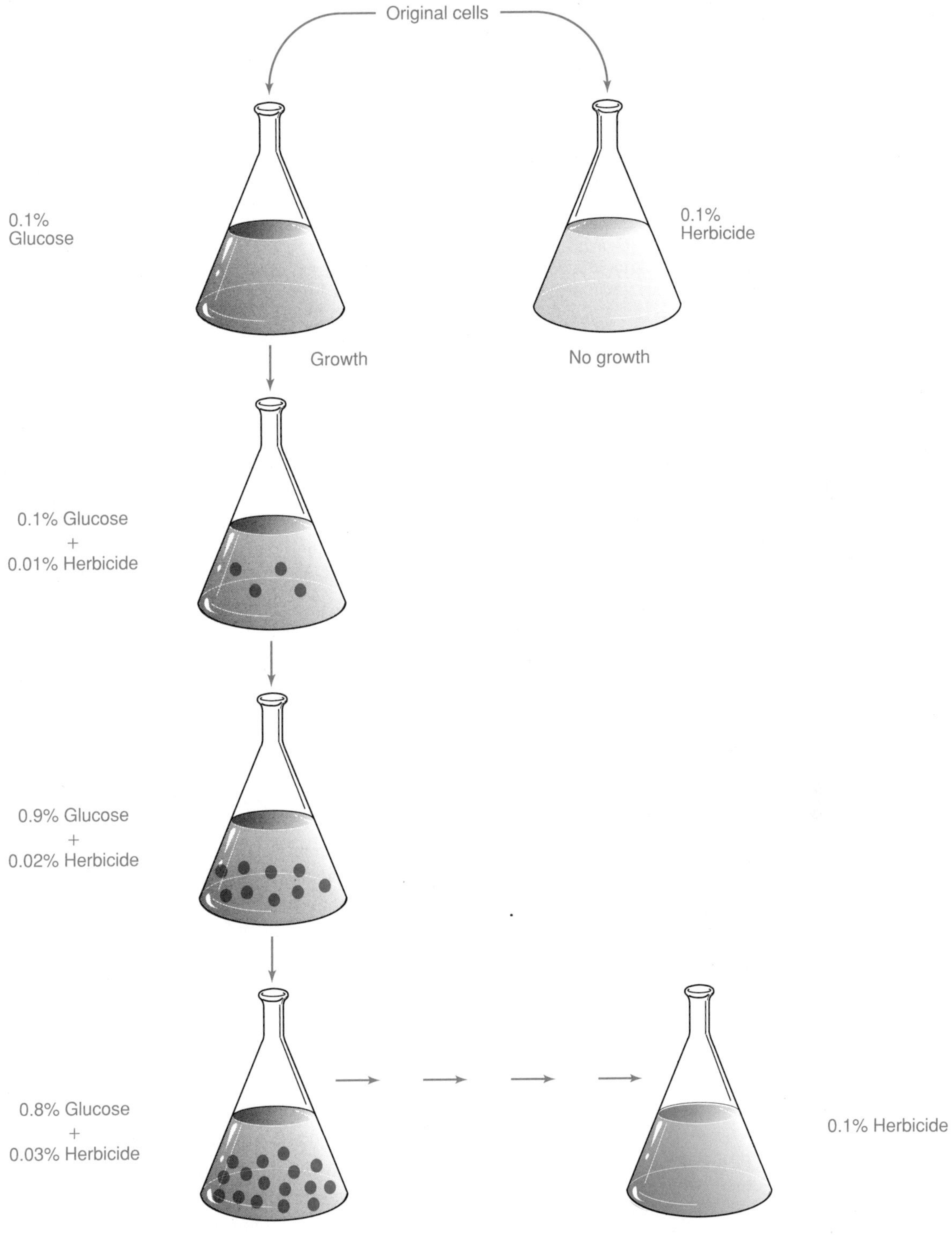

FIGURE 15.1

Experiment in directed evolution. The initial culture does not contain cells capable of metabolizing the target compound (in color), even though this compound is present in the medium. If the culture is transferred to fresh medium at regular intervals and progressively provided with more target compound, an increasing selective advantage is provided to spontaneously mutated cells. Eventually a pure culture of cells containing one or more mutations with the ability to degrade the target compound is obtained.

APPLICATIONS BOX

Immobilized Cells and Enzymes

Sometimes, industrial production of a particular chemical requires only one simple conversion of substrate to product. Instead of adding the substrate to the culture and later purifying away the product, it may be possible to use immobilized cells or enzymes. The basic principle of both technologies is the same. The cells or purified enzyme molecules can be adsorbed by an insoluble support, covalently bonded to an insoluble support, or entrapped within a gel matrix. The essence of the technique is to create a situation in which substrate can be passed over or around the enzyme or cells in a continuous flow. This method provides good control over the reaction parameters and minimizes the cost of the reaction vessel.

In the case of immobilized cells, removal of the cells from their supports has shown that at least some cells remain viable even after treatment. The catalytic activity of immobilized cells does decay, but at least 50% of the activity often can be maintained for several weeks and, in exceptional cases, for more than a year. Similar stability has been reported for immobilized enzymes.

active research projects are underway to clone the necessary information for synthesis of a product into plasmids that can then be placed into bacteria or fungi and used in the manufacturing process. Rennin, the enzyme used for cheese production, provides an example of this situation. The enzyme is found naturally in calf stomach, but its supply is quite variable due to the dependence of cattle raising on economic conditions. Presently, the enzyme has been cloned successfully in the fungus *Aspergillus* and is secreted in its active form. Previous cloning attempts in bacteria and yeasts gave poor yields and/or a denatured form of rennin. It is anticipated that most industrial needs for rennin will eventually be met by the genetically engineered product.

Many other enzymes, primarily those used in scientific research for DNA sequencing, DNA splicing, and general molecular biology, are now produced from organisms carrying cloned DNA. These enzymes in their naturally occurring states were quite expensive to purify originally; therefore, they were good commercial candidates for cloning. Enzymes readily obtainable on a large scale are less likely to repay the developmental costs of a manufacturer.

Fermentations for Chemical Production

Microbial cultures have been used with varying degrees of success to produce a wide variety of chemicals. Basically, an organism that grows on an inexpensive substrate while producing the desired chemical must be found, and then, if possible, it must be modified genetically so that the biosynthetic reactions of interest become unregulated. Under appropriate conditions, it is possible to get as much as 10 to 25% of the mass of a cell as a single product.

In the early days of microbiology, it was not always possible to find one organism to carry out an entire biosynthetic sequence needed for a particular molecule. Instead, one organism was used to synthesize an appropriate intermediate, and then a second organism converted the intermediate into the final product. With contemporary microbiology, gene splicing can produce a single organism to carry out the desired process.

Amino Acid Production

Amino acids can be used for different purposes in food. Sometimes they serve as supplements as, for example, in the case of corn whose protein is deficient in lysine. When corn is part of a varied diet, the fact that it is deficient in an essential amino acid usually poses no problem. For people subsisting primarily or entirely on corn, however, the corn should be supplemented with lysine to provide complete nutrition in this diet. Amino acids also can be used as part of a flavoring process. The sodium salt of glutamic acid (monosodium glutamate) has been found to enhance the flavor of various meats. Monosodium glutamate is used commonly in oriental cooking and can be added as well to chicken during the processing stage to bring out its flavor. The amino acid proline is an important raw material in the pharmaceutical industry.

Use of amino acids as food supplements requires that they be readily available in quantity and sufficiently inexpensive so as to maintain the basic price of such foods. In order to achieve these ends, large-scale fermentations are essential. The various aromatic amino acids (tryptophan, tyrosine, and phenylalanine) are often prepared in this way. Regulatory patterns in the synthesis of aromatic amino acids are shown in figure 15.2. More biochemical details are illustrated in figure 5.6. Note that there are several feedback loops present;

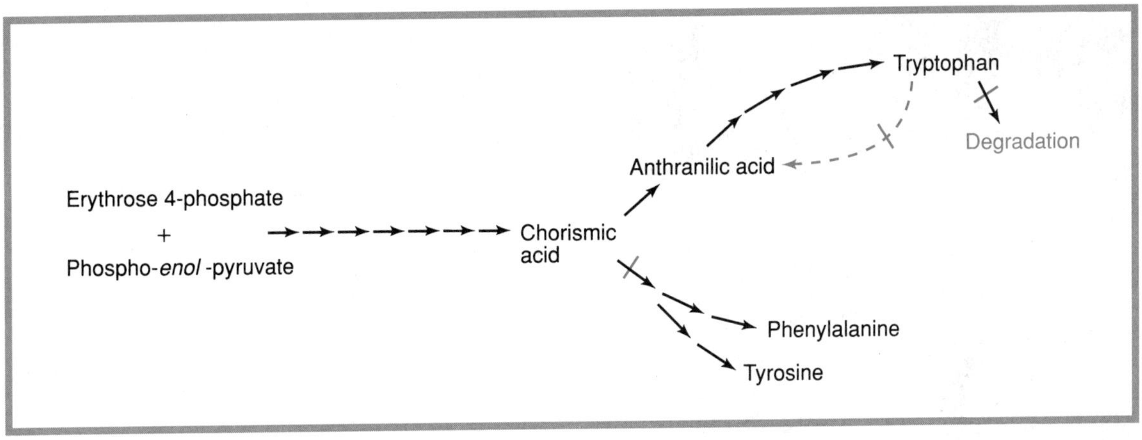

FIGURE 15.2
Regulatory patterns in the synthesis of aromatic amino acids. In order to have tryptophan overproduced, its feedback mechanism (in color) must be blocked. Yield can be improved by interruption of the tryptophan degradative pathway and the phenylalanine-tyrosine biosynthetic pathway.

TABLE 15.1 Some amino acids that can be produced on a commercial scale by microorganisms

Amino Acid	Producing Organism(s)	Carbon Source
L-tryptophan	*Serratia marcescens; Escherichia coli*	Ethanol
L-valine	*Acinetobacter calcoaceticum*	Ethanol
L-glutamic acid	*Brevibacterium flavum; Corynebacterium* sp.	Cellulose Molasses

therefore, if the concentration of any one of the aromatic amino acids is excessive, there is a proportional reduction in its biosynthesis.

The problem facing the bacterial geneticist is one of deregulating the pathway and eliminating secondary pathways. The slashes in figure 15.2 indicate the places at which mutations might have advantageous effects. The general pattern is to block synthesis of the unwanted aromatic amino acids so that all precursors go only to the desired product. In addition, all allosteric enzymes need to be mutated so that they no longer respond to feedback inhibition. When this is done, the yield of amino acid from a *Saccharomyces cerevisiae* culture can be as great as 4.8 nmol/min per mg of protein, a rate that is almost 5,000-fold greater than the wild-type strain. Some additional examples of amino acids that can be produced by microbial activity are listed in table 15.1.

An interesting prospect for obtaining large quantities of proline for the pharmaceutical industry is a **photobioreactor.** This device is basically a series of transparent tubes through which a culture of a photosynthetic microorganism is circulated. For example, consider the green alga *Chlorella.* The alga obtains its energy from sunlight and its carbon from CO_2, minimizing the costs of raw materials to the necessary salts and water. Like many green plants, *Chlorella* responds to osmotic changes in its environment by adjusting its intracellular concentration of proline. Therefore, if the medium circulating in the pipes has a high NaCl concentration, *Chlorella* will synthesize enough proline to prevent water loss from the cytoplasm (it will raise the osmotic strength of its cytoplasm). Under optimum conditions, the quantity of proline synthesized can be as much as 25% of the cell weight. Before commercial production is possible, however, the size of the photobioreactor must be scaled up to the point at which the proline produced is cost-competitive with the proline obtained from other sources.

Microbial Insecticides

Some members of the genus *Bacillus* produce antibiotic compounds during sporulation that are active against other bacteria (chapter 13). Other bacilli also

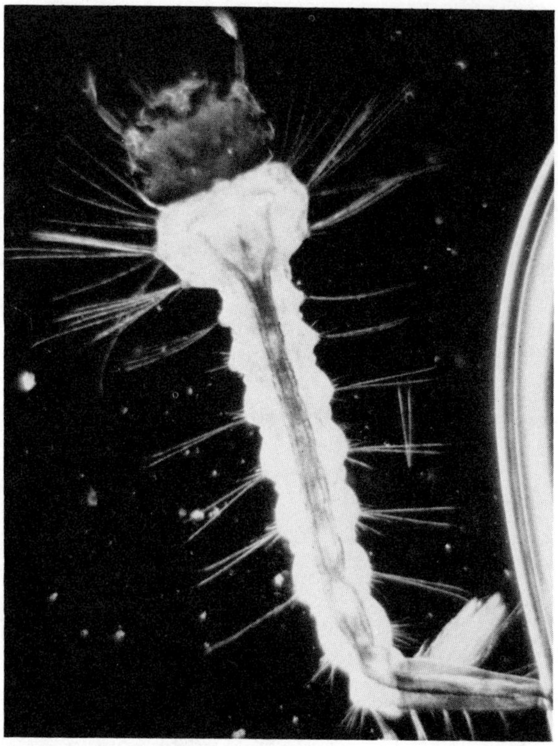

A

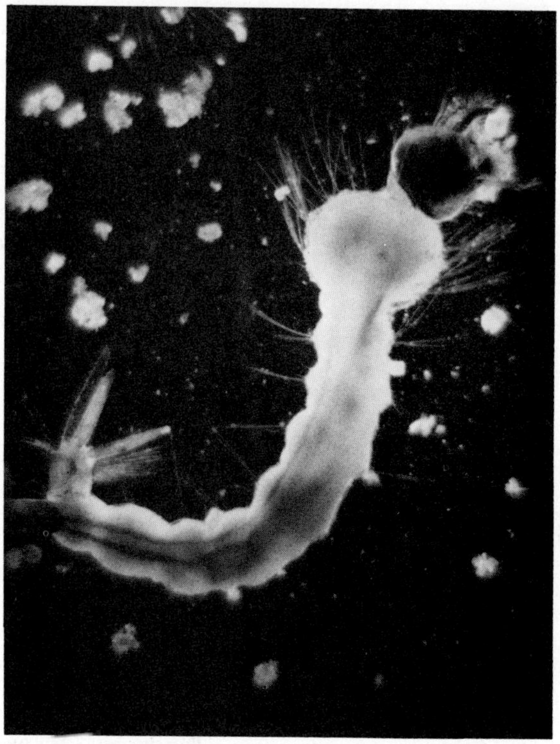

B

FIGURE 15.3

Effect of *Bacillus sphaericus* crystalline toxins. (**A**) A third instar (larva) of the mosquito *Culex quinquefasciatus*. (**B**) A similar larva 48 hours after treatment with the toxin. Endospores associated with the toxin have germinated, and bacteria have filled the larval body.

synthesize special antibiotic compounds during sporulation, but these are active only against various arthropods in the orders *Lepidoptera* and *Diptera*. *Bacillus thuringiensis* variety *israelensis* is the best-known organism that produces a crystal containing an insecticidal protein during sporulation. The protein is highly toxic to the larval stages of various moths and butterflies. It has selective toxicity because the crystal only dissolves and releases the toxic protein in an alkaline environment. The insect gut has the necessary highly alkaline pH, and it is there that the first effects of the toxic protein are seen. The protein's mode of action is not known, but its toxin rapidly causes changes throughout the larva that result in larval death. The crude crystal preparation is sold under various trade names for use against gypsy moth larvae, tomato hornworms, and cotton boll weevils, for example.

There are many diseases transmitted by mosquitoes; therefore, the World Health Organization is sponsoring extensive research into the possibility of using both *B. thuringiensis* and *B. sphaericus* as controlling agents. The latter organism seems to be more specific for mosquito larvae than does *B. thuringiensis*. Figure 15.3A–B illustrates how effective the concentrated toxin can be.

Other insecticidal treatments depend on infection of the host by an appropriate parasite that can be bacterial, viral, fungal, or protozoan. One such treatment that has been licensed for many years is the use of *Bacillus papilliae* against the Japanese beetle. For effective treatment, the bacterium has to be applied to large areas at one time, for example, to an entire golf course. This necessitates large numbers of bacteria, which leads to an awkward complication. *B. papilliae* will not grow on artificial media; therefore, the bacteria are injected into beetle larvae and raised in this way. While this practice undoubtedly is inconvenient, the treatment itself is quite effective. Once introduced into an environment, the bacteria remain in the soil cycling through each new generation of beetle larvae. As a result of their

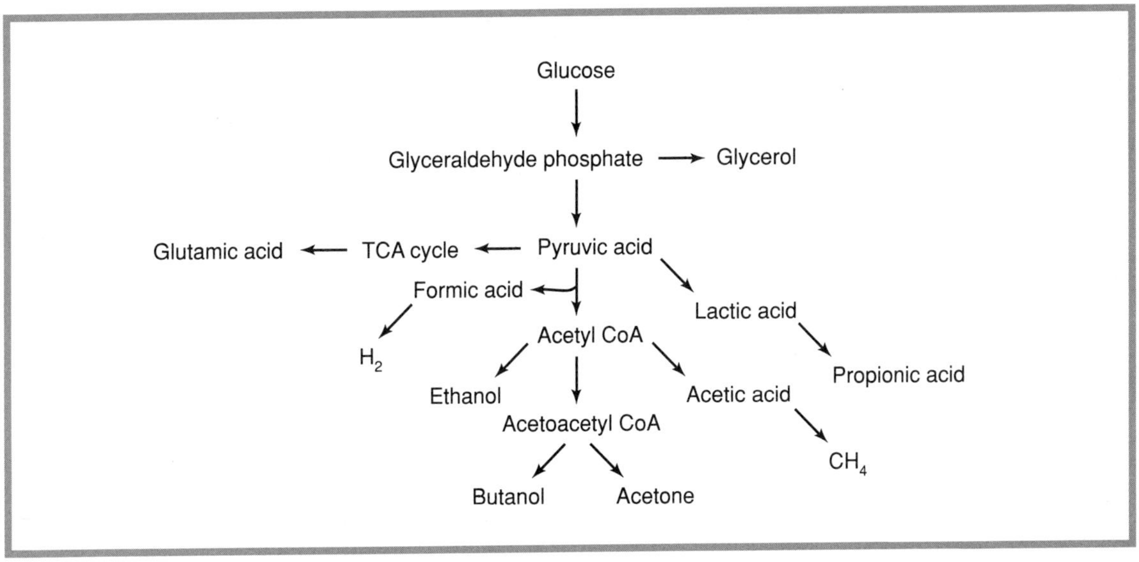

FIGURE 15.4

Biochemical pathways of industrial importance. More complete biochemistry is shown in figure 4.16.

persistence in the environment, treated areas in Connecticut have remained essentially free of Japanese beetles for more than 30 years.

Production of Organic Chemicals

The tremendous variety of chemicals that can be produced during fermentation of glucose (see figure 15.4) includes many that have commercial value. The careful choice of an organism to use for a particular fermentation process can result in the production of significant quantities of a desirable chemical. During the First World War, for example, German scientists took advantage of the EMP pathway to produce the glycerol needed for explosives manufacture.

Glycerol fermentation can be carried out by various yeasts, such as members of *Saccharomyces* or *Torulopsis*. From the cleavage of fructose 1,6-diphosphate, glyceraldehyde 3-phosphate (GAP) and dihydroxyacetone phosphate (DHAP) result, both of which are interconvertible (figure 15.5). GAP is converted stepwise to pyruvate and then to acetaldehyde. NADH + H^+ formed during this process is used to reduce acetaldehyde to ethanol. If sulfite ions are present, they preferentially combine with acetaldehyde and do not allow it to react with the reduced NAD. The cells must reoxidize NAD in order to continue metabolism. They do so by reducing DHAP to glycerol phosphate. The latter molecule then has its phosphate group removed and is secreted into the surrounding medium as a waste product.

Another well-known microbial process that also has seen wartime use is acetone-butanol fermentation. Like glycerol, these two chemicals are of considerable significance as feedstocks for other syntheses, particularly for those leading to various types of plastics such as nitrocellulose. The organism used for fermentation is *Clostridium acetobutylicum,* and the reactions are diagrammed in figure 15.4. As might be expected from the solvent properties of the products, the concentration obtained from fermentation is not very high, approximately 2%. The major products are butanol, acetone, and ethanol produced in an approximate ratio of 6:3:1, respectively.

As long as oil prices remain low, acetone-butanol fermentation is not likely to become cost-competitive. However, because world consumption of oil is steadily increasing and because oil is not a renewable resource on anything but a geologic time scale, it is likely that acetone-butanol fermentation will again come into its own at some future date.

Production of Chemicals Usable for Energy

The use of ethanol as a motor fuel additive is discussed in the preceding chapter. Ethanol is not the only example of an energy-rich chemical that can be produced

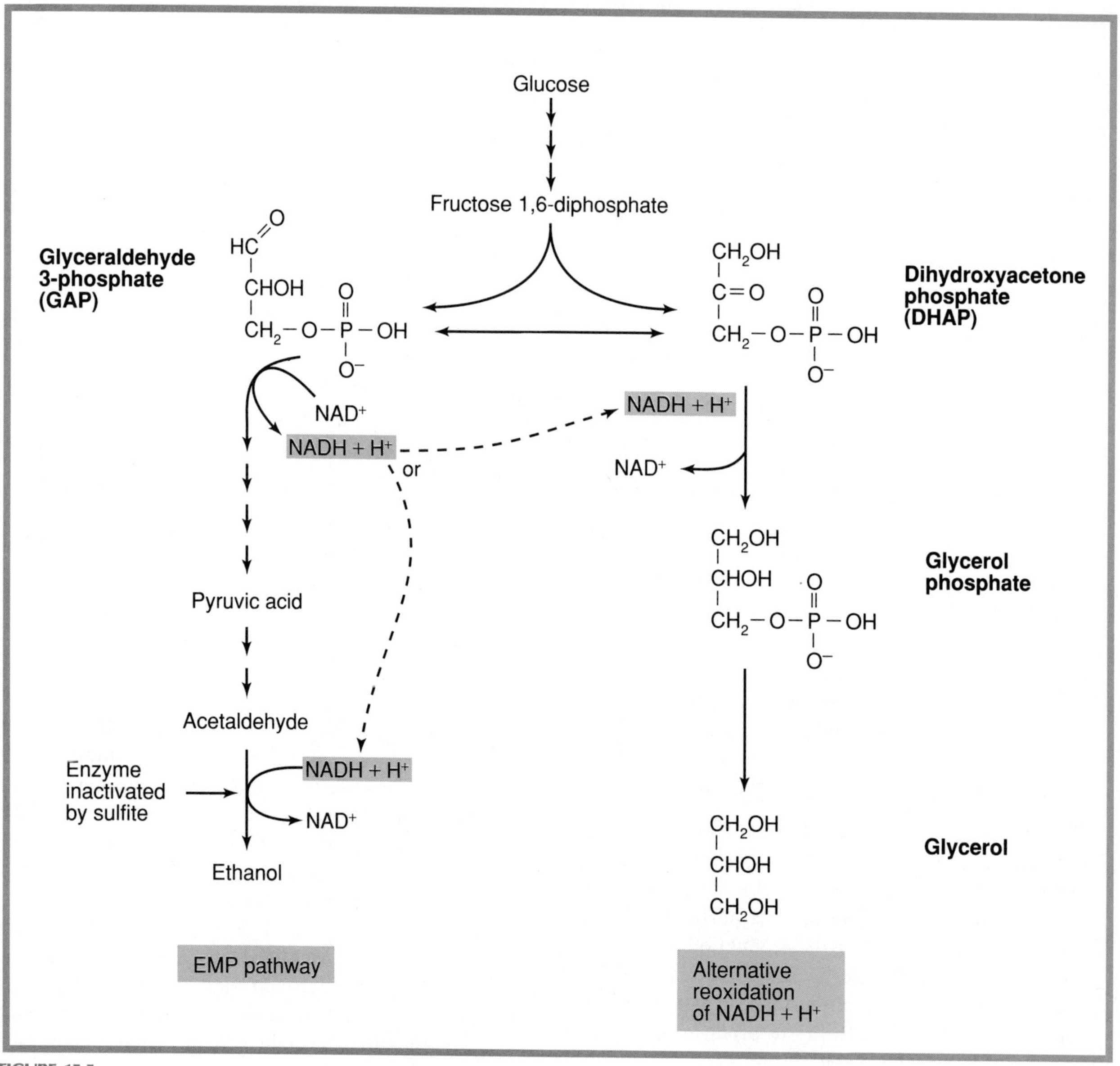

FIGURE 15.5

Microbial production of glycerol. GAP is normally converted to acetaldehyde and thence to ethanol. If sulfite ions block the final step, $NADH + H^+$ is available to reduce DHAP to glycerol.

by microbial action, however. Methane gas is an important chemical that can be used to augment existing supplies of natural gas. Methanogenic bacteria found in genera such as *Methanobacterium, Methanococcus,* and *Methanosarcina* have all been shown to be useful commercially. They are all members of the unusual taxonomic group, the *Archaebacteria.*

There are two general reactions used by methanogenic bacteria to produce methane. Some organisms like *Methanosarcina barkeri* carry out a simple conversion of acetate:

Equation 15.1

$$CH_3COOH + H_2O \rightarrow H_2CO_3 + CH_4$$

APPLICATIONS BOX

Biogas Production

Biogas is the product of an anaerobic digester and consists primarily of a mixture of methane and carbon dioxide. The People's Republic of China has had considerable success with small biogas generators that provide a suitable source of heat for a single rural household. In essence, they are small fermentation tanks fed on a mixture of about 15% human feces, 35% animal feces mixed with straw, and 50% water. Once in operation, the generators need to be refed about every 10 days. A reasonable estimate of the amount of gas produced per day is 1.5 m^3. Further information can be found in the text by Doelle and Hedén that is referenced in the suggestions for further reading at the end of this chapter.

Other methanogens like *Methanobacterium bryantii* carry out a slightly different reaction:

Equation 15.2

$$4H_2 + CO_2 \rightarrow CH_4 + 2H_2O$$

The source of carbohydrate for these reactions generally is a mix of decaying organic matter, such as that found in sewage sludge or in a landfill site. In fact, the best yields have been reported when a broad mixture of plant and organic waste is used as a feedstock. Commercial operators usually set up digesters in areas in which large amounts of material can be obtained easily. One very large plant was established near a cattle feed lot because a steady supply of manure could be assured.

The energy yield from methane-synthesizing reactions is quite low, about one-fourth that produced by the conversion of glucose to lactic acid. Thus, growth of methanogens is quite slow. The material to be digested requires 8 to 30 days in the anaerobic digester for maximum yield.

Methane production from garbage can be obtained in a two-step bioconversion process, as is done with sewage treatment. Initially, garbage is converted to a slurry and inoculated with *Saccharomyces* sp. After several days, the yeasts have converted the usable carbohydrates to ethanol and somewhat broken down the fats and proteins present. One or more methanogenic bacteria are now introduced and allowed to grow on the alcohol and fatty acids as substrates, converting them into methane gas.

If the formation of methane is suppressed by appropriate measures, it is possible to obtain higher alkanes, such as octane, from a fermentation mixture. These alkanes can provide the basis for production of an artificial gasoline.

Another type of energy-rich compound that can result from bacterial metabolism is hydrogen gas. This product of anaerobic growth comes from the following reaction or one similar to it.

Equation 15.3

$$HCOOH \rightarrow CO_2 + H_2$$

The reaction is catalyzed by the enzyme formic-hydrogen lyase. Formic acid itself is a fairly common product in the anaerobic metabolism of pyruvate (figure 15.4). One engineering analysis has indicated that the processing of 4 million liters of wastewater can produce at least 30,000 m^3 of hydrogen gas. At present, however, the efficiency of the conversion process is not high enough to make production of hydrogen gas cost-competitive with conventional technology.

Use of Microorganisms in Bioassays

Certain microorganisms require specific chemicals or growth factors in their media. Without these chemicals, no growth is possible. Such organisms can be used to measure the presence and concentration of a required chemical because the amount of growth that occurs will be proportional to the amount of required chemical present. Use of microorganisms to test for the presence and amount of a specific chemical is called a **bioassay.** Recently, bioassays have been extended to include measurement of properties other than chemical concentration.

Simple Quantitative Analysis

An organism with complex growth requirements may necessitate a rather complicated medium, such as the defined *Pseudomonas* medium described in table 3.1.

In this instance, several different vitamins, for example, thiamine, must be present for growth of a particular organism. If the amount of thiamine added is limited to nonsaturating levels, it is possible to make the amount of bacterial growth proportional to the amount of thiamine present.

This principle has been applied on a commercial scale to the assay of several different vitamins, as in the case of vitamin B_{12}. *Lactobacillus leichmanii* has an absolute requirement for this vitamin and can be used as part of an inexpensive assay to detect its presence and concentration. The amount of cell mass that can accumulate in a medium containing an excess of all nutrients except vitamin B_{12} is directly proportional to the amount of vitamin present.

Mutagenicity Testing

When a new chemical goes into production, it is important to assess the potential hazards that may be associated with its use. Some hazards are obvious (flammability, toxicity) but others are more subtle. Of particular note in this regard are chemicals that possibly may be **carcinogens** or cancer-causing substances. Most testing for carcinogenicity (the ability to cause cancer) used to be performed on test animals that were administered large doses of a chemical and then observed for any increased incidence of cancer. Although this type of testing is still the most realistic, it is expensive, time-consuming, and emotionally unpopular. The number of new chemicals synthesized every year is steadily increasing, and conventional testing methods are not rapid enough to keep pace or provide suitable warnings about potential hazards.

In the mid-1970s Bruce Ames proposed an alternative testing method. He pointed out that known carcinogens seemed to fall into two groups: those that resembled steroids and interacted with receptors on the cell surface, and those that seemed to interact directly with DNA molecules within cells. He further observed that the carcinogenic DNA-interacting chemicals were also good mutagens, although the converse was not necessarily true.

Ames therefore suggested that an inexpensive way to carry out preliminary screening of potentially hazardous chemicals was to see if they would act as mutagens for bacteria. If they proved mutagenic, they then should be targeted for additional testing. If they did not and if they did not resemble steroids, there was no urgent necessity to test them further for carcinogenicity.

The methods used in Ames's testing process are quite straightforward (figure 15.6). A strain of *Salmonella typhimurium* mutant for the production of the amino acid histidine (*his*$^-$) is spread on the surface of an agar plate that lacks histidine. Of the millions of bacteria spread on the plate, a few will spontaneously **revert** or back mutate so that they once again produce histidine and grow. They will form colonies, but the rest of the cells will not (figure 15.6). The number of revertants is quite predictable.

A series of plates containing different concentrations of the material to be tested is spread with a lawn of cells and incubated for several days. At the end of that time, the plates are examined, and the number of revertant colonies determined for each concentration of chemical tested. If the reversion frequency is significantly above background, the chemical undergoing testing must have caused the change and, therefore, must be a mutagen.

Some chemicals are **promutagens** in that they are not in themselves mutagens but can be converted into mutagens by normal metabolic processes. In order to allow for this sort of effect, Ames tested his chemicals in their original form and also after they had been treated with a rat liver extract, because the human liver is generally the site in which unusual chemicals are broken down. A mutagenic result for either the intact chemical or its metabolites warrants further testing. The full-scale test requires several different bacterial strains to test for all possible types of mutagenic activities.

The Ames test has proved to be quick, reproducible, and accurate. Careful validation tests suggest that the *S. typhimurium* system identifies about 84% of the known mutagens tested. A similar *Escherichia coli* system identifies about 91% of the mutagens. The test also allows for the relative risks presented by different chemicals to be quantitated roughly by calculating the number of revertants induced per milligram of chemical tested. This calculation permits regulatory agencies to target those chemicals that appear to present the greatest hazard to the general public.

Modified versions of the Ames test have been developed for use with other bacteria, with induction of prophages, and with animal cell cultures. The principles underlying all these tests are identical. In one particularly interesting variation, dark mutants of the luminescent bacterium *Photobacterium leiognathi* are used. If mutations are induced, the bacteria once again become luminescent and can be detected easily. Inves-

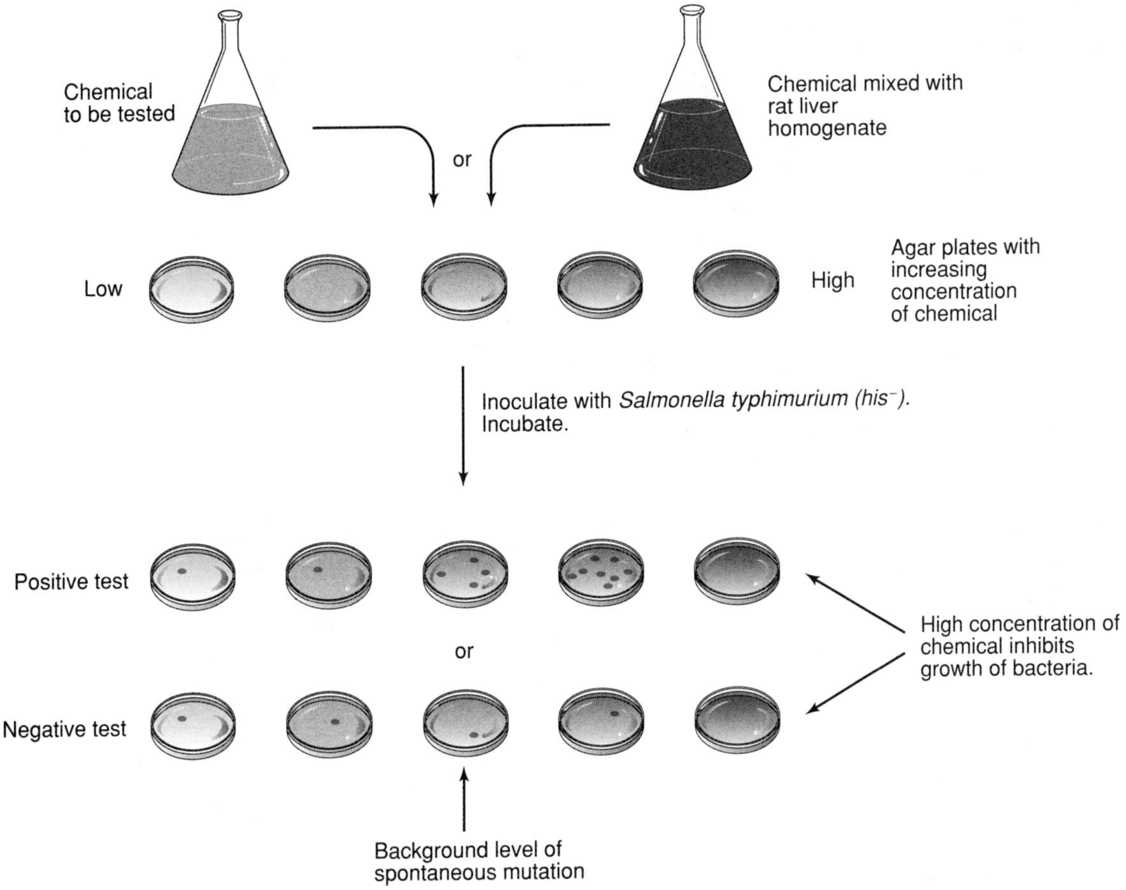

FIGURE 15.6

Ames testing. Chemicals are tested in their original form or after they have been exposed to the enzymes contained in a rat liver extract S9. The chemicals may or may not be toxic to bacterial cells, but even if they are, there will be an appropriate concentration at which killing is minimal. Therefore, plates are prepared with varying concentrations of the chemical to be tested and inoculated with tester bacteria. Among the unkilled cells, revertants will arise and grow on the selective medium used to plate the bacteria. These revertants will be positioned randomly. If either form of the chemical is mutagenic, significantly greater numbers of revertant colonies will be seen on plates containing it than on control plates lacking the chemical.

tigators who carried out experiments with dark mutants estimate that the sensitivity of the system is 100-fold greater than when *S. typhimurium* is used.

Biosensors

Generally speaking, a **biosensor** is a combination of a pH or ion-specific electrode with immobilized bacterial cells. The metabolism of the immobilized cells is proportional to the concentration of a compound in a liquid flowing through or past the sensor, just as in the case of quantitative bioassays. The metabolic reactions then affect the concentration of a second compound, whose concentration is measured by the electrode. Electric output is then an indirect measure of the concentration of the first compound.

Consider the specific case of a *Nitrobacter* sp. The immobilized cells are capable of carrying out the following reaction:

Equation 15.4

$$2NO_2^- + O_2 \rightarrow 2NO_3^-$$

If liquid containing dissolved oxygen is passed over the bacteria, nothing will happen until nitrite is introduced into the liquid. At that time, the nitrite will be oxidized to nitrate, and the level of free oxygen will be reduced. The amount of reduction can be measured by a standard oxygen electrode, an instrument that emits an electric current proportional to the amount of oxygen present. Its reading is inversely correlated with the amount of nitrite introduced.

Similar reactions are possible with other substances, and other types of biosensors can be imagined. Depending on the nature of the metabolic reaction, biosensors can be very general or quite specific. The levels of pollutants in river water can be checked by means of immobilized enzymes. The enzymes will function normally unless inhibited by toxic substances in the water. For example, ATPase is inhibited by vanadium ions, lipase by chlorinated phenols, and ribonuclease by malathion.

The property of bioluminescence has also been used as a tool for analytical work. Bioluminescent light emission results from the oxidation of the flavoprotein luciferin by the enzyme luciferase. It is most commonly seen in the firefly, but bioluminescence also occurs in bacteria. Under appropriate conditions, quantitating the emitted light serves to measure the progress of another reaction. In one instance, the following reaction was catalyzed by the enzyme glycerol dehydrogenase:

Equation 15.5
glycerol + $NADP^+ \longrightarrow$ dihydroxyacetone phosphate + NADPH + H^+

Reduced NADPH was then used to regenerate bacterial luciferin, and the amount of light emitted by the continuing reaction was proportional to the extent of the dehydrogenase reaction. Special instruments called scintillation counters can be used to quantify the emitted light. These counters are used normally to detect fluorescence caused by the decay of radioactive atoms.

Leaching of Ores

Use of thiobacilli to remove sulfur from coal prior to burning is presented in chapter 12 to exemplify how microorganisms can help to remove chemical elements that create environmental problems. The same principles also can be applied to the extraction of desirable elements from low-grade ores. The continuing demand for various metals required by sundry industries has led to the mining of ore deposits that formerly were considered uneconomic. What is needed is a method of concentrating the desired element so that the concentrate can be fed into conventional refining processes.

One area in which microbiologists have been particularly successful is in the case of copper mines. Copper occurs in various insoluble forms, usually combined with iron or sulfur. If a copper compound could be made soluble, simple leaching or extraction of the ore with water would result in a suitable concentrated product. One way in which copper can be converted to a soluble form is to treat it with an acid:

Equation 15.6

$$2CuFeS_2 + 8\tfrac{1}{2}O_2 + H_2SO_4 \longrightarrow 2CuSO_4 + Fe_2(SO_2)_3 + H_2O$$

The various thiobacilli are, of course, particularly well-suited to the production of sulfuric acid (equation 12.2). Hence, if the water used to leach the ore contains members of *Thiobacillus* and some elemental sulfur (usually present in the ores), the result will be the biogenic production of sulfuric acid and the conversion of copper into soluble copper sulfate. This is obviously a comparatively slow process, but the water collected at the bottom of the ore stack can be piped to the top and run through again and again until the concentration of copper reaches desired levels. A water supply to replace evaporative losses and a means of pumping water are the only expenses involved.

A similar process involving *Thiobacillus ferrooxidans* can be used to extract uranium from its ores. Uranium occurs as an insoluble oxide and is converted to a soluble sulfate by ferric sulfate. The latter compound can be derived by bacterial metabolism from the naturally occurring iron pyrite found in uranium ore.

Genetic Engineering of Plants

There are two weak points in the use of *Bacillus thuringiensis* protein toxins to prevent insects from attacking plants: these insecticidal proteins do not persist on plant leaves for a significant time, and the bacteria themselves do not grow significantly on the leaves. Therefore, with DNA splicing methods, considerable research has been carried out to insert the gene coding for *B. thuringiensis* protein toxin into the plant genome. Ideally, the plant could then produce its own protective toxin during its growth.

In the work already carried out, the Ti plasmid has been used as a vector. Chapter 10 describes how all Ti plasmids studied to date induce tumors by transferring a small part of their DNA (T-DNA) into plant cells. Once inside a plant cell, T-DNA is inserted into one or a few sites on the plant chromosomes. Within broad limits, T-DNA does not appear to have a maximum size, which means that even relatively long DNA sequences can be inserted into it and carried into the nucleus of a plant cell. This procedure has been carried out with a bacterial insecticidal protein gene by itself, with a shortened form carrying only the 5′ toxic end of the gene, and with the gene fused to the 3′ end of a plant gene. Only the shortened and fused forms of the gene produced active toxin in a plant, and those plants were toxic for selected insect pests. Organisms that express a gene(s) from another unrelated organism are called **transgenic.**

Some plant-microorganism relationships outlined in chapter 12 are of great economic importance, particularly those involving symbiotic nitrogen fixation. Nitrogenous fertilizers are quite expensive, and the ability to grow crops of legumes without additional fertilizer can aid in reducing the cost of those vegetables. With the advent of serious genetic engineering, major research efforts have begun to add DNA coding for nitrogen fixation to the genome of plants, primarily to corn.

Several different problems must be solved as a part of this research. First, there is the question of what to use as a source of DNA. Rhizobia, being symbiotic nitrogen fixers, would complicate the process because they require symbiotic triggers in order to stimulate their natural regulatory system. Although these regulatory elements can be removed, the consensus is that it is easier to begin with a free-living nitrogen-fixing organism instead. There are many such organisms, but the genus *Klebsiella,* a member of the enteric group of bacteria, has captured primary interest. It is easy to grow and has a genetic system related to that of *E. coli;* therefore, standard procedures can be used in *Klebsiella* manipulation. The genetics of its system is quite complex because at least 12 different *nif* (nitrogen fixation) genes are required for function.

The next problem concerns what vector to use in this endeavor. Once again, attention is focused on the tumor-inducing plasmids found in *Agrobacterium* spp. Unfortunately, there is a problem associated with the use of Ti plasmids as vectors. The *Agrobacterium* plasmid infects dicotyledonous plants (having two leaves within the seed) but not monocotyledonous plants. Because the major interest in genetic engineering of nitrogen fixation revolves around cereal grains, which are monocots, a substantial barrier remains before a product like nitrogen-fixing corn can be developed. Intensive research is underway to establish other types of DNA delivery systems in plants.

Even if genetic engineers are successful in introducing *nif* genes into a plant, they are not certain that plant cells will be able to express the genetic information. Expression requires the presence of both plant promoters and appropriate regulators to allow transcription of new DNA. If by chance the new DNA was inserted in a region of the plant DNA that is not being used in cellular reproduction, there could be no change in the plant phenotype.

Moreover, there is some concern about energy requirements. Nitrogen fixation is a very energy-intensive process; therefore, genetically engineered nitrogen fixation would represent a significant drain on the resources of a cell. Plant physiologists are not in agreement as to whether plant cells have sufficient energy reserves to accommodate the additional metabolic pathway without stunting their growth (and hence reducing crop yield).

One type of genetic engineering in plants that is close to fruition has to do with herbicidal resistance. At the present time, most herbicides must be applied either before a crop has sprouted or after the crop is tall enough so that the herbicide will make contact with weeds, not with the plants themselves. If the plants contain enzymes that are resistant to the effects of the herbicide, it can be applied whenever necessary.

The herbicide glyphosate inhibits the synthesis of aromatic amino acids. *Salmonella typhimurium* is not sensitive to that inhibition; therefore, if the appropriate piece of its DNA is placed into a vector, any plant expressing that DNA will also be resistant to the herbicide. Once again, the Ti plasmid is used as a vector in this process, and it has been shown that tobacco plants (a natural host for Ti plasmids) can become resistant to glyphosate when infected with the spliced plasmid. Other approaches, including the use of mutant plant cells, are also being tried, and some experimental products are undergoing field trials.

Culture Storage and Maintenance

Storage of Cultures in the Laboratory

Although bacterial endospores may be able to survive for very long periods of time, the typical vegetative cell has a much shorter life span in the absence of cell growth. The general growth curve shown in figure 8.4B predicts that once the stationary growth phase has subsided, the death rate of a culture is logarithmic. Clearly, a scientist wishing to preserve a bacterium for future study or a company whose process is based on the metabolic activities of a particular bacterium cannot allow a culture to remain unattended for too long a time, or else the culture may die out. At the same time, the task of transferring each culture to fresh medium on a regular basis quickly becomes burdensome. For a few bacteria and most fungi and algae, there is no alternative. Excellent methods of preservation exist for the vast majority of prokaryotes, however.

The basic strategy involves a system that prevents bacteria from growing and producing additional waste products while simultaneously protecting them from harmful environmental influences. Oxygen, temperature, and degradative enzymes can pose the principal problems in maintaining cultures for extended periods of time. Oxygen can acquire extra energy and be converted into a variety of reactive forms that can damage cellular components (chapter 13). High temperatures can denature proteins and affect membrane fluidity, thereby killing the cells as well. Various autolytic enzymes present in the cells will continue to function and may eventually destroy enough cellular constituents to make the cells inviable. One obvious solution, therefore, is some form of cold storage that will slow the rate of all reactions and protect cell proteins.

Simply storing some (but not all) bacteria in a refrigerator extends their viability from a week or two to several months. Eventually, however, the oxygen effects and the degradative enzymes will decimate the cell population, and the culture will no longer be viable. Covering the culture with a layer of material, such as mineral oil, that does not allow the free diffusion of oxygen serves to greatly extend its life. Cultures kept cold and protected from oxygen can remain viable for a year or more, although degradative enzymes will continue to function slowly. One can simply reach through the layer of oil to obtain a ready portion of the culture, an advantage to this storage method.

A related method of storage is to add some form of antifreeze, such as glycerol or dimethyl sulfoxide, to a culture to prevent the formation of ice crystals within its cells when the culture is frozen. Various temperatures have been used, some as low as that of liquid nitrogen. The colder the temperature, the slower the oxidation and degradation reactions, and the longer the culture is likely to survive. A life span of several years is obtained easily, and cultures can still be used to obtain inocula if needed. They can be partially thawed or a piece of bacteria-containing ice can be chipped off.

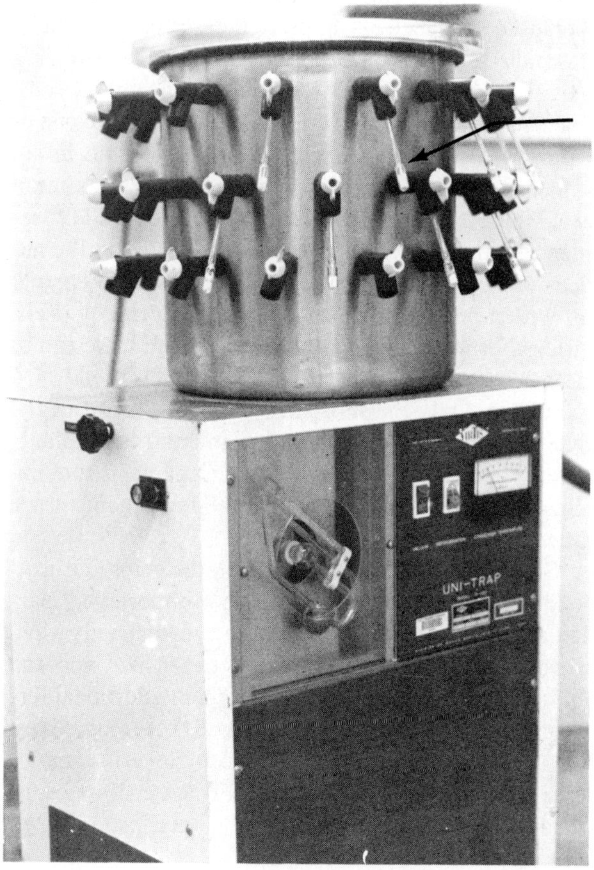

FIGURE 15.7

Lyophilization of bacteria. A small lyophilizer that can freeze-dry many samples simultaneously. The arrow points to a lyophilization vial. If stored properly after sealing with a gas torch, the culture contained in the vial will remain viable for decades.

Lyophilization is a process that provides culture storage for the longest time. A culture is placed in a small glass vial, quickly frozen in a dry ice bath, and then put under vacuum until all water in the culture has sublimed (gone directly from solid to vapor). A gas torch is used to melt the vial in order to seal it permanently (figure 15.7). As long as the vial is protected from heat, there is little reason to expect cellular degradation. Even the internal enzymes of a cell are unable to function in a dry state. All oxygen has been removed from the vial, and a proper seal prevents additional molecules from entering. The life span of a lyophilized culture generally is measured in decades.

Lyophilization is, however, not a panacea. It does not work with all bacteria. Moreover, the treatment has been shown to damage the cells somewhat, and this damage is mutagenic to varying degrees. Nevertheless, it is the most effective preservation method for economic, long-term storage. Lyophils of bacteria prepared in the 1920s have been broken open after 60 years of storage, and the bacteria have been successfully reconstituted with water. In many cases, the best method for preventing damage during reconstitution is to avoid adding any nutrients for the first 24 hours. This allows time for cell enzymes to reactivate and repair any cellular damage before new growth proceeds. Generally it is true that growing cells are more susceptible to all kinds of damage than nongrowing cells.

Culture Collections

The discoverer of a new organism is expected to make it available to others who are interested in its properties. Initially, this can be done by the individual investigator. However, for a new species name to be validly published, a culture of the organism must be deposited in some publicly accessible collection facility. A **type culture** is the culture on which the original species description was founded. It provides the basis for comparisons made by other taxonomists and is critical for the continuity of the science. Type culture collection facilities are located in major countries worldwide. In the United States, the American Type Culture Collection (ATCC) is located in Rockville, Maryland. All type cultures in the collection are available for purchase, and cultures obtained from the organization can be identified by their ATCC number.

Organisms isolated from nature can be used for industrial purposes, deposited with type culture collections, and freely distributed. They cannot be patented. Many companies maintain special cultures of particularly useful microorganisms as trade secrets rather than share them with other manufacturers. If DNA splicing is used to construct an organism that has properties unlike any found in nature, that organism is patentable.

In order to obtain a patent on a microorganism, it is necessary for the applicant to deposit a culture of the appropriate organism(s) necessary to carry out the process with some recognized agency. Once again, the American Type Culture Collection fulfills this role. Distribution of these cultures can be restricted, depending on the status of the patent application. Once a patent has been issued, the organism is available for purchase but its use cannot infringe on the patent.

Summary

Applied microbiology is not limited to the production of food but also includes processes of industrial significance. Many microorganisms produce commercially valuable substances as waste products during fermentation. Others can be forced to secrete the desired product by a mutation blocking a particular metabolic pathway. With an appropriate choice of organism and culture conditions, such chemicals as acetone, butanol, ethanol, octane, various amino acids, and methane gas can be produced. Whether a process is commercially viable depends on the current economic conditions, particularly on the price of the petroleum that constitutes an alternative source for many chemicals.

Microorganisms can play an analytical role in industry as well. They can be used to assess the mutagenic, and possible carcinogenic, potential of newly developed chemicals in variations of the Ames test. The presence or absence of particular nutrients can be assayed with microorganisms whose metabolism responds in a quantitative way to the compound being studied. Examples of assays presented in this chapter include vitamin B_{12} concentration, glycerol concentration, and nitrite concentration, but many others are possible.

Plant-microorganism associations have been shown to be very important in natural ecosystems. They can facilitate the growth of a number of economically useful plants. Some genetic engineers experiment with splicing an organism's DNA coding for certain functions into a plant chromosome. One vector that seems to hold great promise for this type of engineering is a modified form of the Ti plasmid. An extra DNA is inserted within T-DNA sequences and carried along during normal infection of the plant.

When an entire industry depends on the use of a few organisms, it is important to maintain cultures in a pure and unchanging state. In this country, the American Type Culture Collection is responsible for maintaining primary cultures. The principal method of storage for bacterial cultures is lyophilization, but storage in liquid nitrogen or other types of freezers is also possible. Some organisms do not survive any freezing method or lyophilization. These must be transferred to fresh medium on a regular basis, or they will die.

Questions for Review and Discussion

1. Why does the metabolism of pyruvic acid seem to play such a pivotal role in industrial microbiology?
2. In what ways might industrial microbiologists contribute to the resolution of another economic crisis caused by a petroleum shortage? Are any of these methods used today? Why or why not? What factors might cause your answer to change?
3. What is the basis for saying that microorganisms or their products can serve the analytical chemist? Give an example of such an activity.
4. List several different ways of preserving a microbial culture. What are the advantages and disadvantages of each?
5. Why are many corporations looking to hire scientists who know how to splice DNA?

Suggestions for Further Reading

Chaudhry, G. R., and S. Chapalamadugu. 1991. Biodegradation of halogenated organic compounds. *Microbiological Reviews* 55:59–79. Applications of gene splicing to biodegradation problems.

Doelle, H. W., and C.-G. Hedén. 1986. *Applied microbiology.* Kluwer Academic Publishers, Hingham, Mass. A collection of monographs on biotechnology as applied to worldwide social problems.

Genetic Engineering News. Mary Ann Liebert, Inc., New York. This monthly publication in newspaper format deals with all aspects of the biotechnology industry. It is a good source for current brief summaries of activities in this field.

Harrison, L. A., and K. P. Flint. 1987. The use of immobilized enzymes to detect river pollutants, pp. 181–194. *In* J. W. Hopton and E. C. Hill (eds.), *Industrial microbiological testing.* Blackwell Scientific Publications, Ltd., Oxford. A speculative plan to use biosensors for screening drinking water samples.

Hewitt, W., and S. Vincent. 1989. *Theory and application of microbiological assay.* Academic Press, Inc., San Diego. Detailed methods for assays of antibiotics and vitamins.

Höfte, H., and H. R. Whiteley. 1989. Insecticidal crystal proteins of *Bacillus thuringiensis. Microbiological Reviews* 53:242–255.

Huang, S. D., C. K. Secor, R. Ascione, and R. M. Zweig. 1985. Hydrogen production by non-photosynthetic bacteria. *International Journal of Hydrogen Energy* 10:227–231. Provides an example of the sorts of economic evaluations that must be done during the commercial development of any microbial process.

Hütter, R., and P. Niederberger. 1983. Amino acid overproduction, pp. 49–59. *In* N. G. Alaeddinoglu, A. L. Demain, and G. Lancini (eds.), *Industrial aspects of biochemistry and genetics.* Plenum Publishing Corp., New York. Presents problems of genetic engineering in yeasts.

Lavi, J. T. 1984. A sensitive kinetic assay for glycerol using bacterial bioluminescence. *Analytical Biochemistry* 139:510–515. An article about an imaginative bioassay.

Leadbetter, E. R., and J. S. Poindexter (eds.). 1985. *Bacteria in nature,* vol. 1. Plenum Publishing Corp., New York.

Monmaney, T. 1985. Yeast at work. *Science* 85(7):30–36. A historical perspective on industrial uses of yeasts in addition to their great potential for the biotechnology industry.

Morgan, D., and T. Monmaney. 1985. The bug catalog. *Science* 85(7):37–41. Capsule summaries of economically significant uses of microorganisms.

Ogata, S. 1980. Bacteriophage contamination in industrial processes. *Biotechnology and Bioengineering* 22(Suppl. 1):177–193. A general review of an obvious potential problem for industrial microbiologists.

Okada, T., I. Karube, and S. Suzuki. 1983. NO_2 sensor which uses immobilized nitrite oxidizing bacteria. *Biotechnology and Bioengineering* 25:1641–1651. An interesting combination of chemistry and microbiology.

Preuss, P. 1985. Industry in ferment. *Science* 85(7):42–46. An interesting discussion of the engineering problems involved in using microorganisms for chemical production.

TeBeest, D. O. (ed.). 1990. *Microbial weed control.* Chapman & Hall, New York.

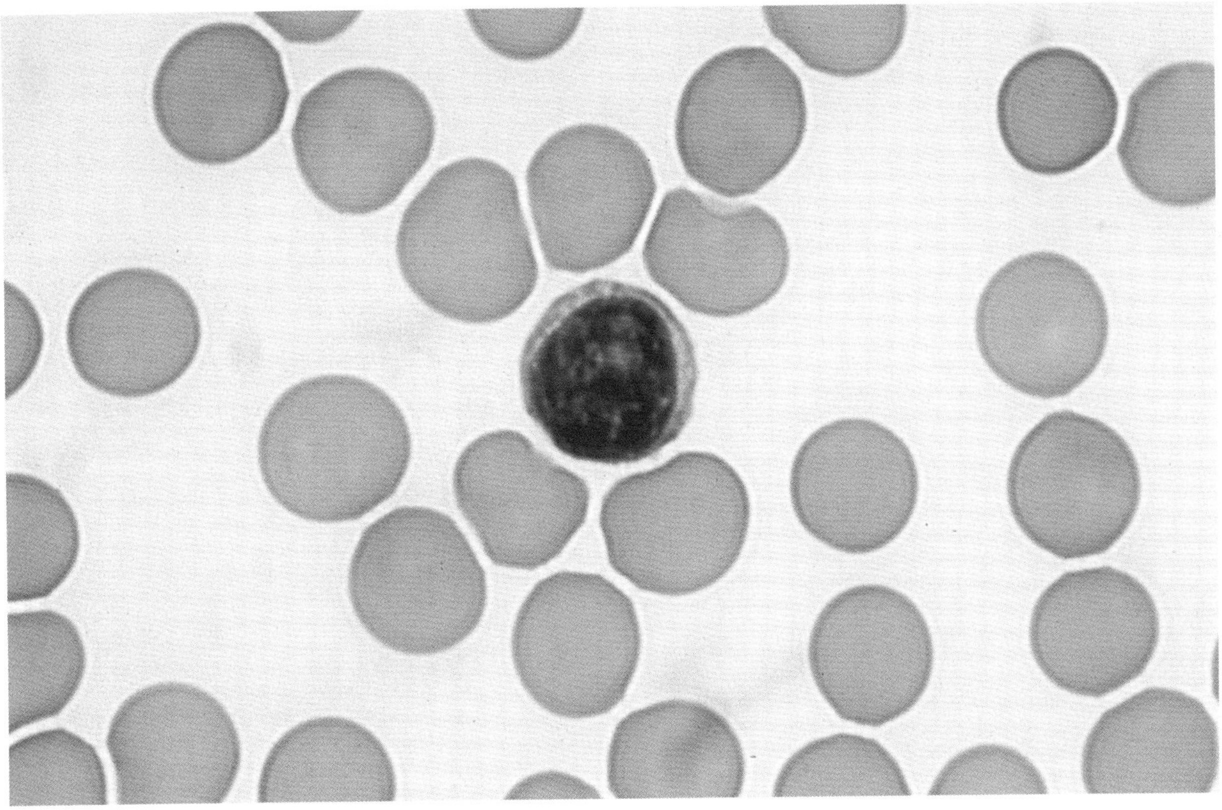

Color plate 7
Lymphocyte.
Courtesy of Valerie Evans, Department of Pathology, University Medical Center, Tucson, Arizona

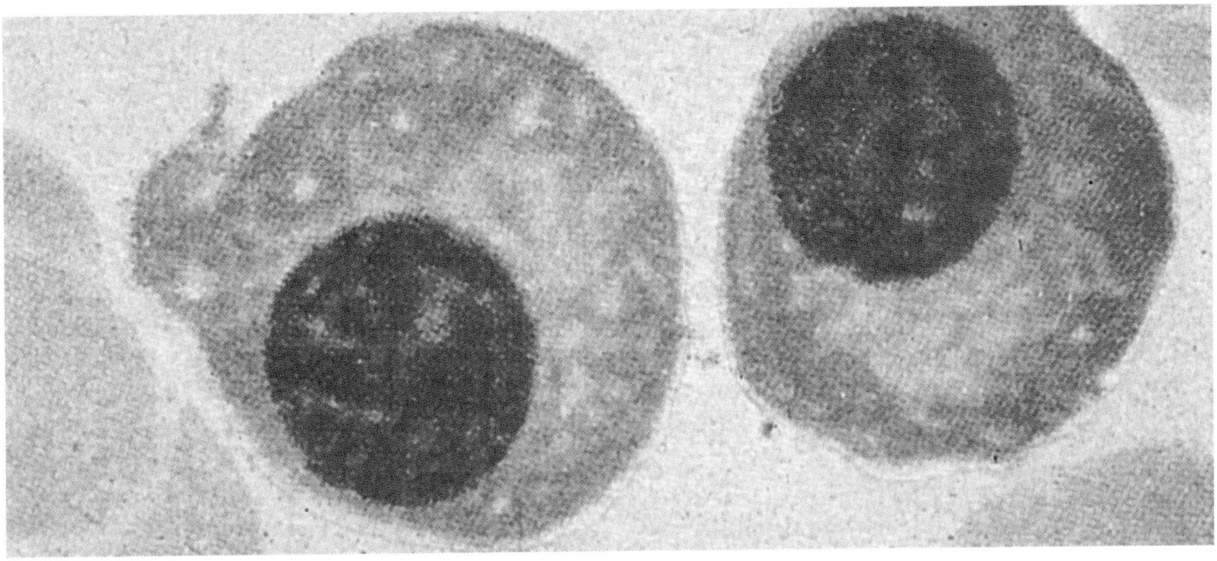

Color plate 8
Plasma cell.
Reproduced from *Sandoz Atlas of Haematology,* 2nd ed. Sandoz Ltd., Basel , Switzerland

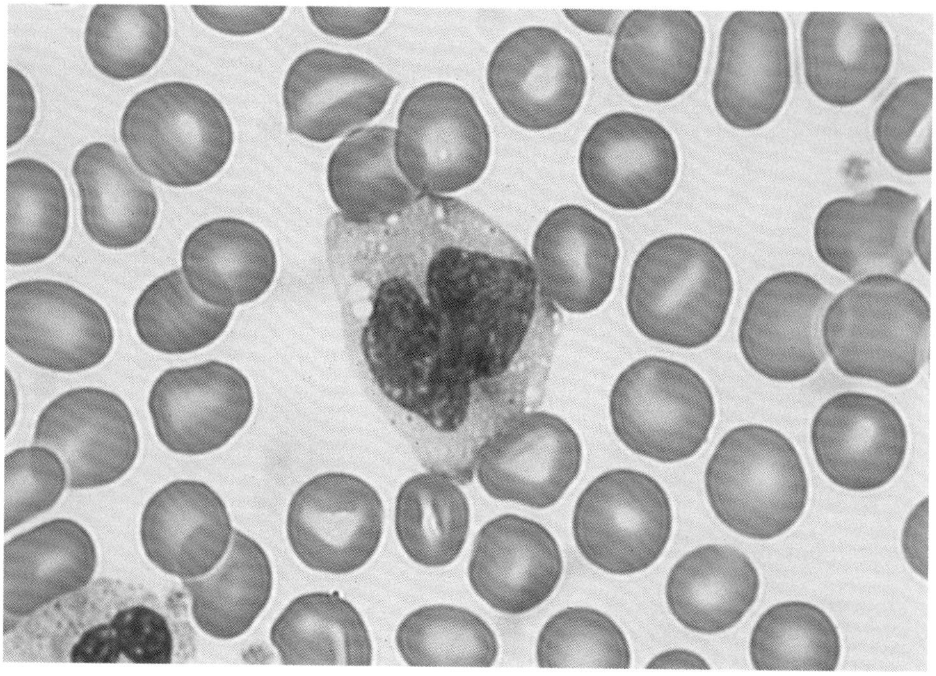

Color plate 9

Monocyte leukocyte. This is a type of phagocytic cell.
Courtesy of Valerie Evans, Department of Pathology, University Medical Center, Tucson, Arizona

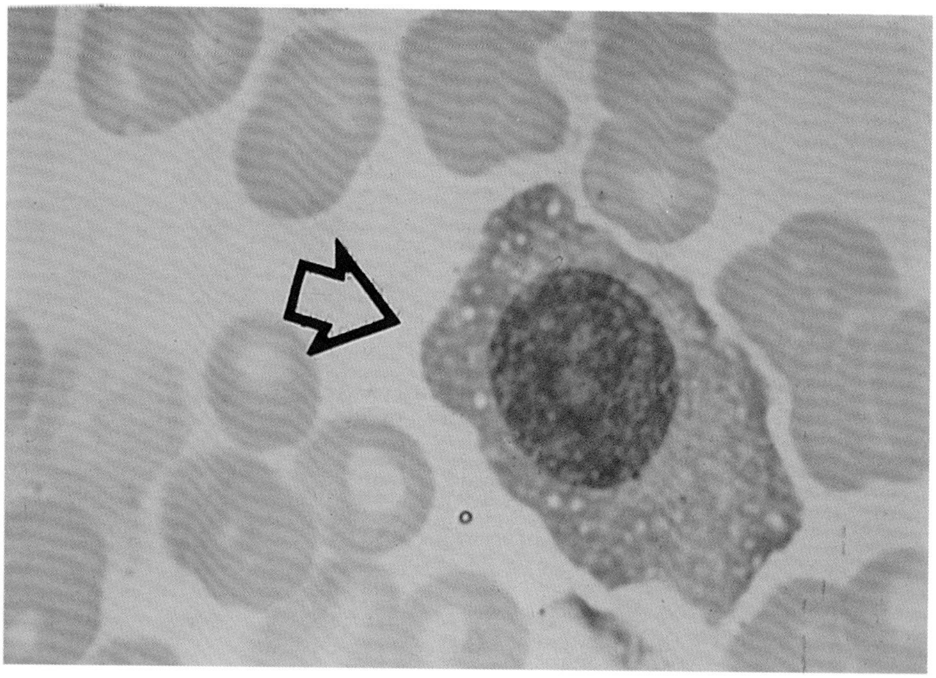

Color plate 10

Tissue macrophage or histiocyte.
Courtesy of the College of American Pathologists

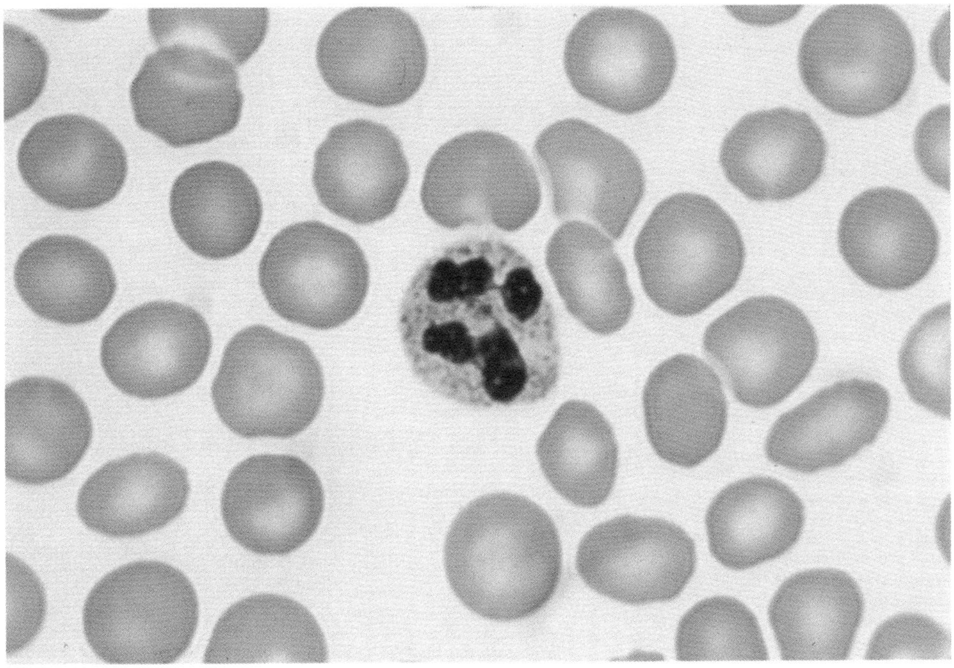

Color plate 11

Neutrophilic leukocyte. This cell is designated as a segmented neutrophil due to the shape of its nucleus. Courtesy of Valerie Evans, Department of Pathology, University Medical Center, Tucson, Arizona

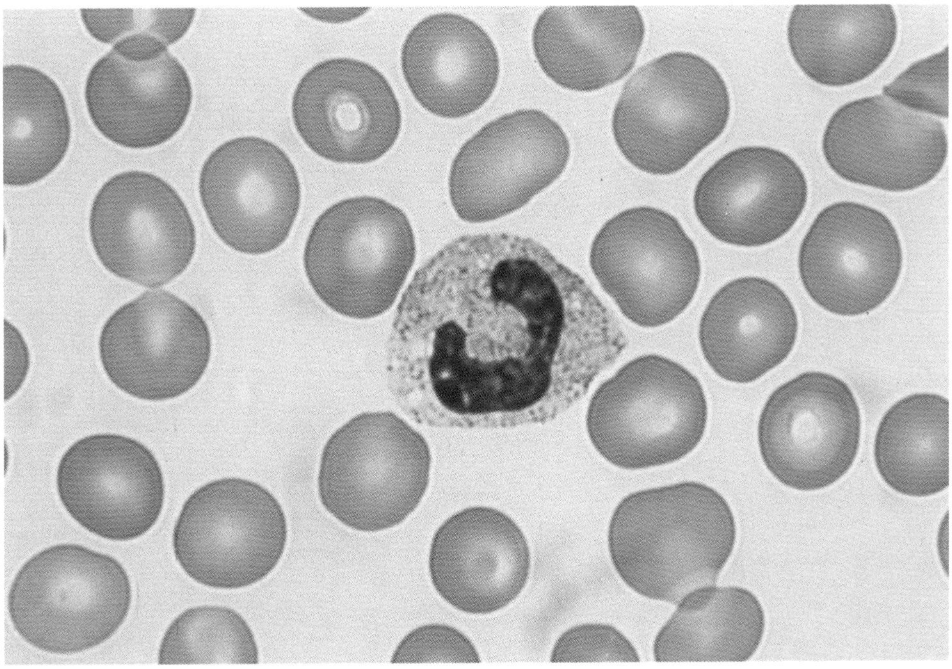

Color plate 12

Neutrophilic leukocyte. This cell is designated as a banded neutrophil due to the shape of its nucleus. Courtesy of Valerie Evans, Department of Pathology, University Medical Center, Tucson, Arizona

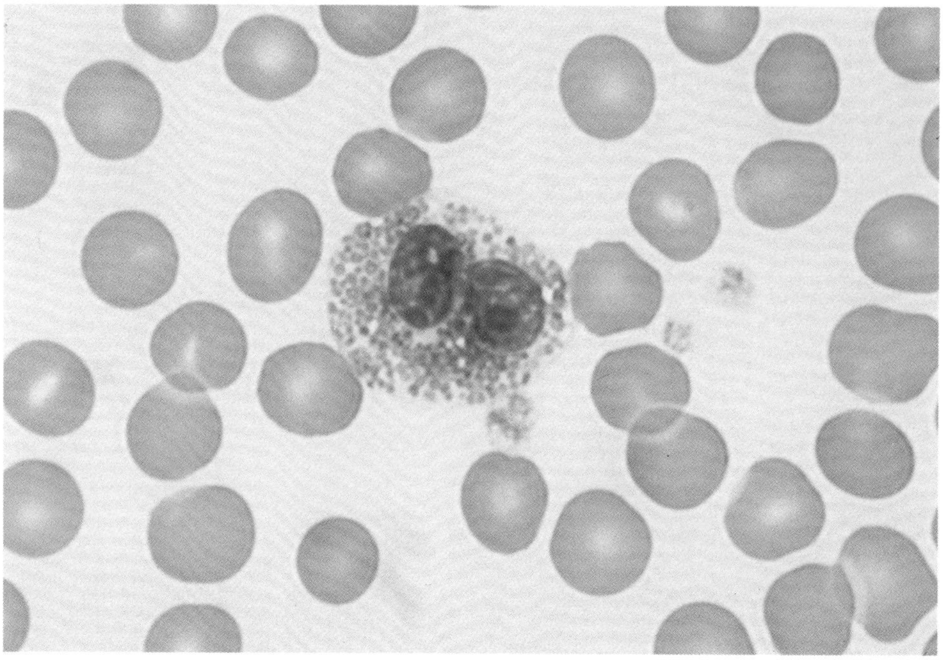

Color plate 13

Eosinophilic leukocyte. This cell is named for the intensely staining granules in its cytoplasm.
Courtesy of Valerie Evans, Department of Pathology, University Medical Center, Tucson, Arizona

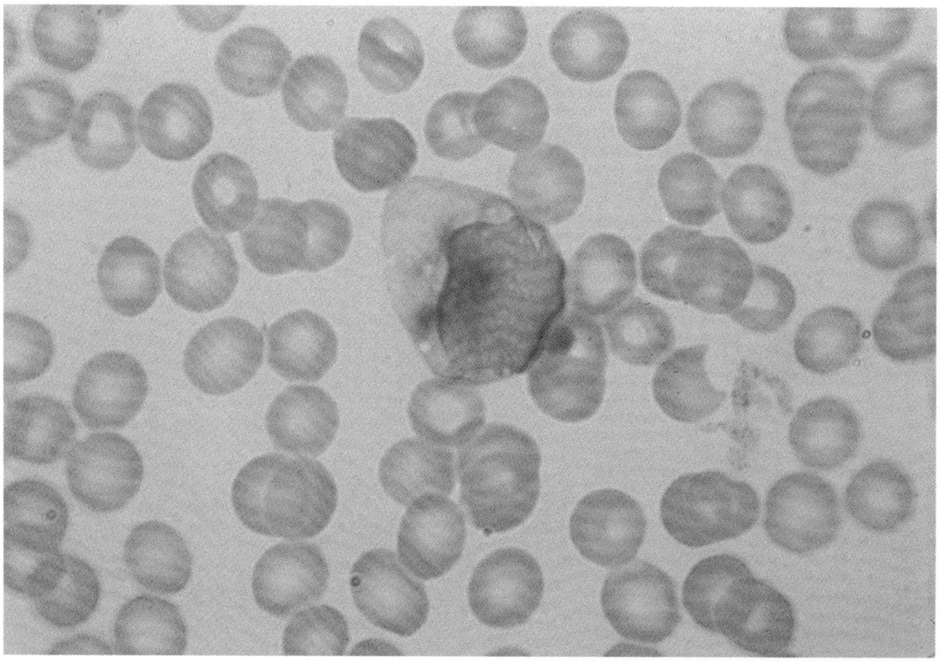

Color plate 14

Reactive lymphocyte or B-memory cell.
Courtesy of Valerie Evans, Department of Pathology, University Medical Center, Tucson, Arizona

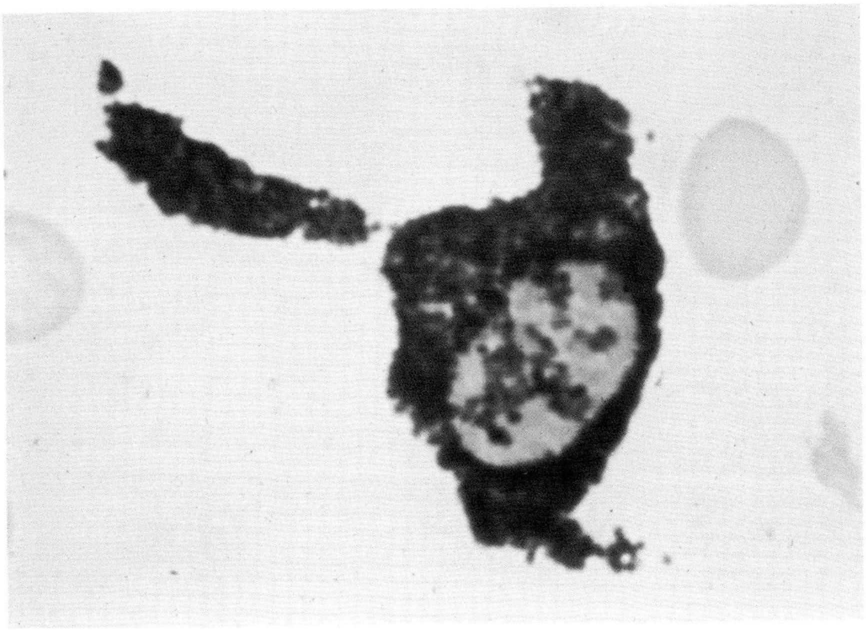

Color plate 15
Mast cell.
Courtesy of the College of American Pathologists

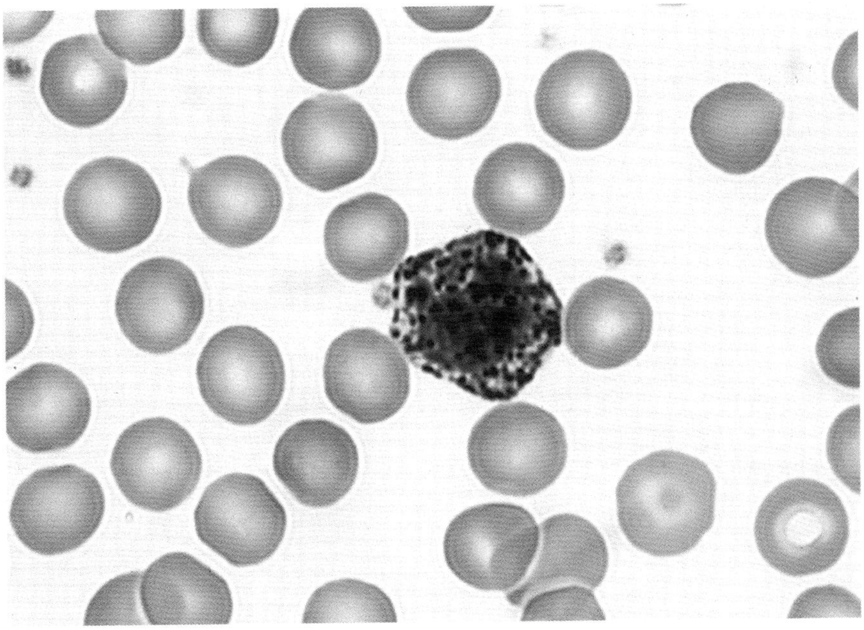

Color plate 16
Basophilic leukocyte. This cell is named according to the staining properties of its cytoplasmic granules.
Courtesy of Valerie Evans, Department of Pathology, University Medical Center, Tucson, Arizona

chapter

17

Immunology

chapter outline

GENERAL GOALS

Be able to explain what antigens and antibodies are and how they are produced. This description should be made both at the cellular and at the molecular level.

Be able to describe autoimmune processes and the differences between humoral immunity and cell-mediated immunity.

Be able to explain the operation of some modern immunoassay techniques and the situations in which they might be used effectively.

Everybody has an intuitive notion of what it means to become immune to a disease-causing agent. This idea is expressed in the phrase "disease of childhood," implying that a person has the disease only in childhood because after having had it, he or she cannot contract it again. The subject of immunology certainly includes the process by which an individual becomes immune to further infection by an organism; however, it encompasses much more than that. The processes by which the body recognizes its various component cells as "self" and defends itself against foreign cells, foreign proteins, and viruses of all types are functions of the immune system as well. Errors in the immune response process can actually cause allergies or autoimmune diseases instead of being beneficial.

Knowledge derived from the study of immunology has a far-reaching impact. The ability to control or prevent errors in the immunologic functions is of obvious significance. Equally valuable is the ability to suppress the normal functioning of the immune system in order to ensure successful organ transplants. In a related vein, the clinical diagnostician has for many years relied on immunologic assays for diagnosis. Finally, the science of immunology has made important contributions to our understanding of molecular biology and the ways in which eukaryotic cells can function.

Introduction to Humoral Immunity

The immune system consists of two discrete, interactive processes: humoral immunity and cell-mediated immunity. Both involve components found within blood but differ in the way in which these components function. The two systems can be distinguished readily by how they are transferred from one individual to another. If immunity can be transferred only by purified cells, it is cell mediated. If immunity can be transferred by blood **plasma,** the fluid portion of blood, it is called **humoral immunity** as a tribute to the Greek physician Hippocrates who considered blood to be one of the four body "humors" (blood, black bile, choler, and phlegm). Components of the immune system are derived from the major lymphoid organs: thymus, spleen, and lymph nodes. Characteristics of the immune system presented in this chapter are seen in vertebrates and do not relate to invertebrates or plants.

Antigens and Antibodies

Fundamental to any discussion of the humoral immune system is the concept of an antigen and an antibody. An **antibody** (abbreviated Ab) is a multimeric (several subunit) protein complex produced by **lymphocytes.** Lymphocytes are small **leukocytes** (also spelled leucocyte) or white blood cells with large nuclei and relatively little cytoplasm (color plate 7). In order for an antibody to be produced in quantity, an appropriate antigen must stimulate a leukocyte to differentiate. The antibody produced by the differentiated cell (plasma cell) binds specifically to the stimulating antigen.

An **antigen** is a comparatively large (molecular weight greater than 1,000) molecule with a complex three-dimensional structure that elicits an antibody response. All antigens stimulate antibody production, and all combine specifically with antibodies produced as a result of that stimulation. In most but not all cases, the antigen bound in an antigen-antibody complex has lost its biological activity. The best antigens are proteins because they provoke the synthesis of more varieties and larger quantities of antibodies. They also exhibit the most elaborate three-dimensional structures of any

APPLICATIONS BOX 1

Evading the Immune System

The immune system is not protective against all pathogenic microorganisms. A good example of this phenomenon is seen in the case of the sleeping sickness disease caused by members of the protozoan genus *Trypanosoma.* The bite of the tsetse fly injects trypanosomes into the blood of humans and cattle. These trypanosomes circulate in the bloodstream and eventually reach the brain, causing death. What has puzzled scientists for years is why the host immune system cannot destroy trypanosomes. The answer lies in the glycoprotein surface that covers the protozoan and has a highly variable composition. Apparently trypanosomes have literally hundreds of genes coding for surface proteins, and they spontaneously change the gene being used at frequent intervals. By the time a primary immune response has occurred, significant numbers of trypanosomes carry entirely different surface proteins; hence, they are not attacked by circulating antibodies. Although some trypanosomes are destroyed, the infection persists. Another primary immune response occurs with respect to the new surface antigen, but by the time it has run its course, even newer surface antigens are in use. For this reason, all attempts to prepare an immunizing treatment effective against mature trypanosomes have failed.

macromolecules. The next best antigens are polysaccharides. They have a more repetitious structure than proteins due to the molecular similarities seen in many of the sugars, but characteristic polysaccharide branching patterns aid in the formation of antigenic structure. Nucleic acids can be antigenic, but they are comparatively weak antigens. Their structure is even more repetitious than polysaccharides, and their number of different subunits is very small. Single-strand nucleic acids and Z-DNA (left-hand helical DNA) tend to make the best antigens among the various types of nucleic acids because both have complex folding patterns. Lipids, whose structure is essentially the same in all cells, have no antigenic properties, although proteins or polysaccharides attached to them can be antigenic.

Location of Antibodies within the Body

Antibody molecules are found on the surface of certain lymphocytes or within the plasma as freely moving molecules. Because blood plasma flows through capillary walls and into tissue spaces (where it is known as lymph), antibody molecules are found in those spaces, in the lymph itself, and in lymph nodes as well. When blood clots, fibrin molecules form a network that traps only red blood cells. The bulk of the plasma outside the clotting network is known as the blood **serum.** Serum, therefore, contains those solutes found in plasma with the exception of clotting factors and cells. The serum fraction of the blood is important to immunologists because it contains most of the antibody complexes and is an excellent and easily obtained source of these molecules. A serologist is a specialist who studies serum and particularly the antibodies within it.

Various proteins within the serum can be separated by a number of different physical techniques. One very common method is **electrophoresis.** Protein molecules are applied to a moistened solid surface across which an electric field is imposed, as can be done with nucleic acids (see chapter 11). The rate at which the proteins move is proportional to their charge and inversely proportional to their size.

A typical electrophoretic profile obtained from human serum is shown in figure 16.1. The major peak indicates albumin, the most prevalent protein in serum. The smaller peaks indicate various kinds of **globulins,** roughly spherical proteins that are insoluble in pure water or in high-salt solutions. They are designated by the Greek letters α, β, γ. Antibody molecules are found primarily in the γ globulin fraction, but some do occur among the β globulins as well. In order to distinguish antibody globulins from other types of globulins, they are designated as immunoglobulins or Ig molecules.

Physical Structure of Immunoglobulins

In the presence of strong reducing agents, like dithiothreitol, and protein denaturing agents, like urea and guanidine hydrochloride, antibodies will separate into

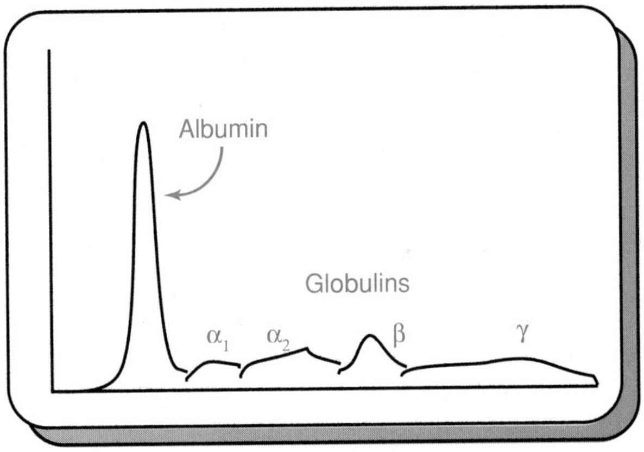

FIGURE 16.1
Electrophoretic profile of serum proteins. An electrode was placed at either side of a strip containing a serum sample, and an electric field was applied to separate the proteins. The densitometer broke the graphic curve to indicate the start of each new peak.
Electrophoretic tracing courtesy of Cathy Downs, Arizona State University.

APPLICATIONS BOX

Collecting Blood Samples

When blood is withdrawn from the veins of an individual, it is usually collected in special evacuated tubes sealed with colored rubber stoppers. The vacuum facilitates entry of the blood into the tube and greatly speeds the collecting process. Tubes sealed with an orange-red stopper contain no extra chemicals, and blood collected in them will clot. This blood is used for serologic tests (those requiring serum). Tubes with caps of other colors contain anticoagulant chemicals, such as heparin or citrate. Blood collected in these tubes does not clot, but cells can be removed by centrifugation to yield the plasma fraction often used for biochemical testing.

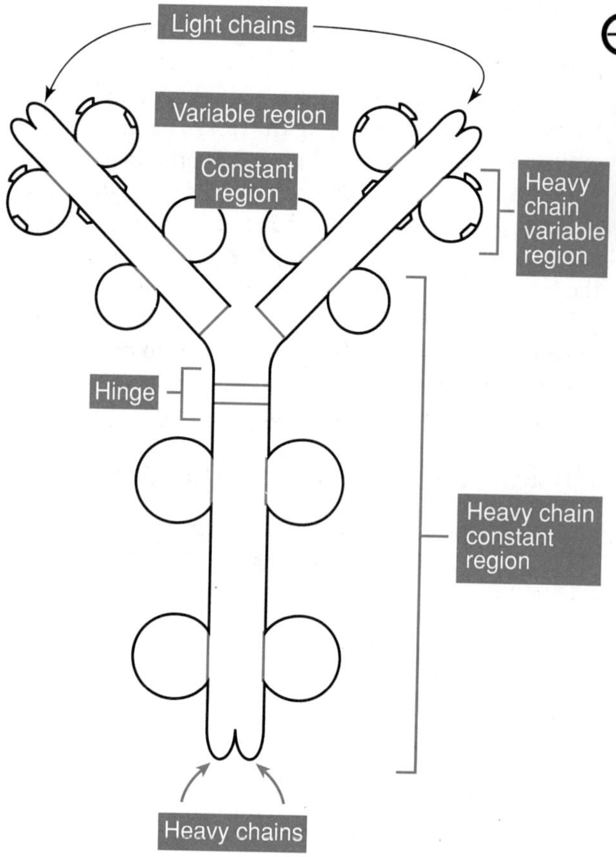

FIGURE 16.2

Typical antibody molecule. It consists of four proteins, two heavy chains and two light chains. Disulfide bonds (in color) formed between cysteine residues stabilize loops within the protein chains and also hold the protein chains together to form the antibody. Each chain has both constant and variable regions. Physical studies of the molecule have shown that the hinge region remains uncoiled and flexible, allowing the "arms" to move closer together or spread further apart. The boxes within the variable regions are the hypervariable regions.

TABLE 16.1 Heavy chain types of human immunoglobulins	
Heavy Chain Type	**Immunoglobulin Type**
γ	IgG
μ	IgM
α	IgA
δ	IgD
ϵ	IgE

their various components. The structure of the most common antibody consists of two identical short proteins composed of about 215 amino acids (light chains) and two identical longer proteins composed of about 450 amino acids (heavy chains). These chains have been shown to associate with one another in a Y-shaped form illustrated in figure 16.2.

The four chains are held together by **disulfide bridges,** chemical bonds linking the SH groups of the amino acid cysteine. These bridges are stable under normal circumstances, but they can be disrupted in the presence of strong reducing agents. Immunoglobulins of this type have two identical antigen-specific binding sites located at the tips of the molecule's Y-shaped arms. The distance between the two binding sites varies because a short "hinge" allows the arms to move through an arc. The arms can point in opposite directions, move quite close together, or maintain any position in between.

Protein chemists have purified the heavy and light chains from many immunoglobulins and determined their amino acid sequences. As expected, immunoglobulins obtained from different animals have differences in their amino acid sequences. There are also several different types of chains within the same individual, however. Light chains can be in either of two forms, designated as lambda (λ) or kappa (κ). Their molecular weight is about 23,000, and they have little or no carbohydrate associated with them. In humans, the κ:λ ratio is about 6:4. Humans also have five different types of heavy chains (shown in table 16.1) with molecular weights ranging from 50,000 to 75,000. All the heavy chains contain significant amounts of carbohydrate and are technically known as glycoproteins. Each type of heavy chain can associate with either a κ or a λ chain. The resulting antibody is named according to the heavy chain that is present.

In both the heavy and light chains, there are regions in which the amino acid sequence is nearly constant within a particular type and others in which substantial variability is present. These regions are designated as constant and variable regions (figure 16.2). The constant regions are always near the carboxyl end of the protein molecules, and the variable regions always involve the amino terminus. Each type of chain (α, δ, ϵ, γ, κ, λ, and μ) has its own unique amino sequence in the constant region, and this is how the individual types can be identified. The variable regions within the heavy and light chains can be subdivided into normally variable areas and hypervariable areas, based on the extent of observed amino acid substitution. Within the constant region of the heavy chain are exactly identical sequences, shown as loops joined by disulfide bridges in figure 16.2.

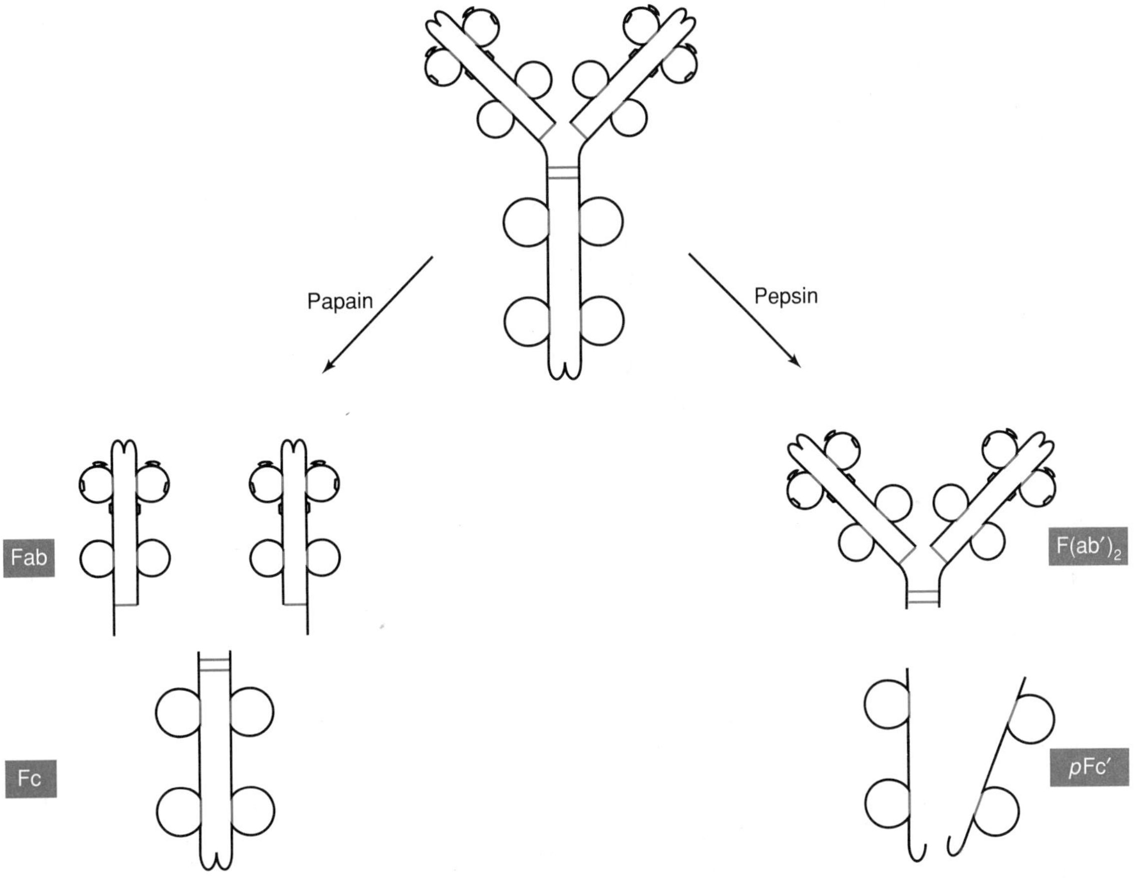

FIGURE 16.3
Protease cuts in immunoglobulin molecules. The enzyme used determines whether the variable regions are obtained separately (Fab) or linked together [F(ab′)$_2$].

The immediate presumption is that the hypervariable regions are the actual recognition (binding) sites for antigens, which, in fact, is the case, as shown by experiments with antibody fragments. If the immunoglobulin molecule is treated with papain, a particular type of protease isolated from papaya, each heavy chain is cut at one specific site in the hinge region (figure 16.3). Of the two types of fragments released, only the variable ends bind to antigens; therefore, these fragments must include the antigen-binding site and are designated as Fab. The fragment containing the constant region is known as Fc, not only because it is constant but because it crystallizes in a cold buffer. The hinge region also allows movement of the Fc region so that the Fc region lies in a plane perpendicular to the plane of the variable regions (dislocation). If the protease used to digest an antibody molecule is the enzyme pepsin, the cut is made below the disulfide bonds, and the two variable regions remain linked. This type of fragment is called F (ab′)$_2$.

The Complement Pathway

When an antigen-antibody response occurs on the surface of a cell, that cell becomes susceptible to attack by a series of proteins that together make up the **complement** (C′) system. This system is composed of a cascade of proteins that functions to destroy any cell whose outer layer has triggered the cascade. There are two pathways for complement activation, the classic and the alternative (figure 16.4).

The classic pathway begins with a pentamer of different C1 subunits, a complex that can be activated by binding to an antigen-antibody complex. When activated, the C1's subunit cleaves C4 protein first and then

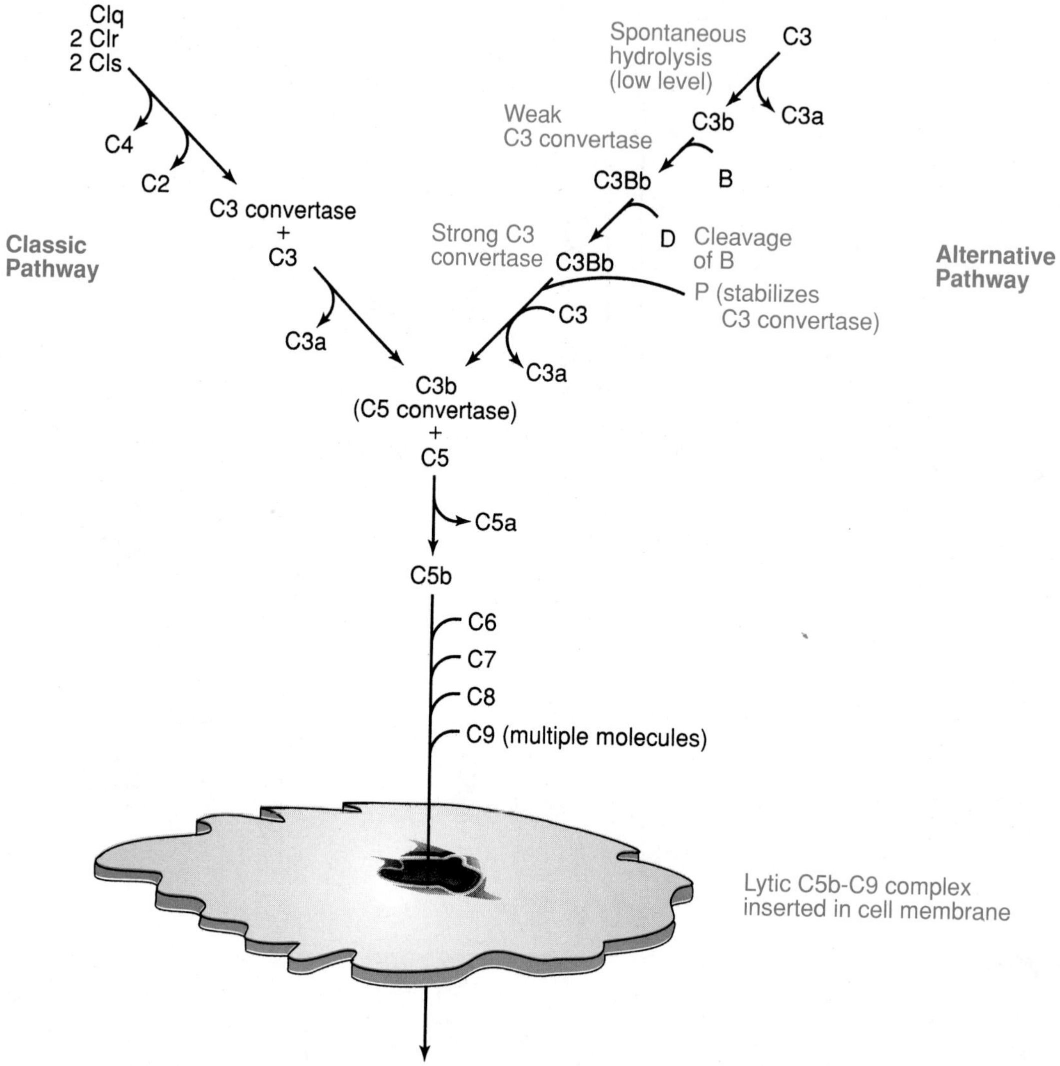

FIGURE 16.4

Complement pathways. The left portion of the diagram depicts the classic complement pathway; the right portion depicts the alternative pathway. Note their convergence at the level of the C5 convertase and the two separate cleavages of C3 in the alternate pathway.

C2. The cleavage products combine to yield a protease holoenzyme known as C3 convertase. This set of reactions normally is inhibited by a serum component, the C1 esterase inhibitor. Antigen-antibody complexes sequester the C1 complex and permit its activation independent of the inhibitor concentration. Localized inhibitor shortages allow C1 activation regardless of the presence of antigen-antibody complexes.

C3 convertase cleaves C3 to give two fragments, only one of which continues in the cascade. The C3b fragment then binds to various receptors located on the cell surface, some of which activate it to become a C5 convertase. This enzyme cleaves C5 and thereby begins the final cascade in which a complex of proteins 5 through 9 is formed. The C5 through C9 protein complex can insert itself into a cell membrane and generate a permanently open pore through which cytoplasmic contents can leak out. As many as 16 molecules of C9 can be added into the complex, and each one results in a larger pore.

In the alternative pathway, cleavage of C3 occurs spontaneously but at low levels. Depending on the type of surface to which the C3b protein fragment binds and on the presence of certain cofactors, production of C3b

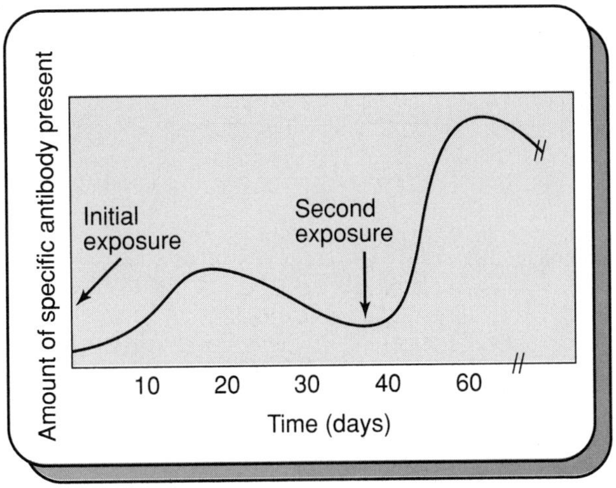

FIGURE 16.5

Time course of an immune response. The initial exposure of the organism to an antigen is the primary response. All subsequent exposures are considered secondary responses.

can be damped or accelerated. If sufficient C3b is formed, C5 convertase results, and the reactions continue as for the classic pathway.

Immunoglobulin Synthesis: The Clonal Selection Hypothesis

Nature of the Immune Response

In any model for immune system function, certain basic observations warrant explanations. Significant quantities of circulating antibodies are not produced at random. They are formed in reaction to the introduction of appropriate antigens into the body. Although polysaccharides, proteins, and nucleic acids of the body are antigenic when introduced into other individuals, they do not normally trigger the production of corresponding antibodies as long as they remain in the body that synthesized them. In other words, the body recognizes what constitutes its "self."

Certain tissues seem to be associated with the immune response. Antibodies are found commonly in blood, lymph, spleen, and lymph nodes. If the thymus gland is removed at birth or absent due to a genetic defect, the immune system never develops properly, and affected individuals are extremely susceptible to infections of all types. Such people normally spend all of their short lives in strict isolation.

The time course of an immune response (figure 16.5) offers significant clues to possible mechanisms of antibody synthesis. During the **primary immune response,** an antigen is introduced for the first time. No antibodies specific for that antigen can be detected for several days. Gradually, the concentration (titer) of antibody molecules rises, peaking after several weeks. The concentration then decays until the system appears to have returned to its original state. This appearance is erroneous, however, for if the antigen is reintroduced, a **secondary immune response** occurs in which the antibody titer rises rapidly to much higher levels than during the primary response and remains at those levels for a longer time. The secondary response is called a memory response as well because it provides evidence of the initiating effect that occurred during the primary immune response. The duration of a memory response is variable, sometimes lasting for only a few months, other times lasting for decades. In some cases, its persistence is thought to be a function of an individual's accidental reexposure to the antigen.

Origin of Immune Cells

A number of different cell types participate in the production of antibodies. Figure 16.6 illustrates some cell lineages involved. Various leukocytes are derived from stem cells produced in the bone marrow. If these stem cells are absent or damaged (as in the case of a person exposed to high levels of radiation), the immune system cannot function. It is sometimes possible to transplant bone marrow from a healthy individual into an affected person and attain normal immune function.

Some leukocytes are ready to function as soon as they differentiate from stem cells, but others require specific maturation in order to function. Of particular interest are the maturation processes of T and B lymphocytes. In the period immediately following birth, **T cells** undergo maturation within the thymus gland. Hence, if the thymus is missing, no T cells are produced. Once T-cell maturation has occurred, the thymus is no longer essential and shrinks in size. **B cells** received their name from the Bursa of Fabricius, an organ in the chicken in which B cells mature. There is no comparable structure in humans, but large numbers of leukocytes are found in the spleen. Although it is thought that B-cell maturation occurs at a number of sites, they primarily mature within the bone marrow. Eventually, some B cells are stimulated to differentiate into **plasma cells** that are active immunoglobulin synthesizers and secreters (color plate 8).

Other marrow-derived leukocytes are also important in the immune response. Monocytes, white blood cells that are somewhat larger than lymphocytes, are released into the blood from the bone marrow and differentiate in various tissues to form long-lived cells

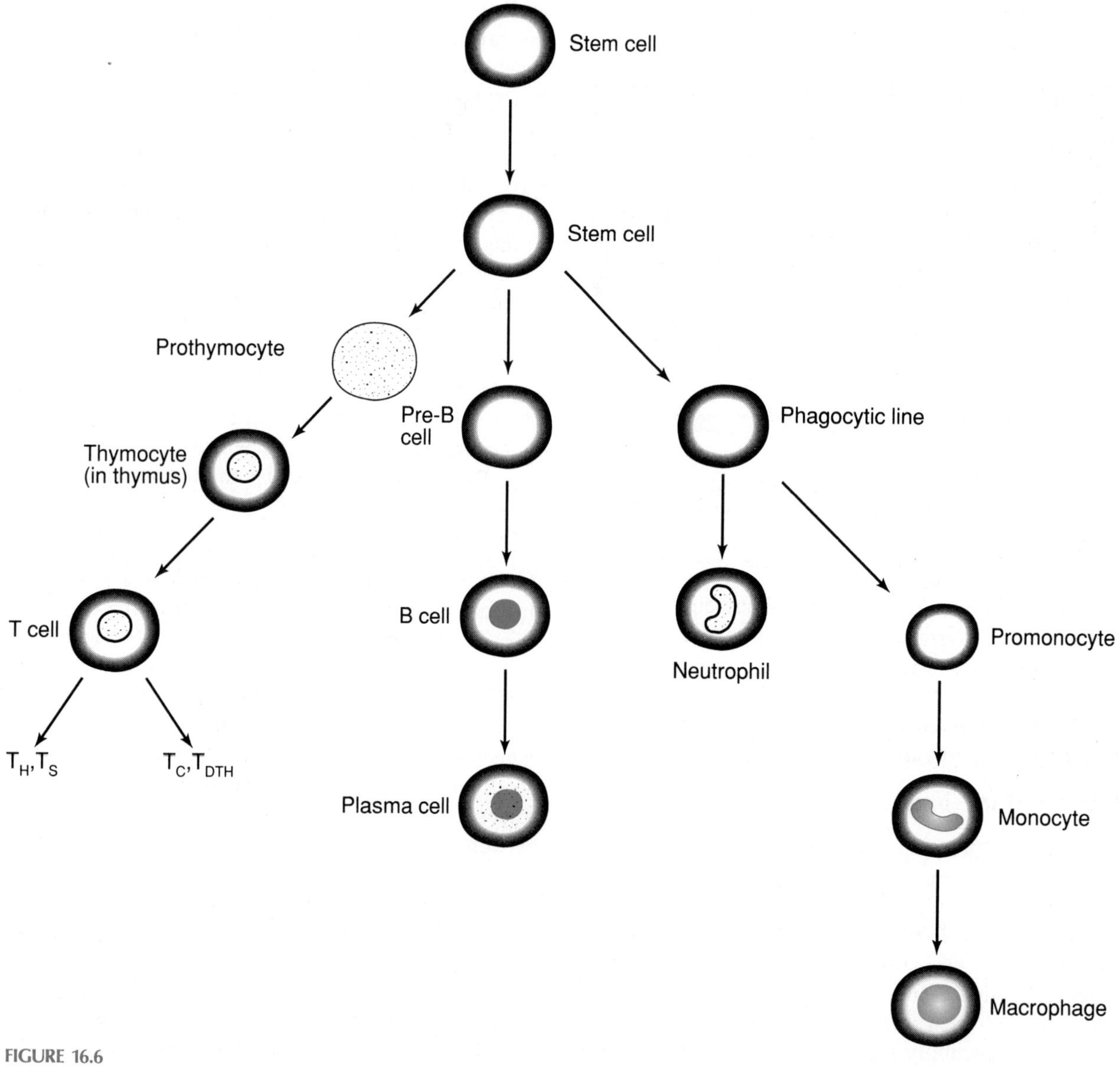

FIGURE 16.6

Differentiation of cells of the immune system. All immune cells depicted in the diagram originate from stem cells in bone marrow. T cells mature within the thymus, while the B cells are thought to undergo their maturation process within the bone marrow itself. Some cells, such as plasma cells, are quite short lived.

called macrophages (color plate 9). **Macrophages** are large, motile, phagocytic cells that engulf particulate matter identified as foreign to the body (color plate 10). Various granulocytes (neutrophils and eosinophils) also function in attacking foreign cells (color plates 11 to 13). They are derived from the same type of leukocytic stem cell but differentiate in the bone marrow itself. Once in the blood, granulocytes survive only for one or two days.

The Clonal Selection Hypothesis

The basic clonal selection hypothesis of Nobel Prize winners Sir MacFarlane Burnet, David Talmage, and Niels Jerne states that all circulating antibodies are synthesized by plasma cells. Control over antibody production is exercised at the level of B-cell differentiation into plasma cells. An antigen triggers this differentiation process, causing a B cell to divide many times and

APPLICATIONS BOX

Bone Marrow Transplants

Reasonable success has been achieved with bone marrow transplants from one individual to another. In some cases, this procedure is necessary because the recipient has developed a tumor in his or her own bone marrow. Following irradiation and chemotherapy to kill all cells, healthy as well as tumorous, the recipient receives the transplant. As with other transplants, the closer the antigenic match between donor and recipient, the greater the likelihood that the bone marrow transplant will be successful.

Many victims of the Chernobyl nuclear disaster lost bone marrow function, and a number of transplants were attempted. Unfortunately, the success rate was not very high.

People with a congenital lack of immune function also are candidates for bone marrow transplants. David, a boy who had spent all of his 12 years of life in strict isolation to protect him from microbial infection, is the most notable example. Finally, a bone marrow transplant was attempted, but the transplant did not take. Unfortunately, David acquired an infection during the procedure and died shortly after the operation.

produce a clone of cells in which all descendant cells are genetically identical to the original. Some of these daughter (clone) cells differentiate into plasma cells that divide a few times at most and live only for a few weeks. During that span, they actively synthesize and secrete immunoglobulins, raising the circulating antibody concentration. As the daughter cells die off, the antibody concentration (titer) falls. The initial burst of plasma cells constitutes the primary immune response, as discussed previously.

The actively dividing B cells are relatively large. During the differentiation process not all B cells become plasma cells. Some return to a form called the small B cell (color plate 14). These cells continue to divide at a slow rate and have a substantial life expectancy, averaging about five years. When exposed to its specific antigen, the appropriate small B cell is stimulated to divide rapidly and differentiate into a plasma cell. In other words, a small B cell is a **memory cell** that "remembers" the original immune response and can repeat it on demand. This ability to remember and repeat on demand constitutes the secondary immune response that is more rapid than the primary due to the existence of a clone of B-memory cells. The antibody titer is higher because of the larger number of plasma cells produced.

The question remains as to why only certain B cells differentiate into plasma cells. Experiments designed to answer that question have also illuminated the roles played by the other types of leukocytes discussed in the previous subsection. If one takes an animal whose immune system has been destroyed by radiation and tries to restore it by injecting purified B cells, its immune system can recover some capacity to produce antibodies. An injection of T cells will restore some cell-mediated immune responses but not the circulating antibody response. Maximum production of circulating antibody is obtained only when a mixture of T and B cells is injected. The conclusion to be drawn is that some T cells are "helping" B cells to recognize the antigen and/or differentiate into plasma cells.

Whenever the secondary immune response is lost or greatly weakened, B-memory cells are thought to have died off to the point at which few or none remain. When an antigen is reintroduced, the differentiation process must begin over again. One way to maintain a secondary response is to "boost" B-cell clones every so often by challenging the organism with the appropriate antigen. The antigen stimulates the B cells to divide and restore their numbers to original levels. The frequency with which a particular immune response needs to be boosted depends on the life expectancy of those B-memory cells. The differences in the longevity of memory cells and the probability of the body's accidental exposure to the antigen that stimulates a specific clone determine whether certain immunizations must be repeated during the life of an individual.

There is one final but critical point to understand about the clonal selection theory. It can be disastrous for an individual to produce immunoglobulins specific for his or her own antigens. At the very least, the binding of antibody molecules to these antigens might prevent specific enzyme functions. At the worst, a true case of **autoimmunity** (an immune reaction against oneself) might result that could be lethal. As a corollary to the clonal selection hypothesis, therefore, all B cells capable of responding to self-antigens must be suppressed or otherwise prevented from functioning, a process that presumably begins during infancy while the immune system develops and continues throughout life.

Antigen Recognition by B Cells

The problem of antigen recognition is not a trivial one. From an evolutionary standpoint, there is no point in stimulating B cells to divide unless they can differentiate into cells that produce the desired antibodies. Any other response is a waste of metabolic energy. Careful examination of B cells shows that they all have actual antibody molecules located within their cell membranes. These molecules are immunoglobulins of the IgM (mostly) or IgD (less frequently) class that are synthesized in the cytoplasm. Then they are specially modified by the addition of a short tail of very hydrophobic amino acids at the carboxyl terminus in order to insert themselves readily into membranes. The orientation of these immunoglobulins is such that their variable regions are exposed on the surface of the cell, and antigens can bind to them. The antibodies on the B-cell surface are the same as those later secreted by plasma cells.

Because the majority of immunoglobulins on the surface of B cells is of the IgM class, it is easy to see why the bulk of antibodies produced during the early stages of the immune response is IgM. As the stimulated clone of B cells matures and prepares to differentiate into plasma cells, there usually is a **class switch** in which the antibody that is produced becomes IgG, IgA, or sometimes IgE. The molecular basis for class switching is considered in the next section, and the functional roles of the different antibody classes are discussed following that.

The nature of the interaction(s) of B and T cells to stimulate differentiation of antibody-producing cells is not entirely understood. There is a gradation in B-cell dependence on T-helper cells. In addition, **lymphokines,** soluble factors from T cells, are necessary for the differentiation of certain B cells. These particular lymphokines are called B-cell differentiation factor (BCDF). Other B cells can react directly with antigen-presenting cells (see following subsection). More information awaits further experimentation.

Antigen Recognition by T Cells

T cells also can bind specifically to antigens. If their cell membranes are examined, however, no immunoglobulins can be found. Instead, there are two protein chains called α and β (molecular weights of 49,000 and 43,000, respectively) that form a complex similar to an antibody. The complex is held together by disulfide bridges and contains both constant and variable regions within the protein chains. Like antibodies, the heavier chain is also a glycoprotein. Nevertheless, the complex is not an antibody, because the DNA sequences of the constant and variable regions of α and β are different from those of the heavy and light antibody chains and because no hypervariable regions are present. Because an appropriate T-cell receptor (α and β complex) can bind the same antigen as an appropriate B cell, the two cells are said to have the same **idiotype,** although the molecules doing the binding are different.

It is not sufficient to have the antigen bound to the surface of the T cell for the stimulation process to occur. The antigen must be presented to the T cell in association with a set of antigens called the **major histocompatability complex** (MHC). The MHC is a group of glycoproteins found on cell surfaces whose composition is characteristic of the individual organism. It can be regarded as the major determinant of self. The term was originally used for the immune system of mice. In humans, human leukocyte antigens (HLA), a comparable term, is often used. Three subsets of MHC antigens are known. Class I MHC antigens are found in the membranes of all nucleated cells. Class II MHC antigens occur only on T and B cells, macrophages, and some dendritic cells (see following paragraph). Antigen presentation must occur in conjunction with class II MHC antigens. Class III MHC antigens include components of the complement system (see discussion on complement fixation).

In order for an antigen to be presented in conjunction with a class II MHC antigen, some normal host cell must become associated with the antigen. An obvious candidate for this function is the macrophage that engulfs foreign antigens. The macrophage then selects portions of these antigens to appear on its cell surface, an activity that provides an automatic MHC-antigen association. The role of presenter cell is not limited to macrophages, however. An alternative group of cells is described as "dendritic," referring to the general shape of the cells. Dendritic cells are nonphagocytic and have quite distinct properties from macrophages, yet they function as presenter cells as well. The exact mechanism by which macrophages and dendritic cells present antigens is not well understood, but it is known that protein antigens enter the cytoplasm and undergo limited proteolysis before they appear on the cell surface in company with class II MHC proteins.

TABLE 16.2 Varieties of T cells and their functions

T-cell Type	Function
Regulators	
T_h	Helper cells; stimulate B-cell functions
T_s	Suppressor cells; inhibit B-cell functions
Effectors	
T_{DTH}	Delayed-type hypersensitivity
T_c	Cytotoxic (cell-killing) cells

TABLE 16.3 Genetic diversity in germline cells

Region[1]		Genes in Human	Genes in Mouse
V_κ		50	200
V_λ		25	2
V_H		100	200
J_κ		5	5 (1 not used)
J_λ		>6	4 (1 not used)
J_H		6	4
D		10–20	10–20
C_λ		>6	3 (1 not used)
C_H	α[2]	2	1
	δ		1
	ϵ	2–3	1
	γ		4
	μ		1

1. The various regions of the antibody complex are indicated in figure 16.7. The capital letters (C, D, J, V) refer to the variable or constant regions, and the subscripts designate the type of chain (κ, λ) or the heavy chain (H) itself. Some uncertainties are present due to the existence of pseudogenes.

2. Heavy chain types.

T-Cell Varieties

T cells can be subdivided according to the functions that they perform when stimulated. These are summarized in table 16.2. There are two basic T-cell groupings, effectors and regulators. T-effector cells are concerned with cell-mediated immunity and are considered later in this chapter. T-regulator cells are concerned with both humoral immunity and cell-mediated immunity.

Helper (T_h) cells and suppressor (T_s) cells differentiate from the same stem cell. A T-helper cell acts as a **mitogen** to assist the antigen in stimulating B-cell clones to divide, and a T-suppressor cell acts to prevent B cells from responding to the antigen stimulus. In neither case is the mechanism truly understood, although it has been hypothesized that T-suppressor cells produce molecules to react with an antigen before it can reach B cells. Normally, there is a balance between helper and suppressor cells. Should T-helper cells predominate, there is the risk of stimulating B-cell clones to release antibodies not needed or wanted. On the other hand, if T-suppressor cells predominate, the ability of B cells to respond to antigenic stimuli is reduced or lost.

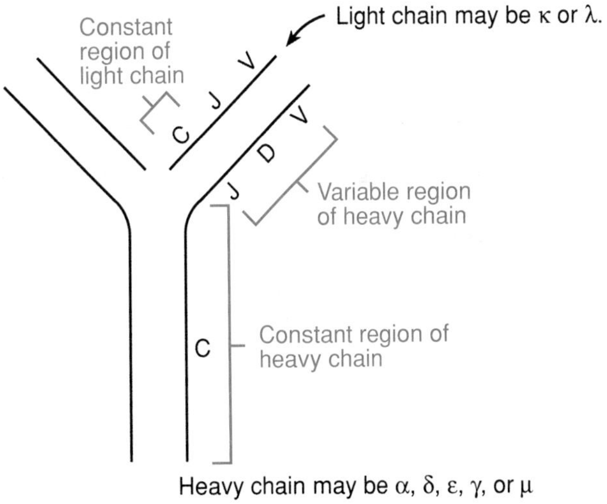

FIGURE 16.7

Genetic view of antibody structure. The light and heavy chain segments are encoded at three and four different chromosomal sites, respectively. Individual letters indicate regions of the protein chains that are encoded separately. These sites must be brought into close proximity in order to synthesize an appropriate mRNA molecule.

Molecular Biology of Clonal Selection

A major challenge for molecular biologists during the 1970s was to account for the incredible diversity of antibodies produced by cells with finite genetic resources. Genetic analyses showed that each individual protein chain of an immunoglobulin is encoded by fusing several genes together. The J or joining region located at the carboxyl end of each variable region, and the D or diversity region that links the J and C regions of the heavy chain (figure 16.7) were both identified on genetic grounds. Table 16.3 details the genetic data with respect to the number of genes coding for constant and variable regions of immunoglobulins found in **germline cells** that have not yet differentiated into the leukocyte precursors.

Genetic mapping experiments have shown that genes coding for the various regions of an immunoglobulin are not contiguous but rather arranged in patterns such as those depicted in figure 16.8. In order for an immunoglobulin molecule to be synthesized, an

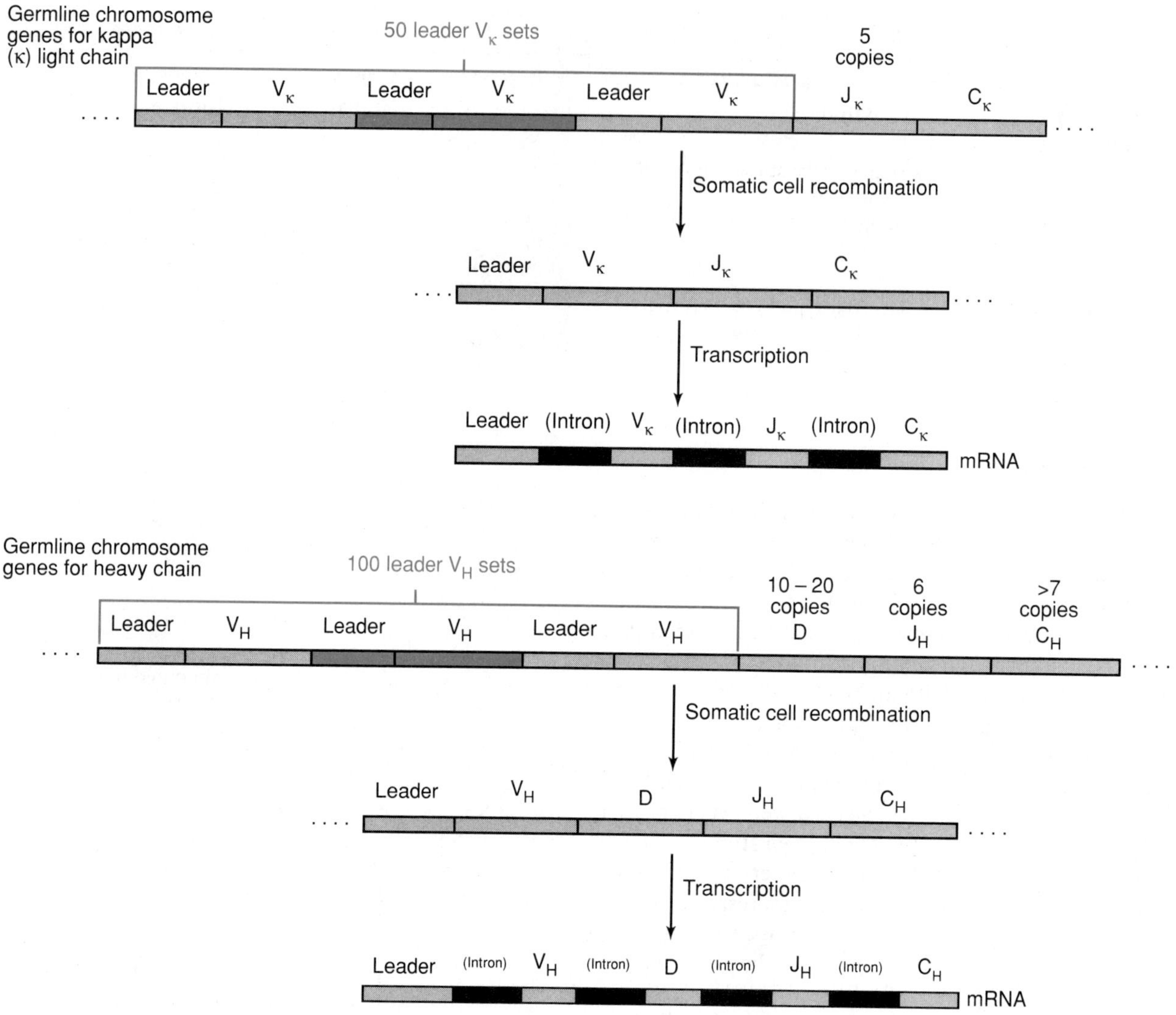

FIGURE 16.8

Possible arrangements of immunoglobulin coding regions determined in germline cells. Examples of the types of rearrangements that can occur are diagrammed also.

mRNA molecule containing one copy of each kind of region is needed. This obviously requires a permanent rearrangement of the germline DNA. Recombination occurring in intergenic regions is thought to be the method of rearrangement, with the number of exchanges required depending on the arrangement desired. Because eukaryotic cells routinely carry out RNA splicing reactions (chapter 11), the coding regions need not be contiguous, but they must be in the correct order. After the mRNA molecule has been synthesized, one or more splicing reactions can be used to generate a translatable molecule that will give rise to an immunoglobulin of the appropriate type (figure 16.9).

The hypervariable regions of immunoglobulin molecules get their altered sequences from **somatic cell mutations,** that is, from changes in the DNA sequence occurring after the cell has differentiated and during mitosis. It has been estimated that 10 or more somatic cell mutations occur in each B-cell line.

Given all the processes that antibody molecules, their precursors, and their associated cells undergo, the possible molecular diversity is staggering. A simple

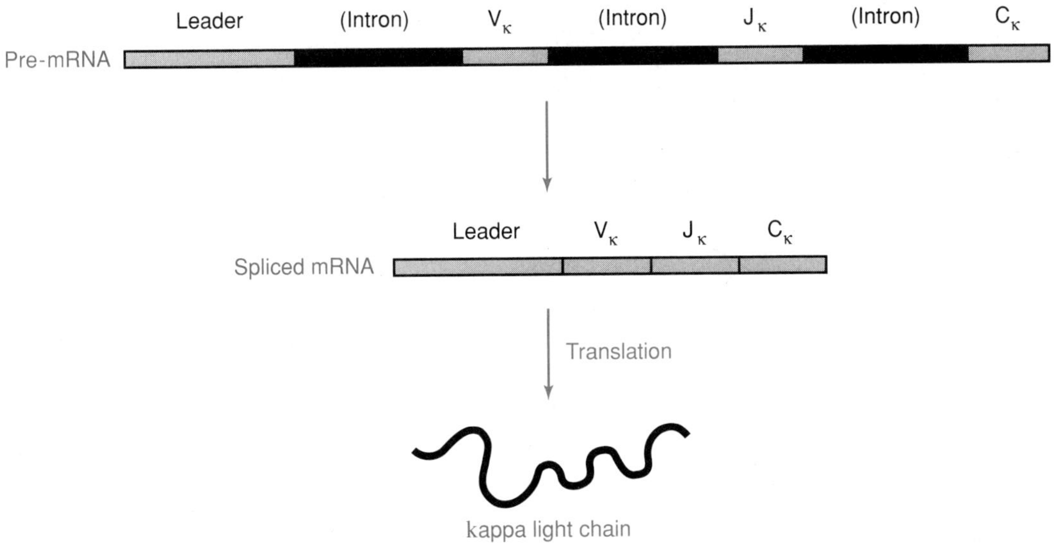

FIGURE 16.9
RNA splicing reactions generate immunoglobulin mRNA molecules that can be translated.

analysis of the ways in which V, J, D, and C regions can be assembled gives more than a million different protein chains. When somatic cell mutations and combinations of heavy and light chains are taken into account, the number of different possible combinations approaches 10^{10}. This number, however, is a fairly certain overestimation.

Several complications can lead to an overestimate in the calculation of potential immunoglobulin diversity. One major difficulty arises with **pseudogenes,** DNA sequences that are nearly identical to known genes but that cannot be transcribed. If any of the various sequences listed in table 16.3 are pseudogenes, the amount of possible diversity is diminished. Another problem lies in the assumption that all possible heavy chains can pair with all possible light chains. There is evidence to indicate that this assumption also may be an oversimplification. Unfortunately, it is not yet possible to formulate rules describing the pairings of heavy and light chains; therefore, the extent to which potential diversity is diminished is unknown.

Serologic Antigen-Antibody Reactions and Immunologic Assays

If mixtures of antigens and their corresponding antibodies are prepared, a **precipitin reaction** can occur in which a macroscopically visible precipitate forms. Formation of a precipitate is highly concentration dependent (figure 16.10). The maximum amount of precipitate is seen when the antigen and antibody are present in roughly equal concentrations. The broad range over which precipitation is seen in a precipitin reaction is referred to as the **zone of equivalence.**

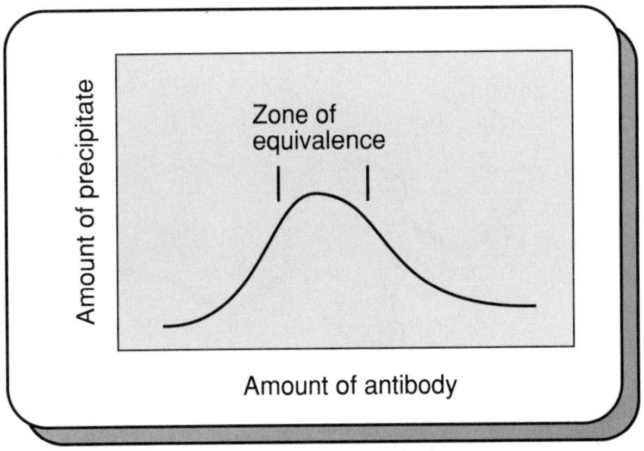

FIGURE 16.10
Stoichiometry of a precipitin reaction. As increasing amounts of antibody are added to tubes containing a constant amount of antigen, a precipitate forms. If the concentration of antibody is too high or too low, however, no precipitate can be visualized.

The existence of a zone of equivalence implies that one antibody molecule must be capable of reacting with more than one antigen. The reasoning is as follows. If an antibody can, in fact, react more than once, there is the possibility of developing a latticework of antigens

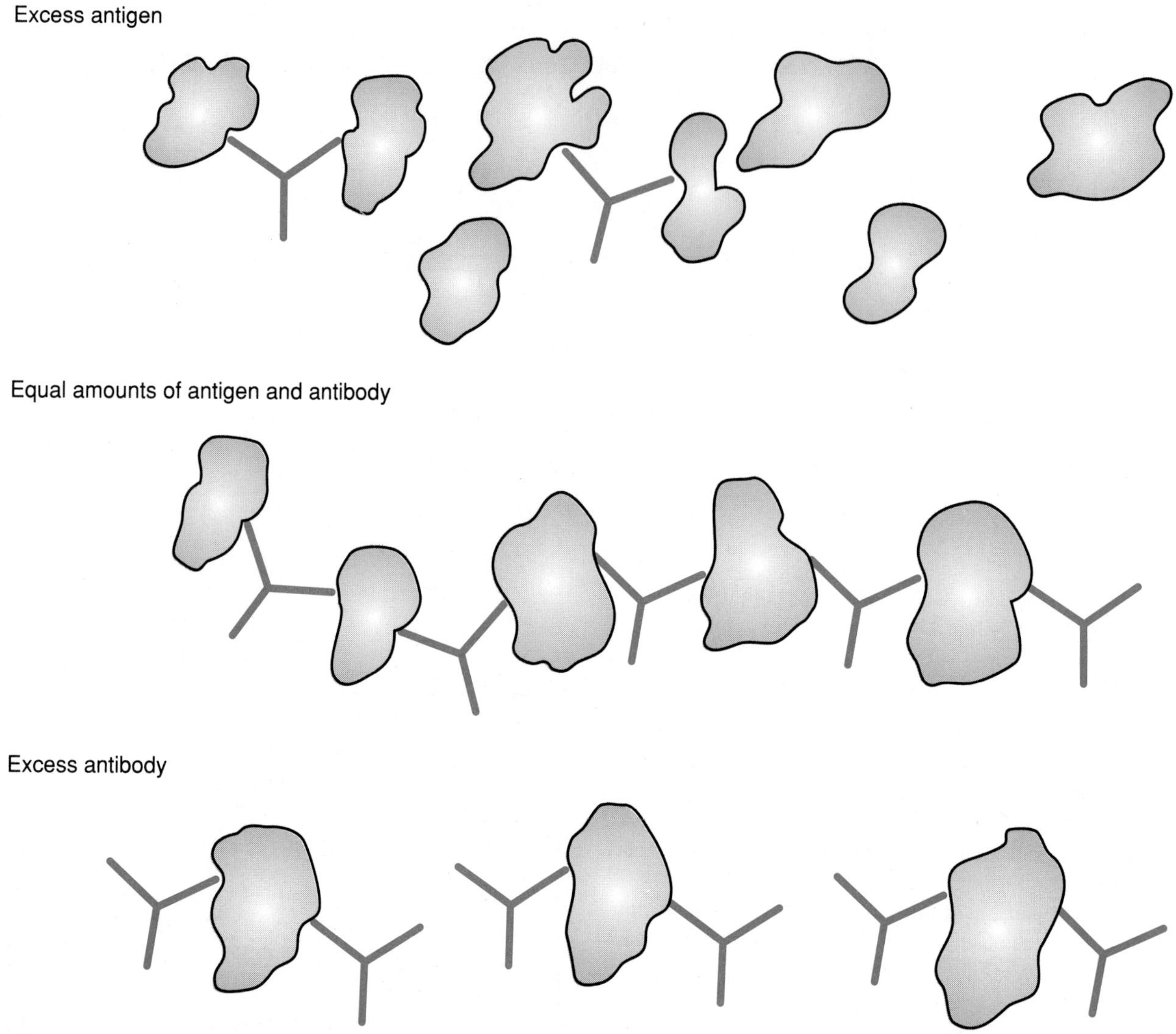

FIGURE 16.11

Zone of equivalence, with bivalent antigens as well as antibodies. When either excess antigen or antibody is present, a molecular chain cannot develop; therefore, no precipitate will form. Only when the molecules are present roughly in equal numbers will long chains develop and precipitate.

and antibodies into a visible precipitate (figure 16.11). When confronted with an excess of antigen, antigen-antibody complexes tend to be short because antibody links in the growing chain are lacking. When antibody is in excess, there will be insufficient antigen links in the growing chain. In either case, therefore, no precipitate will form. Because maximum precipitation occurs with equal numbers of antigen and antibody molecules, antibodies must have two and only two binding sites for antigens. If there were three binding sites, maximum precipitation would occur only with more antigen than antibody.

Note that there is nothing to prevent an antibody from binding to a pair of antigens on the same structure. For example, if several cells have the same antigenic proteins in their cell membranes, an antibody can bind so as to join two cells together or to bridge two antigens on the surface of the same cell (figure 16.12). In the latter case, no precipitate will form.

Because an immunoprecipitate is formed by electrostatic bonding between the various components, it is not a permanent chemical union. The precipitate can be redissolved with solutions containing substances that

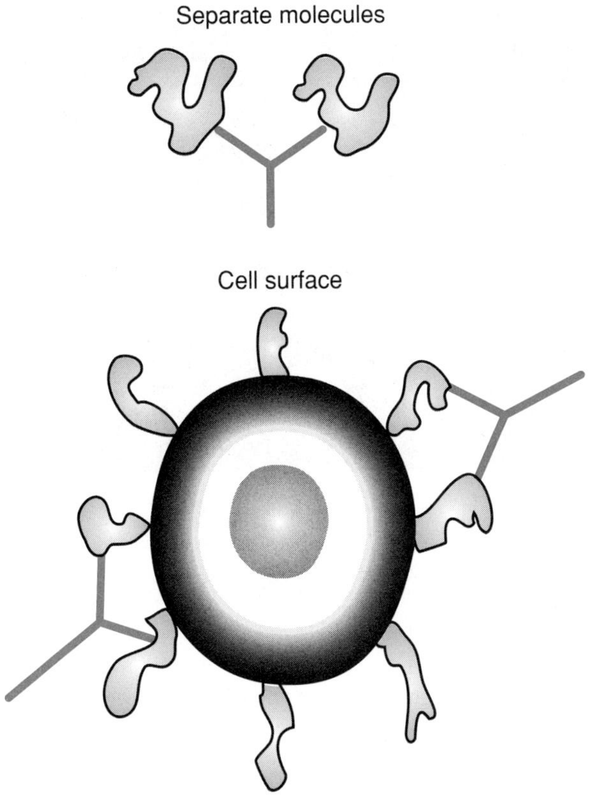

FIGURE 16.12
Antibodies form a bridge between two antigens. These antigens may be separate particles or two entities on the surface of a large structure like a cell.

can neutralize the electrostatic interactions. A buffer containing a high concentration of sodium chloride or the addition of excess antigen to the complex so as to separate the antibodies into smaller clumps can neutralize an electrostatic precipitate.

The amount of antibody present in a serum is often characterized by its **titer,** the most diluted serum that can still cause visible clumping of a standard amount of antigen. The serum containing the antibodies is subjected to a progressive twofold dilution (one-half, one-fourth, one-eighth, etc.). Each dilution is added to a separate tube containing a constant quantity of antigen. The dilution of the last tube to clump is the titer of antibodies in the serum (for example, 1:256). Because of the subjective nature of the test, an antibody titer provides only a relative estimate of the quantity of immunoglobulin present in the serum.

An immunoglobulin is characterized by its idiotype, the shape of the antigen-binding region of the molecular complex. The size of the domain involved in an idiotype can be determined if the subunits of an antigen, such as a carbohydrate, are removed gradually until the molecule no longer reacts with the antibody. The minimum size of carbohydrate molecule that will react with an antibody has been found to have six or seven glucose residues, which corresponds to a space about $3.4 \times 1.2 \times 0.7$ nm.

A unit of seven glucose molecules is too small to stimulate actual production of an antibody even if it is capable of reacting with that same antibody after it has been synthesized. The term **hapten** refers to a molecule that possesses enough structure to react with an immunoglobulin but is of insufficient size to stimulate its production. A hapten can be a discrete molecule like penicillin or aspirin or a fragment of a larger molecule.

Although an antigen-antibody reaction can be described as specific, there are varying degrees of specificity possible. If the original antigen is a complex protein, there may be few or no other proteins of similar structure. On the other hand, if the antigen is a fairly common form of enzyme, such as a dehydrogenase, and the active site is an antigenic determinant, the antibody produced may have a general specificity for dehydrogenases. A **cross-reaction** occurs when an immunoglobulin reacts to some extent with an antigen that is not the original, stimulating antigen.

A powerful new technology based on immune reactions has developed in the last few decades. Very small amounts of antigenic substances can be detected in a variety of situations. Sometimes, the interest is in studying antigens on a cell surface. Other times, the interest is more analytical or diagnostic, attempting to identify certain cell types or locate them within a larger population of cells. The following subsections present a few examples of how this technology is carried out.

Phage Neutralization Assays

It is possible to attach small antigens or haptens to a bacteriophage without altering its infectivity. If a specific antibody later reacts with the attached molecules, however, all infectivity is lost. Therefore, an extremely sensitive assay to detect the presence of a particular antibody can be set up by measuring loss of bacteriophage infectivity. It has been estimated that the amount of antibody produced by a single cell can be detected in this manner.

Ouchterlony Double-Diffusion Test

Antigens and antibodies are generally small enough to diffuse through a liquid medium or a gel at a reasonable pace. A well-known example of this process is illustrated in the **Ouchterlony double-diffusion test** of Orjan Ouchterlony (figure 16.13A–B). Antigens and antibodies are placed in small holes within an agar layer. Both diffuse out in all directions. As the antigens and antibodies diffuse toward one another, there will eventually be a point at which they are in equal concentration (zone of equivalence). If the idiotype is correct for the antigen, a visible precipitate will form, its curved shape reflecting the shape of the interacting concentration gradients. If the antigens do not react or cross-react with the antibody, nothing will happen.

If the antigens in two adjacent wells are quite different and a mixture of appropriate antibodies is provided in the middle well, two overlapping precipitin bands are seen. This is a reaction of nonidentity. If the two antigens are the same, both react with the same antibody molecules and the precipitin bands become continuous. This is a reaction of identity. It is also possible to have a reaction of partial identity in which one of the precipitin bands is the normal size, but the other band stops at the point at which the two bands touch (figure 16.13A–B). A partial identity reaction indicates that at least one of the antigens provided is actually a mixture of antigens. One in the mixture is the same as that in the adjacent well, and their precipitin bands give an identity reaction. Another of the antigens in the mixture, however, is unique and gives an overlapping band of nonidentity.

A mixture of antigens does not necessarily imply that there is more than one type of molecule present. Large molecules like proteins and carbohydrates can be thought of as having domains of unique three-dimensional structure on the surface of their folded configurations (figure 16.14). During antigen presentation, the molecule is cleaved so that each surface represents a different antigen as far as the immune system is concerned. Each antigenic site should stimulate the production of its own unique antibody; therefore, each separate domain of the original macromolecule represents a separate **antigenic determinant.**

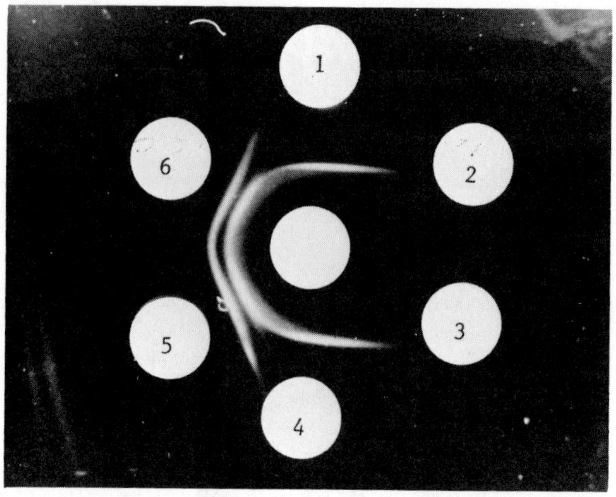

A

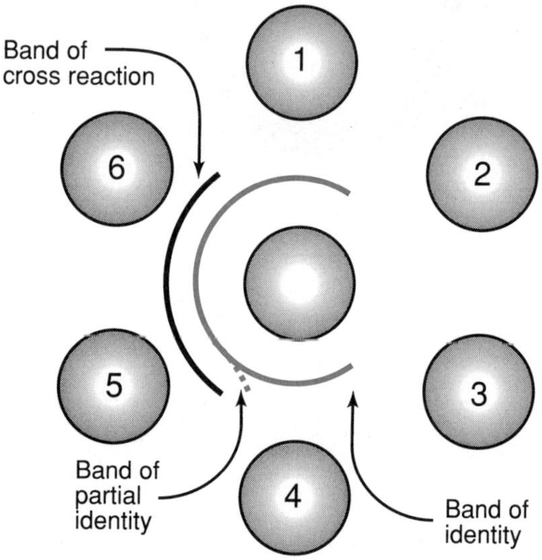

B

FIGURE 16.13

Ouchterlony double-diffusion test. (**A**) In this example, an agar layer was placed on a glass slide, following which small wells were cut into it in a symmetrical pattern. A sample of antibodies prepared against the major antigen of *Blastomyces dermatidis* was placed in the center well, and various antigen preparations were placed in the outer wells. Wells 1 and 4 contained pure antigen corresponding to the antibody and gave single bands. In wells 2 and 3 were extracts of *Histoplasma capsulatum,* a fungus that does not produce an antigen reactive against the antibody. Extracts from *B. dermatidis* cells were placed in wells 5 and 6. These extracts contained the orginal antigen corresponding to wells 1 and 4 (a band of identity) in addition to two other antigens that cross-reacted with the antibody. There was a slight reaction of partial identity at the inner band of well 5. (**B**) The cartoon summarizes these results.

Complement Fixation

When antigens and antibodies react, the complex binds the initiator proteins for the complement cascade. The bound complement proteins are said to be fixed and cannot participate in any future reactions. One long-

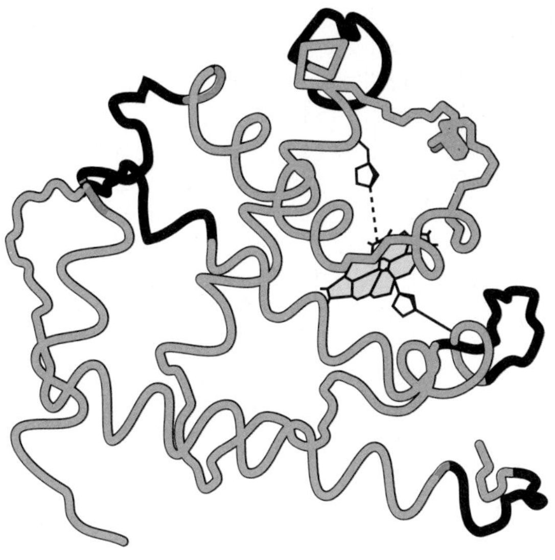

FIGURE 16.14

A single molecule may have several different antigenic sites, and each site can trigger the production of a corresponding antibody. Shown is the structure of the myoglobin molecule. Dark areas are individual antigenic sites. Colored areas, each corresponding only to one amino acid residue, may or may not be included in the adjacent antigenic site depending on the test used. It has been shown that the mottled portions of the molecule are not antigenically active in the intact molecule.

Reprinted with permission from *Immunochemistry* 12:423–438 by M. Z. Atassi. Copyright © 1975 Pergamon Press, Inc., Elmsford, NY.

standing immunologic test that is still sometimes used to detect trace amounts of antigen or antibody is the **complement fixation test.** The test consists of two distinct antigen-antibody reactions performed in sequence (figure 16.15). The first reaction uses serum whose complement has been inactivated by heat and replaced with a measured amount of complement from another serum. This is actually the reaction of interest. If it occurs, the limited amount of complement protein present is fixed, and none is available for the second reaction. The second reaction always occurs and is usually carried out between red blood cells from sheep and a specific antibody. If the complement has been fixed, the red blood cells will not lyse even though an antigen-antibody complex is present on their surfaces (positive test, the first reaction did occur). If the complement has not been fixed, it will bind to and lyse the red blood cells (negative test, the first reaction did not occur).

Monoclonal Antibodies

The clonal selection hypothesis predicts that for every unique antibody there is a corresponding clone of cells, all of which produce only that antibody. If clone cells could be isolated in tissue culture, they would continue their antibody production, but instead of those antibodies commingling with thousands of others (polyclonal antibodies), they would be all alone and could be purified easily. Such antibodies would be called **monoclonal antibodies** because they would all be derived from a single clone of cells. Unfortunately, as discussed in chapter 9, normal cells do not survive for long periods of time in tissue culture, and the concept as stated is not practical. Modern technology has, however, provided alternative means to accomplish the same result.

Tumor cells will grow indefinitely in tissue culture, and **myeloma cells** (cancerous plasma cells) may even secrete antibodies. Unfortunately, tumor cells produce antibodies for which they were originally programmed, and those usually are not the antibodies wanted. If appropriately rearranged DNA from a B cell could be placed into a tumor cell, that cell would secrete the desired antibody, and an ideal solution to growing monoclonal antibodies in tissue culture would result.

There is no good way to move an entire chromosome from one cell into another, but it is possible to force two different cells to fuse their membranes and become one huge tetraploid cell in a process called **cell fusion** (figure 16.16). Fusion can be forced by high concentrations of polyethylene glycol or by an appropriately directed electric field. Animal cells do not tolerate tetraploidy well, and chromosomes are discarded spontaneously during cell division until a reasonable approximation of a diploid state is reached. If by chance a chromosome from a myeloma cell that coded for immunoglobulin variable regions is lost and replaced by a chromosome from an antibody-producing cell of a different type, a new tumor cell (**hybridoma**) is produced that secretes an entirely different antibody.

From a practical standpoint, the probability of a successful fusion can be improved by a careful choice of cells. Usually, multiple myeloma cells that are not secreting antibodies and spleen cells from an animal preimmunized with an antigen that should stimulate production of the desired antibody are the ideal fusion choices. The idea behind using spleen cells from an immunized animal is that if the donor animal is undergoing a secondary or primary response when the

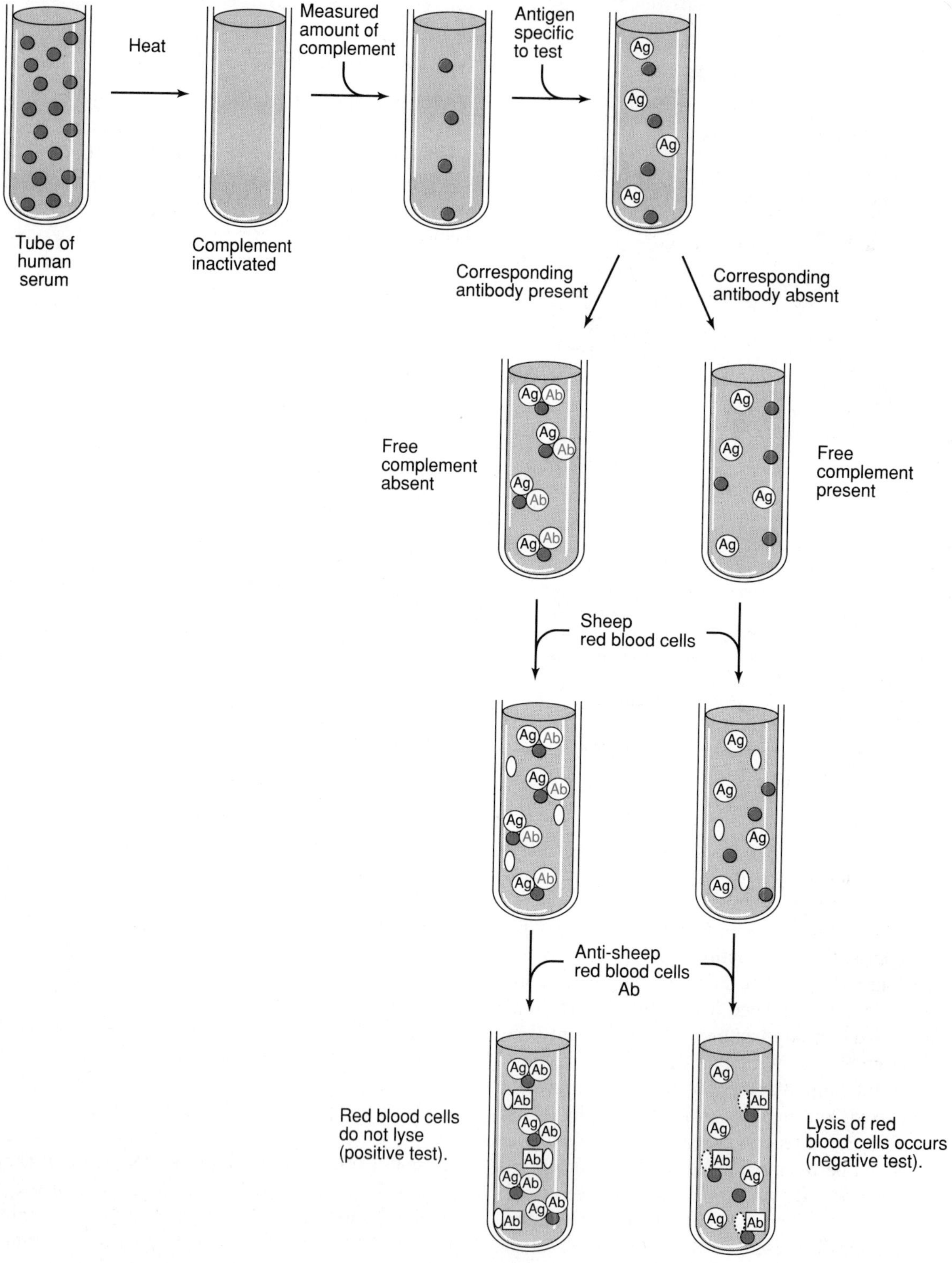

FIGURE 16.15

Complement fixation test. The initial reaction, if it occurs, does not generate sufficient antigen-antibody complexes to form a visible precipitate. It does, however, bind the available complement. The second reaction always occurs but results in lysis of the red blood cells only if complement is available; therefore, cell lysis indicates that the initial reaction did not occur.

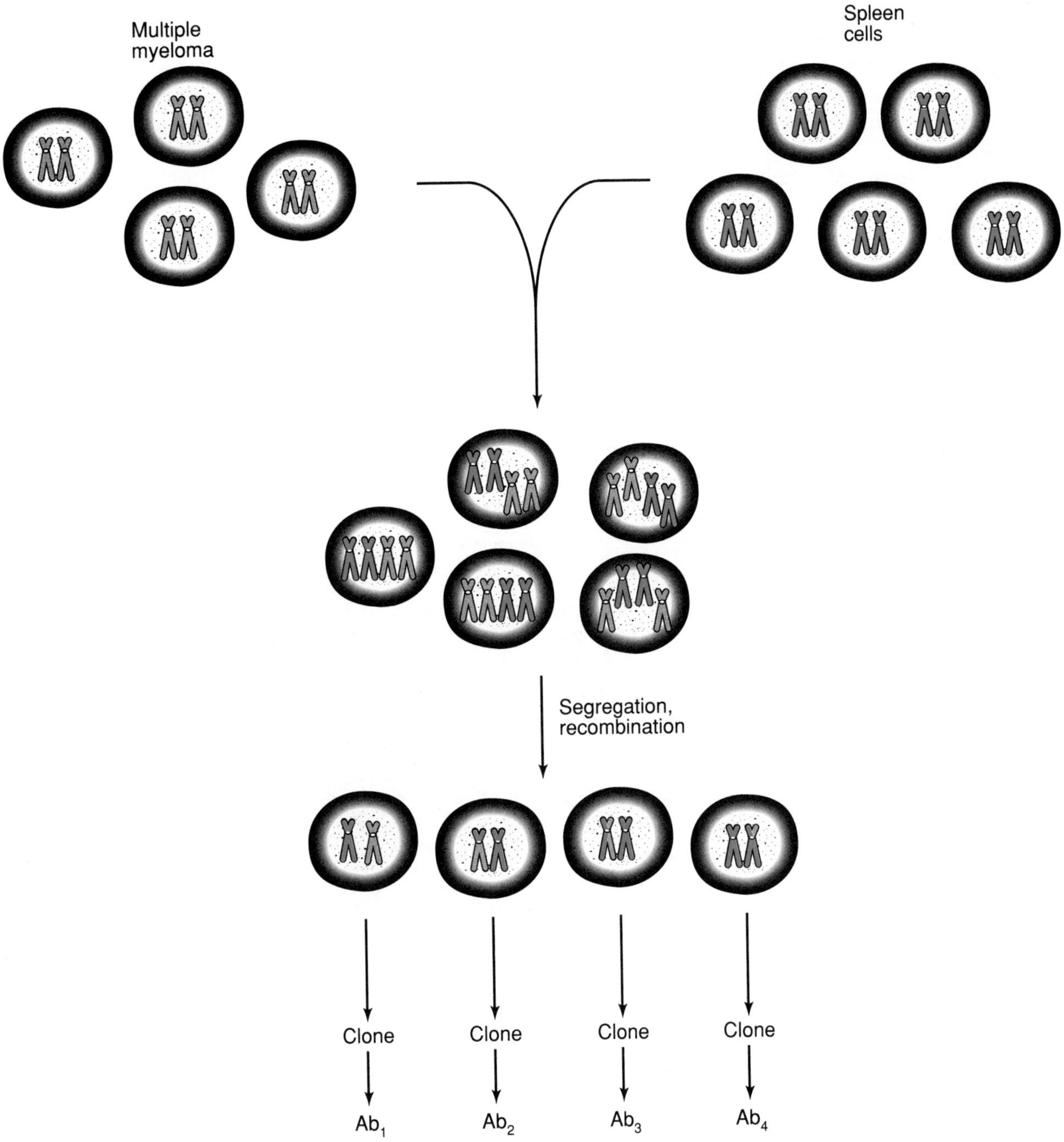

FIGURE 16.16

Preparation of monoclonal antibodies. Fusion of multiple myeloma cells with spleen cells from a preimmunized animal will generate new types of antibody producing cells. Each new cell is cloned separately and examined to see if it produces a desirable antibody.

cells are removed, there will be relatively large numbers of B cells with appropriately rearranged chromosomes present in the total spleen population. The increased number will improve the chances of getting a clone that secretes an antibody that recognizes the target antigen.

Each hybridoma cell is cloned separately and the tissue culture fluid tested for the presence of antibodies against the target antigen. It is to be expected that antibodies with differing affinities for the antigen will be obtained, depending on the antigenic site presented to the original T and B cell. The usual choice for additional work is the antibody with the highest affinity for the antigen. Monoclonal antibodies can be used in any of the procedures described to confer specificity on the assay system, but they are particularly useful for assays done with labeled antibodies and for the ELISA tests discussed in the remainder of this section. Both require more or less purified antibodies.

Assays with Labeled Antibodies

An antigen-antibody reaction can be detected easily if the concentration of the two components is great enough to give a visible amount of precipitate, which is all very well and good for some situations but not useful when one is looking for rare events. For example, suppose that you wish to detect a rare disease-causing bacterium within a large population of other bacteria. If there is an antigen unique to the rare cell, an antibody against that antigen will be able to tag it. Because of the very rarity of the cells, formation of a visible precipitate cannot be guaranteed. What is needed is a method of finding the antibody-antigen complex that is independent of precipitate formation so that a correct diagnosis can be made.

The complement fixation test can be used, but several other methods also are available now. Antibodies can be made radioactive and be detected with some type of radiation-measuring device. Antibodies also can be modified chemically, with fluorescent or enzymatically active compounds attached to them.

In a typical fluorescent antibody test, a sample of bacteria is dried onto the surface of a slide. An antibody preparation to which a fluorescent molecule has been conjugated is added to the cells and allowed to stand for several minutes. The slide is then washed well to remove any unreacted antibody and examined under dark-field conditions with an ultraviolet microscope (figure 16.17). The only cells that will appear bright are those whose surface antigens reacted with the fluorescent antibody molecules and retained them when the slide was washed.

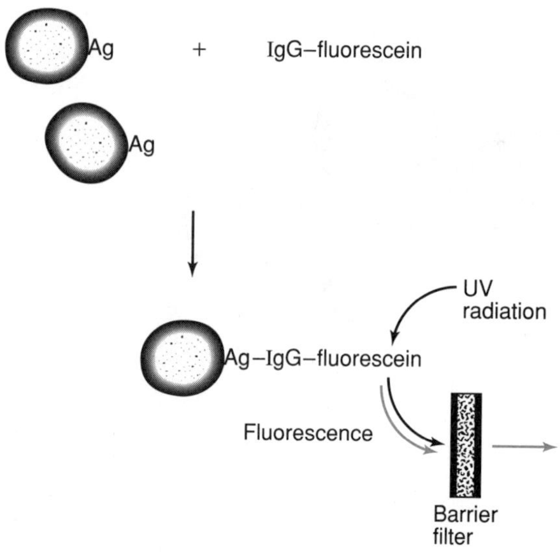

A

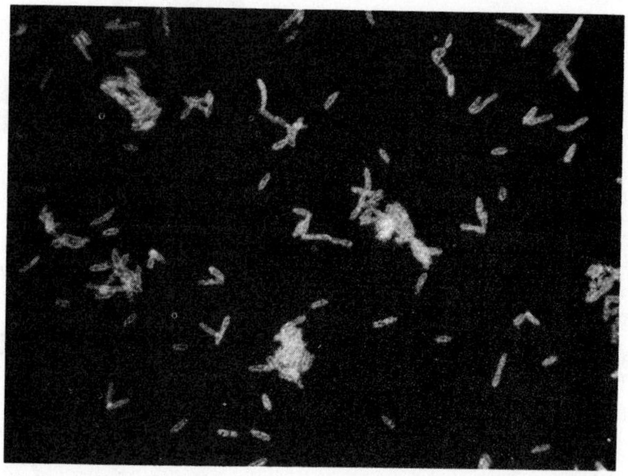

B

FIGURE 16.17

Immunofluorescence. (A) Cells are reacted with antibodies, washed, and examined under UV radiation. A barrier filter prevents all UV rays from reaching the eye of the observer but does allow visible light to pass. (B) Cells carrying the target antigen will appear bright against a dark field. *Legionella pneumophila* seen by direct fluorescence assay.

(B) Courtesy of Lauren Roberts, Department of Microbiology, Arizona State University.

Radioimmunoassays work in similar ways except that a radioisotope detector is used to identify the antibody in question. The type of detector depends on the type of radioactive label employed. One of the most

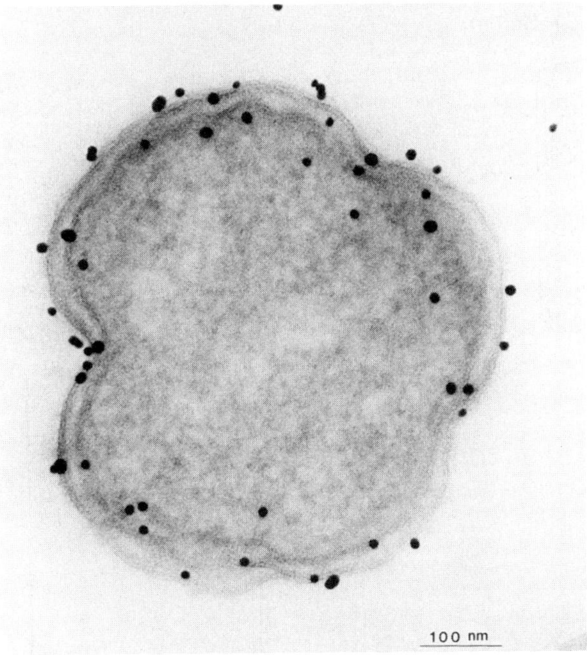

FIGURE 16.18

Immunoelectron microscopy. Antibodies were prepared against subunit 1 of the cytochrome *d* terminal oxidase complex. After the antibody was reacted with a section of *Escherichia coli,* protein A linked to gold atoms was added. Protein A reacts specifically with immunoglobulin molecules and, therefore, attaches a gold atom wherever an antibody molecule is located. When the bacterial sections are viewed under the electron microscope, the gold atoms appear as dark dots and show that the antigen is on both surfaces of the membrane.

Electron micrograph courtesy of James R. Swafford, Brandeis University.

commonly used radioactive labels for antibodies is ^{125}I, and its detection requires a gamma ray detector. The Western blot, analogous to the Southern blots discussed in chapter 11, is one radioimmunoassay used. Proteins are immobilized on a solid surface and allowed to react with radioactive antibodies. A piece of photographic film placed on the surface will be exposed by the radioactive material as it decays. The dark spots on the negative indicate the places on the solid surface where particular proteins are located.

In the case of electron microscopic work, use is often made of antibodies that have been conjugated with heavy metals. The presence of the heavy metal in a sample can be seen as a dark spot in the micrograph. At that spot, it is inferred that an antigen and an antibody have reacted (see figure 16.18).

ELISA Tests

The term *ELISA* is an acronym for enzyme-linked immunosorbant assay. It is a sort of antibody sandwich technique conveniently used to detect small quantities of antibody in a serum sample. In this type of test, generally two antibody preparations are used. For example, if it is important to know whether a person has been exposed to a particular antigen, such as the human immunodeficiency virus, one approach is to discover whether that person has circulating antibodies specific for an antigen characteristic of that virus. One common method used in this determination is illustrated in figure 16.19.

The virus or one of its antigens is affixed permanently to the inner wall of a small container. Simple washing will not dislodge it. Serum collected from a patient is added to the container and allowed to stand for some period of time. If any antigen-specific antibodies are present, they will bind to the virus. The unreacted antibodies then are washed out of the container. At this point, antibodies synthesized by an animal, such as a goat, that had been immunized with human IgG are added to the container. The human IgG served as antigens in the goat to stimulate the production of goat antihuman antibody. Some idiotype-specific antibodies will have been produced, but the vast majority of human IgG-specific antibodies (anti-antibodies) will recognize the constant regions, because all IgG molecules possess those same few constant regions. If the patient's IgG molecules are present in the container, the second antibody will react and not be washed out.

The presence or absence of the second antibody is detected by a simple enzyme assay. The second antibody has been attached covalently to an enzyme, like a phosphatase or peroxidase, that can convert a colorless compound into a colored one. If the substrate is added to the container and no color develops, the patient did not have antibodies against the antigen tested. The converse is true if color appears. Notice that the specificity of the test lies in the original antigen used and not in the second antibody; therefore, one enzyme-linked antibody preparation can be used for a variety of tests.

A possible variation of the test consists of using monoclonal antibodies prepared against a specific antigen and then directly linked to an enzyme. This procedure eliminates the need for introduction of a second antibody but is justified only when a great many tests of the same type must be performed.

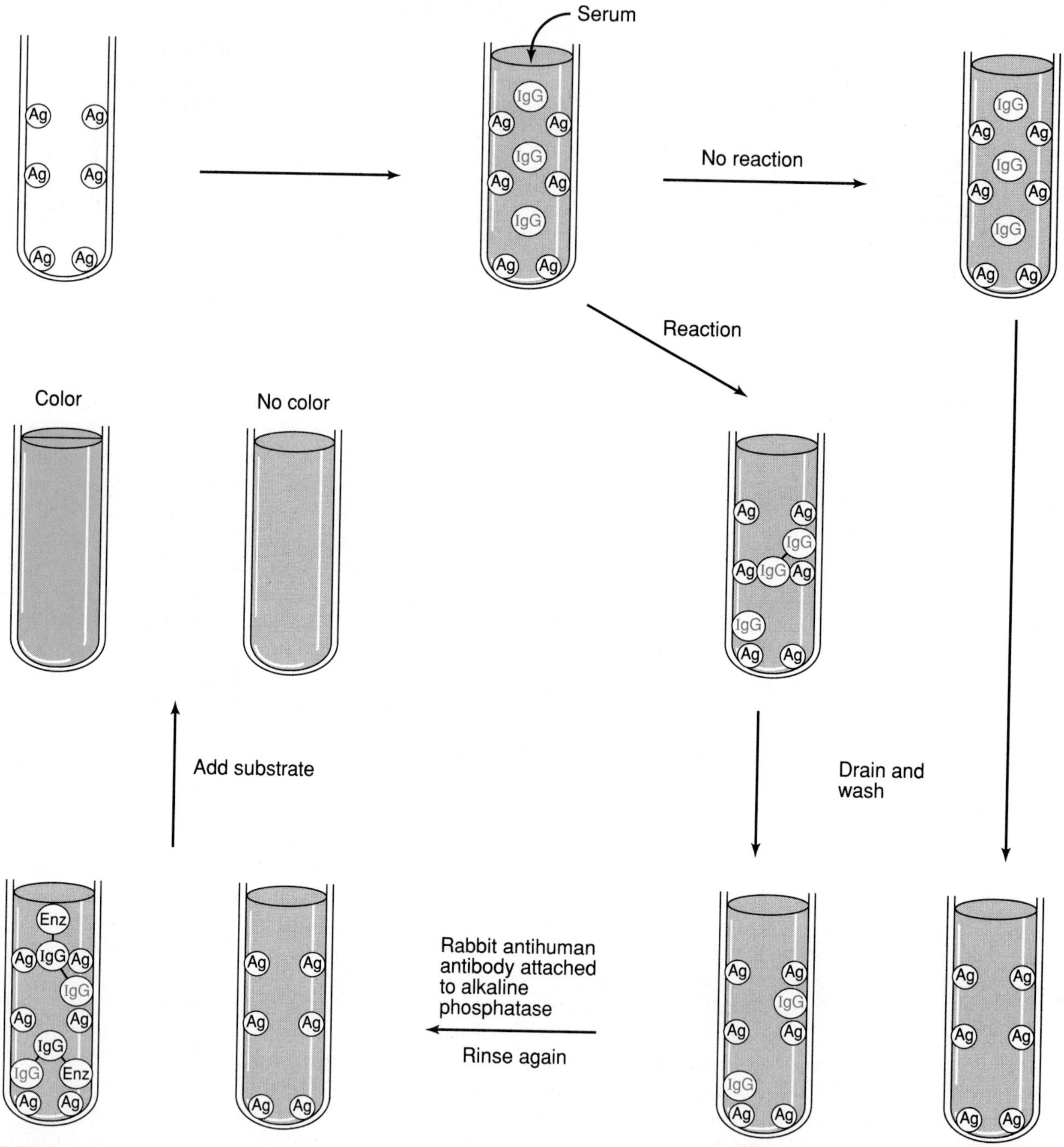

FIGURE 16.19

Example of one type of ELISA test. A known antigen is attached permanently to the walls of a container. Patient serum is added and allowed to react. After washing, a second antibody specific for the constant regions of human IgG and carrying an attached enzyme is added. If the second antibody remains in the container after washing, its attached enzyme will react with an appropriate substrate when added and convert the substrate to a colored compound.

APPLICATIONS BOX

An ELISA Test for Herpesvirus

Herpesviral infections are on the rise along with other sexually transmitted diseases. When this virus is passed from mother to newborn infant, about 50 to 70% of the children die. Correct diagnosis of any herpes infections prior to delivery is necessary, therefore. In one commercially available test, rabbit polyclonal antibody is used to trap the herpesviral antigen. Monoclonal antibodies prepared against a variety of herpes antigens and conjugated to biotin are then added to the mixture. In the final step, the enzyme horseradish peroxidase conjugated to the bacterial protein streptavidin is added and binds to the biotin. Color develops after addition of the substrate if the test is positive.

The assay is reported to detect virus in about 95% of the samples shown to have virus by tissue culture assay. The specificity of the assay is 100%, meaning that all positive samples contain herpesvirus. The time required for the ELISA test is about four hours, whereas the traditional tissue culture assay method requires about seven days.

Cell-Mediated Immunity

Most of the preceding discussion in this chapter concerns the operation of humoral immunity. The complementary immune response is described as **cell-mediated immunity** that, as the name implies, involves entire cells rather than freely circulating proteins. Some of these cells also are members of the T-cell group, namely, cytotoxic T cells (T_c) and delayed-type hypersensitivity cells (T_{DTH}). Macrophages also play an important role in cell-mediated immunity.

The function of cells within this system is to destroy other cells that do not conform with the normal pattern of antigens in an individual. In other words, cell-mediated immunity acts as an **immune surveillance** system. In fulfilling this role, the system can proceed by itself or interact with components of the humoral immune system. Appearance of nonself-antigens on the surface of a cell activates cells of the immune surveillance system. These nonstandard antigens can be produced by viral or bacterial proteins that have been inserted into the plasma membrane of an infected cell, or they can result from the transformation of a normal cell into a tumor cell.

Of course, foreign antigens will stimulate T-cell receptors. Depending on which type of T cell is present, either the humoral or the cell-mediated system will respond. For the moment, consider the effect only on T_c and T_{DTH} cells. As discussed at the beginning of this chapter, antigen stimulation results in the production of compounds that mediate reactions of the immune system. Certain lymphokine proteins can be released from stimulated T cells to affect the growth and differentiation of other cells. These lymphokines include interleukin-1, -2, and -3 (stimulators of other lymphocytes), γ-interferon, and a variety of proteins that affect macrophages. Interleukins are mitogens, chemicals that stimulate other T cells to divide. Interferon acts as previously described (see chapter 11).

The sundry proteins that affect macrophages fall into two groups: those that act as chemotactic attractants and facilitate the migratory process, and those that inhibit the chemotactic response of macrophages once they have arrived at the site of antigenic activity. When these chemicals are produced, the macrophages migrate toward the antigen-containing region, and a substantial population of these cells is maintained within that region as long as the antigenic stimulus continues. The γ-interferon produced by T cells acts on the macrophages, giving them an "angry" appearance. Their cell membranes become ruffled, and phagocytic processes seem to be stimulated (see discussion on phagocytosis at end of chapter).

When stimulated, T_c cells, and to some extent macrophages as well, bind to a cell carrying a foreign antigen and cause it to lyse; however, macrophages and some T_c cells require foreign antigen stimulation before they become cytolytic. The remainder of the T_c cells, called natural killer (NK) cells, require interferon stimulation, not antibody stimulation, to produce their cytolytic effects. NK cells are thought to play the primary role in the surveillance for tumor cells or cells infected with viruses or bacteria.

Like B cells, T cells also must develop a tolerance of self-antigens. If T-helper (T_h) cells do not respond to an antigen, most B cells will be incapable of a response. It has been proposed that tolerance might be due to suppressor cell activity or to blocking of antigen-binding sites on the cell surface. The current hypothesis holds that tolerance is caused by a disappearance of surface receptors from selected cells.

If receptors reappear on the surface of lymphocytes, an autoimmune disease will result. Symptoms of autoimmunity depend on the antigen specificity of reactivated T cells. Examples of autoimmune diseases are listed in table 16.4. Somewhat similar effects are sometimes seen in **immune complex diseases** in which the continuing presence of antigen-antibody complexes

TABLE 16.4. Examples of autoimmune diseases and their effects

Disease[1]	Immune Reaction
Rheumatoid arthritis	Antibodies against membranes lining joints
Systemic lupus erythematosis	Antibodies against blood components, cells, etc.
Hashimoto's disease	Antibodies against thyroglobulin and thyroid cells
Multiple sclerosis	Antibodies against glial cells and myelinated nerve cells

1. The mechanisms that trigger these diseases are not known.

leads to tissue damage. For example, glomerulonephritis (inflammation of the glomeruli in the kidney) can result from the deposition of IgG, complement proteins, and antigens of unknown type in the glomeruli.

When the cell-mediated immune response occurs, changes characteristic of a hypersensitivity (allergic or extreme) reaction occur in the affected tissue and lead to an **inflammatory response.** Symptoms of this response include general vasodilation, resulting in increased blood supply to and reddening of the affected area. Added circulation also makes the area warmer to touch. Increased capillary permeability leads to an accumulation of lymph in the affected area because lymph cannot flow through its ducts at the same rate at which it is now entering the tissue; therefore, the affected tissue swells. If a large number of foreign antigen-bearing cells is present, extensive tissue destruction can occur from T_c cell activity. The area in which the immune reaction occurs may become an open sore, a usual fate of tissue transplanted into an individual not related to the donor. These changes are called **delayed-type hypersensitivity** because the allergic response does not become apparent until several hours following transplantation. In the case of an organ transplant, the time lag may be on the order of a week.

Physiological Role of the Immune System

The immune system does not exist in isolation. It is integrated into the rest of the body's functions and serves an important protective role not only by directly providing immunity but also by stimulating phagocytosis.

Humoral Immunity and Immunization

Foreign antigens can enter the body via the gastrointestinal tract, the respiratory tract, or the broken epithelial layers that might result from an injury or an insect bite. If the immune system is going to act defensively, it must be able to contend with antigens in the food we eat and the air we breathe. For example, sometimes antigens will stimulate an immune response in the intestine before they are degraded by digestive enzymes. This mechanism is thought to account for the presence of circulating antibodies against blood-group antigens in persons who have never had a transfusion. Such an immune response can occur without any overt signs, but it may have serious consequences in the case of food allergies.

Most acquired immunity to pathogens, however, results from a person's exposure to the pathogen, usually in the form of contracting the disease. During the course of the disease, the proper clones are stimulated so that antibodies against the pathogen are produced in quantity. If the individual is exposed again, rapid production of IgG will destroy the invading organism. This process is the reason why certain diseases are diseases of childhood, in which a person who has the disease once is most unlikely to contract it a second time.

An invading microorganism can be assumed to carry on its surface characteristic antigens that should stimulate a normal humoral response. As B-cell clones are stimulated and plasma cells are formed, the specific antibody titer in the blood should rise. Eventually some immunoglobulins thus produced will arrive at the site of infection to react with the surface antigens of the invader. If this is a primary immune response, it may take several weeks for the immunoglobulins to develop and attack these antigens.

To acquire humoral immunity without the inconvenience of having contracted the disease is possible in many but not in all instances of commonly encountered diseases. The causative microorganism must be killed or weakened, or a critical antigen must be isolated from it. An inoculum can be prepared from one of the three and injected into an individual to stimulate a primary immune response. If the antibodies stimulated can inactivate the microorganism or its critical antigen, the body's secondary response mechanism will protect an individual from contracting the disease at a later time. Duration of protection lasts the lifetime of the memory-

APPLICATIONS BOX

Organ Transplants and the Immune System

Recent successes in the field of organ transplantation have been due to the ability of attending physicians to suppress the cell-mediated immune response. Various types of immunosuppressive drugs have been used in the past, often anticancer drugs that kill rapidly dividing cells. Unfortunately, anticancer drugs in particular have many undesirable side effects. The fungal compound cyclosporin A, originally intended as an antibiotic, specifically inhibits the normal functions of T and dendritic cells. Appropriate levels of cyclosporin A permit some humoral immunity to protect a patient against disease, but the drug strongly suppresses T_C and T_{DTH} cells. In some cases in fact, cyclosporin A can work too well. A badly burned youth was grafted with skin from a variety of donors as a temporary expedient. Cyclosporin A so thoroughly suppressed his immune response that the grafts could not be removed, leaving the individual with a mosaic epidermis.

cell clone. Antibodies that do not inactivate the microorganism or its antigen will not protect an individual even if there are high antibody titers.

Sometimes, as a particular disease is controlled and becomes less common in a population or community, acquired immunity to it does not seem to last long. This observation suggests that continued exposure of adults to children with the disease may serve to boost memory-cell clones constantly. It may be necessary, therefore, to provide additional immunizations to adults to maintain their protective levels of circulating antibodies.

In terms of technique, injected antigens often are not sufficient in themselves to stimulate the immune system adequately. It is generally necessary to use an **adjuvant,** a preparation that is highly stimulatory to the immune system and increases its general reactivity. Adjuvants are usually in the form of gels and often serve to maintain a relatively high local concentration of antigen. Freund's adjuvant, the best-known one, is based on a water and oil mixture. The complete adjuvant contains extracts of *Mycobacterium tuberculosis* and is too irritating for use in humans. Presently, the search for new and better types of adjuvants to use with genetically engineered vaccines (discussed in chapter 18) is quite active.

Types of Antibodies Produced

Circulating IgG and IgD antibodies account for much of the body's immune activity. They, or their IgM precursors, can be found throughout blood plasma and lymph. From these distribution channels, they can arrive quickly at any injured site. Unlike IgM and IgG antibodies, IgD antibodies are quite short lived (two to three days in the plasma); hence very little is known about them.

IgG molecules have another important property as well. They are the only antibodies capable of passing the placental barrier and entering fetal circulation. They are also found in colostrum (first secretions of the lactating mother) and can be absorbed through the intestinal lining. Maternal IgG molecules provide **passive immunity** to the newborn, an immunity that does not result from the baby's own metabolic activity. A newborn's immune system has not yet been stimulated sufficiently to provide protection against most microorganisms, but maternal antibodies persist for some weeks in the baby's system and render a decreasing immunity to whatever antigens the mother had encountered. Maternal antibodies gradually are replaced by the child's own antibodies resulting from clonal stimulation to yield **active immunity,** an immunity due to an individual's own metabolism.

IgA antibodies are known as secretory antibodies because they are found mainly in body secretions: mucus, tears, saliva, and sweat. The antibody is actually a dimer of two IgA complexes joined by a short J chain. As IgA molecules coat invading microorganisms, they prevent these organisms from binding to the epithelium and setting up an infection.

IgE antibodies are significant for their role in the development of allergic reactions. Nearly everyone has had an allergic reaction at some time, and those who have not usually know someone who has had such a reaction. Symptoms of an allergic reaction should be very familiar, therefore. Typically, the affected area reddens and swells due to a release of chemical messengers (see section on cell-mediated immunity). If swelling occurs in the sinuses, nasal congestion and accompanying symptoms can develop. If the reaction occurs in the lower respiratory tract (lungs or bronchi), extreme difficulty in breathing and other symptoms of asthma can occur.

Phagocytosis and the Immune Response

Phagocytosis augments and supplements the humoral immune response because the process involves the physical removal of offending antigens from the general circulation. In most but not all cases, these antigens are later destroyed by cytoplasmic enzymes of phagocytic cells. The major phagocytic cells are neutrophils or polymorphonuclear leukocytes (PMNs) and macrophages. Both cell types are chemotactic, responding to phospholipids, peptides, and the C5 fragment of the complement system. Both can take up particulate matter by either phagocytosis or pinocytosis.

Presence of a capsule on a bacterial cell tends to prevent phagocytic attack, probably because the bacterial cell is less likely to stick to the phagocyte. Phagocytic activities against encapsulated bacteria can be stimulated by **opsonization,** the binding of IgG molecules to the target cell. On the phagocytic cell surface are receptors for the Fc region of IgG and for the C3b fragment of the complement system. Neutrophils also have receptors for IgA Fc regions. Phagocytosis is best stimulated by the simultaneous presentation of Fc and C3b regions in which the C3b serves to promote attachment to the phagocyte and the IgG Fc serves as the stimulus for the phagocytic process itself. It has been estimated that a PMN surrounded by 100 bacteria will kill an average of 23 of them before succumbing itself to lack of oxygen and accumulated waste products. A mixture of bacteria and inactivated leukocytes, known as pus, may form after certain infections.

Humoral Immunity and the Allergic Response

The varying symptoms of an immediate allergy response are a function of two different cells, mast cells derived from connective tissue, and leukocytes called basophils (color plates 15 and 16). Both cells contain granules within their cytoplasm and bind IgE antibodies to their cell membranes. When an antigen bridges two antibodies on the surface of one of these cells, degranulation occurs and various mediators of additional responses (table 16.5) are released into the immediate area of the cell. The same reaction can be triggered by the C3a fragment of the complement system. Some of these mediators are present as part of the granules and some are synthesized in response to the antigen stimulus. For the average person with allergies, the most significant of these substances is histamine. Its action leads to breathing difficulty from bronchoconstriction and reduced blood pressure from vasodilation. Various antihistamine preparations are available to reduce allergic symptoms, although they do not interfere with the immune reaction itself. Leukocytes called eosinophils help to moderate the effects of repeated allergic reactions by releasing enzymes to degrade histamine and other mediators of the allergic reaction (color plate 13).

TABLE 16.5 Important compounds released by mast and basophil cells after antigen stimulation

Released Compound	Action
Histamine, leukotriene	Vasodilator, bronchoconstrictor
Heparin (from mast cells only)	Anticoagulant
Prostaglandins	Vasodilator, platelet aggregator
Platelet-activating factor	Platelet aggregator
Eosinophil chemotactic factor (ECF) and neutrophil chemotactic factor (NCF)	Chemotactic attractant for leukocytes (eosinophils and neutrophils)

Note: Mast and basophil cells release other compounds as well. They and those presented in this table have additional actions and effects beyond those listed.

Degranulation of IgE-carrying cells triggers an inflammatory reaction (described previously). In a presensitized person, however, changes occur very rapidly, and the IgE-mediated reaction is called an immediate-type hypersensitivity. In normal individuals, immediate-type hypersensitivity is not remarkable. Nevertheless, some persons seem to be particularly prone to synthesis of IgE antibodies, and in them the appropriate antigen can stimulate a massive release of the mediators listed in table 16.5. The result is **anaphylactic shock,** a condition in which vasodilators cause the blood pressure to fall drastically and various bronchoconstrictors cause the air passage to occlude. A victim can die of asphyxiation within minutes. If he or she is treated successfully, normal blood pressure and respiration will return after about 30 minutes, and no further problems can be expected until the person is exposed to the antigen again. Any compound, such as C3a, that can trigger anaphylaxis is known as an **anaphylatoxin.**

Treatment for anaphylactic shock demands immediate administration of a vasoconstrictor-bronchodilator like adrenaline. In the long term, a sensitive individual must either avoid all exposure to the suspect antigens (which may include things like penicillin, bee stings, and household dusts) or embark on a series of **desensitization shots.** The desensitization treatment consists of injections of extremely small doses of the appropriate antigen(s) administered to the affected individual at regular time intervals. The hope is to stimulate production of IgG rather than IgE antibodies. If IgG is produced in sufficient quantity, it will serve as a **blocking antibody** (one that reacts with the antigens before they can react with the IgE). As in the case of the neutralization assay, an initial antigen-antibody reaction prevents further metabolic reactions. The treatment, although not always successful, offers substantial benefits when it does work. The risk, of course, is that the treated individual will experience anaphylactic shock during the injection process. For this reason, patients are monitored carefully during treatment, and the antigen dose is increased very slowly.

Other Immune Responses

Cells with attached immunoglobulins are targets for the complement system. During fixation of complement proteins, some proteolytic reactions occur that release defined fragments derived from them. C3a and C5a fragments function as anaphylatoxins, and their presence leads to the characteristic reddening and swelling seen at an infection site. In some but not in all cases, when complement proteins have completely bound to antigen-antibody complexes, there is lysis of the invading cell. In other cases, the cell-mediated immune response takes over.

Antibodies against viral surface antigens can act to prevent further spread of the virus by covering the specific attachment sites on the virion that are needed to recognize the appropriate host cells. Immunoglobulins also form cross bridges between virions and cause clumping, which makes the immune complex more susceptible to phagocytosis.

Summary

The immune system acts as a major body defense against foreign substances and cells. It functions by means of molecular complexes (antibodies) whose three-dimensional shapes correspond to portions of the foreign antigens that stimulate their production. Antibodies (immunoglobulins) are composed of four protein chains, two light and two heavy, with the two heavy chains paired in the middle and flanked by the light chains at one end. The entire structure is held together by disulfide bonds. One end of the complex has a constant amino acid sequence. At the other end are two identical variable regions. Within the variable regions are smaller hypervariable regions that serve as antigen-recognition sites. Antibodies come in various classes, IgA, IgD, IgE, IgG, and IgM, depending on the heavy chain used, with each having its own special type of function. Antibodies are found in the cell membrane of B lymphocytes and as secretory products of plasma cells. Similar structures occur on the surface of T cells. All are the result of complex DNA rearrangements within the chromosomes of their producing cells.

The immune system functions by means of various leukocytes, products of stem cells generated by the bone marrow. Antibody response requires both T and B lymphocytes. Both cells must recognize the same antigen using specific recognition molecules in their cell membranes. Then T-helper cells stimulate B cells to divide and differentiate into plasma cells and B-memory cells. Plasma cells actively secrete antibody molecules into the blood plasma, and memory cells serve to give a more rapid antibody response on an individual's reexposure to the same antigen. Another form of T cell, the T-suppressor cell, can act to prevent B-cell response to an antigen. This immunologic tolerance is necessary so that the body does not attack itself.

Cell-mediated immunity does not involve free antibody molecules but rather the actions of entire cells that result in an inflammatory response. This type of immunity is primarily a mechanism for destroying invading cells, virus-infected cells, and tumor cells. Abnormal antigens must be presented to cytotoxic T cells in company with the major histocompatibility complex antigens of the body. Cells carrying abnormal antigens are then destroyed, either by T cells or by macrophages that they attract to the affected area.

Many special assays have been developed with antigen-antibody reactions. Pure antibodies can be obtained in large quantity from cultures of hybridoma cells, fusions of tumor and B cells. Antibody molecules can be labeled with radioactive substances or with nonradioactive chemicals that either fluoresce or catalyze a particular chemical reaction. Sometimes the amount of complement proteins bound by antigen-antibody complexes can be used as an assay. These methods permit the detection of antigen-antibody complexes that are too few to form a visible precipitate. In many cases, one specially prepared antibody can be used in a variety of tests.

Questions for Review and Discussion

1. Describe the primary and secondary immune responses in terms of the clonal selection hypothesis. Why does the immune system not normally attack the cells of its own body?
2. Summarize the properties of T and B lymphocytes. How are they similar and different? What happens to an individual who lacks T cells? B cells?
3. Draw the molecular structure of a typical IgG molecule and indicate the various regions into which the molecule is subdivided. Draw a typical chromosome map for the regions coding for the light chain of IgG. How must this chromosome be rearranged in order to synthesize a light chain?
4. Explain the inflammatory response and its role in body defense. What cells are involved? What kinds of things can go wrong with this response?
5. Explain how an ELISA test works. What conclusions can be drawn from the fact that quite large molecules can be attached to antibodies with no loss of function?

Suggestions for Further Reading

Benjamini, E., and S. Leskowitz. 1988. *Immunology: A short course.* Alan R. Liss, Inc., New York. A textbook for students in the health professions.

Burton, D. R. 1990. Antibody: The flexible adaptor molecule. *Trends in Biochemical Sciences* 15:64–69. A summary of structural information about IgG, IgM, and IgE and the implications for antibody function.

Herberman, R. B., M. J. Bruda, W. Domzig, R. Fagnani, R. H. Goldfarb, H. T. Holden, J. R. Ortaldo, C. W. Reynolds, C. Riccardi, A. Santoni, B. M. Stadler, D. Taramelli, T. Timonen, and L. Varesio. 1982. Immunoregulation involving macrophages and natural killer cells, pp. 139–166. *In* L. N. Ruben and M. E. Gershwin (eds.), *Immune regulation: Evolutionary and biological significance.* Marcel Dekker, Inc., New York. This research report has a good introduction to the subject and is not too complex, except for the many abbreviations.

Kindt, T. J., and J. D. Capra. 1984. *The antibody enigma.* Plenum Publishing Corp., New York. A very readable discussion of antibodies written by a biochemist and a serologist. The text presents material first from one viewpoint and then from the other.

Kinet, J.-P. 1989. Antibody-cell interactions: Fc receptors. *Cell* 57:351–354. A discussion of the ways in which antibodies such as IgE bind to cell surfaces.

Marrack, P., and J. Kappler. 1986. The T cell and its receptor. *Scientific American* 254(2):36–45. Extended coverage of much of the material from this chapter written for the nonspecialist.

Reynolds, C. W., and R. H. Wiltrout (eds.). 1989. *Functions of the natural immune system.* Plenum Publishing Corp., New York. A somewhat technical survey of the nonimmune leukocytes.

Roitt, I. 1988. *Essential immunology,* 6th ed. Blackwell Scientific Publications Ltd., Oxford.

Smith, K. A. 1990. Interleukin-2. *Scientific American* 262(3):50–57. A brief summary of the history and function of this immunologic hormone.

Sunshine, G. H., A. A. Czitrom, S. Edwards, M. Feldmann, and D. R. Katz. 1984. Antigen presentation by dendritic cells, pp. 23–28. *In* A. Ü. Müftüoglu and N. Barlas (eds.), *Recent advances in immunology.* Plenum Publishing Corp., New York.

chapter

17

Host-Parasite Relationships

chapter outline

GENERAL GOALS

Be able to describe some interactions between normal microorganisms and pathogenic invaders.

Be able to describe the various means by which microorganisms can evade normal defenses.

Be able to describe the mechanisms by which microbial production of toxins can have profound impacts on the well-being of their host.

In the mind of the nonscientist, the symbiotic relationship most often associated with bacteria is a parasitic relationship in which one organism assumes the role of host, and the other the role of parasite. The parasite lives in or on the host and, in the process of carrying out its life cycle, harms its host. Any suboptimal physiological state of an organism is considered as a **disease** state; therefore, most parasites are **pathogens** or disease-causing organisms.

Teleologically, a successful parasite is one that can set up a long-term relationship with its host. This allows the parasite to grow steadily without the necessity of searching for a new host constantly. Such a relationship is possible only when the parasite does not damage the host too severely. Long-term parasitic relationships require what is in essence cooperation (often involuntary) between the host and the parasite. Host defense mechanisms and parasitic countermeasures are introduced in this chapter. In order to provide personal interest in the discussion that ensues, examples of human parasitisms are highlighted. Similar situations occur with other mammals.

Normal Human Microflora

All epithelial surfaces exposed to the environment, which includes the skin, oral and nasal passages, reproductive tract, and gastrointestinal tract, usually are occupied by assorted microorganisms known as **normal microflora** (table 17.1). Presence of these organisms constitutes the natural symbiotic state of the body; therefore, microorganisms have important roles to play in the health of the host. In addition to the possibility of cooperative nutrition between intestinal bacteria and humans in whom these bacteria may supply vitamins (chapter 12), the normal flora occupy various body sites that can serve as niches for other organisms as well.

Normal flora are generally well adapted to their interactions with a host, while substitutes may not coexist as well. By occupying the various passages and organs that they do, normal microorganisms can prevent attachment of pathogens. When antimicrobial therapy is administered, the normal microflora are of particular concern. Because antimicrobials are not very specific in their action, they may disrupt the normal flora of the individual being treated. Organisms not normally present then may invade the body (see discussion on *Clostridium difficile* in chapter 19).

Other environmental influences also can have an impact on the body's resistance to disease. A particularly good example is the case of the intestinal flora of an infant. An infant's diet strongly influences the bacteria present (table 17.2). Mother's milk tends to suppress the growth of all but members of the genus *Bifidobacterium* and exhibits other protective effects as well. For instance, *Clostridium difficile* often is present in healthy infants but can cause serious problems in adults and in some other infants.

Persons who have difficulty in digesting lactose often are unable to tolerate milk or foods containing it because of their inability to degrade lactose. Such persons may be helped if they drink milk containing a small quantity of *Lactobacillus acidophilus*. Although this organism does not become a permanent resident of the intestines, it can be shown that *L. acidophilus* causes a transient change in the microbial flora of the gut. While in temporary residence, the lactobacilli can provide lactase activity necessary for lactose degradation.

TABLE 17.1 Examples of organisms commonly found in or on the human body

Location	Organism(s)
Skin	*Staphylococcus epidermidis*
	Propionibacterium acnes
	Lactobacillus spp.
	Clostridium perfringens
	Aerobic corynebacteria
Oral cavity	*Staphylococcus epidermidis*
	Streptococcus mutans
	Streptococcus mittior
	Streptococcus sanguis
	Streptococcus salivarius
	Veillonella spp.
	Lactobacillus spp.
	Actinomyces spp.
	Bacteroides spp.
	Treponema spp.
	Anaerobic cocci
Nasopharynx	*Staphylococcus epidermidis*
	Staphylococcus aureus
	Haemophilus spp.
	Branhamella spp.
	Aerobic corynebacteria
Large intestine	*Bacteroides* spp.
	Fusobacterium spp.
	Peptostreptococcus spp.
	Eubacterium spp.
	Streptococcus group D
	Escherichia spp.
Vagina and cervix	*Lactobacillus* spp.
	Bacteroides spp.
	Peptostreptococcus spp.
	Staphylococcus epidermidis
	Streptococcus group D
	Aerobic corynebacteria

APPLICATIONS BOX

Persistent Abnormal Microflora

The organisms listed in table 17.1 are not necessarily found in every person, and sometimes healthy individuals may have unusual organisms not listed in the table. For example, there are documented instances of otherwise healthy persons who have accidentally acquired organisms in their large intestine that metabolize carbohydrates via an ethanolic fermentation. When such people consume foods high in carbohydrates, sufficient ethanol can be produced to cause intoxication.

TABLE 17.2 Changes in the intestinal bacteria of an infant as a result of changes in diet

Diet	Age of Infant	Bacteria Present
Breast milk or formula	0–2 weeks	*Escherichia coli* *Streptococcus faecalis* *Bifidobacterium* spp. *Clostridium* spp.[1] *Bacteroides* spp.[1]
Breast milk	2 weeks until solid food given	*Bifidobacterium* spp. (nearly 100%)
Formula	2 weeks until solid food given	*Escherichia coli* *Streptococcus faecalis* *Bifidobacterium* spp. *Clostridium* spp.[1] *Bacteroides* spp.[1]
Solid food	Weaning	Adult flora

1. Rare organisms.

Thus, food as well as environmental conditions can influence the growth of microorganisms in or on the body. Perturbations of the normal environment may result in skin eruptions, digestive upsets, or other distressing symptoms. Often, the only real cure is to restore normal flora so that they can displace the pathogens.

Obstacles to Successful Colonization by Parasites

Mechanical Barriers

From an anatomic standpoint, the human body is basically a tube within a tube in that the gastrointestinal tract and its contents can be considered external to the organism. The protective aspect of this arrangement stems from the intact layer of epithelium that normally covers all external body surfaces. Epithelial cells are coated with hyaluronic acid and supported structurally by the connective tissue layers underneath that incorporate chondroitin sulfate. Some of the internal rigidity of connective tissue is derived from polysaccharides incorporating *N*-acetylneuraminic acid.

Epithelial cell layers generally are refractory to infection by most microorganisms. Those comprising the skin are covered by layers of dead cells that obviously cannot serve as hosts for intracellular parasites. The skin prevents routine entrance of microorganisms by serving as a mechanical barrier. The protective aspect of the epithelium lining the digestive tube is somewhat different because these cells are all living and maintained in a moist environment suitable for the growth of many microorganisms. Associated with these epithelial layers of the gut are numerous mucus-secreting cells that function to coat the epithelial lining so that microorganisms cannot easily penetrate the epithelium itself.

Natural openings through epithelial layers have built-in obstacles to invasion. The eyes, for instance, are bathed in tears that contain various antimicrobial compounds such as lysozyme. The ears are sealed from external environmental influences by the tympanic membrane. Nasal hairs initially filter pathogens invading the respiratory tract, and then layers of ciliated epithelium within the bronchi take over. Cilia beat in an upward direction so that mucus, and any microorganisms trapped within it, are moved up into the throat and swallowed. The acidic pH of the stomach serves as a formidable barrier to most microorganisms. In the intestinal tract, bile salts secreted by the liver are powerful detergents that disrupt many different types of cells.

If an opening appears within an epithelial layer as the result of a cut or other mechanical damage, blood clots over the opening. When prothrombin molecules are cleaved by a protease released by platelets, a blood clot begins to form. The resulting thrombin molecules then cleave fibrinogen to generate fibrin. Fibrin forms a network of fibers in which red blood cells and platelets are trapped. The clot completely seals over the wound and prevents any further microbial invasion.

Sequestration of Iron

All microorganisms require iron for growth. Any organism that has large amounts of iron available is a better potential host than one that has relatively little free iron. It has been demonstrated that people who eat a diet rich in iron are more likely to become infected by bacteria or fungi than equally healthy individuals whose diets are low in iron. Obviously, there is a strong selective pressure for minimizing free iron in the body.

If the amount of iron in blood plasma is assayed, the total iron concentration is about 2×10^{-5} M, but the concentration of free iron is about 10^{-18} M. The difference results from the production of special iron-binding proteins, transferrin in blood plasma and lactoferrin in secretions. These proteins act as chelating agents of iron, binding strongly and specifically to free ferric ion and making it unavailable to invading microorganisms. During a fever response (see next section), the amount of available iron is reduced even further because intestinal absorption of iron and erythropoiesis (formation of red blood cells) are suppressed. Severe infections can result in a 50% reduction in plasma iron levels.

Nonspecific, Systemic Responses to Infection

When the body is invaded by a foreign organism, many interrelated responses are activated that, for the most part, involve various segments of the immune system, specifically those aspects discussed in chapter 16. Familiarity with that material is necessary in understanding the following discussion.

Fever Response

A **fever** is said to be present when body temperature is significantly above normal. For most people, normal is 37°C (98.7°F), and temperatures above 38°C (100.4°F) are considered unusual events. Children can tolerate high body temperatures longer than adults and tend to have fevers more frequently. **Pyrogens** are substances that induce a fever, and many different kinds of substances have pyrogenic properties. The outer membranes of gram-negative bacteria are a very common source of pyrogens.

Increased body temperature is a defense mechanism in the sense that normal body temperature is at the upper limit of the mesophilic temperature range; therefore a rise from the normal can impair the ability of an invading organism to grow properly and allow time for other defense mechanisms to come into play. In the case of *Flavobacterium meningosepticum,* it can be shown that many strains are incapable of growth at 38°C. An elevated body temperature also has a significant impact on the availability of ferric iron in the blood.

Humoral Immune Response

Foreign antigens stimulate the humoral immune response, and with the passage of time, circulating IgM and IgG antibody molecules should accumulate in the blood. Antibody molecules can prevent normal function of the antigenic sites to which they bind and serve to activate the complement and phagocytic systems. If a pathogen is highly antigenic, persons who already have undergone the primary response because of previous exposure to the organism are much less at risk of infection. Reduction of risk is the basis for routine immunizations against certain diseases.

Cell-Mediated Response

The cell-mediated immune response has an extensive arsenal of weapons that should be sufficient to ward off invading organisms. Cytotoxic T cells are thought to be capable of binding to the constant regions of complexed IgG molecules and killing foreign cells by degranulation of the T cell. Additional cytotoxic effects result from the presence of phagocytic cells that can release lysozyme to attack the bacterial cell wall, proteinases, interferon, and complement proteins. All these activities are useful particularly in the case of an invading cell that is too large to be engulfed in the phagocytotic process.

Once inside a phagocytic cell, an invading organism or virus is contained within a **phagosome,** a membrane-bound vesicle. Under normal conditions, the phagocytic cell produces metabolic factors to destroy the phagocytosed entity. Some factors require oxygen for the killing effects, others do not (see table 17.3). Most of these products are familiar, but a few need further explanation.

TABLE 17.3 Antimicrobial factors produced by phagocytes

Factor	Macrophages	Neutrophils
Oxygen independent		
Lysozyme	+[1]	+
Acidic pH	+	+
Cationic proteins	−[2]	+
Lactoferrin	−	+
Oxygen dependent		
Superoxide	+	+
Hydrogen peroxide	+	+
Hydroxyl radical	+	+
Hypohalite	+/−	+
Singlet oxygen	+	+

1. Plus symbol refers to presence of factor in that type of cell.

2. Minus symbol refers to absence of factor in that type of cell.

Chelation of iron by lactoferrin should inhibit further microbial metabolism. **Lysosomes** are small, membrane-bound vesicles (often containing particulate matter) found throughout the cytoplasm. After a phagosome has formed, lysosomes fuse with it to release their contents into a composite phagolysosome. Fusion of lysosomal vesicles with the phagocytic vesicle generates an acidic pH.

The same trigger that stimulates phagocytosis also stimulates a **respiratory burst** in the phagocyte. Oxygen consumption increases substantially and various reactive forms of oxygen are produced (figure 17.1). In addition, the lysosomal vesicles supply a myeloperoxidase-halide system that uses hydrogen peroxide to generate hypochlorite:

Equation 17.1

$$H_2O_2 + Cl^- \xrightarrow{\text{myeloperoxidase}} H_2O + OCl^-$$

The hypochlorite can then react further to generate singlet oxygen:

Equation 17.2

$$OCl^- + H_2O_2 \rightarrow H_20 + {}^1O_2 + Cl^-$$

The cytoplasm of the phagocytic cell is protected from these various reactive forms of oxygen by superoxide dismutase, catalase, etc. (see chapter 13). Protection is not absolute, however, and the respiratory burst also

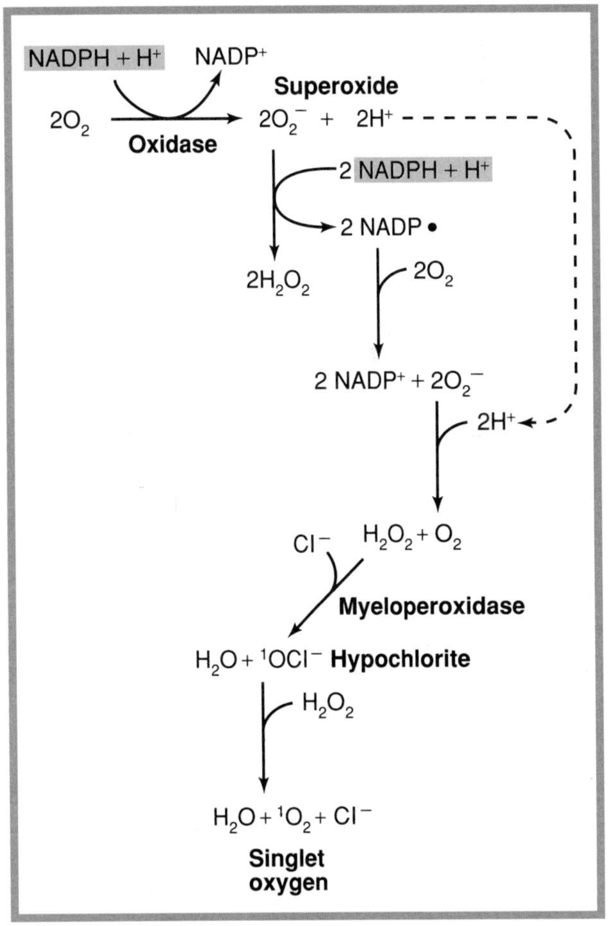

FIGURE 17.1
Formation of active oxygen products during the respiratory burst of a phagocytic cell.

causes the majority of tissue damage that occurs during an inflammatory reaction. Chronic granulomatous disease is characterized by an inability of phagocytes to generate the oxidative burst. Nevertheless, such cells are still capable of some antibacterial activity resulting from nonoxygen-dependent pathways.

Phagocytic cells, as well as some others, can be shown to produce low-molecular-weight peptides (29 to 34 amino acids) called defensins. These molecules act to create anion-specific channels in lipid bilayers and have been shown to have antibacterial, antifungal, antiviral, and cytotoxic properties.

Invasive Properties of Pathogens

The foregoing material makes it sound as though the human body has an impregnable set of defenses suitable for all contingencies. Obviously, this is not true because nobody would ever be sick. Instead, parasitic organisms have evolved mechanisms that serve to counteract, to a greater or lesser degree, one or more host defense mechanisms.

To begin with, parasites must attach themselves to the body. Thus, the surface structures of a microorganism become very important. If a parasite cannot hold on for a reasonable period of time, its chance of infecting a host will be lost. A parasite's motility also plays a significant role in that a motile parasite has a better chance of getting close enough to cell surface receptors to allow its adhesive mechanisms to function.

Parasites can live on the surface of the body or within it. External parasites are usually **opportunistic** in that they do not actively create a niche for themselves, rather they take advantage of some special situation. There may be an excess of nutrients on the skin or in the intestinal tract because of hormonal changes at puberty or loss of some normal flora. In the latter case, normal amounts of nutrients are available, but competition for those nutrients is reduced. Generally, surface infections can be managed by improved cleanliness and topical application of antimicrobial compounds.

The really significant issue concerns true internal pathogens, those organisms that enter body tissues and cause disease. **Virulence** has been defined as the relative capacity of any microorganism to overcome host defenses. Parasites that are both good at entering the body and growing well after they arrive are considered to be highly virulent or pathogenic. In some cases in which microorganisms synthesize poisonous substances, their growth capabilities become less significant. Each of these properties is discussed separately in this section.

Entering the Body

Invasiveness refers to a parasite's ability to penetrate epithelial layers and enter body tissues while resisting various host cell defenses. Although it is possible for opportunists (accidental pathogens) that have no invasive ability to cause significant damage if they enter the body through a wound, they are less common than invasive or deliberate pathogens.

Invasive protozoa generally gain entry into the body by mechanical means. Their comparatively large size precludes their using some strategies employed by bacteria. Fungi, being somewhat smaller, can force hyphae into or between cells (for example, haustoria). Specific cell surface receptors used by a virus can provide an initial anchorage.

Bacteria invade body tissue by locating their target and sustaining their position on the tissue surface while they work to gain entry. There is evidence to indicate that many pathogens also are chemotactic and respond to attractants released by eukaryotic cells. Although motility is not a prerequisite for pathogenicity, motile *Vibrio cholerae* are more pathogenic than their nonmotile relatives. Successful pathogens produce one or more **adhesins** that serve to anchor a bacterium in place once it has chosen an appropriate site. Type 1 fimbriae (common pili) are examples of adhesins. Many bacteria grown on laboratory media fail to produce adhesins, and this lack may account for their greatly diminished pathogenicity.

In the next invasive step, a pathogen simply moves between the cells of an impeding tissue layer, its progress being aided by enzymes like those listed in table 17.4. Hydrolytic enzymes can break down such barriers to pathogenic movement as hyaluronic acid between cells, fibrin proteins holding a clot together, or the neuraminic acid component of some polysaccharides. Hemolysins are a group of proteins that cause lysis of red blood cells. In addition to liberating potentially toxic degradative products of hemoglobin, these proteins help producing bacteria to penetrate blood clots. Hemolytic enzymes usually are subdivided into functional groups designated by Greek letters. One such grouping for *Staphylococcus* is shown in table 17.5.

Streptococci also can produce hemolysins. Their products usually are subdivided according to the visual appearance of colonies grown on blood agar (standard

TABLE 17.4	Bacterial enzymes that promote invasion
Hydrolytic enzymes	
Hyaluronidase (spreading factor)	
Neuraminidase	
DNAase	
Proteinase	
Hemolysins	
Coagulase	

TABLE 17.5 Hemolysins produced by members of the genus *Staphylococcus*

Hemolysin	Mode of Action
α (α-toxin)	Makes pores in membranes
β (phospholipase C)	Attacks sphingomyelins
γ	Mechanism unknown
δ	Mechansim unknown

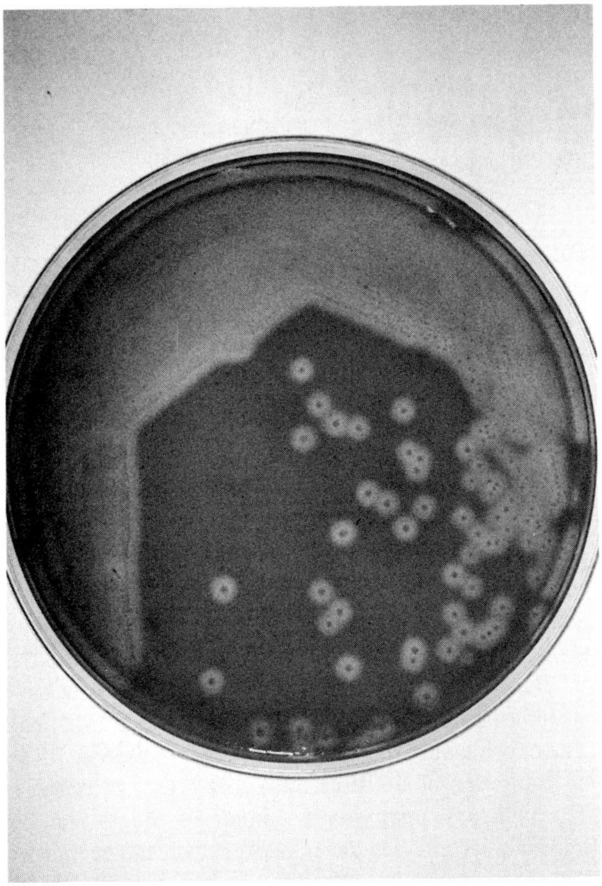

A

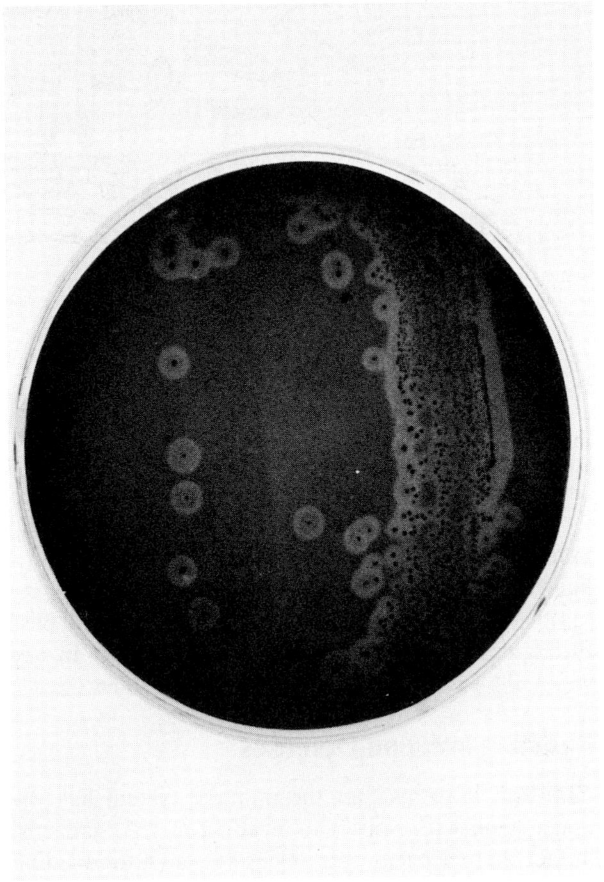

B

FIGURE 17.2

Hemolysin production by streptococci grown on blood agar plates. (**A**) α-hemolysis (partial clearing) produced by *Streptococcus salivarius*. (**B**) β-hemolysis (complete clearing) produced by *Streptococcus pyogenes*.

(A) © G. W. Willis, M.D./BPS. (B) © Eddie Chan/Visuals Unlimited

bacterial nutrients plus sheep red blood cells), as shown in figure 17.2A–B). β-hemolysis is the most potent because red blood cells and hemoglobin are completely destroyed, leaving a broad, clear zone surrounding the colony. The proteins responsible are streptolysins O and S. Streptolysin O is inactivated by free oxygen and therefore works only when the colonies are surrounded by agar. Partial destruction of red blood cells is designated as α-hemolysis. Nonhemolytic streptococci are called γ-streptococci.

Sometimes a pathogen invades the cytoplasm of an adjacent eukaryotic cell. The pathogen can accomplish host cell invasion by stimulating normal endocytotic or phagocytotic mechanisms, leaving the invader encased

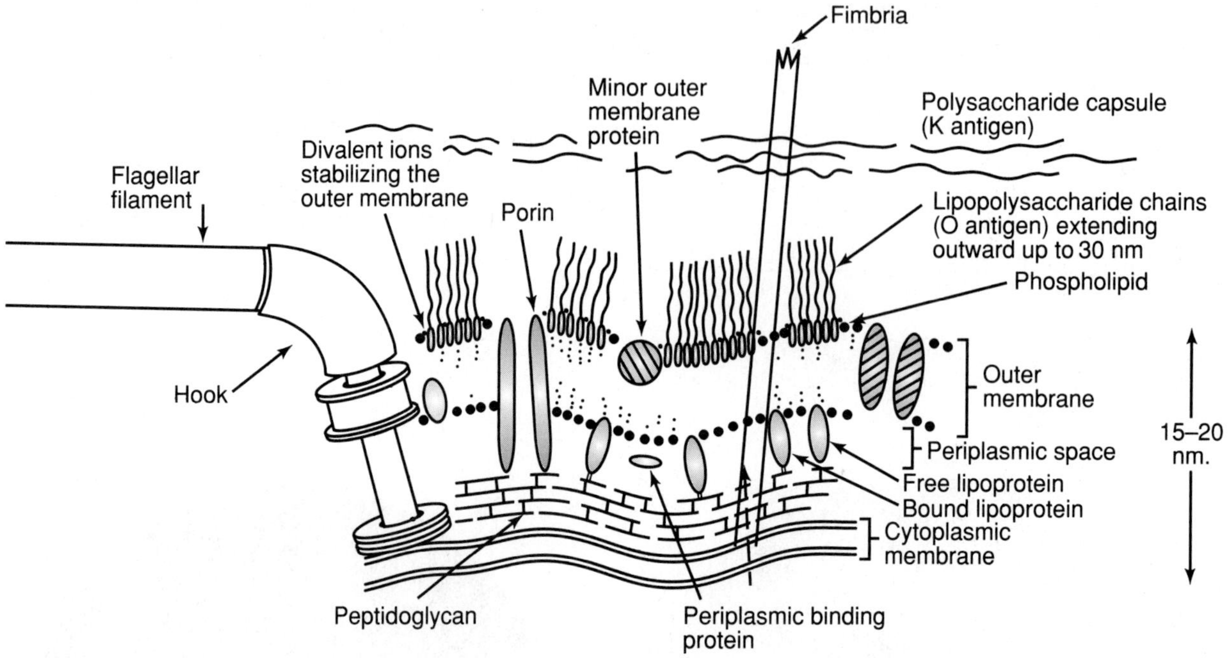

FIGURE 17.3

Diagram of the surface layers of a typical gram-negative cell. Note the multiple-layer effect of the polysaccharide antigens O and K. In addition to these polysaccharide antigens, bacterial surface antigens may include the fimbrial proteins, the flagellin subunits, and the outer membrane proteins.

From S. M. Hammond, P. A. Lambert, and A. N. Rycroft, *The Bacterial Cell Surface.* Copyright © 1984 Croom-Helm, London. Reprinted by permission.

in a membrane-bound cytoplasmic vesicle. Many salmonellae are capable of transcytosis by which they actually move across the cytoplasm of a host cell and pass through the cell membrane on the other side. In one study, the entire process took about four hours.

Resisting Immune Defenses

Although it seems that the immune system has the upper hand with respect to invading microorganisms, in fact there are many strategies available for evading it, most of which involve the nature of the invader's cell surface. A typical gram-negative cell surface is illustrated in figure 17.3. Notice the O and K antigens, the flagella, the fimbriae (common pili), and the outer membrane proteins. One simple mechanism a pathogen can employ for evading the humoral response is to place different antigens on its cell surface. *Salmonella* spp. have the ability to vary their flagellar composition so as to present different surface antigens at different times. Other organisms can vary their fimbriae.

The importance of the pathogen's cell surface can be seen in the case of *Streptococcus pneumoniae,* an organism that can be isolated either with or without a polysaccharide capsule. As is pointed out in chapter 10, organisms surrounded by a capsule are lethal when injected into mice, while those without a capsule cannot set up an infection.

Surveys have shown that all virulent bacteria have capsules, but not all encapsulated bacteria are virulent. The capsule apparently serves several functions. It can be composed of multiple antigenic types, so previous exposure to *S. pneumoniae* cannot guarantee that the immune system has developed antibodies specific for the correct capsular antigens present at a subsequent time. Furthermore, the capsule mechanically prevents some nonspecific phagocytotic processes of macrophages. On the other hand, if antibodies are bound to capsular antigens, opsonization and consequent phagocytosis occurs. Most virulent organisms are surrounded by capsules that are relatively weak antigens, presumably as a result of the immune system's continued selective pressure. In *Neisseria gonorrheae,* for example, a major component of the capsule is sialic acid, a derivative of neuraminic acid that also occurs on the surface of human cells. Therefore, the nonprotein portion of *N. gonorrheae* capsules can be antigenic only at the risk of autoimmune disease.

Other components of the bacterial cell surface can be shown to protect the cells from the attack of complement. There are some 200 antigenically distinct forms of O antigen, yet pathogenic bacteria tend to have only about 20 different types. Those cells possessing the appropriate O antigens are resistant to complement. The alternative pathway fails to be activated because the antigens that do bind C1 are located so far within the capsular structure that phagocytic cells cannot reach them. The classic pathway seems to be inactive due to the presence of excess inhibitors or enhanced binding of existing inhibitors. Some organisms prevent the attack of complement by releasing molecules that activate the complement cascade away from the cell surface, thereby reducing the amount of material available to attack the cell.

Many bacteria can produce proteins that directly intercept immunoglobulin molecules. An IgA protease has been isolated from *Streptococcus* spp. that has as its substrate the specific type of IgA (IgA1) that comprises 90% of serum IgA but only about 50% of IgA secreted by mucous membranes. This protease cleaves the polypeptide chains in the hinge region of the IgA antibody to release Fab and Fc. Some *Staphylococcus aureus* strains synthesize a different sort of substrate called protein A that binds strongly and specifically to the Fc region of the antibody molecule and inhibits antibody function.

Some staphylococci produce coagulase, an enzyme that can activate prothrombin. The staphylothrombin that is formed serves to cleave fibrinogen to fibrin in the usual way to yield a blood clot. Although this activity may seem somewhat strange because the body normally uses clots to wall off infection, in fact it has a practical basis. A clot can work both ways; therefore, antibodies, phagocytes, and other products of the immune system may not be able to penetrate an infection site that has been walled off by a clot.

If it is assumed that the immune system is able to respond to the presence of an invading organism, there still can be problems. If a parasite manages to invade a normal cell, the immune system will be unable to act directly against it. If the intracellular parasite stimulates the production of new antigens on the host cell surface, various cytolytic activities of the immune system can eliminate both the parasite and the infected cell. In the case of the protozoan parasite *Plasmodium,* the parasite apparently can stimulate deliberate phagocytosis by red blood cells in order to enter the cytoplasm of its host. Reoviruses have a similar mode of entry into phagocytic cells, and they depend on lysosomal enzymes to carry out their uncoating.

Even if a parasite is phagocytosed, it may not be destroyed. Although phagocytic cells usually engulf invading bacteria and kill them by oxidation or other methods, other outcomes are possible. Many bacteria have developed ways to neutralize or minimize the harmful effects of phagocytosis. Some of their strategies are obvious, such as overproduction of superoxide dismutase and catalase by the bacterium *Toxoplasma* to minimize the effect of the respiratory burst in phagocytic cells.

Other organisms, for example, the protozoan *Trypanosoma,* bind to macrophages and are taken up by phagocytosis. Trypanosomes, however, cause the phagosomal membrane to lyse before lysosomal fusion can take place, thereby they escape into the cytoplasm in which they begin to divide. *Chlamydia psittaci,* on the other hand, produces a substance that prevents the fusion of lysosomes with the phagosome. Mycobacteria and some others seem to produce substances of an unknown sort that allow them to survive the environment of the phagolysosome. *Bordetella* spp. produce a very active adenylate cyclase that can disrupt normal phagocytic cell function. *Salmonella typhimurium* carrying a 100-Kb plasmid makes specific factors that protect against the activity of defensins.

Other bacteria like *Mycoplasma* spp. bind to the surface of phagocytic cells but are not taken in. Instead, they remain attached to the host cell surface and continue to reproduce from that location. Apparently they carry a protective protein on their surface because protease treatment results in normal phagocytosis. Protein M from group A streptococci inhibits chemotaxis of phagocytes and also prevents the complement C3b fragment from coating the cell surface. The few patches of C3b that do form on the cell surface have no opsonizing activity.

Acquiring Necessary Iron for Growth

One common host response to parasitic invasion is the sequestration of iron, discussed at the beginning of this chapter. It then follows that the parasite must respond with countermeasures to extract iron from the host if it is to grow in its new location. A good hemolysin, for example, will lyse red blood cells and release their contents, including a substantial amount of iron contained in hemoglobin.

Microorganisms can also produce siderophores, chelating agents that serve to bind ferric iron more strongly than do host molecules. The two most common types of siderophores are diagrammed in figure 17.4. Hydroxamate siderophores are most prevalent among

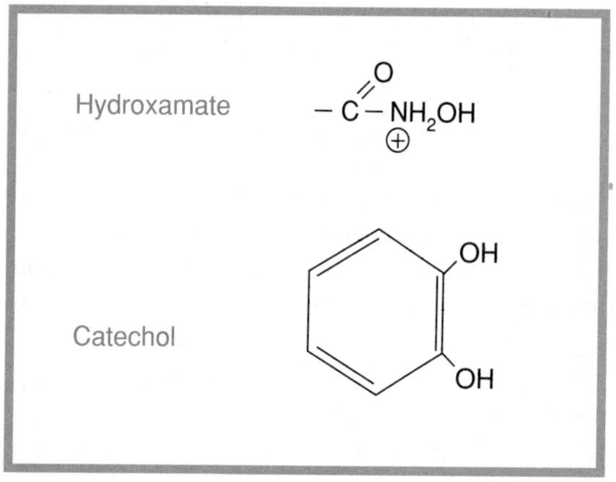

FIGURE 17.4

Common types of siderophores used by pathogenic microorganisms to acquire ferric iron. Fungi mainly produce the hydroxamates, while the phenolates, catechol being an example, are primarily the products of bacteria.

fungi, while phenolate siderophores are most frequent among bacteria. Certain plasmids also can code for iron chelating agents and thus contribute to virulence.

Toxigenicity of Pathogens

The overall virulence of an organism is the sum of its invasiveness and the damage that it does when inside the body. A significant contributor to virulence is the production of one or more microbial toxins (figure 17.5). A microbial **toxin** is a high-molecular-weight antigenic substance that disrupts normal physiological processes. In other words, it is a poison. Thus, a person who is intoxicated is one who is being poisoned. In common usage, the term *toxin* refers to all kinds of poisonous molecules, and intoxicated often is used to refer to someone who has consumed excess quantities of ethyl alcohol, a known poison when taken in large doses. It has been suggested that toxins alter the immediate environment of the producer cell so that the organism can more readily establish an infection.

Bacterial cells can produce exotoxins and endotoxins. **Exotoxins** are true toxins in the sense defined. They are protein molecules released from the bacterial cell during growth that can promote damaging effects in parts of the body remote from the site of infection. By way of contrast, **endotoxins** are complexes of lipopolysaccharides released from the outer membrane of gram-negative cells when they die. They too may have effects beyond the immediate site of infection, but in general, they are less toxic by weight than exotoxins.

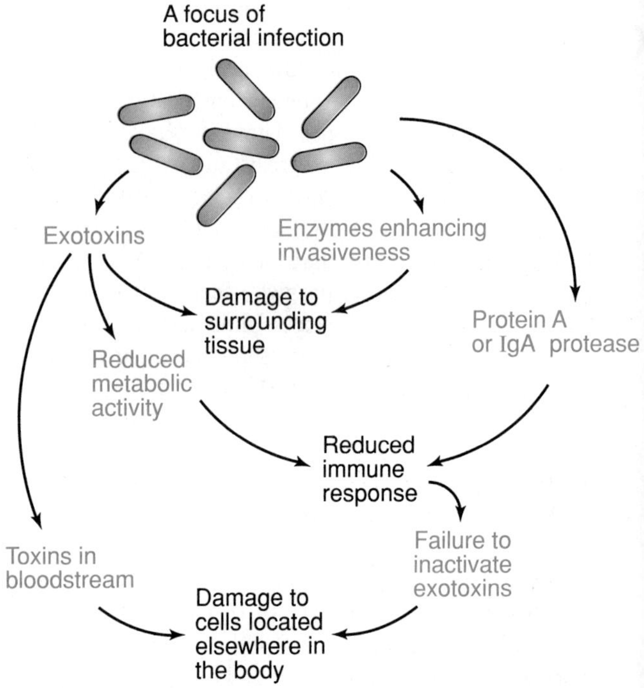

FIGURE 17.5

Pathogenesis encompasses the possible interactions between microbial exotoxins and auxiliary factors such as enzymes enhancing invasiveness (proteases, lipases, and nucleases).

Filamentous terrestrial fungi are known to produce an array of toxic products called **mycotoxins.** These are products of secondary metabolism as are antibiotics, and like fungal antibiotics, they display a wider range of structures than those of bacteria. Viruses too may be able to damage host cells via some of their structural components. At the present time, however, there is no firm evidence that toxicity plays a major role in the pathogenesis (establishing an infection) of protozoa, viruses, and fungi.

Bacterial Exotoxins

Numerous bacterial species produce significant exotoxins, and some of the more important ones are listed in table 17.6. Their names are often derived from their physiological target (figure 17.6). The most notorious are neurotoxins and enterotoxins. The former induce dysfunction of portions of the nervous system, while the latter cause severe intestinal upset, diarrhea, etc. It is important to note the distinction between an enterotoxin, a protein exotoxin affecting the gastrointestinal tract, and an endotoxin, the toxic portion of the gram-negative outer membrane.

TABLE 17.6 Examples of bacterial exotoxins and their modes of action

Toxin	Producing Organism	Mode of Action
Diphtheria toxin[1] (≤100 ng injected)	*Corynebacterium diphtheriae*	Inhibits protein synthesis
Botulism toxin[1] (1 ng)[2]	*Clostridium botulinum*	Induces muscle paralysis beginning in limbs
Erythrogenic toxin[1]	*Streptococcus pyogenes*	Causes a skin rash interpreted as a generalized hypersensitivity reaction
Tetanus toxin (<2.5 ng)[2]	*Clostridium tetani*	Blocks inhibition of spinal motor nerves; causes muscle contraction
Cholera toxin	*Vibrio cholerae*	Activates adenylate cyclase; causes massive loss of fluid
Enterotoxin	*Escherichia coli*	Is similar to cholera toxin but less intense

Source: Data taken from Gill, "Bacterial Toxins: A Table of Lethal Amounts," *Microbiological Reviews* 46:86–94, 1982.

1. This toxin is the result of lysogenic conversion.

2. This value is the lethal dose for humans.

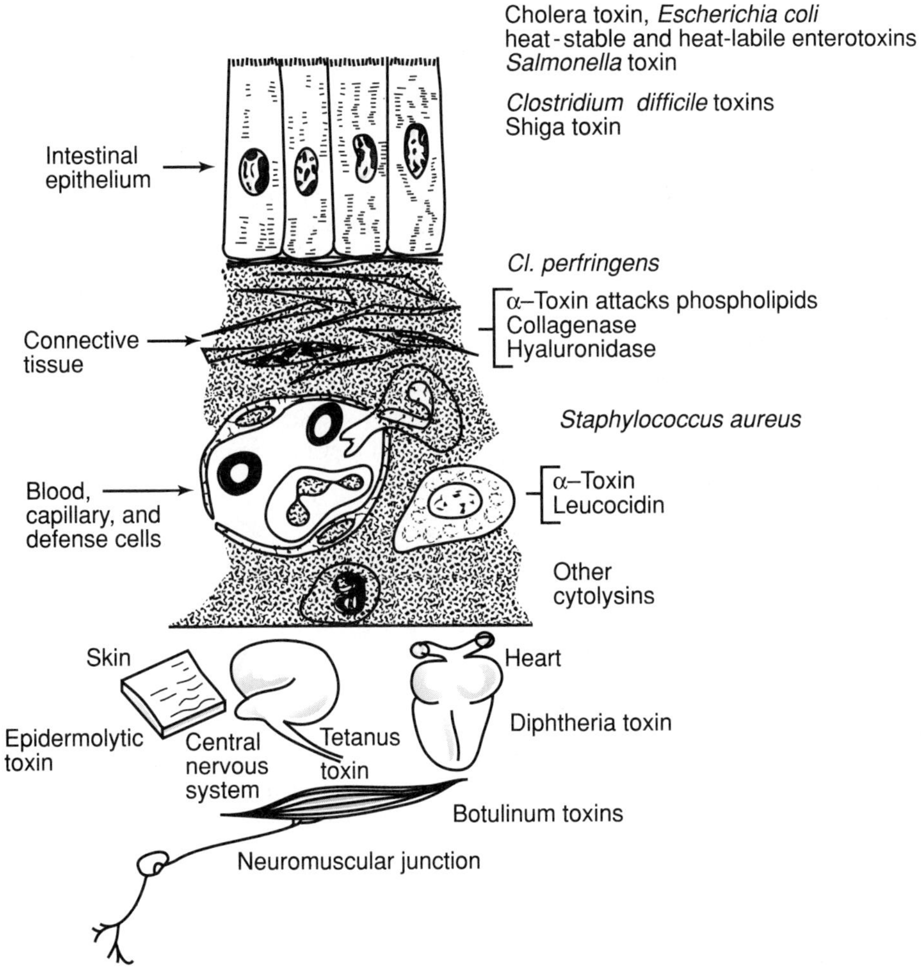

FIGURE 17.6

Diagrammatic representation of bacterial exotoxins that damage the host at epithelial surfaces, subepithelial sites, and tissues or organs distant from the focus of infection.

From J. P. Arbuthnott, "Host Damage from Bacterial Toxins," in *Philosophical Transactions of the Royal Society*, London, B303, 149–165 (1983). Reprinted by permission.

Some bacteria are not really responsible for their exotoxins, rather one of their prophages is. Diphtheria, botulism, and erythrogenic toxins are coded by prophage DNA (table 17.6). In a lysogenic conversion process, infection and lysogenization of a nontoxigenic bacterium converts it into a toxin producer. The specific details of diseases caused by these toxins are highlighted in chapter 19. In this context, however, note that the severity of this type of disease is controlled by the amount of toxin produced. The actual bacterial infection often is harmless.

Bacterial Endotoxins

The bacterial endotoxin can be thought of as a piece of outer membrane such as the one diagrammed in figures 17.3 and 17.7. The lipid A portion of the complex is the point of actual attachment to the outer membrane, with the extended polysaccharide chain of the O antigen attached to it. The way in which the fragment of outer membrane was generated determines whether some protein is also present in the complex. The polysaccharide and protein (if any) are not without toxic activity, but it can be shown that the majority of toxicity lies with the lipid A moiety.

The mechanisms yielding the endotoxin complex are somewhat uncertain. It seems likely that normal growth processes occasionally might release membrane fragments as might cellulolytic processes of the immune response. What is known is that phagocytic cells that engulf gram-negative bacteria later degrade them and form endotoxic fragments of membrane. Some fragments are retained within the phagocytic cells, and some are released into the surrounding fluid (lymph or plasma).

The effects of endotoxin are many and varied. It binds to specific receptors and causes release of second messenger compounds like interleukin-2 or tumor necrosis factor. Symptoms include elevated temperature, hypotension (low blood pressure), and intravascular coagulation. Both the regular and alternate pathways for the activation of complement are triggered. These effects of endotoxin are ones normally associated with the defense against invading parasites. The magnitude of the response is too great, however, and substantial damage to the host, including death, may occur. In one study, it has been estimated that one-third of untreated infections with a gram-negative bacterium result in death due to endotoxin production.

Activation of complement poses a problem in that it triggers a number of other immune responses reminiscent of immediate hypersensitivity (figure 17.8). Monocytes are activated and produce thromboplastin that, in large quantities, causes the blood to coagulate. Thromboplastin also stimulates the production of lysosomal granules that may damage cell membranes. The endotoxin itself causes the cells lining the blood vessels to roll up, exposing their inner lining to bacteria. It also causes cells to produce plasminogen activating factor that indirectly triggers a loss of fibrin molecules and increases bleeding. From all of the preceding information, it is easy to see why the presence of endotoxin results in severe circulatory problems that can lead to shock.

APPLICATIONS BOX

Therapeutic Use of a Bacterial Exotoxin

The potential hazards from exposure to bacterial exotoxins are great, but in one instance a practical use for a toxin has been found. In persons who develop a condition characterized by persistent twitching of a muscle or muscles, the problem can be minor or seriously debilitating, depending on the location of the twitch. If, for example, muscle contractions occur in the muscles surrounding the eye, it becomes difficult if not impossible for an individual to use that eye. The botulism neurotoxin has the effect of paralyzing muscle action. If minute doses of this neurotoxin are carefully injected directly into the affected eye muscle, a person can receive significant relief. When the injections are administered properly, there is little risk of a generalized effect on the patient, although there is always the chance that the immune system will be stimulated to produce antibodies to inactivate the neurotoxin.

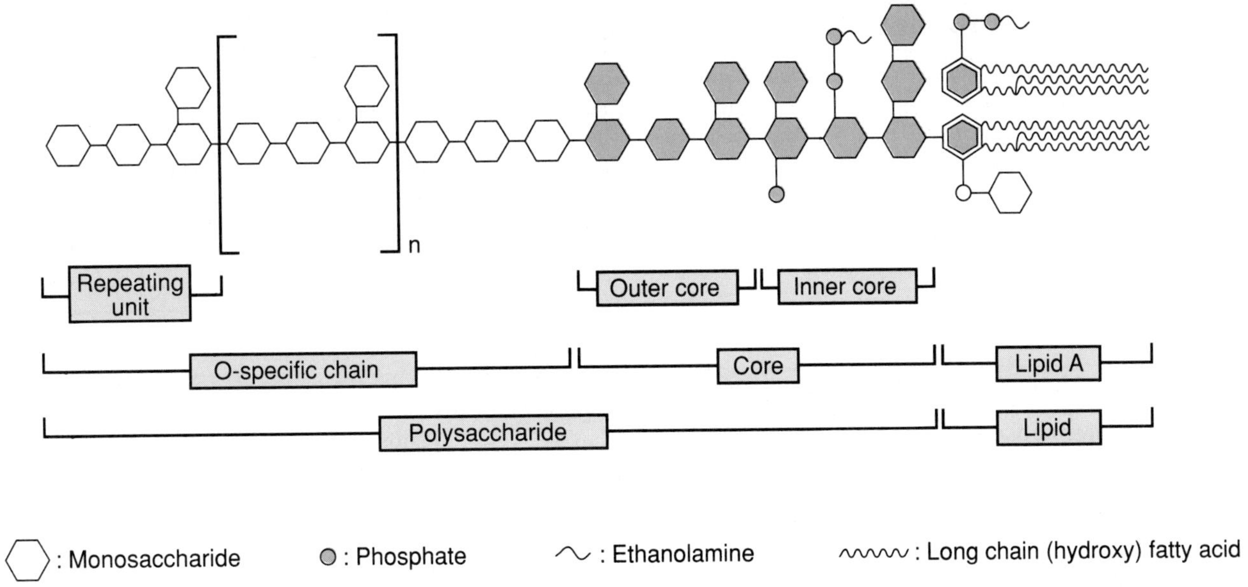

FIGURE 17.7
General structure of endotoxin. The lipid A portion of the complex is inserted into the outer membrane. The long polysaccharide chain is the O antigen.

From E. T. H. Rietschel, et al., "Newer Aspects of the Chemical Structure and Biological Activity of Bacterial Endotoxins," in *Progress in Clinical and Biological Research,* Vol. 189:31. Copyright © 1985 John Wiley & Sons, Inc., New York, NY. Reprinted by permission of John Wiley & Sons, Inc.

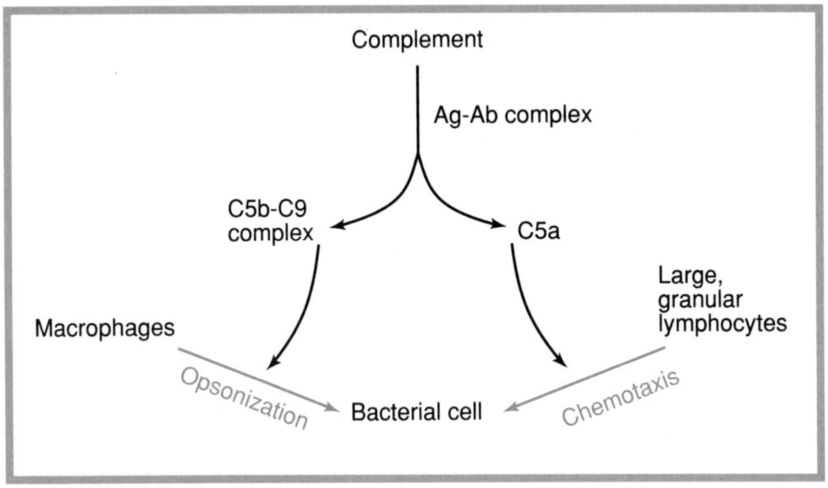

FIGURE 17.8
Activation of the complement pathway and its consequences for a bacterial cell.

APPLICATIONS BOX

Drastic Effects of Endotoxin

The potential severity of a gram-negative bacterial infection should not be underrated. Recently, a university student infected by *Neisseria meningitidis* presented initial symptoms of influenza (high fever, lethargy, etc.). By the afternoon of the same day, a painful purplish rash had developed over her body due to massive bleeding triggered by bacterial invasion of the circulatory system. Despite antibiotic treatment, the student died within a few hours. The probable site of initial infection was the nasopharynx, or possibly the meninges of the brain although the student presented no central nervous system symptoms. Once the bacteria broke through the blood vessels and into the circulatory system, a large inoculum was present and the disease progressed in its typical, rapid fashion.

TABLE 17.7 Examples of mycotoxins

Precursor Intermediate	Mycotoxin	Producer Organism
Tetraketide	Patulin	*Penicillium* spp.
Heptaketide	Alternariol	*Alternaria* spp.
Octaketide	Secalonic acid	*Aspergillus, Claviceps, Phoma* spp.
Decaketide	Aflatoxin	*Aspergillus* spp.

Mycotoxin Effects

Mycotoxins are derived from the same fundamental biochemical intermediates as antibiotics and can be grouped according to their precursor molecules (table 17.7). Basically, polyketides are intermediates from fatty acid biosynthesis (figure 5.8) in which the keto groups derived from the addition of acetyl CoA have not been reduced. They can be subclassified according to the length of the chain, with the minimum being 4 acetyl CoA units and the longest common chain being 10. Formation of the mycotoxin patulin is shown in figure 17.9.

Another group of mycotoxins, the trichothecenes, is derived from mevalonic acid formed by the condensation of three molecules of acetyl CoA. Figure 17.10 depicts the essential features of the trichothecene structure as well as the structures of other mycotoxins. Members of the genus *Fusarium* often are producers of trichothecenes.

Some mycotoxins are simple cyclic polypeptides. Others, like ergotamine, are elaborate polycyclic structures derived from amino acids (see figure 17.10). Other toxins are derived from a combination of mevalonic acid and amino acids.

A person usually acquires a primary mycotoxicosis (poisoning by a fungal toxin) by consuming a food on which the producer fungus has grown. If mycotoxicosis occurs in a basically asymptomatic animal slaughtered for consumption, the mycotoxin is passed up the food chain and a secondary mycotoxicosis results.

Mycotoxicotic effects can be extremely varied. Some mycotoxins specifically affect a particular organ system such as the kidneys. Others promote more general effects. Some have been shown to be carcinogenic or teratogenic (cause birth defects). More details about mycotoxic disease can be found in chapter 19.

Summary

A successful parasitism requires the interaction of host and parasite. If the parasite is too damaging to the host, it will kill the host and search for a new one. At the same time, however, the parasite must resist host attempts to destroy it or prevent it from acquiring a suitable niche. Mammalian hosts use several different methods to protect themselves against parasites. They have varied immune responses to foreign antigens as well as a series of mechanical barriers to prevent parasitic access to inner body tissues. Because iron is an essential element for microbial growth, lactoferrin and transferrin, both naturally occurring molecules, chelate iron and thereby keep its available concentration low. This response is exaggerated during a fever. Even the intestine, which is technically outside the body, is

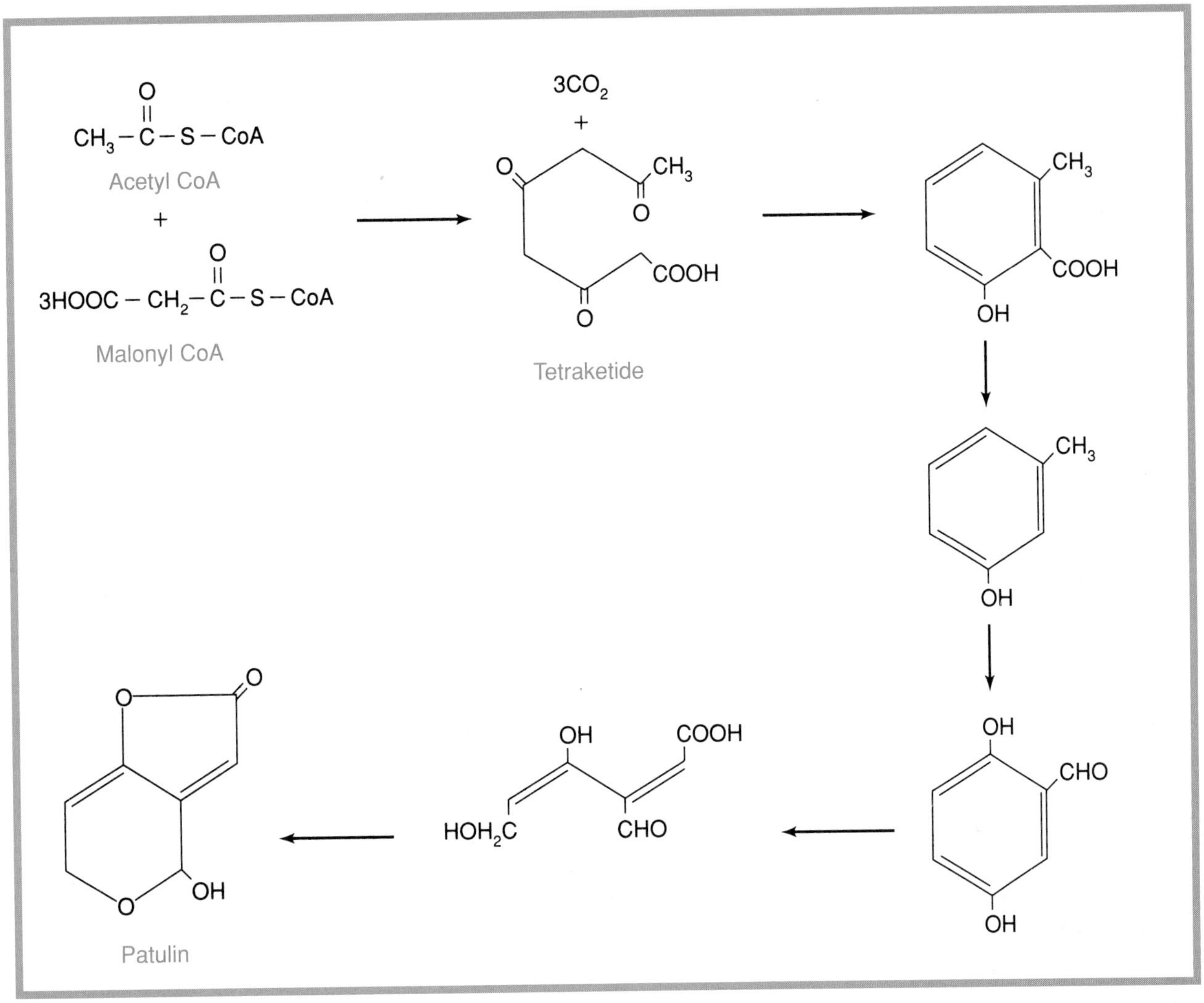

FIGURE 17.9

Formation of patulin from a tetraketide.

APPLICATIONS BOX

Mycotoxins and Food

It is comparatively easy for mycotoxins to be present in food. Storage of grains and other staples under damp conditions provides a good medium in which fungi are able to grow. Peanuts are particularly susceptible to aflatoxin production if improperly stored, but other foods also can be affected. In a compilation of data reported in 1976, 35% of peanut samples and 6% of rye samples contained some aflatoxin. Oat samples contained none. A recent U.S. government report has suggested that the potential cancer-causing effects of aflatoxins in food may be greater than many environmental chemicals to which we are exposed, even though the latter receive more publicity. Along similar lines, stored grains can become contaminated with ergot as a result of the growth of *Claviceps purpurea*. When eaten, the contaminated grains can cause a burning sensation in the extremities, a condition known historically as Saint Anthony's fire. There were numerous outbreaks of ergotism during the fifteenth and sixteenth centuries, and thousands of deaths were reported.

Aflatoxin B_1 from *Aspergillus flavus* or *A. parasiticus*

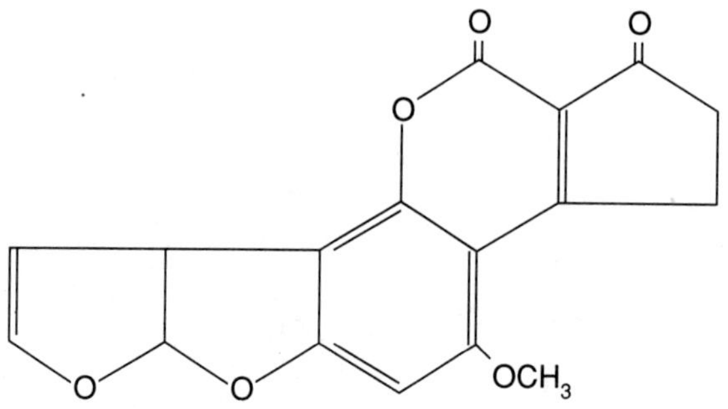

Deoxynivalenol, a trichothecene from *Fusarium* sp.

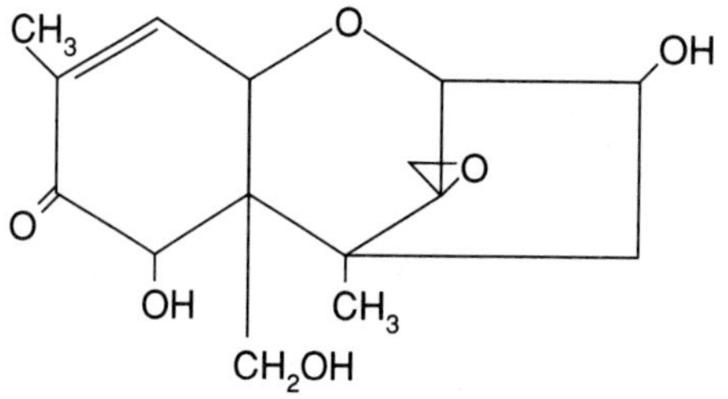

Ochratoxin A from *A. ochraceus*

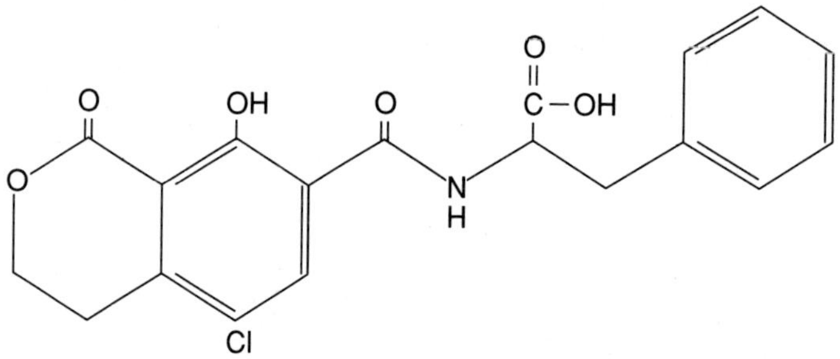

FIGURE 17.10

Essential structures of some mycotoxins.

protected to some extent by the highly acidic pH of the stomach and by some secretory antibodies.

For a parasite, the ability to attach to a host, resist eviction, and maintain growth means success. Attachment mechanisms are not always well understood, particularly in the case of bacteria. Protozoan parasites often have large mechanical "suckers" that can be used as anchors. To augment its invasive power, a parasite can synthesize various enzymes to attack critical portions of epithelial tissue that bar its entry into the body. The virulence of an organism depends on the amount of damage it inflicts after establishing an infection. In turn, virulence often is dependent on the amount of parasitic growth. The level of virulence results from the combined synthesis of various special products. Leukocidin, coagulase, and IgA protease all serve to protect a parasite against various host cell mechanisms. Siderophores compete for available iron with the chelating agents of a host. Of particular concern are those parasites that produce toxic substances because their effects on the host can be disproportionate to the actual extent of the infection.

Intracellular parasites are especially resistant to many host defenses. They manage to survive by subverting normal phagocytotic and cellulolytic processes of the cell-mediated immune system. By virtue of their location within a host cell, intracellular pathogens are protected from other body defenses and are rapidly disseminated throughout the body.

Questions for Review and Discussion

1. What are the distinctions between invasive organisms and virulent organisms? What is meant by the term *accidental pathogen,* and what does this concept tell us about the possibility of a totally harmless bacterium?
2. Give some examples of molecules that contribute to parasitic invasiveness and virulence. How does each molecule function? Which of these molecules takes advantage of normal biochemical reactions of host cells and converts them to the parasite's advantage?
3. How do intracellular parasites survive host defenses? Why is it not possible for all parasites to become intracellular parasites? What strategies might a host employ to eliminate intracellular parasites?
4. Explain the difference between an exotoxin and an endotoxin and give an example of each. What are their toxic effects?
5. In what ways are bacterial toxins and mycotoxins similar? In what ways are they different? Give some examples of mycotoxins.
6. Give some examples of the ways in which the bacterial cell surface affects the ability of the host immune system to interact with the pathogen.
7. Discuss the essential role of iron in parasitic metabolism. Why must it also be present in the mammalian host? What strategies are used by host and by parasite in their competition for this essential nutrient?

Suggestions for Further Reading

Baggiolini, M., and M. P. Wymann. 1990. Turning on the respiratory burst. *Trends in Biochemical Sciences* 15:69–72. A consideration of the signal-transduction mechanisms involved in activating the respiratory burst.

Finlay, B. B., and S. Falkow. 1989. Common themes in microbial pathogenicity. *Microbiological Reviews* 53:210–230. A good general discussion from one of the main laboratories in the field.

Groisman, E. A., and M. H. Saier, Jr. 1990. *Salmonella* virulence: New clues to intramacrophage survival. *Trends in Biochemical Sciences* 15:30–33. A more detailed survey of this phenomenon than is possible in this chapter.

Hentges, D. J. (ed.). 1983. *Human intestinal microflora in health and disease.* Academic Press, Inc., New York. Of particular interest are the chapters on normal flora, flora of infants, and production of intestinal mutagens.

Holder, I. A. (ed.). 1985. *Bacterial enzymes and virulence.* CRC Press, Inc., Boca Raton, Florida. A collection of seven short reviews dealing with aspects of countering host defenses.

Joiner, K. A. 1988. Complement evasion by bacteria and parasites. *Annual Review of Microbiology* 42:201–230. A somewhat technical article covering broader issues than this chapter presents.

Jones, T. C., and H. Masur. 1980. Survival of *Toxoplasma gondii* and other microbes in cytoplasmic vacuoles, pp. 157–164. *In* H. Van den Bosch (ed.), *The host-invader interplay.* Elsevier/North-Holland Publishing Co., Amsterdam. Although the title implies that the text concentrates on one organism, this chapter in fact provides a very nice, short review of how microorganisms evade destruction.

Lehrer, R. I., T. Ganz, and M. E. Selsted. 1990. Defensins: Natural peptide antibiotics from neutrophils. *ASM News* 56:315–318.

Leslie, R. G. Q. 1984. Evaluation of phagocyte function, pp. 77–110. *In* W. G. Reeves (ed.), *Recent developments in clinical immunology.* Elsevier/North-Holland Publishing Co., Amsterdam. The first half of this chapter offers an excellent overview of phagocytes.

Ogata, R. T. 1983. Factors determining bacterial pathogenicity. *Clinical Physiology and Biochemistry* 1:145–159. An excellent review with vocabulary well within the range of readers of this text.

Smith, H. 1984. Extension of consideration of the role of toxins in pathogenicity from bacteria to fungi, protozoa, and viruses, pp. 1–12. *In* J. E. Alouf, F. F. Fehrenbach, J. H. Freer, and J. Feljcszwicz (eds.), *Bacerial protein toxins.* Academic Press, Ltd., London.

Smith, J. E., and M. O. Moss. 1985. *Mycotoxins: Formation, analysis and significance.* John Wiley and Sons, Inc., Chichester. A short, not too technical account of fungal toxins affecting humans and animals.

chapter

18

Principles of Epidemiology and Public Health

chapter outline

■

GENERAL GOALS

Be able to describe how an epidemic disease spreads and how it can be controlled.
Be able to describe various methods for treating diseases and some advantages and disadvantages of each.
Be able to explain how immunization represents an important element in the protection of a society from a particular disease.
Be able to discuss the rights and responsibilities of those who for one reason or another cannot be immunized.

The science of microbiology is most often associated with disease processes and disease-causing organisms, even though it encompasses many other facets. This is not necessarily an injustice because public health and medicine in the twentieth century have experienced major successes and remarkable improvements. Consider, for example, the statistics listed in table 18.1 for the years 1882 and 1983.

This chapter presents the ways in which public health specialists seek to understand and control the spread of disease. The use of vaccines, the principles underlying their manufacture, and the risk-benefit analysis for evaluating the potential of any given treatment are highlighted.

Communicable Diseases

If disease is defined as a deviation from the normal homeostatic state of a particular species, there are several different ways in which a disease state can arise. It can result from some inborn biochemical or genetic error present at birth or arise from some mutation at a later time. Although such an error may be catastrophic for the individual, as is the case with an inborn anomaly like phenylketonuria, it does not present a risk to those associated with the individual. A **communicable disease,** on the other hand, can be spread from one individual to another and, therefore, represents a risk not only to the afflicted person but also to the population at large. The causative organism of a communicable disease is known as its **etiologic agent.**

Sources of Communicable Diseases

Communicable diseases are always present somewhere in the world, although they can be absent from a particular subpopulation. There is an expected level of activity for a given disease, a sort of background level always present unless circumstances change dramatically. For any particular disease, the background level can be quite low or rather high. The incidence of certain diseases in the United States is listed in table 18.2. It is easy to see the low frequencies of some and the high frequencies of others.

TABLE 18.1 Mortality and morbidity (illness) in New York City

Disease	Mortality in 1882[1]	Morbidity in 1983[2]
Smallpox	269	0
Measles (rubeola)	912	87
Diphtheria	1,521	0
Scarlet fever	2,070	Not reportable
Typhus	66	0
Malaria	533	45

1. Data on mortality are taken from 1983 *Scientific American* 248(3):14.

2. Data on morbidity are taken from the Centers for Disease Control annual summary of 1983. Morbidity is the state of being diseased or ill; therefore, these totals do not just represent fatalities.

TABLE 18.2 Incidence of some diseases whose occurrence must be reported to the Centers for Disease Control

Disease	Cases in 1989	Cases in 1990	Mean Number of Cases between 1984–89
AIDS	33,722	41,129	15,420
Cholera	0	9	7
Diphtheria	3	4	2
Gonorrhea in civilians	733,151	664,159	820,739
Malaria	1,277	1,185	1,083
Measles (rubeola)	18,193	26,510	6,156
Plague	4	2	15
Syphilis	44,540	48,363	33,904
Toxic shock syndrome	400	293	407
Rabies in animals	4,724	4,219	5,112
Rabies in humans	1	1	1

Sources: *Morbidity and Mortality Weekly Report* 38:89, 38(54):3, 53, 1990; and *Morbidity and Mortality Weekly Report* 39:941, 1991.

There are many specialized terms applied to infectious diseases. The speed with which a disease spreads is reflected in its **incubation period,** the amount of time that elapses between a person's exposure to a pathogen and the first appearance of symptoms. **Infectious dose** refers to the minimum number of viral particles or microbial cells necessary to establish a disease state in a normal, healthy individual. An **epidemic** occurs when a disease is present in a population at levels statistically higher than expected. The key to identifying an epidemic is not, however, the absolute number of cases but rather the number of cases compared to those expected due to normal statistical fluctuations (figure 18.1). In terms of the diseases presented in table 18.2, an additional 50 cases of plague in one year would constitute a major outbreak, but 8,500 additional cases of gonorrhea would hardly be noticed. An epidemic usually is quantitated on a local or regional basis. A **pandemic** is the equivalent term used to describe an outbreak of disease on a worldwide basis. If a particular disease has reached epidemic proportions in the wildlife of an area but has not significantly affected the human population, it is referred to as an **epizootic.** A disease that is **endemic** is always present in a particular population.

Individuals infected with an organism that can cause a communicable disease constitute a **reservoir of infection** for that disease if they can infect others. As long as there are active cases of the disease, there remains the potential to precipitate another epidemic. The reservoir for a disease may be members of the same species or members of a different species. For example, the influenza virus can infect both humans and swine. In

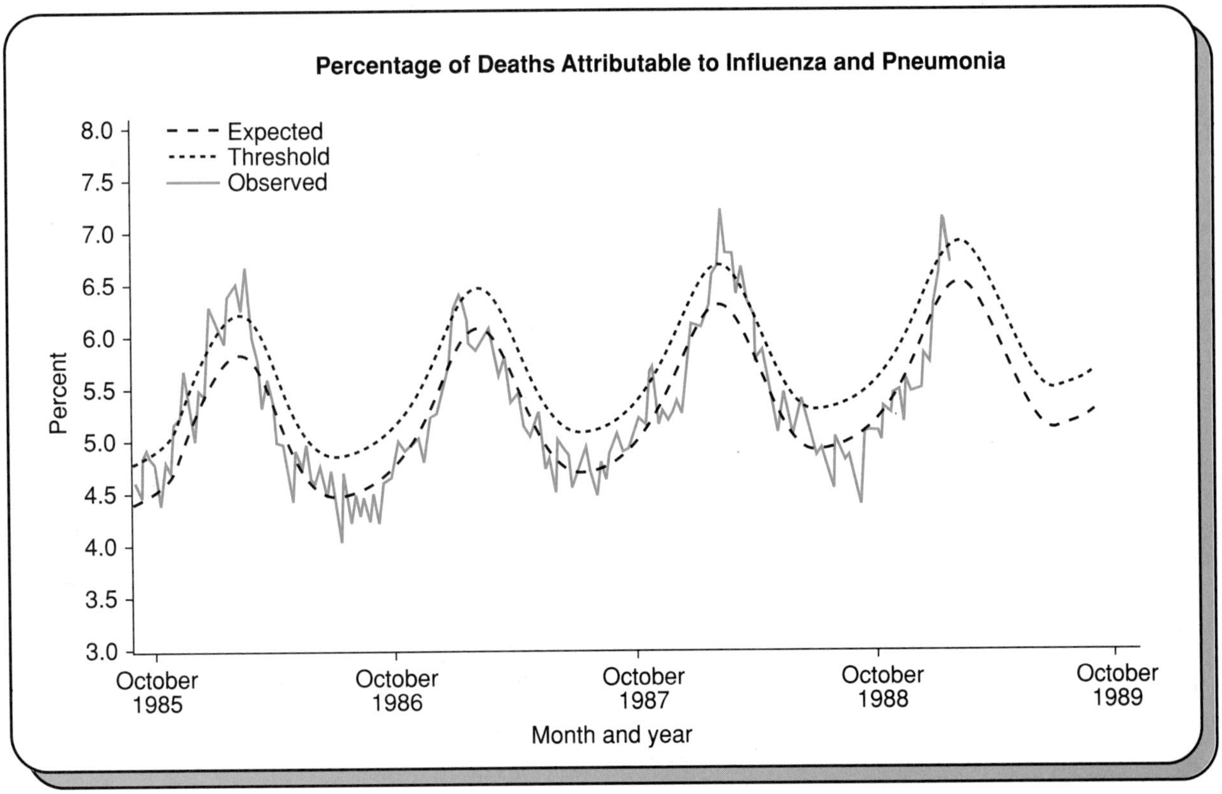

FIGURE 18.1

Identification of an epidemic. As an endemic disease, influenza is always present in the population and can be expected to cause a certain number of deaths either directly or indirectly by predisposing a person to pneumonia. The disease is a seasonal one, with most cases occurring during the winter months. Based on years of collecting and evaluating statistics, the Centers for Disease Control can predict the expected percentage of total deaths that will result from influenza and pneumonia at any given time (long dashed line). There is a certain margin for error in the calculation, the upper limit of which is indicated by the threshold value (short dashed line). The colored line shows the actual (observed) percentage of deaths attributed to influenza and pneumonia. Whenever the observed value rises above the threshold value, the percentage of deaths attributable to the two diseases is significantly greater than expected statistically, and an epidemic is in progress.

Source: Adapted from *Morbidity and Mortality Weekly Report* 38:97, 1989.

the so-called swine flu episode in 1976, swine served as a reservoir with respect to the human population.

A person who is a carrier of an infectious agent also is a reservoir. A **carrier** state results when the immune system prevents a parasite from growing to the point at which it will cause symptoms, but the system cannot kill the organism outright. Although an organism grows slowly in a carrier and presents no obvious symptoms, the carrier can transmit it to others or release it into the surrounding environment. New infections easily can result from contact with those parasites. The best-known example of a disease spread by carriers is typhoid fever (see chapter 19).

Not all reservoirs consist of infected host organisms, however. It is equally possible for inanimate objects to be reservoirs. A municipal water supply that contains sufficient bacteria of the appropriate type to cause a diarrheal illness is an obvious example. The bacteria in question may not actually reproduce in the water, but as long as they continually are reintroduced into the water by fecal contamination, the water supply will remain a disease reservoir. Bacteria that are natural inhabitants of the soil also can cause an infection when introduced into a wound. In this more subtle example, the bacteria are reproducing, but the infection actually interrupts their life cycle.

Modes of Transmission of Communicable Diseases

Communicable diseases can be spread by various means. What is involved in all cases is some mechanism for delivering the pathogen from its original host to the appropriate **portal of entry** of a potential new host. The portal of entry is the site through which a microorganism can establish an infection. The respiratory tract, genital area, and intestinal tract are examples. Each pathogen has a characteristic portal(s) of entry. The portal of entry for an accidental pathogen is a wound site.

Obviously, direct contact with the infection or infected area constitutes an easy way to spread a disease. If a pathogen has lodged in the respiratory tract, **aerosols** (fine droplets of liquid or fine granules of powder suspended in air) produced by sneezing or coughing easily can transmit the infection to others. Sexually transmitted diseases are spread by direct contact with mucous membranes of the genital tract or oral cavity.

If an infectious organism within the body is located outside the respiratory tract, the aerosol from a person having the disease will not necessarily be the mode of its transmission. In this case, the usual route of transmission is via contact with some body secretion or waste product. As documented in chapter 19, feces are an excellent transmitter of a number of pathogens. For this reason, houseflies are potential sources of infection because they frequently land on feces as well as on food. Urine from a person with a bladder infection also can transmit disease. Pus produced by an infected wound often contains infectious organisms.

Occasionally, it is possible for an infectious microorganism to survive for some period of time on environmental surfaces. Persons touching those surfaces might become infected if they should happen to contaminate the appropriate portal of entry on their own body. The term **fomite** is applied to an inanimate object that can transmit a disease.

A spore-forming bacterium may produce a similar effect due to its ability to survive in inhospitable environments. *Bacillus anthracis* grows in soils and sporulates readily. Inhaling an aerosol of infected dust is sufficient to cause an anthrax infection. A biological warfare test carried out during World War II on a remote island off the coast of England left the island heavily contaminated with *B. anthracis* spores. After 40 years, this island is still uninhabitable.

Sometimes a **vector** transmits a pathogen. A vector is a biting or sucking animal that can harbor a pathogen and, in the course of its normal feeding activities, contaminate the wound that it inflicts. A vector can simply transport the pathogen or participate in its life cycle (compare Rocky Mountain spotted fever with malaria in chapter 19). Note that this solves the problem of pathogenic invasion rather nicely for some organisms. The pathogen does not have to be at all invasive because the vector provides its portal of entry. Although many types of vectors are known, the most common are such members of the phylum Arthropoda as fleas, lice, and ticks.

Technology also is responsible for transmission of infections. Blood transfusions or tissue transplants from infected donors can transmit disease-causing microorganisms of all types. At the present time, the AIDS epidemic strikes heavily at intravenous drug abusers who share needles with one another. So strong is this link that one experimental program is being used to test whether AIDS transmission can be reduced if addicts are provided with free, sterile needles.

TABLE 18.3 Normal route of administration and side effects of some antimicrobial agents

Antimicrobial Agent	Route(s) of Administration	Major Side Effects
Streptomycin	Oral, IV[1]	May cause deafness
Penicillin	Oral, IV	May cause allergic reactions
Chloramphenicol	Oral	May cause aplastic anemia
Neomycin	Topical	May be toxic to kidneys
Erythromycin	Oral, IV	May cause gastrointestinal disturbances
Sulfamethoxazole	Oral	May cause loss of blood cells and platelets
Cefazolin	IV	May cause hypersensitivity reactions, reduction in blood cell counts
Nitrofurantoin	Oral, IV	May cause hypersensitivities, reduction in blood cell counts, nervous system dysfunction

1. IV stands for intravenous administration.

Antimicrobial Treatments for Communicable Diseases

Antimicrobial agents constitute a big business. Worldwide consumption of antibiotics in 1980 was estimated at 5.4 g per person per year. It is, therefore, extremely important that antimicrobials be used carefully and logically.

Therapeutic Uses of Antimicrobial Agents

For a bacterial infection, the treatment of choice usually is an antimicrobial compound. Fungal diseases generally are treated only when the symptoms are particularly severe because antifungal agents are quite toxic. Drug toxicity is a consequence of inherently similar metabolism among eukaryotes; that is, the target of an antimicrobial agent is present on both the host and the pathogen. An appropriate antimicrobial agent must affect microbial cells at a concentration significantly below that at which it causes severely toxic effects in the host. The **therapeutic ratio** is determined by dividing the agent's minimum inhibitory concentration by its toxic threshold; the higher the ratio, the more effective is the antimicrobial treatment.

The minimum inhibitory concentration of an antimicrobial as determined in the laboratory is not necessarily a concentration that the body can achieve. Some compounds are insoluble in water and, therefore, provide only low concentrations in the blood. Others are not absorbed well by the intestines and only can be administered intravenously. Still others are so toxic that they can be used externally but not internally. Examples of antimicrobial agents and their properties are exhibited in table 18.3.

Most side effects (effects on the host rather than on the parasite) of antimicrobial agents tend to involve specific organ systems. Some agents like penicillin and cefazolin are haptens that may cause a hypersensitivity reaction. Many antimicrobials are nephrotoxic (poisonous to the kidneys) as they are being filtered from the blood. The antimicrobial dose, therefore, must be carefully adjusted on the basis of body mass so that a therapeutic but minimally toxic level is maintained in the blood.

Occasionally, side effects can be lethal, as in the case of chloramphenicol. In a minority of individuals to whom chloramphenicol is administered, an irreversible aplastic anemia develops in which bone marrow cells cease production of various blood components. Afflicted individuals eventually die from a lack of red blood cells (anemia). Because of this risk, chloramphenicol is used only sparingly, chiefly in cases of typhoid fever when, in the opinion of the attending physician, an individual's risk of death from the disease is greater than his or her probability of developing aplastic anemia (see chapter 19).

Antimicrobial Resistance during Therapy

Widespread use of antimicrobial agents has provided an excellent example of evolution in action. As new antimicrobials are introduced, populations of microorganisms resistant to them appear. The issue is basically one of probability. Every time a microorganism is exposed to an antimicrobial, there is a finite probability that it will already be resistant to that agent due to chance mutation. For particularly useful antimicrobials, this probability is low, at least 10^{-8} or less. In any case, the more microorganisms that are exposed to the antimicrobial selection, the greater the chance that at

TABLE 18.4 Plasmid-determined resistance in enteric bacteria

Antibacterial Drug	Year First Administered	Year R Plasmid Detected
Sulfonamides	1936	1959
Streptomycin	1948	1959
Tetracyclines	1949	1959
Chloramphenicol	1950	1959
Neomycin	1954	1963
Ampicillin	1962	1965
Gentamicin	1964	1972
Trimethoprim	1968	1977
Phosphonomycin	1969	1977
Amikacin	1974	1974

Source: Data from N. Datta, "Plasmids of Enteric Bacteria," pp. 487–496. In *Antimicrobial Drug Resistance*, edited by L. E. Bryan. Copyright © 1984 Academic Press, Orlando, Florida.

least one will be spontaneously resistant and survive treatment so as to give rise to a population of resistant microorganisms. In order to minimize the survival of resistant organisms, certain precautions must be taken.

Once the decision is made to use a specific antimicrobial treatment, it is essential that the appropriate dose be administered to achieve therapeutic levels. Moreover, these levels must be maintained long enough to ensure that all pathogens have been killed or sufficiently weakened in order for normal immune mechanisms to eliminate them. This generally requires 7 to 10 days or more, depending on the ease with which an antimicrobial agent can reach the site of infection. The bladder, for example, has relatively poor circulation and hence is not easily infiltrated by antimicrobial molecules circulating in the blood. On the other hand, if the kidneys quickly removed an antimicrobial compound from the blood, its concentration in the urine would rise, and the bladder could be treated in that manner.

If treatment is of short duration, there is the chance that some microorganisms will have been exposed only to moderate antimicrobial concentrations. In this case, it is relatively easy to find cells that are resistant to moderate levels of the antimicrobial agent. The effect of the brief treatment is to enrich the microbial population for these cells. If the infection recurs and the same treatment is given, it will have less effect than the first time the treatment was administered. Eventually, if partial treatment is repeated often enough, it is reasonable to assume that a culture completely resistant to therapeutic levels of the antimicrobial agent will be produced.

Bacterial development of antimicrobial resistance is particularly troublesome when that resistance is actually encoded on a conjugative R plasmid. In this case, moderate levels of an antimicrobial agent serve only to select for those cells carrying or receiving the plasmid. Soon the entire microbial population in an infected individual may be carrying the plasmid and will resist conventional treatment. Moreover, R plasmids frequently confer resistance to more than one antimicrobial agent, further complicating matters. On the other hand, if an antimicrobial agent is administered at the correct therapeutic level for the appropriate period of time, the initially sensitive cells will be killed before they have a chance to acquire the plasmid. The few resistant cells present in the population then can be eliminated by the immune system.

These precautions in treatment are very important because the time lag between the introduction of an antimicrobial on the market and the first appearance of an R plasmid conferring resistance to it seems to be decreasing (table 18.4). Although the reasons for reduced time lags are varied, failure to administer antimicrobials properly can only make the situation worse.

Nontherapeutic Uses of Antimicrobials

Interestingly, and perhaps unfortunately, low levels of antimicrobial agents introduced into certain animal feeds cause the animal to gain weight more rapidly. They do not seem to improve the general health of the animals, however, and the mechanism(s) by which they work is unknown. Antimicrobials used for this purpose are not particularly expensive, and their use results in cheaper food production than would otherwise be possible. The practice of using antimicrobials in animal feed is widespread in the United States. Although the quantity used per animal is small, the total consumption is enormous and represents a major portion of antimicrobial sales in this country. There are, therefore, compelling economic reasons for continued use of antimicrobials in animal feed.

APPLICATIONS BOX

Antimicrobials and Public Policy

The World Health Organization (WHO), sponsored by the United Nations, has recognized that antimicrobial agents can be maintained as useful tools only if the selective pressures on general microbial populations are minimized. The organization has provided guidelines to establish the proper use of these drugs and avoid their abuse. Each hospital under the auspices of WHO must establish a written policy to govern the use of antimicrobial agents and to restrict staff access to new agents in order to avoid indiscriminate use.

Unfortunately, WHO has assumed that the medical establishment has sole control of antimicrobial use. In many third world countries, however, purchase and use of antimicrobial agents do not require a doctor's prescription. These drugs are available over the counter like aspirin is. Inappropriate use of antimicrobial agents in this situation is potentially enormous. Formal restrictions on antimicrobial availability are unlikely to be useful, however, because most of the population in these countries has limited resources to pay for drugs and limited access to professional health care. Therefore, guidelines to require prescriptions for antimicrobial drugs may be difficult to implement and counterproductive in terms of overall public health.

The public health problem arising from the use of antimicrobial agents in fattening animals is that there are also those that are useful therapeutically. It has long been suspected that an animal's prolonged exposure to antimicrobials would cause bacteria normally present in or on it to become resistant to the antimicrobials used. One laboratory study has shown that exposure of bacterial chemostat cultures to 1/10 the minimum inhibitory concentration of certain antimicrobials is sufficient to enrich the population for antimicrobial-resistant bacterial cells. This and other studies have led to fears that antimicrobial-resistant bacteria might cause serious diseases within the human population.

One bacterium very prevalent among animals consumed by humans is *Salmonella*. A recent study of outbreaks of *Salmonella* infections that occurred during a 12-year period supported some concerns about the use of antimicrobials in animal feed. There were 52 outbreaks identified, and the **case:fatality ratio** (percentage of cases resulting in death) was 4.2 for antibiotic-resistant organisms but only 0.2 for antibiotic-sensitive bacteria. Moreover, of the 38 outbreaks whose source could be identified, animals consumed by humans were the source of 11 out of 16 antibiotic-resistant outbreaks and only 6 out of 13 antibiotic-sensitive outbreaks. Although not conclusive, these data suggest that there may be risks to society from continued use of antibiotics in animal feed.

Reducing the Development of Antimicrobial Resistance

Despite the psychological benefits of being able to "cure" a disease, oftentimes the only really appropriate treatment is either no treatment at all or treatment to alleviate symptoms. This situation occurs when symptoms of the disease are not particularly severe and/or when the pathogen is not sensitive to any known treatment. Often viral diseases can be treated only in a symptomatic sense. There are very few chemicals that will interfere specifically with the viral life cycle because a virus uses the metabolic pathways of its host cells. Treatments for the common cold have no real effect on the viral agent. They merely relieve the nasal congestion and discomfort accompanying a cold. Nevertheless, antimicrobial agents are the backbone of modern health care. The necessity is to use them appropriately.

For example, a severe viral respiratory infection (not treatable with antimicrobials) can disrupt normal respiratory function sufficiently to allow a bacterial **secondary infection.** Under these circumstances, it is entirely appropriate to administer an antimicrobial agent to prevent the development of bacterial pneumonia.

Antimicrobial agents also can lose their effectiveness if epidemic levels of a disease continue for too long a period. Table 18.2 indicates that the number of gonorrheal cases has been at very high levels for years. Originally, the disease was treated easily with penicillin, but within the last decade, penicillinase-producing *Neisseira gonorrheae* (PPNG) has evolved to the point at which it is isolated commonly. More than 20% of the isolates are resistant to penicillin and/or tetracycline. Other antimicrobials, such as spectinomycin, have been employed to control PPNG, but as they have been used more frequently the *Neisseria* organisms have begun to develop resistances to them. Already significant numbers of *Neisseria* strains resistant to spectinomycin have been reported, and its usefulness obviously will be limited.

Problems such as these do not mean that antimicrobial agents are no longer useful. They should be taken as a warning, however, because careful consideration must be given to the use of antimicrobials only in appropriate situations and only in appropriate ways.

Methods for Preventing Disease Transmission

When bacterial viruses were first discovered, it was thought that they might play a role in the treatment of disease by seeking out pathogenic organisms within the body and destroying them. In fact, the novel *Arrowsmith* by Sinclair Lewis is based on just such a premise. However, early attempts to treat diseases with viruses failed. Recently, the concept has been revived, and phages isolated from *Escherichia coli* that cause diarrheal diseases in young livestock have been used successfully to limit or halt the spread of such diseases. Unlike antimicrobial agents, phages must be given at the onset of symptoms or earlier, at which time they are more effective than conventional antimicrobial therapy. Presumably, once the infection is established, the viruses cannot be administered in sufficient quantity to infect all bacterial cells involved.

In a sense, an epidemic is analogous to the growth curve of a bacterial culture. The nutrients in this case are susceptible individuals. New cases of a disease result when susceptible individuals become infected with the etiologic agent; therefore, the graph representing the total reported cases increases logarithmically (see the figure in the applications box in chapter 8 on bacterial growth curves). Eventually, all susceptible individuals have contracted the disease, and the number of cases levels off. That point in time represents a critical period. If new susceptible individuals are introduced, the epidemic will again spread. A **quarantine** or enforced isolation of infected individuals is often instituted to keep diseased individuals away from susceptible ones in order to abate the spread of an epidemic.

Note also that the timing of a particular antimicrobial treatment can have a remarkable influence on its perceived impact. An effective treatment administered early will cause the epidemic curve to level off sooner than it would if the pathogens were left to follow their own course. Contrariwise, any treatment at all will appear effective if it is administered just before the epidemic curve is about to level off of its own accord. Hence, because the epidemic stopped shortly after the treatment was applied does not provide sufficient information to judge the efficacy of the treatment. In addition, one must know whether or not there were any susceptible individuals left to be exposed to the disease after treatment had been applied.

Acquired Immunity to Disease

Over the centuries, people noticed the natural phenomenon of acquired immunity, but it was not until after the work of Louis Pasteur and Robert Koch that a scientific understanding developed.

The Concept of Vaccination

The first serious attempts to induce an immune response in a person who had not contracted a particular disease seem to have been made with the smallpox virus (variola virus). Initially, some people attempted to find others with a mild case of the disease and infect themselves by rubbing exudate from the sores of an afflicted person into their own skin or by ingesting the crusts of sores, a process called variolation. The hope, not always realized, was that one mild case followed by an unusual route of infection would beget another mild case. The ever present risk of contracting a virulent form of the disease made variolation a dubious procedure, however.

In 1796 Edward Jenner tried a different approach. He had noticed that milkmaids rarely if ever got smallpox. Instead, they usually contracted a much milder disease known as cowpox. Jenner reasoned that perhaps their exposure to the cowpox virus conferred on them a lasting immunity to smallpox. In terms of our present understanding, the assumption was that the cowpox virus has antigens in common with the smallpox virus, and antibodies formed against those antigens will cross-react with the smallpox virus and inactivate it as well.

Initially, the virus was administered as a droplet of liquid on the skin and forced into it with a series of needle punctures. Under normal circumstances, a small hypersensitivity reaction developed at that site in a few days, indicating that an infection had stimulated the immune system. The virus used was referred to as variola vacciniae (smallpox of the cow). Immunization against smallpox has continued until recent times and conferred immunity for some 3 to 7 years. To maintain immunity, reimmunization was required about every 3 years.

Edward Jenner was not the only person to advocate immunization against smallpox, but he was the most active and most public in this quest; therefore, he is most remembered. He and others like him set the stage for the work of Koch and Pasteur by showing that immunity can be induced deliberately and safely. Pasteur extended Jenner's work by asserting that an **attenuated** or weakened pathogen is able to stimulate the immune

APPLICATIONS BOX

Polio Vaccine

When the Salk vaccine (killed virus) was first introduced, one of the early batches of vaccine prepared by a California laboratory was not inactivated properly. The virus used for vaccine preparation had been the normal live pathogen, and consequently the improperly prepared vaccine caused a large number of cases of paralytic polio. Other batches of the vaccine from different manufacturers caused no particular problems. Therefore, the problem was identified to be in the manufacturing phase rather than in the design of the process. The improperly prepared vaccine was recalled, and the immunization programs continued across the United States.

Albert Sabin later prepared an attenuated live viral vaccine that is the one in current use. Occasionally, it too causes a rare case of paralytic polio if a recently immunized person transmits the virus through contact with a nonimmunized individual. Presumably, the infected individual has a defective immune system or the virus has lost its attenuation by passage through a human host.

response before succumbing to the body's defenses but without precipitating the disease-causing process. He demonstrated the success of this technique with *Bacillus anthracis,* the causative agent of anthrax, a serious disease of sheep.

Pasteur called his immunization process **vaccination** in honor of the pioneering work of Jenner. The organism or derivative of it used in the process is referred to as a vaccine. Similarly, the virus used for smallpox immunizations is designated as the vaccinia virus (see chapter 9), although there is some dispute as to whether it is a separate virus or a highly attenuated form of variola.

Generally, it is thought that the best immune response results from exposure to a living but attenuated microorganism. Microbial growth under suboptimum conditions or culture storage for a period of time results in attenuation. Pasteur had success in attenuating *B. anthracis* by growing it at 42 to 43° C, a temperature nearly outside the growth range of that mesophilic bacterium. Sometimes many passages (cycles of growth) on laboratory medium will suffice to reduce the pathogenicity of a bacterium. In the case of the rabies virus, Pasteur found that aging the viral extract for several weeks is sufficient to prevent it from causing the disease.

Potential Problems with Vaccines

Naturally there are some risks to the use of live microorganisms for immunization. If the attenuation process has been improperly carried out, the disease state may result. Then too, there is always the risk of a spontaneous mutation occurring to restore the virulence of the immunizing organism. For these reasons, killed vaccines also have been prepared. The main problem in preparing a killed vaccine is to destroy the pathogenicity of the organism or virus without damaging its antigenicity. Thus, certain treatments that denature proteins (for example, heat sterilization) are unsuitable for vaccine preparation. One commonly used method is to treat a culture with formaldehyde or glutaraldehyde, both of which cross-link proteins and render them enzymatically inactive although the proteins do retain their natural shape. Killed vaccines are superior to live vaccines because of their storage properties. In the final analysis, the decision to use a live or killed vaccine depends on the relative antigenicities of the preparations, their potential for side effects, and the available storage conditions (for example, lack of refrigeration).

Occasionally, immunization results in a metabolic effect that actually potentiates a subsequent infection (see the article by Oehen and coworkers in suggestions for further reading). Both circulating antibodies and T cells have been reported to produce the effect in cases of infection with viruses causing dengue fever, measles, AIDS, and lymphocytic choriomeningitis.

One way to completely sidestep the issue of whether to use a live vaccine or a killed vaccine is with DNA splicing. If the pathogen has one or only a few prominent, stable protein antigens, DNA coding for those antigens can be cloned and used to genetically transform a nonpathogenic organism. If cloned DNA is associated with the appropriate promoter and ribosomal binding site, the transformants can synthesize the pathogen's antigen in large quantities. The antigen can be purified and used to immunize susceptible individuals. The risk associated with such a vaccine is substantially less than that associated with a live vaccine, but it is still not zero. Two vaccines that have been prepared by DNA splicing are a vaccine for animals against hoof and mouth disease and a vaccine for humans against hepatitis B virus.

APPLICATIONS BOX

Vaccine Technology

There is considerable enthusiasm about using new technologies to develop vaccines that have the capacity to produce fewer side effects than those presently being used. One interesting idea is the anti-idiotypic vaccine. Consider that disease-protective antibodies specifically bind to a particular antigenic determinant from the pathogen whose three-dimensional shape is constant. If the Fab portion of the protective antibody (the idiotypic portion of the molecule) is used as an antigen in a laboratory animal, it will induce the production of an antibody (an anti-idiotypic molecule) whose shape is similar to the original antigenic determinant of the pathogen. In theory, if the Fab portion of the anti-idiotypic antibody is injected into an animal, its shape will induce the animal to produce the same antibody whose production would be stimulated by its direct exposure to the pathogen. If successful, an anti-idiotypic vaccine would confer immunity to a disease without the necessity of ever exposing an individual to a pathogen or its products.

TABLE 18.5 Vaccines used routinely in the United States

Vaccine[1]	Antigen Used	Reimmunization[2]
Diphtheria	Diphtheria toxoid	Every 10 years after initial 3 doses
Influenza	Attenuated virus	Yearly
Measles (rubeola)	Attenuated virus	Once during middle school years is suggested
Mumps	Attenuated virus	None
Pertussis (whooping cough)	Attenuated bacteria	Not after age 7
Pneumonia	Capsular polysaccharides	None
Polio	Attenuated virus	None after initial 3 doses
Tetanus	Tetanus toxoid	Every 10 years after initial 3 doses

1. All are recommended for administration to children.

2. Reimmunization times refer to adults over 18 years of age. If no reimmunization time is indicated, adults are considered to be immune.

As is discussed in chapter 17, some organisms are feared more for their toxin production than for the actual damage caused by the invasion of the microorganism. The strict anaerobe *Clostridium tetani* provides a good example. In such an instance, an immunization is prepared against the toxic molecule(s) rather than against the bacterium itself. To inject the toxin itself, however, clearly is very dangerous; therefore, a **toxoid** or inactivated toxin is used instead. The principles for preparing a toxoid are similar to those for preparing a killed vaccine. Formaldehyde often is the agent used in this preparation process. A purified toxoid preparation obviously presents no risk of infection, although a trace of toxic activity possibly might remain in the event of improper preparation.

Some vaccines commercially available in the United States are listed in table 18.5. The Centers for Disease Control coordinate recommendations on their use.

Public Health Programs

The previous discussion focuses on microorganisms and diseases from the point of view of the affected individual. The public health practitioner, although not indifferent to individual suffering, focuses on the population as a whole and considers not only how to cure an afflicted person but how to protect the rest of society from contracting the same disease. The key aspect of public health is to prevent problems before they start rather than to treat them afterward. **Epidemiologists** are public health workers who focus on each epidemic that occurs. They try to identify the source of the epidemic and provide recommendations to the public health community on how to prevent a recurrence.

Identifying the Source of a Disease

In order to identify the nature of an outbreak of a particular disease, the first task that a microbiologist faces is to identify the organism and its source. In this instance, identification implies more than just supplying a genus and species name for the organism. The species must be subdivided into identifiable groups so that it is possible to say with some certainty that either all cases have been caused by the same organism or by more than one subgroup of the same species.

One very common method of characterizing bacteria is by means of their phage sensitivity. For example, staphylococci can be subdivided into groups by their sensitivity to 23 different phages. Only certain phages will lyse a particular *Staphylococcus* organism, based on the lysogenic state of the bacterium. Remember that just as a prophage can provide superinfection immunity, so it can also make the cell immune to infection by closely related viruses. R. C. Lancefield set up a means of categorizing β-hemolytic streptococci based on their reactions with certain antisera (in other words, based on their surface antigens). The groups are designated by the letters A through O. Other possible methods for bacterial identification are described in chapter 20.

Once a disease-causing organism has been identified as the cause of an outbreak, the epidemiologist must then locate the source of the microorganism. In order to do this, it is necessary to obtain a history from each person who has contracted the disease so that its course can be checked for factors in common with other patients. A **case definition** or statement of the symptoms must be prepared to assist an epidemiologist in deciding which individuals have been affected by the disease during this outbreak. This may present no problem if the symptomology is unique, but if the disease is of a more general nature, such as a diarrheal illness, it is almost certain that some cases will be missed and others not really part of the outbreak will be included.

When a common factor(s) has been identified, a statistical test is applied to calculate the probability that the cases attributed to this common factor(s) did not occur by chance. An example of an epidemiologic analysis is shown in figure 18.2. Cases of a disease that result from one particular factor, such as contaminated milk, are considered to be a common source outbreak.

Preventing the Spread of a Disease

Sometimes a certain disease is quite rare at a particular time or in a particular locale, but at other times or in other places it may be endemic or always present at significant levels. The strategy used to prevent the spread of endemic diseases is slightly different from that used to prevent the spread of epidemics. Stopping an epidemic requires treatment procedures like those discussed in chapter 17 and earlier in this chapter. Prevention of endemic diseases is based on the principles of cost-benefit analysis. In the logic behind the analysis, there is no medical treatment that is totally 100% without risk. No matter how careful a practitioner is or how cautious a pharmaceutical manufacturer may be, a few people always will experience side effects from a particular procedure. Although it is not usually possible to identify these individuals, it is possible to estimate their number statistically. One significant problem for health professionals is that the present tort system used for civil lawsuits does not always recognize the possibility that preventive measures prepared and administered carefully may nevertheless cause harm in some cases.

To take a specific example, years of experience in administering smallpox vaccinations have revealed that for every 1 million primary (first-time) vaccinations given, roughly 1 person was expected to die, another 10 to have serious side effects and require hospitalization, and another 1,000 to experience various skin eruptions beyond the normal localized reaction. The risk from the disease itself varied with the particular infecting strain. Variola major was the classic smallpox virus and normally had an average case:fatality rate of 15 to 45%. Most of the time, smallpox was caused by a strain of variola major that had a case:fatality rate of less than 10%, but occasional mutants had a case:fatality rate of more than 90%. Variola minor was a variant form that occurred in Africa and had a case: fatality rate of less than 1%.

Because smallpox was a disease that could be found worldwide at one time, there was a substantial risk of death to any given individual at any time, far greater than the risk of dying from vaccination. India reported over 1 million cases of smallpox (231,000 deaths) in an epidemic in 1944. The United States had a small epidemic of 49,000 cases (173 deaths) in 1930. Because of the individual risk, all countries that were in a position to do so required smallpox vaccinations as a way of protecting the populace.

In contrast, plague is a very low-level, endemic bacterial disease in the United States. In any given year, there are only 10 to 20 cases in the entire country. Even though a good vaccine is available, it is not administered to the public at large because the risk from the vaccine is considered to be greater than the risk from the disease itself.

The general public health policy becomes one of evaluating the incidence of the disease and the efficacy of the vaccine. If disease transmission can be halted by simple means short of vaccination, that is the appropriate way to proceed. Naturally, if specific segments of a population are at particular risk of contracting a disease due to occupational hazards, for example, these individuals would receive immunization regardless. The plague vaccine is routinely administered to military personnel but not to the civilian population.

Epidemiologic Notes and Reports

Salmonella heidelberg Outbreak at a Convention—New Mexico

Of approximately 1,000 persons attending a convention October 6-8, 1985, in Santa Fe, New Mexico, 91 reported a diarrheal illness with onset of symptoms between 10 A.M., October 7, and 11P.M., October 12. *Salmonella heidelberg*, sensitive to all antibiotics tested, was isolated from the stools of five attendees. Three persons were hospitalized. The ill attendees reported spending over $11,000 on medical costs and lost 117 days of work.

A telephone survey of 76 convention attendees living in New Mexico showed that, of four meals consumed at the convention, only the breakfast of October 7 was significantly associated with illness ($p < 0.002$).

In a subsequent mail survey of the approximately 550 convention attendees who ate the breakfast, the only food significantly associated with illness among the 60% who responded was eggs. All of 91 ill attendees ate the eggs, compared with 189 (92%) of 206 well attendees ($p = 0.01$). Eggs served at the meal were not available for culture; other eggs from the same distributor were culture-negative for *Salmonella*. The eggs had been cracked and stored in tall 2-gallon containers in a walk-in refrigerator the evening before the breakfast. They were then cooked in batches in a steamer in the morning. Several attendees commented that the eggs seemed "runny."

Of the staff who worked at the breakfast, three reported illness compatible with salmonellosis with onset during the same period as the conventioneers, and all three had eaten the eggs. *S. heidelberg* was isolated from the stools of two staff members who did not handle food but had eaten the eggs.

Reported by P Weisse, E Libbey, MD, St. Vincent's Hospital, Santa Fe, L Nims, MS, P Gutierrez, MS, New Mexico Scientific Laboratory Div, T Madrid, MPA, N Weber, MS, C Voorhees, V Crocco, C Hules, S Hill, Environmental Improvement Div, TM Ray, R Gurule, F Ortiz, District II Health Office, M Eidson, DVM, CM Sewell, DrPH, S Castle, MPH, P Hayes, Office of Epidemiology, HF Hull, MD, State Epidemiologist, New Mexico Health and Environment Dept; Div of Field Svcs, Epidemiology Program Office, CDC.

Editorial Note: In the 1960s, eggs were responsible for a large proportion of salmonellosis outbreaks. With improvements in egg processing and quality control, egg-related outbreaks decreased dramatically in the 1970s *(1)*. However, as this outbreak illustrates, egg-related illness remains an important public health concern. Pathogens may proliferate in eggs or in other food refrigerated in large containers, since the center of the container may be inadequately cooled *(2)*. In this outbreak, the fact that many well attendees also ate eggs suggests that only some egg containers were contaminated, that only some eggs were cooked sufficiently to kill the bacteria, or that susceptibility to infection may have varied among the attendees.

For the 10-year period 1973–1982, 11 outbreaks of salmonellosis due to eggs were reported to CDC's Foodborne Disease Surveillance System. Of the 307 ill people in these outbreaks, 45 (15%) were hospitalized, and nine (3%) died *(3)*. *S. heidelberg* has been frequently associated with poultry, accounting for 29% of *Salmonella* isolates from poultry submitted to the U.S. Department of Agriculture in 1982 *(4)*.

References

1. Cohen ML, Blake PA. Trends in foodborne salmonellosis outbreaks: 1963-1975. J Food Protection 1977; 40: 798-800.
2. Bryan FL. Foodborne diseases in the United States associated with meat and poultry. J Food Protection 1980; 43: 140-50.
3. CDC. Unpublished data.
4. CDC. 1982 *Salmonella* surveillance. Atlanta, Georgia: Centers for Disease Control.

FIGURE 18.2

Epidemiologic analysis of the outbreak of a disease.

Source: As appeared in *Morbidity and Mortality Weekly Report* 35:91, 1986.

When immunization is considered inappropriate, waterborne diseases can be eliminated by proper sewage and water treatment (covered in chapter 12). Food-borne diseases also can be prevented by proper preservation and storage techniques (outlined in chapter 14). If a disease is spread by a vector, it may be possible to eliminate the vector.

Environmental monitoring often is helpful to the epidemiologist in verifying that a disease outbreak is unlikely. For example, traps can be set to check for the presence of a specific vector such as a mosquito. In the case of waterborne diseases, other measures can be employed.

To take a specific example, cholera is spread by fecal contamination of water, generally as a result of absent or inadequate sewage treatment. Outbreaks of cholera often are associated with the consumption of shellfish because shellfish take in large quantities of water, retaining particulate material as food and discarding the water. Unfortunately, this filter feeding process results in the accumulation of *Vibrio cholerae* in shellfish, and the bacteria are not rapidly destroyed. Because shellfish often are eaten raw, the unwary consumer can sometimes receive a large inoculum of bacteria in a single meal.

The size of the inoculum is of some significance because clinical disease may not result from from a minimal exposure. Not all organisms are equally infectious. For instance, an inoculum of 180 cells of fully virulent *Shigella,* the organism that causes dysentery, is infectious for humans, but a dose of about 10^8 cells is required to initiate a *V. cholerae* infection.

In the United States, careful enforcement of water treatment standards has helped to prevent the spread of cholera and other diseases resulting from fecal contamination. In areas in which fecal contamination is reported, a ban on harvesting shellfish usually is imposed. Even so, scattered cases of cholera are reported in the United States each year, often along the Gulf or Atlantic coasts, and similar patterns of disease transmission are seen in the case of the other diarrheal diseases. Other organisms can survive for several weeks in the water supply even if they do not grow significantly; therefore, they can extend the period of risk following fecal contamination. The obvious precaution to take with respect to a water supply is to check it regularly for fecal contamination.

A potential problem in a water supply can be with organisms present at a comparatively low level, perhaps only 1,000 per milliliter, that are capable of causing a disease particularly in individuals who have other medical problems. Water-quality testers often check for an **indicator organism,** one usually abundant in fecal material and easy to culture in the laboratory but not necessarily a disease-causing organism. Presence of significant numbers of an indicator organism means that potentially dangerous levels of fecal contamination are present in the supply. The two most commonly used indicator organisms are *Escherichia coli* in the United States and *Streptococcus faecalis* (also known as *Enterococcus faecalis*) in Europe. Actual determination of the quality of drinking water is based on the number of cells morphologically and biochemically resembling *E. coli* present in the sample. The number thus obtained is called the **coliform count.** Various selective and differential media, such as MacConkey agar or eosin methylene blue agar, are used to select for coliforms present in water. Figure 18.3 illustrates a water-quality testing procedure.

When a disease can be spread by direct contact between humans or when the disease has a reservoir of significant size in close physical proximity to a human population, other measures are necessary. Initially, a major effort is undertaken to immunize the general population. This effort, of course, requires the production of an effective vaccine. Vaccine production is not necessarily straightforward. There may be no good way to attenuate an organism, or there may be many different strains in the microbial population so that it is difficult to decide which ones are appropriate to use. The pneumonia vaccine contains polysaccharide capsular antigens from 23 different strains of *Streptococcus pneumoniae.* The yearly flu vaccine generally provides immunity against the three most prevalent forms of influenza observed in the preceding year.

The customary means of implementing a program to immunize the general population with a new vaccine is to encourage the passage of laws requiring that school children be immunized before they begin classes. Such laws must of necessity allow for exemption of persons whose religious beliefs are against immunization or whose immune systems are too deficient to respond properly to a vaccine. Members of the latter group can be severely endangered if they are immunized, particularly if the vaccine is based on an attenuated microorganism. Even when exemptions are allowed, disease transmission within the school system is reduced significantly due to the phenomenon of herd immunity.

The term **herd immunity** pertains to a situation in which the vast majority of a population is immune to a particular pathogen, but some susceptible individuals are still present. As long as susceptible individuals are

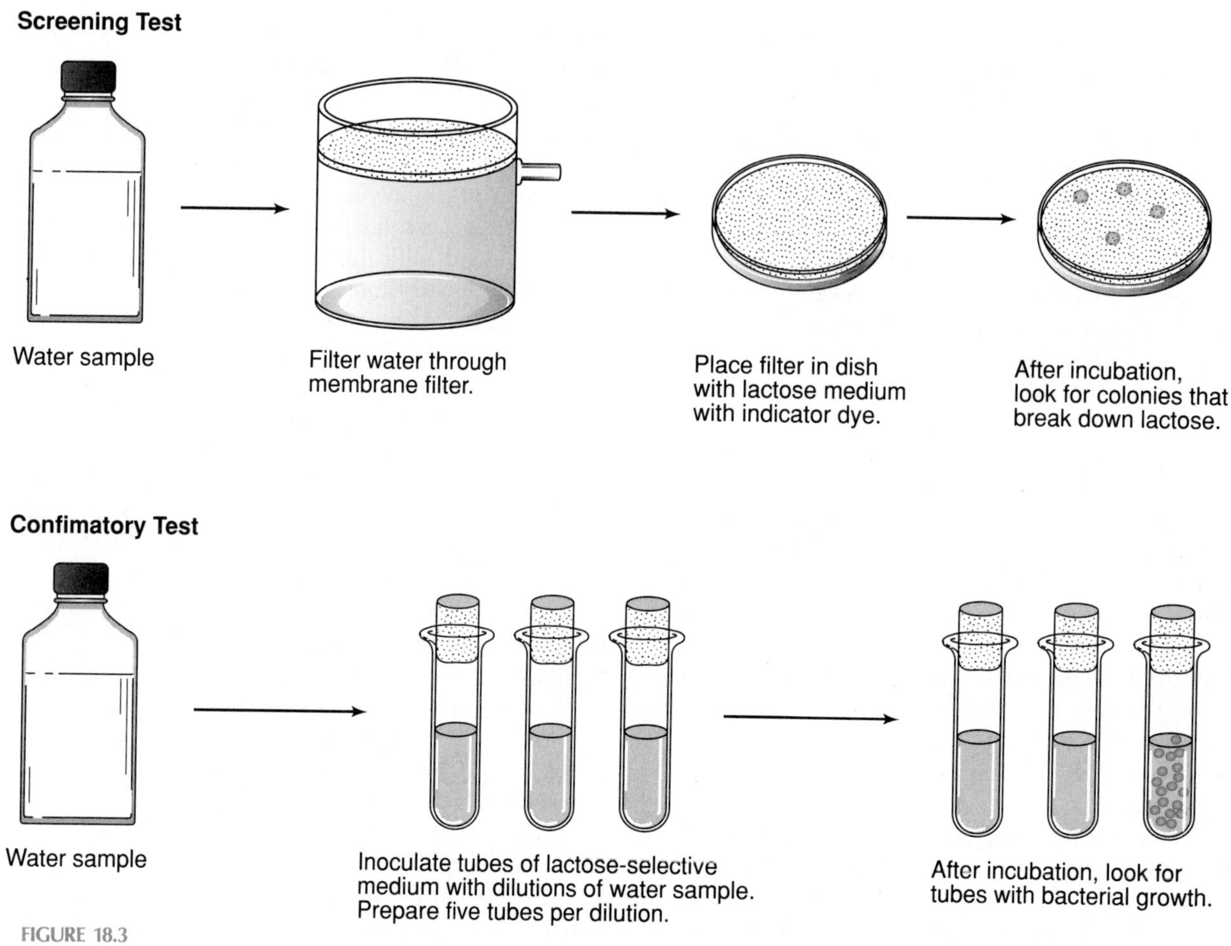

FIGURE 18.3

Water-quality testing procedure. The indicator organism, *Escherichia coli,* will ferment lactose to acid and gas and can grow on a selective medium containing bile salts. The initial test is performed using a membrane filter with pores small enough to retain bacteria from the water that is filtered through it. If the colonies that grow on the filter resemble *E. coli* after the filter is placed on the lactose medium, a confirmatory test is necessary. Various dilutions of the water sample are added to tubes of selective lactose medium. If growth is observed in the tubes, the presence of coliforms in the water sample is confirmed. Statistical tables are available that allow scientists to estimate the most probable number (MPN) of coliforms in the original water sample based on the number of tubes that have growth and the original dilutions prepared.

dispersed randomly throughout the population, they are surrounded by members of the "herd" who are immune. Thus, their risk of infection is greatly lowered, and the herd effect provides them with the same protection as an immunization does. Immune individuals will not transmit a disease to susceptible ones, although those who have been immunized recently might be "shedding" the attenuated organism. Obviously, the critical factor is the percentage of the population that must be immune in order to ensure an operative herd immunity. Most estimates place this value between 70 and 80%, which leads to an interesting sociological problem. A few isolated people who decide not to be immunized have little effect on the population or themselves. If, however, the majority of the population decides not to be immunized, a disease can be expected to occur at its former frequency.

At the present time, measles (rubeola) is a disease for which immunization is being encouraged with a view to eradicating the disease completely. An effective vaccine has been available for the past 20 years, and much of the population is now immune without ever having contracted the disease. Although the incidence of measles in the United States had plummeted, it is now rising again (see table 18.2 and figure 18.4). Scattered outbreaks of the disease still occur, and particular attention is focused on the school-age population that seems to harbor the bulk of susceptible individuals at this time.

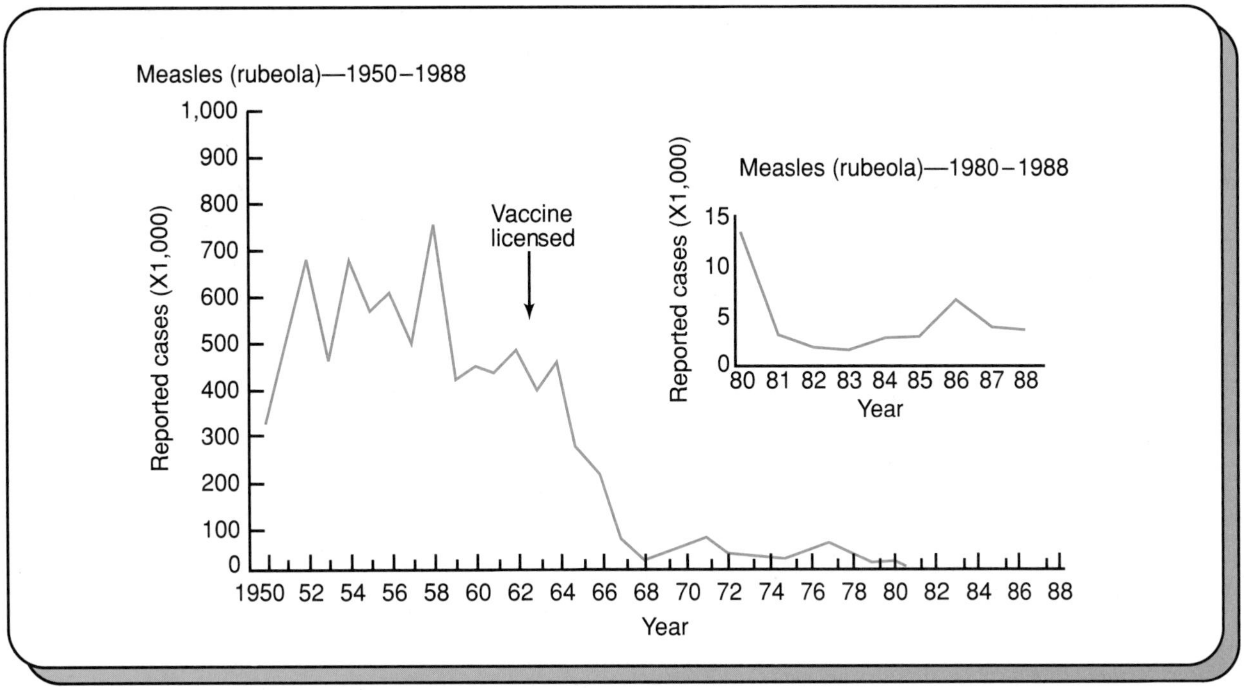

FIGURE 18.4
Incidence of measles (rubeola) in the United States.
Source: *Morbidity and Mortality Weekly Report* 37(54):28.

The immunity conferred by childhood immunization does not seem to confer protection as long as had been expected; therefore, a second immunization has been recommended during the teenage years. The Centers for Disease Control had as their goal the reduction of measles to the status of a rare disease by 1990, but this has not been achieved.

Even if a population is essentially free of a disease such as measles, there is always the risk that it may be imported by someone traveling from a foreign country. Many outbreaks of measles at the present time can be traced to foreign sources. Once again, smallpox provides a good example of what is likely to happen in the future. At the time when smallpox had essentially been eliminated from the United States population but was still present in certain countries (1960s), the risk to a nontraveler became very low. The immunization requirement was changed so that persons staying at home were not required to be vaccinated, but anyone traveling to a country reporting cases of smallpox had to have proof of immunization in order to reenter the United States. It is likely that immunization with the measles vaccine will be required under these circumstances as well.

When susceptible individuals congregate together, as in the case of a religious community that refuses vaccination or an economically impoverished area that cannot afford it, herd immunity is not a consideration. If a disease happens to be introduced into such a population, it will spread rapidly. Immediate immunization is the solution, preferably of the entire community; however, if that is not possible, all people in the immediately surrounding area should be immunized and the infected community quarantined. If the quarantined area is bounded by an immune population, the disease cannot spread further and will eventually die out.

Public health officials employ a similar tactic with a serious outbreak of a disease to which people are not immune. In the event of an exposure to plague (for example, of hospital personnel), all persons exposed to the infected patient are immunized and/or treated with antibiotics as soon as the diagnosis has been confirmed. When the incubation period is lengthy, prompt immunization coupled with close scrutiny for overt symptoms of infection generally prevents any transmission beyond the initial case.

APPLICATIONS BOX

The Swine Flu Scare

The decision to embark on a large-scale immunization campaign is not a trivial one because it is a virtual certainty that potentially hazardous side effects will occur. A particularly good example of what can go wrong can be illustrated by the swine flu immunization program for 1976. At that time, it was feared that a major antigenic shift had occurred in the influenza virus that would leave large numbers of persons susceptible to the disease and generate a larger than usual influenza epidemic. A new vaccine formulation was prepared and heavily publicized. After the program was well underway, there were numerous reports of Reye syndrome, a rare paralysis of uncertain origin occurring in people who had been administered the flu vaccine. A **syndrome** is a unique collection of nonspecific symptoms that when taken together constitute a recognizable disease state.

Although no statistical correlation could be established between the flu vaccine and occurrence of Reye syndrome, the immunization program was greatly reduced in scope because of the publicity and resultant outcry. When the anticipated epidemic did not occur, additional recriminations were forthcoming. A further account of this affair can be found in the book by Neustadt and Fineberg listed in the suggestions for further reading at the end of this chapter.

When the time lag between exposure to a pathogen and the appearance of the symptoms is quite short, there may not be time for the primary response to occur after vaccination. In some instances, there may be no vaccine available. In either of these situations, it may be better to use preformed antibodies that provide passive immunity. Passive immunization is most often used when symptoms are caused by an exotoxin. The antibody preparation administered to the threatened population is called an **antitoxin.** If an animal immunization procedure is available, serum obtained from an immunized animal can be used. Otherwise, pooled human immunoglobulins (γ globulin) can be employed. Immunoglobulins injected to provide the immunity will last for several weeks before being destroyed by a recipient's immune system. Active immunization cannot be attempted until the passively acquired antibodies have been removed.

There are risks to the frequent use of passive immunity. A recipient normally is stimulated to produce antibodies against the constant regions of injected IgG molecules. Reexposure to the same type of IgG usually gives an allergic reaction known as serum sickness.

Smallpox, an Example of Public Health in Action

Smallpox has been a serious scourge throughout the world for all of recorded history. Nevertheless, it has now been completely eradicated from the natural population. Eradication is possible only in certain special circumstances in which the disease reservoirs also can be treated. In the case of smallpox, this was relatively easy because humans are the only known hosts. Hence, if there are no human cases, there is no disease.

WHO is charged with coordinating reports and activities of various national public health agencies. By the 1960s, it had become apparent that the worldwide incidence of smallpox was at an all-time low, and the disease was concentrated primarily in underdeveloped countries. WHO proposed a campaign to eradicate smallpox with a concerted effort to immunize everyone in all areas in which the disease was reported. The campaign would be funded primarily by industrialized nations. After the plan was adopted by the member nations, public health teams were trained, equipped, and sent out into the field to immunize entire populations that were still threatened.

Goodfield's article, referenced at the end of this chapter, chronicles the activities of the public health teams in Bangladesh, the last country on earth that reported cases of smallpox. Briefly, the strategy was to examine every person in the country in order to identify cases of smallpox. Whenever a case was found, all immediate family and neighbors were immunized at once. It was found that 50% immunity in this population gave sufficient herd immunity to prevent transmission of the disease.

Public health workers had to fight not only the disease but also the apathy and fears of the people to whom vaccination was a mysterious rite. As the number of cases diminished, substantial monetary rewards were offered to anyone who reported a confirmed case of smallpox. Finally, a period of time greater than the incubation period of the disease elapsed with no additional cases reported. It was assumed that the disease was eradicated, but global surveillance was maintained for another year. When no new cases appeared, WHO declared smallpox completely eradicated, and vacci-

APPLICATIONS BOX

AIDS, A Disease Out of Control

Unlike the rest of the diseases discussed in this chapter, acquired immune deficiency syndrome (AIDS) is an epidemiologist's nightmare. The virus is spread primarily by sexual contact or intravenous drug use, and many people have resisted behavioral modifications that would provide a greater margin of safety. As a consequence, the worldwide numbers of AIDS cases increases yearly with no signs of leveling off. Furthermore, to this date, there is no effective treatment once symptoms appear; therefore, the number of fatalities also is increasing steadily. Until the chain of disease transmission is interrupted by behavioral changes and/or by discovery of an effective treatment, there is little likelihood of any change in the increasing incidence of AIDS worldwide.

nation requirements were dropped by most countries. There have been a few scattered reports of smallpox since then, but careful examination has always revealed a misdiagnosis, usually of chicken pox. Some stocks of the virus are still maintained. In the event that the disease were to start up again, it would be possible to reproduce the vaccine from these stocks.

Summary

Public health problems are somewhat different from those of individual health. The focus is on what is good for the majority of society, and it is recognized that all medical treatments entail some risk to the individual. Diseases are categorized in two ways: by their modes of transmission and by the availability of vaccines to immunize against them. Transmission of a disease can be interrupted by good levels of immunity (active or passive) within a population or by careful attention to activities like proper purification of water and preparation of food. The history of smallpox provides the only known example of a disease that went from being a major problem to being eradicated completely.

A disease can be present at a characteristic endemic level within a society. Significant local increases in numbers of cases from this level constitute an epidemic. A worldwide epidemic is a pandemic. Diseases can be spread by direct contact between individuals, by aerosols, by contaminated food or water, by vectors, and by fomites. An animal reservoir is a nonhuman species that can sustain a microbial infection and later transmit it to humans. A lack of susceptible individuals (induced immunity), quarantining of susceptible individuals, or antimicrobial therapy can halt an epidemic.

Antimicrobial agents are extremely powerful tools for the treatment of disease, but they can be subjected to misuse. Overuse of antibiotics or failure to achieve therapeutic levels during treatment often results in development of resistant organisms within affected individuals. If resistance is due to the presence of an R plasmid, the problem is particularly acute because many of those plasmids are also conjugative and can move from one cell to another.

Questions for Review and Discussion

1. What is cost-benefit analysis as applied to public health? How does the principle of herd immunity enter into the calculation?
2. Why is it that some immunizations need to be repeated at regular intervals while others do not? Why is it that the average citizen does not receive all possible immunizations but only a selected few?
3. What would be the possible responses of public health officials to an epidemic of a mild disease like tonsillitis? How might the responses be different if the disease were more serious, like plague?
4. What is the appropriate way to administer an antibiotic? What can go wrong if antibiotic use is not carefully monitored? Why does administration of an antibiotic apply a selective pressure to a bacterial population?

Suggestions for Further Reading

Allman, W. F. 1984. Drugs in feed: Fatter cattle, fitter bacteria. *Science* 84(10):16.

Behbehani, A. M. 1983. The smallpox story: Life and death of an old disease. *Microbiological Reviews* 47:455–509.

Bittle, J. L., and F. L. Murphy (eds.). 1989. *Vaccine biotechnology.* Academic Press, Inc., San Diego. Additional details about new approaches to vaccine technology can be found in this text.

Cliff, A., and P. Haggett. 1984. Island epidemics. *Scientific American* 250(5) 138–147. A description of the spreading patterns of measles epidemics in Iceland.

Fuller, J. G. 1974. *Fever! The hunt for a new killer virus.* E. P. Dutton, New York. A very exciting retelling of a major epidemiologic problem involving an African hemorrhagic fever virus that was accidentally imported into the United States.

Germanier, R. (ed.). 1984. *Bacterial vaccines.* Academic Press, Inc., Orlando, Fla. A good source of specific details on the vaccines most commonly used in the United States.

Ginsberg, H., F. Brown, R. A. Lerner, and R. M. Chanock. 1988. *Vaccines 88. New chemical and genetic approaches to vaccination: Prevention of AIDS and other viral, bacterial, and parasitic diseases.* Cold Spring Harbor Laboratory, Cold Spring Harbor, N.Y. A collection of technical papers with emphasis on AIDS and malaria.

Goodfield, J. 1985. The last days of smallpox. *Science* 85(8):58–66. An account of the efforts to eradicate smallpox in Bangladesh.

Holmberg, S. D., J. G. Wells, and M. L. Cohen. 1984. Animal-to-man transmission of antimicrobial-resistant *Salmonella:* Investigations of U.S. outbreaks, 1971–1983. *Science* 225:833–835.

Neustadt, R. E., and H. Fineberg. 1983. *The epidemic that never was. Policy-making and the swine flu affair.* Random House, New York. An account of the decision-making process that includes commentary by the officials themselves.

Oehen, S., H. Hengartner, and R. M. Zinkernagel. 1991. Vaccination for disease. *Science* 251:195–198. This article discusses a possible role for T cells in potentiating subsequent infection.

Reynolds, C. W., and R. H. Wiltrout (eds.). 1989. *Functions of the natural immune system.* Plenum Publishing Corp., New York. Chapters 3 through 8 concern the control of microbial infections.

Smith, H. W., and M. B. Huggins. 1983. Effectiveness of phages in treating experimental *Escherichia coli* diarrhea in calves, piglets and lambs. *Journal of General Microbiology* 129:2659–2675.

chapter

10

Some Dread (and Not So Dread) Diseases

chapter outline

■

GENERAL GOALS

Be able to describe some important diseases, how they are transmitted, and how their modes of transmission can be interrupted.

Be able to provide examples of public health programs that have aided communities in achieving full control of a target disease and programs that have yet to achieve their goals.

This chapter builds on the principles underlying epidemiology and disease control discussed in the previous chapter. Specific examples are presented to illustrate the mechanisms used to apply those principles to control some diseases common within the United States. The reasons why a few diseases are rare in this country but common elsewhere also are considered. Although some attention is given within this chapter to methods for treating diseases, major emphasis is on the epidemiology and public health aspects of these diseases. Because a single organism can sometimes cause several different kinds of problems in various parts of the body, all manifestations of each organism are discussed as a unit. Diseases are not considered and discussed according to the organ system(s) affected, as generally is the case in most microbiology texts. Table 19.1 summarizes the diseases covered in this chapter. Their etiologic agents and the specific organ systems that these diseases affect also are listed in the table. Statistics on the frequencies of the diseases presented throughout the chapter are taken from various publications of the Centers for Disease Control (CDC).

Bacterial Diseases

The bacterial diseases discussed in this section are alphabetized according to the name of the species that causes the disease.

Bordetella pertussis. Pertussis, better known as whooping cough, is primarily a disease of young children. The infection begins in the upper respiratory tract with symptoms resembling a cold. As the *Bordetella* bacteria grow along the surface of the epithelium lining the throat, the infection moves toward the lower respiratory tract and the bronchi. At this stage of the disease, some 3 weeks after the initial infection, a severe cough develops. Coughing can occur paroxysmally with a distinctive whoop that signifies a gasp for breath between coughs. If the infection continues down into the bronchi, mucous plugs can develop to obstruct the airway.

Mild cases of pertussis require no particular treatment. More severe cases usually require antibiotic therapy against *B. pertussis* and possibly against secondary invaders of the lungs that can give rise to pneumonia (fluid in the lungs). Any toxins produced by the bacteria have only local effects and generally are not troublesome.

Pertussis is of concern not so much because every case is serious but because of its very high attack rate. In a group of susceptible children, the attack rate can be as high as 90%, with significant mortality among infants. Adults, on the other hand, are hardly affected. For these reasons, early immunization (generally at 2 months of age) is recommended. By the age of 7 a child is usually immune to the disease by exposure if not by vaccination, and no further immunizations are required.

Immunizations are given as a DPT shot (diphtheria, pertussis, tetanus) that has come under severe criticism in recent years. The problem is that children experience significantly greater side effects from the pertussis vaccine than from most other vaccines in common use (about one in 140,000). Pertussis vaccine seems to cause convulsions in some children, and parent protests have led to an impressive drop in pertussis immunization as well as to a significant increase in the observed number of cases of pertussis in the United States. The median number of pertussis cases for the years 1984–1988 was 3,379, but the number of cases in 1990 was 4,188, a generally upward trend. If this increase continues, pertussis may once again become a serious disease in certain groups of children. In a study conducted during 1984–1985, hospitalization was required in 41% of all cases and in 74% of all cases under the age of 6 months.

In an attempt to reduce the number of serious side effects from the pertussis vaccine, recommendations for administration of the vaccine have been reassessed. **Contraindications** (signs that the vaccine should not be administered) now include any history of convulsions in a child. The hope is to provide a good herd immunity in which unimmunized children can be protected from contracting pertussis within a given population. In addition, substantial efforts are being devoted to the development of a genetically engineered vaccine that might be able to reduce some of the side effects. No success in this effort has been reported to date, however.

Borrelia burgdorferi. This spirochete that causes Lyme disease or Lyme borreliosis is named after the town in Connecticut in which it was first identified. The disease is transmitted by the bite of deer ticks of the genus *Ixodes,* in which it sets up a systemic infection, and probably is passed to humans and other animals through tick saliva. It appears that ticks are able to transmit the infection to their offspring through their eggs. In humans, a systemic infection results that can affect a variety of organ systems including the central nervous system, heart, and liver. Large numbers of bacteria are never present. Basic symptoms are fever, headache, muscle pain, a circular rash, and sometimes

Table 19.1. Diseases discussed in this chapter	
Disease	**Etiologic Agent**[1]
Diarrheal Diseases	
Cholera	*Vibrio cholerae*
Dysentery	*Shigella dysenteriae*
Giardiasis	*Giardia lamblia* (P)
Other Intestinal Infections	
Infant botulism	*Clostridium botulinum*
Pseudomembranous colitis	*Clostridium difficile*
Exotoxin-Related Diseases	
food poisonings	
Nonspecific food poisonings	*Clostridium perfringens, Staphylococcus aureus*
Aflatoxicosis	*Aspergillus flavus* (F)
Botulism	*Clostridium botulinum*
Cholera	*Vibrio cholerae*
Ergotism	*Claviceps purpurea* (F)
nonspecific infections	
Diphtheria	*Corynebacterium diphtheriae*
Scarlet fever	*Streptococcus pyogenes*
Tetanus	*Clostridium tetani*
Toxic shock syndrome	*Staphylococcus aureus*
Nervous System Infections	
Creutzfeld-Jakob disease	Prion (unconventional virus)
Meningitis	*Haemophilus influenzae, Neisseria meningitidis*
Polio	Polio virus
Rabies	Rabies virus
Skin Eruptions	
Boils	*Staphylococcus aureus*
Candidiasis	*Candida albicans* (F)
Chicken pox, shingles	Herpes zoster virus
Cold sores	Herpes simplex virus, type 1
Leprosy	*Mycobacterium leprae*
Measles (rubeola)	Measles virus
Rubella	Rubella virus
Respiratory Tract Infections	
upper respiratory tract	
Common cold	Rhinovirus
Infectious mononucleosis	Epstein-Barr virus
Influenza	Influenza virus
Pertussis (whooping cough)	*Bordetella pertussis*
lower respiratory tract	
Legionellosis	*Legionella pneumophila*
Pneumonic plague	*Yersinia pestis*
Tuberculosis	*Mycobacterium tuberculosis*
Valley fevers	*Histoplasma capsulatum* (F), *Coccidioides immitis* (F)

Continued on next page

fatigue and weight loss. Early treatment with tetracycline or penicillin provides a cure, while delayed treatment often is ineffective because many symptoms appear to result from proinflammatory molecules released by the host in response to the infection.

State health departments in conjunction with the CDC have not yet required local physicians to report the occurrence of Lyme disease; therefore, exact statistics are difficult to find. In 1988 there were 4,572 cases reported to the CDC, but this is probably an

Table 19.1. Continued

Systemic Infections	
AIDS	Human immunodeficiency virus
Bubonic plague	*Yersinia pestis*
Gas gangrene	*Clostridium perfringens*
Hepatitis	Hepatitis viruses
Lyme disease	*Borrelia burgdorferi*
Malaria	*Plasmodium* (various species) (P)
Rheumatic fever, puerperal fever, acute glomerulonephritis	*Streptococcus pyogenes*
Rocky Mountain spotted fever	*Rickettsia rickettsiae*
Typhoid fever	*Salmonella typhi*
Typhus	*Rickettsia prowazekii*
Sexually Transmitted Diseases	
AIDS	Human immunodeficiency virus
Genital herpes	Herpes virus
Gonorrhea	*Neisseria gonorrhoeae*
Syphilis	*Treponema pallidum*
Vector-Transmitted Diseases	
Bubonic plague	*Yersinia pestis*
Lyme disease	*Borrelia burgdorferi*
Malaria	*Plasmodium* (various species) (P)
Rocky Mountain spotted fever	*Rickettsia rickettsiae*
Typhus	*Rickettsia prowazekii*

1. A letter following an agent indicates the type of microorganism: protozoan (P) and fungus (F). Agents not otherwise designated are bacteria.

APPLICATIONS BOX

Compensation for Vaccine Side Effects

Prior to 1988, persons suffering severe reactions to immunizations had to either finance their medical care themselves or sue a manufacturer for negligence. As discussed in chapter 18, properly manufactured vaccine will have significant side effects in a small number of cases. The Centers for Disease Control estimate that approximately 20 million children are immunized against various diseases in any given year. About 500 of these children have serious reactions to the vaccine, and about 75 die. In 1988 the Congress established a National Vaccine Injury Compensation Program financed by a surcharge on every vaccine dose (for example, $4.56 for DPT, $0.29 for polio). Any person whose side effects from an immunization meet specific criteria can receive compensation determined by the U.S. claims court judge.

underestimation. Cases have been reported worldwide, and the disease has been known in Europe for decades under a variety of names. In the United States, outbreaks have been reported from all states but primarily from New Jersey to Massachusetts, Wisconsin, and Minnesota (figure 19.1A). There is no vaccine available at the present time, and prevention consists of protective clothing and insect repellent to ward off tick bites. Figure 19.1B shows a sample information card being distributed in Wisconsin to encourage people to take proper precautions while in endemic areas.

Clostridium botulinum. Botulism in adults results from the ingestion of a neurotoxin produced by the *C. botulinum* bacterium. In the normal course of events, the organism causes no bacterial infection because it is incapable of establishing itself in the adult digestive tract. During processing, food can be contaminated by dirt containing *C. botulinum* endospores. If foods are improperly canned, the endospores can germinate to form a bacterial culture. Comparatively little growth is required in order for the bacteria to produce lethal amounts of toxin; therefore, the taste and/or smell of

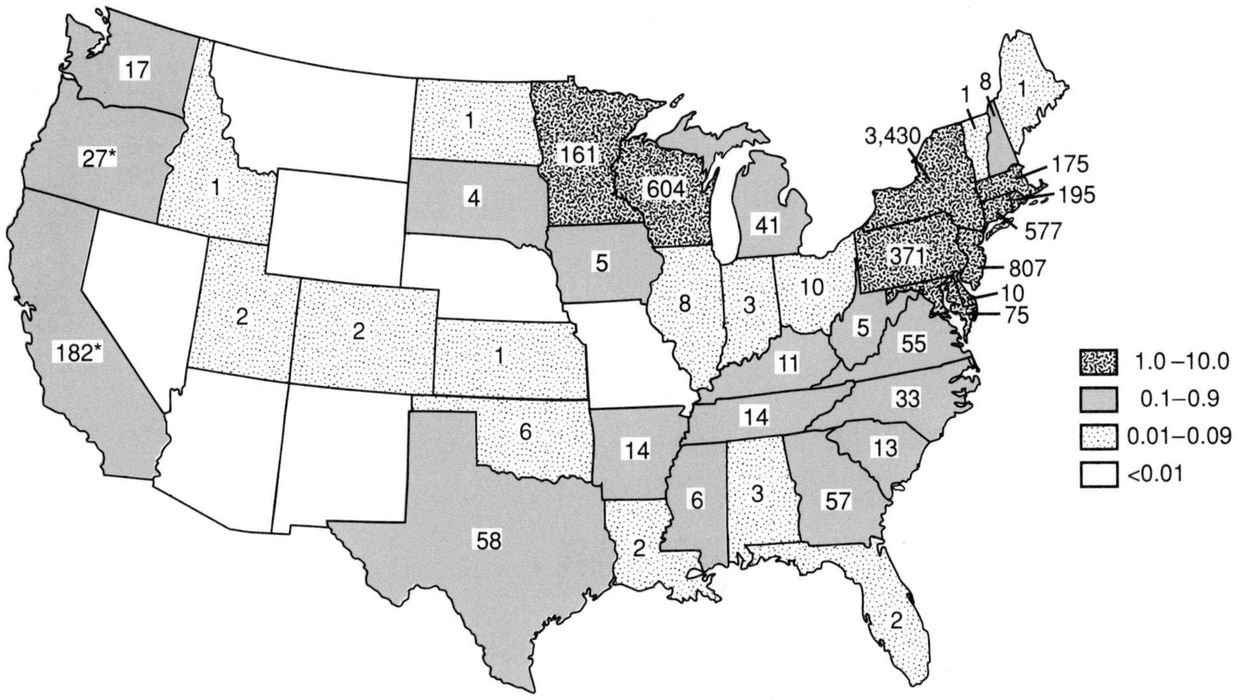

*Data for Oregon and California are for 1987 only.

A

Continued on next page

FIGURE 19.1

Lyme disease. (**A**) Map showing the number and average incidence rates of Lyme disease cases per 100,000 persons reported for the years 1987–1988.

(A) Source: *Morbidity and Mortality Weekly Report* 38:669, 1989.

the food is not necessarily altered in the process. The bacteria's ability to produce the toxin is sometimes the result of a prophage, making botulism an example of lysogenic conversion. Because some speciation of the genus *Clostridium* is based on toxin production, changing the prophage present in a cell may change the species to which the bacterium is assigned.

For the most part, commercially canned products present no problems with respect to botulism. The number of cans of food sold by the average supermarket in the course of a year attests to the safety of commercially prepared canned goods. The same, unfortunately, is not necessarily true for home canned products. When prepared properly, canned foods are sterile or have such a low pH that *C. botulinum* cannot grow. Home canners, however, may not follow recipe directions strictly when cooking and sealing foods to be canned. If proper sterilization temperatures are not achieved, it is quite likely that clostridia will grow.

In order to protect the home canner, recipes always recommend that canned foods be boiled for 10 minutes before serving because the botulism toxin is **heat labile** (has its activity destroyed by high temperatures). The final cooking step will compensate for any errors made during preparation. The average consumer is protected in restaurants because home-canned products cannot be served by law. In addition, commercial food processing plants are inspected regularly. These regulations are successful because there were only 21 cases of food-borne botulism in the United States in 1990. Botulism is treated with an antitoxin prepared from the serum of an immunized laboratory animal, which provides passive immunity.

A variant form of botulism that was recognized only in the 1970s is infant botulism. For reasons that are not entirely known, *C. botulinum* can become a member of the intestinal flora in some infants. The level of risk seems to be increased if the infant is fed honey prior to

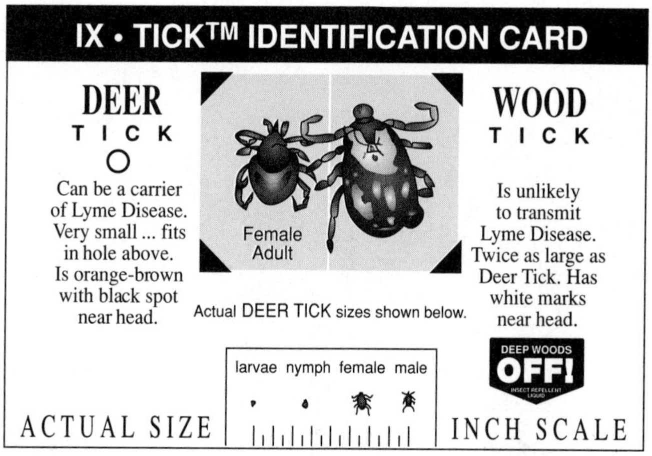

LYME DISEASE INFORMATION DEEP WOODS OFF!

EXPOSURE: Lyme disease is a bacterial infection transmitted by the bite of a deer tick, found on all types of vegetation and animals, especially in the woods. The deer tick is most active in the spring and fall, and present throughout the summer.
PRECAUTION: Wear long sleeves and pants. An insect repellent such as Deep Woods OFF! can help repel ticks. Brush off clothing and pets before going inside. Check yourself and pets carefully for ticks. Remove attached tick at once with fine-jaw tweezers by grasping tick's head as close to the skin as possible and gently pulling straight out. Be careful not to squeeze tick's body as this may cause it to inject fluid into you. Wash bite area and apply antiseptic.
SYMPTOMS: A ring-shaped rash may occur within 4-20 days. Symptoms may include fever, chills, headache, stiffness in joints, weakness and fatigue. Common symptoms of Lyme Disease may mimic arthritis. In some cases there are no symptoms.
DIAGNOSIS: Contact your doctor if you suspect Lyme Disease, or its symptoms.
TREATMENT: If detected early, Lyme Disease is usually treatable with antibiotics.
©1988 **GUNDERSEN MEDICAL FOUNDATION, LTD.,** 1836 South Ave., LaCrosse, WI 54601
OFF! and Deep Woods OFF! are S.C. Johnson & Son, Inc.'s trademarks for insect repellents.
IX • TICK is a trademark of Gundersen Medical Foundation, Ltd.

B

FIGURE 19.1 CONTINUED

(**B**) Small card offering protective advice is distributed widely in Wisconsin and Minnesota in an attempt to stop the spread of the disease.

(B) From the Gundersen Medical Foundation, Ltd., La Crosse, WI. Copyright © 1988. Reprinted with permission.

the age of 1 year. The amount of toxin produced is not lethal generally, but it is sufficient to induce neurologic symptoms, including lethargy and poor feeding. Symptoms can be relieved by appropriate supportive therapy. An enema is sometimes used to eliminate the clostridial toxin from the intestine. Antibiotics and antitoxins do not show clear-cut effects. In 1990 there were 58 cases of infant botulism in the United States.

Clostridium difficile. Pseudomembranous enterocolitis, the disease caused by *C. difficile,* fundamentally results from modern antibiotic therapy. Usually, microflora of the colon prevent the bacterium from residing there. Because antibiotic therapy decimates the normal microbial flora in some patients, this organism can establish itself in the colon. It produces a heat-stable enterotoxin that causes diarrhea and irritates the epithelial lining of the intestinal tract. The areas of irritation, visually distinct from normal epithelium, are described as plaques even though they have nothing to do with a viral infection.

Persons susceptible to enterocolitis should use antibiotics with care. Generally, if they must be on antibiotic therapy, they are encouraged to eat bacteria-containing foods, such as yogurt, that provide competitors for the clostridia. This tactic often prevents a reoccurrence of the disease.

Clostridium perfringens. This organism is responsible for several serious conditions. In the case of food that has been improperly canned, *C. perfringens* can grow and produce an enterotoxin along with copious quantities of gas. Food cooked but not sterilized and then held at a temperature that allows spore germination also induces clostridial growth. Food poisoning is caused by the enterotoxin and results in a self-limiting diarrhea.

More severe problems can arise if *C. perfringens* happens to infect a wound. Several toxins in addition to enterotoxin are produced by the organism, and all are potentially lethal. The worst possible situation is one in which the organism infects muscle tissue. Toxin production induces extensive tissue death, an effect exacerbated by the physical damage wrought by the gas produced during bacterial growth. Nutrients released by the damaged tissue fuel bacterial growth. This condition is known as gas gangrene. In the days before antibiotics, amputation of the affected limb was the only treatment. Today's treatment includes antibiotic therapy and sometimes hyperbaric (higher than normal pressure) oxygen in an attempt to directly oxygenate tissues whose blood supply has been damaged. If oxygenation is successful, *C. perfringens* cells will stop growing.

Careful **debridement** (cleaning and dressing) of wounds coupled with appropriate antibiotics prevents gas gangrene. Tissue in which blood supply has been irreversibly damaged must be removed to prevent growth of *C. perfringens*.

Clostridium tetani. Tetanus is another disease that has virtually disappeared in the United States. Only 60 cases were reported in 1990. It, like botulism, results from the effects of a neurotoxin rather than from a disseminated bacterial infection. The usual portal of entry is a wound, often a puncture wound. Extensive tissue damage does not occur because the organisms need relatively little growing time in order to cause death.

Tetanus neurotoxin blocks the inhibitory mechanism of the central nervous system that relaxes muscles after stimulation. The word *tetany* means the fully contracted state of a muscle. Because all muscles in the body are susceptible to the toxin, a person suffering from tetanus becomes immobile as his or her muscles work against one another. In a lying position, the spine is arched and only the back of the head and the heels of the feet touch the surface of a bed. Death occurs from exhaustion or suffocation.

Routine immunization with tetanus toxoid prevents the disease. In children, this is administered in combination with diphtheria and pertussis vaccines (DPT shots). For adults, diphtheria and tetanus are combined (DT shots). Failure to reimmunize against tetanus is common with an estimated 92% of adult cases occurring in underimmunized persons. Tetanus immunizations last for at least 10 years but are given routinely after someone suffers a serious wound. Should disease symptoms appear, passive immunization with tetanus antitoxin (antibodies against the tetanus toxin produced in a laboratory animal) is the appropriate treatment. Tetanus antitoxin prevents the development of additional symptoms but may not eliminate the effects of the toxin already bound to muscle cells.

Corynebacterium diphtheriae. Diphtheria primarily results from toxin production. *C. diptheriae* can occur in two forms, toxin producers (tox^+) and toxin nonproducers (tox^-), with the distinction being whether or not it is a lysogen for β phage. The toxin gene is located on the prophage, and its transcription and translation are triggered by an iron deficiency in the bacterium. The toxin is an ADP-ribosyl transferase that modifies EF2, an elongation factor (protein) required for mammalian ribosomal protein synthesis, and thereby blocks protein synthesis. A single diphtheria toxin molecule is fatal to the cell it inhabits.

The presence of *C. diphtheriae* within the body does not necessarily indicate that a person will contract the disease. The organism is often observed within the upper respiratory tract of humans who exhibit no apparent symptoms. Presumably, these bacteria are of the tox^- variety. When a toxin-producing bacterium precipitates an infection, the lining of the throat becomes inflamed. The accumulated products of the inflammatory reaction form a pseudomembrane along the throat lining. The pseudomembrane can, if sufficiently extensive, mechanically block the air passage, and suffocation results. Toxic effects of the infection are generalized and include tissue destruction and bleeding of the heart, lungs, and kidneys. Eventually paralysis may result.

As with the other diseases in which toxins play a significant role, prompt administration of an antitoxin is essential. Once the threat of additional toxicity is removed, the bacterial infection easily can be treated with administration of antibiotics like penicillin or tetracycline. An excellent immunization is available, usually combined with tetanus and pertussis vaccines. Consequently, the number of cases of diphtheria in the United States is extremely low (4 in 1990).

Haemophilus influenzae. The species name of *H. influenzae* is a complete misnomer and exemplifies careless science. Because this bacterium often was found associated with cases of influenza, it was named before anyone attempted to fulfill Koch's postulates in identifying the etiologic agent. In point of fact, *H. influenzae* causes meningitis, primarily in children. The meninges are the membranes surrounding the brain and spinal cord, and meningitis is an inflammation of those membranes. The organism invades the meninges from the upper respiratory tract. Septicemia (invasion of the bloodstream by an organism) can result in particularly severe cases.

The infection is treated with ampicillin or with ampicillin and chloramphenicol in conjunction because some R plasmids found in *H. influenzae* are known to code for β-lactamase. A vaccine against *H. influenzae* type B was introduced in 1985 and is now recommended for use in all children.

Legionella pneumophila. In 1976 in Philadelphia, an American Legion convention was held. During that week-long meeting, 182 attendees became ill with a pneumonia-like disease that could not be attributed to any etiologic agent known at that time. Twenty-nine people died. The CDC spent over six months studying possible bacterial, viral, or toxic chemical diseases. At last, a bacterium grew on a charcoal medium originally intended to isolate *Rickettsia* spp. The bacterium was the first example of a hitherto unknown group of organisms that tend to grow in freshwater. The organisms are found particularly in cooling water towers, and the Philadelphia outbreak, as well as several others, was traced to that source. The disease is not always severe, and it can be treated with antibiotics. No vaccine has been prepared against this organism because the median number of cases observed in the United States in the last five years is only 832. The risk to any particular individual is generally small therefore. Most preventive measures involve careful decontamination of water circulated through cooling towers and engineering modifications to ensure that air intakes within buildings are not located near cooling towers because spray from these towers can enter these intake systems.

Legionella spp. grow so slowly in culture that it has been suggested that they grow in association with other organisms. In vitro experiments have shown that bacterial concentrations are much higher when various free-living amoebae are present. A suggested host for *Legionella* is the amoeba *Hartmannelli vermiformis*.

Mycobacterium bovis. Like its more famous relative, *M. tuberculosis,* this organism can cause a serious lung disease. The bacterium can infect cows and humans. Prior to pasteurization, milk was an important source of mycobacterial infections. For a description of the course of the disease and its treatment, refer to the subsection on *M. tuberculosis*.

Mycobacterium leprae. This organism is responsible for leprosy, a disease that is much misunderstood. Its psychological connotations are so great that it is often referred to as Hansen's disease after the Norwegian doctor who pioneered research on it. *M. leprae* does not grow on artificial medium; therefore, less is known about the organism than about many other comparable etiologic agents. The organism does infect the nine-banded armadillo as well as humans, making some experimentation possible.

A close study of the disease shows that it is not particularly contagious unless there is prolonged, close contact with sores on an infected individual. The organism does not invade the body but relies on some mechanical injury to the skin to provide its portal of entry. In the lepromatous phase, phagocytosis of the bacteria by macrophages occurs with concomitant tissue damage. Often, peripheral nerves are damaged resulting in sensory loss; however just as much damage can result from unwitting injury due to lack of sensation to pain. Patients often cut or burn themselves badly before realizing that anything is wrong. In the tuberculoid stage, the infection is more or less walled off as tuberculosis is.

Leprosy is not a common disease in the United States. Only 203 cases were reported in 1990. The disease is a chronic one with a very long incubation period, generally on the order of years. Although sufferers used to be confined to leprosaria in Louisiana or Hawaii, the disease was not often transferred from the patients to the medical staff. Indeed, the length of the incubation period makes it very difficult to trace the route of infection. Appropriate chemotherapy can control the disease completely, but any tissue damage that may have already occurred cannot be reversed.

In light of the long incubation period and slow progression of the disease, it seems strange that leprosy has acquired such a ferocious reputation. Two explanations have been put forward. One suggests that the original form of leprosy was much more virulent, but now the organism has moderated its metabolism to establish a better host-parasite relationship. The other proposes that the early reports confused a particularly virulent form of syphilis with leprosy.

Mycobacterium tuberculosis. Tuberculosis is a well-known disease that has been linked to poverty and overcrowding. The disease appears in many forms of which the familiar lung infection is the most common manifestation. Transmission of the organism is normally by way of aerosol inhalation, either from dust containing dried sputum or from an infected patient. When the mycobacteria invade the lungs, the tissues become inflamed, initiating a cell-mediated immune response.

During the cell-mediated response, macrophages become hypersensitive to the mycobacteria and form spheres of cells about the bacteria wherever they occur. These spheres differentiate into characteristic lumps called **tubercles.** If the bacteria continue to grow, tubercles can enlarge and their centers liquefy, resulting in the formation of a cavity visible by X-ray examination. If the lung infection is not curtailed, the bacteria can spread into the large network of adjacent blood vessels and thence to the rest of the body. Once into the circulatory system, they can infect and destroy essentially any organ of the body.

Exposure to tuberculosis can be detected by means of a skin test in which an extract of partially purified mycobacterial proteins is injected into the dermal layer of the skin. If a cell-mediated immune response has previously occurred, a localized hypersensitivity reaction will be seen, giving a positive result. If a skin test is positive, a lung X ray is necessary to determine whether an active disease state is present. Tuberculosis is treated by long-term therapy with the antimicrobial chemicals isoniazid and rifampin. *M. tuberculosis* grows very slowly, and the tubercles tend to protect the bacteria from the effects of antibiotics. A year's treatment generally is required to be certain of a cure.

Prompt treatment of infected individuals and alleviation of crowded, unsanitary conditions are the normal means of disease prevention. The bacille Calmette-Guerin (BCG), an attenuated strain of *M. bovis* isolated in France, is thought to confer at least some immunity to *M. tuberculosis.* It is often used to protect laboratory workers who experiment with the bacterium in the United States, but BCG is not in general use. Unfortunately, administration of the BCG vaccine results in a positive tuberculin skin test; therefore, it becomes impossible to know whether an individual has been exposed to the actual tuberculosis organism. There were 23,720 cases of tuberculosis reported to the United States in 1990, many of them traced to AIDS cases. AIDS depresses cell-mediated immunity, the system that attempts to wall off the tuberculosis infection in the lungs. When that process is less efficient, affected individuals who would normally be asymptomatic because of a rapid immune response develop full-blown symptoms.

At least nine other species of mycobacteria can infect humans, although they have been estimated to cause less than 15% of the total tuberculosis cases reported. Among them are the extremely similar *M. avium* and *M. intracellulare. M. avium* causes tuberculosis in humans, swine, and birds, while *M. intracellulare* is much less pathogenic. *M. kansasii* also causes tuberculosis in humans and, together with *M. avium,* is responsible for most cases of tuberculosis that are not due to *M. tuberculosis.*

Neisseria gonorrhoeae. Gonorrhea is a sexually transmitted disease (STD) that affects the mucosa of the reproductive tracts of men and women. The bacteria are phagocytized by epithelial cells lining the reproductive tract and grow inside these cells. Eventually the organism ruptures the cells, thereby penetrating the epithelial barrier. The inflammatory response evoked leads to production of a vaginal or penile discharge, the primary symptom of gonorrhea. Symptoms are more obvious in males; therefore, males are more likely to seek treatment. Potential consequences of the disease are more problematic in females, however, because the bacteria can spread up the reproductive tract to the fallopian tubes. The fibrosis that is a consequence of infection can result in sterility in such cases. Both males and females occasionally can develop a septicemia from the disease.

Normally, *N. gonorrhoeae* infections are treated with one of the penicillin derivatives such as ampicillin. However, epidemic levels of gonorrheal cases have occurred for many years now (about 673,000 cases in 1990 alone), and R-plasmid-mediated antibiotic resistance has been documented many times. Penicillinase-producing *Neisseria gonorrhoeae* (PPNG) causes significant problems in certain areas of the country. The second antibiotic of choice in treatment is spectinomycin, but already there have been reports that organisms resistant to this antibiotic also have been isolated. The real solution, therefore, must be prevention rather than treatment.

A vaccine against *N. gonorrhoeae* is presently undergoing testing to determine its efficacy and safety. If all tests are satisfactory and if the general population can be persuaded to use the vaccine, the gonorrheal epidemic can be stopped. Unfortunately, the public health problems in coping with sexually transmitted diseases are often compounded by varying moral philosophies.

One innocent victim of gonorrhea can be the newborn. If a woman infected with gonorrhea gives birth vaginally, the baby is coated with significant numbers of the bacteria. The disease is particularly serious with respect to the eyes because bacterial growth in them results in permanent blindness. Normal prenatal care includes testing for gonorrhea, but every state also requires that the eyes of all newborns be treated with an appropriate antibacterial substance as a preventive measure. An antibiotic like penicillin or erythromycin or a solution of silver nitrate is used. In cases in which gonorrhea has been diagnosed, the infant must be treated for a possible systemic infection as well.

Neisseria meningitidis. This organism is prevalent in the upper respiratory tract. Carriers can trigger an outbreak whenever overcrowded conditions predominate. Formerly, meningococcal meningitis was a major problem in the military services particularly because the mortality rate is about 85% in untreated cases.

The development of a vaccine containing the polysaccharide capsular antigens has greatly reduced meningococcal meningitis. In the military, immunization against the organism is required. In 1990 there were only 2,349 cases reported in the United States.

Pseudomonas aeruginosa. This organism and other members of the genus *Pseudomonas* are the most common opportunistic pathogens in humans. A particular problem can arise in burn patients who have lost much of their epithelial protection and are easy prey to these organisms. R plasmids are common within this genus; therefore, pseudomonas infections are difficult to treat. With respect to burn patients, preventive treatment consists of skin grafts from cadavers or the use of some sort of artificial skin to guard against infections.

Salmonella typhi. Typhoid fever is potentially as serious as cholera and has additional complications. *S. typhi* is a member of the enteric group of bacteria and is transmitted by fecally contaminated food and water. In 1990 there were only 503 cases of the disease in the United States.

S. typhi grows well in the intestine and usually invades the intestinal mucosa. As it does, the organism induces a comparatively high fever that can last for weeks. Septicemia often results from the bacteria penetrating the intestinal lining. The bacteria are resistant to killing after phagocytosis; therefore, macrophages serve to further disseminate the infection as they move through the body. Macrophages also tend to protect the salmonellae from the effects of antibiotics. Chloramphenicol is the most commonly used antibiotic in the treatment of typhoid fever, despite its attendant hazards (see chapter 18). If the infection is not controlled, bacterial invasive activities can cause hemorrhage or perforation of the intestinal tract. Thus, the risk from the infection is greater than the risk from the antibiotic therapy.

S. typhi is very good at establishing a carrier state. A carrier remains infectious, particularly if employed in a food-handling area, and there is no reliable treatment for a carrier. For these reasons, local public health officials often require routine cultures on persons who are hired to work in restaurants to ensure that they are not salmonellae carriers. The most famous case in this regard was Mary Mallon of New York City (Typhoid Mary) who, in the early twentieth century, was responsible for several typhoid fever outbreaks. She refused to give up her occupation as a cook and was eventually quarantined to protect the rest of the public.

Salmonellae are widely distributed in nature and are common on farms. They contaminate the surfaces of poultry and have been isolated from the inside of eggs in the northeastern United States. Some outbreaks of salmonellosis have been traced to contaminated milk supplies. Infections by species other than *S. typhi* do not usually have drastic effects on an individual. Normally, their symptoms are the result of enterotoxin production, and their infections are self-limited. Deaths have been reported from various *Salmonella* infections, however, particularly among the elderly.

Shigella dysenteriae. Dysentery is an acute diarrheal disease resulting from an infection by an enterotoxin-producing organism. One of the most common bacteria associated with this disease is *S. dysenteriae.* Its pathogenicity is highly correlated with its ability to invade the intestinal wall. When the bacteria are fully virulent, a human infection can be initiated by as few as 180 cells. The traditional statement about enteric diseases being transmitted by "fingers, feces, food, and flies" is very appropriate. In severe cases, a high fever leaves the affected person in a debilitated state, and mortality rates as high as 20% have been reported.

Shigella readily develops antibiotic resistance, primarily through the presence of conjugative R plasmids. Treatment, therefore, requires careful laboratory screening to be certain that an appropriate antibiotic is chosen. There is no vaccine in general use, although preliminary trials have been made. Persons living in a particular area quickly acquire a moderate resistance to locally endemic strains of *Shigella,* a finding that may account for tales of "traveler's diarrhea."

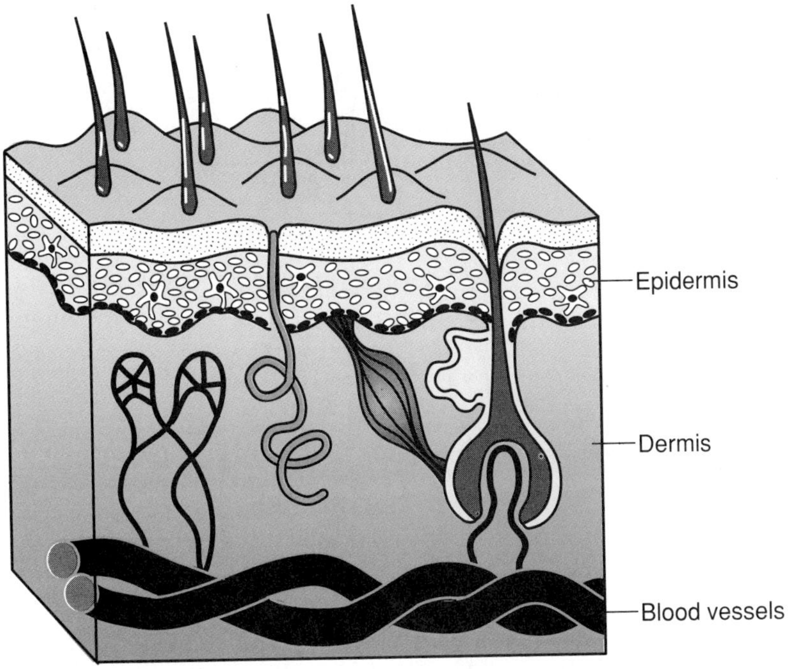

FIGURE 19.2

Anatomy of a hair follicle. *Staphylococcus aureus* can initiate infections either within the sebaceous glands or at the base of the hair follicle near the blood vessel.

Staphylococcus aureus. This golden-colored organism can cause a variety of health problems, particularly in a hospital environment. *S. aureus* is remarkably hardy, tolerating 7% NaCl in its growth medium and surviving for weeks in dried pus. As noted in chapter 18, the organism produces a wide range of virulence factors that enhance its ability to set up an infection.

S. aureus is found on the skin surface of some 30% of the adult population. It normally causes few if any problems for a healthy individual. Injections of the organism under the epithelium are generally safe as well. If an irritating agent such as a suture (material used for sewing up a wound) is present, however, *S. aureus* grows very well and can cause major problems. Sometimes the bacterium will establish an infection in a pore of the skin, in a sebaceous gland, or at the base of a hair follicle (figure 19.2). Pus develops at the center of the infection and forms a hard, swollen area called a boil. If more than one boil develops, they may coalesce into a larger structure called a carbuncle. Boils or carbuncles can be dangerous if the pressure from the infection itself or from misguided attempts to expel the pus by squeezing drives the bacteria down into the dermis because septicemia can result.

S. aureus produces a heat-stable enterotoxin and can cause serious outbreaks of food poisoning. The usual route of infection is through food prepared by someone with an active staphylococcal infection on the skin. If that food is later left at moderate temperatures that allow bacterial growth before serving, an individual can become infected. Pastries, foods containing dairy products, and some meats particularly are susceptible to staphylococcal contamination. Education is the key to disease prevention. Food preparers must be alert for open sores on their hands and instructed to always refrigerate food after it has been prepared. Food handlers also should be screened for possible carrier status. Staphylococcal food poisoning is common during the picnic season when foods are carried long distances, often with minimum refrigeration. Subsequent cooking does not destroy the toxin.

Hospitals have a difficult time with *S. aureus* because it readily acquires resistance to those antibiotics commonly used. To counteract this effect, most hospitals have a regular program to change the spectrum of antibiotics used routinely in order to keep *S. aureus* infections at minimum levels. In the case of boils, it is often necessary to lance or open the site of infection and drain the general area because staphylococcal products can disrupt local circulation.

A rare disease caused by *S. aureus* infection is toxic shock syndrome in which the organism produces a lethal toxin. The nature of the toxin is not well understood. The disease was originally described primarily with respect to women's use of menstrual tampons, but it can occur in males and in females independently of menstruation. Whenever anyone has an *S. aureus* infection, he or she apparently is at risk for developing toxic shock syndrome. Although not receiving the publicity it did when first reported in 1978, the disease is still present at endemic levels because 293 cases were reported in 1990, the majority of which were associated with menstruation. One report has suggested that toxin production may be stimulated by chelation of magnesium ions in the vicinity of the infection. Those tampons most likely to be associated with the disease were also the best chelators.

Streptococcus pyogenes. Several major health problems are caused by β-hemolytic streptococci, of which *S. pyogenes* is the most common. As detailed in chapter 18, the organism produces many virulence factors that can aid in establishing the disease. An infection of the upper respiratory tract, a "strep throat," is the typical streptococcal disease. For most people, the problem resolves itself within a week or so, but some serious complications can develop (figure 19.3). Streptococci can establish themselves in areas adjacent to the respiratory tract, such as the tonsils or the middle ear. They can invade and infect organs elsewhere in the body as well.

If the infecting *Streptococcus* is a lysogen of the appropriate type, it produces an erythrogenic toxin that can precipitate in susceptible children a whole body rash known as scarlet fever. The mode of action is uncertain, but a generalized hypersensitivity reaction is suspected. There have been recent reports of serious complications, including death, in cases of scarlet fever. These observations represent a dramatic change from the experience of the last few decades, and their meaning is not yet clear.

Persons diagnosed with a strep throat may wonder why their physicians recall them for treatment after the infection has subsided. The answer lies in the complications that sometimes occur. After a period of time sufficient for a primary immune response, significant levels of antistreptococcal antibodies can be detected. A few individuals seem to overproduce these antibodies. In some cases the antistreptococcal antibodies

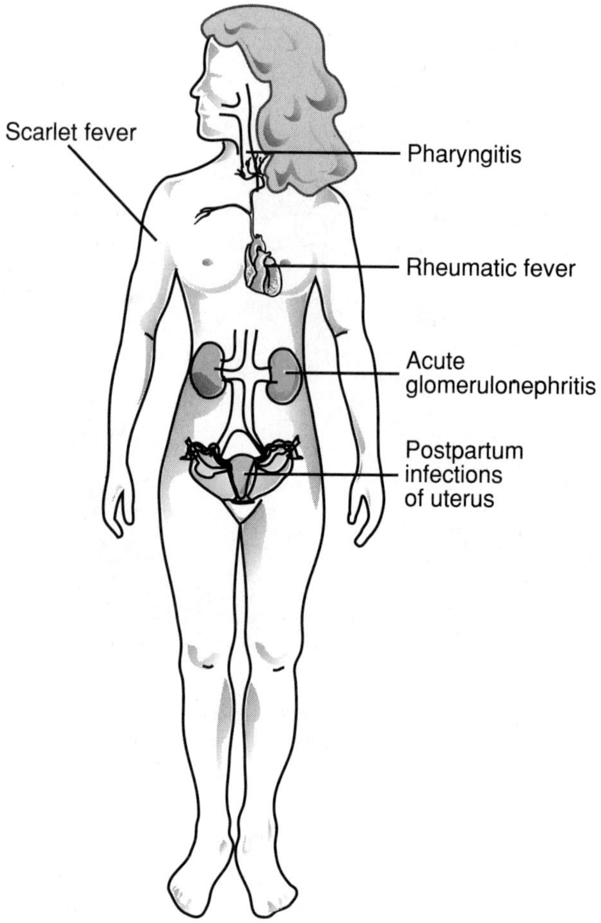

FIGURE 19.3

Possible effects of a streptococcal infection.

can cross-react with normal body tissue and stimulate typical cytotoxic responses. Symptoms presented depend on the affected tissue. The two most common diseases seen following a streptococcal infection are rheumatic fever, in which the heart valves are affected, and acute glomerulonephritis, in which the kidneys are attacked.

Antibiotics are particularly effective against *S. pyogenes,* and resistant strains rarely are isolated. Thus, penicillin is and has been the treatment of choice for many years. Those persons who have had rheumatic fever or acute glomerulonephritis normally are maintained on low levels of penicillin at all times to prevent them from acquiring a streptococcal infection that might trigger a secondary immune response. A secondary immune response would probably do a great deal more damage to tissues than a primary response.

S. pyogenes is quite capable of infecting various types of wounds and is the etiologic agent for puerperal fever, a serious disease that can affect postpartum women. Apparently, attendants at a delivery can inadvertently infect any wounds resulting from the delivery process if they have an active respiratory tract infection or have been in recent contact with someone else who has. Puerperal fever was severe particularly in the days before the germ theory of disease was generally accepted. This disease stimulated Ignaz Semmelweiss to propose that physicians could infect their patients (chapter 1). Modern antiseptic deliveries and antibiotics have greatly reduced the incidence of the disease.

Rickettsia prowazekii. This obligate intracellular parasite causes typhus, a disease transmitted by the human body louse *Pedunculus humanus.* Lice are infected by feeding on infected humans. During the two or three weeks that infected lice survive before they succumb to the disease, they transmit it to other hosts via their feces. The bite of the louse itches severely, and any scratching action rubs the rickettsia into the wound. The bacteria multiply in the lining of host blood vessels, gradually spreading throughout the body. They cause hemorrhages, meningitis, heart problems, and kidney failure. A rash results from superficial bleeding.

Typhus has been a scourge of humankind for centuries but is easily controlled by eliminating body lice. The book *Rats, Lice, and History,* referenced at the end of this chapter, provides some details. The discovery of the insecticide DDT was a major step in preventing the spread of typhus at the end of World War II. Unfortunately, DDT itself is very toxic; therefore, other insecticides have had to be substituted. Tetracycline and chloramphenicol are the antibiotics of choice for treatment of typhus. Vaccines are available that reduce the severity of the disease.

Rickettsia rickettsiae. At the present time, the preponderance of the approximately 650 cases per year of Rocky Mountain spotted fever, caused by *R. rickettsiae,* occurs in the Carolinas. The disease is transmitted by another arthropod vector, the tick. In this instance, the tick survives the infection and continues to infect new hosts for long periods of time. Moreover, an infected female tick can infect the eggs that she lays. Symptoms of Rocky Mountain spotted fever are quite similar to those of typhus except that the rash begins around the wrists and ankles. The untreated disease has a mortality rate of at least 25%. The organism is susceptible to the same antibiotics as *R. prowazekii,* giving an average case:fatality ratio of 4%. No vaccine is available.

Treponema pallidum. This spirochete causes syphilis, a sexually transmitted disease with considerable impact on people worldwide. The organism is difficult to study because it does not grow on laboratory media. Probably it is a microaerophile. Syphilis is transmitted by direct contact and manifests itself in three stages. Mucous membranes of the reproductive tract or oral cavity make contact with a sore containing the bacteria.

During the primary stage, a small ulcerative lesion called a chancre develops at the site of actual infection. The time lag between infection and appearance of the chancre is highly variable, ranging from 10 to 60 days. Although this lesion is infectious, it heals spontaneously. The afflicted individual often assumes that no further action is necessary following this healing process, whereas in actuality serious risks remain.

During the secondary syphilitic stage that occurs some 2 to 12 weeks after the chancre first appears, a generalized rash and local ulcers develop on the skin as a consequence of the disseminated treponemal infection. All surface lesions are highly contagious and can form anywhere on the body. Most treponemal organisms eventually are eliminated by the cell-mediated immune system, but some tissue damage can result from immune activity. Rosebury has suggested that the secondary stage of syphilis may have been confused with leprosy in Biblical times (see the reference to his book in the suggestions for further reading at the end of this chapter).

In pregnant women, *T. pallidum* can readily cross the placental barrier and infect the fetus, resulting in congenital syphilis. A child may be stillborn or survive for a period of weeks or years before succumbing. A routine prenatal blood test for syphilis is done on all pregnant women. If the disease is detected, treatment can begin in time to protect the fetus. In recent years, the number of congenital syphilis cases has increased dramatically (685 cases in 1990). In roughly 48% of all cases, the mother received no prenatal care.

Tertiary syphilis, the third stage of the disease, occurs in only a minority of cases, but it is extremely severe. Generally, the time lag is on the order of years after the secondary stage. Symptoms of tertiary syphilis are indeterminate and depend on the affected organ system, leading to the designation of the disease as the "great imitator." Persistence of bacteria within one or more organs after the major immune response has occurred gives rise to symptoms. At some later time, another hypersensitivity reaction can occur that results in massive destruction of affected tissue. If treponemes are found in the brain, madness results. Other symptoms can include blindness, deafness, and heart disease.

Penicillin therapy is the treatment of choice for the primary and secondary stages of syphilis. Penicilin is used against tertiary syphilis as well, but it is not as effective because the major problem is the immune response, not the growth of the organism. Best results are obtained if the disease is treated prior to the appearance of tertiary symptoms.

Syphilis is still common in the United States, with 48,363 cases reported in 1990. This is an increase from 35,000 in 1987. A generally rising trend is noted, and women using prostitution to support drug habits are thought to be the principal cause. There is some risk to medical and dental personnel who are in direct contact with individuals in the secondary stage of syphilis. Appropriate precautions should be taken whenever any personnel are exposed to persons having suspect lesions.

Vibrio cholerae. Cholera is one of the most-studied diarrheal diseases and is caused by the ingestion of *Vibrio cholerae.* The CDC requires notification of cholera cases but classifies the disease as one of low frequency. Generally fewer than 100 cases per year are reported. During the first four months of 1991, eight domestic cholera cases were reported, but an epidemic centered in Peru had caused over 1,400 deaths. The organism does not invade but attaches to the intestinal lining and releases its toxin. The toxin affects adenylate cyclase and stimulates massive release of fluid from the intestinal walls into the intestinal lumen. The fluid volume can be immense. A case in which a patient voided as much as 20 liters of fluid per day has been documented. Typically, a cholera patient is described as having a "rice water stool" because the feces are composed primarily of water.

The sheer volume of water lost during the course of the disease is both a help and a hindrance. It is a help in the sense that the vibrios cannot remain attached to the intestinal epithelium for long periods of time; therefore, the disease is self-limiting. The hindrance stems from the dehydrating effect of excess water loss. Along with its secretion-stimulating effect, the cholera toxin inhibits normal absorptive activities of the intestinal epithelium. It is not possible for an untreated patient to ingest sufficient liquid by mouth to compensate for the liquid loss, and death results from dehydration.

Until a few years ago, the necessary therapy for cholera consisted of intravenous fluid administration to maintain hydration. The problem with that technique was that cholera is most prevalent in countries that cannot afford such a technological treatment. The World Health Organization has announced the development of a balanced salts and glucose formula that can be taken by mouth and will allow the uptake of water by the intestinal lining. As long as pure water is available to use with the formula, economic cholera treatment is now possible worldwide. On the other hand, the primary source of the disease is water contaminated with fecal material; therefore, improved treatment may be difficult due to lack of a clean water supply.

Yersinia pestis. The disease caused by *Y. pestis* is plague, the stuff of history and folklore. Known formerly as *Pasteurella pestis,* the bacterium caused the "black death" that swept across Asia and into Europe between 1347 and 1350, killing 25 to 30% of the population. It is spread by an arthropod vector, any of several fleas commonly found on rodents. Fleas feed on the blood of rodents infected with *Yersinia* and acquire the bacterium. The organism grows in the gut of the flea and produces sufficient numbers of cells to block its digestive tract. When the flea subsequently attempts to feed on a human host, it wounds its victim and regurgitates the ingested blood as well as its own bacteria, thereby infecting its host.

Typically, the bacteria multiply in the tissues and lymph of a human host. As lymph flows through its ducts, the bacteria accumulate in the lymph nodes, primarily in the groin, and localized swellings occur as a result of the inflammatory reaction induced by the bacteria. A swollen lymph node is called a **bubo,** and this form of plague is known as the bubonic plague. If the immune response in the lymph nodes is insufficient, septicemia can result and the infection will be disseminated throughout the body. Symptoms can include shock and high fever. Very large numbers of bacteria can accumulate, and localized bleeding and clotting can result. Lack of oxygenated blood confers the characteristic black color of plague victims. The incubation period of the disease can be from one to six days, depending on the size of the bacterial inoculum. If left untreated, bubonic plague has a case:fatality ratio of 50% or more. The injection of 10 bacterial cells into a mouse can be fatal.

One possible outcome of septicemia is pneumonic plague in which the infection establishes itself in the lungs. Pneumonic plague is a particularly dangerous condition because fluid in the lungs becomes heavily contaminated with bacteria, and sputum and aerosols from the respiratory tract are very infectious. A person inhaling the bacteria quickly develops a similar lung

APPLICATIONS BOX

Applied Bacterial Taxonomy

Taxonomy is often considered to be a dull subject, but it can have important practical consequences. Taxonomists try to group organisms based on criteria like those presented in chapter 20. Normally, pathogenicity is not a criterion used. If pathogenicity is ignored, however, it is difficult to distinguish between *Y. pestis* and *Y. pseudotuberculosis,* an organism that is not particularly pathogenic. Therefore, the international governing body for bacterial taxonomy has proposed the elimination of the species *Y. pestis* and the incorporation of both organisms under the species name *Y. pseudotuberculosis.* The proposal has engendered a huge outcry from the medical community on logical grounds because a person working with *Y. pestis* is much more likely to use very careful laboratory technique than a person who thinks that the organism is a less pathogenic one.

infection; therefore, transmission of the disease becomes independent of the arthropod vector. Large amounts of bloody sputum are produced in the final stages of this disease. Pneumonic plague is the most virulent form (nearly 100% of untreated cases are fatal) and the most likely to cause epidemics such as the one that devastated Europe. Death normally occurs in three to four days. Treatment after 12 to 15 hours of fever is rarely effective. Any patient diagnosed with plague is kept in respiratory isolation (all staff and visitors must wear masks) until the infection is controlled.

Treatment is with antibiotics like streptomycin or chloramphenicol. A vaccine is available to immunize persons known to be at risk, such as members of the armed forces, persons exposed to wild animals, etc. Booster doses must be given at one- to two-year intervals. The number of human cases in the United States is comparatively small. There were 2 cases in 1990. Worldwide there were 770 cases of plague in 1989, mostly in Vietnam and Madagascar. In 1984 there were 31 cases reported to the CDC, of which 19% were fatal.

Viral Diseases

In this section, the viral diseases are alphabetized according to the name of the disease as opposed to the name of the virus.

Acquired immune deficiency syndrome (AIDS). AIDS was first identified in 1981, although in retrospect some cases of the disease had been observed earlier. It is caused by a retrovirus, human immunodeficiency virus (HIV), also known as human T-cell lymphotrophic virus III (HTLV III) or lymphadenopathy-associated virus (LAV). The case definition presently outlined by the CDC includes individuals who have a demonstrable HIV infection in addition to any infection from an opportunistic pathogen such as *Pneumocystis carinii* pneumonia (infection by a protozoan parasite), disseminated histoplasmosis (a fungal disease discussed in the following section), extrapulmonary tuberculosis, or Kaposi's sarcoma (unknown etiology but probably sexually transmitted). All these ailments result from a loss of immune function and are the most commonly observed illnesses among AIDS patients. Loss of normal immune response due to a failure of T-lymphocyte function is the primary characteristic of AIDS. Death results from a continuing series of infections that eventually overwhelm what remains of the body's immune system.

The number of cases of the disease has increased rapidly, in part due to its spread and in part due to better diagnosis. In 1990 in the United States, there were 41,129 cases of AIDS reported compared with 34,340 reported in 1989 and a median of 13,405 cases reported between 1984 and 1988. These statistics are frightening, but the general populace has been alarmed even more by the fact that the outcome of HIV infections in people in whom symptoms have appeared has been invariably fatal. The case:fatality ratio for the disease is about 50%, primarily due to the large number of new cases being reported. As of the middle of 1991, a total of 179,136 AIDS cases had been reported, among which there were more than 113,000 fatalities.

Much of the public hysteria about AIDS stems from our lack of specific knowledge about the disease and its transmission. The virus can be found in various body secretions, such as semen, tears, and occasionally saliva. It also is concentrated in T4 lymphocytes. Table 19.2 lists some risk factors that have been identified for AIDS infection. It is apparent that intimate contact is necessary for disease transmission. Note that AIDS is readily transmitted by heterosexual activity; therefore, some public attitudes toward persons infected with HIV have to be revised. A rising number of cases of AIDS are due to transmission from mother to fetus. Congenital AIDS is one of the greatest tragedies in public health at the present time. Children with AIDS survive for only a few years.

APPLICATIONS BOX

AIDS Worldwide

Although the problem of AIDS in the United States is alarming because roughly 1% of the population is infected with HIV, in many areas of the world AIDS is a catastrophe. In Romania 10% of 10,657 children living in orphanages or public hospitals have tested positive for HIV infection.

Much of the continent of Africa is in dire straits. In the last decade, over 400,000 AIDS cases have been reported with an estimated 3.5 million more cases believed to exist. Roughly 16% of those cases are believed to be children under the age of 5. An additional million children are not infected but have mothers who are. They can expect to be orphaned within the next few years. One report from Uganda indicated that 20% of the rural adult population is infected. Another report from the Ivory Coast indicated that 41% of male cadavers and 31% of female cadavers tested positive for HIV infection, making AIDS the leading cause of death. The AIDS epidemic in Africa has the potential to do as much damage as the plague epidemics did in Europe.

TABLE 19.2 Risk factors for 16,458 cases of AIDS reported in the United States

Risk Factor	Total Cases before 1985 (%)	Total Cases in 1988 (%)
Adult Males		
Homosexual/bisexual	69	63
Intravenous drug user	15	20
Homosexual/bisexual intravenous drug user	10	7
Hemophiliac	1	1
Born outside U.S. in a country reporting heterosexual AIDS transmission	3	1
Transfusion recipient	1	2
Heterosexual contact	<1	1
No identified risk or still under study	2	4
Adult Females		
Intravenous drug user	58	53
Hemophiliac	<1	<1
Heterosexual contact	16	26
Born outside U.S. in a country reporting heterosexual AIDS transmission	8	3
Transfusion recipient	8	10
No identified risk or still under study	10	8
Children		
Mother with/at risk for AIDS infection		
Intravenous drug user	45	40
Risky sexual partner	11	21
Born outside U.S.	22	7
Transfusion recipient	11	11
Hemophiliac	4	7
No identified risk	6	6

Source: *Morbidity and Mortality Weekly Report* 38(S–4):19, 1989.

There are few reported cases of AIDS among families of AIDS patients or among health personnel who may have been working with them. This fact, in turn, suggests that the disease is not particularly contagious. Health personnel who have contracted AIDS appear to have done so through accidental contamination with blood or blood products. For this reason, it is considered prudent to assume that in any health care setting all blood samples are contaminated with HIV unless specifically proved otherwise. None of 64 household contacts with pediatric AIDS patients exposed for an average of 21 months had any symptoms or produced any antibody against HIV.

Transmission of AIDS by contaminated blood products has been greatly reduced since the introduction of monoclonal antibody tests for the AIDS virus (Western blotting or ELISA tests; see chapter 16). The ability to identify the presence of the virus in the blood also has added to the confusion, however. There is at the present time no certain indication that every individual who tests positive for anti-HIV antibodies will come down with the disease, because many individuals with no symptoms do have antibodies against the virus. By the same token, the finding of the virus in asymptomatic individuals may mean that the incubation period for the disease is quite lengthy. Only the passage of time will provide the answer.

Some individuals with apparently mild cases of AIDS have been identified, and the existence of such cases provides hope that AIDS is not necessarily a terminal disease. More importantly, such cases may provide a testing ground for new drugs that might be able to halt or reverse the course of the disease. To date, the only drug certified for use in treating AIDS is AZT (azidothymidine), a thymine analog that specifically inhibits the RNA-dependent DNA polymerase required for retroviral function.

Chicken pox. Chicken pox, also called varicella, is chiefly a disease of childhood. No vaccine is available, and adults have more difficulty with the disease than do children. The virus, a member of the herpesviral group, is spread via aerosols. Within two weeks after entering the respiratory tract, it is disseminated via the blood to surface tissues. A rash characterized by small red lumps designated as **papules** or pocks appears and is accompanied by a fever. Although the rash is unsightly, it is not particularly painful and resolves itself in about a week. The papules can become infected by bacteria.

In some persons who have had chicken pox, the virus may go into a latent state in sensory ganglia. At a later time, possibly years later, it can be reactivated. Reactivation causes a painful inflammatory reaction of local sensory nerves. Inflammation occurs in broad red patches on the body surface corresponding to the distribution of the ends of one or more spinal nerves. From this distribution it is assumed that the virus has traveled along nerve fibers to reach its new location. The disease, called shingles or herpes zoster, is extremely uncomfortable and painful. Paralysis and meningitis are possible complications of herpes zoster.

Common cold. The common cold is so common because it is not caused by a single virus but rather by several families of viruses, the most common of which are the rhinoviruses. Rhinoviruses are picornaviruses, and more than 100 antigenically distinct, biologically related rhinoviruses have been identified. Similar symptomology can be produced by immunologically unrelated viruses of the coronaviral or coxsackieviral families. The disease consists of an acute inflammation of the upper respiratory tract that is not accompanied by a fever. Despite its name, there is no indication that exposure to low temperatures or deliberate chilling predisposes an individual to a cold. Transmission seems to occur by aerosol inhalation.

The immunologic unrelatedness of the causative viruses means that production of a successful vaccine against the common cold is unlikely. Although it is true that vaccines against multiple serotypes have been made, none has been attempted on a scale that would be required to protect against all viruses that can cause the common cold. Infection with a particular virus does confer immunity that can last for several years; therefore, adults tend to be more resistant to colds than younger individuals.

Genital herpes. Details of the herpes simplex viral life cycle are presented in chapter 9. Type 1 virus primarily promotes oral and ocular infections, while type 2 virus primarily promotes genital infections. Both viruses are found in active and latent states and are infectious only during the active phase in which overt symptoms are not necessarily displayed. The latent state involves various ganglia and results from subviral particles traveling up axons to neuron cell bodies. At irregular intervals, viral particles travel down axons and reinitiate an active viral infection. The infection spreads by contact of mucous membranes with a sore containing the herpes virus. Work is progressing on the development of a vaccine. At the present time, no cure is possible, although certain preparations can reduce the severity of the symptoms. For example, acyclovir apparently prevents reactivation of latent virus, especially following the first viral exposure.

The tendency of all herpesviruses to gravitate to nervous tissue is manifest in herpes simplex. Adults with type 1 infections run some risk of developing severe meningitis that can cause permanent brain damage. Infants born to mothers with active herpes simplex type 2 infections develop a massive infection in which the virus replicates in nearly every organ of the body, including the brain. Survivors often have severe brain damage. For this reason, pregnant women with genital herpes infections must deliver by cesarian section.

Hepatitis. Hepatitis is an inflammatory disease of the liver that can be caused by different viruses: hepatitis A virus, hepatitis B virus, and several non-A non-

B viruses. Little is known regarding non-A non-B viruses although a biotechnological company has reported that hepatitis C virus, a recently identified togavirus, is transmitted through contaminated blood. By comparison, hepatitis A and B viruses are fairly well characterized.

The A virus is a small RNA virus that is comparatively resistant to common disinfectants. Food contaminated by feces transmits the virus, which is a frequent problem in restaurants and day-care centers. The course of the infection is similar to that of polio, with the virus colonizing the intestine, moving into the circulatory system and thence into the liver. A vaccine is presently undergoing testing, but for now, prevention of the disease requires good sanitary practices, especially by food handlers and day-care workers.

The B virus is a DNA virus primarily transmitted by transfusions or exposure to contaminated blood or blood products. As such, hepatitis B has always been an occupational risk for health care personnel. It can cause both acute and chronic infections. Chronic infections can result in hepatocellular carcinoma after a 30- to 50-year incubation period. Infants are readily infected by carrier mothers during childbirth. Both an inactivated viral and a genetically engineered vaccine are available for those persons whose job or life-style places them at an increased risk of infection.

Hepatitis is still a very common problem in the United States even with the hepatitis B vaccine. In 1990 there were 28,919 cases of hepatitis A, 19,939 cases of hepatitis B, and 2,773 cases of non-A non-B hepatitis. These numbers have been slowly decreasing over the preceding six years. Worldwide there are an estimated 300 million cases of chronic hepatitis. In southeast Asia and tropical Africa, at least 10% of the population are thought to have chronic infections. Proposals to require immunization against hepatitis B in children are under active consideration.

Infectious mononucleosis. The Epstein-Barr (EB) virus is another member of the herpes group of viruses and follows a similar pattern in infecting its host. Initial infection of the nasopharynx leads to a latent infection of a large number of B lymphocytes in the tonsils and adenoids. An extensive immune response develops from the latent infection and is characterized by a proliferation of monocytes. As the immune response progresses, symptoms characteristic of infectious mononucleosis appear, including painful sore throat and extreme fatigue. The immune response and the disease last for several weeks. Gradually, the immune system and body energy levels return to normal. Even after all symptoms have disappeared, however, latently infected B lymphocytes can be detected, and infectious virus appears in the nasopharynx at irregular intervals.

Influenza. The influenza viruses are members of the orthomyxoviral group and are found to display a variety of surface antigens. There are three types of influenza viruses: A, B, and C. Type A viruses are subdivided further according to their H (hemagglutinin) and N (neuraminidase) surface antigens. The general pattern of the disease is the same in all cases. As winter begins and people spend more time indoors in comparatively close contact, the virus is transmitted via aerosols. The neuraminidase activity present in the viral coat acts to neutralize the antiviral effects of mucoproteins (neuraminic acids) in the respiratory tract. The virus primarily invades the ciliated epithelial cells lining the respiratory tract. Because these cells are involved in protecting the lungs against infection, secondary infections are a frequent complication of influenza. Influenza itself is characterized by fever, chills, and a generalized aching, especially of the muscles.

In healthy adults, the disease runs its course in some three to seven days. Its effects are more serious in the very young, in the very old, and in those with preexisting respiratory problems. For these reasons, each year an influenza vaccine is formulated from inactivated viruses to immunize against those three viral strains (two type A and one type B) thought most likely to present problems in the coming year. In the United States, this decision is made in February to allow time for preparation of the vaccine for the following year. The immunity conferred by the vaccine declines rapidly, and yearly revaccination is necessary. Massive outbreaks can arise in the event of a major antigenic shift, something that has not occurred since the 1950s (the last pandemic occurred in 1957). The influenza pandemic of 1918–1919 was the most serious ever recorded, killing an estimated 20 million people worldwide. A major factor in that pandemic was secondary bacterial infections that can now be better controlled by the use of antibiotics.

As in the case with nearly all viral diseases, there is no specific treatment for influenza. When the yearly outbreak begins, those persons who are not immunized can derive some protection from amantadine, a substance that prevents uncoating of the type A virus. It is apparently of little value once the infection has begun.

Measles. There are actually two forms of measles, morbilli (also called rubeola) caused by a paramyxovirus and rubella (sometimes called German measles) caused by a togavirus. Current CDC policy uses the word *measles* to refer to morbilli, and this policy is employed in this text. Both viruses affect children and young adults, infecting the respiratory tract via aerosols. Eventual viremia leads to a generalized papular rash. Unfortunately, a person becomes infective before the rash appears so that epidemic transfer can occur rapidly. For most people, measles of either type is not a major problem, with rubella causing milder symptoms than measles. Serious complications are nevertheless possible.

Rubella is known to cause severe birth defects in babies whose mothers contract the disease during the first trimester of the pregnancy (congenital rubella syndrome). Those infants that survive can have a broad range of problems including deafness, cataracts, heart anomalies, and microcephaly. Measles has always been a more serious disease than rubella and, in some cases, can lead to pneumonia, middle ear infection, and encephalomyelitis (inflammation of the myelin sheath of the brain). Encephalomyelitis occurs about once in every 2,000 cases and has a mortality rate of 10%. Permanent damage occurs in many survivors.

For all the reasons discussed, a measles vaccine has been a high priority of health officials. The first vaccine developed did not induce a reliable immunity, but the newer, live viral vaccines seem to provide a more lasting immunity. The number of measles cases in 1990 was 26,520, and the number of rubella cases was 1,093. Although these numbers are low compared to prevaccine days, they in fact represent a major epidemic. The median number of measles cases between 1984 and 1988 was only 3,065. For rubella the median number of cases for the same time period was 530, and congenital rubella decreased from 62 cases in 1969 to 2 in 1985.

At the present time, the population most susceptible to contracting measles seems to be of school age because the majority in this age group have received either no vaccine or the less effective one. Public health officials have targeted this group for a major immunization campaign. Recently, a second immunization to be given during the early teenage years has been suggested in order to protect any person in whom immunity is fading or in whom no immunity was induced previously.

Polio virus. Polio virus is a picornavirus transmitted by fecal contamination. When ingested, the virus infects the intestinal mucosa and oral cavity. A viremia develops, and the virus migrates to other tissues, including fat layers and lymphoid tissue of the gut. For most individuals, the progress of the disease stops at this point, but in some cases, viremia persists and results in viral penetration of the central nervous system. Once in the nervous tissue, the virus continues to replicate, killing its host cells and causing paralytic polio, a familiar disease before the development of vaccines. Paralytic symptoms depend on the nervous tissue destroyed.

One unusual aspect of polio epidemiology is that paralytic polio is primarily a disease of modern society in that it is far more frequent in well-developed countries with good sanitation than in less-developed countries. The virus is found all over the world, but in most countries with poor sanitation, even children of 5 years are already immune to its effects. They apparently are infected early in life while they still have some resistance from their mothers, and they develop few or no symptoms. Children in countries with good sanitation are not exposed to the virus until later in their lives, at which time it is far more likely to cause the paralytic disease.

Two types of vaccines have been developed, a killed vaccine and an attenuated live viral vaccine. The killed vaccine was first developed by Jonas Salk. It has the advantage of being easy to store and transport but needs to be administered by injection and boosted at regular intervals. The live viral vaccine of Albert Sabin, on the other hand, induces an actual infection in the gut and maintains itself there. Fecal contamination thus inadvertently immunizes those people not actually receiving the vaccine. Moreover, the live vaccine can be administered orally, a very desirable feature for children. However, it does not transport well and is poorly suited for use in third world countries in which refrigeration is intermittent or lacking.

A dilemma has arisen in the United States with respect to paralytic polio. Most years, there are only a few cases observed (five confirmed in 1989), but these cases are nearly always the result of a nonimmunized person becoming infected with a virulent mutant of the virus used in the live polio vaccine. In other words, the vaccine itself may be maintaining the low level of polio cases in this country. The article by Hinman, referenced at the end of this chapter, provides more background about this dilemma.

Rabies. Rabies is another disease that has been known for centuries, and its reputation has not been exaggerated. A rhabdovirus found in the saliva of infected humans and other mammals causes the infection that is typically transmitted by a bite, although inhalation of aerosols released by a large population of bats is a possible mode of transmission as well. The length of the incubation period depends on the site of infection as well as on the quantity of virus transferred.

In about 30 to 50% of all infections, the virus migrates along the peripheral nerves to the brain and spinal cord and induces encephalitis, an inflammation of the brain that results in extensive tissue degradation. Irritability followed by spastic contractions of facial muscles are the presenting symptoms. Facial contractions result in hydrophobia, a condition in which the sight of water or other liquids stimulates further contractions. Eventually, damage to the central nervous system becomes irreversible, and death intervenes. The elapsed time from first appearance of symptoms to death is about three to four weeks.

Although it is no longer correct to say that rabies is fatal in all cases, the number of humans known to have survived rabies after symptoms have appeared is probably less than 10 in all of recorded history. Nonfatal infections have occurred in recent years thanks to massive support therapy. Because of the inevitable death associated with the disease it is no wonder that one of Louis Pasteur's early projects was the development of a rabies vaccine. Pasteur found that if the immune system can be stimulated into a primary response before the appearance of symptoms, the disease can be prevented. Pasteur's definitive treatment consisted of intraperitoneal injections (in peritoneal cavity of gut) of extracts from rabid rabbit brain, starting with extracts from brain aged for several weeks and gradually progressing to extracts from fresh brains. The total number of intraperitoneal injections ranged from 15 to 20. As vaccine technology improved, the virus was inactivated by phenol, but the basic methodology of injection remained unchanged.

The rabies vaccine prepared from brain tissue suffered from one serious drawback. Occasionally, it stimulated an allergic response that cross-reacted with antigens of the host brain. The vaccine used today is prepared in France from a virus grown on human diploid cell cultures. The number of required doses of this vaccine is only about five, a substantial improvement over the previous vaccine.

Rabies is endemic in the United States and at near epidemic levels in portions of Mexico. It can be found in many wild animals, but particularly in skunks, bats, raccoons, and foxes. There are no vaccines licensed for use with wild animals, but very good vaccines are available for cats and dogs. The Wistar institute has a genetically engineered vaccine for wild animals based on the vaccinia virus that is in the early trial stage. In Europe a genetically engineered vaccine has been administered through a special bait particularly attractive to foxes. It was successful in halting an epizootic. There were 4,219 cases of animal rabies reported to the CDC in 1990 which include both domestic and wild animals. Most municipalities have laws requiring rabies immunization for dogs but not necessarily for cats. Consequently, in the last few years, cats have accounted for the majority of the rabies cases in domestic pets (table 19.3). Of significance to the farmer is the fact that cattle are very susceptible to rabies and often are infected by contact with rabid wild animals.

Prevention of rabies requires that domestic pets be immunized. The same is not true for most people, although those working with animals (veterinarians, animal caretakers, etc.) should receive a **prophylactic** (preexposure treatment) series of rabies shots. A person who has been preimmunized should still receive the normal series of injections if exposed to rabies. For the vast majority of people, the best prevention is to avoid contact with bats and other wild animals that might carry rabies.

Several tragic cases of rabies have occurred in persons receiving corneal transplants from donors who had died of atypical rabies. Diagnosis of rabies can be done easily at autopsy; therefore, a new rule has been instituted to prevent the use of tissue transplants from cadavers unless the cause of death is clearly established not to be rabies.

TABLE 19.3 Reported rabies in the United States by animal type[a]

Year	Dogs	Cats	Farm Animals	Foxes	Skunks	Bats	Raccoons	Other Animals	Humans	Total
1953	5,688	538	1,118	1,033	319	8	40	79	14	8,837
1965	412	289	625	1,038	1,582	484	99	54	1	4,584
1975	129	104	200	276	1,226	514	192	31	3	2,675
1986	95	166	255	207	2,379	788	1576	85	0	5,551

Source: [a]Adapted from *Morbidity and Mortality Weekly Report* 36(3S):16S (1987).

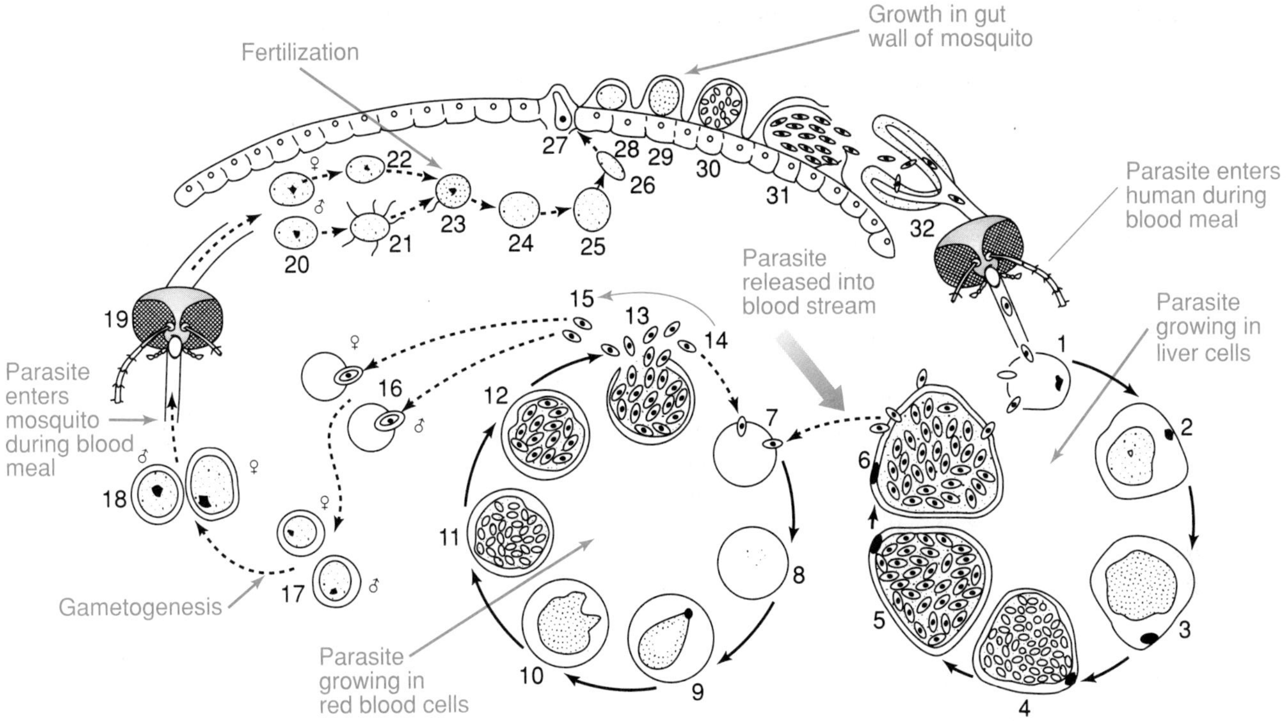

FIGURE 19.4

Life cycle of *Plasmodium* sp.

From Marvin C. Meyer and O. Wilford Olsen, *Essentials of Parasitology*, 3d edition. Copyright © 1980 Wm. C. Brown Publishers, Dubuque, Iowa. All Rights Reserved. Reprinted by permission.

Diseases Caused by Prions (Unconventional Viruses)

Prions are filtrable infectious agents composed of protein instead of nucleic acid (chapter 9). They seem to cause a family of degenerative brain diseases characterized by slow onset and inevitable fatality. The best studied of these diseases is kuru, a disease identified in Papua New Guinea among tribes that have been reported to have practiced ritual cannibalism. Extracts from the brains of affected individuals can be shown to be highly infectious in chimpanzees. The disease has nearly disappeared now that tribal elders have banned risky practices, like removal of brains from corpses.

Closer to home, there is a disease very similar to kuru called Creutzfeld-Jakob disease. It too is characterized by progressive brain degeneration. A recent cluster of Creutzfeld-Jakob cases involving persons who had received injections of human pituitary extract (containing growth hormone) suggests that this disease may be transmitted in the same manner as kuru is. Continuing studies seem to bear out this assumption.

Protozoan Diseases

Giardiasis. Giardiasis is another diarrheal disease caused by the drinking of fecally contaminated water. If *Giardia lamblia* cysts are present, they activate and release the trophozoite form that attaches to the walls of the small intestine. Once in place, the protozoan causes a persistent diarrhea and can, in sufficient numbers, block normal absorption of food materials. At the present time, the disease is endemic in the Rocky Mountain area. Prevention includes proper treatment of a water source whether from a lake serving a city or from a mountain stream. Giardiasis is treated with antimicrobials like quinacrine.

Malaria. Members of the *Plasmodium* genus cause malaria. At least five species are known to infect humans, but only two pose major problems (*P. vivax* and *P. falciparum*). The protozoans are spread by *Anopheles* mosquitoes, arthropod vectors that form an essential part of the protozoan life cycle (figure 19.4).

The female mosquito (the male does not feed on humans) may in the course of a blood meal receive the gametocytes of the *Plasmodium* organism. Gametocytes complete their differentiation into gametes in the midgut of the mosquito. After fertilization, sporozoites develop that eventually migrate to its salivary glands. During a blood meal, the mosquito uses its saliva as an anticoagulant. In this way, an infected mosquito inoculates its host with the *Plasmodium* organism. Sporozoites migrate first to the host liver and differentiate into merozoites that are eventually released to infect red blood cells. Merozoites form trophozoites in red blood cells that, in turn, produce more merozoites. Eventually, differentiation into gametocytes occurs, and the life cycle is repeated.

Infection of new red blood cells by merozoites occurs as a result of lysis of existing red blood cells. The cellular debris is pyrogenic and causes the characteristic fever of malaria. The cycle of infection and release of merozoites becomes synchronized so that the fever recurs on a regular basis, every 48 hours for *P. falciparum* and *P. vivax* (tertian malaria) and every 72 hours for *P. malariae* (quartan malaria). The accumulated degradative products of the red blood cells can cause brain or kidney damage.

Prevention of the disease is by avoidance of mosquitoes. Insect control programs have helped many urban areas to greatly reduce the potential for malaria. Malaria was formerly a serious problem in the United States, particularly along the Gulf Coast, but all the 1,044 cases reported in 1988 were diagnosed in individuals who had contracted the disease outside the country. However, among the 1,232 cases reported in 1989 were several outbreaks of malaria that occurred among migrant farm workers in California. Apparently, persons returning from Mexico brought *Plasmodium* organisms with them and infected California's local mosquito populations. Local transmission then occurred among individuals camped along rivers and streams.

Normally, travelers to areas in which malaria is endemic are urged to take chloroquine prophylactically, although resistant strains of *Plasmodium* are now known. Chloroquine kills plasmodia as they enter the body and prevents establishment of an infection. Because the drug is rather slow acting, it must be taken before departure and continued for several weeks after return. If malaria develops regardless, higher doses of chloroquine can be used in conjunction with primaquine.

Fungal Diseases

Aflatoxicosis. Aflatoxicosis is a type of food poisoning induced by aflatoxins that are synthesized by *Aspergillus flavus* and *A. parasiticus.* The fungi are known to grow on stored food such as cottonseeds (animal feed), peanuts, rice, and maize. Metabolites of the aflatoxins are poisonous to the liver and extremely carcinogenic. For this reason, the permissible levels of aflatoxins in grain and nut stores are very low.

Complex interactions in the food chain sometimes come into play. For example, cottonseed contaminated with aflatoxin can be used as cattle feed. There is concern that aflatoxins might pass from the cow into her milk. Thus, humans could be affected by a contaminated product not intended for human consumption. Generally, the secret to prevention is proper storage of susceptible foods under sufficiently dry conditions in which the fungi are unable to grow.

Candidiasis. *Candida albicans* is an opportunistic dimorphic fungus. Surface infections by this organism primarily yield yeast cells, while deeper infections yield actual hyphae. *C. albicans* is found commonly among the normal oral, genital, and intestinal microflora. Its invasive activities can be triggered by malignancies, immunosuppression, or antibiotic destruction of normal flora. Surface infections can be treated with one of the polyene antibiotics, and amphotericin B is used for persistent deeper infections.

Ergotism. Ergotism is another type of food poisoning known since the Middle Ages (formerly called St. Anthony's fire). A mycotoxin produced by *Claviceps purpurea* is the causative agent. The fungus is a common plant pathogen that grows on cereal grains, particularly on rye. When consumed, it can induce various neurologic problems including convulsions or destroy the circulation to a limb with consequent gangrene. Good farming and storage procedures as in the case of aflatoxicosis, are preventive measures.

Valley fevers. There are two fungi that cause very similar diseases named after river valleys. *Histoplasma capsulatum* is found commonly along the Ohio and Mississippi river valleys, and *Coccidioides immitis,* first identified in the San Joaquin valley of California, is found in Arizona, New Mexico, Texas, and northern Mexico as well. Both fungi are dimorphic soil organisms. The hyphal forms occur in the soil and release their spores. Inhalation of a dust and spore mixture normally infects the vast majority (as much as 80%) of humans living in these areas. In the usual course of events, the fungal spores enter the lungs and germinate to yield the yeast form of the fungus. The resulting infection induces symptoms resembling a cold and is limited by the normal cell-mediated immune response, but not until some scarring of the lung tissue has occurred. Persons who have been exposed to the disease can be identified by skin tests similar to the tuberculin skin test.

In some individuals, the immune response apparently is insufficient to control the fungus, and a disease similar to tuberculosis results. As in the case of tuberculosis, lung lesions can eventually break and allow invasion of the vascular system. In this event, the fungi will be disseminated throughout the body. Prior to the advent of amphotericin B, such infections were always fatal. Amphotericin B controls histoplasmosis well and coccidiomycosis moderately well. At the present time, vaccines are not available for either disease.

Summary

The real triumphs of modern public health practices can perhaps best be seen in the context of the diseases discussed in this chapter. Despite some really serious consequences, these diseases are by and large of comparatively little significance in the United States today. For those diseases for which the general population is at risk, vaccines often are available and their use mandated before students enter the public school system. In some cases, vaccines are available but held in reserve until actual exposure occurs. These diseases are not a threat to the average person. Where vaccines are not available, transmission of the infecting organism must be interrupted.

Disease transmission can occur in many ways, but two stand out. Fecal contamination is the mode of transmission of a wide variety of diseases, not all of which are diarrheal. Arthropod vectors present significant hazards to many populations, and their control would effect a significant improvement in public health. In the case of AIDS, modification of various behaviors regarding sexual practices and intravenous drug abuse is necessary to prevent its spread.

Questions for Review and Discussion

1. Give four examples of diseases transmitted by different means. How might a public health worker try to break the transmission pattern for each disease?
2. Give an example of a vaccine whose use is recommended for all people. Give an example of a vaccine whose use is restricted. Why are both vaccines not used all the time?
3. As you look over the lists of diseases presented in this chapter, you might notice that some can be treated more readily than others. Why is this? What problems arise in treating certain types of diseases? Does future research offer any hope of improvement?
4. What diseases might be spread by contact with saliva or mucus? Do you think that the present fad of spitting in public places poses a health hazard? Why?
5. Give some examples of situations in which antibiotics are not totally effective, and explain what the problem is in each case.

Suggestions for Further Reading

Davis, B. D., R. Dulbecco, H. N. Eisen, and H. S. Ginsberg. 1988. *Microbiology: Including immunology and molecular genetics,* 4th ed. Harper & Row, Publishers, Inc. Hagerstown, MD. A textbook intended as an introduction to microbiology for medical students and others with similar interests.

Fox, J. L. 1989. Rabies vaccine field test—pending. *ASM News* 55:355–357. This article provides some insight into the problems of obtaining regulatory approval for the use of genetically engineered vaccines.

Hinman, A. R., J. P. Koplan, and W. A. Orenstein. 1988. Live or inactivated poliomyelitis vaccine: An analysis of benefits and risks. *The American Journal of Public Health* 78:291–295.

Morbidity and Mortality Weekly Report. This short but informative summary of public health activities in the United States is prepared by the Centers for Disease Control and distributed by the Massachusetts Medical Society, CSPO Box 9120, Waltham, MA 02254–9120. The price is about the same as that for a monthly magazine. The terminology used is not beyond the scope of the student who has completed this book.

Radetsky, R. 1985. The rise and (maybe not the) fall of toxic shock syndrome. *Science* 85 6(1):73–79.

Rosebury, T. 1971. *Microbes and morals: The strange story of venereal disease.* Viking Press, New York. A survey of the common sexually transmitted diseases with emphasis on syphilis written by a scientist who spent most of his life studying them. A most interesting book for the scientist or layperson.

Stevens, J. G. 1989. Human herpes viruses: A consideration of the latent state. *Microbiological Reviews* 53:318–332. This is a good summary of current information about the physiology and biochemistry of latent herpes viruses. It includes material on all herpes viruses discussed in this chapter.

Szczepanski, A., and J. L. Benach. 1991. Lyme borreliosis: Host responses to *Borrelia burgdorferi*. *Microbiological Reviews* 55:21–34. This article reviews the immune responses of the host and their contribution to disease symptoms.

Tiollais, P., and M. A. Buendia. 1991. Hepatitis B virus. *Scientific American* 264(4):116–123. This article reviews some of the medical problems caused by this virus as well as the unusual molecular biology of its replication.

Zinsser, H. 1963. *Rats, lice, and history*. Little, Brown & Co., Boston. A biography that, after 12 preliminary chapters indispensable for the preparation of the layperson, deals with the life history of typhus fever. The title says it all.

part

4

MICROBIAL TAXONOMY

Taxonomy (or systematics) is an important discipline within biology in which we attempt to organize our knowledge about organisms so as to permit the discovery of new relationships. It also is valuable to students because taxonomic divisions are like a form of shorthand. When a taxonomic group is specified for an organism, certain other traits are implied automatically. The final chapter of this book provides an overview of the generally accepted taxonomic organization for microorganisms. A more detailed exposition of prokaryotic taxonomy can be found in appendix A.

chapter

21

Taxonomy of Microorganisms

chapter outline

■

GENERAL GOALS

Be able to describe some methods used in classic taxonomy as well as the newer methods used in molecular taxonomy.

Be able to discuss some possible ways in which a species can be defined.

Be able to describe the general taxonomic classifications of microorganisms.

This chapter considers **taxonomy,** the science of classifying organisms into related groups. Although a general taxonomic outline for all three kingdoms of microorganisms is included in this chapter, major emphasis is placed on strategies used to classify microorganisms and on what those classifications imply for the evolution of organisms. Detailed bacterial taxonomy can be found in appendix A. In order to understand the following discussion, familiarize yourself with the five-kingdom classification of Whittaker presented in table 2.2.

Taxonomic Survey of Eukaryotic Microorganisms

Typical eukaryotic microorganisms are shown in color plates 2 and 3. It is easy to see that although they have many similarities, there are fundamental differences among them reflected by their classification into two separate kingdoms.

The Kingdom *Fungi*

The basic structure of nearly all fungi is the hypha, the tubular structure of protoplasm shown in figure 2.13A–C. Like the taxonomic organization of plants and animals, the taxonomic organization of fungi is grounded primarily in their methods of reproduction, both sexual and asexual. The most elementary type of reproduction is the simple fragmentation of a mycelial mat into hyphal pieces. If the fragmentation is such that individual cells result, the cells are assigned special names that generally include the suffix "spore." If the cells are essentially unmodified, they are called **arthrospores** (figure 20.1A). On the other hand, if they develop a thickened cell wall as they separate from the hypha, the cells are designated as **chlamydospores.** Although chlamydospores are relatively more resistant to environmental changes than are arthrospores, they are not nearly so resistant to environmental influences as bacterial endospores.

Because both arthrospores and chlamydospores are products of mitotic cell divisions, they are elements of **asexual reproduction.** More elaborate asexual spores can be produced in a manner that keeps them attached to the hyphae (figure 20.1B–D). If they are contained in sacs they are called **sporangiospores,** whereas if they develop as separate cells on the ends or sides of the hyphae they are called **conidia** (singular conidium). Occasionally, asexual reproduction occurs by budding instead of by the conventional process of cell cleavage. Budding is most frequently encountered in certain types of fungi called yeasts.

Sexual reproduction in the fungi is a somewhat different process than it is in animals. Meiosis, or reduction division, is still a distinguishing characteristic and occurs on special reproductive hyphae (as distinguished from the vegetative hyphae that are used to increase the mass of a fungus). The haploid products of meiosis are not necessarily gametes, however. **Gametes** are specialized cells that fuse together to form a diploid zygote or fertilized cell from which the new individual is produced. Among many eukaryotic microorganisms, the result of meiosis is the production of haploid cells that give rise to individuals with their own life cycles. Specialized cells within these haploid individuals can then differentiate to form gametes that fuse to start a new diploid cell. Alternatively, the haploid cells may fuse, but the nuclei do not. This process gives rise to a single cell with two different haploid nuclei. Such a cell is called a **heterokaryon.**

The differences in sexual cycles of fungi can be used to develop a taxonomic schema (table 20.1). One commonly used approach is to break the kingdom *Fungi* into two divisions, the *Mastigomycotina* (fungi with centrioles) and the *Amastigomycotina* (fungi without centrioles). The *Mastigomycotina* are the "lower" fungi characterized by coenocytic hyphae (lack any partial cross walls) that form various types of asexual and sexual spores. Included within this group are all organisms that form sporangiospores. One member of this division belongs to the genus *Phytophthora* and is depicted in figure 20.2.

The "higher" fungi are grouped into four phyla *(Zygomycotina, Ascomycotina, Basidiomycotina, Deuteromycotina)* that are distinguished on the basis of their mechanism of sexual reproduction (or lack of it). In the *Zygomycotina,* sexual reproduction results when specialized hyphae (**gametangia**) from different mating types fuse to produce a **zygospore.** A typical member of this subdivision is *Rhizopus,* the common bread mold (figure 20.1B).

A typical member of the *Ascomycotina* is *Neurospora,* the organism familiar to generations of geneticists. Ascomycetes produce a saclike structure called an ascus in which the products of meiosis (ascospores) can be found arrayed (figure 20.3).

In the *Basidiomycotina* are found the familiar mushrooms that "sprout" after a heavy rain. The visible structure is actually only a small portion of the body

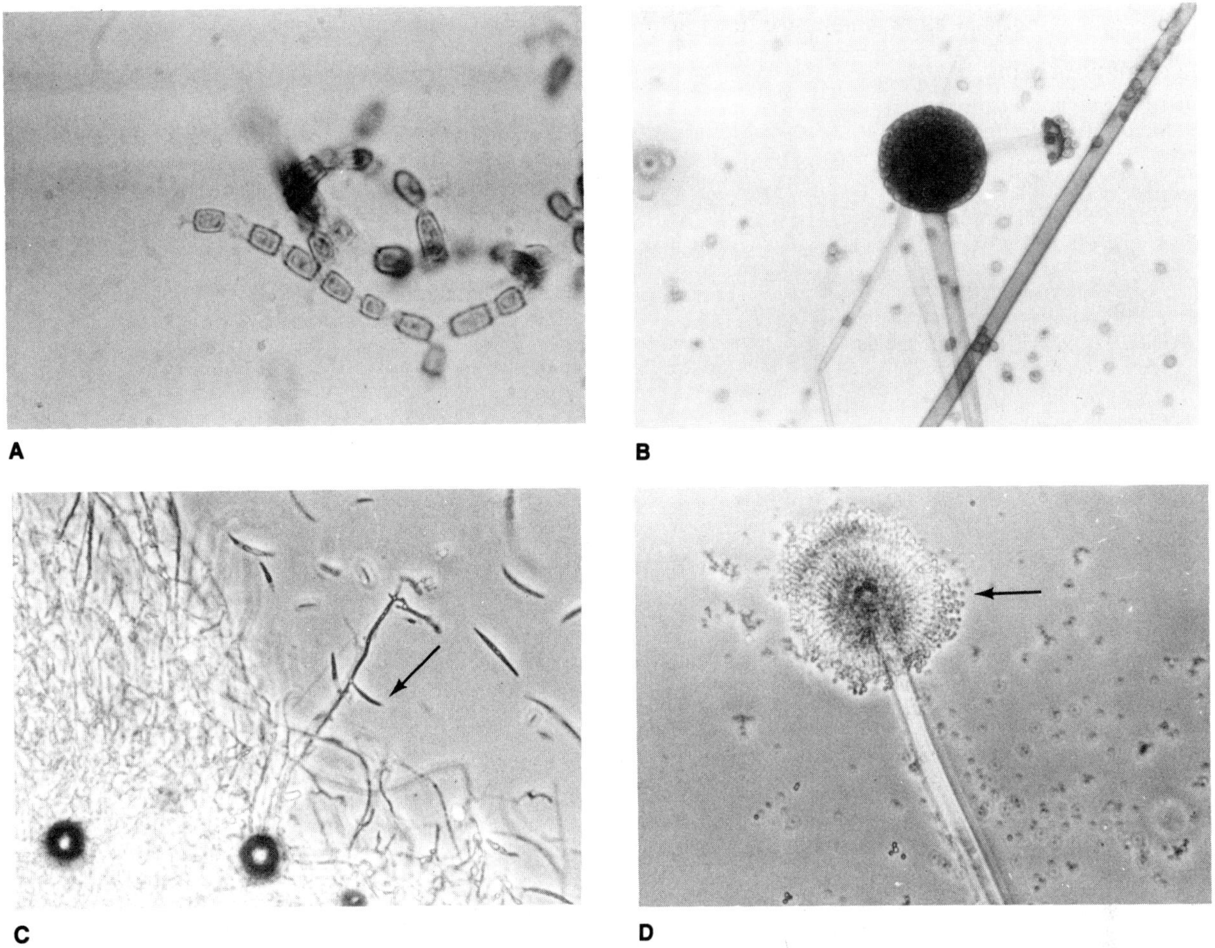

FIGURE 20.1

Asexual fungal spores. (**A**) Arthroconidia of *Coccidoides immitis.* The spaces between the arthrospores are empty or vacuolated cells. (**B**) Sporangiospores of *Rhizopus* sp., a member of the phylum *Zygomycotina.* The spores are contained within a large sac or sproangium. (**C**) A conidium (arrow) projecting from the side of a *Fusarium* sp. hypha. (**D**) Chains of conidia (arrow) on the surface of an *Aspergillus* sp. vesicle.

(A) Courtesy of the Centers for Disease Control, Atlanta, Georgia.

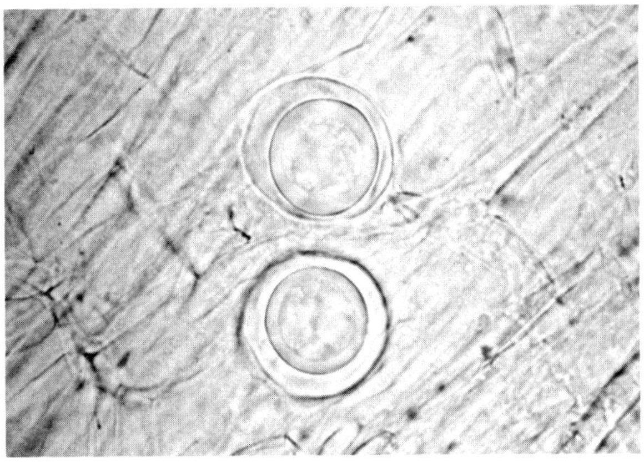

FIGURE 20.2

Phytophthora, a plant pathogen and lower fungus.

© D. A. Glawe/BPS

of the fungus, the rest of which is located underground. The vegetative hyphae proliferate underground until hyphae from two different mating types touch one another. The cells then fuse to form a heterokaryon that grows and divides to produce the body of the mushroom (figure 20.4A–B). When a heterokaryotic cell divides, a clamp connection forms between the two daughter cells to allow passage of one haploid nucleus (figure 20.5). The other haploid nucleus is transmitted by normal mitosis. As the mushroom cap forms, the nuclei within the heterokaryon fuse to give a single diploid nucleus that then goes through the meiotic process to give rise to special cells called **basidia** that bear the haploid basidiospores (figure 20.4B). The mushroom most commonly found in stores is *Agaricus brunescens.*

TABLE 20.1 A classification scheme for fungi

Taxonomic Schema	Characteristics
Division *Eumycota* (true fungi)	
Subdivision *Mastigomycotina*	Have centrioles
Phylum *Mastigomycotina*	Have motile, flagellated stage (zoospores or motile gametes)
Subdivision *Amastigomycotina*	Do not have centrioles
Phylum *Zygomycotina*	Produce zygospores
Phylum *Ascomycotina*	Produce ascospores
Phylum *Basidiomycotina*	Produce basidiospores
Phylum *Deuteromycotina*	Have no sexual stage
Division *Myxomycota* (slime molds)	
Myxomycetes	True slime molds
Acrasiomycetes	Cellular slime molds
Plasmodiophoromycetes	Endoparasitic slime molds

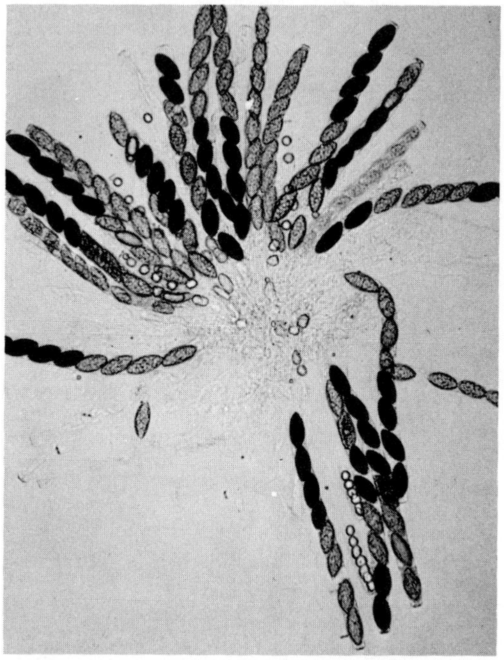

FIGURE 20.3

Neurospora, perhaps best known as the organism used in early work on the nature of meiosis.

© Omikron/Photo Researchers

A

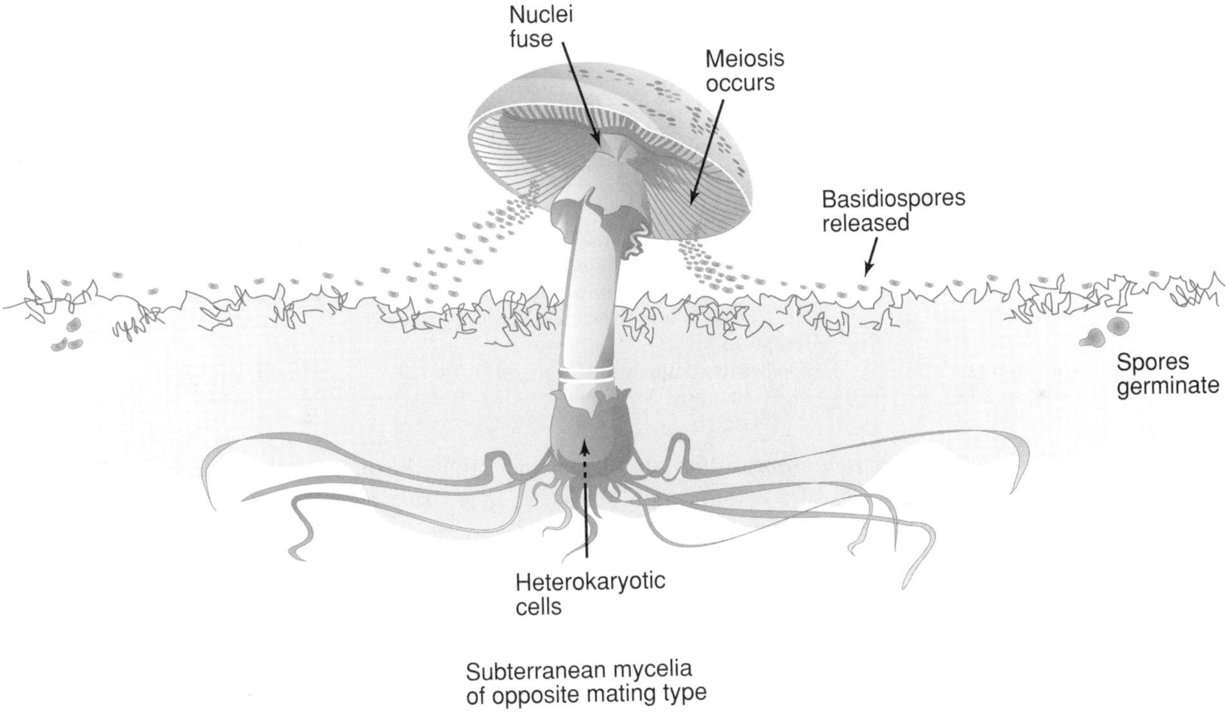

B

FIGURE 20.4
(**A**) A mushroom. (**B**) Diagram of a typical life cycle.

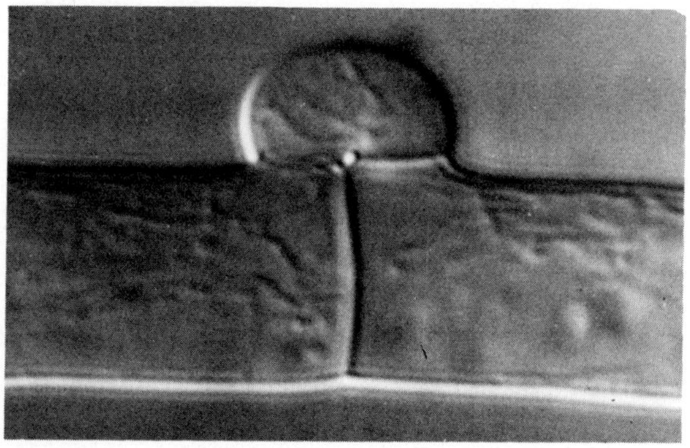

FIGURE 20.5

A clamp connection seen in *Sclerotium rolfsii.* During mitosis one of the extra nuclei in a dividing heterokaryotic cell can migrate to a daughter cell through this connection.

Micrograph courtesy of Dr. Robert Roberson, Department of Botany, Arizona State University.

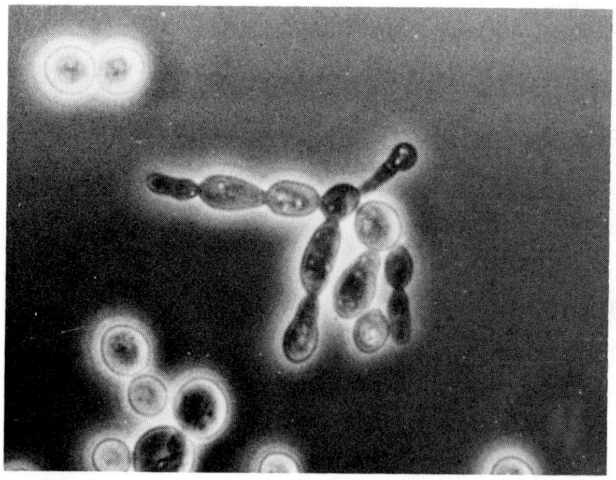

FIGURE 20.6

A pseudohypha of *Saccharomyces cerevisiae.*

TABLE 20.2. A classification scheme for the slime molds

Class	Characteristics	Example Organism
Acrasiomycetes	Amoebae with no cell walls; do not lose their identity within the plasmodium	*Dictyostelium discoideum*
Myxomycetes	So-called true slime molds in which the plasmodium is a multinucleate mass of protoplasm	*Physarum polycephalum*
Plasmodiophoromycetes	Endoparasitic on plants, algae, and fungi	*Polymyxa graminis*

Members of the subdivision *Deuteromycotina* are distinguished by their apparent lack of a sexual stage. Although a few organisms with sexual stages have been left in *Deuteromycotina* for taxonomic convenience, in general the sexual process does not occur or occurs so infrequently that it has not yet been observed. In some cases, this failure of sexual reproduction is the result of heterothallism (see chapter 10). Pure fungal cultures do not contain the two mating types necessary for sexual reproduction. The *Deuteromycotina* includes most of the fungi that cause disease in humans.

The previously discussed members of the fungal kingdom all have the filamentous appearance usually described as "mold." Not all fungi are molds, however, some are yeasts. A yeast exists either as single cells or as single cells aggregated loosely into **pseudohyphae** (figure 20.6). Most yeasts reproduce by budding, but a few reproduce by the cell fission process typical of the rest of the fungi. In fact, bacteria were originally classified as schizomycetes or fission fungi. Yeasts can be found in any of the "higher" fungal groups, but the yeasts of commercial value are ascomycetes.

Some fungi have the property of being **dimorphic;** that is, they spend part of the time as a typical hyphal fungus and the rest of the time as a yeast. An example of such a fungus is *Coccidioides immitis* that causes the respiratory disease coccidiomycosis (chapter 19). The fungus grows in the soil as hyphae but converts into a yeast form in the lungs of a susceptible host (figure 20.7A–C).

The slime molds are discussed briefly in chapter 2 as organisms with intermediate properties between protozoa and fungi. Their classification is primarily morphologic (table 20.2). One group of organisms formerly classified with the slime molds is now grouped with the protists.

The Kingdom *Protista*

The protists include both phototrophs and chemotrophs and are, therefore, metabolically diverse. Prior to the five-kingdom hypothesis, they were variously classified as algae, protozoa, and occasionally fungi.

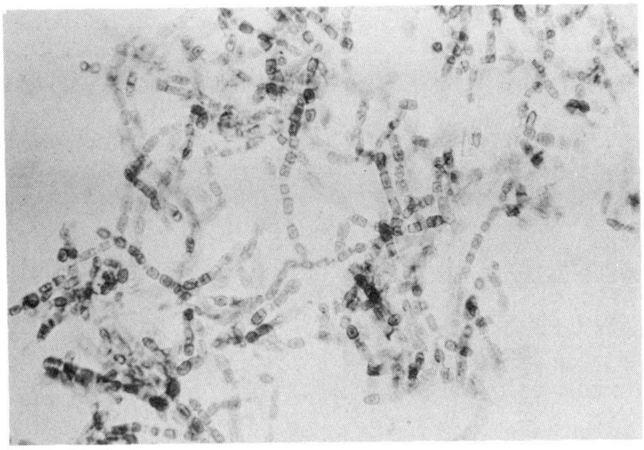

A

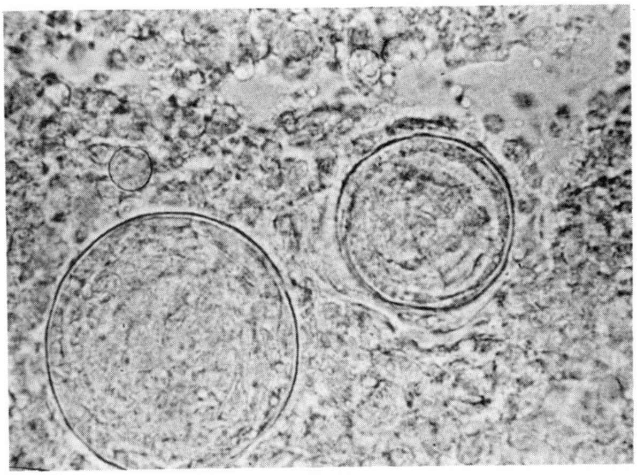

B

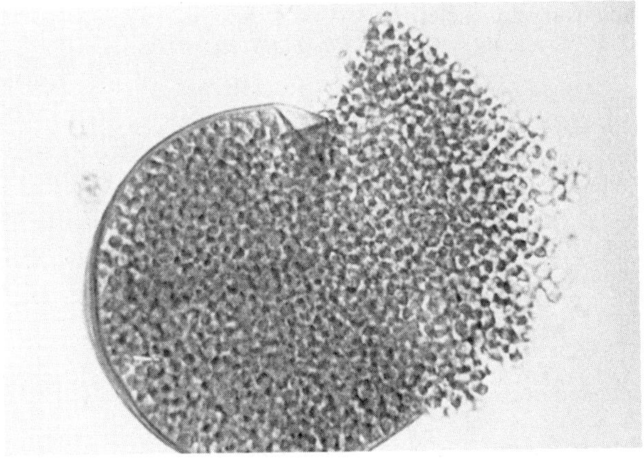

C

FIGURE 20.7

Coccidioides immitis, a dimorphic fungus. (**A**) The filamentous form of the fungus as it occurs in soil. (**B**) A spherule develops from individual cells growing in lung tissue. (**C**) An isolated spherule that has broken open to show the endospores inside it. Each endospore can give rise to another spherule.

(A) Courtesy of the Centers for Disease Control, Atlanta, Georgia. (B–C) Courtesy of Lauren Roberts, Department of Microbiology, Arizona State University.

Protists are a heterogeneous group of organisms ranging from unicellular to multicellular and from strictly microscopic to plantlike structures tens of meters in length. For these reasons, some biologists place the larger algae into a separate group, considering them to be in the kingdom *Plantae.* In this chapter, however, all algae are taken as a single group of eukaryotic organisms that lack true tissue specialization and have only a thalloid structure if multicellular. The prokaryotic algae (cyanobacteria, blue-green algae) are considered bacterial and discussed later. The various phyla assigned to the protists are presented in table 20.3, and some micrographs of representative protists are presented in figures 2.15 and 20.8A–D.

Some clues as to the historical nature of the organisms in any given phylum can be obtained from the suffix used. A name ending with "phyta" is indicative of algae, while a name ending with "mycota" indicates fungal origins. It is important to remember, however, that most phyla have exceptional members, such as organisms that appear identical to photosynthesizers except for a lack of chloroplasts or organisms that are photosynthetic only by virtue of endosymbiotic algae.

In alternative classification systems, some phyla listed in table 20.3 are combined. *Chrysophyta* also may include *Xanthophyta* and *Bacillariophyta. Ciliophora* may include *Stephanopognomorpha.* The phyla *Apicomplexa, Microspora, Acetospora,* and *Myxospora* were formerly grouped together as the *Sporozoa.*

The protists formerly categorized as algae exhibit four different modes of sexual reproduction that can be classified according to the site at which the meiotic events occur (table 20.4). The first (gametic) is typical of multicellular eukaryotes in which haploid gametes are formed that fuse to yield a new zygote. In this mode, there is no persistent haploid stage. The second (zygotic) is the reverse of the first. Haploid organisms differentiate into gametes that fuse to yield the diploid zygote. The zygote immediately undergoes meiosis to begin the haploid phase again.

The third or somatic mode combines the first two. There are two continuing types of cells, haploid and diploid. Each can reproduce itself mitotically. Occasionally, the haploid cells are triggered to differentiate into gametes that fuse to start a new diploid cycle. Sometimes during vegetative growth, the diploid cells are stimulated to undergo meiosis to yield not gametes or spores but the haploid cell type. Haploid cells that can differentiate to give gametes are from the **gametophyte generation**, while cells that can undergo meiosis

TABLE 20.3 Major phyla of protists

Phylum	Characteristics[1]	Common Example (Genus)
Originally Considered as Algae		
Dinophyta (dinoflagellates)	Two dissimilar flagella; chlorophyll c_2	*Gonyaulax*
Chlorophyta (green algae)	Chlorophyll *b;* store starch; have cell walls	*Volvox*
Euglenophyta (euglenoids)	Chlorophyll *b;* have no cell wall; highly motile; grow as photoautotrophs or chemoheterotrophs	*Euglena*
Chrysophyta (golden algae)	Chlorophylls c_1 and c_2; golden chloroplasts	———
Raphidophyta	Chlorophyll *c;* oil as storage product; yellow-green chloroplasts	———
Bacillariophyta (diatoms)	Cell wall is elaborate silica boxes; chlorophylls c_1 and c_2	*Fragilaria*
Phaeophyta (yellow or brown algae)	Chlorophyll c_1 and c_2 but not *b;* size ranges from unicellular to the giant kelp	*Macrocystis*
Xanthophyta	Biflagellate zoospores, chlorophylls c_1 and c_2	*Vaucheria*
Eumastigmatophyta	Coccoid; reproduce by zoospores; single hairy anterior flagellum	———
Haptophyta	Unicellular; has spiral-shaped filiform appendage (haptonema); chlorophylls c_1 and c_2	———
Cryptophyta	Chlorophyll c_2; generally unicellular; reproduce by longitudinal binary fission	*Cryptomonas*
Rhodophyta (red algae)	Have phycobilins as accessory photosynthetic pigments	*Porphyra*
Originally Considered as Protozoa		
Kinetoplasta	Kinetoplast (large concentration of fibrillar protein and DNA) in mitochondrion	*Trypanosoma*
Ciliophora	Motile by cilia arranged in rows; have two kinds of nuclei	*Paramecium*
Stephanopognomorpha	Like *Ciliophora* but have homokaryotic nuclei	———
Apicomplexa	Parasites with mitochondria; life cycle in several stages	*Plasmodium*
Microspora	Parasites with unicellular spores	———
Acetospora	Complex uni- or multicellular spores differentiating in plasmodial mass	———
Myxozoa	Parasites growing as plasmodia within animals	———
Sarcodina	Motile by pseudopodia	*Amoeba*
Choanoflagellata	Unicells similar to sponges; have collar of microvilli at base of flagellum	———
Originally Considered as Fungi		
Phycomycota	Asexual zoospores with single anterior flagellum	———
Chytridiomycota	Asexual zoospores with single posterior flagellum	*Blastocladiella*
Originally Considered as Slime Molds		
Labyrinthomorpha	Produces hyaline cytoplasmic material on which cells of spindle or rhizoid shape glide	———

Note: In addition to those listed, there are three phyla of endobionts (protists living within another organism). For additional details, see Barnes, R. S. K. (ed.). 1984. *A synoptic classification of living organisms.* Sinauer Associates, Sunderland, Mass.

1. Only the predominant characteristics for each phylum are listed. In many cases, the common name (given in parentheses under phylum) is useful in distinguishing members of the phylum. All phototrophs have chlorophyll *a.* Other types of chlorophyll molecules are listed if they are present.

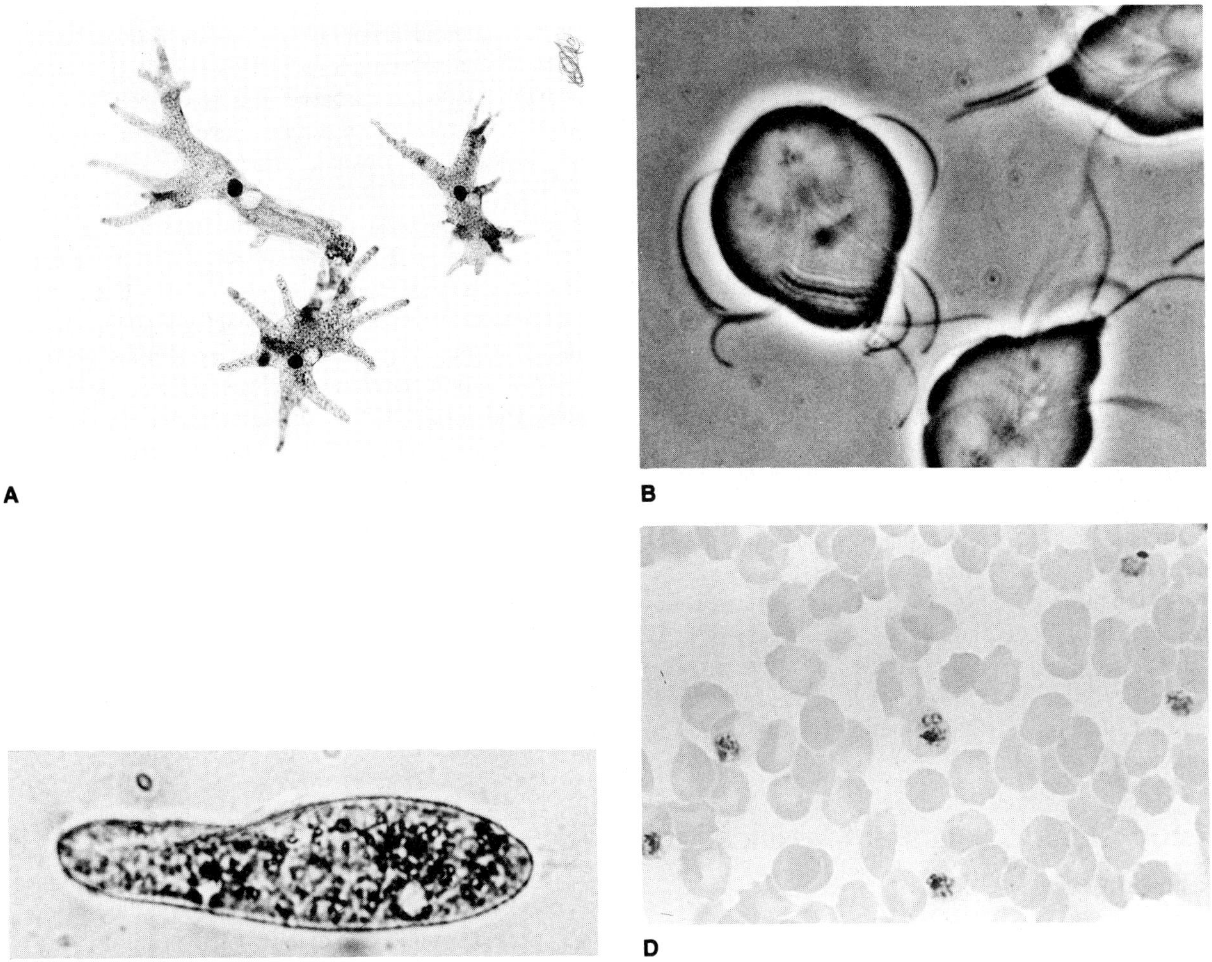

A B C D

FIGURE 20.8

Examples of protists. (**A**) *Amoeba proteus.* (**B**) *Giardia lamblia.* (**C**) *Paramecium aurelia.* (**D**) *Plasmodium falciparum.*

(A & C) © Carolina Biological Supply. (B) © George J. Wilder/Visuals Unlimited. (D) G. W. Willis, M.D./BPS

TABLE 20.4 Modes of sexual reproduction observed in the eukaryotic algae.

Sexual Mode	Meiotic Occurrence	Example (Genus)
Gametic	During gamete formation	*Acetabularia*
Zygotic	As zygote germinates	*Chlamydomonas*
Somatic	During vegetative growth without forming spores or gametes	*Prasiola*
Sporic	During spore formation	*Porphyra*

to yield gametophytes are from the **sporophyte generation.** Gametophytes and sporophytes can have the same or different cell morphology.

Sporic, the final mode of sexual reproduction, is one in which meiosis occurs during spore formation. Germination of a spore yields a haploid cell that eventually differentiates into a gamete.

Most other protists lack obvious sexual processes; therefore, the taxonomic schema developed has relied on other traits. For the chemotrophs, a relatively simple classification is based on their modes of locomotion in the case of free-living forms or on their modes of reproduction in the case of parasitic forms (table 20.3). The morphology of the locomotor apparatus is of great importance even among the phototrophs. For example, members of *Haptophyta* and *Cryptophyta* both have two flagella. Cryptophytes, however, have hairy flagella while haptophytes have naked flagella. Sometimes the spores themselves have flagella, making them zoospores.

Certain specialized structures also can be seen within some protists. Frequently, these involve the cells' DNA. A kinetosome is composed of fibrous strands of protein associated with specific, nongenomic DNA. Heterokaryons have two types of nuclei in one cell, a macronucleus and a micronucleus.

Among the phototrophs, the *Rhodophyta* are considered to be the most similar structurally to prokaryotic cyanobacteria. They have no chloroplast endoplasmic reticulum and no flagella. Their mode of fatty acid biosynthesis is also similar. At the other extreme are multicellular algae, *Phaeophyta* and some members of *Chlorophyta,* that sometimes are assigned to the kingdom *Plantae.*

Theory Behind Bacterial Classification

Definition of a Species

Bacterial taxonomy is based on the same system of binomial nomenclature that was originally proposed by Linnaeus. Thus, there is an appropriate genus and species name for every bacterium. The difficulty arises, however, with respect to what constitutes a species. The situation is well defined in the kingdom *Animalia* in which different species are reproductively isolated physiologically and/or genetically. Each species, therefore, represents a separate breeding group. Matters are not quite so clear-cut in the kingdom *Plantae,* however. There is a strong tendency for plants of what are considered to be different but closely related species to cross-pollinate one another, resulting in the production of hybrids. These hybrids usually have difficulty in reproducing by meiosis and are sterile, unless they also double their chromosome number to give each chromosome a homolog for pairing. Such polyploid offspring are fertile and, depending on the definition used, may or may not constitute a new species.

Fungi and protists represent a grouping between animals and plants on the one hand and monerans on the other. Their comparatively small size has rendered them more difficult to study, and hence their taxonomy has been grounded not only on modes of sexual reproduction but also on other properties as well. In contrast, moneran taxonomy is based almost entirely on properties other than those directly involved in reproduction, sexual or asexual, which has led to some difficulties.

Because bacteria do not reproduce by mitosis or meiosis and do not have sexual processes in the usual sense of the word, there has always been uncertainty and disagreement as to which bacteria should be included within a particular species. Philosophically, bacteria within the same species should be closely related, and species within the same genus should have less extensive similarities, but how is relatedness to be defined and measured? The following subsections introduce some ways in which bacterial relationships are tested and evaluated. The question of what constitutes a species is reconsidered at the end of this section.

Physical and Biochemical Description of Bacteria

Chapters 2 through 4 present some methods used in describing bacteria. Cells can be described by their size and shape as well as by their type of general metabolism (chemotrophic, facultatively anaerobic, etc.). Various stains provide information about the chemical composition of the cell wall, and the electron microscope permits a general examination of the cell surface, cell envelope profile, and cytoplasm. The ability to synthesize essential metabolites can be characteristic. The problem that arises, however, is determining which of these traits is the most significant in assigning the organism to the correct species.

At various times, that question has been answered differently. Historically, the Gram reaction has been of fundamental importance, and in nearly all taxonomic schemas it is used as an indicator of cell wall structure. Nevertheless, certain bacteria present problems. In some instances, the gram-negative reaction is not always meaningful. Some bacteria have no cell walls or atypical cell walls. For other bacteria, the Gram reaction is variable, being gram positive during one phase of the growth cycle and gram negative during another. The genus *Bacillus* provides a good example because its members have cells that tend to stain gram negative in the process of forming spores, while normally growing *Bacillus* cells are strongly gram positive. Perhaps the strangest example, however, is in the genus *Arthrobacter,* whose members are coccoid in shape and stain gram positive when in the stationary phase of growth, but they are rod shaped and frequently gram negative

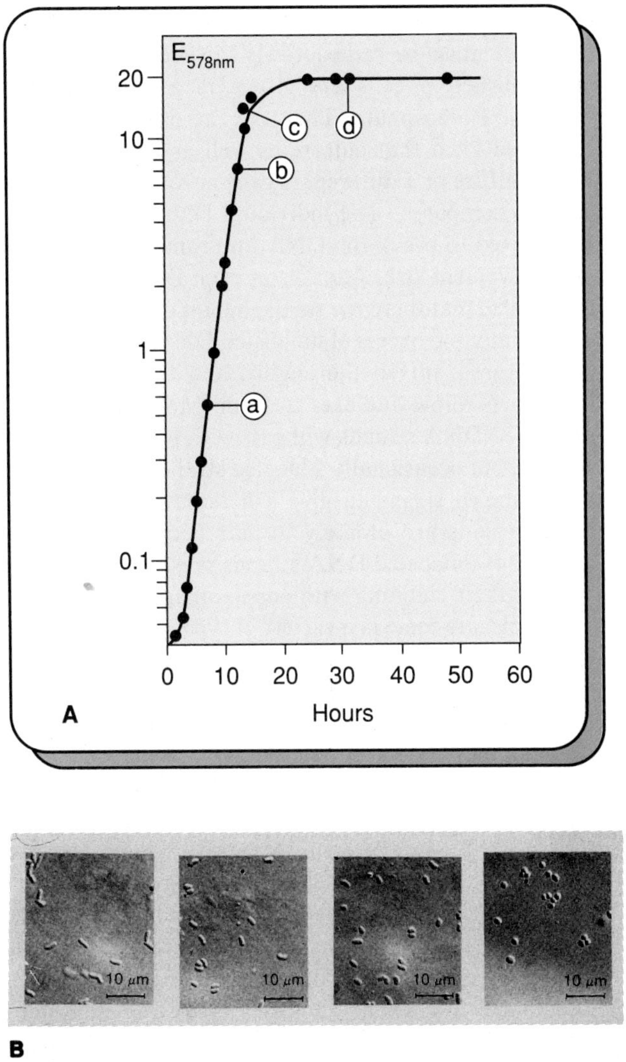

FIGURE 20.9

Influence of growth phases on cell morphology in *Arthrobacter crystallopoietes.* (**A**) Growth curve for the culture. Samples were taken at the indicated points. (**B**) Light micrographs of the culture samples. Note the change in cell shape.

(**B**) Reprinted from Faller, A. H., and K. H. Schleifer. 1981. Effects of growth phase and oxygen supply on the cytochrome composition and morphology of *Arthrobacter crystallopoietes. Current Microbiology* 6:253–258.

unless very carefully stained when in the exponential phase of growth (figure 20.9A–B). It is thus extremely important that the culture conditions employed be specified whenever the results of an experimental procedure are described. Electron microscopic examination of cell walls also is helpful.

Beveridge has analyzed the results of modified Gram stain procedures in the transmission electron microscope and has subdivided the gram-variable organisms into two broad groups (see his article referenced in the suggestions for further reading at the end of this chapter). Members of *Actinomyces, Arthrobacter, Corynebacterium, Mycobacterium,* and *Propionibacterium* tend to become gram negative during the exponential growth phase, and the number of cells staining gram negative can be as high as 40% as these cells go into the stationary phase. The common features of the cells that stain gram negative are initiation of septum formation and a generally weakened cell wall. Members of *Bacillus, Clostridium,* and *Butyrivibrio* also tend to become gram negative as they age so that older cultures are almost entirely gram negative even though in lag and early exponential phases they are clearly gram positive. The effect is most pronounced during the transition to stationary phase and again is associated with a weakening of the cell wall.

Organisms that seem identical except for one fairly obvious trait are commonly encountered. For example, members of the genera *Escherichia* and *Shigella* show no real differences except that *Escherichia* can use the sugar lactose as a carbon source and *Shigella* cannot. *Clostridium botulinum* cannot be distinguished from *C. sporogenes* on any morphologic or biochemical criteria, only on the basis of toxin production. *C. botulinum* toxin comes in distinct varieties with similar effects and is encoded by any one of several temperate phages. Disagreement easily can arise in such cases as to whether the difference is significant enough to warrant the creation of two separate genera or species. The two sides in such a taxonomic disagreement are commonly referred to as "lumpers" and "splitters," depending on their choice in the dispute.

As such dilemmas became more frequent, a new school of thought developed in which it was held that no trait could be considered more important than any other. Instead, the possible results from any test are assigned a numeric score, and an organism is characterized from as many test results as possible, generally at least 50 and sometimes several hundred. A computer is used to compile patterns of similarity and develop a matrix consisting of a list of strains tested and a list of test results. The observed patterns are used to propose taxonomic relatedness. This system is called numeric taxonomy or, perhaps more appropriately, computer-assisted taxonomy.

However, even the proponents of numeric taxonomy caution that its main validity is statistical and that the system reflects only similarity in phenotype. On the other hand, if taxonomic relatedness has any real meaning, it is in terms of genotype, the actual genetic code used by the organisms under study. Proponents of numeric taxonomy presume that organisms having diverged recently in evolutionary time will show more similarities in their DNA and RNA sequences than organisms having diverged in the remote past. The present trend is to develop measures of relatedness that focus on the actual genetic code contained within the organism, therefore.

DNA Similarity Analysis

It would be a tremendous amount of work to determine the complete nucleic acid sequence of a single organism; therefore, it is clearly out of the question for the large number of bacteria that need classification. Instead, various measures are used to reflect the genetic composition of a particular portion of the DNA sequence that can be determined easily, and these limited measures can be compared between organisms. If two organisms have the same genetic information, their base composition, that is, the percentage of DNA bases that are thymine, cytosine, guanine, and adenine, should be similar. Because cytosine is always paired to guanine, base composition usually is expressed as the percentage of bases that are either guanine or cytosine and is abbreviated as %(G+C). Bacteria can have a %(G+C) ranging from about 35% on the low end to more than 70% on the high end. A similarity in the %(G+C) is considered suggestive of a relationship between two organisms, but it has been shown that dissimilar organisms sometimes have similar base compositions.

Additional information about the relatedness of two organisms can be obtained from DNA hybridization studies. The double-helical structure of DNA is not maintained under all conditions, and if the temperature or the pH is high enough, the DNA is **denatured** and separated into its individual strands. When the temperature or pH is lowered once again, the DNA strands will attempt to renature (reassociate) or return to their original double-strand state of maximum hydrogen bonding and minimum thermodynamic energy. This state can be attained only when the base sequences of the two strands are exactly complementary. Intermediate states are possible, however; therefore, two strands that are not perfectly complementary can form a reasonably stable duplex (a **heteroduplex** or duplex made up of strands from different DNA molecules). The probability of finding such a duplex is proportional to the degree of base sequence homology. Thus, a measurement of the amount of such duplexes provides an indication of the similarity of two DNA molecules. An example of a procedure for measuring homology is shown in figure 20.10.

For DNA hybridization studies, the DNA of one organism must be radioactively labeled. This can be accomplished if cells are grown in a medium containing ^{32}P phosphate. Then the radioactive DNA is extracted from that culture as well as from an unlabeled culture of a different organism to which the first is to be compared. The individual DNA preparations are treated to break the DNA into random fragments of a convenient size. Aliquots of each DNA are mixed together so that there is a small amount of labeled DNA and a manyfold excess of unlabeled DNA. The mixture is denatured, mixed thoroughly, and then slowly renatured to allow duplexes to form. Most of the time, unlabeled DNA strands will pair with unlabeled DNA strands, but occasionally a labeled strand will pair with an unlabeled strand to give a heteroduplex. Two labeled strands are unlikely to pair because the large number of unlabeled DNA strands present will tend to prevent their colliding with one another.

Nucleases come in a variety of types, and one called S1 nuclease can break down single-strand DNA while leaving duplex DNA untouched. After the DNA mixture obtained by reannealing has been treated with the enzyme, the amount of heteroduplex DNA remaining can be ascertained by passing the reannealed DNA through an hydroxyapatite gel packed into a glass tube (column). Hydroxyapatite specifically binds duplex DNA while allowing single-strand fragments and individual nucleotides to pass through. The percentage of the radioactive label that attaches to the gel is taken as the percentage of similarity between the DNA sequences of the two organisms. Note that this percentage is a measure of overall homology and not of any specific sequence.

The most recent development along the lines of DNA homology analysis is the use of specific DNA fragments to recognize particular species or genera. The basic idea is to identify a comparatively short region of DNA that is specific to the group to be identified and then to use that piece of DNA as a probe in a Southern or Northern blot reaction (see chapter 11). If there is strong binding of the probe, the unknown DNA is homologous. If there is no binding, the unknown DNA belongs to a different group. The degree of stringency of the test can be controlled if the conditions of the hybridization reaction are varied; for example, a higher temperature would require a better match of bases. Probe technology is useful particularly in identifying rare bacteria from environmental or other complex sources.

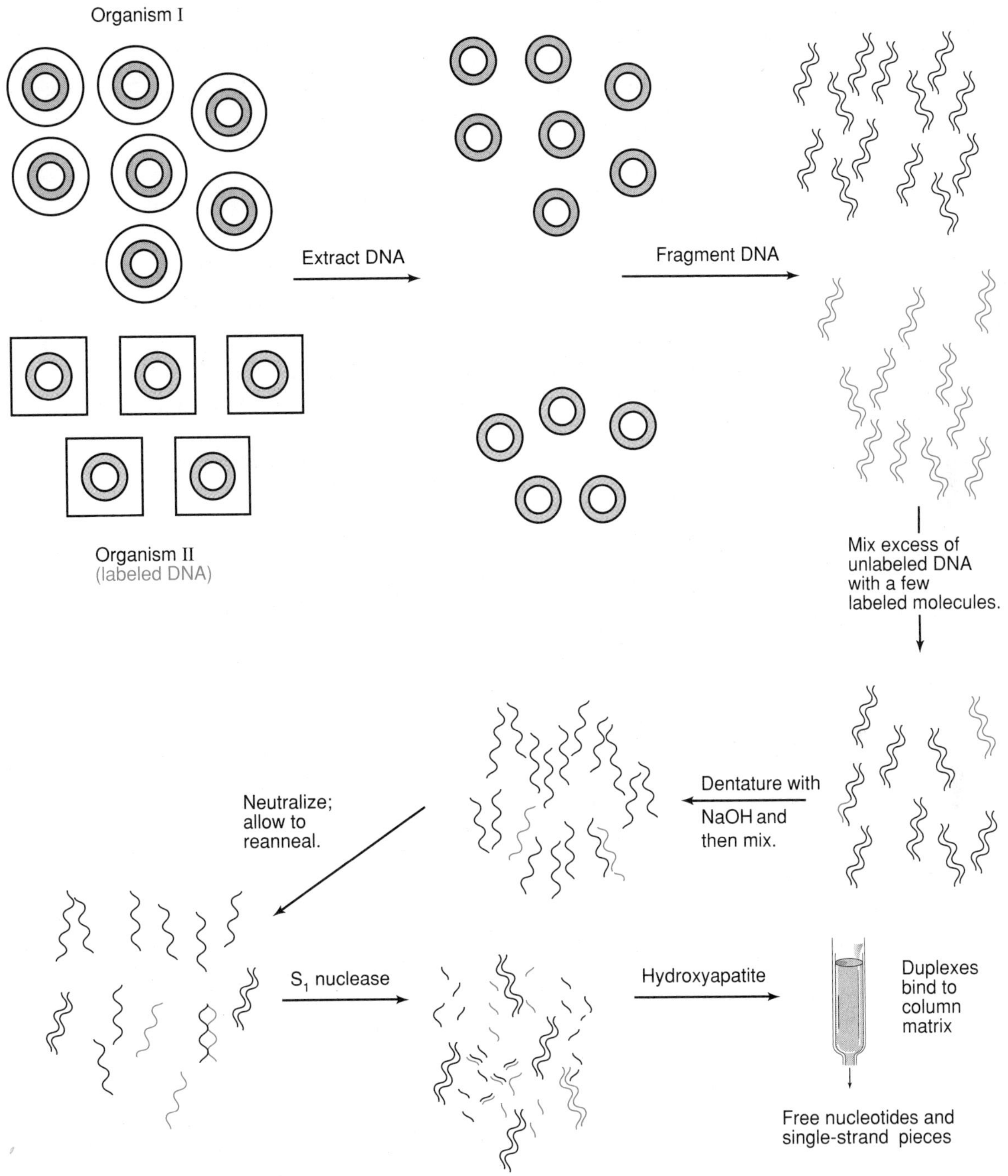

FIGURE 20.10

DNA hybridization analysis. Circular DNA molecules found in bacterial cells are extracted and broken into smaller, linear pieces. Then labeled (in color) and unlabeled DNA fragments are mixed and denatured at a high pH. When the pH is lowered, reannealing occurs and heteroduplexes may form. DNA not incorporated into duplexes is degraded by S1 nuclease. At this point, the degraded and duplex DNA are passed over an hydroxyapatite column. Only duplex DNA is retained by this column, and the percentage of the radioactive label found in the duplexes indicates the amount of homology between the two original types of DNA molecules.

In cases in which DNA homology testing is not yet available, it is possible to focus on cellular proteins as an indirect measure of the genetic code of bacteria. Many bacteria produce proteins that carry out the same function. It is reasonable to assume that if bacteria are very closely related, the amino acid sequences of their proteins also will be very similar. The question then becomes how to compare the proteins. One way is to determine the amino acid sequence for each of the proteins directly, but this process is very time-consuming and has been done for relatively few organisms for any given protein.

With electrophoresis, it is possible for proteins to migrate through a gel of polyacrylamide at a rate determined by their charge or molecular weight. In one common method, a rectangular slab of gel is used, and two separate electrophoretic runs are performed at right angles to one another. During the first run, the separation is based on charge; during the second, it is based on molecular weight. The result is a pattern of protein spots that is characteristic of the organism from which the proteins were extracted (figure 20.11A–B). Comparisons of the positions and intensities of the various spots can yield useful taxonomic information.

Another simple method involves the comparison of the three-dimensional structure of proteins and the assumption that it reflects their amino acid sequence. This comparison can be made by the use of antibodies, taking advantage of the observation that an antibody specific for one antigen can "cross-react" with another molecule of sufficiently similar structure, as seen in figure 16.14. The extent of the reaction is taken to indicate the degree of similarity. For example, table 20.5 shows data obtained from a group of organisms, all of which produce an ATPase. When an antibody fully reacts with the protein, enzyme activity is lost; therefore, it is easy to measure the relative reactions between the proteins. Based on the data shown, it can be concluded that *Micrococcus luteus* and *M. roseus* are related closely to each other, but *M. varians* is much less similar. *Nocardia rhodochrous* shows no similarity at all even though other characteristics of the organism suggest a relationship.

The real advance in modern taxonomy has come about due to the discovery of methods to determine the base sequence of nucleic acids comparatively easily and quickly, thereby allowing a direct assessment of genetic similarity. The actual method by which the sequence is determined is not important, but the results are. Biologists in general are in the middle of a literal information explosion of nucleic acid sequence data. For

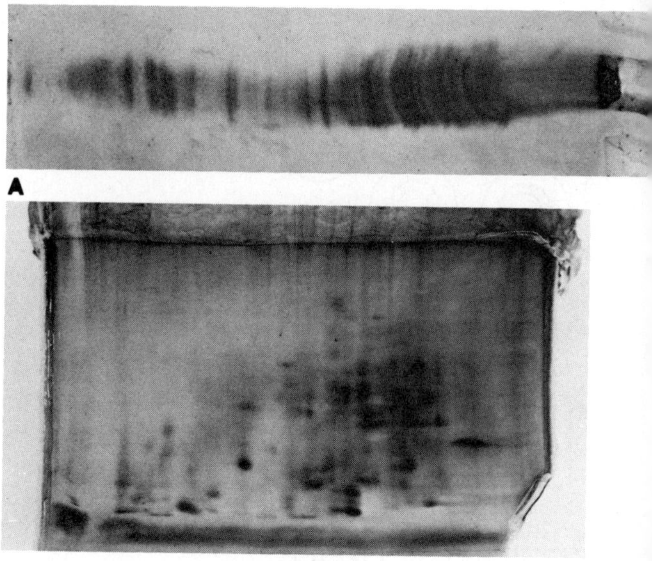

A

B

FIGURE 20.11

Two-dimensional protein electrophoresis. (**A**) This acrylamide gel has a pH gradient caused by a superimposed electric field. The pH is 10 at the left and 3 at the right. Proteins migrate within the pH gradient to their isoelectric point, the pH at which their net molecular charge is zero. They have been stained with Coomassie blue dye in order to be visualized. A gel similar to that shown in A has been placed atop another acrylamide gel, the proteins denatured by the detergent sodium dodecyl sulfate, and an electric field applied. Under these conditions, proteins migrate according to molecular weight, with the smallest proteins moving farthest. The bands at the left are molecular weight size standards. The spots on the remainder of the gel represent individual proteins from the isoelectric focusing gel. They have been stained with silver.

Gels courtesy of Dr. Pearl H. Lin, Department of Microbiology, Arizona State University.

ease of comparison, however, taxonomists have tended to concentrate on a few nucleic acid sequences that are common to all bacteria, namely the ribosomal RNA molecules.

Carl Woese and his collaborators have exploited this line of research by developing a large body of information regarding the base sequences of 5S ribosomal RNA and, more recently, 16S and 23S ribosomal RNA as well. The comparisons of sequence data from different organisms are thought to be useful particularly because the sequences of the ribosomal RNAs tend to be highly conserved during evolutionary processes in order to maintain the normal functioning of the ribosomes. Therefore, large differences in the base sequences are taken to mean that the organisms in question are not closely related; that is, they presum-

TABLE 20.5 Immunologic cross-reactivity among different bacteria

Bacterium	% Inhibition by Antibody	%(G+C)
Micrococcus luteus	100	73.5
M. luteus (a different isolate)[1]	89	72.4
M. roseus	84	73.5
M. varians	35	72.4
Streptomyces coelicolor	30	65.3
Nocardia rhodochrous[2]	0	70.4
Bacillus subtilis	0	43.0
Sporosarcina ureae	0	42.9

Source: Data taken from T. L. Whiteside, A. J. De Siervo, and M. R. Salton, "Use of Antibody to Membrane Adenosine Triphosphatase in the Study of Bacterial Relationships," *Journal of Bacteriology* 105:957–967.

Note: Antibody was prepared against the protein ATPase produced by the bacterium *Micrococcus lysodeikticus* and then used to treat similar proteins produced by other bacteria. A reaction between antibody and enzyme was assumed to be reflected in a loss of enzyme activity; therefore, the greater the inhibition produced by the antibody, the more reaction occurred and the more similar the proteins were taken to be.

1. Classified in the genus *Sarcina* at that time.

2. Classified in the genus *Micrococcus* at that time.

TABLE 20.6 Binary comparisons of 16S ribosomal RNA catalogs for several bacteria

	Source of Comparison DNA			
	M. luteus	***M. roseus***	***S. ureae***	***B. brevis***
Source of DNA	Bases in Common[1]			
Micrococcus luteus	—	411 (.73)[2]	211 (.35)	178 (.31)
Micrococcus roseus	411 (.73)	—	192 (.32)	165 (.29)
Sporosarcina ureae	211 (.35)	192 (.32)	—	319 (.52)
Bacillus brevis	178 (.31)	165 (.29)	319 (.52)	—

Source: Data from E. Stackebrandt and C. R. Woese, "A Phylogenetic Dissection of the Family Micrococcaceae," *Current Microbiology* 2:317–322.

1. Fragments of 16S RNA from various bacteria are sequenced and catalogued. Comparisons then are made between the catalogs for pairs of organisms. Data values indicate the number of bases in sequences of six bases or longer common to each pair of catalogs.

2. The numbers in parentheses are similarity coefficients for each pair of organisms. A higher value for the similarity coefficient indicates greater similarity. A diagram of the relationships suggested by the similarity coefficients is presented in figure 20.12.

ably diverged from a common ancestor at a point in the distant past. Examples of the sorts of data obtained are shown in table 20.6 and figure 20.12. It is easy to see that some organisms appear to be very similar in their sequences, while others are quite different. This has profound implications for bacterial taxonomy that are discussed in the following subsection.

Redefinition of a Bacterial Species

The information provided by DNA homology analysis has made possible actual quantitative expressions of what constitutes a bacterial species. Patrick Grimont proposed that two organisms are in the same species if they exhibit >80% reassociation of their DNA in the S1 nuclease protection assay at the optimum temperature and if the resulting heteroduplexes show no more than a 5° C temperature spread during denaturation (regions having fewer hydrogen bonds will denature more readily, suggesting base mismatches). Contrariwise, the organisms are in different species if the reassociation is <60% and the temperature spread is >7° C. Organisms whose reassociation values fall between 60 and 80% must be evaluated on additional properties such as morphology, metabolism, etc.

From criteria like these, several interesting anomalies in classic taxonomy become evident. All *Shigella* species except serogroup 13 of *S. boydii* become indistinguishable from *Escherichia coli.* All members of the genus *Salmonella* are grouped into a single species with several subspecies evident. *Yersinia pestis* and *Y. pseudotuberculosis* become the same species (see chapter 19). *Neisseria gonorrhoeae, N. meningitidis, N. lactamica,* and *N. polysaccharicum* become subspecies of

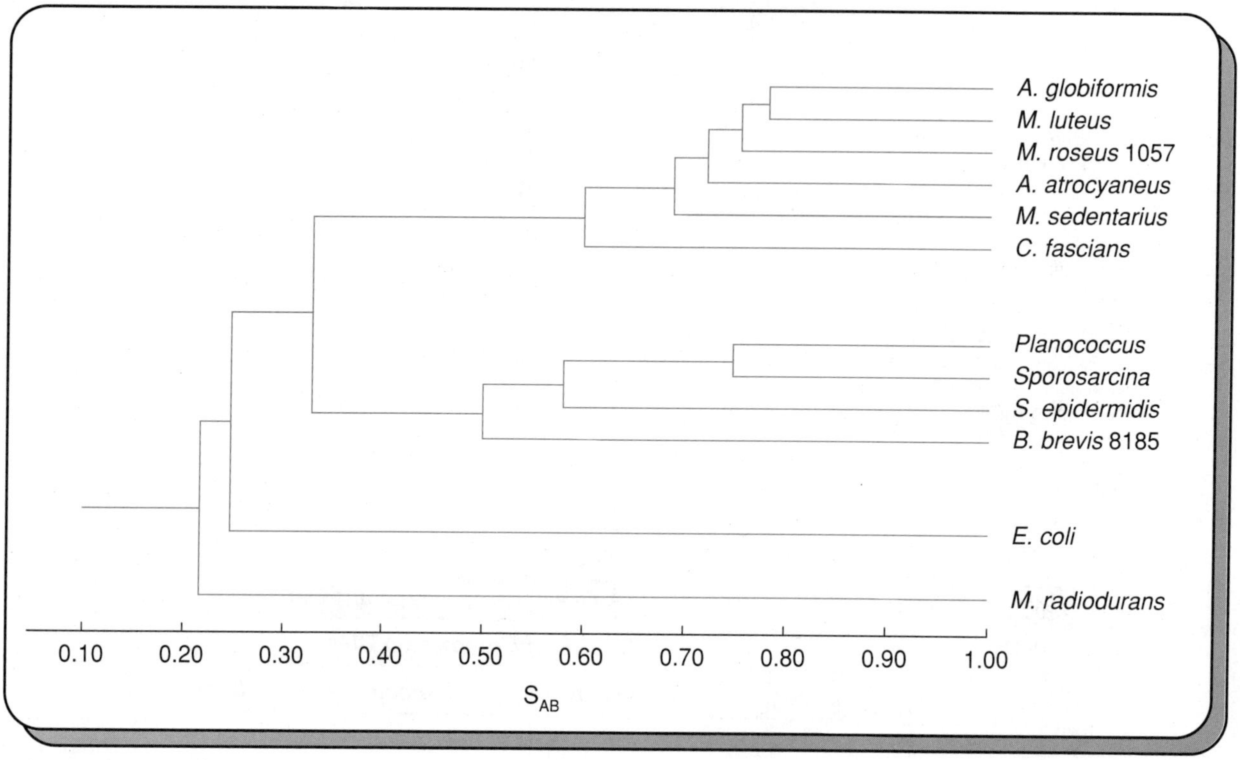

FIGURE 20.12

Relatedness tree or dendrogram. The scale across the bottom of the diagram is a measure of the similarity between a present-day organism and a hypothetical progenitor organism located just before a branch point. For example, in table 20.7 the similarity coefficient for the pair *Bacillus brevis—Sporosarcina ureae* is 0.52, and on the dendrogram the branch point that connects the two organisms has a value of 0.5.

From R. Stackebrandt and C. R. Woese, "A Phylogenetic Dissection of the Family Micrococcaceae," in *Current Microbiology* 2.317–322. Copyright © 1979 Springer-Verlag, Heidelberg, Germany. Reprinted by permission.

a single species. On the other hand, *Clostridium botulinum* becomes subdivided into several additional species.

Whether these taxonomic changes actually will be incorporated into the nomenclature remains to be seen. There are several possible difficulties. In some cases, such as with *E. coli* and *Shigella* spp., medical terminology is so firmly established that the proposed changes are unlikely to be accepted. In other cases, such as within the *Yersinia* genus, one organism causes an extremely serious disease while the other does not. Maintaining the present nomenclature serves the useful purpose of reminding laboratory personnel of the great potential danger from *Y. pestis*. Nevertheless, changes like those proposed for members of *Salmonella* may well be adopted. All salmonellae can cause disease, and medical identification has emphasized testing for surface antigens, with each new antigenic combination constituting a "species." Under the new proposal, there would be only one species and some subspecies, all of which could still be characterized by their antigenic structure.

Evolutionary Studies on Bacteria

There have long been questions not only about how bacteria are interrelated but also about how they relate to eukaryotic organisms. There are three logical possibilities. First, they may not be related at all. Similarities in metabolism and various other parameters argue against this idea, however. Second, it has been suggested that all eukaryotic cells are derived from the same lines of descent as present-day bacteria. Third, although bacteria and eukaryotes are not directly related, they possibly might share a common ancestor in the remote past. The second and third possibilities are more difficult to distinguish between and to discount.

Woese and his collaborators have expanded their taxonomic work to encompass a variety of organisms from all areas of the biological world. Using their cat-

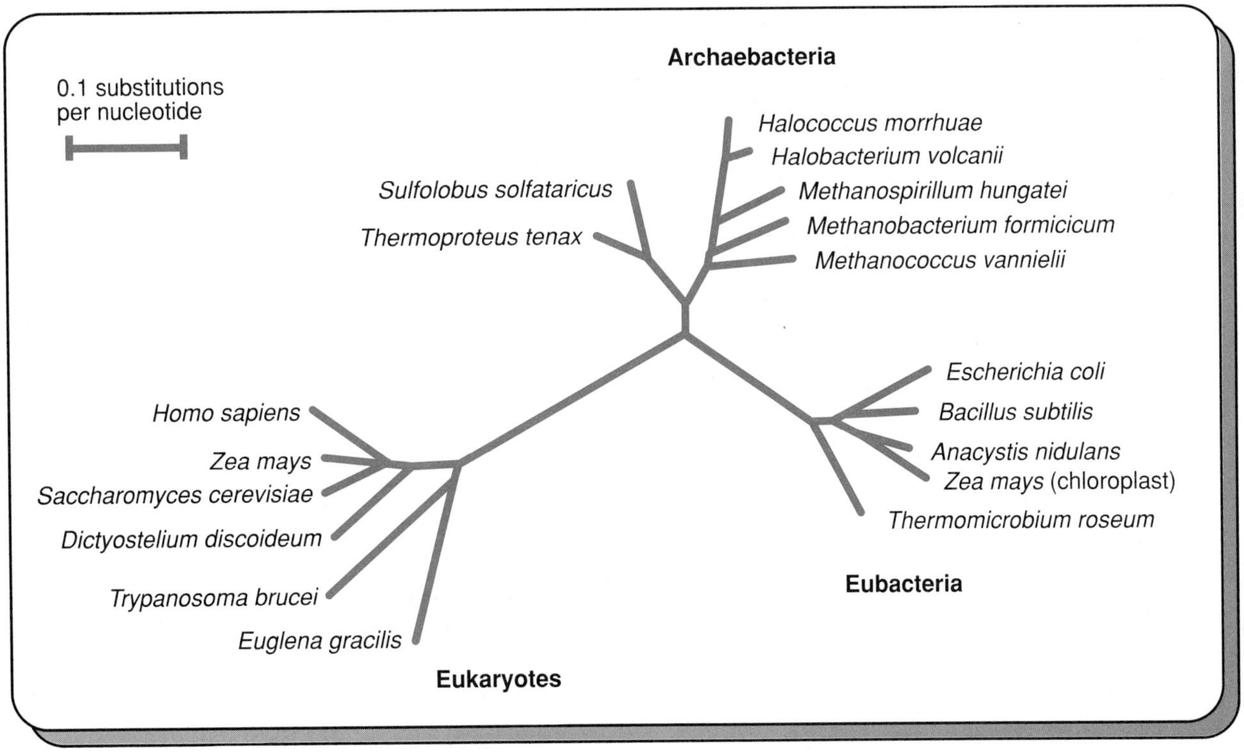

FIGURE 20.13

Universal phylogenetic tree determined from rRNA sequence comparisons. There are three primary kingdoms. The branching order within each kingdom is correct only to a first approximation. The hypothetical common ancestor is indicated at that point at which the three major lines intersect.

From G. J. Olsen and C. R. Woese, "A Brief Note Concerning Archaebacterial Phylogeny." Reproduced by permission of the National Research Council of Canada from the *Canadian Journal of Microbiology 35*, pp. 119–123, 1989.

alogs of rRNA sequences, they have been able to develop **dendrograms,** treelike structures that show the relationships between organisms. The distance between branches of a dendrogram is a measure of lack of similarity between corresponding sequences. An example of a dendrogram can be seen in figure 20.12. As more information accumulates, the complexity of the diagram that can be produced increases. Eventually, complete phylogenetic trees can be developed in an attempt to express the relationships between major groups of organisms. A phylogenetic tree developed by Woese and his collaborators is depicted in figure 20.13. The statistical analyses required for the graphic illustration of a phylogenetic tree are not trivial and the methodology is often disputed. An example of such a dispute can be found in Olsen and Woese's article referenced in figure 20.13.

If we assume for the moment that the phylogenetic tree presented by Woese and his collaborators is correct, there are a number of striking features. First, there are three major groups of organisms rather than two as originally was suspected. In addition to prokaryotes and eukaryotes, there also are archaebacteria. These bacteria differ from eubacteria in terms of metabolism, cell structure, and DNA sequence. The short stem on the phylogenetic tree suggests that they have diverged little from the primordial ancestor and, as such, they are more closely related to the eukaryotes than are the traditional prokaryotes. Note also that although the cytoplasmic rRNA of *Zea mays* (corn) fits in well with other eukaryotes, the rRNA sequences of its chloroplast ribosomes are closely related to the present-day cyanobacterium *Anacystis nidulans*. This observation supports the endosymbiont theory for the origin of chloroplasts.

Present State of Bacterial Taxonomy

In order to understand bacterial taxa, it is important to bear in mind two facts. First, there are very stringent international rules regarding the naming of any organism that are enforced by appropriate professional

TABLE 20.7	Proposed higher taxa for bacteria

Kingdom *Prokaryotae*
  Division I. *Gracilicutes* (gram-negative, typical bacteria)
    Class I. *Scotobacteria* (nonphotosynthetic bacteria)
    Class II. *Anoxyphotobacteria* (nonoxygenic photosynthetic bacteria)
    Class III. *Oxyphotobacteria* (oxygenic photosynthetic bacteria)
  Division II. *Firmicutes* (gram-positive, typical bacteria)
    Class I. *Firmibacteria* (simple, gram-positive bacteria)
    Class II. *Thallobacteria* (branching, filamentous bacteria)
  Division III. *Tenericutes* (bacteria lacking a cell wall)
    Class I. *Mollicutes*
  Division IV. *Mendosicutes* (bacteria with defective cell walls or lacking peptidoglycan)
    Class I. *Archaebacteria*

Source: Data from R. G. E. Murray, "The Higher Taxa for Bacteria, or, a Place for Everything?" pp. 2329–2332 in *Bergey's Manual of Systematic Bacteriology*. Copyright © Williams and Wilkins, Baltimore, MD.

bodies. International commissions have been created to adjudicate disputes about whether new organisms have been identified and named correctly. Second, there are no equivalent rules for the creation of categories above the level of genus; therefore, any orders, families, etc. that have been created are for the convenience of microbiologists. To the extent that they are useful, the names are preserved. If they serve no useful function, they are changed or discarded. The structured presentation that follows should be considered as tentative and subject to revision as more knowledge becomes available.

The complexities of the decision-making process in bacterial taxonomy have meant that no one person has maintained an authoritative position in all aspects of the field. Instead, the state of knowledge has been codified every decade or so by a committee that has written a manual for use by all bacteriologists. The first such committee was headed by Dr. John H. Bergey, and seven subsequent editions have been titled *Bergey's Manual of Determinative Bacteriology.*

In 1984, however, the Bergey's Manual Trust began publication of a four-volume set of *Bergey's Manual of Systematic Bacteriology.* This event marked the beginning of a new era in bacterial taxonomy and a complete revamping of the taxonomic groupings and nomenclature at taxonomic levels above the genus. The *Manual of Systematic Bacteriology* does not present bacterial taxonomy in quite the same manner as that used for eukaryotic taxonomy. That is to say, attempts to produce higher taxa (families, orders, phyla, etc.) for the kingdom *Monera* have been difficult, because general agreement has not always been possible on the criteria to be used to define these groups.

Four divisions of bacteria have been proposed based on the possible cell wall characteristics of bacteria. The division *Gracilicutes* encompasses all organisms with conventional cell walls and gram-negative staining reaction. The gram-positive organisms are placed into the division *Firmicutes.* Those organisms with no cell wall at all are members of the division *Tenericutes,* and those organisms with defective cell walls or cell walls of unusual composition are members of the division *Mendosicutes.* This arrangement is summarized in table 20.7.

The manual includes 33 sections, each of which consists of one or more closely related genera. Those sections having numerous genera usually are subdivided into families. Occasionally, one or more orders is specified, depending on previous custom. An annotated table of contents for the manual can be found in appendix A, and a much abbreviated version is presented in table 20.8.

TABLE 20.8 The groups of bacteria listed in *Bergey's Manual of Systematic Bacteriology*

Volume 1

Section 1. The spirochetes
Section 2. Aerobic/microaerophilic, motile, helical/vibrioid, gram-negative bacteria
Section 3. Nonmotile (or rarely motile), gram-negative, curved bacteria
Section 4. Gram-negative aerobic rods and cocci
Section 5. Facultatively anaerobic gram-negative rods
Section 6. Anaerobic gram-negative straight, curved, and helical rods
Section 7. Dissimilatory sulfate- or sulfur-reducing bacteria
Section 8. Anaerobic gram-negative cocci
Section 9. The rickettsias and chlamydias
Section 10. The mycoplasmas
Section 11. Endosymbionts

Volume 2

Section 12. Gram-positive cocci
Section 13. Endospore-forming gram-positive rods and cocci
Section 14. Regular, nonsporing, gram-positive rods
Section 15. Irregular, nonsporing, gram-positive rods
Section 16. The mycobacteria
Section 17. Nocardioforms

Volume 3

Section 18. Anoxygenic phototrophic bacteria
Section 19. Oxygenic photosynthetic bacteria
Section 20. Aerobic chemolithotrophic bacteria and associated organisms
Section 21. Budding and/or appendaged bacteria
Section 22. Sheathed bacteria
Section 23. Nonphotosynthetic, nonfruiting gliding bacteria
Section 24. Fruiting gliding bacteria: The myxobacteria
Section 25. Archaebacteria (archaeobacteria)

Volume 4

Section 26. Nocardioform actinomycetes
Section 27. Actinomycetes with multilocular sporangia
Section 28. Actinoplanetes
Section 29. Streptomycetes and related genera
Section 30. Maduromycetes
Section 31. *Thermomonospora* and related genera
Section 32. *Thermoactinomyces*
Section 33. Other genera

APPLICATIONS BOX

Alternative Views of Bacterial Taxonomy

DNA and RNA sequence information continues to have an impact on bacterial taxonomy, although most of the information has not yet been incorporated into *Bergey's Manual*. Major changes are likely to occur before the next edition is prepared; however, it is difficult to predict exactly which proposals will stand the test of time.

For example, the genus *Streptococcus* can be shown to be heterogeneous, and it has been proposed that the present streptococcal organisms should be distributed among three genera: *Streptococcus, Lactococcus,* and *Enterococcus*. In terms of food microbiology, the organism known as *Streptococcus lactis* would receive the new name of *Lactococcus lactis*. *Streptococcus faecalis*, a common fecal organism, would be renamed *Enterococcus faecalis*.

Higher taxa also have been targeted for possible change. A rearrangement has been proposed for division I (gram-negative) bacteria described in table 20.7. The separate class for Oxyphotobacteria would be maintained, but all other gram-negative bacteria grouped with the Anoxyphotobacteria and the Scotobacteria would be entered into a new class called *Proteobacteria*. Such an arrangement would be more reflective of 16S rRNA sequence data.

Summary

Traditionally, taxonomic schema have always been extremely useful in broad outline but tend to break down in detail. Organisms never seem to fit neatly into the categories devised by biologists. Nowhere is this more true than with microorganisms. To define a species can sometimes be difficult, especially if the organism only reproduces asexually. Historically, the classification of microorganisms has been based on morphologic and metabolic characteristics. Increasingly, however, direct sequencing of selected nucleic acid segments is being used to provide data for studies of relatedness. Once DNA (or RNA) homology has been measured, it is possible to apply quantitative criteria to define a species or a genus. With respect to prokaryotes, DNA and RNA sequence analysis sometimes has supported the more traditional methods of classification. In other circumstances, major rearrangements may be appropriate. Whether any rearrangements will be adopted will not be known for some time. The major catalog of bacterial taxonomy is *Bergey's Manual of Systematic Bacteriology.* Over the course of five years, four volumes have appeared; therefore, the manual is not completely consistent. The kingdom *Monera* is divided into 33 sections in the manual based primarily on morphology and/or physiology, although some nucleic acid sequence data have been incorporated.

A second exciting aspect of the study of taxonomy is its implications for evolutionary biology. With nucleic acid sequence data, it has been possible to prepare phylogenetic trees in an attempt to show the relationships of widely diverse organisms. The biggest surprise resulting from this type of analysis has been the observation that there are three major groups of organisms (eubacteria, archaebacteria, and eukaryotes) rather than two. As more sequences are analyzed, additional anomalies may be uncovered.

Questions for Review and Discussion

1. How would a biologist have defined a species in the year 1900? How would he or she define it now? What do you think might happen to the definition over the next 50 years?
2. What sort of metabolic traits could be used as data for numeric taxonomy? Do you agree with the assumption that no particular trait should be considered as more important than any other trait? Why?
3. Give an example of a DNA homology assay. Does your example make a direct or an indirect comparison of the sequence involved?
4. Why do you think that some organisms whose rRNA sequences have been tested seem to diverge faster than other organisms? What kinds of metabolic differences might be present? In terms of Darwinian theory, what might happen to an organism that never diverged, whose rRNA sequences were unchanged over millenia?
5. At minimum, what properties do you think ought to be reported by a scientist who claims to have discovered a new bacterium?

Suggestions for Further Reading

Beveridge, T. J. 1990. Mechanism of Gram variability in select bacteria. *Journal of Bacteriology* 172:1609–1620. A report of a modified Gram stain procedure that allows examination of the results with the electron microscope.

Doolittle, W. F. 1989. Bacterial evolution. *Canadian Journal of Microbiology* 34:547–551. A brief discussion concerning the implications of modern taxonomic analysis and phylogenetic trees.

Holt, J. G. (ed. in chief). 1984–1989. *Bergey's Manual of Systematic Bacteriology.* Williams & Wilkins Co., Baltimore. This is the definitive work in four volumes. A set of essays on bacterial taxonomy, methods, and evaluation is included at the beginning of each volume and is well worth reading.

Krieg, N. R. 1988. Bacterial classification: An overview. *Canadian Journal of Microbiology* 34:536–540. An excellent review in a special issue in which many articles deal with taxonomic topics.

Margulis, L., and K. V. Schwartz. 1982. *Five kingdoms. An illustrated guide to the phyla of life on earth.* W. H. Freeman, San Francisco. A lovely book that provides at least one example of every phylum in each biological kingdom. It is filled with marvelous examples of biological diversity.

Starr, M. P., H. Stolp, H. G. Trüper, A. Balows, and H. G. Schlegel. 1981. *The prokaryotes. A handbook on habitats, isolation, and identification of bacteria.* Springer-Verlag KG, Berlin. This is an excellent overview of the kingdom *Monera* that not only includes some taxonomy but also provides other types of information.

Woese, C. R. 1987. Bacterial evolution. *Microbiological Reviews* 51:221–271. A discussion of evolution including some early work from the time of Darwin. This is a good source of information about rRNA sequence analysis.

appendix

A

A Survey of *Bergey's Manual of Systematic Bacteriology*

The material presented in this appendix is basically an annotated table of contents of the four volumes of *Bergey's Manual of Systematic Bacteriology.* Each section outlined in *Bergey's Manual* is introduced briefly and one or two commonly encountered examples of the organisms that comprise each section are presented. This material is not intended for memorization but rather as a reference tool.

Volume 1

Section 1. The spirochetes

Order I. *Spirochaetales*
Family I. *Spirochaetaceae*
Family II. *Leptospiraceae*
Other Organisms

The group comprising section 1 are gram-negative, nonphotosynthetic organisms with an easily discernible spiral shape (figure A.1). The organisms are distinguished from other spiral-shaped bacteria by the extreme flexibility of their cells and by the presence of one or more **axial fibrils,** flagella that are located within an outer sheath surrounding the cell.

The *Spirochaetaceae* include the genera *Spirochaeta, Cristispira, Treponema,* and *Borrelia,* while the *Leptospiraceae* are differentiated by cells with bent or hooked ends and consist only of the genus *Leptospira* whose members cause a relatively common disease of animals, leptospirosis. The most frequently discussed member of the *Spirochaetaceae* is *Treponema pallidum,* the causative organism of syphilis. However, members of the various genera are widely distributed and most do not cause disease.

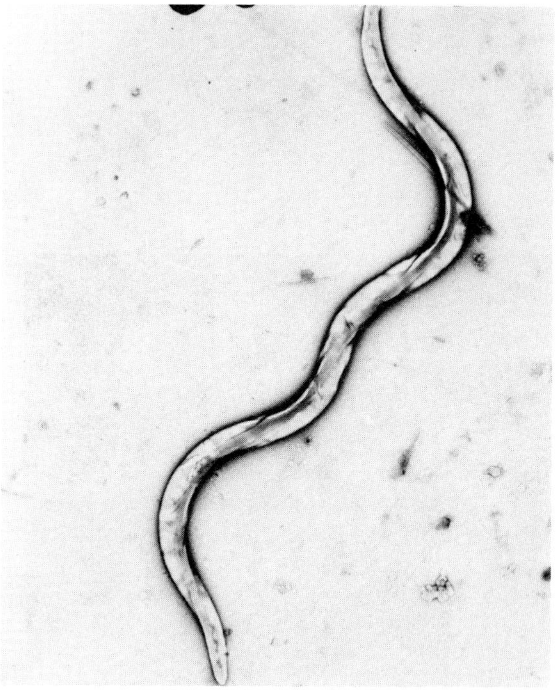

FIGURE A.1

Spirochaeta sp., a spirochete.

Electron micrograph by Dr. Jean M. Schmidt, Department of Microbiology, Arizona State University.

Section 2. Aerobic/microaerophilic, motile, helical/vibrioid, gram-negative bacteria

As the characteristics of this group imply, section 2 comprises a catch-all group of organisms that are not strikingly similar. The term *vibrioid* refers to a cell shape that is less than a complete helical turn (figure 12.10). Unlike the members of section 1, however, no axial fibrils are present and the cells are more rigid in nature. They move with a twisting motion by means of tufts of polar flagella. The genera found in section 2 include *Aquaspirillum, Spirillum, Azospirillum, Oceanospirillum, Campylobacter, Bdellovibrio,* and *Vampirovibrio.* As the genus names indicate, most member organisms are encountered commonly in aquatic environments. Members of the genus *Bdellovibrio* are unusual in that many are parasitic on other bacteria, penetrating through the cell wall and digesting the host cell to provide energy and nutrients for their reproduction.

Section 3. Nonmotile (or rarely motile), gram-negative, curved bacteria

Family I. *Spirosomaceae*
Other Genera

Members of section 3 are widespread in nature with habitats ranging from salt-cured hams to freshwater ponds, but they have not been studied extensively. Their principal claim to fame is that the cells frequently are ring shaped (figure A.2A). The most commonly encountered genera are *Microcyclus, Spirosoma, Flectobacillus, Meniscus,* and *Runella.*

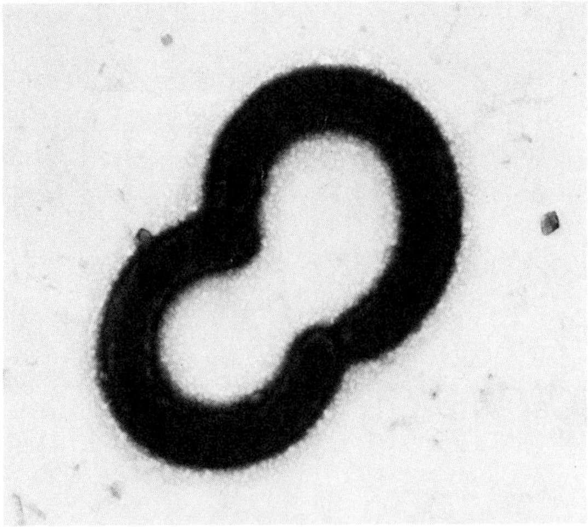

A

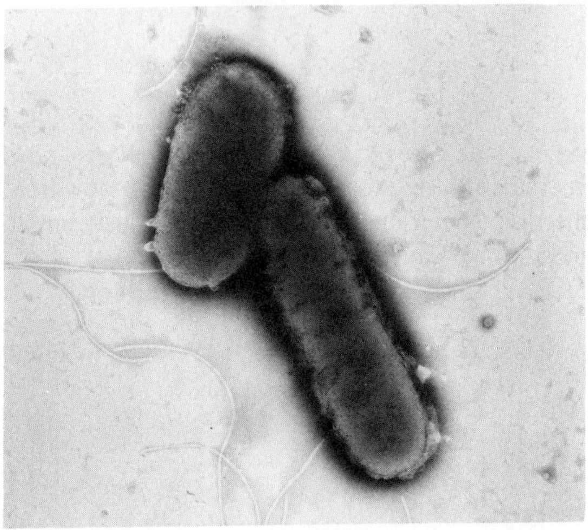

B

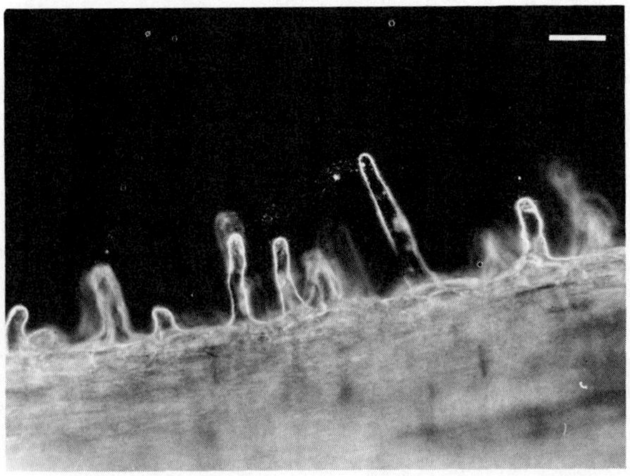

C

FIGURE A.2

(**A**) *Flectobacillus,* a ring-shaped nonmotile organism.
(**B**) *Pseudomonas aeruginosa,* a gram-negative aerobic bacterium.
(**C**) *Rhizobium leguminosarum* (formerly *R. trifolii*) attached to a root hair.

(A) Electron micrograph courtesy of Dr. Jean M. Schmidt, Department of Microbiology, Arizona State University. (B) © G. Musil/Visuals Unlimited. (C) Reprinted from Dazzo, F. B., M. R. Urbano, and W. J. Brill. 1979. Transient appearance of lectin receptors on *Rhizobium trifolii. Current Microbiology* 2:15–20.

TABLE A.1 Gram-negative aerobic rods and cocci

Family	Representative Genus	Characteristic
Pseudomonadaceae	*Pseudomonas, Xanthomonas*	Nutritionally versatile group
Azotobacteraceae	*Azotobacter*	Free-living nitrogen fixers
Rhizobiaceae	*Rhizobium*	Symbiotic nitrogen fixers
Methylococcaceae	*Methylococcus*	Methane oxidizers
Halobacteriaceae[1]	*Halobacterium*	Require high osmotic strength
Acetobacteriaceae	*Acetobacter*	Acetic acid producers
Legionellaceae	*Legionella*	Cause Legionnaires disease
Neisseriaceae	*Neisseria*	Parasitic in human and animal hosts

Note: Section 4 of *Bergey's Manual* is so large that is has been subdivided for convenience in visualizing relationships. There are 16 genera that still remain unassigned, however. Included among them are *Alcaligenes, Bordetella, Brucella,* and *Francisella.*

1. This family is also classified in section 25, which is probably a more appropriate assignment.

TABLE A.2 Facultatively anaerobic gram-negative rods

Family	Representative Genus	Characteristic[1]
Enterobacteriaceae	*Escherichia, Salmonella*	Do not have cytochrome oxidase; do have catalase
Vibrionaceae	*Vibrio*	Have cytochrome oxidase and catalase
Pasteurellaceae	*Haemophilus*	Have cytochrome oxidase and catalase

1. Cytochrome oxidase is an enzyme of the respiratory pathways (see chapter 5), and catalase serves to protect the cell from hydrogen peroxide (see chapter 13).

Section 4. Gram-negative aerobic rods and cocci

Family I. *Pseudomonadaceae*
Family II. *Azotobacteraceae*
Family III. *Rhizobiaceae*
Family IV. *Methylococcaceae*
Family V. *Halobacteriaceae*
Family VI. *Acetobacteraceae*
Family VII. *Legionellaceae*
Family VIII. *Neisseriaceae*
Other Genera

Section 4 encompasses an enormous group of organisms that are very important medically, environmentally, and industrially. They are discussed extensively in chapters 12 and 14. Table A.1 shows the general characteristics for each family. Figure A.2B–C depicts two typical members of this section.

Section 5. Facultatively anaerobic gram-negative rods

Family I. *Enterobacteriaceae*
Family II. *Vibrionaceae*
Family III. *Pasteurellaceae*
Other Genera

The diversity of organisms in section 5 is not as great as that in section 4, but the number of known genera is equally large due primarily to the number of disease-causing organisms included in this group. Family properties are listed in table A.2. The principal family is the *Enterobacteriaceae* (**enteric bacteria**) that has 20 different genera assigned to it. As the name implies, the members of this family are primarily the inhabitants of the intestinal tracts of animals. *Escherichia coli* is a member of this group (figures 2.3, 3.10B) as is *Enterobacter* (figure 2.8B).

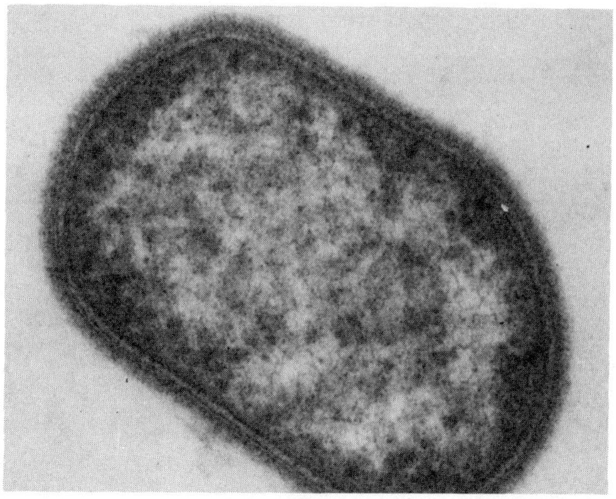

FIGURE A.3

Bacteroides gingivalis isolated from an oral infection of a human.

Reprinted from Okuda, K., J. Slots, R. J. Genco. 1981. *Bacteroides gingivalis, Bacteroides asaccharolyticus,* and *Bacteroides melaninogenicus* subspecies: Cell surface morphology and adherence to erythrocytes and human buccal epithelial cells. *Current Microbiology* 6:7–12.

Section 6. Anaerobic gram-negative straight, curved, and helical rods

Family I. *Bacteroidaceae*

There are 13 genera within the one family in section 6. The most commonly encountered member of this group is the genus *Bacteroides* whose species make up the bulk of the bacteria in fecal material. Comparatively little is known about the bacteria in this section because of their sensitivity to oxygen in the air. An electron micrograph of *Bacteroides gingivalis* can be seen in figure A.3.

Section 7. Dissimilatory sulfate- and sulfur-reducing bacteria

The chemotrophic bacteria classified in section 7 obtain their energy for growth by oxidizing carbon-containing molecules and concomitantly by reducing either sulfate or sulfur. The term *dissimilatory* refers to the fact that although sulfides can be used in metabolic reactions, in this case, they are a waste product and are not used further by the cell. This is an example of anaerobic respiration, a topic discussed in greater detail in chapter 5. There are seven genera in this section and no family grouping. Figure A.4A presents an electron micrograph of *Desulfovibrio,* a relatively common member of section 7.

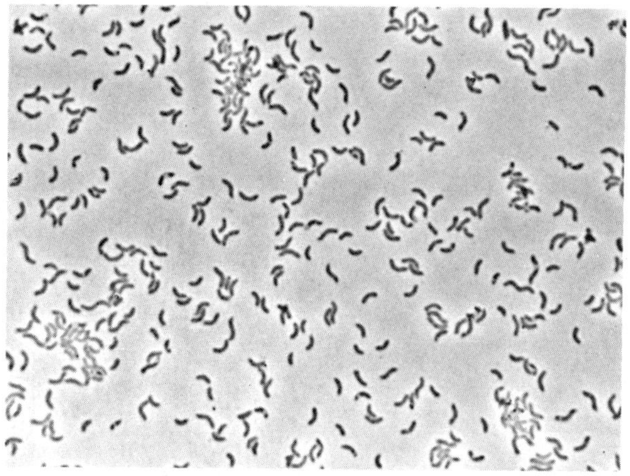

A

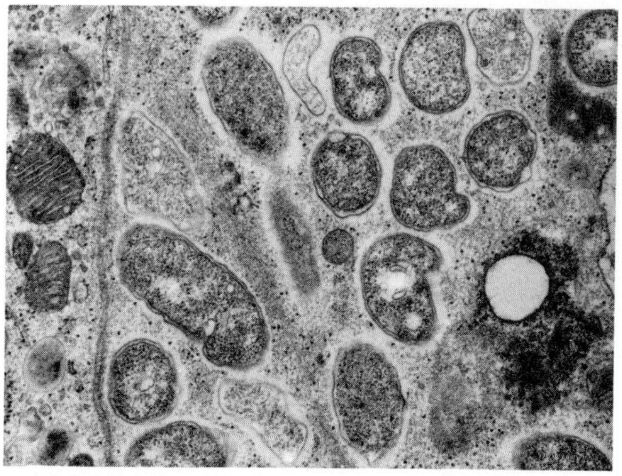

B

FIGURE A.4

(**A**) *Desulfovibrio vulgaris.* (**B**) *Rickettsia prowazekii*

(A) © F. Widdel/Visuals Unlimited. (B) Burgdorfer.

Section 8. Anaerobic gram-negative cocci

Family I. *Veillonellaceae*

There are only three genera within section 8: *Veillonella, Acidaminococcus,* and *Megasphaera.* Unlike the members of section 7, these bacteria are fermentative.

Section 9. The rickettsias and chlamydias

Order I. *Rickettsiales*
  Family I. *Rickettsiaceae*
  Family II. *Bartonellaceae*
  Family III. *Anaplasmataceae*
Order II. *Chlamydiales*
  Family I. *Chlamydiaceae*

TABLE A.3 The rickettsias and chlamydias

Family	Tribe[1]	Characteristic(s)
	Order I. *Rickettsiales*	
Rickettsiaceae	*Rickettsieae*	Live in arthropods; may infect vertebrates
	Ehrlichieae	Infect mammals other than humans
	Wolbachieae	Symbiotes of arthropods
Bartonellaceae		Live in erythrocytes of vertebrates; have cell wall
Anaplasmataceae		Enclosed in a membrane; live in erythrocytes of vertebrates
	Order II. *Chlamydiales*	
Chlamydiaceae	Genus *Chlamydia*	Energy parasites

1. A tribe is a classification level intermediate between the family and the genus. In all cases, there is a genus corresponding to the name of each tribe (for example, in *Ehrlichiaeae* can be found the genus *Ehrlichia*).

TABLE A.4. The mycoplasmas

Family	Representative Genus	Characteristic
Mycoplasmataceae	*Mycoplasma*	Require sterols for growth
Acholeplasmataceae	*Acholeplasma*	Sterols not required
Spiroplasmataceae	*Spiroplasma*	Helical forms during some portion of life cycle

Note: There also are many mycoplasma-like organisms that infect plants and vertebrates; however, they are not included in this list.

The organisms grouped in section 9 are unusual in the sense that many are obligate intracellular parasites, that is, they cannot be cultured except inside living cells. As such, they are all of medical importance and have been extensively studied taxonomically. Their classification is presented in table A.3. The rickettsias cause diseases such as typhus and Rocky Mountain spotted fever, while the chlamydias are responsible for parrot fever. A photograph of these organisms can be seen in figure A.4B.

Chlamydias cannot be grown outside of a living cell, and they lack the enzymes necessary for the production and storage of ATP. As such, they are considered to be energy parasites on their host cells. Rickettsias, on the other hand, can live intracellularly but still provide their own enzymes for ATP production. Indeed, many rickettsias have more of a mutualistic rather than parasitic relationship with their hosts, and some can even be grown in pure culture on artificial media.

Section 10. The mycoplasmas

Division. *Tenericutes*
- Class I. *Mollicutes*
  - Order I. *Mycoplasmatales*
    - Family I. *Mycoplasmataceae*
    - Family II. *Acholeplasmataceae*
    - Family III. *Spiroplasmataceae*
    - Other Genera

The organisms grouped in section 10 lack a cell wall. The three families of this section are summarized in table A.4. A sample organism is depicted in figure A.5A–B. Mycoplasmas are the smallest of the free-living organisms, and because they have no cell wall, they essentially are living membrane vesicles. They frequently produce diseases in humans and other animals as well as in plants. In the tissue culture laboratory, they are a notable scourge.

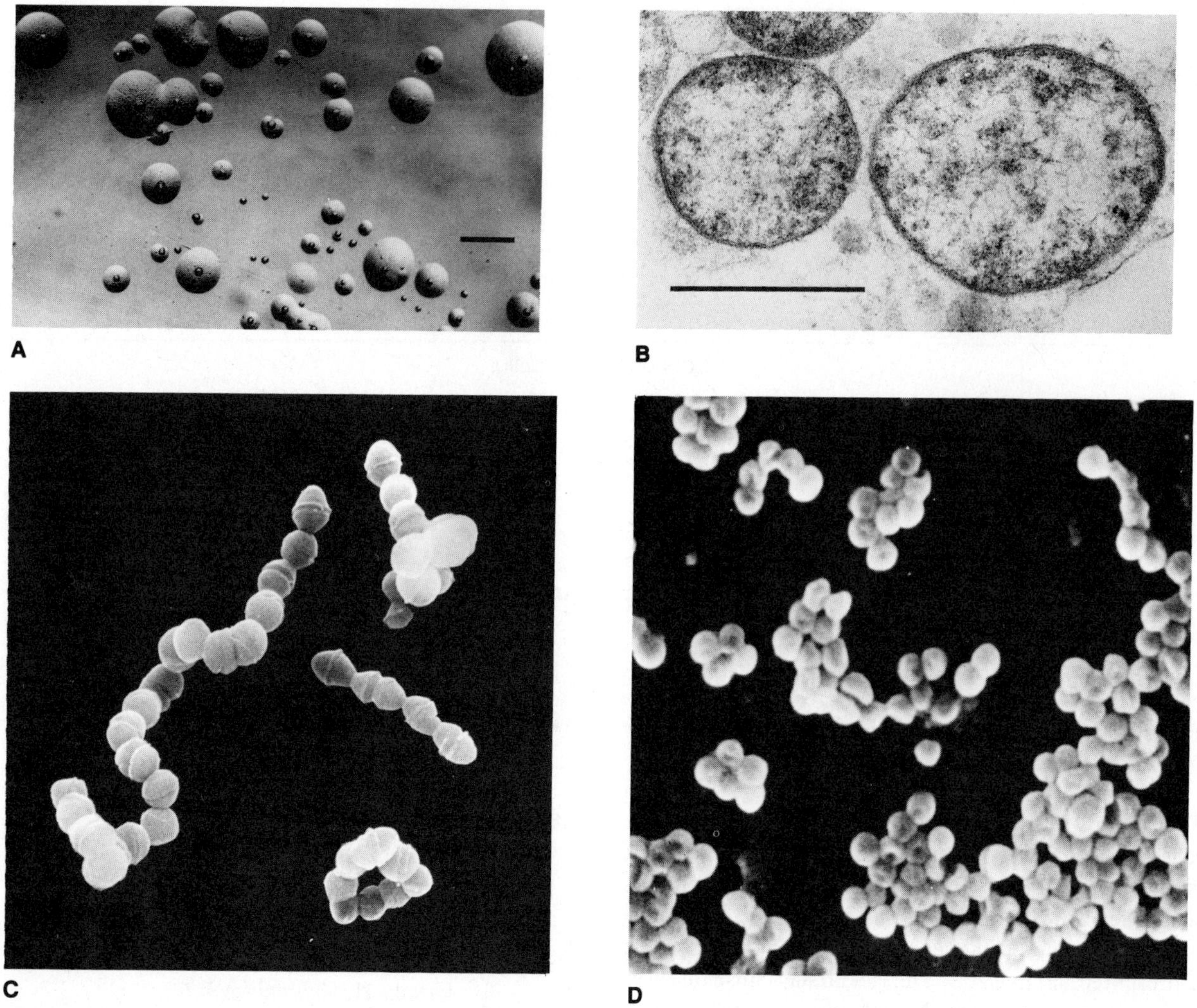

FIGURE A.5

(**A**) Colonies of an *Acholeplasma* sp. (**B**) Thin section of an *Acholeplasma* sp. (**C**) *Streptococcus* sp., a gram-positive coccus that forms chains of cells. (**D**) *Staphylococcus* sp., a gram-positive coccus that forms clusters of cells.

(A–B) Reprinted from Eden-Green, S. J., and J. G. Tully. 1979. Isolation of *Acholeplasma* spp. from coconut palms affected by lethal yellowing disease in Jamaica. *Current Microbiology* 2:311–316. (C-D) © David M. Phillips/Visuals Unlimited.

Section 11. Endosymbionts

I. Endosymbionts of protozoa
II. Endosymbionts of insects
III. Endosymbionts of fungi and invertebrates other than arthropods

This group of organisms is an entirely new category in bacterial taxonomy and reflects a growing appreciation of the fact that bacteria are often found in more or less permanent association with eukaryotic organisms, both unicellular and multicellular. A well-known example is *Blattabacterium cuenoti,* an organism that lives in the fat body, ovaries, and embryos of the cockroach and seems to provide essential trace nutrients to its host. Uninfected cockroaches can be raised, but only on a special, complex diet. The bacteria in section 11 have not yet been grown on artificial media.

TABLE A.5. Gram-positive cocci		
Cocci	**Typical Genus**	**Characteristic**
Family *Micrococcaceae*	*Micrococcus, Staphylococcus*	Cells in clusters
Family *Deinococcaceae*	*Deinococcus*	Resistant to large doses of radiation
Other organisms (> 12 genera)	*Leuconostoc*	Lenticular cells
	Streptococcus	Round cells in chains

Volume 2

Section 12. Gram-positive cocci

Family I. *Micrococcaceae*
Family II. *Deinococcaceae*
Other Genera

Section 12 also comprises a heterogeneous grouping of organisms based on a single pair of traits. As a result, the organisms are not necessarily closely related. At the present time, there are more genera not assigned to a family than genera that are (see table A.5). Among the genera grouped in this section are several familiar names: *Streptococcus,* whose members include both pathogens and food fermenting organisms (figure A.5C); *Staphylococcus,* whose members are frequently pathogenic (figure A.5D); and *Leuconostoc,* whose members are important in food and industrial fermentations (see chapter 14). Note that the prefix "strepto" indicates an organism that tends to form chains after cell division, while "staphylo" indicates that clusters of cells are formed.

Section 13. Endospore-forming gram-positive rods and cocci

There are only six genera in section 13, and two are very common, namely, *Bacillus* (figure 2.8) and *Clostridium.* All members grouped in this section produce heat-resistant endospores (discussed in chapter 7); therefore, their principal differences lie in their cell shape and/or physiology. For example, a member of the genus *Bacillus* forms spores under aerobic conditions, while a member of *Clostridium* forms spores only under anaerobic conditions. Otherwise, as figure A.6A–B illustrates, they are very similar in appearance.

Section 14. Regular, nonsporing, gram-positive rods

Only seven genera are grouped in section 14, the most common of which is *Lactobacillus,* an organism widely used in food fermentations (see chapter 14). Another genus sometimes in the news is *Listeria,* which can occasionally be found in unpasteurized milk and can induce abortions or perinatal infections in pregnant women who consume milk or milk products contaminated with *Listeria* organisms. The distinction between the members of this section and those of section 15 has to do with the uniformity of the cell shape (compare the representative cells in Figure A.6C).

Section 15. Irregular, nonsporing, gram-positive rods

The irregularity of form of the organisms grouped in section 15 is exemplified by members of the genus *Corynebacterium.* These organisms produce complicated multicellular structures because they fail to separate completely after cell division. This failure, known as **snapping division,** leaves the cells attached at their corners, as shown in figure A.6D.

Section 15 contains both aerobic organisms such as *Corynebacterium* and *Arthrobacter* (figure 20.9A–B) and anaerobic organisms such as *Actinomyces* and *Propionibacterium.* The genus *Corynebacterium* is subdivided into two groups, the animal and saprophytic corynebacteria as opposed to the plant corynebacteria. The two groups differ with respect to their %(G+C) content and their cell wall composition. The most notable animal corynebacterium is *Corynebacterium diphtheriae* that causes diphtheria.

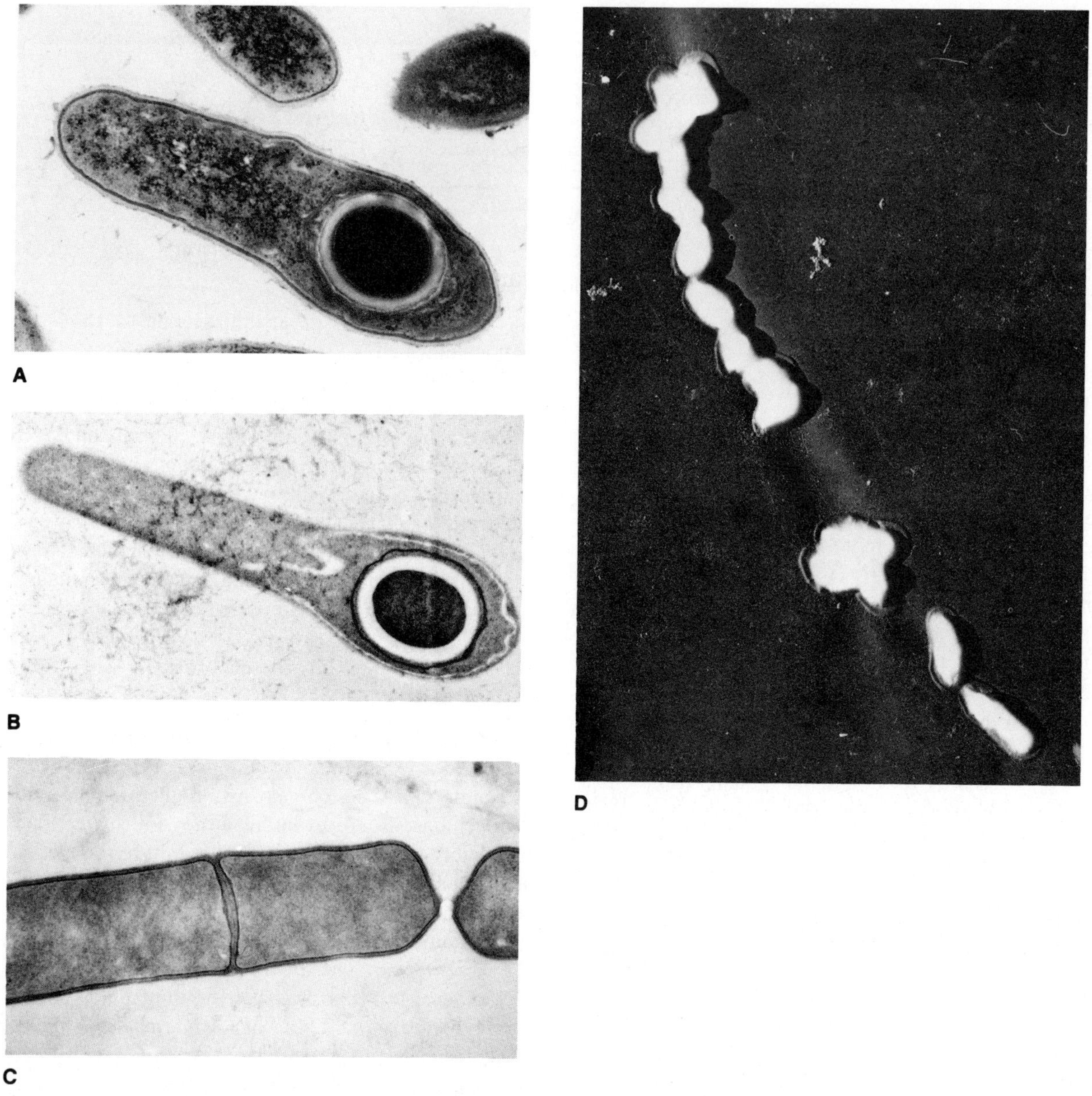

FIGURE A.6

Two endospore-forming bacteria, (**A**) *Bacillus* sp. and (**B**) *Clostridium sporogenes.* Note the similarity of cell shape to (**C**) *Lactobacillus,* which does not form spores. (**D**) Cell division in a *Corynebacterium* sp.

(A) Micrograph courtesy of Dr. Joseph Madden. (B–C) Dr. T. J. Beveridge, Dept. of Microbiology, University of Guelph, Ontario/BPS. (D) © Fred E. Hossler/Visuals Unlimited.

TABLE A.6. Anoxygenic photosynthetic bacteria

Group and Family	Typical Genus	Characteristic(s)
Purple bacteria		Pigments in cell membrane or intracytoplasmic structures linked to membrane
Chromatiaceae	*Chromatium*	Purple sulfur bacteria; use sulfide in metabolism and deposit sulfur grains internally
Ectothiorhodospiraceae	*Ectothiorhodospira*	Photoautotrophs and/or photoorganotrophs; use sulfide as e^- donor and deposit sulfur grains externally
Purple nonsulfur bacteria	*Rhodospirillum*	Photoorganotrophs by preference; can use hydrogen as e^- donor; do not tolerate high levels of sulfide
Green bacteria		Green pigments in chlorosomes
Green sulfur bacteria	*Chlorobium*	Obligate phototrophs; deposit sulfur grains externally
Multicellular filamentous green bacteria	*Chloroflexus*	Gliding motility; flexible cells

Section 16. The mycobacteria

Family *Mycobacteriaceae*

There is only one genus in section 16, *Mycobacterium.* The organisms in this genus are acid alcohol fast, weakly gram positive, and fundamentally unicellular. An example can be seen in color plate 1C. *Mycobacterium* organisms are very slow growing and can be either pathogenic or saprophytic. *Mycobacterium tuberculosis* causes tuberculosis and is a major health problem worldwide.

Section 17. Nocardioforms

Nocardioforms are gram-positive bacteria that are frequently acid alcohol fast or at least partially so. They are distinguished from mycobacteria by their production of a mycelium that later fragments into round- and rod-shaped cells. As the name of section 17 implies, the principal genus is *Nocardia,* which is comprised of saprophytes as well as plant and animal pathogens. During the five-year course of the manual's publication, it was decided to include these organisms also in section 26.

Volume 3

Section 18. Anoxygenic phototrophic bacteria

I. Purple bacteria
  Family I. *Chromatiaceae*
  Family II. *Ectothiorhodospiraceae*
  Purple sulfur bacteria
II. Green bacteria
  Green sulfur bacteria
  Multicellular filamentous green bacteria
III. Genera incertae sedis (included for convenience but may not belong)

As the name of section 18 implies, these gram-negative organisms do not produce oxygen during photosynthesis. Such metabolic processes can be expected to resemble those occurring during the initial evolution of life on earth; therefore, they are of considerable biochemical interest. Table A.6 presents the general classification scheme for this section, and some sample organisms can be seen in figure 4.5A–B.

Section 19. Oxygenic photosynthetic bacteria

Group I. Cyanobacteria
  Subsection I. Order *Chroococcales*
  Subsection II. Order *Pleurocapsales*
  Subsection III. Order *Oscillatoriales*
  Subsection IV. Order *Nostocales*
    Family I. *Nostocaceae*
    Family II. *Scytonemataceae*
    Family III. *Rivulariaceae*
  Subsection V. Order *Stigonematales*
Group II. Order *Prochlorales*
    Family I. *Prochloraceae*
    Other Taxa

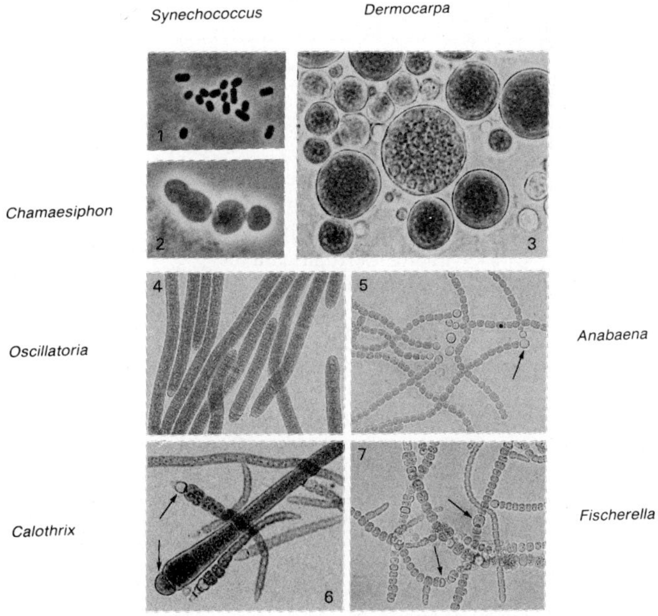

A

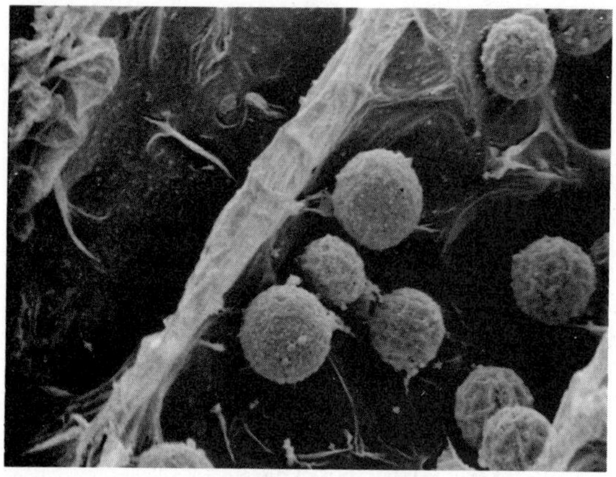

B

FIGURE A.7

(**A**) Some oxygenic phototrophs: (1) *Synechococcus,* (2) *Chamaesiphon,* (3) *Dermocarpa,* (4) *Oscillatoria,* (5) *Anabaena,* (6) *Calothrix,* (7) *Fischerella.* (**B**) *Prochloron* sp.

(A) Reprinted from Stanier, R. Y., and G. Cohen-Bazire. 1977. Phototrophic Prokaryotes: The cyanobacteria. *Annual Review of Microbiology* 31:225–274. (B) Electron micrograph courtesy of Lanna Cheng, University of California.

TABLE A.7 Taxonomy of the cyanobacteria

Subsection	Order	Typical Genus	Characteristic(s)
I	*Chroococcales*	*Synechococcus*	Unicellular; divide by binary fission or budding; coccoid to rod shaped
II	*Pleurocapales*	*Xenococcus*	Reproduce by formation of multiple baeocytes by internal fission of a single cell
III	*Oscillatoriales*	*Spirulina, Oscillatoria*	Filamentous; fission in one plane; have trichomes but no heterocysts or akinetes
IV	*Nostocales*	*Nostoc, Anabaena*	Same as subsection III but can make heterocysts
V	*Stigonematales*	*Chlorogloeopsis*	Multiserate trichomes or branches; fission in more than one plane

This is the first time that cyanobacteria have actually been integrated into *Bergey's Manual.* In the 8th edition, they were a separate division; prior to that time, they were not included with bacteria at all and were called blue-green algae. Roger Stanier forcefully argued, however, that while their photosynthetic systems resembled those of eukaryotic organisms, all their other attributes were those of prokaryotes. His thesis now is generally accepted by bacteriologists (although not necessarily by all biologists).

Section 19 is divided into two groups, the cyanobacteria (Figure A.7A) and the order *Prochlorales,* with the cyanobacteria being far more numerous. The general properties of cyanobacteria are summarized in table A.7. There are several unusual structural features in certain cyanobacteria. Baeocytes are small cells produced internally by multiple fission of a single parental cell. **Trichomes,** a term taken from algae, are nonsheathed chains of cells. Heterocysts are discussed in chapter 12 in connection with nitrogen fixation.

TABLE A.8 Aerobic chemolithothrophic bacteria

Classification	Typical Genus	Characteristic
Nitrifying bacteria		Have reduced energy source; nitrogen-containing compounds
	Nitrobacter	Oxidize nitrate
	Nitrosomonas	Oxidize ammonium
Colorless sulfur bacteria		Use reduced sulfur compounds for energy
	Thiobacillus	Oxidize hydrogen sulfide or sulfite
Obligately chemolithotrophic hydrogen bacteria	*Hydrogenobacter*	Use hydrogen gas for energy source
Iron- and manganese-oxidizing and/or depositing bacteria	*Siderocapsa*	Oxidize or deposit metals
Magnetotactic bacteria	*Aquaspirillum*	Respond to magnetic field

Prochloron is the only genus included in group II (figure A.7B). The photosynthetic system of the members of *Prochloron* is intermediate in composition between those of the cyanobacteria and the eukaryotic cells. None of these organisms have been grown in culture, and the only known examples are found as extracellular symbionts associated with marine invertebrates. The recently identified *Prochlorothrix* is included as "other taxa."

Section 20. Aerobic chemolithotrophic bacteria and associated organisms

I. Nitrifying Bacteria
  Family *Nitrobacteraceae*
    Nitrate-oxidizing bacteria
    Ammonia-oxidizing bacteria
II. Colorless sulfur bacteria
III. Obligately chemolithotrophic hydrogen bacteria
IV. Iron- and manganese-oxidizing and/or depositing bacteria
  Family *Siderocapsaceae*
V. Magnetotactic bacteria

As suggested by the group name, these bacteria all derive their energy from oxidizing inorganic materials such as nitrogen and sulfur compounds. Members of the genera *Nitrobacter, Nitrococcus,* and *Nitrospina* oxidize nitrite to nitrate (see figure A.8A and chapter 12). Some unusual organisms assigned to section 20 are the so-called magnetotactic bacteria that contain small crystals of magnetite enabling them to orient themselves with respect to the earth's magnetic field. It has been hypothesized that this ability would enable mud-dwelling bacteria in temperate latitudes to stay at the bottom of lakes and ponds. The general taxonomic scheme of section 20 is summarized in table A.8.

Section 21. Budding and/or appendaged bacteria

I. Prosthecate bacteria
  A. Budding bacteria
    1. Buds produced at tip of prostheca
    2. Buds produced on cell surface
  B. Bacteria that divide by binary transverse fission
II. Nonprosthecate bacteria
  A. Budding bacteria
    1. Lack peptidoglycan
    2. Contain peptidoglycan
  B. Nonbudding, stalked bacteria
  C. Other bacteria
    1. Nonspinate bacteria
    2. Spinate bacteria

Although bacteria do not have appendages in the sense of having arms or legs, they frequently do have stalks protruding from the cell body proper. In **prosthecate** bacteria, the stalk is cellular (contains cytoplasm), while in **nonprosthecate** bacteria, the stalk is composed of filaments that do not contain any cytoplasm. Some examples of this morphologically diverse group are presented in figures A.8B–C and 1.14C–D. Table A.9 summarizes the taxonomic subdivisions of this section.

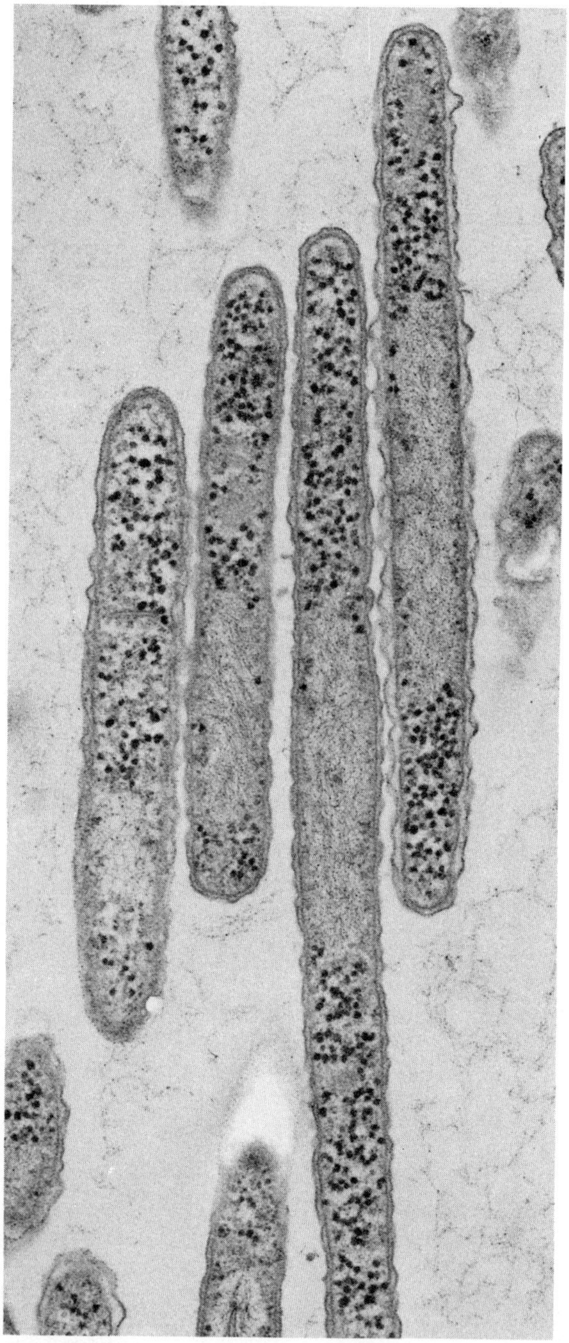

A

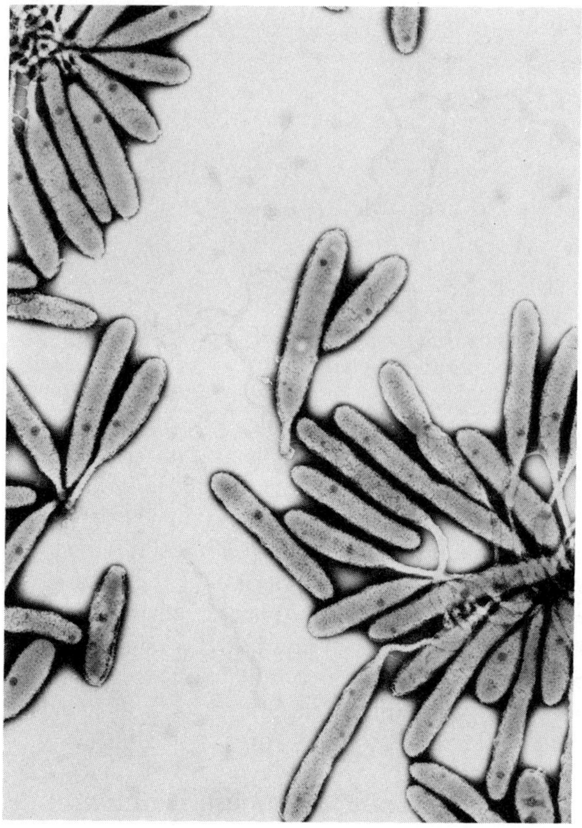

B

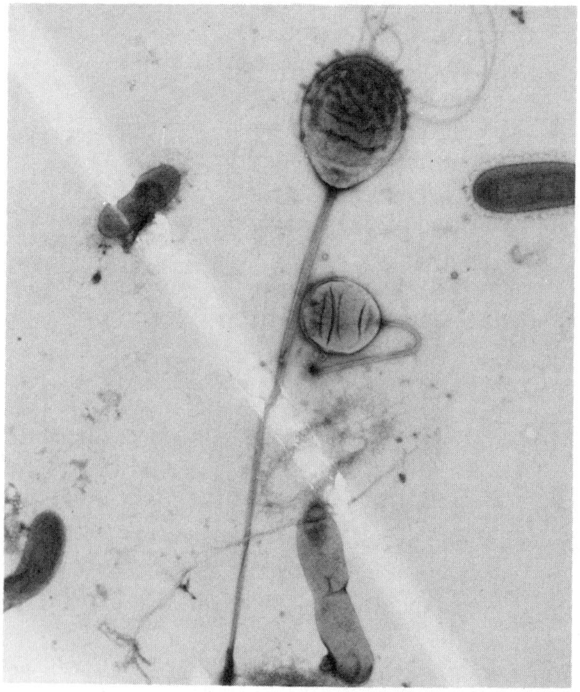

C

FIGURE A.8

(**A**) *Nitrospina gracilis,* a nitrite-oxidizing bacterium. The dark granules are glycogen. (**B**) *Caulobacter bacteroides,* a prosthecate bacteria having a cellular stalk. (**C**) Members of the *Blastocaulis-Planctomyces* group with noncellular stalks.

(A) Reprinted from Watson, S. W., and J. B. Waterbury. 1971. Characteristics of two marine nitrite-oxidizing bacteria: *Nitrospina gracilis* nov. gen. nov. sp. and *Nitrococcus mobilis* nov. gen. nov. sp. *Archiv für Mikrobiologie* 77:203–230. (B–C) Electron micrographs courtesy of Dr. Jean M. Schmidt, Department of Microbiology, Arizona State University.

TABLE A.9	Budding and/or appendaged bacteria

I. Prosthecate bacteria
  A. Budding bacteria
    1. Buds at tip of prostheca *(Hyphomicrobium)*
    2. Buds on cell surface *(Prosthecomicrobium)*
  B. Binary transverse fission *(Caulobacter)*
II. Nonprosthecate bacteria
  A. Budding bacteria
    1. No peptidoglycan *(Planctomyces)*
    2. Have peptidoglycan *(Blastobacter)*
  B. Nonbudding, stalked bacteria *(Gallionella)*
  C. Other bacteria
    1. Nonspinate *(Seliberia)*
    2. Spinate (pericellular nonprosthecate bacteria usually catalogued in other sections)

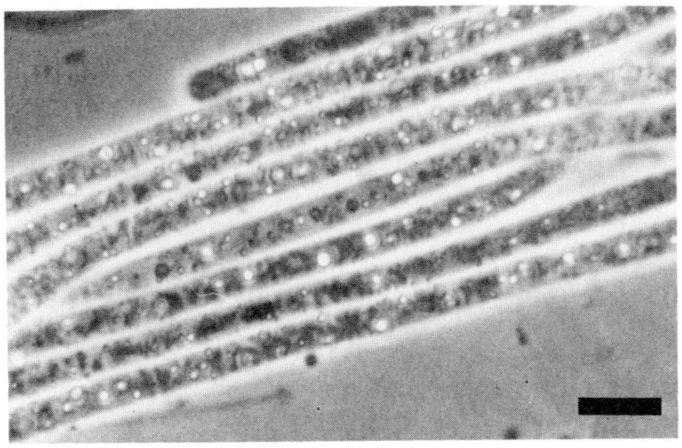

A

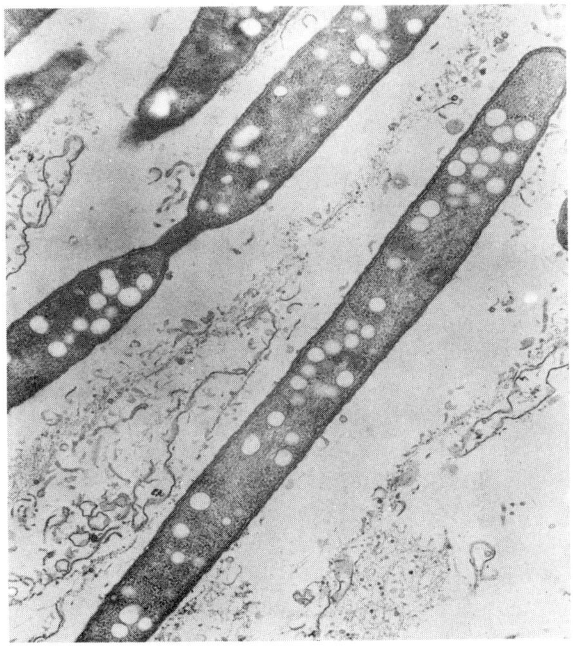

B

FIGURE A.9

(A) The sheathed bacterium *Beggiatoa* sp. **(B)** *Myxococcus* sp.

(A) Reprinted from Strohl, W. R., and J. M. Larkin. 1978. Cell division and trichome breakage in *Beggiatoa*. *Current Microbiology* 1:151 155.
(B) © BPS.

Section 22. Sheathed bacteria

Sheathed bacteria form chains of cells surrounded by a tubular sheathed structure. Sometimes the chains of cells are parallel to one another within the sheath and are designated as trichomes. Sheathed bacteria are all obligate aerobes, but they do not necessarily require a high oxygen tension. As a result, they can be found in soil, sewage treatment facilities, etc. They frequently deposit either ferric hydroxide or manganese dioxide in the sheath walls, both of which add to the rigidity of the structure. The most commonly encountered genera of sheathed bacteria are *Sphaerotilus* and *Leptothrix*, both of which are gram negative.

Section 23. Nonphotosynthetic, nonfruiting gliding bacteria

Order I. *Cytophagales*
  Family I. *Cytophagaceae*
  Other Genera
Order II. *Lysobacterales*
  Family I. *Lysobacteraceae*
Order III. *Beggiatoales*
  Family I. *Beggiatoaceae*
Other families and genera
  Family *Simonsiellaceae*
  Family *Pelonemataceae*
  Other genera

These gram-negative bacteria comprising section 23 are distinguished by their method of motility and by an absence of stalks and spores, such as are seen in organisms grouped in section 21. They are widely distributed in nature, and their complex taxonomic relationships are of interest primarily to the specialist. Two genera encountered by the beginning student are *Beggiatoa* (figures A.9A, 2.1C, and 2.18), the genus with the largest known prokaryotic cells, and *Cytophaga*, the most commonly encountered genus of gliding bacteria.

Section 24. Fruiting, gliding bacteria

Order *Myxococcales*
Family I. *Myxococcaceae*
Family II. *Archangiaceae*
Family III. *Cystobacteraceae*
Family IV. *Polyangiaceae*

These gram-negative bacteria resemble the cellular slime molds in life-style and function. They are all assigned to the order *Myxococcales* and are subdivided into four families based on the nature of the fruiting bodies produced, the shape of the spores, and the habitats. The organisms occur in the soil, on dead and decaying matter, and sometimes on living trees. They are frequently capable of degrading cellulose. The genus *Myxococcus* is a good example of organisms that fall into this classification (figure A.9B).

Section 25. Archaebacteria (archaeobacteria)

Group I. Methanogenic archaebacteria
Order I. *Methanobacteriales*
Family I. *Methanobacteriaceae*
Family II. *Methanothermaceae*
Order II. *Methanococcales*
Family *Methanococcaceae*
Order III. *Methanomicrobiales*
Family I. *Methanomicrobiaceae*
Family II. *Methanosarcinaceae*
Other taxa
Group II. Archaebacterial sulfate reducers
Order *Archaeoglobales*
Family *Archaeoglobaceae*
Group III. Extremely halophilic archaebacteria
Order *Halobacteriales*
Family *Halobacteriaceae*
Group IV. Archaebacteria without cell walls
Group V. Extremely thermophilic S^0 metabolizers
Order I. *Thermococcales*
Order II. *Thermoproteales*
Family I. *Thermoproteaceae*
Family II. *Desulfurococcaceae*
Other bacteria
Order III. *Sulfolobales*
Family *Sulfolobaceae*

Archaebacteria are a fascinating group of organisms that represents a new subdivision within *Bergey's Manual.* It was originally proposed by Carl Woese and his coworkers who showed that the nucleic acid sequences of certain bacteria were unrelated to the main group of bacteria. If anything, archaebacterial sequences are more closely related to those sequences found in eukaryotic cells. Moreover, the physiology and biochemistry of these organisms tended to suggest a very different form of cellular organization from that found in eubacteria, hence the name "archaebacteria" is intended to imply a sort of living fossil. It is spelled in the manual as "archae*o*bacteria," but this spelling has not taken hold among contemporary workers in the field.

Because this is a new taxonomic grouping, the status of the various family names is uncertain. Table A.10 presents the current arrangement. Archaebacteria are predominantly methane producers, sulfur oxidizers, and heat- and/or salt-tolerant organisms. Figure A.10A provides an example of a methane-producing member of the group. Note that halobacteria also are listed in volume 1, section 4. It is anticipated that after this edition of the manual, the halobacterial group will no longer be cross-listed.

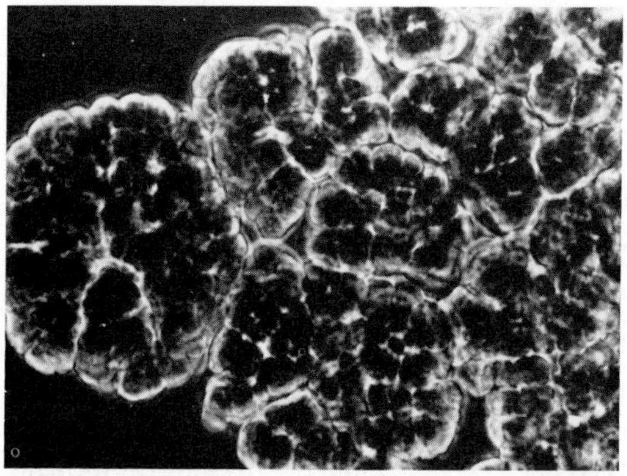
A

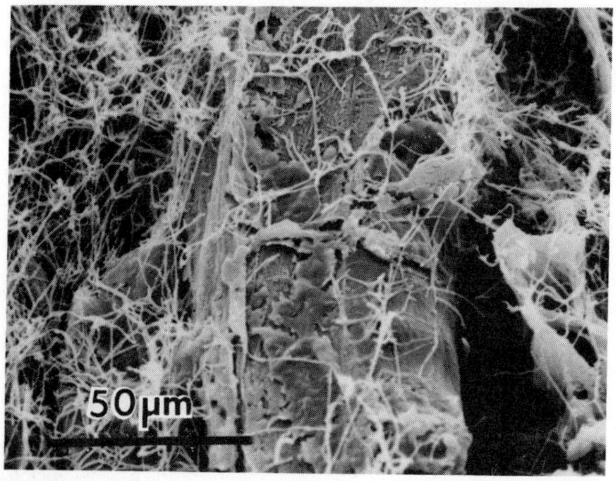

B

TABLE A.10 Archaebacterial groups		
Order	**Family**	**Characteristic(s)**
Methanogenic Archaebacteria		Rod shaped; catabolize $H_2 + CO_2$ or formate to methane; cell wall has pseudomurein
Methanobacteriales	*Methanobacteriaceae*	Little or no growth at temperatures $> 70°$ C
	Methanothermaceae	No growth below 60° C; optimum growth temperature $>70°$ C
Methanococcales	*Methanococcaceae*	Coccoid; cell walls composed of protein
Methanomicrobiales		Catabolize methyl groups as well as $CO_2 + H_2$ or sometimes formate
	Methanomicrobiaceae	Oxidize H_2, alcohols; reduce CO_2 to CH_4
	Methanosarcinaceae	Reduce CO_2 with H_2; ferment methanol
Other taxa	*Methanoplanaceae*	Unassigned methanogens
Archaebacterial Sulfate Reducers		
Archaeoglobales	*Archaeoglobaceae*	Cocci; cell envelope has glycoproteins
Extremely Halophilic Archaebacteria		
Halobacteriales	*Halobacteriaceae*	Require at least 1.5 M NaCl for growth; have ether-linked phospholipids; are not sensitive to penicillin or chloramphenicol
Archaebacteria without Cell Walls		
No order or family; the only genus is *Thermoplasma,* an obligate thermoacidophile.		
Extremely Thermophilic S^0 Metabolizers		
Thermococcales	*Thermococcaceae*	Strictly anaerobic cocci; obligate heterotrophs; sulfur reduced to H_2S; optimum temperature for growth is 88–100° C
Thermoproteales		Heterotrophs that use sulfur respiration
	Thermoproteaceae	True branching rods or filaments
	Desulfurococcaceae	Coccoid
Other Bacteria		
Sulfolobales	*Sulfolobaceae*	Extreme thermoacidophiles

Note: These unusual organisms are similar only in their ribosomal RNA sequences rather than in a particular life-style. They are, however, capable of growing under anaerobic conditions; therefore, they might possibly have been present during the early stages of the evolution of the earth's atmosphere.

◂◂◂ FIGURE A.10

(A) *Methanococcus mazei,* a methane-producing archaebacterium. These young cells are arranged in packets. **(B)** *Streptomyces flavovirens* after five weeks of growth on the inner bark of a Douglas fir.

(A) Reprinted from Mah, R. A., 1980. Isolation and characterization of *Methanococcus mazei. Current Microbiology* 3:321–326. (B) Reprinted from Sutherland, J. B., R. A. Blanchette, D. L. Crawford, and A. L. Pometto III. 1979. Breakdown of Douglas-fir phloem by a lignocellulose-degrading *Streptomyces. Current Microbiology* 2:123–126.

Volume 4

The content of volume 4 is devoted entirely to funguslike bacteria. Some confusion arises, however, because historically these bacteria have been referred to as the actinomycetes even though the genus *Actinomyces* is listed under section 15. This usage of the term is unfortunate but unlikely to change; therefore, you should be particularly careful not to confuse the genus with the group.

Section 26. Nocardioform actinomycetes

This is a slightly expanded group of section 17.

Section 27. Actinomycetes with multilocular sporangia

The principal genus in section 27 is *Geodermatophilus,* whose members live only as animal parasites and grow by means of branching filaments. A genus of some interest is *Frankia* whose members form a root nodule symbiotic relationship with nonleguminous plants similar to the symbiosis *Rhizobium* spp. form with legumes.

Section 28. Actinoplanetes

Members of section 28 produce spores within sporangia. The most typical genus is *Actinoplanes,* whose members are soil bacteria.

Section 29. Streptomycetes and related genera

This group of organisms is extremely important to industrial microbiology because the type genus *Streptomyces* is the source of numerous antibiotics. Streptomycetes are primarily soil organisms. A sample organism can be seen in Figure A.10B.

Section 30. Maduromycetes

Maduromycetes are all aerobic, gram-positive organisms with branched asporogenous vegetative mycelia and aerial hyphae bearing spores. Members of the genus *Actinomadura* are found in soil, and some tropical species are pathogenic.

Section 31. *Thermomonospora* and related genera

Section 31 comprises an artificial grouping to accommodate several different genera. Members of *Thermomonospora* are found in composts and other hot areas.

Section 32. *Thermoactinomyces*

Section 32 accommodates the genus of the same name. Organisms of *Thermoactinomyces* are found in such places as molding stacks of hay.

Section 33. Other genera

The organisms included in section 33 have been identified so recently that there is insufficient information to place them in any other category. An example is the soil organism *Glycomyces.*

Suggestions for Further Reading

Holt, J.G. (ed. in chief). 1984–1989. *Bergey's manual of systematic bacteriology.* Williams & Wilkins Co., Baltimore. This is the definitive work in four volumes. A set of essays on bacterial taxonomy, methods, and evaluation is included at the beginning of each volume and is well worth reading.

Sandine, W. E. 1988. New nomenclature of the nonrod-shaped lactic acid bacteria. *Biochimie* 70:519–522. A review of the recently proposed changes that includes some dendrograms on which the changes were based.

Stackebrandt, E. 1988. Phylogenetic relationships vs. phenotypic diversity: How to achieve a phylogenetic classification system of the eubacteria. *Canadian Journal of Microbiology* 34:552–556. This article considers the problem of arranging higher taxa for the most common bacteria.

Stackebrandt, E., R. G. E. Murray, and H. G. Trüper. 1988. Proteobacteria classis nov., a name for the phylogenetic taxon that includes the "purple bacteria and their relatives." *International Journal of Systematic Bacteriology* 38:321–325. A not particularly difficult example of an article published in the major taxonomic journal for bacteriologists.

Starr, M. P., H. Stolp, H. G. Trüper, A. Balows, and H. G. Schlegel, 1981. *The prokaryotes. A handbook on habitats, isolation, and identification of bacteria.* Springer-Verlag KG, Berlin. This is an excellent overview of the kingdom *Monera.* The text includes some taxonomy and also provides other types of information.

appendix

B

Answers to Study Questions

Chapter 1

1. A dark-field microscope will allow you to determine cell shape quickly and inexpensively. A bright-field microscope works well with stained or naturally pigmented specimens. A phase-contrast microscope allows you to view internal structures in transparent objects. Transmission electron microscopy provides high-resolution images of thin specimens. Scanning electron microscopy provides good resolution of specimen surfaces.
2. Briefly boil a mixture of spores and cells. If boiling is of the correct duration, the cells will die, but the spores still will be viable. As long as nutrients are not added, the spores should remain as spores.
3. If you are exposed to appropriate body fluids of someone who has a disease, you can contract it. Following this, you can transmit the disease to other people in the same way, but treatment(s) that kills disease-causing cells will prevent further transmission. It is also possible to have a pure preparation of a disease-causing organism. When that organism is administered in suitable quantity, the disease results. Once again, treatment(s) that kills cells will prevent the disease.

Chapter 2

1. The basic cellular organization is similar: Cells contain cytoplasm and are surrounded by a plasma membrane. The fundamental biosynthetic processes also are similar although they differ in detail. For example, both eukaryotes and prokaryotes have ribosomes, but their sizes are different. Both have DNA as their genetic material.
2. The distinction between algae and protozoa was often difficult to maintain in these unicellular organisms. Grouping fungi with plants was awkward because fungi are not photosynthetic. Bacteria differ in detail from eukaryotes; for example, they lack a nucleus.
3. Ascertain if the organism is eukaryotic or prokaryotic. If prokaryotic, the organism is a moneran. If eukaryotic, determine if it is photosynthetic. If it is a eukaryote, true plants have tissue differentiation and algae do not. Animals are nonphotosynthetic, display movement during at least part of their life cycles, and have tissue differentiation. The kingdom *Protista* was created precisely to avoid having to determine whether an organism is an alga or a protozoan.
4. Positive effects: organisms used to prepare foods like beer, wine, bread, yogurt, etc. Negative effects: organisms that spoil food, cause diseases, and attack materials from which buildings are constructed.

5. A microorganism can be unicellular or multicellular. It lacks true tissue differentiation if it is multicellular. The requirement for cellular organization rules out all viruses, viroids, and prions.
6. Gram-positive cells have a characteristically thick peptidoglycan layer. Gram-negative cells can have a thin peptidoglycan layer, a cell wall without true peptidoglycan, or no cell wall at all. Sometimes, electron microscopic observation will be sufficient to identify the actual state of affairs. Otherwise, biochemical analysis is needed.

Chapter 3

1. Any medium must supply all the essential elements for life (C, H, O, N, P, S, Mg, K, etc.). It also must provide any special nutrient chemicals that the cells cannot synthesize with their own metabolism. Unless the cell is a phototroph, the medium also must contain molecules that can be metabolized to provide energy for growth.
2. Unless pure cultures are employed for routine observations, a scientist can never be certain which organism is causing which effect. On the other hand, if pure cultures are used, all synergistic effects (requiring two or more organisms for their production) will be missed.
3. The cell walls of both gram-negative and gram-positive organisms contain peptidoglycan. However, gram-negative cells have only a single layer of peptidoglycan, while gram-positive cells have multiple layers. A gram-negative cell wall has an outer membrane based on lipid A. The different Gram stain reactions result from the relative ease of extraction of the dye complex from a thin peptidoglycan layer. Archaebacteria have cell walls based on proteins, pseudomurein, or other types of polysaccharides. Eukaryotic cells have cell walls based on cellulose (plants) or chitin (fungi).
4. Both capsules and outer membranes should limit access to the plasma membrane; however, a capsule fundamentally has a general retarding effect, while an outer membrane is truly semipermeable like the cell membrane. Therefore, an outer membrane is more likely to have specific effects on cell accessibility.
5. Bacteria can be seen to run or tumble when observed under a light microscope. By varying the relative amounts of time spent in these activities, a cell can move gradually in a desirable direction. At the biochemical level, bacteria can rotate their flagella in a clockwise or counterclockwise direction. Clockwise rotation of the flagella initiates the tumbling reaction. Reversal of the motor occurs in the presence of specific phosphorylated chemotaxis proteins.
6. Substances can be transported into a cell by simple diffusion, facilitated diffusion, symport, or active transport. All diffusion processes are driven by simple chemical gradients of the substance being transported. In symport reactions, a chemical gradient for one molecule is used to supply the energy necessary to transport another type of molecule against its own concentration gradient. In active transport, ATP or a molecule synthesized from ATP is involved in these reactions. Transport processes that do not involve chemical modification of the transported molecule are fundamentally reversible.
7. There is no particular reason to expect archaebacteria to behave differently from eubacteria, at least in broad outline. The same types of experiments used with eubacteria also can be applied to archaebacteria.

Chapter 4

1. The Embden-Meyerhof-Parnas (EMP) pathway converts glucose to pyruvate yielding NADH + H^+ as well as ATP. This is the most common route to the TCA cycle, but the waste products formed from pyruvate often are acidic. Alternative pathways exist therefore. The hexose monophosphate (HMP) pathway has the advantage of yielding ribose and erythrose that can be used for various biosynthetic purposes as well as sugars that can be fed into the EMP pathway. The HMP pathway can synthesize ribose from glucose without a decarboxylation reaction, thereby

conserving reduced carbon. The Entner-Doudoroff pathway is an alternative route to pyruvate. The phosphoketolase pathways are common in fermentative organisms and provide an alternative route to the same products. Along the way, they generate useful sugar intermediates.

2. Chemiosmotic theory states that the movement of an ion through a pore in a semipermeable membrane can be coupled to the synthesis of ATP via an ATP synthase complex located with one end in the membrane and the other in the cytoplasm. The energy to drive the reaction comes from the movement of the ion along its charge gradient. The reaction is reversible, and if ATP is hydrolyzed, the energy released in the reaction can be used to pump an ion from the cytoplasm to the exterior of the cell. Although most often described in terms of H^+ ions, there is no intrinsic reason why other ions like Na^+ cannot be used for chemiosmotic processes as well.
3. Electron transport is sequential oxidation and reduction of a chain of molecules located within a semipermeable membrane. During the sequence of reactions, protons can be carried along and left on the exterior of the membrane. Depending on the difference in the oxidoreduction potentials of the beginning and ending components of the electron transport chain, more than one proton may be transported. Ultimately, the electron being transported is used to reduce a final molecule to form a characteristic waste product.
4. Some chemotrophs and all phototrophs can use electron transport to store metabolic energy. The chemiosmotic principles are essentially the same in both cases, only the energy inputs are different. Fermentative chemotrophs use substrate level phosphorylation to synthesize ATP. Their ion gradients across the cell membrane are generated by ATP hydrolysis or reverse symport, not electron transport.
5. Anoxygenic photosynthetic bacteria carry out a photosynthetic pathway that is, in essence, the same as photosystem I in plants. Some bacteria can operate their photosystem in a noncyclic fashion to generate reduced NADP for biosynthetic purposes. Plants rarely do this.
6. Pyruvate can be converted into substances like lactic acid, ethanol, acetic acid, formic acid, and 2,3-butanediol. The general purpose of these conversions is to reoxidize the reduced enzyme cofactors synthesized during formation of pyruvate. Some more complex pathways, like the one to propionic acid, also can generate extra ATP molecules.
7. The simplest description of cells is to call them aerobic (capable of growing exposed to air), anaerobic (not capable of growing exposed to air), or facultative (capable of growing in the presence or absence of air). A microaerophilic cell requires oxygen but at reduced partial pressure. There are two types of facultative cells, those that never use oxygen under any circumstances and are strictly fermentative (obligate anaerobes, aerotolerant anaerobes) and those that use oxygen when it is available (oxybionts). Respiratory cells can be found among all these types except the obligate anaerobes because oxygen is not the only possible terminal electron acceptor.

Chapter 5

1. 2–Oxoglutaric acid is used to synthesize glutamate and glutamine, both of which act as donors in transamination reactions used to synthesize other amino acids. It is produced from glucose by the combined action of the EMP pathway and the first portion of the TCA cycle. Acetate feeds into the TCA cycle as acetyl CoA, and from that point on, the reactions are identical with those for glucose.
2. Both types of carbon fixation reactions require CO_2 as a substrate. However, the Calvin reactions are cyclic, the compounds initially carboxylated being regenerated as part of the cycle. There is no cycle during reductive carboxylation, and there must be a constant input of molecules to be carboxylated. Methane oxidizers are unlikely to use any of these reactions because they are producing CO_2 as a waste product.
3. A cell might use nitrogen fixation, transamination from existing amino acids, or reductive amination reactions with ammonium ion as a substrate. Use of preexisting amino

acids is the simplest solution, but ammonium ion is very common in nature.

4. Yes. All other members are derived from it. An example is methionine.
5. Transamination reactions involve the transfer of amino groups in a reaction with no energy input. Reductive amination reactions require the input of energy in the form of reduced enzyme cofactors.
6. Phosphoenolpyruvate (PEP) is part of the EMP pathway; oxaloacetate and 2-oxoglutarate are part of the TCA cycle; erythrose 4-phosphate is part of the HMP pathway. Most pathways begin with components from the EMP pathway. The next most common is the TCA cycle.
7. NAD and FAD contain adenosine diphosphate (adenine dinucleotide).

Chapter 6

1. Discontinuous DNA replication resolved the lack of a 3′ to 5′ DNA polymerase activity.
2. A cell wall must be synthesized in such a manner that the load-bearing structure is never broken. A cell membrane is not under stress; therefore, lipids can be inserted in just about any part of a membrane. Cell wall material, however, must be inserted into nonloaded strands that later have a load applied to them.
3. Nucleic acid synthesis requires the equivalent of one ATP molecule for every base added and additional ATP molecules to provide the supercoils. Every amino acid added to a protein requires the equivalent of an ATP molecule and an additional GTP to translocate the ribosome to the next codon. Input of another GTP is required to activate elongation factor Tu. Each sugar in a polysaccharide requires the equivalent of an ATP for its addition.
4. Topoisomerases are enzymes that add or remove supercoils from DNA molecules. They are necessary for DNA replication because strand separation adds supertwists to a DNA molecule, and newly synthesized DNA lacks supertwists.

Chapter 7

1. With transcriptional control, the synthesis of mRNA is regulated; with translational control, the synthesis of proteins is regulated. Translational control normally is mediated through the protein products of translation. An excess of product binds to mRNA and prevents further ribosomal attachment. Transcriptional control is mediated through a variety of molecules including sigma factors, repressors, and activators.
2. Positive control of the lactose operon is exercised by cAMP and CRP. They attach near promoter 1 and improve its ability to bind RNA polymerase. Negative control is maintained by the allosteric protein called the lactose repressor. In one allosteric state, the lactose repressor binds to the operator region of the promoter and blocks RNA polymerase binding. In the other state, it binds to the inducer. The sensor is the phosphorylation state of one component of the phosphotransferase system used to transport glucose and other molecules into the cell. Promoter 2 maintains a background level of transcription of the operon.
3. If the need for conversion of the substrate to the inducer is eliminated, the need for promoter 2 thereby is eliminated. Then the requirement for activation of promoter 1 can be made absolute.
4. An allosteric protein is one that has several stable configurations. It can be frozen in one particular state by the binding of a small allosteric effector molecule. One example is given in answer 2.
5. With feedback inhibition, a cell can give a more prompt response to overproduction of a molecule than with inhibition of transcription. After all, unless translational control is imposed, translation of existing mRNA molecules will continue for some time.

Chapter 8

1. Bacteria can initiate a new round of DNA replication before the old one has finished. In rapidly growing cells, new DNA helices being synthesized will not actually segregate into

separate daughter cells until two or three cell divisions later.

2. A differential medium contains substances that change its appearance when affected by the growth of certain bacteria. A pH indicator that changes color in the presence of acid-producing bacteria is an example. A selective medium is one that prevents the growth of most organisms. A medium containing an antibiotic is selective. An enrichment medium allows some organisms to grow better than others, but nearly all organisms can grow. A medium with only trace quantities of ammonium ion and amino acids is suitable for enrichment because nitrogen-fixing organisms will grow better in it. However, nearly all organisms can grow at least a little bit.
3. A closed culture receives no input of fresh nutrients and cannot dispose of its waste products. An open culture both receives and disposes. In a closed culture, nutrient limitation means fewer cells per milliliter when stationary growth phase is achieved than in an open culture. In an open system, the limiting nutrient can be added continuously in small quantities. The result is continuous slow growth of the cultured organisms.
4. Some parameters to be controlled are temperature, oxygen, osmotic strength of the medium, pH, and atmospheric pressure. The suffix *philic* means that an organism prefers a certain set of conditions; for example, a thermophilic organism grows only at high temperatures.
5. It is not possible to measure the true minimum growth temperature because the time between cell divisions would be infinite at this temperature.

Chapter 9

1. Unusual modes of replication include rolling circle replication of DNA, all types of RNA replication, and retroreplication (RNA making DNA). Each of these methods of replication solves a particular problem of making a type of nucleic acid not normally present in a host cell. Rolling circle replication gives single strands of DNA, RNA replication provides more RNA, etc.
2. A primary cell culture has just been explanted from an organism. A diploid cell strain has been growing in culture long enough to accumulate some abnormalities, but the chromosomes are still grossly normal. An established cell line has gross chromosomal abnormalities and is transformed. An established cell line is the easiest culture to maintain, but the cells within least resemble the normal organism from which the culture was started. The converse is true for primary cultures.
3. A tumor is a mass of tissue that has lost normal growth regulation and continues to divide long after growth should have ceased. It consists of transformed cells. A virus that can cause transformation is the retrovirus called Rous sarcoma virus.
4. Intracellular growth of a virus is best studied by some variation of the premature lysis experiment. To distinguish host from viral proteins, label the host cell with radioactive amino acids and then wash the cells well before they are infected. Preexisting proteins will be labeled, proteins synthesized later will not be. Some ambiguity will arise, depending on the amount of labeled amino acid inside the cell at the time of infection.
5. Viroids are independent RNA molecules found in plants. Virusoids are extra pieces of RNA within an RNA virus. Prions are protein molecules that can stimulate transcription of the normally repressed DNA genes that code for them. Prions are not particularly like viruses, although viroids and virusoids are.

Chapter 10

1. Plasmids can promote transfer of DNA from one organism to another. They also can contribute to the metabolic capabilities of the host cell. Sometimes they can move themselves to other cells by conjugation. If small enough, they can be moved by transduction. Genetic transformation is also a possibility, although transformation is not very efficient with plasmids.
2. Temperate bacteriophages that have nonintegrated prophages are very similar to plasmids. Those with integrated prophages resemble plasmids like F. Lytic viruses are

plasmidlike only for the time that they are actually replicating within a host cell. Normal plasmids are found inside viral coats only when generalized transduction is occurring.

3. Conjugation. Specialized transduction. Specialized transduction is greatly limited in size by the necessity of preserving large portions of the viral genome.
4. Transposons undermine the genetic stability of an organism because they can change their position more or less at random. For this same reason, they are very exciting because DNA contained within a transposon can be inserted into unusual locations.
5. There is no definitive answer. You must assess, answer yes or no, and defend your position.

Chapter 11

1. DNA splicing allows any DNA sequence to be mixed with any other sequence. Theoretically any combination of genetic traits is possible. Major concerns are: development of antibiotic resistances in pathogenic bacteria, displacement of normal organisms by genetically engineered ones, and Frankensteinian monsters of all sorts. Safety regulations require biological and physical containment according to the Centers for Disease Control guidelines for working with pathogenic microorganisms.
2. There are three types of restriction endonucleases. All recognize a specific sequence of bases. Type I cuts at an arbitrary sequence 1,000 or more bases away. Type II cuts at defined points within the recognition sequence so that the base sequence of the cut ends can be predicted. Type III cuts at a definite site not within the recognition sequence so that the base sequence of the cut ends cannot be predicted. Only enzymes that generate predictable ends are useful for genetic engineering.
3. A vector is a self-replicating DNA molecule used to provide replicational functions for cloned DNA. In addition, it may or may not provide the means to transcribe and translate the cloned DNA. It is sometimes useful to have extra cloning sites within the vector or to regenerate the original cloning site so that restriction endonucleases can be used to remove the cloned DNA.
4. Genetically engineered products are human interferon, human insulin, human growth hormone, tissue plasminogen activating factor, numerous restriction endonucleases, and other enzymes. Any protein that is in large demand and/or commands a high price.

Chapter 12

1. Symbionts within cells: phagocytosed cyanobacterium in a protozoan. Symbionts within an organism: rumen of cattle. Symbionts on exterior: mycorrhiza. Adaptations: the saliva of ruminants has no digestive enzymes. A special organ (the rumen) is provided to house microorganisms and encourage their growth. The behavior of calves is such that they will ingest a good dose of bacteria for their rumens before they are weaned.
2. Agriculturally important symbioses include the animal rumen, the legume root nodule, and an edible mycorrhiza. Symbiosis in the termite gut is important in a negative sense.
3. Describe figures 12.3 and 12.4 in words.
4. Pollution-removing processes are those in which material is either degraded to provide food for organisms or oxidized by exposure to the atmosphere. Pollution problems have increased greatly due to large population increases worldwide. Waste material is being generated faster than it can break down spontaneously. Microbiologists can help by finding ways to accelerate decomposition processes.
5. Disease-causing organisms and viruses are eliminated through competition with free-living microorganisms, through exposure to oxygen and sunlight, and, as a final safeguard, through chlorination of sewage effluent and sterilization of sludge.
6. BOD is the amount of oxygen necessary for microorganisms to oxidize the usable material in a water sample over a defined period of time. A low BOD means relatively little growth of microorganisms and, therefore, clear water in lakes and streams. Increased BOD means more growth and more turbidity. If the BOD exceeds the available oxygen, the

water will become anaerobic and fermentations will begin. BOD can be increased by any organic waste or by inorganic substances that can be oxidized by chemolithotrophs using oxygen as a terminal electron acceptor.

Chapter 13

1. Tyndallization: slow, but relatively low temperatures. Autoclaving: fast, but relatively high temperatures, may get some nutrient decomposition. Radiation: fast but expensive. Only the first two methods are normally used for liquid media. Radiation is used for plasticware because the other methods of sterilization will melt plastics normally used in petri dishes.
2. Disinfectant: chlorine bleach, used to treat countertops. Antiseptic: Mercurochrome, used to treat surface wounds. Sanitizer: detergent used to clean bathrooms.
3. Antibiotics are organic chemicals produced by living organisms that kill or inhibit the growth of other, nonproducer organisms. Some are suitable for internal use, but others are used only externally in humans or other animals. Some can be used both ways. Others that are too toxic for use on the body can function as disinfectants or antiseptics.
4. A synthetic antimicrobial is one produced by an organic chemist rather than by a microorganism. A semisynthetic antimicrobial is an antibiotic that has been chemically modified. In terms of function, there is no essential difference between synthetic antimicrobials and antibiotics.
5. Time-dose-rate relationships are inverse. If a dose is increased, less time is required to effect the same killing. If a dose is decreased, more time is required.
6. Enveloped virus: nearly any treatment that will dissolve the lipid envelope, such as an alcohol. Nonenveloped virus: any treatment (phenol for example) that will attack the capsid proteins and denature them. Bacteria: an antibiotic. Bacterial endospore: autoclaving. Protozoan: boiling. Fungus: boiling. Antibiotics and alcohols can be used on the surface of a living host. Only certain antibiotics can be used internally.

Chapter 14

1. See tables 14.5, 14.6, and 14.7.
2. Milk is allowed to sour and clump. The curds are collected free from the whey. An unripened cheese consists only of curds. A ripened cheese is allowed to stand for some period of time while microorganisms grow and contribute flavoring compounds to it. Soft cheeses are surface ripened while hard cheeses are ripened by organisms growing within them.
3. A source of carbohydrate is fermented, usually by *Saccharomyces cerevisiae.* Within a week to ten days, the alcohol concentration can reach 12 to 15%. Wine is fermented longer than beer because of its higher alcohol concentration. Normally, beer is fermented at lower temperatures and is carbonated. The carbohydrates used in wine and beer fermentations are different. Distilled alcoholic beverages include brandy, whiskey, vodka, etc.
4. Drying removes the water needed for growth of microorganisms. Freezing prevents growth of microorganisms by shutting down all metabolic functions. Canning sterilizes the food product. On a routine basis, canning offers the cheapest and longest lasting method of food storage.

Chapter 15

1. Pyruvic acid is the most common end product of fermentative metabolism of glucose (see chapter 4 for details).
2. Microorganisms are capable of producing fermentative waste products that can serve as feedstocks for petroleum-based products like plastics. The waste products include acetone, butanol, and some higher fatty acids and alkanes. Few of these methods are used today owing to the relatively low prices of petroleum. Any political event that cuts off oil supplies from the Middle East would signal a resurgence of interest in these processes.
3. Analytical chemists are interested in biosensors. These are devices based on microbial metabolism that measure the concentration of a particular molecule indirectly. For example, *Nitrosomonas* organisms oxidize ammonium ion to nitrite

using oxygen; therefore, measurements of dissolved oxygen in a culture are an indirect indication of the amount of ammonium ion available.

4. Microbial cultures can be preserved by refrigeration, freezing, lyophilization, or anaerobic storage. Lyophilization is the most cumbersome but also potentially the longest lasting method. No method works with all organisms. Refrigerated cultures are subject to mechanical failures.
5. DNA splicing allows scientists to create microorganisms that produce a particular product in large quantity. All kinds of proteins are suitable candidates for having their genes cloned.

Chapter 16

1. During the primary response, new antigens are engulfed by macrophages, and pieces of them are presented to B and T cells. Certain B cells, assisted by T-helper cells, are stimulated by the reaction of the antigen pieces with the antibodies present on the surface of the B cells. They divide and differentiate to yield plasma cells that actively secrete antibodies and small B cells that serve as memory cells. Eventually the clone of plasma cells dies off. During a secondary response, the antigen stimulates preexisting memory cells to divide and differentiate rapidly into plasma cells. Consequently, the secondary response is faster and of greater magnitude. T-suppressor cells prevent the activation of B cells whose antibodies would react with the body's own antigens.
2. Both T and B cells are small, round lymphocytes derived from bone marrow. They mature in different organs within the body. B cells have actual antibody molecules on their surface, while T cells have antibody-like molecules on their surface. Only B cells actually differentiate into cells that secrete antibody molecules. A person without B cells would have cell-mediated immunity but not humoral immunity. A person without T cells would lack cell-mediated immunity and have greatly depressed humoral immunity.
3. See figures 16.8 and 16.9.
4. The inflammatory response is triggered by the cell-mediated immune system. Capillaries dilate, fluid and phagocytic cells accumulate in the tissue. Within the fluid are samples of all the antibodies in general circulation. Killing of invading cells by cytotoxic T cells or the complement system can release toxic molecules that aggravate the process. Most problems with this response can be traced to the presence of IgE antibodies that can trigger the degranulation of mast cells and a massive release of histamine. In small doses, histamine precipitates allergic symptoms. In larger doses, anaphylactic shock results.
5. An ELISA test is an antibody sandwich technique. An antibody specific for a particular antigen is added to a sample. If the antigen is present, the antibody molecules bind and stay bound during washing. If not, the antibody molecules are washed away. Following this, a second antibody that specifically binds to the constant region of the first antibody is added. The second antibody has an enzyme attached to its constant region where it will not affect binding to the first antibody. The test is completed by addition of a chromogenic enzyme substrate. If the second antibody is present, the substrate is cleaved and a colored compound produced. The color indicates that the first antibody was bound; therefore, the antigen in question is present. If there is no color, the antigen is not present.

Chapter 17

1. An invasive organism is one that can enter the body without much difficulty. It does not necessarily cause a disease upon entry, however. A virulent organism is one that can readily cause a disease and often is invasive as well. An accidental pathogen is a noninvasive organism that can cause disease if it enters the body through a wound; therefore, a microorganism that is safe to a person with intact epithelium may be very dangerous to a person with an injury.
2. Bacterial enzymes that promote invasion are hydrolytic enzymes, hemolysins, and coagulase. Toxins, capsules, chelating agents,

and protein A promote virulence. Coagulase converts a normal host defense reaction into one that protects the invading bacterium.

3. Intracellular parasites must survive the respiratory burst following phagocytosis. Sometimes they produce high levels of detoxifying enzymes or cause premature lysis of the phagosome. Elimination of intracellular parasites usually involves destruction of the parasitized cell.
4. An exotoxin is a poisonous protein molecule synthesized and released by a bacterium. It can produce effects distant from the site of infection. Tetanus toxin is an example that causes generalized muscle contraction. An endotoxin is a poisonous portion of a bacterial cell. It is released only after the cell is destroyed. An example is lipid A from the gram-negative cell wall that can induce generalized shock within the body.
5. Mycotoxins are produced by fungi. Normally they affect the body only when they are ingested. Bacterial toxins can be released as a result of an actual infection, or they can be ingested with food.
6. Capsules can prevent phagocytosis of bacteria. Some surface proteins can bind antibodies and prevent their specific reaction. Polysaccharide chains can act to keep antibodies at a distance from a cell to prevent activated complement from attacking the cell.
7. All organisms must have iron for growth. Both host and parasite attempt to chelate iron and keep it for their respective uses.

Chapter 18

1. Cost-benefit analysis means weighing the possible adverse reactions to a particular medical treatment against the possible benefits. In the case of herd immunity, for example, a vaccine is administered to as many people as possible. There is always some risk of side effects, but they should be less frequent than the disease was at the beginning of the program. If the percentage of immunized persons is high enough, the few remaining susceptible individuals will have substantial protection against the disease because the persons surrounding them cannot contract the disease and, therefore, cannot transmit it.
2. The frequency with which an immunization needs to be repeated is a function of the life span of the plasma cells that secrete the pertinent antibodies. Different clones survive for different lengths of time. In addition, persons are often accidentally exposed to a disease-causing organism in the course of their daily lives, and each accidental exposure serves to restimulate the clone. Therefore, immunizations that need to be given frequently have short-lived clones, and/or an individual has little chance of being accidentally exposed to the antigen.
3. If the disease is a mild one with few side effects, the risks from immunization are likely to outweigh the benefits. For a potentially lethal disease, the converse is true.
4. The selection of an antibiotic ideally should be based on susceptibility tests, and the antibiotic should be administered at a proper dose for the recommended length of time, regardless of how fast the disease symptoms dissipate. Common failings include discontinuation of an antibiotic as soon as the patient feels better or administration of an antibiotic in a situation in which it is ineffective as, for example, against a virus. Selective pressure means that a particular treatment forces a population of cells to change or die. A mild pressure (as with suboptimum doses of an antibiotic) allows more time for mutations to occur. When mutation does occur, the antibiotic is no longer effective.

Chapter 19

1. Typhoid fever is transmitted by contaminated water and prevented by proper treatment of sewage. Bubonic plague is transmitted by infected fleas and can be prevented by insecticides to kill the fleas and other products to control infected rodents. Tuberculosis is transmitted via inhalation of aerosols or direct contact with sputum, and a reduction in crowded conditions can help prevent its spread. Active cases can be treated with antibiotics, and spitting in public places should be discouraged. Syphilis is transmitted by direct sexual contact or by direct contact by any of the ulcerative lesions with any mucous membrane. Preventive measures

include monogamy and use of condoms. Active cases can be treated with antibiotics.

2. The measles vaccine is recommended for everyone; the vaccine for plague is not. The risk of death from plague is very small because there are so few cases in the United States. The risk of death or serious injury from measles is much greater because there are more cases every year.
3. It is much easier to treat diseases caused by prokaryotes than diseases caused by eukaryotes because the metabolic differences between the two types of organisms allow antibiotics to be targeted to one specific group. It is also much easier to treat diseases in which the causative organism can be grown in culture than diseases in which it cannot. Research is greatly simplified if an animal model system for the disease can be found. The production of genetically engineered vaccines and anti-idiotypic vaccines offers some interesting opportunities for the future.
4. Tuberculosis, strep throat, meningitis, diphtheria, and other diseases affecting the upper and lower respiratory tracts.
5. Antibiotics are not always effective against gonorrhea because of penicillinase producers. Valley fevers are difficult to treat because few antibiotics are effective against fungi and most have serious side effects for the patient. Most viral diseases cannot be inhibited specifically by antibiotics owing to their parasitic nature.

Chapter 20

1. The first bacteria to be speciated were characterized by basic morphology and life-style. At the present time, species are more rigidly defined in terms of nucleic acid sequence homology.
2. All observable traits have at one time or another been used as taxonomic criteria.
3. Direct comparison. Sequence the 5S ribosomal RNA.
4. The rate of diversion will depend, among other things, on the rate of cell division. Rapidly growing bacteria will replicate their DNA more often and, therefore, have a greater probability of introducing mutations into their DNA. An organism that had a mutation rate of zero would be totally unadaptable. If conditions changed, it would not be able to cope and would soon be outcompeted.
5. %(G+C), cell size, shape, habitat, optimum growth conditions, type of cell wall, and primary energy metabolism among others.

glossary

a

accessory pigment a colored molecule found in a photosynthesizer that is not directly involved in photosynthesis but can nevertheless absorb light energy and pass it to the photochemical reaction center. This pigment is a component of the photosynthetic antenna system.
active immunity resistance to a disease resulting from circulating antibody molecules that are synthesized by the individual's own cells.
active site that portion of a folded protein chain (enzyme) to which a substrate molecule binds and is converted to the product.
active transport a process in which metabolic energy is required to move a substance across a cell membrane because the substance is moving against its own concentration gradient.
acyl carrier protein a protein that serves as a carrier of the intermediates during the biosynthesis of a fatty acid.
adhesin a substance that anchors a bacterial cell to a particular site.
adjuvant a preparation that enhances an immune response to an antigen. It is usually some sort of gel or emulsion that acts to maintain the antigen at relatively high local concentration.
aerobic having oxygen present.
aerobic digester a mechanical device to promote the aerobic growth of microorganisms on water passed through a sewage treatment plant.
aerosol fine droplets of liquid or fine particles suspended in a gas.
agar a complex polysaccharide based on agarobiose and used for solidifying microbiological media.
algal bloom a sudden burst of growth by one or a few algae that imparts a characteristic color to a body of water.
allosteric effector a small molecule that binds to an allosteric protein and causes a shift in the equilibrium between the various configurations of the protein.
allosteric protein a protein that can fold into more than one stable three-dimensional configuration. The configurations are in equilibrium with one another and their stability can be affected by the binding of small molecules.
ammonification the process of converting amino nitrogen from organic molecules into free ammonium ions.
anabolic reaction a biosynthetic reaction that builds up a molecule rather than disassembles it.
anaerobic lacking oxygen.
anaphylactic shock severe drop in blood pressure and bronchoconstriction resulting from the rapid degranulation of mast and basophil cells after a person is exposed to an anaphylatoxin.
anaphylatoxin a molecule that can stimulate an anaphylactic response.
anaplerotic pathway an alternative series of biochemical reactions that acts to bypass a portion of a normal biochemical cycle in order to supply extra quantities of a particular compound. This replenishing activity allows other metabolic pathways to remove intermediate compounds from the cycle without destroying its function.
antibiotic a substance produced by one organism that kills or inhibits the growth of other, nonproducer organisms.
antibody a complex of protein molecules produced by a B lymphocyte or its descendants that reacts specifically with a molecule or portion of a molecule that stimulated the original B cell.
anticodon a region or loop on a tRNA molecule that forms hydrogen bonds with a codon on an mRNA molecule during translation.
antigen any substance or portion of a molecule that is capable of stimulating division and differentiation of B lymphocytes from which antibodies specific to the antigen are produced.
antigenic determinant a domain within a larger molecule that can stimulate production of an antibody.
antiporter system system by which a cell can transport two substances in opposite directions across a membrane.
antiseptic a chemical that can be used on the skin to prevent infection of a wound. It is too toxic to be used internally.
antiseptic surgery an operation in which the surgeon uses various chemicals and/or antibiotics prophylactically to prevent subsequent infection of a wound.
antitoxin a preformed antibody preparation that inactivates a toxin and can be used to provide passive immunity to that toxin.
apoenzyme a protein(s) that has catalytic activity but not all the substrate specificity of a holoenzyme complex.
aquifer a stratum of rock that contains water.
arthrospore an essentially unmodified cell released from a mycelial mat as a method of asexual reproduction in certain fungi.
ascus a sac containing sexual spores that is seen in fungi called ascomycetes.
aseptic lacking contaminating microorganisms. Aseptic technique is used, for example, to maintain culture purity during inoculation of medium or to prevent infection of a surgical wound.
asexual reproduction new individuals are produced from a parent organism by a process that does not involve meiosis.
assimilatory nitrate reduction conversion of nitrate to ammonium ion for biosynthetic purposes.

attenuation (1) A process of operon regulation in which transcription is allowed to begin but is followed by early termination of mRNA before the protein coding sequences are copied. (2) In vaccine production, an attenuated microorganism is so weakened that it can no longer cause a disease, although it is still capable of stimulating the immune system.
autoclave an instrument used for sterilization consisting of a pressure vessel into which a sterilizing agent (usually steam) is introduced. Contact between the sterilant and the material to the sterilized is maintained for the appropriate length of time, and then the sterilant is removed before the vessel can be opened safely.
autogenous regulation a situation in which a substance, usually a protein, controls the rate of its own synthesis. In most cases, the substance acts to prevent transcription of its gene or translation of its mRNA.
autoimmunity the condition in which an organism produces antibodies that react with its own antigens.
autolysin an enzyme that can degrade peptide cross bridges or glycan chains of peptidoglycan, reshaping the cell wall. In excess, it can cause sufficient damage to a cell wall that osmotic pressure will destroy the producing cell.
autoradiography the process of exposing a photographic emulsion to the decay products of radioactive atoms. The silver grains in the emulsion form an image of the radioactive substance.
autotroph an organism that can reduce carbon dioxide in order to form the carbohydrates it needs to make polysaccharides, amino acids, etc. It does not require any other source of carbon for metabolism.
axial fibrils flagella that are located within an outer sheath surrounding gram-negative nonphotosynthetic bacteria.

b

bactericidal agent a treatment or substance that promptly kills bacteria. Treated cells are no longer capable of reproduction, even if treatment is discontinued.
bacteriocin a chemical or protein complex synthesized by a bacterium that kills nonproducing bacteria.
bacteriophage a virus that infects a bacterial cell.
bacteriorhodopsin a chromoprotein synthesized by *Halobacterium halobium* under anaerobic conditions; it functions as a photosynthetic proton pump.
bacteriostatic agent a treatment or substance that inhibits the growth of bacteria but does not immediately kill their cells. Although the cells may eventually die after a period of days or weeks, the cells normally resume growth when the bacteriostatic agent is removed.
bacteroid a prokaryotic cell found in the cytoplasm of a eukaryote that has lost its typical appearance and its ability to live independently.
barophile an organism that will grow only under conditions of high pressure, as in the lower strata of an ocean.
basidium an elongated cell found in certain fungi (basidiomycetes) on which sexual spores are produced.
B cells lymphocytes that, in the chicken, mature in the Bursa of Fabricius. In humans, there is no comparable structure, but B cells do exist and presumably mature in the lymphoid tissues of the intestinal tract.
binary fission a process of cell division in which a single, large cell creates a new septum at its approximate midline and splits to give two daughter cells. The process does not involve mitosis or meiosis.
bioassay a test in which a living organism is used to measure a parameter quantitatively.
biochemical oxygen demand the amount of oxygen required in a body of water to allow all types of organisms to respire using the available nutrients. It is a measure of the amount of oxidizable carbon present.
biodegradable refers to a compound that can be used by certain organisms as part of their normal metabolism so that it does not accumulate in the environment.
biogenic produced by a living organism.
biosensor a measuring instrument consisting of immobilized microbial cells and a pH or ion-specific electrode. The level of cell metabolism depends on the concentration of a specific compound (the first compound). As microbial cells metabolize, they affect the concentration of the substance (the second compound) measured by the electrode and generate a current proportional to the concentration of the original (first) compound.
biosphere that portion of our planet that contains living organisms.
blanch to heat a vegetable prior to freezing it in order to inactivate autolytic enzymes.
blocking antibody IgG molecules that react with an antigen before it can stimulate other functions of the immune system. An antibody stimulated by desensitization injections used in the treatment of allergies is an example.
bubo a swollen lymph node.
buffer a substance that resists changes in the pH of a solution by combining with free hydrogen ions as the pH drops or by releasing hydrogen ions as the pH rises.
burst size the average number of viral particles released per infected cell.

c

Calvin cycle a series of chemical reactions that allow a cell to take carbon dioxide from the atmosphere, reduce it, and incorporate it into glyceraldehyde phosphate. It is also called the reductive pentose phosphate cycle.
capneic refers to an organism that requires a partial pressure of carbon dioxide greater than what is normal in order to grow well.
capsid the protein coat of a virus.
capsule an organized layer, usually of a gelatinous polysaccharide, that surrounds a bacterial cell and serves a protective function (see also glycocalyx).
carcinogen a substance that can cause cancer.
carrier (of a disease) a person who exhibits no symptoms but nevertheless is infected with an etiologic agent. A carrier is a potential source of infection for all other members of a population.
carrier protein a protein that spans a cell membrane and facilitates the movement of one or more substances across it.
case definition an epidemiologic term that refers to the collection of symptoms taken to be characteristic of a particular disease. It is used to decide which individuals have been affected by the disease under study.
case:fatality ratio the percentage of cases of a disease that are fatal. The ratio can be affected by the availability of treatment procedures.
catabolism metabolic reactions involved in the degradation of a substance so as to yield energy and/or substrates for biosynthesis.
catabolite repression reduction in the amount of transcription from an operon when glucose is present (also known as glucose effect).
cDNA see complementary DNA.
cell fusion the process by which two cells are forced to join their plasma membranes and merge their cytoplasmic contents. The resulting composite cell is unstable genetically and discards chromosomes more or less at random to achieve a more normal number.
cell-mediated immunity an immune response based on the actions of cells rather than on individual, circulating protein molecules.
chelator a compound to which metallic ions, especially divalent ions, bind strongly; however, the union is reversible. The complex of chelator and metal normally is soluble so that the metal is more readily available for use by growing cells.
chemiosmotic theory any process involving movement of solutes across a semipermeable membrane in which energy is used to create a gradient of chemical or electric potential or the gradient is used as a source of energy to power another reaction.
chemolithotroph a chemotrophic organism that oxidizes inorganic molecules.
chemoorganotroph a chemotrophic organism that oxidizes organic molecules.

chemostat a culture vessel that permits the continuous addition of fresh culture medium and the removal of a corresponding volume of culture and cells. A chemostat permits the steady-state culture of an organism or group of organisms.
chemotaxis movement of a cell in response to a concentration gradient of a chemical. The chemical can be a repellent or an attractant.
chemotherapy the use of chemicals to kill or inhibit the growth of disease-causing microorganisms while they are in the body. The chemicals used must do less damage to the host than to the parasite.
chemotroph an organism that derives the energy required for growth and metabolism from the oxidation of organic or inorganic chemical compounds.
chlamydospore a cell with a thickened wall released by hyphal fragmentation as part of asexual reproduction in certain fungi.
chlorosome a membranous vesicle attached to the inner surface of the cell membrane in *Chlorobium.* It serves to increase the membrane surface area for photosynthesis.
chromosome a DNA molecule that contains genetic information essential for cell function and is transmitted to all progeny cells. Chromosomes in eukaryotic cells have firmly attached proteins giving rise to a nucleosome structure, while those in prokaryotes have less firmly bound proteins and no obvious nucleosome structure.
class switch a change in the heavy chain of an antibody molecule being produced. Initially, most antibodies are produced as IgM and later switched to IgG, IgA, or IgE.
clone a group of cells all descended from one original cell.
cloned DNA DNA that has been spliced into a vector and then introduced into a host cell that is allowed to grow. All cells of the culture carry the cloned DNA, usually as an extrachromosomal element.
coccobacillus a bacterial cell shaped like a cylinder that is only slightly longer than it is round. A cell that, under the light microscope, is difficult to distinguish from a true coccus.
coccus a bacterial cell that is shaped like a sphere.
coenocytic a continuous filament of cells in which there are multiple nuclei but no cross walls. This term is used normally to describe certain fungal hyphae.
coliform count the number of *Escherichia coli* and cells of similar morphology present in a sample.
colony a macroscopically visible mound of cells on a solid surface. All cells in a colony descend from one original cell.
commensalism a symbiotic relationship in which two or more organisms live in close physical association.
commercially sterile refers to a product that has no organisms detectable by standard methods or that has so few organisms as to present no threat to long-term storage. In the latter case, the product is often processed so that growth of the surviving organisms is inhibited.
communicable disease an abnormal physiological state caused by an organism that can be transmitted from an infected individual to noninfected individuals.
competence the ability of a bacterial cell to transport fragments of DNA into its cytoplasm for the purpose of recombination. Some cells are competent at all times but others, only when they are nutritionally stressed.
competitive inhibition when two molecules, one of which is not an enzyme substrate, compete to bind at the active site of an enzyme. The observed level of enzyme activity depends on the relative ratios of substrate and inhibitor.
complement a protein cascade that, when activated by antibodies bound to a cell's surface, results in the destruction of the triggering cell.
complement fixation test an immunologic test for the presence of an antigen-antibody complex that actually measures the binding of the complement proteins rather than the complex itself.
complementary a strand of nucleic acid whose chemical polarity and base sequence are such that it can form a double helix with the single-strand nucleic acid being studied.
complementary DNA (cDNA) a DNA molecule that has been prepared using reverse transcriptase with a single strand of RNA as a template.
conditional mutation a change in a DNA base sequence that results in a phenotypic change only under particular conditions, such as higher than normal temperature.
congeners flavoring compounds such as ethyl esters of fatty acids that are found in brandy.
conidium asexual fungal spores that remain attached to the sides of hyphae from which they were produced.
constitutive enzyme a catalytic protein molecule whose biosynthesis is unregulated.
constructive interference when two light rays interact to reinforce one another and appear brighter to the eye than either ray alone.
consumer an organism in an ecosystem that must eat another organism or its products in order to obtain certain molecules that it cannot synthesize for itself. A particular consumer organism can be a producer with respect to another consumer.
contact inhibition normal animal cells growing in tissue culture continue to divide until their edges touch other cells or the walls of their container. At this time, further cell division ceases. Cells susceptible to contact inhibition grow in monolayers in culture.
contraindications signs or symptoms indicating that a particular medical treatment or procedure should not be used.
cross-reaction a complex formation between an antibody and an antigen not used to stimulate the production of the antibody originally. The degree of cross-reaction indicates the extent of similarity between different antigens.
culture organisms, usually unicellular, growing in a discrete container (for example, a culture of bacteria). By extension, the term is used also to describe preparations of cells taken from a plant or animal that are grown as separate entities in the laboratory.
curds the coagulated protein from milk.

d

deaminate to remove an amino group.
death point the dose of a particular antimicrobial treatment that is necessary to kill all cells in a standard amount of time, usually in 10 minutes.
death time the time required to kill all cells in a culture when a particular dose of an antimicrobial treatment is applied.
debridement the cleansing of a wound and the removal of any damaged tissue to promote healing and prevent potential infections.
decimal reduction time the amount of time required for a specific dose of a bactericidal treatment to reduce the number of viable cells or spores of a particular organism tenfold.
decomposers microorganisms that are secondary consumers, breaking down complex molecules in the course of their metabolic activities and, thereby, initiating nutrient recycling.
delayed-type hypersensitivity a localized inflammation or allergic reaction caused by substances released by cells of the immune system that have been triggered by a person's exposure to an antigen.
denaturation a process in which either high temperature or high pH is used to break the hydrogen bonds that hold a DNA double helix together so that single strands are released. A similar process involving the application of chaotropic agents also is used to unfold a protein molecule so that its function is lost.
dendrograms treelike structures that show the relationships between organisms. The distance between branches of a dendrogram is a measure of lack of similarity between corresponding sequences.
denitrification conversion of nitrate ion to nitrogen oxide or free nitrogen gas. Nitrate ion is used as a terminal electron acceptor in this process.
desensitization shot injection of an allergen with the intent of stimulating the production of IgG to act as a blocking antibody.
destructive interference when two light rays interact so as to cancel one another. The eye perceives no light.
diauxie the two growth phases observed when bacteria are cultured on a mixture of glucose and another sugar.

differential medium a mixture of nutrients and other chemicals in which many organisms grow but only certain organisms produce a visible effect, such as a color change, on the medium.

differential stain a staining procedure that yields cells stained in either of two colors depending on the particular molecular architecture of the cell.

diffusion movement of molecules from an area of higher concentration to an area of lower concentration. In thermodynamic terms, the chemical activity of the molecules is proportional to their concentration, and the movement is down the gradient of chemical activity.

dimorphic fungus a eukaryotic microorganism that can exist in either a filamentous form or a yeast form, depending on growth conditions.

diploid a cell that has two copies of each chromosome.

diploid cell strain cells of a tissue culture that grow extensively and can be transferred many times. These cells have some chromosomal abnormalities but are still fundamentally normal. They cannot be transferred indefinitely, however.

directed evolution a change or series of changes in the phenotype of an organism that is forced by artificially imposed selection.

disease a suboptimum physiological state of an organism.

disinfectant a chemical that kills infectious microorganisms on environmental surfaces. It is too harsh to be used on living tissue.

dissimilatory nitrate reduction the use of nitrate as a terminal electron acceptor. The reduced nitrogen compounds are waste products that are not used for biosynthesis.

disulfide bridge a sulfur-sulfur chemical bond formed between two cysteine residues located in the same or different protein chains. Disulfide bridges hold protein molecules in particular configurations.

docking protein a protein found in the cell membrane that serves as an attachment site for ribosomes synthesizing proteins destined for export.

e

eclipse period that period in the infectious cycle of a bacteriophage during which an infected cell can be disrupted without releasing any infectious phage particles. It is the period from the entry of viral DNA into the cell until the assembly of the first virus.

electrophoresis a technique used to separate macromolecules according to their size and charge. An electric field is applied to molecules located in some sort of solution or gel. The more highly charged molecules move faster than the uncharged ones, and the larger molecules generally move more slowly than the smaller ones.

Embden-Meyerhof-Parnas pathway one possible catabolic route for bacteria to metabolize glucose. This pathway often is referred to as "glycolysis" and includes the splitting of fructose diphosphate into two 3–carbon molecules that are oxidized to pyruvate via phosphoenolpyruvate.

endemic disease one that is always present at significant levels within a population.

endocytosis a generalized phagocytic process in which viral particles are enclosed in a portion of the cell membrane and brought into the cytoplasm of eukaryotic cells.

endosymbiont an organism (usually prokaryotic) that lives within the cytoplasm of another organism (usually eukaryotic).

endotoxin a high-molecular-weight poison that is a structural component of its producing cell.

enrichment medium a mixture of nutrients and other chemicals that allows most organisms to grow slowly and only a few organisms to grow well. The latter group will become more prevalent in the population.

enteric bacteria members of the family *Enterobacteriaceae* that occur in the digestive tracts of animals.

Entner-Doudoroff pathway an alternative catabolic pathway for glucose commonly found in *Pseudomonas*. Glucose is first oxidized to 5–phosphogluconate and then converted to pyruvate.

envelope a lipid bilayer, derived from a previous host cell, that surrounds a virus.

epidemic refers to a disease that is present in a population at levels significantly higher than expected.

epidemiologist a scientist who studies the causes of diseases and how they are transmitted with the intention of trying to eliminate the disease from a population.

episome a plasmid that can exist readily in either an autonomous or integrated state.

epizootic an epidemic among a wild animal population.

established cell line cells of a tissue culture that have undergone chromosomal rearrangements and other DNA modifications so as to adapt to growth in the laboratory. These cells can be maintained indefinitely provided that they are transferred regularly.

etiologic agent an organism that causes a communicable disease.

eubacterial the "true" bacteria as opposed to the archaebacteria.

eukaryote (also eucaryote) a cell that possesses a true nucleus separated from the cytoplasm by a true unit membrane. It also contains various subcellular organelles bounded by unit membranes.

exon portion of a gene that codes for part of a protein. If a gene has more than one exon, they are separated by intervening sequences (introns) and made contiguous by mRNA splicing.

exonuclease an enzyme that breaks the phosphodiester bond holding the last nucleotide of a linear nucleic acid.

exotoxin a poisonous protein produced by a bacterium and released into the surrounding medium.

expression vector a cloning vector that not only provides replicational functions but also provides a promoter and ribosomal binding site to allow expression of the cloned gene.

extrachromosomal element a DNA sequence that is capable of dissociating itself from the chromosome. It can be a plasmid capable of self-replication or a transposon that can move itself to different locations on the same or different molecules.

f

facilitated diffusion movement of a substance across a cell membrane mediated by a carrier protein embedded within the membrane. Without the carrier protein, diffusion alone proceeds at a slow rate or not at all. If there is no concentration gradient, there is no net movement of molecules across the membrane.

facultative anaerobe an organism that can grow in the presence or absence of oxygen. It may or may not use oxygen as part of its metabolism (see also oxybiontic and obligate anaerobe).

feedback inhibition a reduction in the activity of the first enzyme in a multienzyme pathway caused by the accumulation of the final product of the pathway.

fermentation in biochemical terms, the oxidation of an organic molecule involving another organic molecule as a terminal electron acceptor instead of involving an electron transport system. Industrial microbiologists often use this term to describe the process of growing a culture of microorganisms in a large vessel.

fever a body temperature that is significantly above normal. In humans, normal body temperature is 37° C. A temperature above 38° C constitutes a fever.

filter mating a conjugation experiment in which donor and recipient cells must be physically held together by trapping them on a microporous membrane.

flagellar hook the bent portion at the base of a bacterial flagellar filament at which the flagellum is connected to the flagellar motor that is embedded in the cell membrane.

flagellum a filamentous projection from the surface of a cell used for locomotion. In eukaryotic cells, the flagella are flexible, whereas in prokaryotic cells they are rigid helices.

fluid mosaic membrane refers to a model of the structure of the cell membrane, first proposed by Singer and Nicholson, in which the membrane is depicted as a dynamic structure with a main body composed of a lipid bilayer that is sufficiently flexible to allow proteins embedded in it to have limited motion, such as the ability to change their tertiary structure.

fluorescence a phenomenon in which a material absorbs energy striking it at one wavelength and re-emits that energy at a different, longer wavelength.
fomite an inanimate object that can carry infectious organisms from one person to another.
F prime a larger than normal F plasmid that includes some bacterial DNA sequences. These bacterial sequences are fully functional and can be transferred during conjugation.
freeze-fracture process by which bacterial cells are suspended as a paste and then frozen to temperatures at which they become brittle. The frozen cells crack along the lines of least resistance (usually lipid bilayers) when struck with a rammer.
fusel oils long-chain alcohols (C_8 to C_{12}) produced during ethanol fermentations.
fusion polypeptide a portion of a protein molecule that has been joined to all or part of another protein from the splicing of their two genetic coding regions together.

g

gametangium a specialized haploid fungal hypha that can fuse with a gametangium of a different mating type to yield a zygospore.
gamete a specialized haploid cell produced by meiotic division that functions to fuse with another gamete to generate a new diploid individual.
gametophyte generation those haploid cells of certain eukaryotic algae that are capable of differentiating into gametes (see also sporophyte generation).
gas vacuoles protein- or lipid monolayer-bound bacterial structures containing a gas.
gel electrophoresis an electric current is used to separate charged molecules embedded in a soft matrix composed of substances like polyacrylamide, agar, and agarose.
generation time the time that elapses between one cell division and the first division of the daughter cells. Normally, the generation time is taken as the average time required for the number of cells in a culture to double.
genetic transformation the genetic transfer process in which DNA from a donor cell diffuses through a liquid medium until it contacts a competent cell. The DNA is then transported into the cell and recombined if it is homologous to the resident DNA.
genome equivalent the amount of DNA equivalent in mass to one complete bacterial chromosome.
genotype the catalog of all genetic information for a cell regardless of whether this information is being expressed.
germline cell a cell of a multicellular organism that has not yet differentiated.
gliding motility motile bacteria that lack flagella possess a gliding motion on a solid surface as opposed to bacteria that possess the typical swimming ability due to their flagella.
globulins roughly spherical proteins that circulate in blood plasma. They are insoluble in pure water or in high-salt solutions.
gluconeogenesis the biosynthesis of the hexose glucose from short-chain carbon molecules.
glucose effect see catabolite repression.
glycocalyx a diffuse polysaccharide associated with the exterior of a bacterial cell. If it is closely adherent to the cell, it becomes a capsule.
glycolipid a molecule comprised of glycerol, two fatty acids esterified to the glycerol, and a sugar esterified at the third position of the glycerol.
glyoxylate bypass an anaplerotic pathway of the TCA cycle that provides extra quantities of malate and succinate to overcome losses that occur when 2-oxoglutarate is used to synthesize amino acids.
gnotobiote a "germfree" animal that is delivered by cesarian section and maintained in a sterile environment all its life. It has no internal or external microorganisms.
gratuitous inducer a compound whose presence will cause the genetic regulatory system of an operon to derepress; however, it is not itself a substrate for the enzymes encoded within the operon.
greenhouse effect the trapping of planetary infrared radiation by waste gases in the atmosphere resulting in a mean temperature rise.
groundwater water located in underground reservoirs and rivers that is replenished by surface water that sinks and drains through the soil.
group translocation a transport process in which a solute is actively transported across the plasma membrane and into the cytoplasm while being modified chemically.
growth rate the number of cell divisions per hour, usually represented by the Greek letter mu (μ).

h

halophile an organism that will only grow in the presence of high-salt concentrations. Extreme halophiles grow in saturated NaCl solutions and require a minimum concentration of about 15% NaCl.
haploid an organism that has only one copy of each chromosome.
hapten a molecule that is too small to stimulate antibody production by itself, but it is antigenic when complexed to a larger molecule. A hapten can react with an antibody even though it cannot stimulate antibody production directly.
haustorium a specialized fungal hypha that inserts itself into a cell of another organism so that the fungus can obtain nutrients.
heat labile a molecule or complex of molecules whose biological activity can be destroyed by raising the temperature to an appropriate level.
helicase an enzyme that unwinds a DNA double helix.
hemagglutination the clumping of red blood cells to form a visible precipitate.
herd immunity if the preponderance of individuals in a population are immune to a particular disease, a nonimmune individual is protected because he or she is surrounded by persons who cannot transmit the disease.
heterocyst a thick-walled cell that results from the differentiation of a normal cyanobacterial cell under conditions of nitrogen starvation. Within the heterocyst, nitrogen fixation occurs.
heteroduplex a double-strand DNA molecule produced by hybridization of single strands from two different DNA molecules.
heterofermentation energy production in the absence of electron transport with more than one type of organic molecule used as a terminal electron acceptor. The result of the fermentation is, therefore, an accumulation of several types of chemicals in the medium.
heterokaryon in certain fungi, a single cell that contains two unfused haploid nuclei. In protists, a cell that has both macro- and micronuclei.
heteroploid cells that have lost some chromosomes and gained extra copies of others; therefore, they are not perfectly diploid, triploid, etc.
heterothallic an adjective used to describe a fungus that mates two haploid cells to form a diploid cell only when two different cell types are present.
heterotroph an organism that procures the carbon needed for growth from organic molecules.
hexose any six-carbon sugar.
hexose monophosphate pathway an alternative pathway for glucose catabolism that yields less ATP than the EMP pathway and provides ribose phosphate for use in nucleic acid biosynthesis.
Hfr cell a bacterial cell carrying an integrated, conjugative plasmid. When the plasmid attempts to transfer itself to another cell, varying amounts of bacterial DNA are transferred also.
holoenzyme a complex of proteins that has complete specificity and enzyme activity. RNA polymerase is an example of a holoenzyme in which some proteins that confer specificity of initiation can be removed without affecting polymerizing ability.
homeostasis a cell's ability to maintain constant internal conditions (pH, ionic concentration, nutrient levels) in the face of changing external conditions.
homofermentation energy production in the absence of electron transport with only one type of organic molecule used as a terminal electron acceptor; therefore, only one type of chemical accumulates in the medium.

homopolymer a macromolecule composed of identical subunits joined together.

homothallic an adjective used to describe a fungus that can mate two haploid cells of the same type to produce a diploid cell.

host range a catalog of organisms that can be infected by a particular virus or serve as recipients for conjugation of a particular plasmid.

humoral immunity the defense mechanism mediated by globular proteins (immunoglobulins, antibodies) circulating in blood plasma.

hybridoma an antibody-producing cell resulting from the fusion of a plasma cell and a multiple myeloma cell. It produces monoclonal antibodies when grown in tissue culture.

hydrophilic liking water. A term applied to a molecule or structure that is charged or that has a charge separation and tends to bind water molecules.

hydrophobic avoiding water. A term applied to a molecule or structure that is uncharged and, therefore, cannot bind water molecules or dissolve in water.

hypha a filament of cells found in fungi. The filament can exhibit true branching in which a Y-shaped cell is formed.

i

icosahedron a regular polyhedron that has 20 faces and 12 corners. The shadow cast by an icosahedron is a regular hexagon.

idiotype the shape of the antigen-binding portion of the molecular complex that comprises an antibody molecule.

immune complex disease a bodily dysfunction caused by the accumulation of antigen-antibody complexes at a particular site. Examples are glomerulonephritis and rheumatoid arthritis.

immune surveillance the removal of cells carrying inappropriate antigens on their surface by various cells of the immune system.

incompatibility the term used to describe a situation in which two different plasmids cannot coexist in the same cell.

incubation period the period of time that elapses between an individual's exposure to a disease-causing organism and the first appearance of symptoms.

indicator bacteria (1) With respect to phages, these are bacteria used as hosts so that plaques will appear on a lawn. (2) With respect to water-quality analysis, these are bacteria whose presence indicates fecal contamination.

inducer a substance that results in the expression of an operon.

inducible operon a unit of transcription that is normally repressed but can be activated by the presence of an appropriate inducer molecule.

induction the turning on of cellular functions that had been turned off.

infection thread a tube formed by root cells of a legume in response to the invasion of *Rhizobium* species. Rhizobia use the tube to invade further into the root before stimulating production of a root nodule.

infectious dose the minimum number of viral particles or microbial cells necessary to establish an infection in a particular host.

inflammatory response a body's reaction to a foreign antigen characterized by reddening and swelling that result from capillary dilation, an accumulation of plasma exudate at the site, and interactions of complement with antigen-antibody complexes.

infusion a nutrient solution prepared by boiling a particular substance in water (for example, a meat infusion).

initiator codon the first codon in an mRNA base sequence that codes for a protein. It is usually, but not always, AUG.

inoculation the process of introducing a microorganism into a culture medium in which it can grow.

insertion sequence a relatively short segment of DNA (less than 1,500 base pairs) that is capable of moving itself from one location to another. Transposons are bounded by pairs of insertion sequences.

integration the process of recombining two circular DNA molecules, one of which is smaller than the other, so as to create a single, large circle. The smaller molecule is said to be integrated into the larger.

interference a vector sum of two or more light waves. If two light waves of the same wavelength and equal amplitude arrive at a particular point in phase, the perceived light is twice as bright as either wave alone. This is constructive interference. On the other hand, if the waves arrive out of phase, destructive interference occurs, and no light can be seen by an observer.

interrupted mating experiment conjugating cells are treated so as to prevent further DNA transfer, and recipient cells are checked to determine which donor genes have been received. By varying the times of interruption, it is possible to construct a crude genetic map.

intron an intervening sequence of bases found in a DNA molecule that does not appear in the final mRNA molecule. It is removed by an RNA-splicing process.

invasiveness the ability of a parasite to penetrate the epithelial tissue layers in order to infect a host internally.

l

β-lactamase enzyme that opens the β-lactam ring found in penicillin and its chemical derivatives; also called penicillinase.

lagging strand the second strand of a DNA duplex to be synthesized. Initially, it is produced as a series of Okazaki fragments that later are linked together.

landfill site a depression in the ground, such as an old quarry, that is filled in with trash and/or garbage.

latent phase the interval at the beginning of a one-step growth experiment during which there is no increase in the number of infectious centers.

lawn a confluent, uniform layer of cells on the surface of an agar medium.

leader region the portion of the 5′ end of an mRNA prior to the first codon. It includes the ribosomal binding site.

leading strand the first strand of a DNA duplex to be synthesized. It is probably produced as one continuous strand.

L form bacterial cells with defective or missing cell walls that biochemically resemble members of several common genera. Their name comes from the Lister Institute in London.

leghemoglobin a special form of hemoglobin synthesized by root nodule cells in leguminous plants. This hemoglobin is used to transport oxygen to nitrogen-fixing bacteroids.

leukocyte (also leucocyte) a white blood cell.

limiting nutrient the nutrient whose rate of addition controls the growth rate of the organism in a chemostat culture.

lipid A the lipid component of the outer leaflet of the outer membrane. It is based on glucosamine instead of glycerol and has fatty acids bonded both to the glucosamine and to each other.

lymphocyte a small white blood cell that has a large, approximately round nucleus and relatively small amount of cytoplasm.

lymphokine a substance that causes a lymphocyte to proliferate and differentiate. The stimulated cell may be the producer or one of its neighbors.

lyophilization freeze-drying of a solution. The solution is frozen at dry ice temperatures and then placed under a high vacuum. Water sublimes out from the cells, preserving their structure.

lysogen a bacterial cell that is harboring a prophage.

lysogenic conversion a prophage codes for synthesis of an exotoxin and causes all of its lysogens to become toxin producers and, therefore, more virulent.

lysosome a membrane-bounded cytoplasmic vesicle that contains hydrolytic enzymes to degrade the contents of phagocytosed particles.

lytic virus one that always causes cell lysis following an infection.

m

macrophage a large, motile, phagocytic leukocyte that functions to engulf foreign particulate matter in the body.

major histocompatability complex a group of antigens found on cell surfaces whose composition is characteristic of the individual organism.

mating aggregate a clump of conjugating donor and recipient cells.
medium a nutrient solution used to grow microorganisms that can be either complex (composed of chemically heterogeneous substances) or defined (composed of specific chemicals).
memory cell a small B lymphocyte that can be stimulated quickly to differentiate into plasma cells. It is the source of the secondary immune response.
mesophile an organism that grows best at moderate temperatures, usually defined as ranging from 10 to 45°C. In another definition, the temperature ranges from 270 to 330 K.
mesosome a narrow channel within the cytoplasm of a bacterial cell formed by the invagination of its cell membrane. The mesosome is seen in electron micrographs of fixed specimens but not in electron micrographs of frozen sections.
metachromatic a color reaction in which an object that has been stained with a dye of one color will nevertheless exhibit a different color.
methyl-accepting chemotaxis proteins membrane proteins that serve as receptors of chemotactic signals in the unmethylated state. The binding of a chemotactic agent to such a protein triggers an adaptation process that results in methylation of the protein.
methyl red test a biochemical test that identifies organisms that carry out a mixed-acid fermentation by the characteristic low pH of the culture medium after the cells have reached stationary phase (see also MRVP test).
microaerophile an organism that grows in the presence of oxygen but only at partial pressures lower than those of normal ranges in air.
minimum inhibitory concentration the least amount of a substance necessary to prevent the growth of a particular microorganism in its immediate vicinity on an agar plate.
mitogen anything that stimulates cell division.
mobile genetic element a piece of DNA, such as a transposon, that is capable of moving from one DNA molecule to another, or a plasmid that is capable of transferring itself from one cell to another.
modification enzyme a protein catalyst that recognizes specific sequences of bases on a DNA molecule. The enzyme makes one or more simple chemical modifications, such as the addition of a methyl group, to one or more bases within the specific sequence.
monoclonal antibodies identical immunoglobulins that are all derived from the same clone of plasma cells.
MRVP test a combination biochemical test for two metabolic pathways from glucose. The MR portion of the test (addition of methyl red pH indicator) is positive if mixed acids are synthesized from pyruvate. The VP (Voges-Proskauer) portion of the test is positive if acetylmethylcarbinol, an intermediate in 2,3–butanediol biosynthesis from pyruvate, is present.
multiple myeloma cells cancerous plasma cells.
mutagen anything that causes a mutation.
mutation any change in the base sequence of a nucleic acid that constitutes the hereditary material of a cell or virus.
mutualism a symbiosis in which both partners benefit from the relationship.
mycelium a fungal structure formed by twisted and interwoven hyphal filaments.
mycorrhiza an association between plant roots and fungi that serves to enhance the ability of the plant to absorb nutrients from the soil.
mycotoxin a poisonous product of a fungus's secondary metabolism.
myxamoebae unicellular forms of a cellular slime mold produced by germinating spores that can later aggregate to form a plasmodium and eventual fruiting body.

n

neutralism a theoretical symbiotic relationship in which neither organism is benefitted or harmed.
niche the position of an organism within the physical or biological community.
nick the loss of a phosphodiester bond in a DNA molecule. In a nicked DNA molecule, all the bases are still present.
nick translation movement of a nick to a different physical location as a result of the action of DNA polymerase I binding to the original nick and replacing some existing bases with new ones. In essence, the nick is "pushed" ahead of the enzyme.
nitrogen fixation the capacity of a few bacteria to reduce gaseous nitrogen (N_2) into ammonium ion.
noncompetitive inhibition a reduction in enzyme activity caused by the binding of an effector molecule away from the active site.
nonprosthecate refers to bacteria without cellular stalks.
normal microflora those microorganisms usually found in or on another organism. Normally they are harmless and sometimes they can even be beneficial.
Northern blotting a technique similar to Southern blotting except that RNA molecules are used (see also Southern blotting).
nuclease an enzyme that degrades nucleic acids.
nucleocapsid the DNA and associated proteins found inside a viral capsid.
nucleoid the nucleuslike structure of a bacterial cell. It consists of DNA together with associated RNA and proteins but lacks the unit membrane that surrounds the nucleus of a true eukaryotic cell.
nucleosome a cylinder of histone proteins around which a double-strand DNA molecule is wrapped in a eukaryotic chromosome.

o

oncogenic a virus, a specific gene product, a chemical, etc. that is capable of causing a tumor.
one-step growth experiment an experiment in which only one cycle of phage infection, reproduction, and cell lysis is permitted.
open reading frame a region of a DNA molecule that could code for a protein but that has not yet been shown to do so. It includes code for a ribosomal binding site, some amino acids, and a translational termination signal.
operon a unit of transcription that can be regulated by appropriate activators and repressors. At minimum, it consists of a promoter, a coding sequence, and a transcriptional terminator.
opportunist a pathogen with no invasive ability that can cause disease if it enters the body through a wound.
opsonization the process in which a cell is made more attractive to phagocytes by binding antibody molecules to its surface.
osmotic pressure the net pressure that must be applied to the solution on the side of a semipermeable membrane with the higher solute concentration in order to prevent the net movement of water molecules from a solution of low concentration to a solution of high concentration. It is assumed that solute molecules cannot pass through the membrane.
Ouchterlony double-diffusion test an immunologic assay in which antigens and antibodies are allowed to diffuse toward one another within an agar or agarose gel. If a reaction occurs between antigens and antibodies, a visible precipitate will form in the zone of equivalence.
oxidase test the application of a solution of N,N, dimethyl-*p*-phenylenediamine to a suspension of bacteria. If a cytochrome *c* oxidase enzyme is present, a colored compound is formed.
oxidoreduction reaction a chemical reaction between two compounds in which one compound is oxidized and the other correspondingly is reduced.

p

pandemic a worldwide epidemic.
papule a small red lump appearing transiently on the skin.
parasitism a symbiotic relationship in which one partner benefits while the other is harmed.
passive immunity immunity that results from the injection of preformed IgG antibodies into

an individual. Passive immunity also results from the transport of IgG antibodies from maternal to fetal circulation. It is of comparatively short duration.
pasteurization the process of heating a food or beverage just to the point at which harmful microorganisms are killed. It is not a sterilizing treatment.
pathogen an organism that can cause a disease.
peptide a short chain of amino acids linked by peptide bonds. It is shorter than either a polypeptide or a protein.
peptidoglycan the material that comprises the majority of bacterial cell walls. It consists of chains of sugars oriented perpendicular to the long axis of a rod-shaped cell and connected by short chains of amino acids running parallel to the long axis.
periplasmic space the potential space or gel layer located between the outer membrane and the peptidoglycan layer in gram-negative bacteria.
phagosome a membrane-bounded cytoplasmic vesicle containing particulate matter that is the end product of phagocytosis.
phenol coefficient an inverse ratio of the amount of a substance necessary to kill certain microorganisms to the amount of phenol needed to accomplish the same task. If the coefficient is greater than 1, the chemical being tested is more lethal than is phenol.
phosphoketolase pathways various routes in which glucose is catabolized by fermentative reactions having keto sugars as intermediates.
photobioreactor a culture vessel designed for growing photosynthetic organisms in order to produce specific chemicals.
photoreactivation repair of pyrimidine dimers in a DNA molecule with an enzyme (photolyase) that uses the energy of blue light to reverse the dimerization process.
photosynthetic antenna a series of light-absorbing pigments surrounding a photochemical reaction center that can capture photons of light and transmit their energy to the reaction center.
phototroph an organism that can convert the energy of sunlight into metabolic energy for all its growth needs.
pili thin, filamentous structures other than flagella found on the surface of bacterial cells. In the pure sense of the term, pili are associated with certain conjugative plasmids and involved in the formation of mating aggregates. Common pili are not associated with conjugative plasmids but are involved in adhesion and are often referred to as fimbriae.
plaque a hole or region of poor growth in an otherwise confluent layer of cells. This hole is caused by the destruction of cells or by the retardation of their growth resulting from a viral infection.
plaque-forming unit a viral particle or infected cell that can infect surrounding cells and lead to the formation of a hole or region of inhibited growth in a lawn of cells.
plasma the fluid portion of the blood.
plasma cell a cell that actively synthesizes and secretes antibody molecules into the blood plasma. It is derived from a B lymphocyte.
plasmid an autonomous DNA molecule in a cell that is capable of self-replication and sometimes of self-transfer as well.
plateau phase the interval at the end of a one-step growth experiment after all infected cells have lysed and the number of infectious centers is no longer increasing.
plus-strand RNA an RNA transcript of a DNA molecule that is a copy of the sense strand of the DNA helix; therefore, it can be translated to give a functional protein. In virology, the genome of a plus-strand RNA virus is also its mRNA.
polymer a large molecule, such as a protein or DNA, composed of identical or nearly identical subunits.
polymerase an enzyme that joins subunit molecules together to make a polymer.
polysome a shortened form of the term *polyribosome*. It refers to an mRNA molecule that is being translated by multiple ribosomes arrayed along its length.
porin an outer membrane protein that forms a channel or pore through the outer membrane and is part of a complex usually comprised of three identical proteins.
portal of entry an opening in the body through which a pathogen can enter.
precipitin reaction a reaction that occurs when antigens and specific antibodies are present in roughly equal concentration so that large insoluble molecular complexes (precipitates) are formed.
premature lysis experiment an experiment in which a culture of cells is infected with a phage, and then the cells are disrupted at various times prior to when the infecting phage normally would cause cell lysis.
primary cell culture cells growing in a laboratory that are identical to cells removed from an organism.
primary immune response production of a circulating antibody in response to an organism's first exposure to an antigen. It requires several weeks for completion and results in the synthesis of comparatively low levels of circulating antibody as well as in the growth of a memory cell clone.
primosome an enzyme complex that lays down a short chain of nucleotides opposite a single strand of DNA to provide a site for DNA polymerase to begin synthesis.
prion a noncellular disease-causing agent that is not a conventional virus. It apparently is a protein molecule normally encoded in the DNA of its host cell, and it can stimulate its own production.
probe a single-strand fragment of DNA or RNA that is used to test for the presence of homologous sequences in Southern or Northern blotting techniques.
producer an organism in an ecosystem that, as part of its normal life cycle, can carry out one or more reactions to synthesize organic molecules for subsequent use by consumers. It serves as a food source for other organisms in the ecosystem.
prokaryote (also procaryote) a cell that lacks a true nucleus. Although the cell itself is bounded by a unit membrane, there are no internal unit membrane-bounded structures such as mitochondria or chloroplasts.
promoter the site on a DNA duplex to which the RNA polymerase holoenzyme complex binds in order to begin transcription.
promutagen a substance that can be converted into a mutagen but is not itself mutagenic.
prophage a DNA molecule of a bacteriophage that has formed a stable relationship with its host cell and replicates at the same rate as its host DNA. It is inherited by both daughter cells at cell division. Physically, a prophage can be a part of the bacterial chromosome or exist as an independent plasmid.
prophylactic refers to a treatment applied in advance of need, such as an immunizing shot.
prosthecate refers to bacteria with a cellular stalk.
protease an enzyme that can break the peptide bond between two amino acids in a protein chain.
proton motive force the potential energy ($\Delta\tilde{\mu}_{H^+}$) available to a cell resulting from the sum of the electric ($\Delta\Psi$) and chemical (ΔpH) gradients across the cell membrane.
protoplast a round bacterial cell without any cell wall that is derived from a gram-positive organism if peptidoglycan synthesis is inhibited.
provirus a quiescent viral genome that replicates in step with a cell's DNA and is passed to all progeny cells.
pseudogene a nontranscribed DNA sequence that is nearly identical to a gene.
pseudohypha a short chain of yeast cells. Unlike a true hypha, the cells of a pseudohypha are demarcated clearly.
psychrophile an organism that grows only at cool temperatures, usually defined as between −5 and +20° C. In another definition, a psychrophile's minimum growth temperature is less than 261 K and maximum temperature is less than 290 K.
psychrotroph an organism that grows at temperatures between those for psychrophiles and mesophiles. The temperature range has been defined from 269 K to 305 K.
pyrogen a substance whose presence in the body causes a fever.

q

quarantine restrictions placed on a person's or animal's movements to prevent the spread of a communicable disease.

r

racemase an enzyme that changes the stereoisomerism of a molecule, for example, by converting an L-amino acid to a D-amino acid.
recombination the process by which two DNA molecules exchange strands so as to create a new molecule(s) carrying a composite of the genetic information contained in the original molecules.
reductive amination a reaction between an ammonium ion and a keto group to yield an amine.
reductive carboxylation alternative methods for carbon dioxide fixation found in various bacteria that lack the enzymes necessary for the Calvin cycle.
replication fork the actual site at which one DNA duplex is becoming two as new complementary DNA strands are being synthesized.
replicative form a double-strand DNA molecule used as a replicational intermediate by a single-strand DNA virus.
repliconation the process of taking the single strand of DNA transferred during bacterial conjugation, circularizing it, synthesizing a complementary strand, and preparing it for self-replication.
repressor a molecule that binds to DNA and prevents transcription of an operon by blocking attachment of the RNA polymerase holoenzyme complex.
reservoir of infection an organism or physical entity like a body of water that can harbor a disease-causing organism and transmit it to susceptible individuals.
resolution the resolution or resolving power of a microscope refers to the ability of that microscope to permit an observer to discern details clearly. Specifically, it is the minimum spacing needed for two objects to be seen as separate and is always expressed as a length measurement.
respiration production of an ion gradient across a membrane using an electron transport chain. Respiration can be aerobic or anaerobic depending on the terminal electron acceptor employed.
respiratory burst a set of metabolic pathways that produce reactive forms of oxygen in order to destroy a cell in a phagocytic vesicle.
restriction endonuclease an enzyme that recognizes and binds to a particular sequence of bases on a DNA molecule. If certain bases within the sequence are not chemically modified in a particular way, the enzyme cleaves phosphodiester bonds within both strands of the DNA molecule.
retroreplication the process in which an RNA molecule is replicated via a DNA intermediate.
reverse transcriptase an RNA-dependent DNA polymerase that makes a DNA copy of an RNA molecule.
reversion a back mutation. A mutation that converts the genetic code for a gene from a mutant form to a wild-type state.
rise phase the interval during the middle of a one-step growth experiment in which phage-infected cells are lysing and the number of infectious centers is increasing with time as viral particles are released.
rod one type of cell shape observed in bacteria. A rod-shaped cell is a cylindrical object longer than its diameter.
rolling circle a mode of DNA replication in which a double-strand DNA molecule is nicked in one strand and then, as one end of the nicked strand is extended by DNA polymerase, a single strand of DNA is displaced from the duplex.
rugose possessing a rough surface.
rumen a special pouchlike organ found in the digestive system of certain animals (ruminants). It serves as a small fermenter, containing microorganisms to digest low-quality food. The animal then uses some of the microorganisms and their waste products as food.

s

sacculus the meshwork of peptidoglycan filaments that covers the surface of most bacterial cells.
sanitizer a chemical that reduces the number of infectious microorganisms on environmental surfaces to acceptable levels but does not necessarily kill them all.
secondary immune response production of a circulating antibody in response to an organism's repeated exposure to a specific antigen. The amount of antibody synthesized and the rate of production are both high as a result of a preexisting clone of memory cells.
secondary infection a microbial infection that occurs as a result of weakened host defenses caused by a primary infection.
sedimentation coefficient the rate at which a particle moves through a solution. A particle with a sedimentation coefficient of 1S travels 10^{-13} cm/s in a gravitational field of 1 g. The sedimentation coefficient is a function of molecular shape and weight.
selective advantage having the ability to grow faster or more efficiently than other, competing organisms.
selective medium a mixture of nutrients and other chemicals that allows growth only of a certain organism or a certain group of organisms.
semiconservative DNA replication the process of synthesizing new DNA by pairing new bases to old ones. The products are two DNA duplexes consisting of an old strand paired to a new one.
sense strand the strand of a double-helical DNA molecule that is complementary to mRNA transcribed from that region.
sepsis an infection caused by growth of microorganisms.
septate possessed of true septa or partitions between cells in a filament. This term is used to describe certain fungal hyphae that have partial cross walls.
serum the portion of blood that remains fluid after clotting has occurred. It contains no cells.
shuttle vector a plasmid used in DNA cloning experiments that is capable of replicating in two or more diverse organisms.
siderophore a molecule synthesized by microorganisms to chelate iron and make it available for microbial growth.
signal peptide a short chain of amino acids found at the amino terminus of proteins destined to be exported across the cell membrane. The chain is hydrophobic and can insert readily in a membrane's lipid bilayer.
single-cell protein a food product derived from unicellular microorganisms.
single-hit kinetics only one interaction with an inactivating agent is necessary to destroy a cell or virus. A graph of the logarithm of survivors as a function of intensity of treatment is a straight line.
site-specific recombination exchange of DNA sequences that occurs only at a specific location(s).
sludge the particulate matter separated from raw sewage and aerobic digestion during the sewage treatment process.
snapping division a form of binary fission in which the rod-shaped daughter cells remain attached by one corner.
somatic cell mutation any change in the DNA sequence of a differentiated cell.
SOS repair a fast DNA repair process invoked when the damage is too great to be repaired promptly by normal processes. The repairs normally are prone to have mistakes.
Southern blotting a technique in which the DNA to be tested is immobilized on a solid support and then exposed to a labeled DNA probe. After washing, the presence of label indicates regions in which homologous DNA segments are located.
spheroplast a round bacterial cell derived from a gram-negative cell if peptidoglycan synthesis is inhibited. It retains portions of the original cell wall, unlike a protoplast.
sporangiospore an asexual reproductive structure found in fungi. The spores are contained within sacs attached to hyphae.
sporangium a cell that is producing a spore.
sporophyte generation those cells of certain eukaryotic algae that are capable of meiosis during vegetative growth without giving rise to spores or gametes (see also gametophyte generation).
steady state a nonequilibrium situation in which various parameters remain constant while material (nutrients, organisms, etc.) passes through continuously.
sterile a condition in which no viable cells of any type, active viruses, functional viroids, etc. are present (see also commercially sterile).
substrate analog a molecule whose three-dimensional structure sufficiently resembles that of a normal enzyme substrate such that

it will bind to an active site but cannot be converted into a product.
substrate level phosphorylation ATP is produced by a reaction in which a phosphorylated compound and ADP react to give a dephosphorylated compound and ATP.
superinfection reinfection of a cell that is already infected with a virus.
symbiosis the growth of two or more organisms in association with one another.
symport the simultaneous transport of a substance against its concentration gradient together with an ion moving in the direction of its own concentration gradient. Thus, the movement of the two solutes is in the same direction.
syndrome a unique collection of nonspecific symptoms that when taken together define a recognizable disease state.

t

taxonomy the science of classifying living things into groups of related organisms.
T cell a thymus-derived lymphocyte of the immune system.
temperate response a viral infection that results in the production of a prophage. The host cell survives because production of more viral particles is supressed; therefore, cell lysis does not occur.
terminal electron acceptor a molecule that receives an electron as a part of an oxidation-reduction reaction and then is eliminated from the cell.
thalloid refers to a structural shape observed in fungi and algae. Although these structures resemble roots, stems, and leaves superficially, there is no true tissue differentiation like that seen in higher plants such as trees.
therapeutic ratio the ratio between the concentration of an antimicrobial agent that will kill or inhibit pathogenic growth and its toxic concentration that will harm a host. The best antimicrobial agents have a high therapeutic ratio.
thermophile an organism that will grow only at high temperatures, usually defined as temperatures greater than 40° C. In another definition, a thermophilic organism has a minimum growth temperature of 290 K, an optimum temperature of 315 K, and a maximum temperature greater than 330 K.
time-dose-rate relationship the effect of an antimicrobial treatment is proportional both to the absolute quantity of the treatment used as well as to the duration of the cells' exposure to the treatment.
time of entry the first instant during a conjugative experiment at which a particular donor trait can be detected in any recipient cell.
tincture an alcohol solution of a particular substance.
tissue culture a preparation of plant or animal cells grown under artificial conditions in a laboratory.
titer the number of molecules or particles of interest present in a solution. The term is most commonly employed with respect to viral particles and antibody molecules. In the case of antibodies, a titer is expressed as the greatest dilution of serum that will still cause clumping of a standard quantity of antigen.
topoisomerase an enzyme that can add or remove supercoils from a covalently closed, circular DNA molecule.
toxin a poisonous substance. A microbial toxin is an antigen of high-molecular weight.
toxoid a toxin that has been treated so as to destroy its poisonous properties but not its antigenic properties. It is used for immunizations.
trace metal any of the metallic elements essential for cell metabolism and growth but required only in very small amounts. A typical concentration of a trace metal might be 10^{-4} M or less.
transamination a chemical reaction in which an α-amino acid and an α-keto acid exchange their respective α substituents. The stereoisomerism of the α-carbon is preserved in the reaction.
transconjugant a cell that has received DNA during conjugation.
transcription the process of copying the information from a region of a double-strand DNA molecule into a single-strand RNA molecule.
transduction a genetic transfer process in which DNA from a donor is carried to a recipient inside a viral capsid.
transfection genetic transformation in which viral DNA is used. A virus-infected cell results.
transformation conversion of normal eukaryotic cells into tumorigenic cells. The cells can be carried in tissue culture indefinitely and will cause a tumor if returned to the organism from which they were taken (see also genetic transformation).
transgenic organism an organism that has undergone genetic engineering so that it expresses one or more genes from an unrelated organism.
transition mutation a change in the base sequence of a nucleic acid resulting from the substitution of a purine for a purine or a pyrimidine for a pyrimidine (for example, adenine can be replaced by guanine).
transition temperature that temperature at which lipids in a cell membrane convert from a gel to a fluid and vice versa.
translation the process of converting the genetic information inherent in the structure of an mRNA molecule into a sequence of amino acids joined by peptide bonds. Translation occurs on a ribosome.
translocation the process of moving a ribosome and nascent peptide chain along mRNA during protein synthesis.
transpeptidase an enzyme that can transfer a preexisting peptide bond from the amino group of one amino acid to the amino group of a different amino acid.
transposition movement of a DNA sequence from one location to a different location on the same molecule or to a different DNA molecule. The movement can be conservative or can include DNA replication.
transposon a DNA sequence that is capable of transposition.
transversion mutation a change in the base sequence of a nucleic acid in which a purine is substituted for a pyrimidine or vice versa (for example, adenine can be replaced by thymine).
tricarboxylic acid cycle a means by which a cell can completely oxidize acetyl groups to water and carbon dioxide with the concomitant production of reduced enzyme cofactors. Operation of the complete cycle requires that electron transport functions to generate ion gradients. Also called the Krebs or citric acid cycle.
trichomes chains or filaments of algal or bacterial cells.
tubercle a lump of tissue formed in the lungs when macrophages surround an invading organism and wall it off. Tubercles are seen in cases of tuberculosis, histoplasmosis, and coccidiomycosis.
tubidimetry a method of estimating the concentration of cells in a culture by measuring the absorbance of the culture.
turbidostat a continuous culture device designed to maintain a culture at a constant turbidity.
type culture a culture that represents a species or patent description. Taxonomic comparisons are made among type species.

u

uncoating the process of removing a viral coat in the cytoplasm of a eukaryotic cell so that the infectious process can continue. Uncoating often occurs inside lysosomes.

v

vaccination the process of introducing a killed or attenuated microorganism or virus into the body in order to stimulate the immune system to produce antibodies against the organism without causing an actual disease state.
vaccine substance used to immunize an animal against a disease.
vector (1) In DNA splicing, a DNA molecule that is self-replicative and can be used as a carrier for the DNA of interest. (2) In epidemiology, a biting or sucking organism

that serves to transport an infectious microorganism or virus from an infected to an uninfected host.

virion an intact viral particle consisting of nucleic acid surrounded by a protein coat.

viroid an uncoated RNA molecule that is parasitic on certain plants.

virulence the relative ability of a parasite to overcome host defenses and cause disease.

virus a noncellular, obligate intracellular parasite. Its genetic material, which may be composed of single- or double-strand DNA or RNA is encased within a protein coat.

Voges-Proskauer test a test for the presence of acetylmethylcarbinol, an intermediate compound in the synthesis of 2,3–butanediol from pyruvate.

W

water activity (a_w) the actual proportion of water molecules that are available to a cell and not bound in spheres of hydration around solute molecules. Pure water has an a_w value of 1.0.

whey the fluid portion of milk that remains after coagulation of milk protein.

wrapping choice a model for the packaging of phage DNA in which it is assumed that any suitably sized piece of DNA can be inserted into a phage, although the probability of nonviral DNA being used mistakenly is low.

Z

zone of equivalence region of maximum precipitate formation between antigens and their specific antibodies. It is seen when the concentrations of antigen and antibody are roughly equal.

zygospore a sexual fungal spore produced by the fusion of gametangia of different mating types.

subject index

Page numbers in **boldface** indicate boldfaced terms that can also be found in the glossary.
Page numbers in *italics* indicate figures and tables.

a

b

c

d

e

f

g

m

n

O

P

t